Library of Congress Control Number: 2017930179

Front jacket image: Zsolt József Simon, *Untitled*, 2012. Porcelain, no glaze, 13.75" × 13.75" × 9.5". Photo: Stephen Papandropoulos.
Jacket flap image: Antoine-Louis Barye, *Rabbit, Ears Raised*, first offered for sale in 1845. Paris, France, bronze, 1.62" × 2.62" × 1". Courtesy of The Walters Art Museum, Acquired by William T. Walters before 1889, CC0, GNU FDL.
Jacket spine image: Yasmin Vinograd, *Memory Figurine #2*, 2015. Silver, iron, 2.2" × 1.9" × 0.7". Photo: Alex Kucherenko.
Back jacket images (clockwise from top): Detail, bas relief bull from the Ishtar Gate, 575 BCE. Babylon, glazed brick, gate 46' × 100'. Photo: © Josep Renalias, Pergamon Museum, Berlin / Wikimedia Commons / CC by 3.0.
David Giulietti, *Moon Curve Earrings*, 2014. Sterling silver, 18K, 22K, diamonds, moonstone, 3" × 1.4" × 0.38". Photo: Hap Sakwa.
Cup Plate, Boston and Sandwich Glass Factory, ca. 1830s. Sandwich, Massachusetts, USA, pressed sapphire glass, 3.5". Courtesy of LACMA, Gift of the Reverend W. Daniel Quattlebaum in memory of Edith Brockett Quattlebaum, www.lacma.org, LACMA public domain.

Type set in Tiffany Lt BT/Josefin Sans
ISBN: 978-0-7643-5338-3
Printed in China

Published by Schiffer Publishing, Ltd.
4880 Lower Valley Road
Atglen, PA 19310
Phone: (610) 593-1777; Fax: (610) 593-2002
E-mail: Info@schifferbooks.com
Web: www.schifferbooks.com

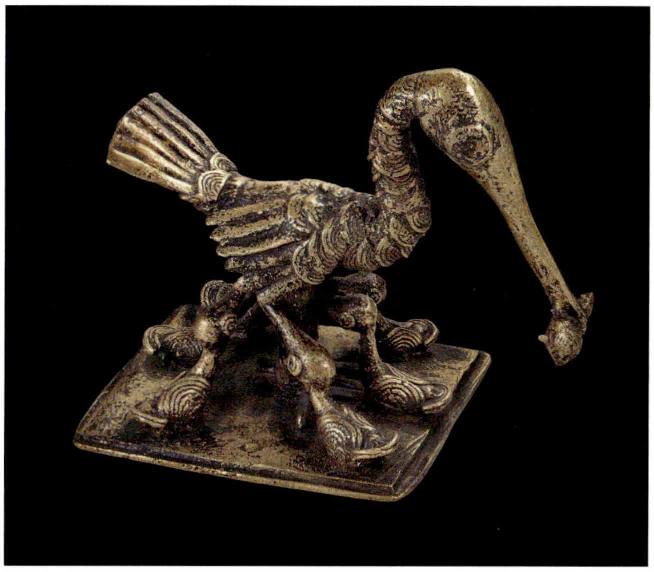

For Ed Vates and
John and Tommy Sterling:

Your understanding of our casting obsession made this book possible and your love and support has transformed our lives.

Akan Gold Weight, Bird and Its Chicks, nineteenth century. Akan culture, Africa, cast brass (0.14 to 0.19 g). Former collection of French botanist and ethnologist Jean Baur, donated to the Museum of Toulouse in 2010.
Photo: © Didier Descouens at the Museum of Toulouse / Wikimedia Commons / CC by 3.0.

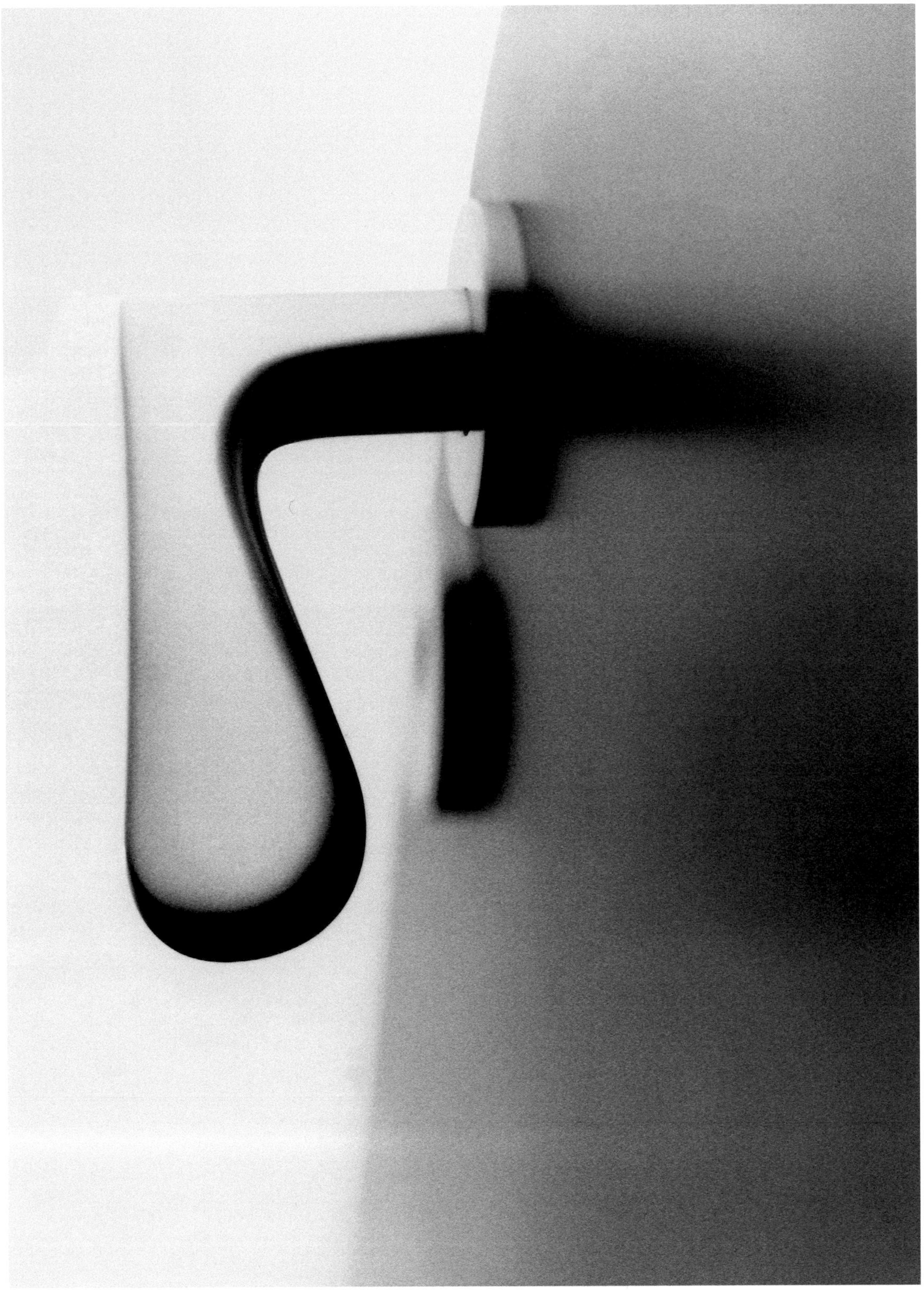

Above Jasper Morrison Ltd, produced by FSB, *Door Handle Series 1144*, Germany, 1990.
Photo: Hans Hansen/FSB.

Opposite page Recumbent Hare, eleventh century. Egypt, bronze, hollow cast, 1.75" × 3.5".
Courtesy of LACMA, Gift of The Ahmanson Foundation, www.lacma.org, LACMA public domain.

There are things known and there are things unknown, and in between are the doors of perception.

—Aldous Huxley

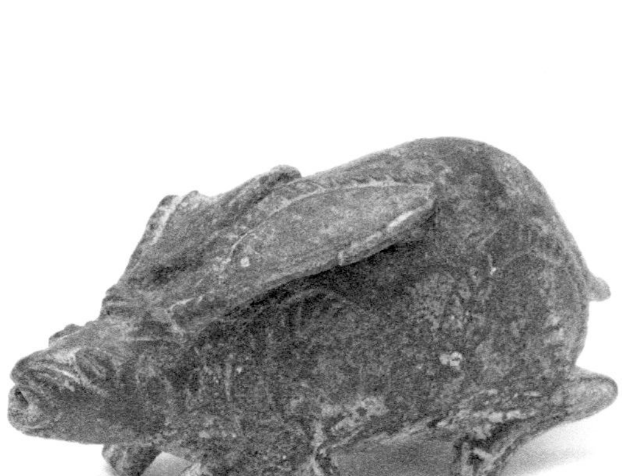

CONTENTS

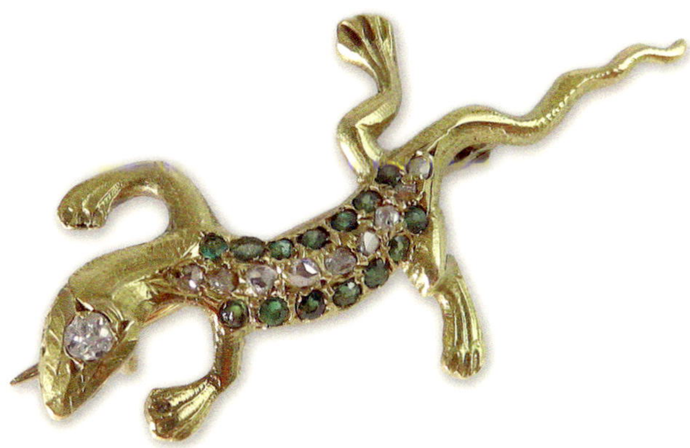

Diamond and Emerald Salamander Brooch, ca. 1880. 18K yellow gold, diamonds, emerald, cast and hand engraved, 1.4" × 0.5" × 0.3". Photo: Susan J. Sifert, Courtesy of Susan J. Sifert, Lind House Antiques.

PREFACE

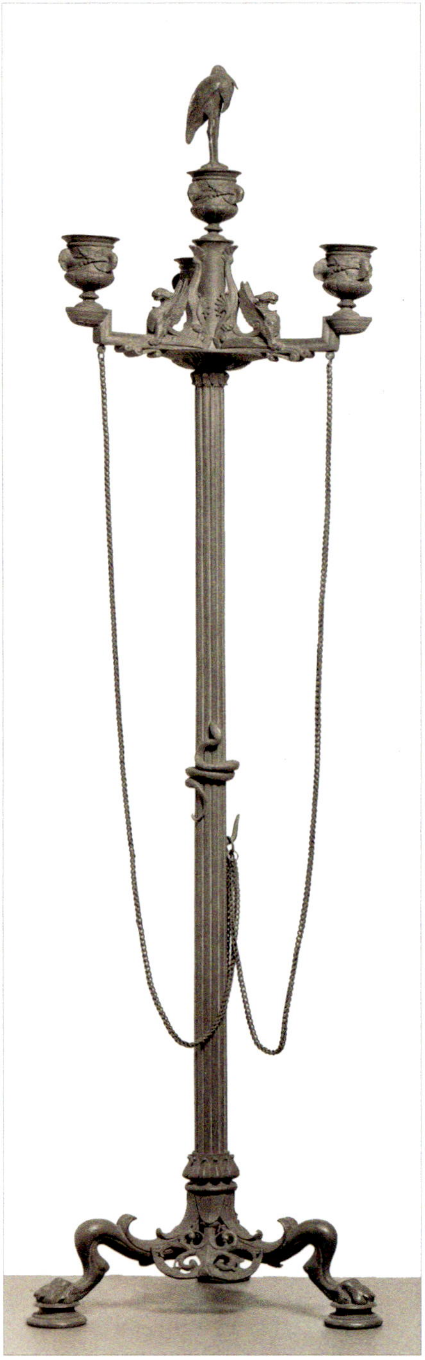

Antoine-Louis Barye, *Candelabra with Three Branches, Chains, and a Stork Finial*, nineteenth century. Paris, France, bronze, 31.8" tall.
Courtesy of The Walters Art Museum, Acquired by William T. or Henry Walters, CC-by-0, GNU FDL.

Genesis

This book developed through a series of conversations about casting, our love of the process, and our unrequited desire for great reading material about it. We have taught casting for many years and wished we had a book for students that showed casting's many strengths in the creation of art and objects. As artists who cast, we have wished for a book that we could look to for inspiration—a place to see the pinnacle of the process in the work we admire most. Many manuals address the technical side of casting and all of its complexities, but we have never seen one that reveres the process, shows its many unique and beautiful applications, and inspires makers to use casting in their own work. Since the book we longed for didn't exist, we decided to write it.

Initially, we planned to write a small book on casting in jewelry and metalsmithing. We are well versed in both the history and the contemporary landscape of our native field, but as our conversations unfolded and we considered the multitude of mediums used in contemporary jewelry making, we started to see the book's horizons expand.

Then came the conundrum of the odd and often inconsistent division between the sculpture and the metalsmithing traditions; the work can be interchangeable at times, but the roots of fine art and craft education are quite different. These divisions have always confounded both of us—but they seemed even more arbitrary as we began to realize that casting was impossible to compartmentalize in any way that made sense. We see merits in all areas of the process and wanted to take an inclusive approach to this book.

Why aren't there any books on the topic of casting as it relates to craft and fine art when it's such a pivotal process? While we see a spectrum of making rather than separate entities, the differences in the cultures of fine art and craft make us suspect our question may actually have two very different answers.

Craft: The Handmade vs. Industry

In general, craftspeople are defined by their materials and processes. It is not uncommon to hear them wax poetic about the importance of the handmade, the traditions associated with their techniques, the unique qualities of their chosen materials, their favorite tools, and their commitment to keeping ways of working alive that might otherwise be lost in the postindustrial age.

We (the authors) come from a jewelry and metalsmithing background. Our field is steeped in pride of process, but for many craftspeople, casting is resented rather than revered. It is taught in almost every academic institution, but can be treated as a necessary evil and is often taught with little know-how or affection. Why is this the case? Even though we use casting very differently in our own studio practices, our passion for the process is matched, and we have never understood the apathy (sometimes verging on distaste) for our favorite way of working. After much thought, we have come to the conclusion that this aversion most likely has to do with the industrial use of casting and our consumer culture.

Some say industry is the nemesis of craft. Mass production floods the world with cheap, poorly-made, disposable "stuff," creating expectations in the public about the value (or lack thereof) of objects. Industry has undeniably put craftspeople in a very tough spot. As a jeweler, how can you compete with mass-produced work being sold for barely your material cost? How can a ceramicist compete with $1 mugs? The world has been oversaturated with manufactured objects since the Industrial Revolution. This glut of mass-production has distanced the public from how things are made; the process, materials, and labor invested in the design and creation of objects have become a mystery to non-makers, leaving craftspeople trying to find their footing.

To further complicate the matter, there is a wide range of quality available in mass-produced cast objects—some are poorly made, but there are also many finely designed and well-produced items that are hard to compete with for a maker with a small studio practice. Craftspeople are therefore challenged to educate their audience about the value of their work. What is the value of pieces that are well considered, unique, finely made, loved, and labored-over?

Industry does some things well and some things poorly, but one of its undeniable strengths is casting. American malls are filled with affordable and functional cast jewelry, glass vases, and ceramic dinnerware. Big-box stores are filled with rotocast- and injection-molded plastic items, from kitchenwares to toys. Industry has mastered casting.

When you consider these ideas, it is easy to see how there is a resentment toward industry (especially casting) within the crafts.

Stepping away from the perspective of the craftsperson, industry's contribution to humanity through casting is undeniable. Take the simple yet indispensable toilet as an example. Modern life as we know it—with basic sanitation—is defined by casting.

Because casting is so linked to industry today, modern craftspeople have a complicated relationship to it. Many makers have therefore chosen to focus on processes that do not lend themselves to industrial applications as easily. It is common at contemporary craft fairs, for example, to hear an exhibitor proudly proclaim that "none of their work is cast." We would argue that this is a myopic thing to say.

Casting is a process like any other. It can be executed and applied well or poorly, it can be the right or the wrong choice in the creation of any given work, but it offers certain undeniable possibilities that can't be achieved any other way. It is our hope to illustrate this point and create a renewed interest in the process by showing the incredible depth and breadth that it has to offer. When we look at a piece of jewelry like *Rapture* by Lin Stanionis, we don't see how casting could be anything but worshipped. The same goes for the works of Zsolt József Simon, Jamie Harris, and David Umemoto. These makers use casting in such a skillful way that their work could never be the same without it.

Fine Art: Idea vs. Object

Turning from craft to fine art, we suspect there are other reasons behind a lack of books on casting. Ever since conceptual art took hold in the mid-twentieth century, there has been an elevation of "idea over object." This value system has affected everything from the education of fine art students, art historians, and critics to what is studied, discussed, and written about in fine art journals and books. Because technique has become a footnote, there are many sculptors graduating with MFAs who are wonderful thinkers, but haven't actually been educated in the creation of objects. Craftsmanship is often seen as secondary and is sometimes dismissed as menial and prosaic. If process is craft's pride, it's fine art's dirty little secret.

We believe, though, that truly exceptional fine artists must understand and value the technical. It's through process that strong, well-developed ideas come to fruition. What would Marc Quinn's work be without an extensive understanding of all aspects of casting? What of the brilliance of Dario Robleto's "alloying" of bizarre, touching, and profound materials to create completely unique casting mediums imbued with meaning?

The most interesting works, whether in the fine or the decorative arts, are those that are well developed in conception and adroitly executed. Ai Weiwei's *Sunflower Seeds* is one such piece. *Sunflower Seeds* is an installation piece made up of 100 million handmade cast porcelain sunflower seeds, each painted and finished individually, and collectively weighing over 150 tons. They are displayed on the floor in a 4.5-inch-deep "carpet" intended to be walked over. The seeds were made by 1,600 Chinese workers in the town of Jingdezhen over the course of two and a half years. The piece comments on mass production and the Western demand for cheap Chinese labor, individuals (represented by the seeds) contributing to (but getting lost in) a huge group, the power of groups, the potential of seeds to grow, and on and on. Sheena Wagstaff, the chief curator at the Tate Modern (one of the venues where the work was displayed), has commented that the work incorporates "an epic sense of scale with an exquisite level of craftsmanship."* This piece in particular comments on so many areas of interest to us—the importance of molds, multiples, industry, craft, fine art, and the fact that the piece is entirely dependent on casting. It touches on so many of the reasons we feel the topic of casting is important for makers of all kinds and the public at large; makers are liberated through an understanding of process and are then able to actualize ideas to their best effect.

Everyone will benefit from thinking about how objects are made, the implications of mass production, appreciating what goes into the creation of the objects we interact with every day, and understanding the value of the handmade. As importantly, we hope we have made a beautiful book that will inspire anyone who picks it up.

We have been stunned by our research into casting; the breadth and depth of its history and impact are too vast to capture in a single book. Even with more than 800 images, we feel we've barely scratched the surface of this behemoth topic.

At the heart of all of this, we are just two people who are passionate about casting. We are drawn to the cast object. There is a kind of magic—almost that of alchemy—when a bit of ephemera is turned into bronze, that has enchanted both of us since the first moment we tried it. We hope to share with you our love of the process, some of our favorite cast works, and how casting has transformed our world with the objects that inhabit it.

Jen Townsend and
Renée Zettle-Sterling

* Faurschou Foundation website, "Ai Weiwei Sunflower Seeds." 20 July 2016. http://www.aiweiweiseeds.com/about-ai-weiweis-sunflower-seeds. This website also offers detailed information about the piece.

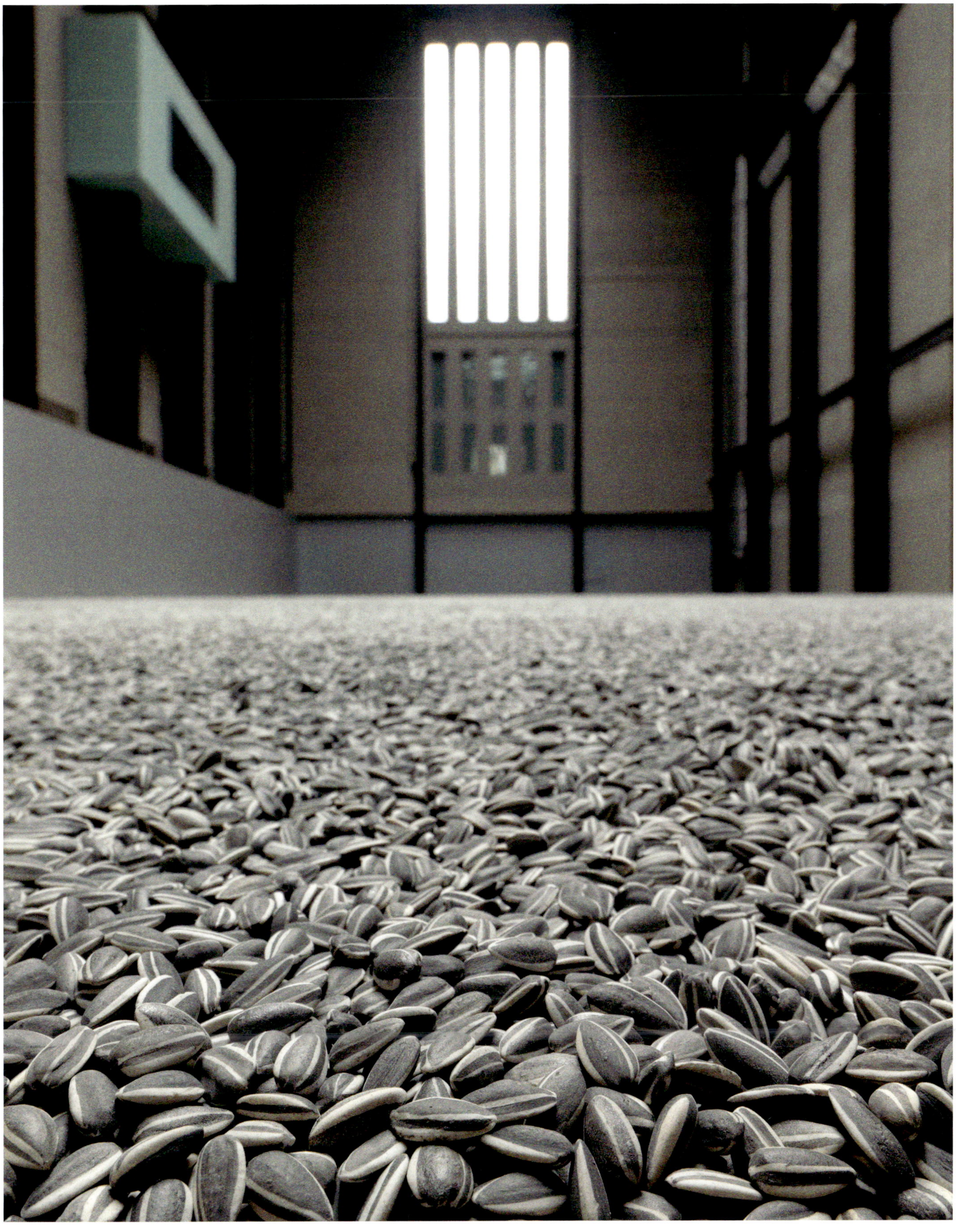

Ai Weiwei, *Sunflower Seeds,* installed at the Tate from October 12, 2010 to May 2, 2011. Porcelain.
Photo: Tate Photography 2010.

PROLOGUE

Jen Townsend

Casting has always been my favorite way to work. Carving and modeling wax line up with the way I envision building forms in my mind—both additive and reductive, organic, fluid, and dynamic. Casting offers me a way to manifest my ideas in all their dimensionality.

I've always had one foot in the academic world and the other in the world of commercial jewelry, and have seen how each realm approaches casting. I have made my living creating custom fine jewelry, but I've always thought in terms of composition, sculptural form, and concept. My passion for art influences every piece I make, whether it's an engagement ring, a piece of art jewelry, or a sculpture.

I first thought about writing this book in graduate school when I was casting bronze, silver, rubber, and plaster. It was one of those seeds that was planted long ago, but had to wait for the right conditions to sprout. Those conditions became favorable in 2012 when Renée and I began discussing our mutual passion for casting.

Renée uses casting in a very different way than I do, but her love for the process matches my own. Her poignant use of casting found objects is thrilling to me because it's so different than the way I have always worked. From 2012 to 2014, we had endless conversations about teaching casting, our own work, and other cast work we admired...our zeal for returning our favorite process to its rightful place as a revered process was building. We bemoaned the lack of inspiring books on the topic, and eventually, I popped the question... "Since the book we both want to own doesn't exist, do you want to write it with me?"

Writing *Cast* has been one of the most grueling and thrilling things I've ever done, but working with Renée has been a delight. I hope our passion for casting is contagious and evident in this, our love letter to our favorite process.

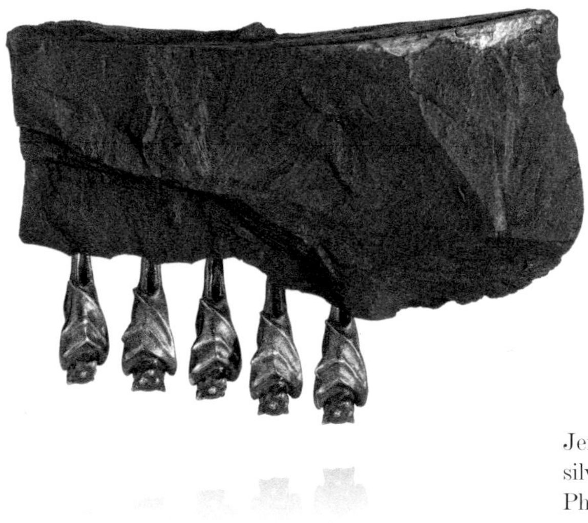

Jen Townsend, *Sleeping Bats*, 2014. Jet, sterling silver, 3" × 2" × 0.5".
Photo: Hank Drew.

Renée Zettle-Sterling

Material culture has always inspired my work. The idea that casting could transform and deepen our relationship with material culture is so profound that I've never stopped casting. It has always excited me that I could take a found object and transform it into something else entirely: a piece of artwork that could express strong emotions and the indefinable. Casting has allowed me to imbue a found object with new meaning while still maintaining its original form and context.

I teach many different courses as a professor, but the most demanding, challenging, and rewarding one is the casting course. Nothing gives me more pleasure than sharing casting techniques and the importance of casting (not only in the jewelry world, but in our daily lives) with my students.

Writing this book has been a learning adventure from start to finish. It was, at times, mentally draining, but conversely, invigorating, hilarious, exciting, and joyful. For me, working with Jen has been the most rewarding part of this process, and even if *Cast* had never gone to print, it would still have been worth it.

Although Jen and I approach casting in very different ways, we both have an undeniable and unalloyed passion for the process. We hope *Cast* will fill the gap in available books that cover casting the way we see it: beyond its technical aspects. It was our hope to write a book that we would want to own and one that would inspire readers to look at this ancient, and still relevant, process in a new light.

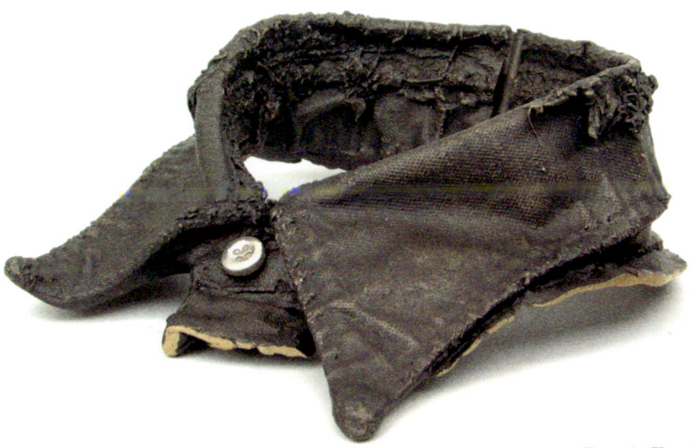

Renée Zettle-Sterling, *Object of Mourning: Impermanence #2* (neckpiece), 2013. Bronze, silver, copper, direct cast of her brother's shirt collar (Tommy Lanigan Zettle: January 14, 1975–May 4, 2002), 3.75" × 5.75" × 7.25". Photo: Renée Zettle-Sterling.

INTRODUCTION

Karen LaMonte, *Kabuki*, 2012. Bronze, 60.23" × 31.88" × 32.67".
Photo: Martin Polak, © 2012 Karen LaMonte.

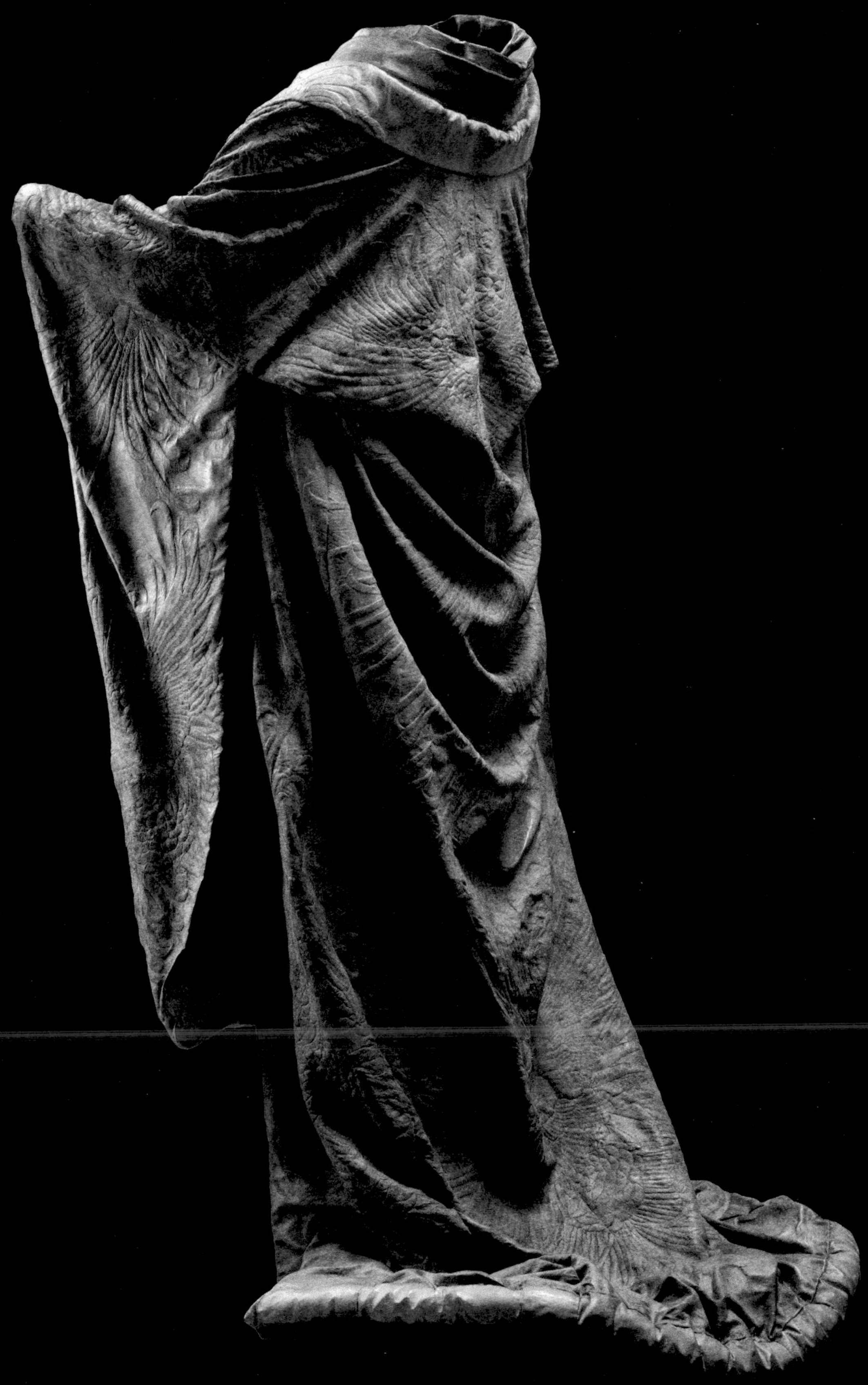

Casting has changed human existence in ways that are hard to overstate. We only have to consider common linguistic references like "breaking the mold," "seamless," and "one-of-a-kind" to see how deeply ingrained the process is in our collective psyche. Although these words roll off the tongue without a second thought, casting's influence is in every corner of our lives. Although it often goes unnoticed, casting has propelled civilization and transformed our world to such an extent that would make it unrecognizable to our ancestors.

Cast examines casting in all its materials and manifestations. Inclusivity is a key philosophy behind this book. There are certain commonalities in approaching design and manufacturing (whether Greco-Roman bronze figure or plastic monobloc chair) that are fascinating food for thought for anyone interested in the process. While *Cast* touches on industry, a special focus is given to the fine and decorative arts.

History and Impact looks at the development of the casting process since its inception and what it has meant to civilization. There have been certain pivotal advancements that have irrevocably changed life as we know it. From cannons to dentures and concrete blocks to rubber shoe soles, casting is everywhere and once seen, cannot be unseen.

Casting as It Relates to the History of Fine and Decorative Arts looks at casting's unique role in our aesthetic history. Suzanne Ramljak introduces the chapter with an insightful look at molds and reproduction and the part they play in art history. The gallery features a survey of fascinating objects from 2000 BCE to the present day.

Creation delves into the act of casting and the mindset of casters. This is an exploration of the way casters think and their practices. Peek into the Caster's Studio, the gallery component of this chapter, features seventeen exceptional artists who have opened their studios to share a glimpse at their varied processes.

Metal, Ceramics, and Glass look at specific mediums. These chapters feature rich and varied historic objects and exciting contemporary examples of work that offer visually stunning and thought-provoking pairings. Joseph Becherer (metal), Ezra Shales (ceramics), and Susie J. Silbert (glass) present finely honed perspectives on each medium. Each chapter features a robust gallery full of diverse and compelling work ranging from the historic and traditional to the contemporary and avant-garde.

Jewelry is its own chapter since it is a peculiar genre that includes a multitude of materials. Jewelry is so diverse, and demonstrates so many ways to use casting, that it demands special consideration. Jen Townsend explores this in her essay, and the gallery has a stunning array of jewelry from the quiet and wearable to the bold, quirky, and conceptual.

(Im)material investigates everything else. Humans have cast just about every material that can be poured, pressed, or melted into molds. Elaine A. King opens the chapter and explores the broadening field of sculpture as the use of alternative materials becomes more and more mainstream. This chapter looks at cast objects throughout history. These objects and their media are fascinating in their differences as well as their commonalities. There are two galleries in this chapter; the first, Gallery: Spectrum, is a survey of objects cast in materials from plaster to plastic; the second, Gallery: Unconventional Approach, features work by artists pushing the boundaries of the process. The galleries are especially rich in this chapter, with astonishing image pairings worthy of contemplation.

The Alchemy of Casting explores the mysterious and spellbinding nature of casting and our primordial attraction to it. The transformative nature of pouring molten bronze into a cavity can be magical at times…preserving the ephemeral or giving form to the void are both profound acts of creation. The second part of this chapter, Meaningful Materials, looks at artists who use the nature of the process to imbue their works with concepts through the casting materials they create, such as Marc Quinn's piece, *Nicholas Grogan – Insulin (Diabetes)*, cast in polymer wax and the drugs that saved the model's life. When artists thoughtfully consider and craft their casting mediums, it adds a conceptual underpinning to the work that's impossible to achieve any other way.

Cast shines a light on all the overlooked ways that casting is present in our lives and investigates our history as well. Contemporary casters are part of a continuum of thousands of years of makers and through the collective history of the process, there is much to learn. This book seeks to inspire makers in every casting discipline through the cross-pollination of ideas, through a reminder of our place in history, and by offering a glimpse at the exciting future of the process.

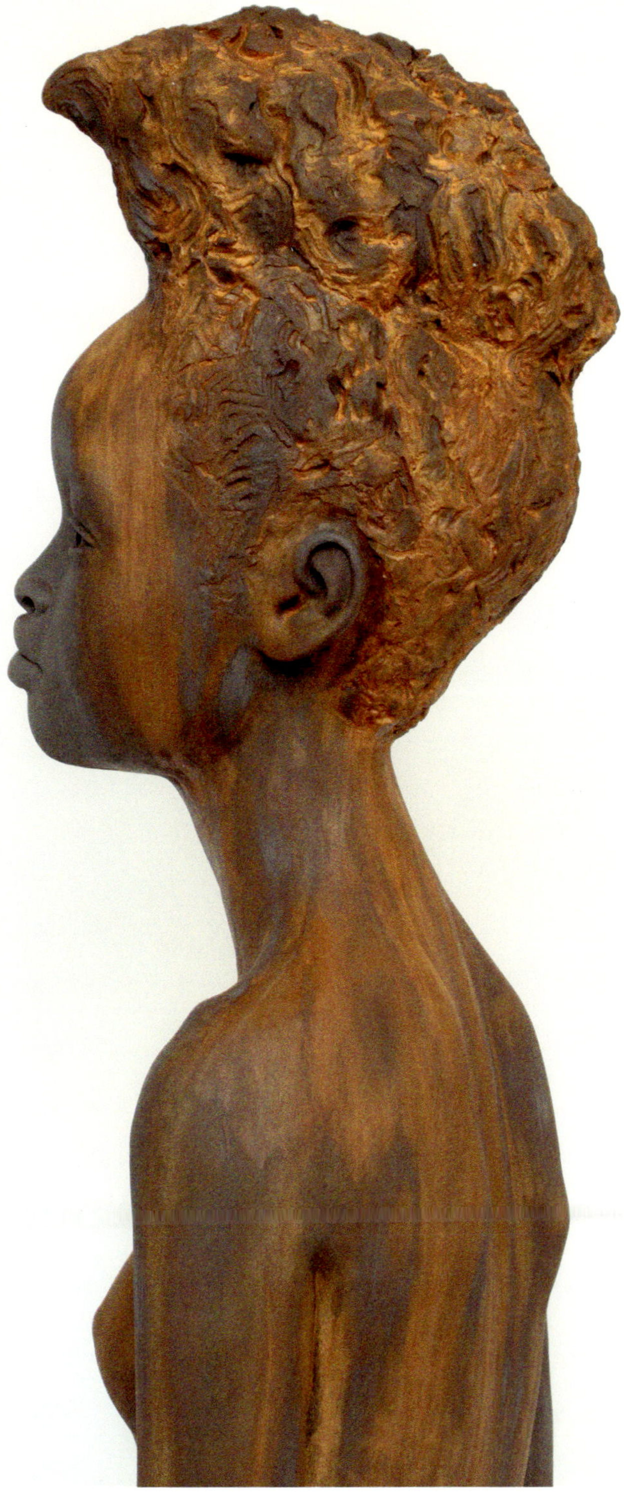

Gosia, *Open Your Eyes–Rust Edition–2/5*, 2014. Polymer, gypsum, and patina, cast and patinated, 9" × 22" × 8". Photo: Gosia, gosia.ca.

HISTORY and IMPACT

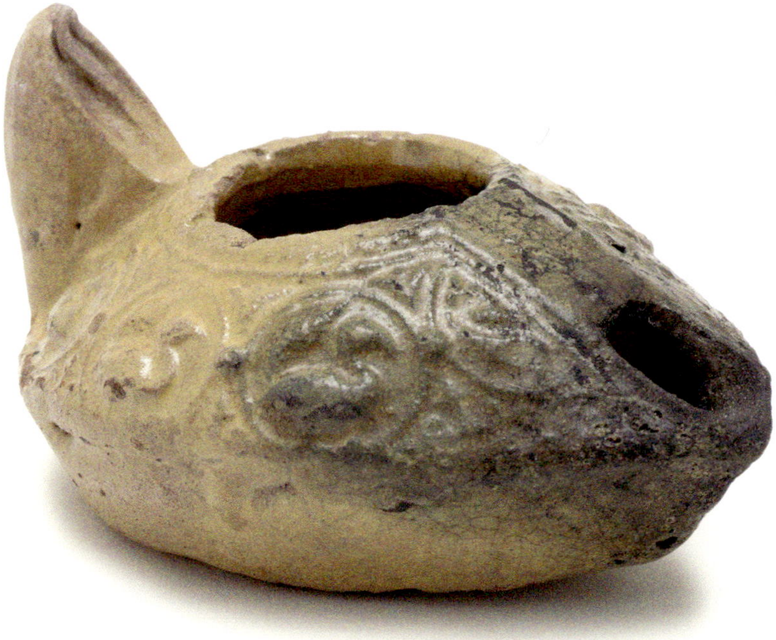

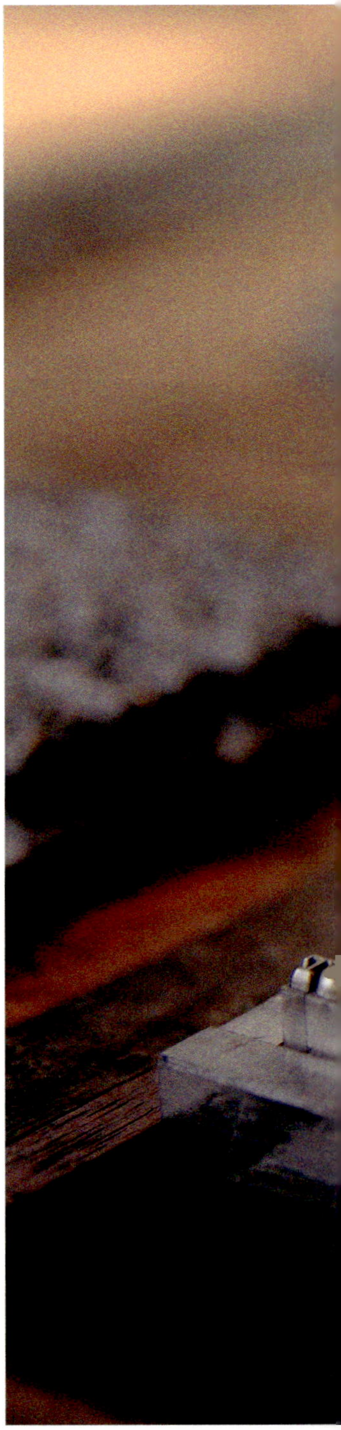

Oil lamp, seventh to eighth century. Egypt, ceramics (earthenware), 1.31" × 3.5".
Courtesy of LACMA, Gift of Camilla Chandler Frost, www.lacma.org, LACMA public domain.

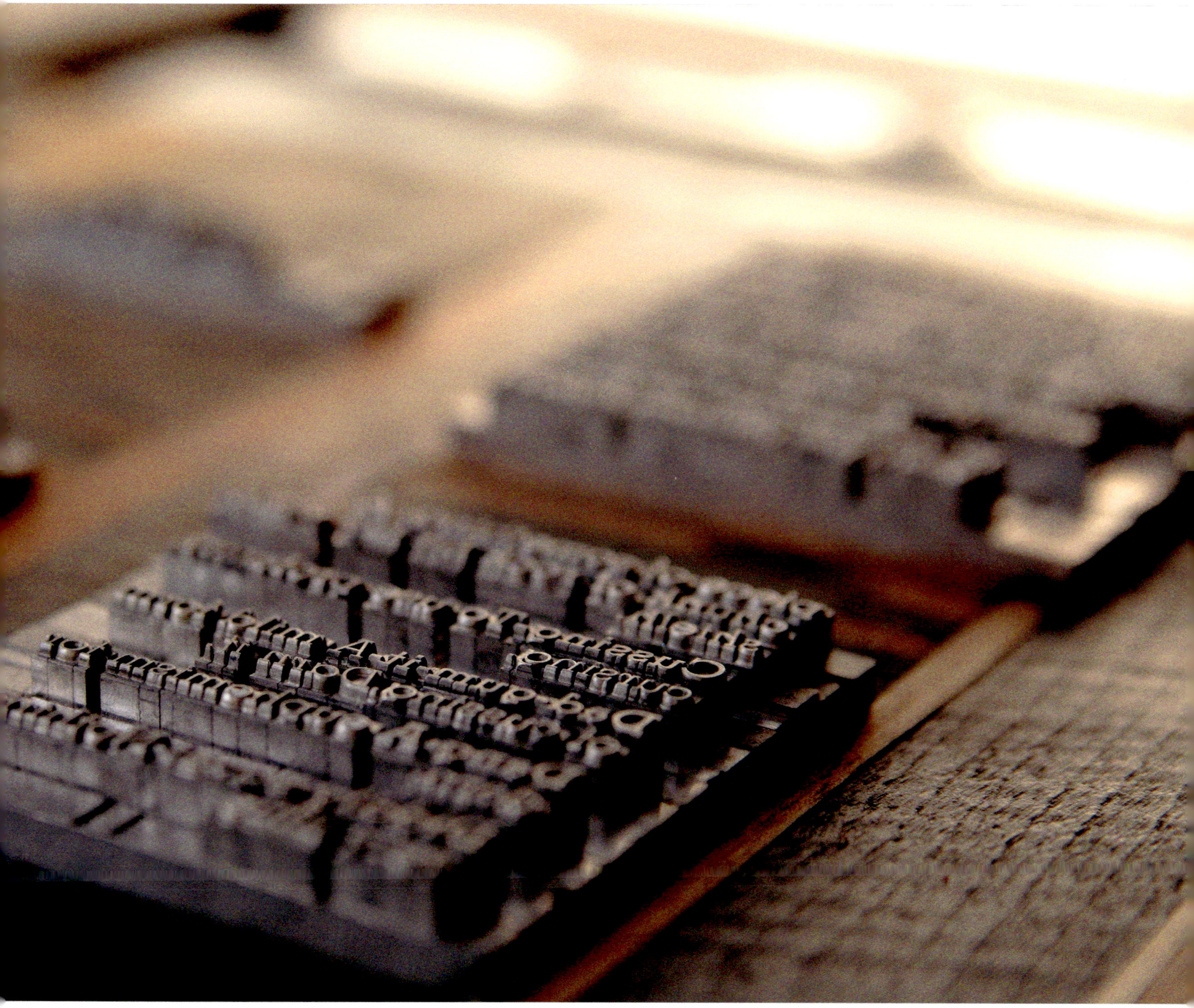

Letterpress metal type secured in a chase before inking. Metal.
Photo: © Alice Day, Depositphotos.

UNDERSTANDING OUR HISTORY

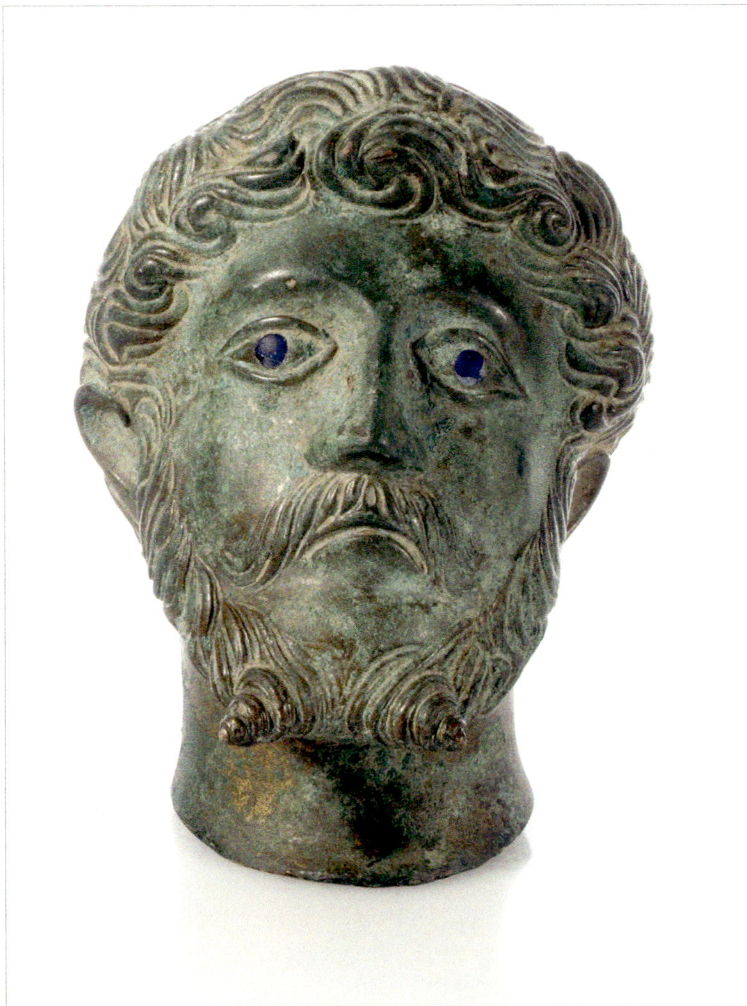

Bust of Marcus Aurelius, 100 to 200 CE. Britain (Romano-British Workshop), primary material: copper alloy, secondary material: glass, 6.3" × 4.75".
Courtesy of Portable Antiquities Scheme / CC-by-2.0 / The British Museum.

Some of the most pivotal objects in human history were cast. The careful work of archaeologists and historians has helped us to understand these objects and their significance. The humble press-molded ceramic oil lamp extended our ancestors' ability to see in the darkness of night. Cast metal type allowed the mass dissemination of information for the first time, enabling literacy, education, and elevated thought. Rulers long dead are made human once more through their bronze likenesses, their power and charisma harnessed and captured in perpetuity.

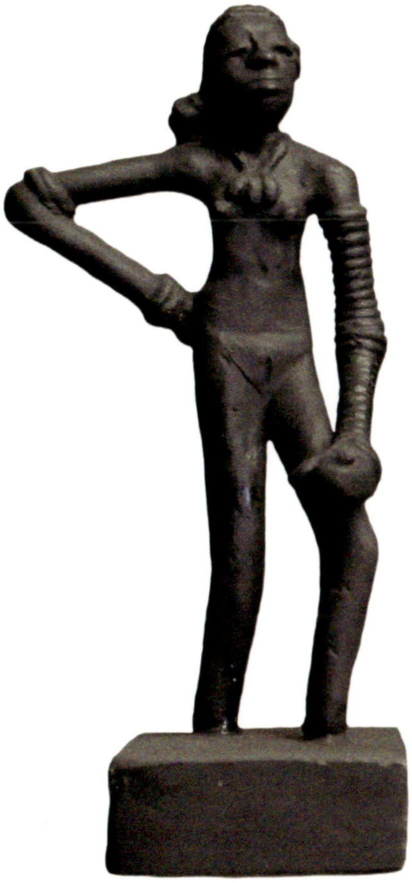

These potent objects speak to cultural shifts with far-reaching implications, but arguably, connecting with our own humanity is equally important. Objects like jewelry remind us of our connection with our ancestors. Whether a Viking arm band worn to war or a portrayal of a young Indian girl covered in bangles, these objects give us a glimpse of life thousands of years ago and remind us that we are part of a continuum...one of living, making...and casting.

Left One of the oldest bronze figures found to date, Dancing Girl exhibits a deep knowledge of both metallurgy and artistry.

Dancing Girl, 2500 BCE. India (found at the Mohenjo-daro site), bronze, 4.1" tall.
Photo: © Joe Ravi / Wikimedia Commons / CC-by-SA-3.0 / Chhatrapati Shivaji Maharaj Vastu Sangrahalaya (formerly Prince of Wales Museum of Western India).

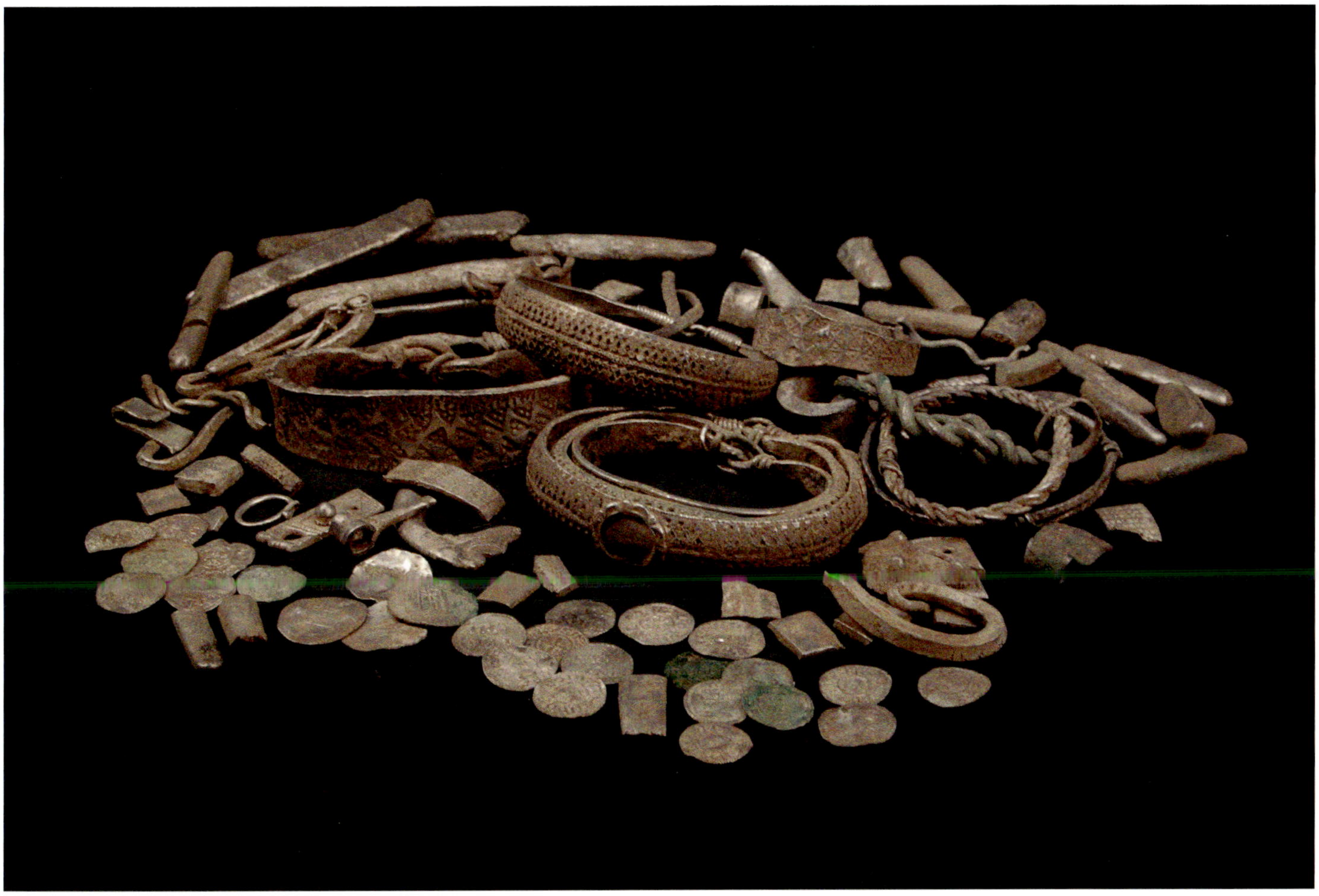

Above One of the largest Viking hoards ever discovered in the UK, the Silverdale Hoard includes over 200 pieces of silver jewelry and coins. Discovered by Darren Webster, a metal detectorist. It was purchased by the Lancashire Museums Service.
The Silverdale Hoard, 900 to 910 CE. Viking, found in Lancashire, UK by Darren Webster, silver.
Courtesy of Portable Antiquities Scheme / CC-by-2.0 / Lancashire Museums.

RELIGION

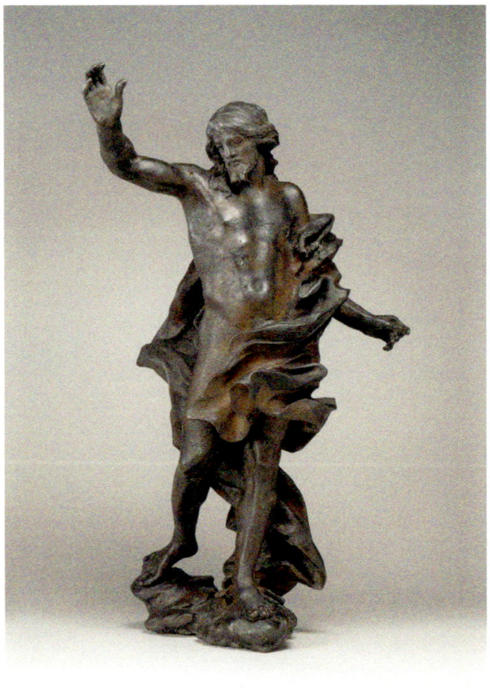

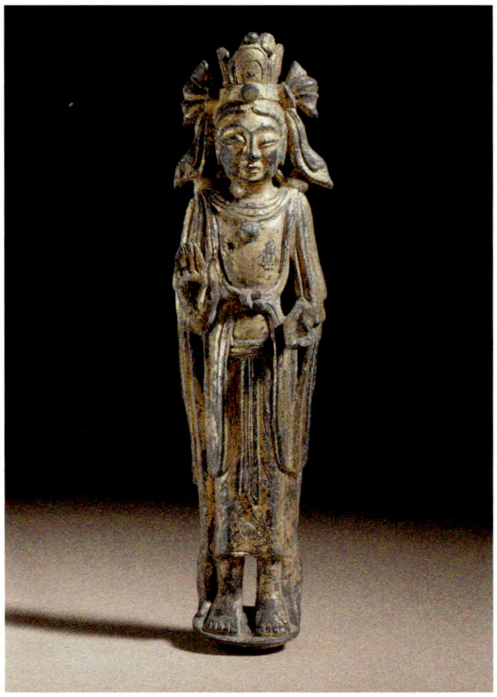

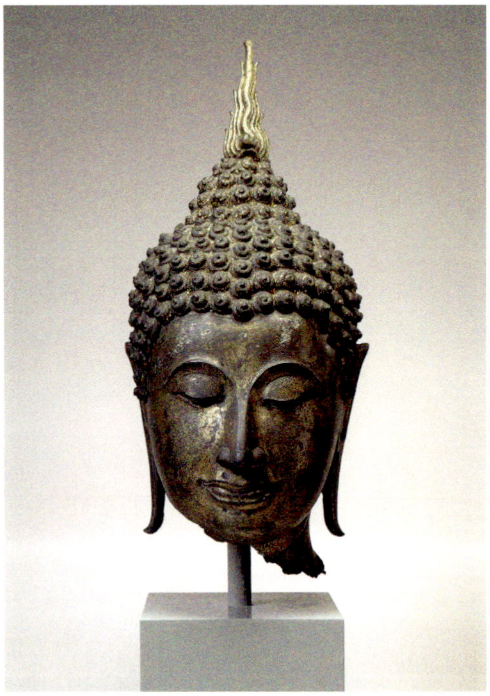

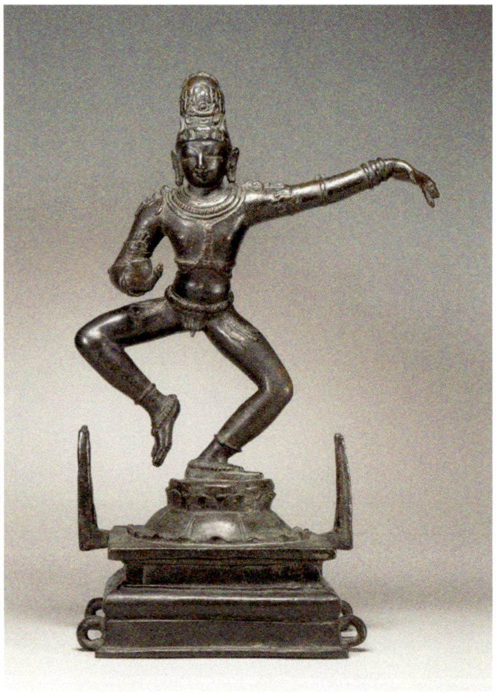

Top left Gian Lorenzo Bernini, *The Risen Christ*, 1673–1674. Bronze, 17.3" tall.
Courtesy of The Walters Art Museum, CC0, GNU.

Top right Probably Avalokiteshvara (Guanyin), the Bodhisattva of Mercy, 591 to 618 CE, Sui dynasty. China, cast bronze with gilding, 6.5" × 1.5" × 1.5". Courtesy of LACMA, Gift of Mr. and Mrs. Eric Lidow, www.lacma.org, LACMA public domain. FDL.

Bottom left Head of Buddha, second half of fourteenth century. Thailand, leaded bronze, traces of gilding, 17.62". Courtesy of The Walters Art Museum, CC-by-0, GNU FDL.

Bottom right Krishna, late tenth century. India (place of origin), bronze.
Courtesy of The Walters Art Museum, CC-by-0.

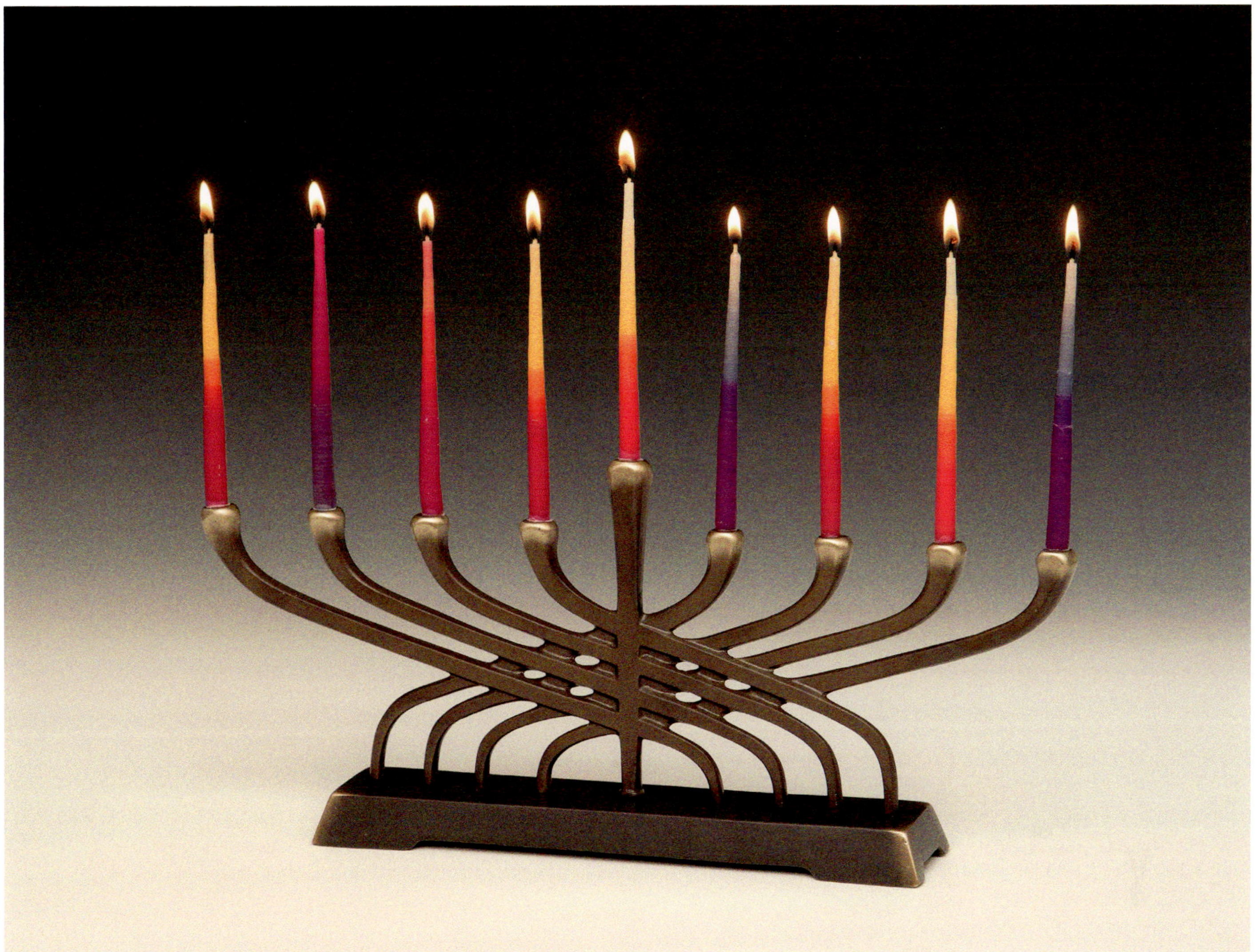

Scott Nelles, *Contemporary Menorah*, 2012. Bronze, 7" × 15" × 2.5". Photo: Don Rutt.

Casting has been used in sacred spaces for thousands of years. Whether for church, temple, or shrine, lovingly made cast objects often play a role in creating the environment for worship. From crude bronze votive figures to Bernini's sophisticated and dynamic *Risen Christ*, people have made a wide range of objects in service of their faith. Many traditions center around images of the divine in human form—Christ, Buddha, Bodhisattva, and Krishna all inspire devotion, piety, and zeal in their millions of followers. Before literacy was common, these figures were used to tell religious stories, educate the faithful, protect and honor relics, and embody the transcendent here on earth.

Religions also have many nonfigurative devotional objects associated with them. Since casting provides a way to produce—and reproduce—objects of all kinds, it's a logical means of production for denominations with millions of followers. Cast objects create a point of access to faith allowing organized religions to expand their following. Devotional objects like menorahs, medals, rosaries, bells, and chalices have helped people to stay connected to their beliefs for thousands of years.

COMMERCE

Cast metal plays a significant role in the history of commerce around the world. Whether in the form of early coins, gold weights, stamps, or today's gold ingots, cast metal has been crucial ever since it was first poured into a mold.

Gold weights, used in Africa for centuries as a standard measure, were cast in a variety of different forms. These were used to ensure fairness in trade and provided a universally agreed upon weight against which other goods could be measured.

Objects like stamps were used to seal everything from documents to casks of wine. The wax seal would assure recipients that letters remained unread or that their goods had not been corrupted. This was an early form of branding and a way people could identify the source of a particular item, allowing vendors to build identifiable businesses.

Casting allowed for the mass production of early coins (they are now die-struck). Currency made wealth portable and easy to quantify. Values became standardized and numeric.

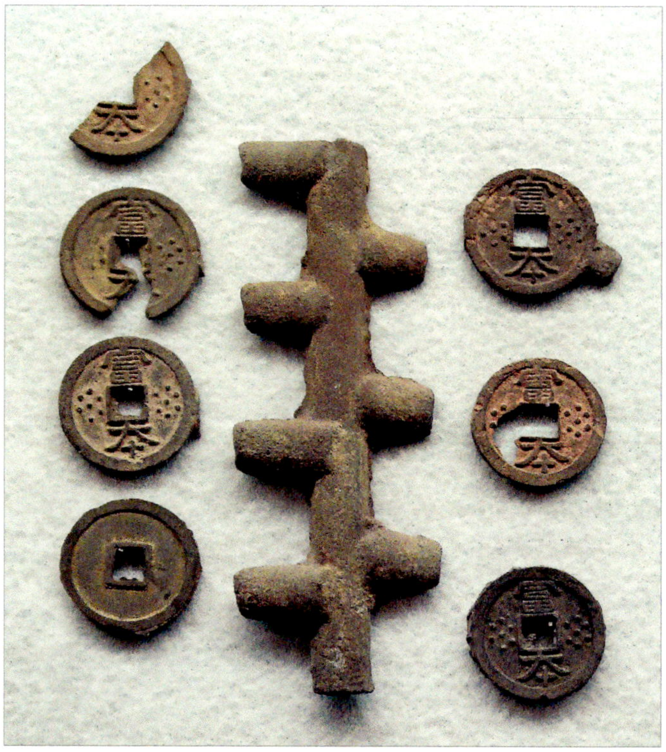

Early Japanese minting: Tomimotosen, end of seventh century. Found in Tobishimaike, alloy of copper and antimony.
Photo: © PHGCOM at the Tokyo Currency Museum / Wikimedia Commons / CC-by-3.0.

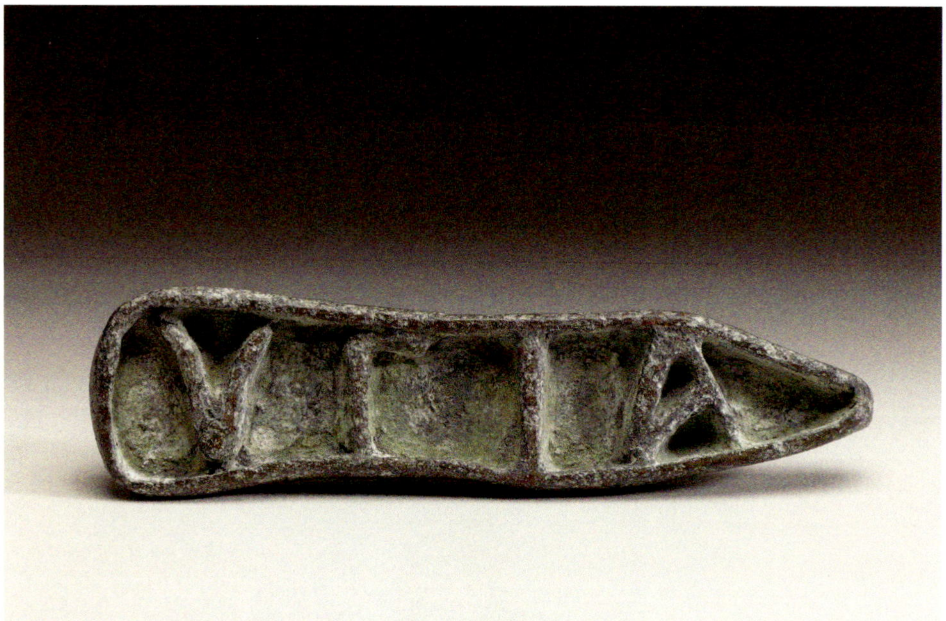

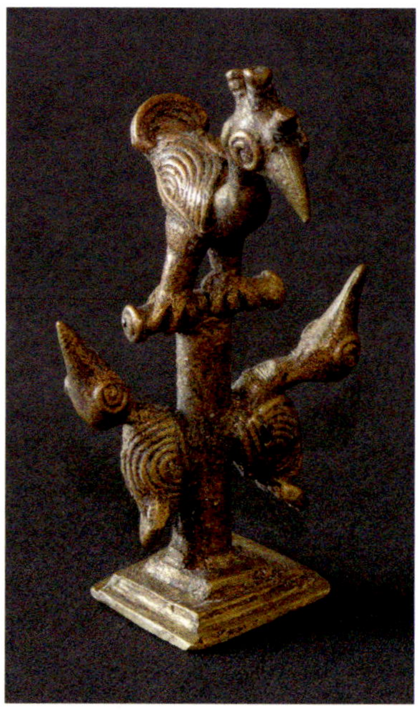

Commercial stamp, sixth to eighth century (Late Antique), Byzantine, bronze, 0.93" × 3.43" × 0.87".
Courtesy of The Walters Art Museum, CC0, GNU FDL.

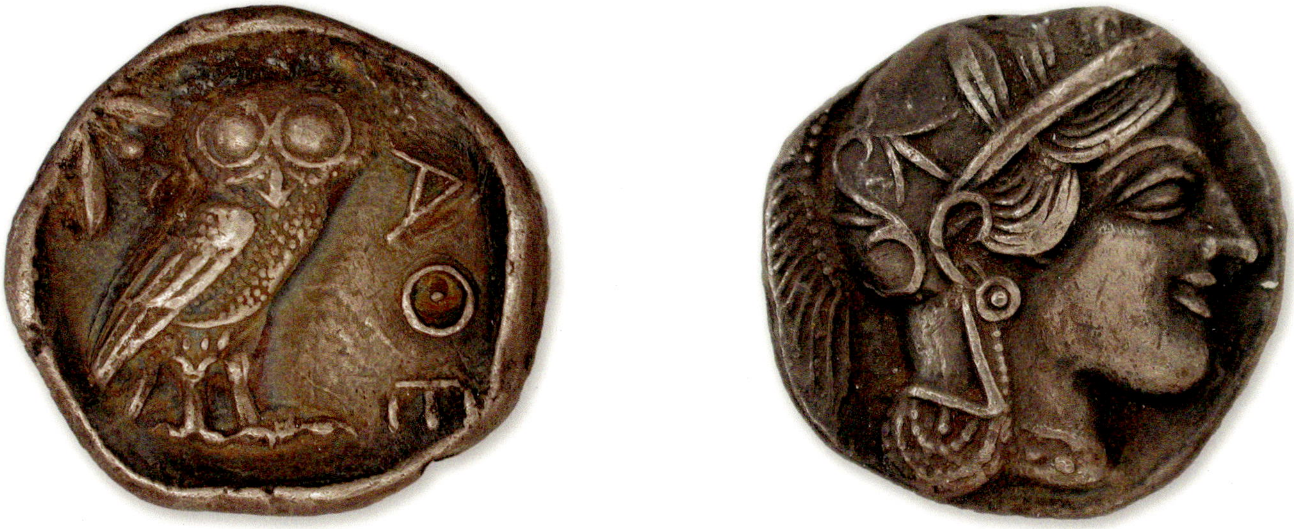

Front and back of Macedonian coin, ca. 450 BCE. Macedonia, silver.
Courtesy of LACMA, The Phil Berg Collection, www.lacma.org, LACMA public domain.

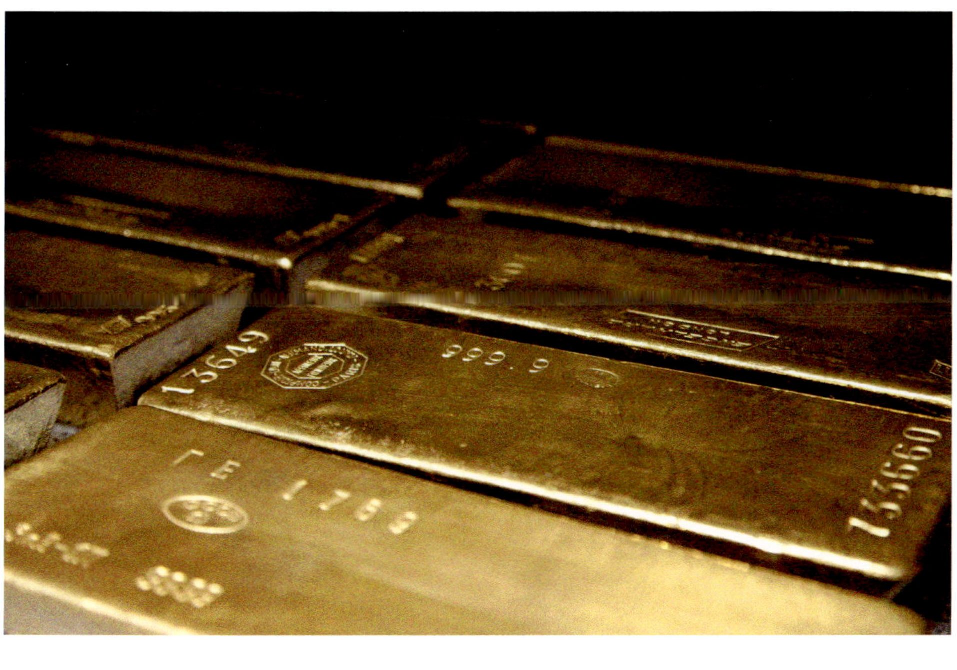

Left Standard 400-oz. gold bars, 0.999% pure gold.
Photo: © Andrzej Barabasz / Wikimedia Commons / CC-by-4.0.

Opposite page, bottom right Akan Gold Weight: Three Birds, nineteenth century. Akan kingdoms of southern Ghana and Ivory Coast, brass. Former collection of French botanist and ethnologist Jean Baur, donated to the Museum of Toulouse in 2010.
Photo: © Pierre Selim at the Museum of Toulouse / Wikimedia Commons / CC-by-3.0.

POWER and PROPAGANDA

Castings made in service of power and propaganda have changed the course of human history, sometimes in tangible (and bloody) ways, but other times, their impact has been more mercurial.

Personal weapons, the earliest expressions of human power, were key in both hunting and self-defense. Cast bronze daggers, axe heads, and arrows (all developments of the Bronze Age) were far more lethal than their Stone Age counterparts. En masse, armed individuals became formidable forces and the scope of tribal struggles expanded, ultimately evolving into states, kingdoms, empires, and the rise of their formal armies.

The Chinese were the leaders in the development of weaponry, creating incredibly sophisticated bronze crossbow triggers, as well as the first cannons, cannonballs, and bullets. These innovations spread to the Middle East and eventually to Europe, changing the face of warfare forever by making battles a matter of mass destruction rather than hand-to-hand combat.

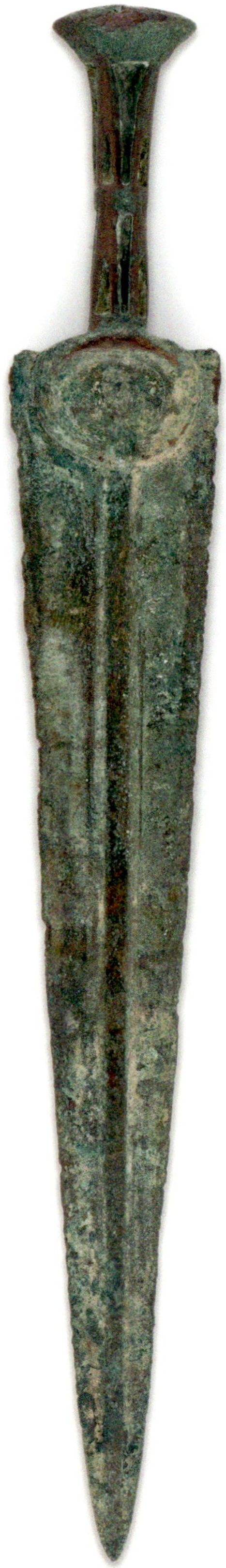

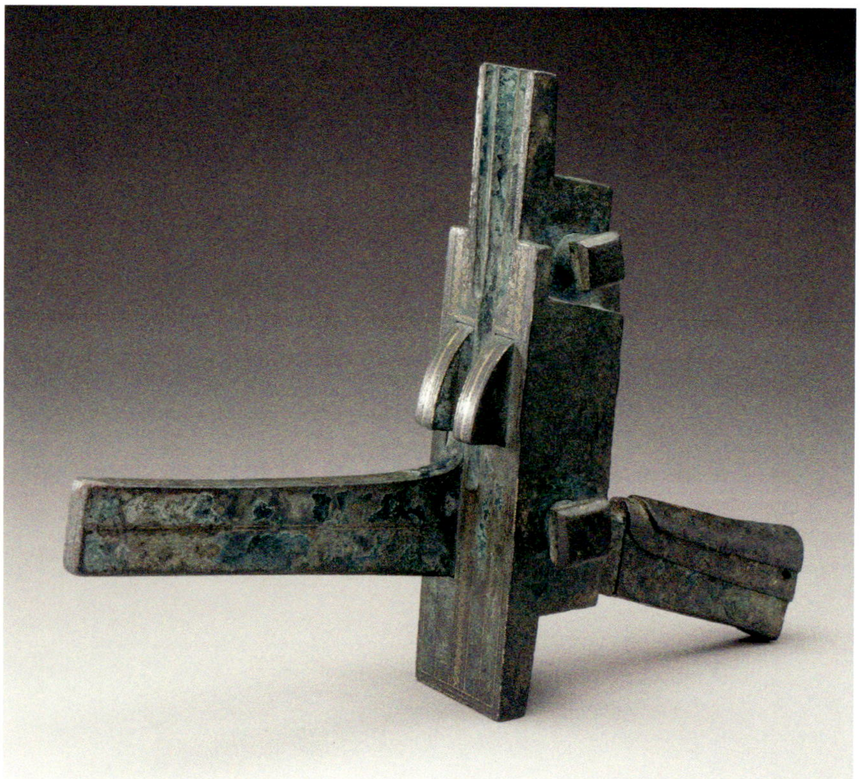

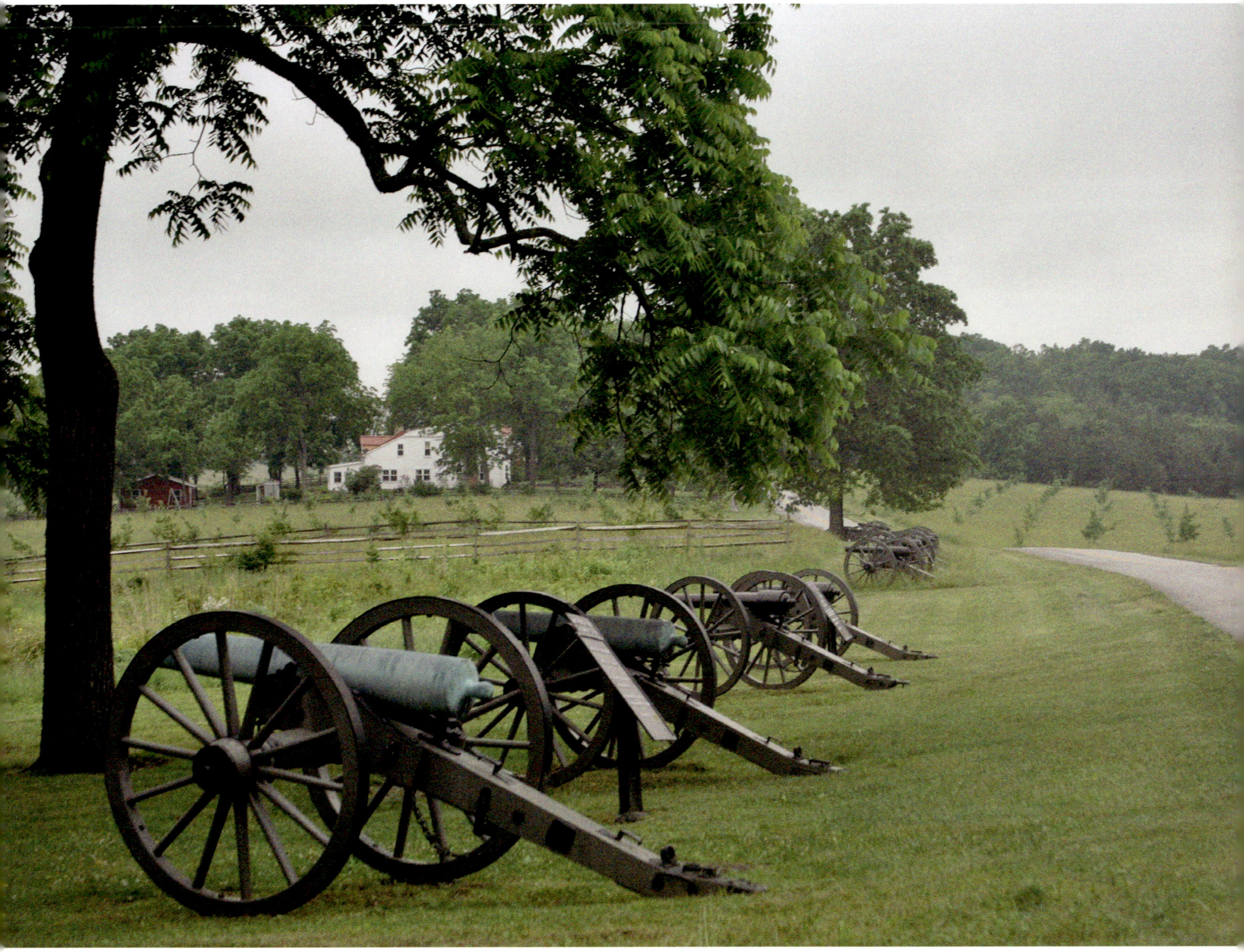

Civil War cannons. Gettysburg, Pennsylvania, USA, original bronze tubes, cast iron replica carriages. Gettysburg National Military Park. Photo: © Dragonfly819 at the Gettysburg National Military Park/ Wikimedia Commons / CC-by-3.0.

OPPOSITE PAGE

Top right Dagger, ca. 1350 to 1000 BCE. Northern Iran, bronze, overall length: 16.25"; hilt length: 3.5"; blade length: 12.5". Courtesy of LACMA, Gift of The Ahmanson Foundation, www.lacma.org, LACMA public domain.

Bottom left Crossbow trigger (Nugong), ca. 206 BCE to 25 CE. China, Western Han dynasty, cast bronze with gold and silver inlay, 9.25" × 7". Courtesy of LACMA, Gift of Mr. and Mrs. Eric Lidow, www.lacma.org, LACMA public domain.

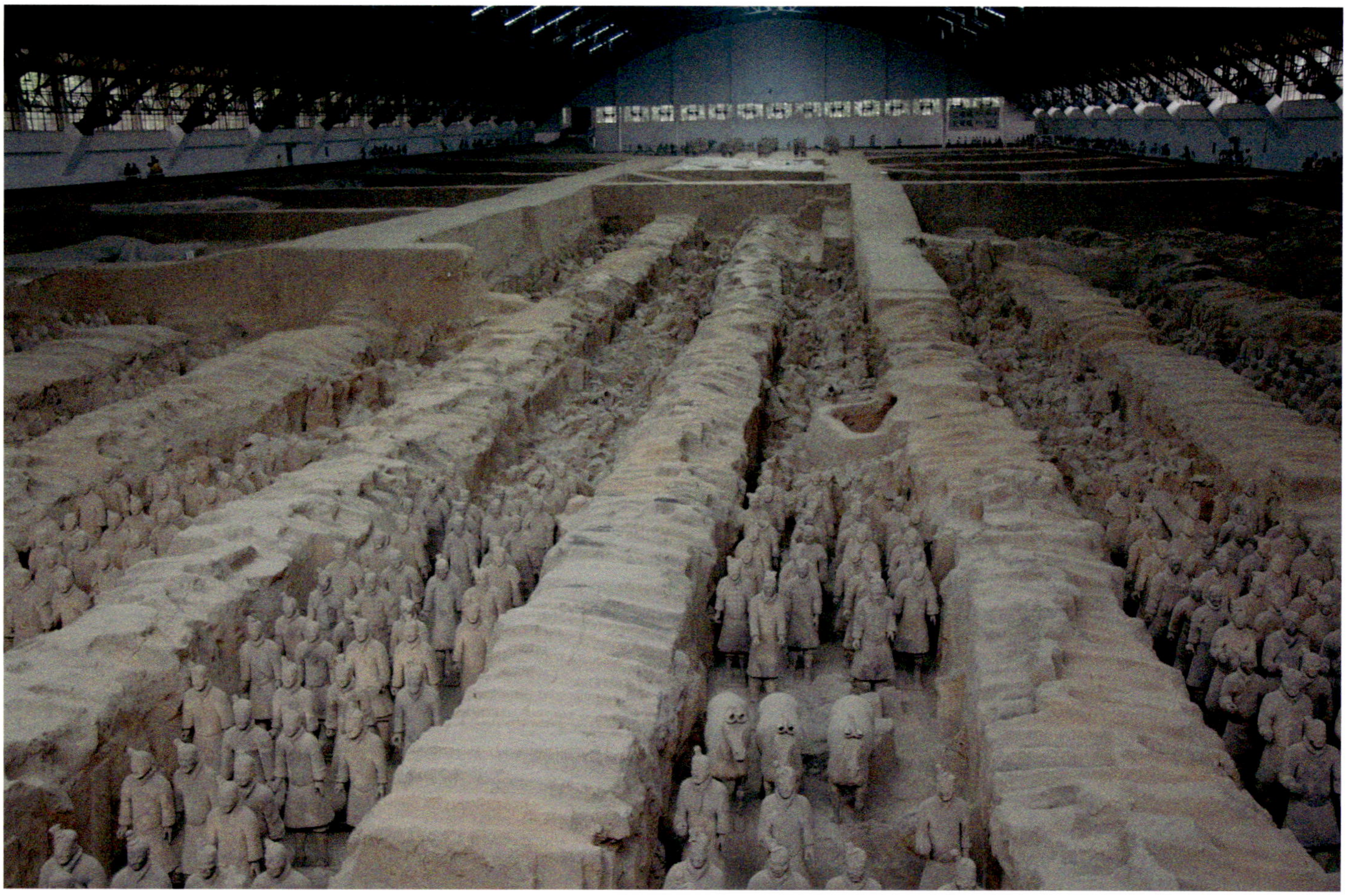

Terracotta Army Pit 1. Work began in 246 BCE. Xi'an, China, terracotta figures, bronze weapons, horses, and chariots.
Photo: © Maros M r a z (Maros) / Wikimedia Commons / CC-BY-SA-2.5.

Munitions (and death tolls) are tangible ways to measure casting's impact on the exercise of power, but propaganda, which works on a psychological level rather than a physical one, is worthy of consideration. While it's harder to measure, the perception of power and the collective buy-in of the masses is what cements the loyalty of an army and a population to its leader.

Displays of power vary in their scale and impact, but two noteworthy examples from history are *The Terracotta Army* (created to guard the mausoleum of Qin Shi Huang, the first Emperor of China) and the statues of Communist dictator Joseph Stalin.

The Terracotta Army consists of thousands of life-size warriors armed with real cast weapons. The bodies of the ceramic warriors were press-molded in multiple parts, assembled, and customized to make each warrior an individual. The (forced) labor required to create it was one credible show of power, and the final product must have been a terrifying and awe-inspiring sight.

On the face of it, freestanding statues of Joseph Stalin that were sprinkled around the Soviet bloc can hardly be compared to the thousands of imposing warriors in the *Terracotta Army*, but these are both masterful approaches to exerting power on a psychological level. Stalin's influence was far-flung and pernicious, and was supported by powerful propaganda at every level, even resulting in Socialist Realism, the state sanctioned style and approach to art making of all kinds. The insidious and lasting nature of propaganda can't be overemphasized; according to a poll conducted by the Carnegie Endowment for International Peace, in 2011 Joseph Stalin had approval ratings of over 45% in parts of Russia.*

Since its earliest days, casting has played a deadly role in both physical and psychological warfare. Whether we are safer or more imperiled by the development of so many weapons is, in large part, determined by who wields them.

* Rachlin, Samuel. "Stalin's Long Shadow." *The New York Times*, 04 Mar. 2013. Web. 28 July 2016. http://www.nytimes.com/2013/03/05/opinion/global/stalins-long-shadow.html?_r=0.

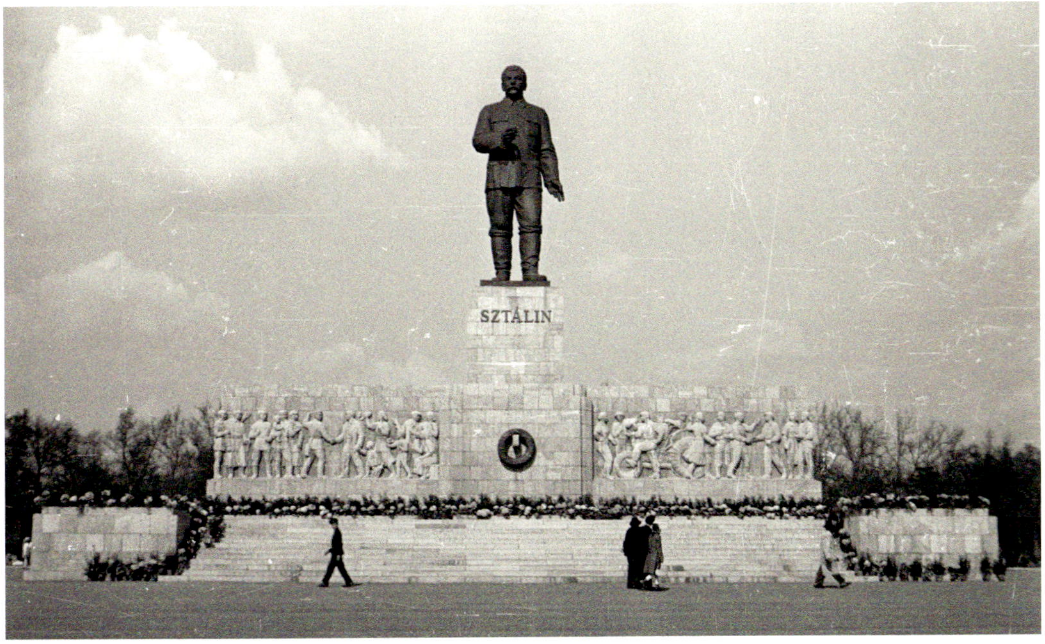

Statue of Stalin in Budapest, 1953. Hungary.
Photo: Gyula Nagy / Wikimedia Commons / CC-BY-SA-3.0.

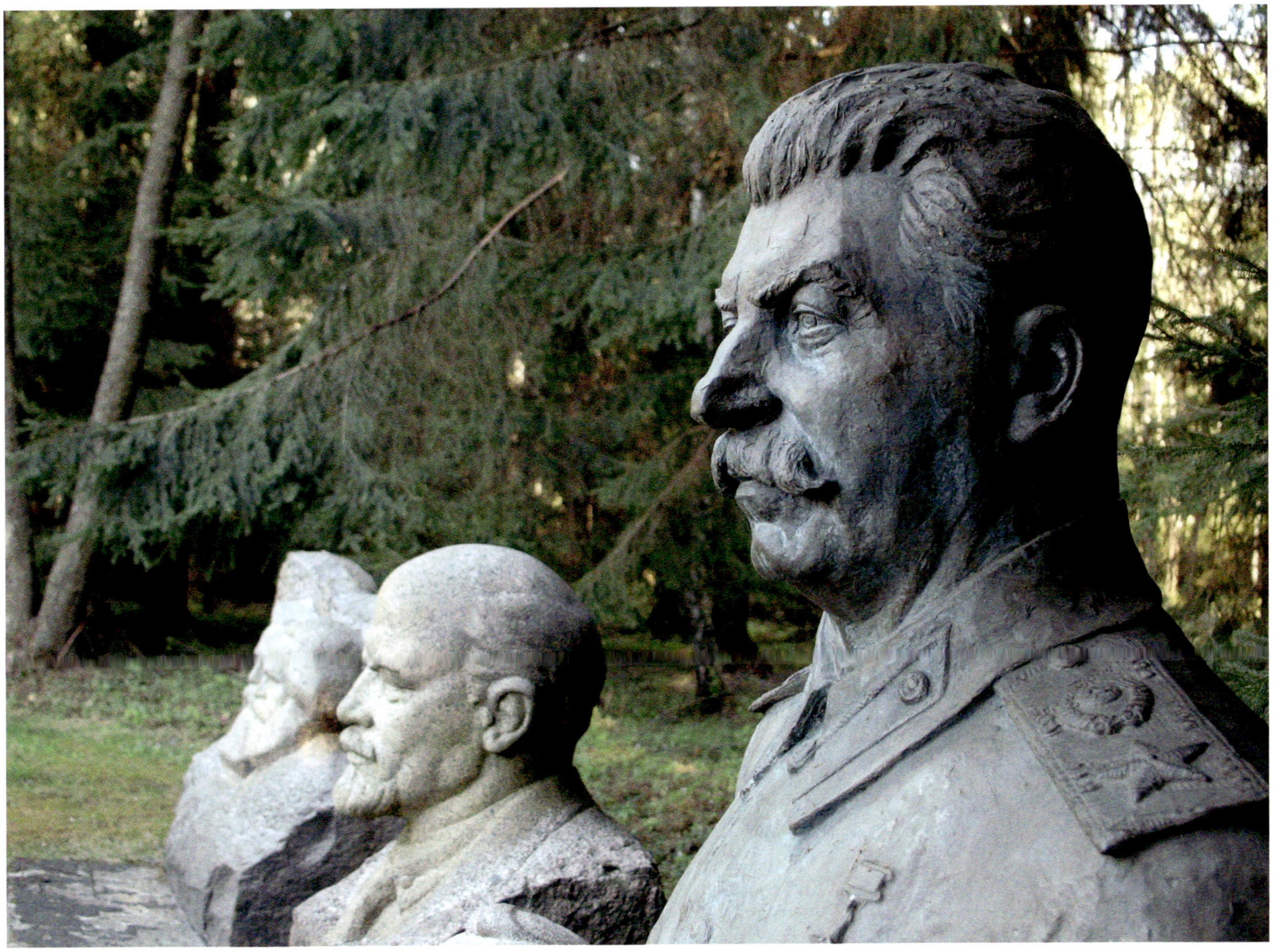

Unknown, busts of Stalin and Lenin, mid- to late twentieth century. Grūtas Park, Lithuania, bronze (serial reproduction), 43.31" × 43.31" × 129.92".
Photo: © Adriao / Wikimedia Commons / CC-by-3.0.

INDUSTRY

Casting is a process that fits hand-in-glove with the goals of industry: it lends itself to mass-production and when output is substantial enough, price per part can be kept at a minimum. While casting has been masterfully executed since the Bronze Age, the Industrial Revolution fundamentally changed the scale of the process. Iron production, the creation of machine tools, advanced mechanization, and the development of the factory system all contributed to a renaissance for casting.

The twentieth century saw a vast expansion of materials and processes. Castable materials like plastics and casting processes like injection molding and rotocasting redefined how objects are produced. From telephones to lawn chairs, cast plastic revolutionized consumer culture around the world.

Large scale industrial casting is now used to manufacture products in materials of all kinds—affordable metal, ceramic, glass, and plastic goods helped make the middle class a reality, which was a seismic shift in the course of human history.

OPPOSITE PAGE

Top to bottom
Plastic telephone production, from *Bell Telephone* magazine, vol. XXVI, published by American Telephone and Telegraph Co., 1947.
Courtesy Prelinger Library collection, via www.internetarchive.org.

Unfired ceramic toilets in a factory setting.
Photo: © junrong, Shutterstock.com.

Monobloc Chairs, mid-twentieth century to present. Plastic/polypropylene.
Photo: © Knut Niehus, Depositphotos.

Many variations of this chair, originally designed by Italian designer Vico Magistretti in 1967, have gone into production and millions of the Monobloc Chairs have been injection molded. They are now found in every corner of the world.

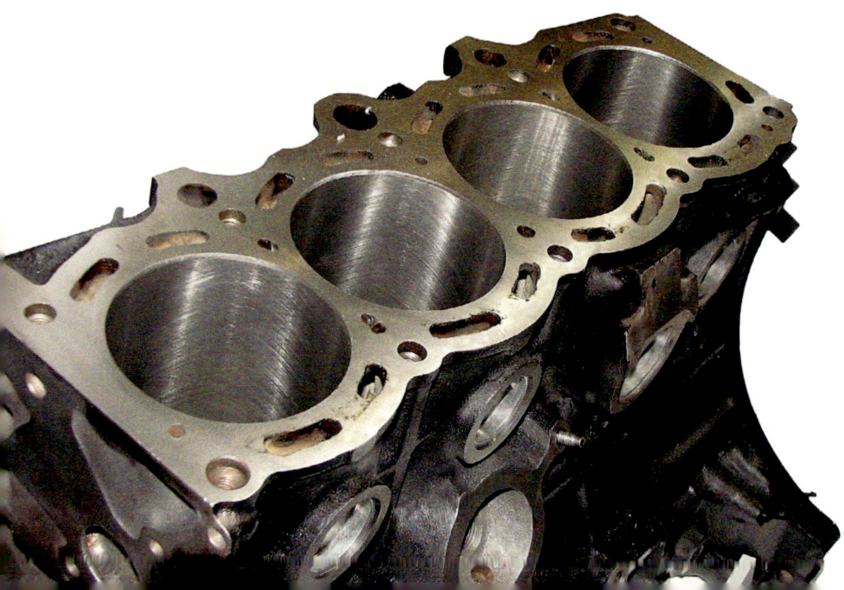

Left Engine block, steel, cast and machined.
Photo: © Vladislav Gajic, Depositphotos.

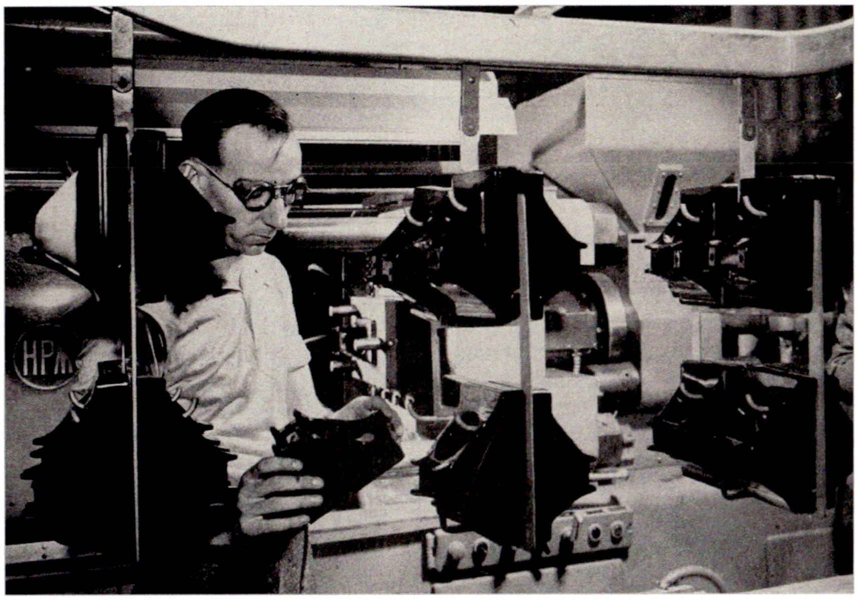

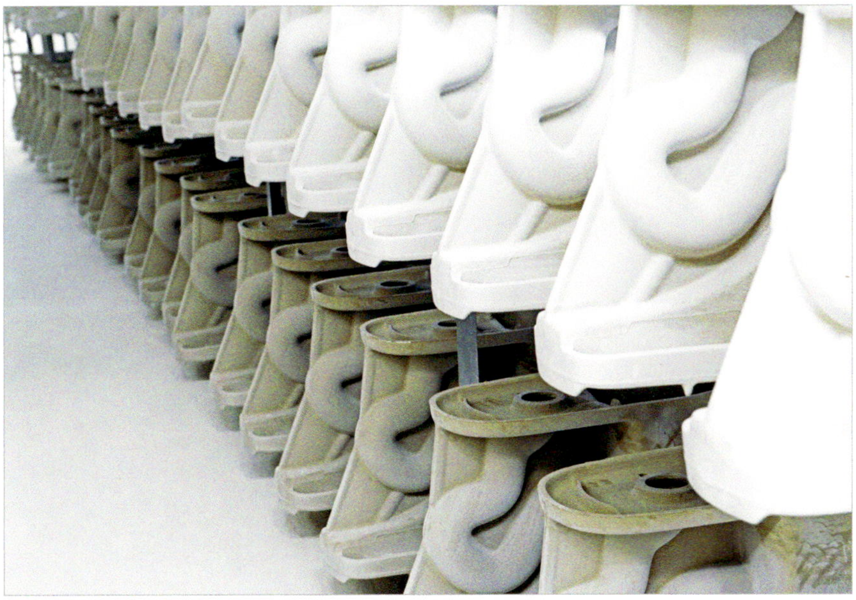

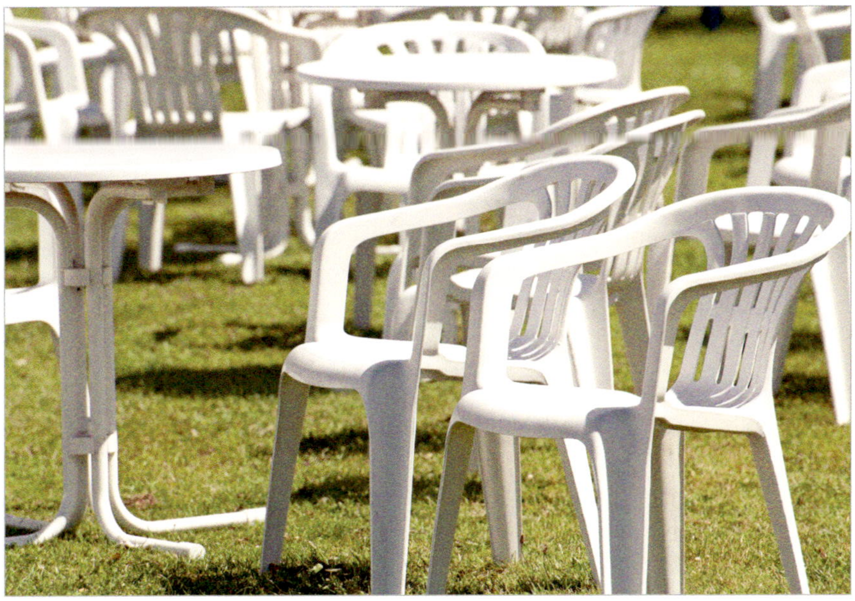

ARCHITECTURE

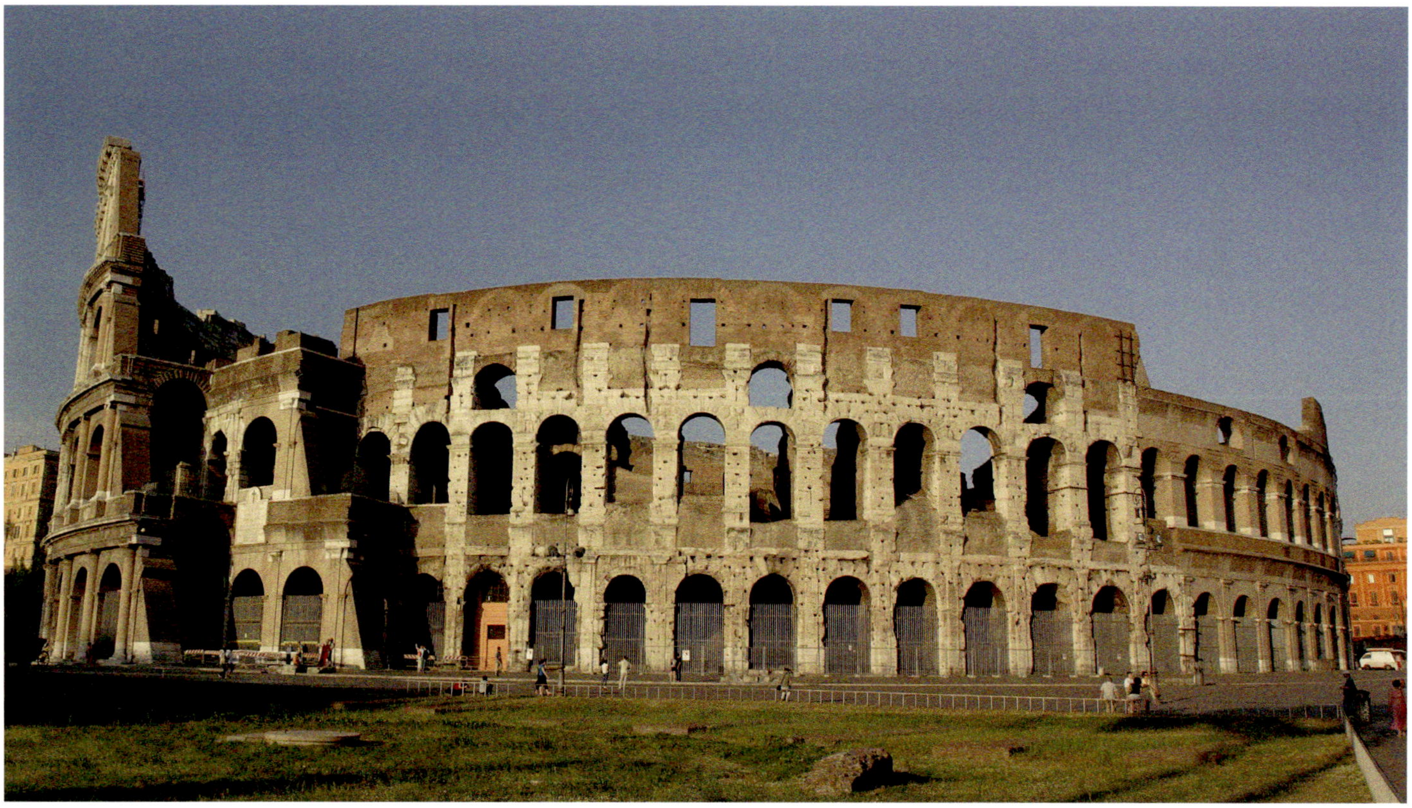

Top Initiated by Vespasian, stewarded by Titus, finished by Domitian, the Colosseum, 80 CE. Rome, concrete, sand, stone, bronze.
Photo: © Yelobird / Wikimedia Commons / CC-by-4.0.

Right These cast and die-struck coins were distributed at the grand opening of the Colosseum in 80 CE and show the significance of this monumental public space.

Colosseum Sestertius, 80 CE. Rome, bronze (cast as blanks and die-struck).
Photo: Joe Jaroch, Colosseo Collection.

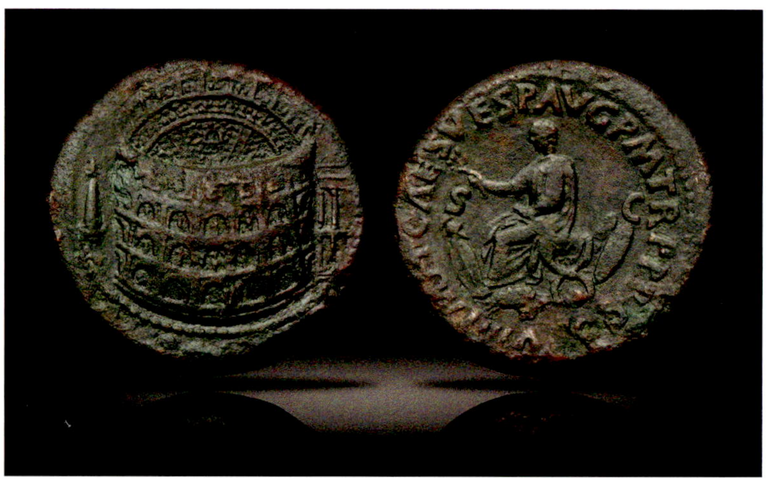

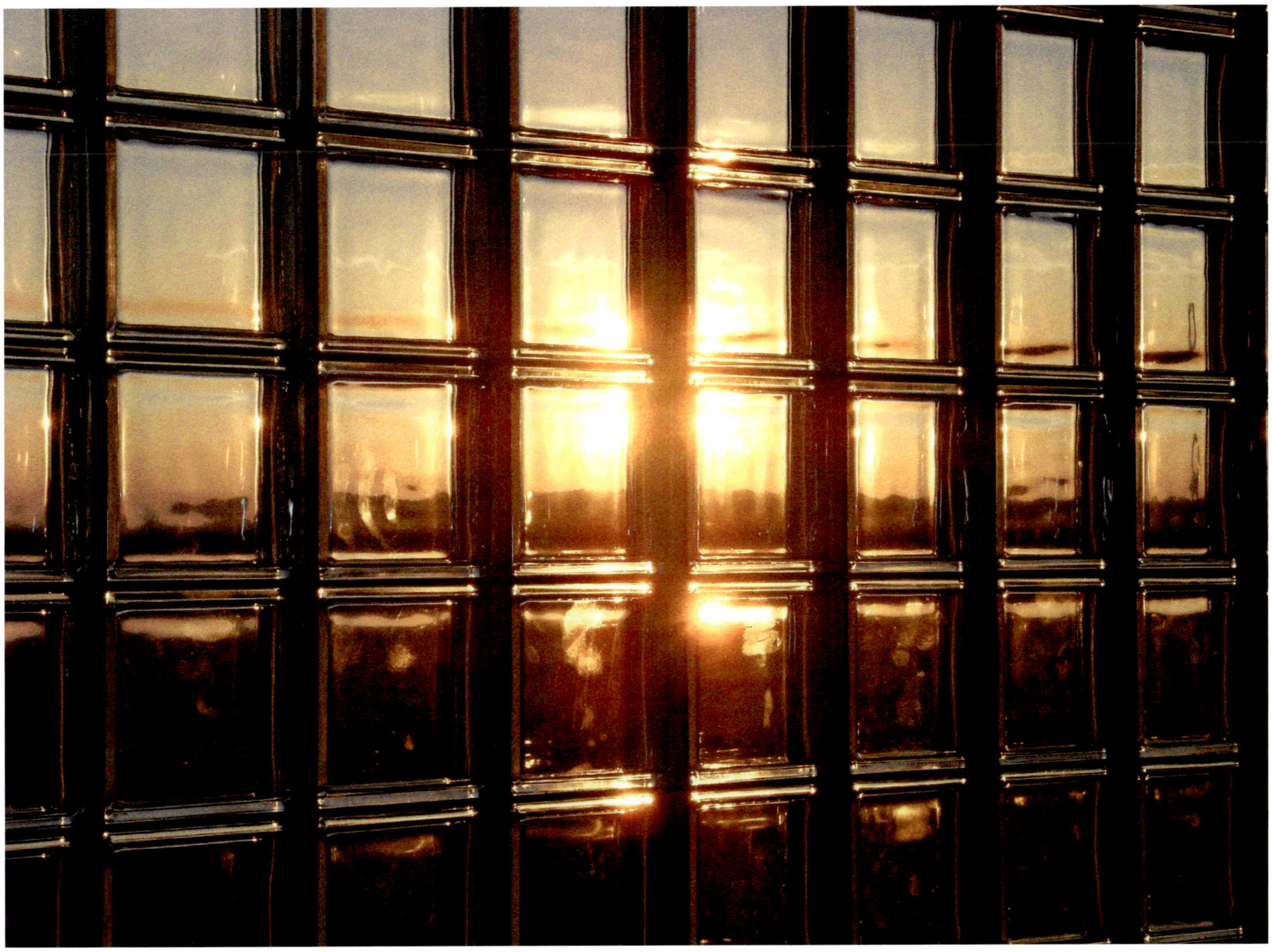

Glass block wall, glass.
Photo: © Mark Ahsmann / Wikimedia Commons / CC-by-3.0, GNU FDL.

Casting and architecture might seem like strange bedfellows, but they actually have a long and intertwined history. Concrete is the most broadly used castable material in architecture and has changed the way we approach building—its versatility is unequalled. It has been used for thousands of years in noteworthy buildings such as the Pantheon and the Colosseum, not to mention the countless other concrete structures that pepper our cities. Completely cast concrete buildings may not be the norm, but almost all modern buildings use concrete in some way...concrete pads and cast concrete blocks are literally the foundation of today's construction. Cast concrete, when polished, can be quite refined and elegant and is now being used for high end interior flooring as well as countertops, steps, and other architectural focal points.

Concrete isn't the only cast-in-place material for flooring and countertops, though. Resin can also be cast in place and works beautifully for a variety of architectural applications. Bricks are cast, and so are tiles, for that matter... when we really begin to examine our surroundings, casting is everywhere.

The other major player when it comes to architecture and casting is glass block. Glass blocks are typically cast in two parts and joined by heating the top edge of each part leaving a space in the center of each unit. Glass blocks were invented in the early twentieth century and became very popular in the pre-World War II Art Deco era. These cast glass blocks are useful because they allow light to pass through but unlike a sheer single pane of window glass, they maintain privacy as well. In addition, they function as structural building blocks that offer some insulation and the ability to create large uninterrupted walls of glass.

Whether it's used in a component form, like brick, tile, and glass block, or cast in place like concrete and resin, casting is fundamental to architecture.

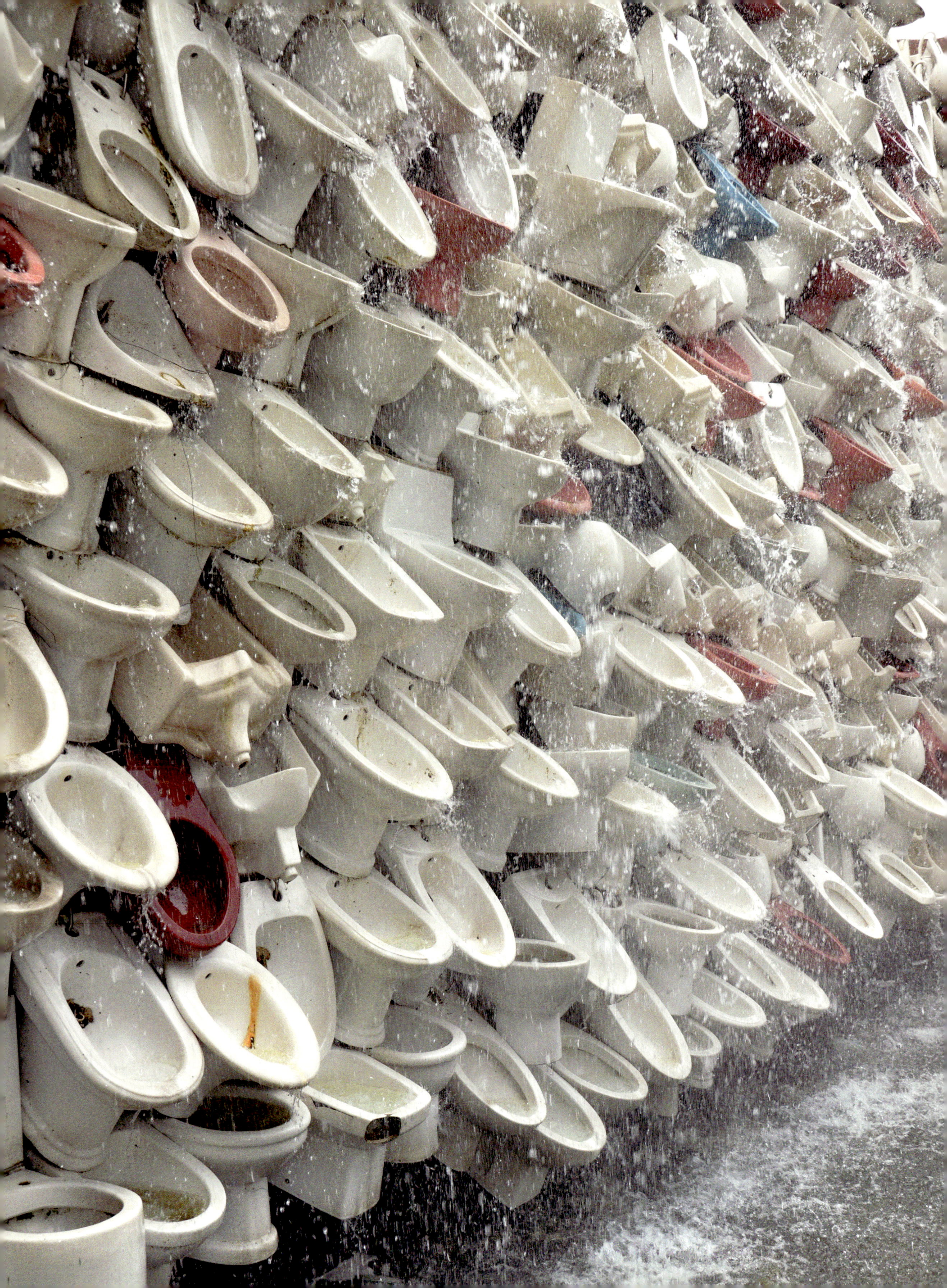

FIXTURES and HARDWARE

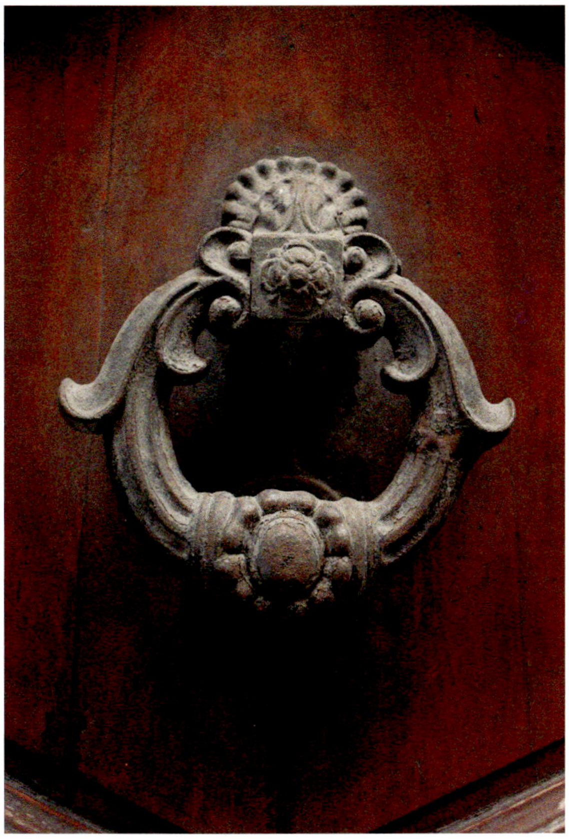

Perhaps casting's most significant contributions to daily life are in the myriad fixtures and bits of hardware that fill our homes, vehicles, and buildings. We take these fixtures for granted, hardly noticing them in our daily lives, but they make our world function. The advent of the cast sink, bathtub, and toilet gave us sanitation in our homes, improving hygiene and health, not to mention the convenience of being rid of chamber pots and privies. These fixtures can be cast in a variety of materials such as ceramic, fiberglass, enameled cast iron, and in recent years, manufacturers have begun casting some in a mixture of powdered stone and resin.

Cast metal radiators were another huge improvement in the quality of life. While many people are replacing them with forced air or baseboard heat today, having heated rooms that didn't require firewood or coal marked another step into modernity and made life far more comfortable, which was a significant contribution.

Along with the momentous improvements of heated buildings with indoor plumbing, cast hardware peppers our homes in the form of knobs, pulls, handles, hinges, and faucets, making everything function seamlessly. Light fixtures are often cast, as are decorative grates, and escutcheon plates; the list gets more obscure from there, but the point is, our homes are filled with castings, down to the cast plastic light switches and switch-plate covers we touch every time we turn the lights on.

We interact with cast objects everyday, and with astounding frequency. The utility of these objects helps define modern life, but this phenomenon certainly isn't unique to the modern era.

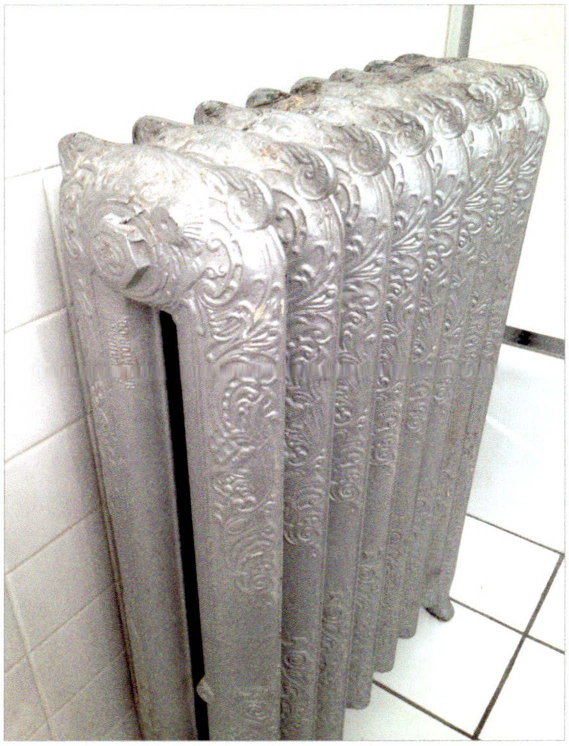

Top Bronze door knocker. Florence, Italy, metal.
Photo: © Toni Sánchez, Depositphotos.

Left Ornamental radiator, late nineteenth century. Germany (?), metal.
Photo: © Frank Vincentz / Wikimedia Commons / CC-by-3.0, GNU FDL.

Opposite page Wall of toilet bowls, porcelain.
Photo: © Yann Song Tang, Depositphotos.

EVERYDAY INTERACTIONS

As soon as makers mastered and refined the casting process, cast objects became staples of everyday life. A simple press-molded ceramic drinking vessel, like this example from the eighth century, was surely exceptionally convenient in the hot, dry climate of the Middle East. Today, we depend on cast objects just as much (if not more) than our ancestors did.

The industrial production of cast plastic in particular has filled our daily lives with utilitarian products. It can be argued that our culture is oversaturated with them, contributing to the American tendency to consider everything disposable. This attitude about the disposability of these products comes with a price. Repair has become an expensive and infrequent act, and considering the low cost and abundance of these everyday objects, replacement makes financial sense, even if it's environmentally destructive.

Setting the murky ethics of our disposable culture aside, there are also huge advantages in the countless and sophisticated products available today... it's no wonder we love products like shock-absorbing running shoes with their cast soles.

The prevalence of cast objects today is not limited to plastic, however—casting has been mastered in every commercially viable material, providing us with metal tools and jewelry, ceramic dinnerware, and countless other products. From toys to tools, cookware to combs, and rings to running shoes, we devour whatever industry dishes up.

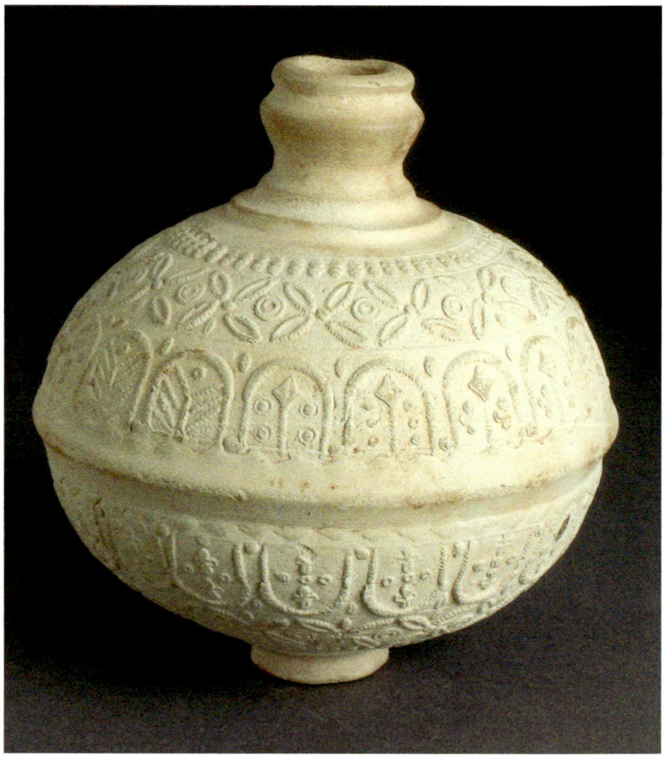

Vessel, first half of eighth century. Iraq, Iran, Syria, or Egypt, ceramics (earthenware) molded, 5.75" × 16.87" (at widest point). Courtesy of LACMA, Gift of Camilla Chandler Frost, www.lacma.org, LACMA public domain.

OPPOSITE PAGE

Top left Engagement ring, 2014. 18K white gold, 1-carat diamond. Photo: © Tomvers / Wikimedia Commons / CC-by-3.0.

Top right Lodge Manufacturing Company, 15" cast iron skillet. Courtesy of The Lodge Manufacturing Company.

Center left Articulated doll, produced 1923–1926, celluloid. Photo: © Bernd Schwabe / Wikimedia Commons / CC-by-3.0.

Center right Plastic comb. Photo: © Ekaterina Palkina, Depositphotos.

Bottom left Shoe soles. Rubber. Photo: © Michael Zwahlen, Depositphotos.

Bottom right Wrenches. Steel. Photo: © ДенисГладкий, Depositphotos.

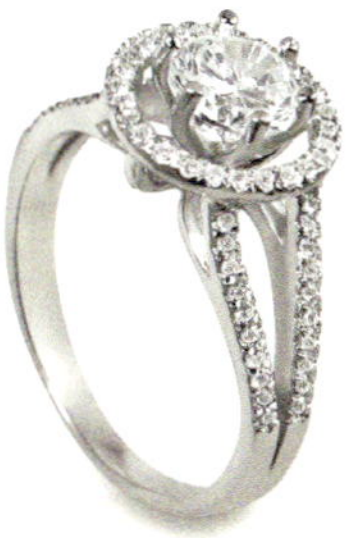

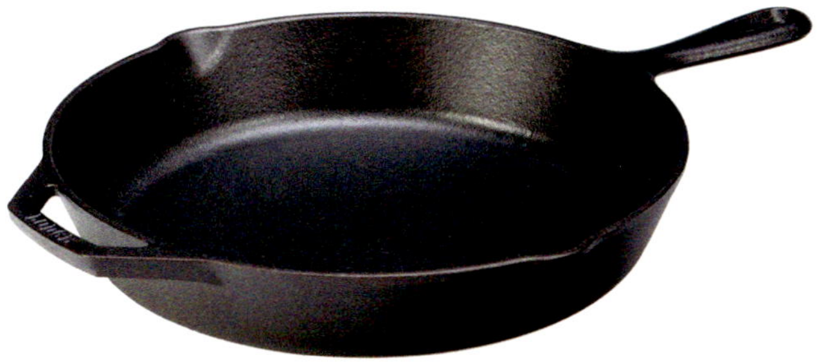

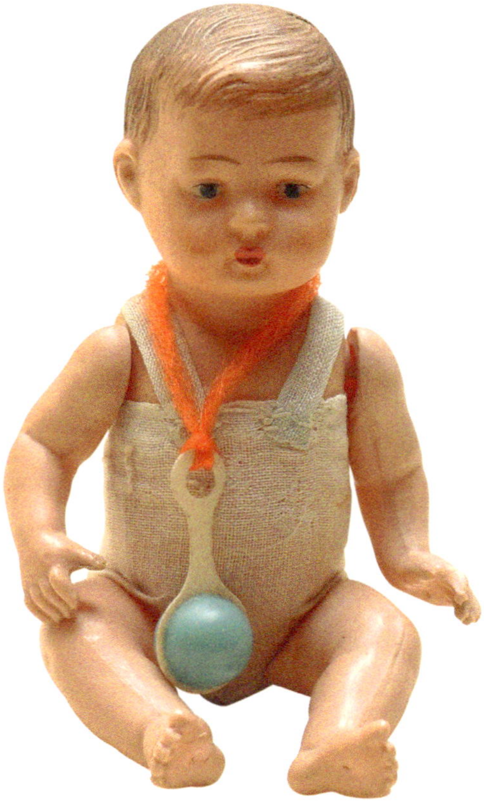

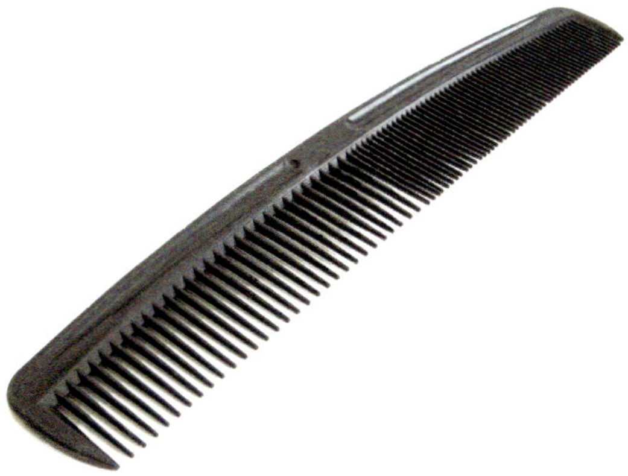

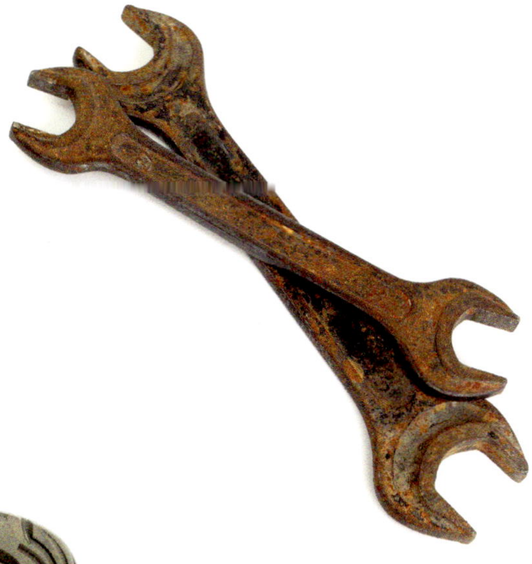

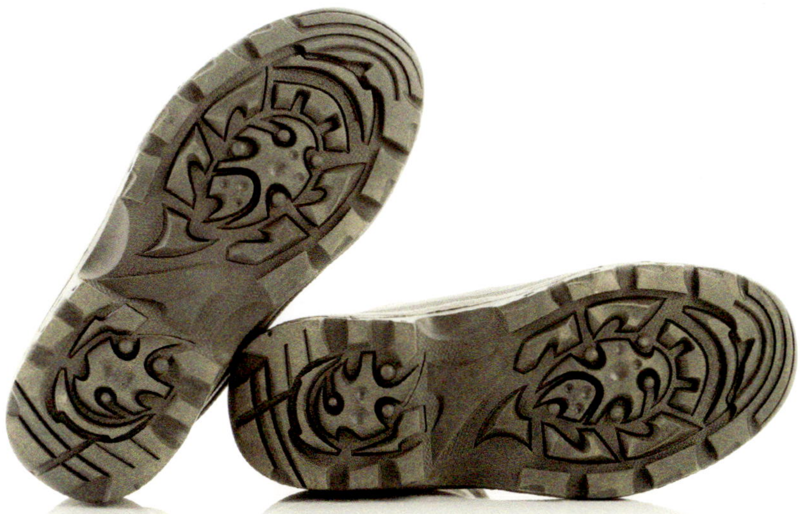

HEALTH

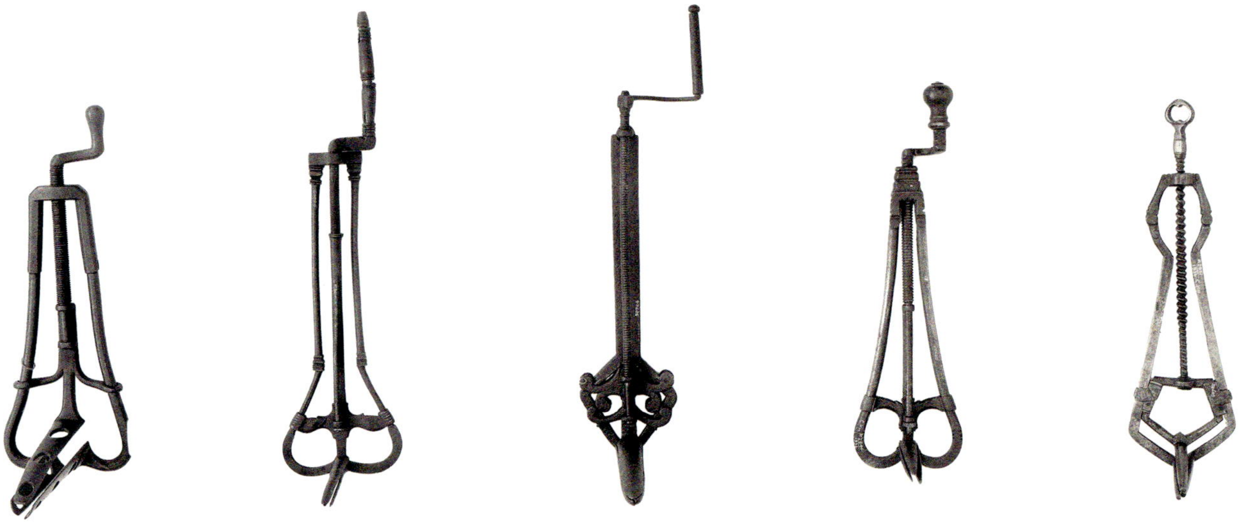

From pills to prosthetics, there is a broad range of applications for casting in medicine. As in every other field, many metal tools are cast. Medical instruments like surgical clamps, hemostats, and speculums, just to name a few, have cast parts. One of casting's strengths is the ease with which organic forms can be produced, so tools with complex handles or unusual shapes are likely candidates for being cast, as are implants such as joint replacements. The exciting developments in three-dimensional (3-D) printing are opening the door to incredible innovations in casting for medicine. Models can be designed using software, printed on a 3-D printer, and those models can be either burned out directly in a lost-wax casting process, or molded and cast using any number of materials. This is an area of growing interest and one that holds great promise.

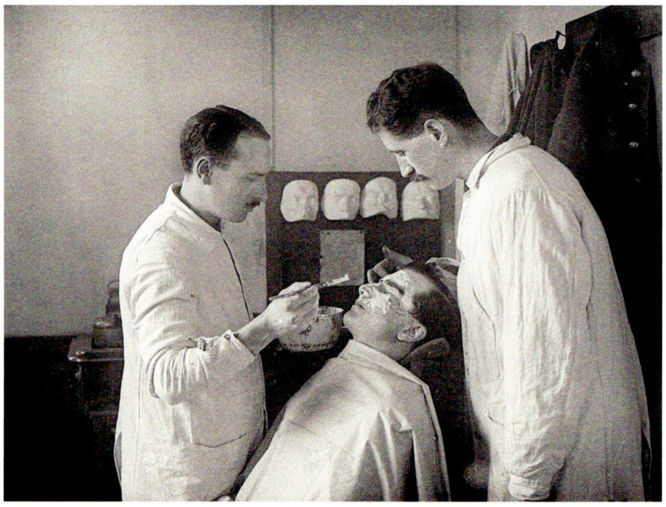

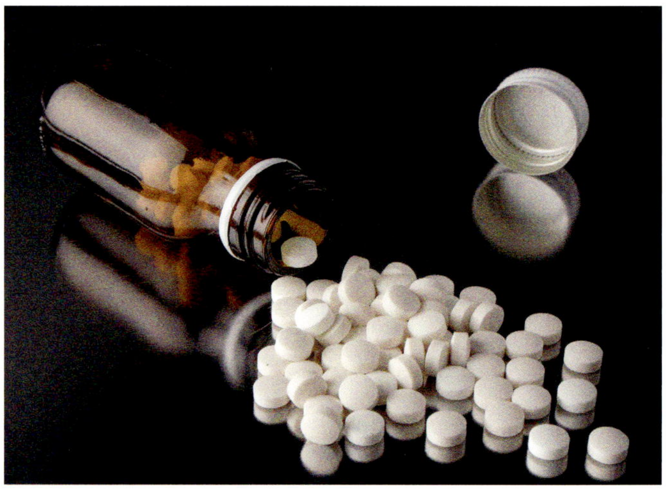

Top to bottom

Five speculums, fourteenth century, fifteenth century, and sixteenth century. Harmonic collection, Wellcome Library, London. Courtesy of Wellcome Images / Wikimedia Commons / CC-by-4.0.

Captain Francis Derwent Wood takes a plaster mold of a First World War soldier's injured face to make a mask that would conceal his injuries.
3rd London General Hospital, plaster.
Photo: Horace Nicholls, Imperial War Museum, Photograph Archive Collection.

Pills.
Photo: © Ari Nousiainen, Depositphotos.

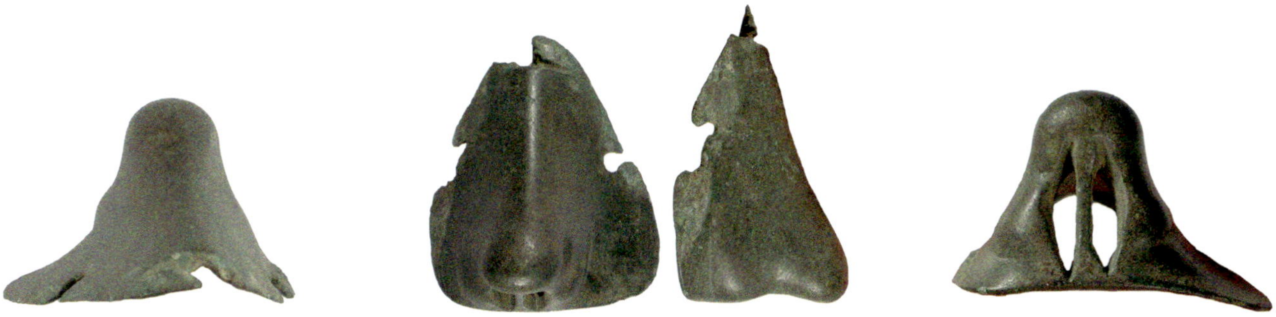

Another area of casting that has a long history and a bright future is prosthetics. Casting is an ideal fit for making a prosthesis or a replacement tooth. Taking a mold of the body's damaged part allows for the creation and alteration of a wax or plaster model and eventually, a cast prosthesis that is customized to the body. Whether a sixteenth- to seventeenth-century bronze nose, a partial face recreation for a World War I soldier, or contemporary dental work, molds and casting can restore function, reduce pain, and help repair the trauma of disfigurement. New materials such as silicone have added translucency and the "feel" of flesh, so that when an experienced maker applies their skills, the prosthesis can be almost impossible to spot. This noble work makes a palpable difference in the lives of those living with an injury or illness, allowing them a return of function and the ability to move through the world with dignity.

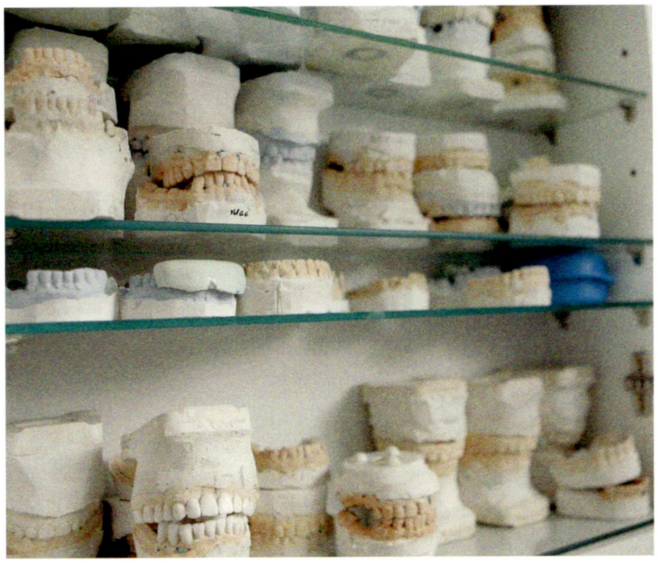

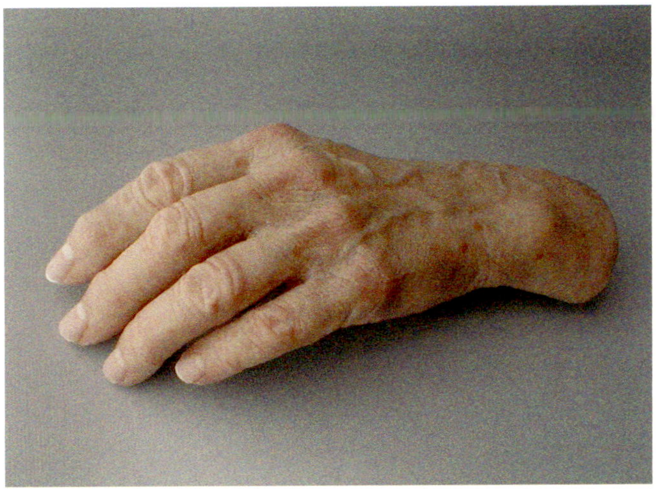

Top to bottom

The Beddingham Nose, sixteenth or seventeenth century. Beddingham, East Sussex, England (place of discovery), bronze.
Courtesy of Portable Antiquities Scheme, CC-by-2.0.

Gypsum dental molds. Gypsum.
Photo: © Alisa Gurinova, Depositphotos.

Sharon Haggerty, MAMS, CCA, custom left hand prosthesis, dorsum (top view), 2008. Two-part silicone, pigments, rayon fiber, 9.5" × 4" × 2.75".
Photo: Sharon Haggerty, Custom Prosthetic, Ltd.

CASTING as it RELATES to the HISTORY of the FINE and DECORATIVE ARTS

In some ways, this entire book examines how casting relates to the fine and decorative arts, but this chapter takes a closer look at the history of that relationship. Suzanne Ramljak's essay Mold Culture: Casting through Western Art History investigates the impact of moldmaking and multiples. Gallery: Old and New pairs historic and contemporary cast works, shining a light on some of the thematic threads that make up the fascinating and varied tapestry that is the history of cast art objects.

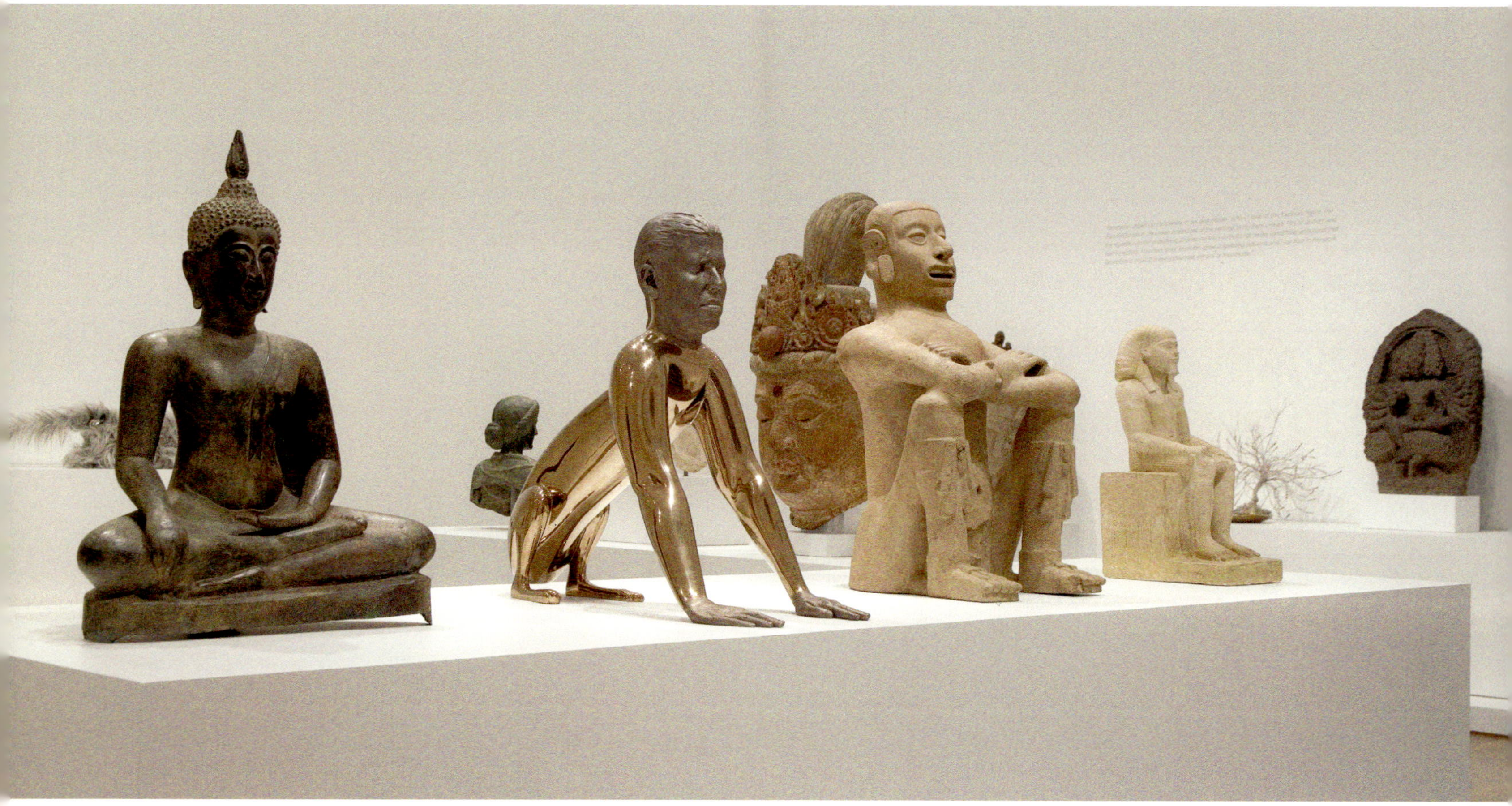

Rona Pondick, *The Metamorphosis of an Object*, Worcester Museum of Art. *Left to right: Seated Buddha in Maravijaya*, late fifteenth to early sixteenth century, Sukhothai period. Thailand, bronze, 31" × 22.5" × 9.5". Rona Pondick, *Dog*, 1998–2001. Yellow stainless steel, edition of 6 + 1AP, 28" × 16.5" × 32". *Seated Male Figure*, 900 to 1200. Mexico (Remojadas), ceramic, 31" × 14.5" × 17.5". *Statuette of Hapidefai*, 2060 to 1780 BCE, Middle Kingdom. Egyptian, limestone, 20.6" × 8.8" × 12".
Photo courtesy of Galerie Thaddaeus Ropac, London/Paris/Salzburg and artist.

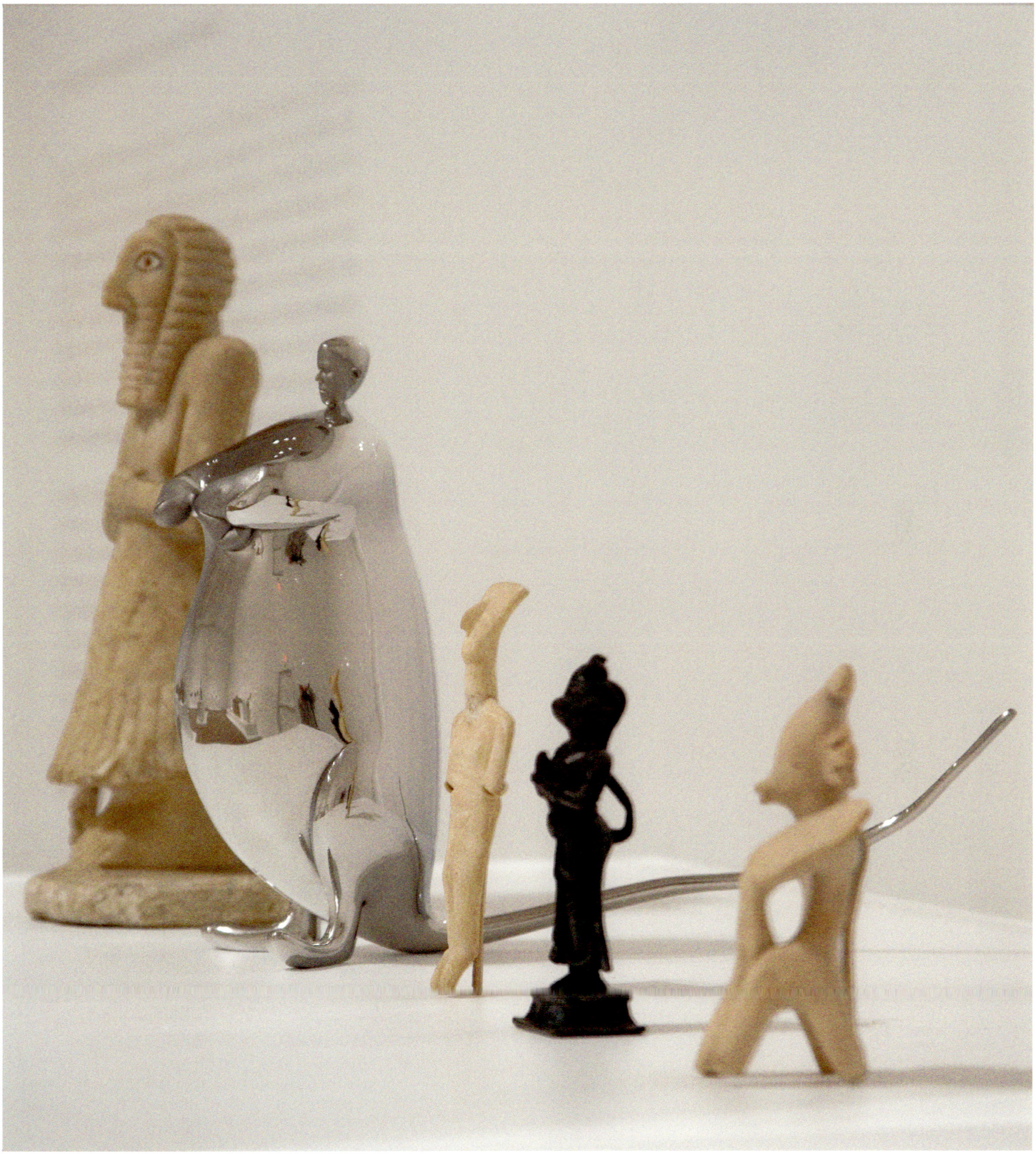

Rona Pondick, *The Metamorphosis of an Object*, Worcester Museum of Art. *Left to right: Statuette of a Man,* 3000 to 2500 BCE. Mesopotamian (Sumer), limestone and shell, 13.8" × 4.8" × 4.8". Rona Pondick, *Muskrat,* 2002–2005. Stainless steel, edition of 3 + 1 AP, 10.5" × 12.5" × 4.6". *Female Votive Figure,* ca. 2400 BCE, Greek (Cyclades Islands), marble, 5.8" × 2.3" × 0.75". *Hanuman,* eighteenth century. South Indian, bronze, 4.8" × 1.8" × 1.3". *Figurine with Four Legs,* 300 BCE to 400 CE Mexico (Nayarit), ceramic, 5" × 2.5" × 1.8".
Photo courtesy of Galerie Thaddaeus Ropac, London/Paris/Salzburg and artist.

MOLD CULTURE
Casting Through Western Art History

by Suzanne Ramljak

For millennia, the technique of casting has proved an efficient means of producing three-dimensional form. It could be argued that casting has helped shape human culture, and it has literally played a formative role within the history of art. The word *cast* has several meanings: to throw or dispense; to cause a magic spell; to allocate acting parts. These various functions—disseminating, captivating, and populating—also apply to cast sculpture throughout Western art history.

The most pertinent definition of *casting* for the purpose of this essay is the creation of form through the use of a mold. Molds are at the core of the casting process, providing a model or die for the transformation of matter. Without a mold, a cast cannot occur. Just as microbial molds are enlisted as "starter cultures" for food production—in the making of cheeses and yogurt, for example—casting molds serve to generate visual culture. Over the centuries, these molds have yielded sculptural models that have, in turn, established artistic norms.

Casting has been employed since antiquity both for creating unique works of art, and for generating multiples. The technique has prompted creative strides and expanded the sculptural repertoire. Bronze casting alone has allowed sculptors to lend strength and permanence to their modeled forms, and most public statues ensconced in parks and piazzas owe their existence to the mighty mold.

A quick jaunt through the history of sculpture reveals the prime contributions that this technology has wrought. While casting was practiced throughout the European Bronze Age (ca. 3200-600 BCE), the resulting statuettes were small and tightly contoured owing to the costliness of bronze and the warping of solid metal. With the cultivation of hollow lost wax casting in Ancient Greece, the scale and range of sculpture was greatly increased, and by the sixth century BCE cast bronzes had reached a high level of sophistication. Through the lost wax method, bronze statues were cast in separate parts then assembled and smoothed at the seams. This process enabled sculptors to create life-size figures in complex poses, exemplified by the extended arms of *Artemision Zeus* (ca. 450 BCE) and the later, more supple *Victorious Youth* (300-100 BCE). The Romans further mastered hollow wax casting, and produced legions of bronze sculptures depicting individuals and deities.

Hollow casting lay dormant during the Middle Ages, as did many artistic practices, buried beneath religious priorities or indifference. Solid bronze casting of small objects was still done, but was rare and expensive. With the renewed interest in Ancient Greece and Rome during the Italian Renaissance, the hollow cast technique was resurrected. This rebirth took the form of Donatello's *David* (ca. 1445), hailed as the first life-size bronze figure created since antiquity. In addition to its scale, the *David* is distinguished by its freestanding nature, unsupported by architecture, an achievement made possible through hollow wax casting. Donatello's finesse with bronze casting is visible in the figure's satiny surface and the warm patina that enhances its sensuality.

Following Donatello's hollow cast revival, the next major episode in the creative evolution of casting occurred in the late nineteenth century, with the sculpture of Auguste Rodin (1840-1917). During this era of experimentation and spurned traditions, Rodin enlisted the time-honored method of casting as an element in his sculptural vocabulary. Often making multiple casts of a figure or body part, Rodin would freely reassemble them, sometimes combining duplicates in the same composition, as in *Les Trois Ombres* (1889). The artist also made casts of antique urns or stelae in his private collection, then united them with cast elements he had sculpted himself. Beyond this, Rodin would unapologetically leave join marks from the separate casting molds exposed and unpolished. This forthright evidence of the casting process within a final work would later prevail in sculpture of the twentieth and twenty-first centuries, most notably in that of artist Rachel Whiteread, whose subject is casting itself.

Equally significant as the role of molds in producing unique works, is their importance in *reproducing* sculptures. Cast reproductions have been immensely influential in the history of art, giving sculptural masterworks a foothold in the larger culture.

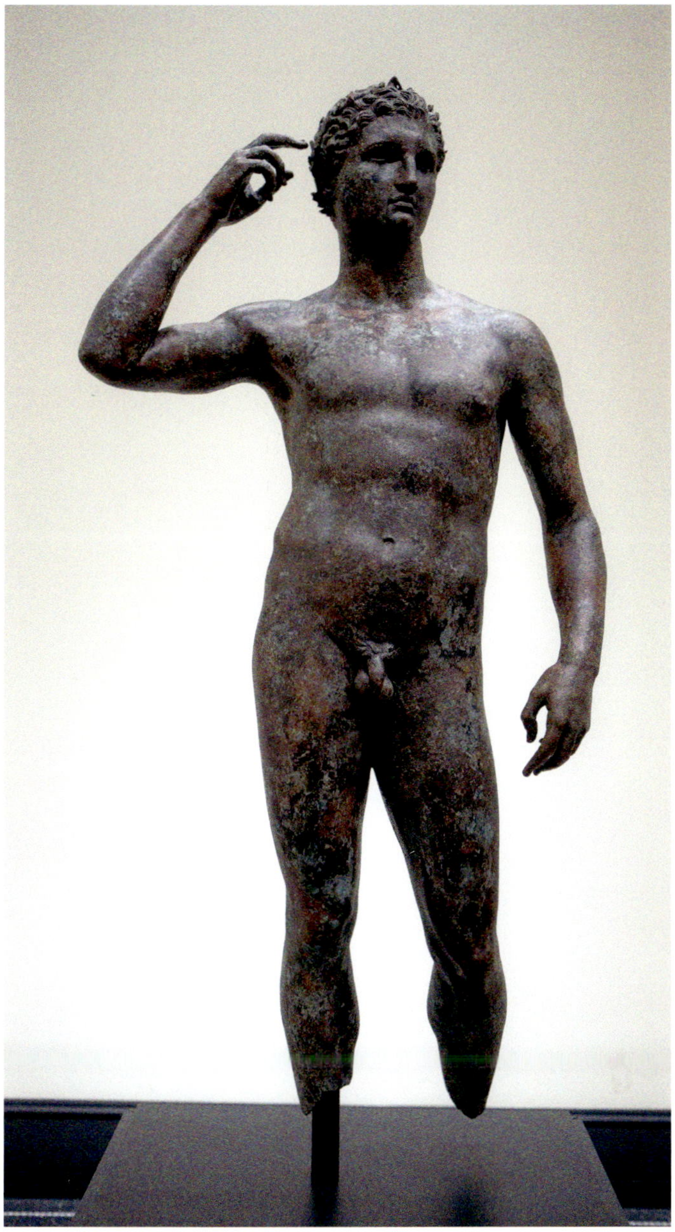

Attributed to Lysippos or school of Lysippos, *Victorious Youth*, 300 to 100 BCE. Greece, bronze with inlaid copper, 59.62" × 27.56" × 11". Photo: Magnus Manske, Collection of The J. Paul Getty Museum / Wikimedia Commons / public domain.

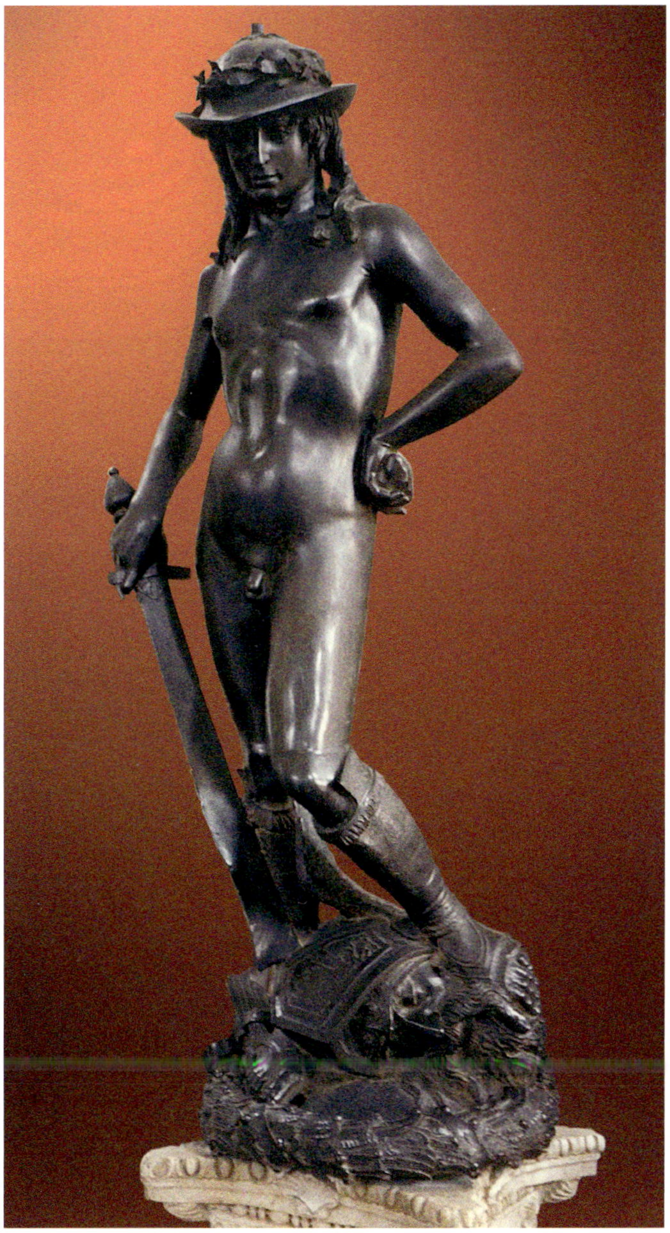

Donatello, *David*, ca. 1440. Florence, Italy, bronze, 62.2". Photo: © Patrick A. Rodgers at the Museo Nazionale del Bargello / Wikimedia Commons / CC-by-2.0.

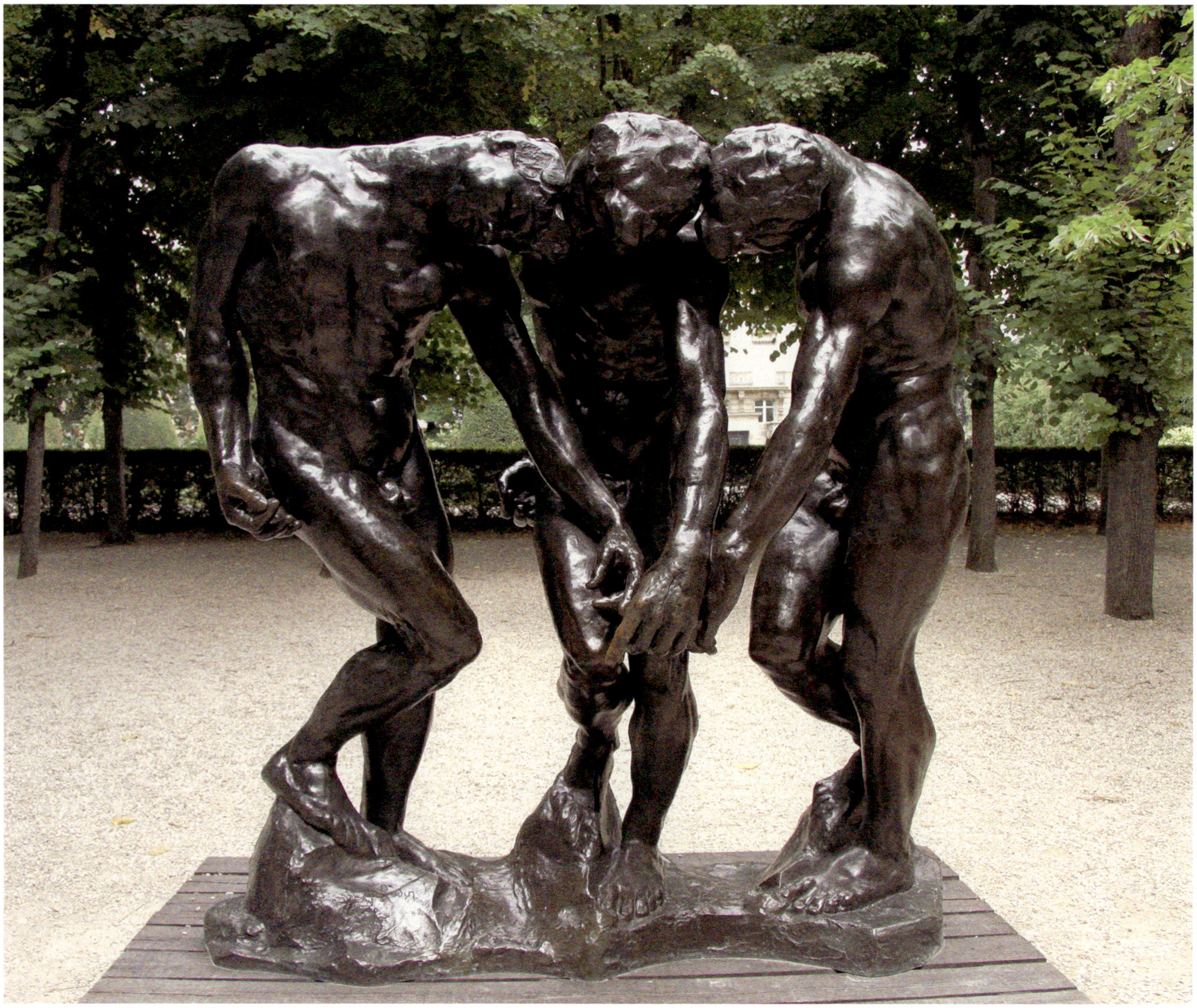

Auguste Rodin, *The Three Shades*, ca. 1880–1885. France, bronze, 38.5" × 36" × 21.37".
Photo: © Piponwa / Wikimedia Commons / CC-by-3.0.

Just as the Greeks excelled at the art of cast sculpture, they also pioneered casting as a means of replicating statuary. Copying was also fundamental to ancient Roman sculptural production, and the Romans established an industry of casts based on Greek prototypes. Not only did they cultivate the casting technique, but they also valued the copies in themselves, and the first known collections of bronze and plaster casts were gathered in the Roman Empire. Through their enamor of replication, the Romans propagated the Greek aesthetic and sired the sculpture canon that would reign for centuries.

With its embrace of Greco-Roman culture, the Italian Renaissance saw a resurgence of casting for the dissemination of ancient works. Collecting of cast reproductions gained sway in this period, with humanists and noblemen acquiring plasters and small bronzes to flaunt their knowledge of the classical past. Sculpture casts also began to populate the studios of Renaissance artists. From the founding of the first art school in 1563, Florence's Academia delle Arti del Disegno, casts became central to art education, providing ideal models for inspiration and emulation.

The establishment in 1648 of the Académie Royale de Peinture et de Sculpture (later Académie des Beaux-Arts) in Paris, codified the art curriculum. Copying from cast replicas became the bedrock of training. The study of casts and ancient sculpture served many purposes; it developed draftsmanship skills, fostered a grasp of human anatomy, and imparted a sense of classical form. As affirmed by art historian Carl Goldstein, "With the ascendancy of the French academy, the antique came to be regarded as the definitive measure, in the literal sense of the word, of beauty and perfection."[1] Every student was required to absorb the cast sculpture pantheon. To facilitate the multi-year program of drawing from casts, each art academy amassed a collection of key copies that typically included the *Apollo Belvedere*, *Venus de Medici*, *Farnese Hercules*, and *Laocoön*.

Along with the centrality of casts in art instruction, the 1600s ushered in the Grand Tour, a social phenomenon that would flourish in the eighteenth and nineteenth centuries. An educational rite for wealthy young men, the Grand Tour provided exposure to great works of antiquity and the Renaissance through travel, with Italy as the prime destination. These excursions doubled as shopping sprees for the purchase of paintings and cast sculpture reproductions destined for stately homes. During the nineteenth century, steam-powered trains made travel less exclusive, allowing the middle class to partake in such cultural pilgrimages.

The growth in tourism went hand in hand with the emergence of the souvenir. A reciprocal relationship arose between venerated masterworks and cast replicas that travelers brought home from their journeys. Tourists eagerly sought out the canonical works they had earlier encountered in reproduction, only to bring back their own copies for display. These souvenir castings took up residence with both aristocrats and the working class, becoming ubiquitous emissaries of high art. Among the "first mass-market commodities" in Europe,[2] souvenirs were popular consumer products, and duly inculcated the public with a knowledge of sculpture history.

While casts were potent agents in art education and tourism, their impact within the museum sphere was even more vital. Following the Louvre's opening in 1793 as Europe's first public museum, a raft of museums appeared across the continent. Cast facsimiles were essential within these new "palaces for the people."[3] Belief in the usefulness of copies in spreading sculptural literacy led to the Convention for Promoting Universal Reproductions of Works of Art, in 1867. The initiative was spearheaded by Victoria and Albert Museum director Henry Cole, who persuaded fifteen European princes to establish a network to make molds and exchange casts between their nation's collections, issuing copies of the finest works of art which each museum possesses."[4]

In the United States, public collections of reproduced antique and Renaissance sculpture began to be assembled in the 1880s. Visitors to art museums at the turn of the century would customarily be greeted by plaster casts, copies of masterworks for those who could not afford a trip abroad. These reproductions were partly put forth in hopes of refining the citizenry and elevating popular taste. Beyond their aesthetic appeal and idealized forms, antique casts performed a moral function, imparting lessons of Western civilization and instilling values of nobility, restraint, and character. In every sense, these cast copies served as role models for an adolescent nation.

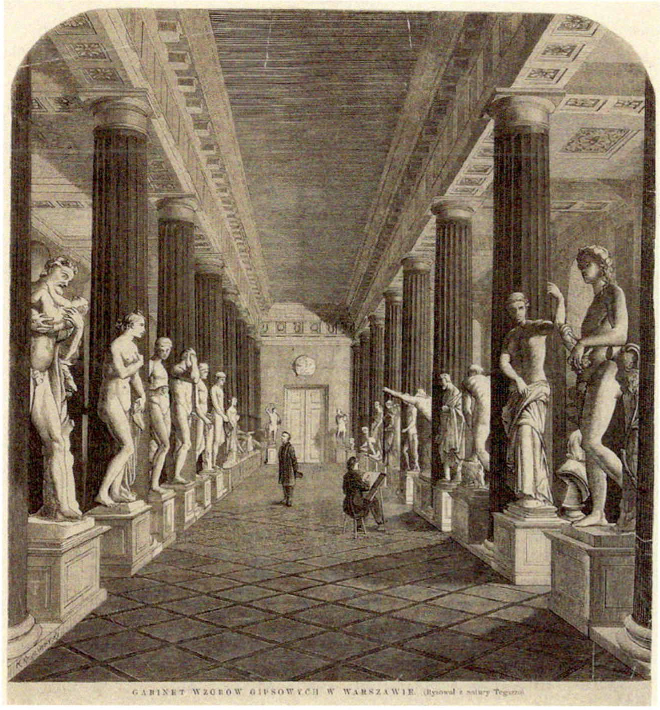

Kazimierz Krzyżanowski, Cabinet of plaster casts of the Warsaw University, 1866. Poland, woodcut on paper.
Courtesy of the National Museum in Warsaw / Wikimedia Commons / public domain.

Within university museums, collections of cast replicas helped to systematize the disciplines of art history and classical archeology. The scholarly study of sculpture casts within academic settings allowed for comparative analysis and connoisseurship, furthering the understanding of stylistic development over time. Cast plaster collections in universities and colleges facilitated the proper practice of art history itself.

As castings entered the cultural bloodstream through museums, art schools, and widespread travel, certain sculptures acquired a lofty standing. This exposure gave birth to the celebrity sculpture, which became famous for being famous. Casting was the mechanism through which such sculptural notoriety was achieved. In the art-historical halls of fame, the *Venus de Milo* provides a case study of this process of ascension through cast replication. Arguably the most reproduced sculpture on Earth, the statue's story is one of cultural saturation and triumph. While relatively new on the historical stage, she rapidly stole the limelight through the power of reproductive technology. Discovered in Greece in 1820, the sculpture was acquired for

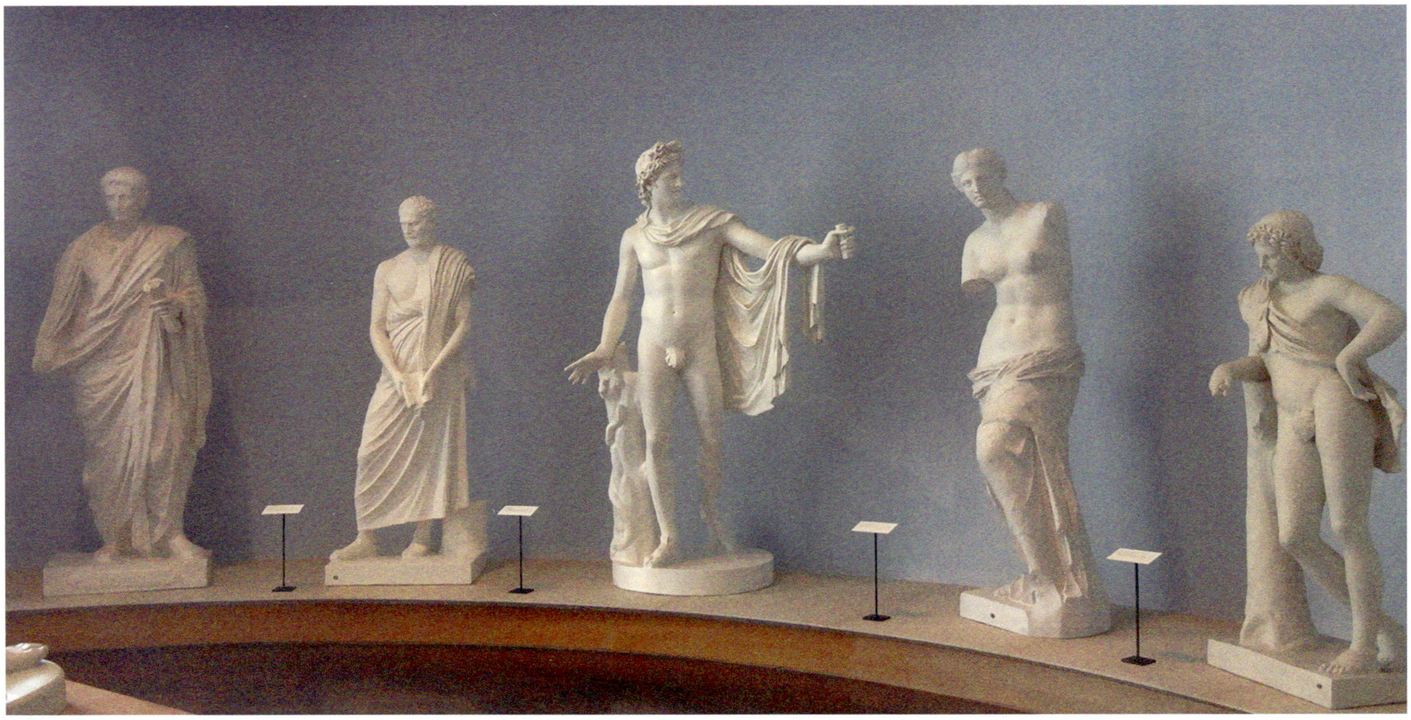

Installation view of the William J. Battle Collection of Plaster Casts, acquired 1894–1923. Plaster.
Photo: Matthew Felix Sun, Collection of the Blanton Museum of Art, University of Texas at Austin.

France and installed in the Louvre in 1821. A mold was soon taken and casts were quickly produced, permitting the *Venus de Milo* to join her plaster counterparts in museums and art schools worldwide.

Beyond just a statue for copying, the *Venus* became the subject of other artists' work, making appearances in Surrealist art, Pop art, and late twentieth-century appropriation art. Perhaps no contemporary artist has depicted the *Venus de Milo* more frequently than Jim Dine, who absorbed the Greek icon into his visual lexicon, transforming and recasting it within his own sculpture. In Dine's art, like that of other artists, the antique statue comes full circle from High Classical masterwork, to cast mass-market product, then back into the fine arts realm. By simultaneously engaging the copy and the original, many contemporary artists—from Andres Serrano to Yayoi Kusama—reflect the larger function of castings as hand-me-down art history. In Larry Jens Anderson's *Analysis of Beauty* we are given a sampler of *Venus de Milo* casts, each with a different formal twist, revealing the inevitable mutations that result when a sculpture is so widely dispersed.

Without casting, or the culture of the mold, works such as the *Venus de Milo* could not have embedded themselves so deeply in the mass psyche. The virulence with which certain sculptures multiply and spread has been likened to a type of contagion. As artist Edward Allington says of these oft-copied works, "There is something almost terrifying about them. Like some form of disease that reproduces itself. It is awe inspiring, something unstoppable."[5] While there may be an unsettling quality about the intractability of such sculptures, there is solace to be taken as well. Familiarity can breed contempt, but it can also offer comfort. Through the shared iconography that casting issues forth, prominent works act as beacons or guideposts in the hazy morass of public imagery. Like friendly faces in the crowd, these recognizable cast characters can put us at ease. Without them, our cultural landscape would be less hospitable and our connection to art history more tenuous. Indeed, these offspring of the mold provide models from the past to help us shape our unformed future.

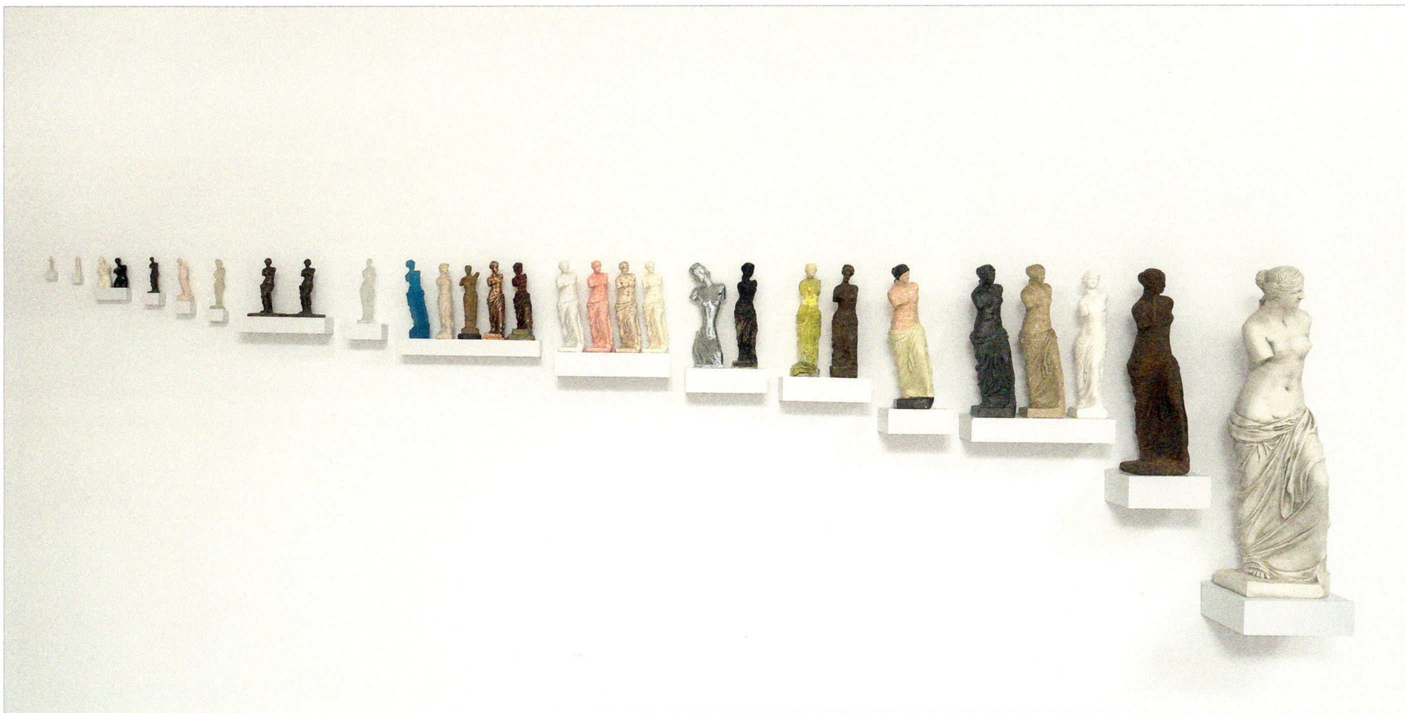

Larry Jens Anderson, *Analysis of Beauty,* 2015. Metals, ceramic, resin, concrete, paint, olive oil in glass jar, plastic, rubber, wood, MDF, glass, 138" × 75" × 9".
Courtesy of the Museum of Contemporary Art of Georgia.

Notes

1. Carl Goldstein, *Teaching Art: Academies and Schools from Vasari to Albers* (Cambridge and New York: Cambridge University Press: 1996), 151.
2. Sarah Benson, "Reproduction, Fragmentation, and Collection: Rome and the Origin of Souvenirs," D. Medina Lasansky and Brian McLaren, eds., *Architecture and Tourism: Perception, Performance and Place* (Oxford: Berg Publishers, 2004), 25.
3. Nathaniel Burt, *Palaces for the People: A Social History of the American Museum* (New York: Little Brown & Co., 1977).
4. Henry Cole, *Memorandum of the International Exchange of Copies of Works of Fine Arts* (London: South Kensington Museum, 1864).
5. Edward Allington, unpublished interview with author, September 23, 1998.

Suzanne Ramljak, a writer, art historian, and curator, is currently editor of *Metalsmith* magazine and curator at the American Federation of Arts. She was formerly editor of *Sculpture* magazine and of *Glass Quarterly,* as well as associate editor of *American Ceramics.*

GALLERY
Old and New

Many threads weave their way through the history of cast art objects. This gallery identifies some of those threads by pairing historic and contemporary works that share a common material as well as other striking similarities. Some pieces seem like siblings, while others are more like doppelgängers, their differences just as compelling as their commonalities. These thought-provoking couplings provide a lens through which we can examine the continuum of casting from antiquity to today.

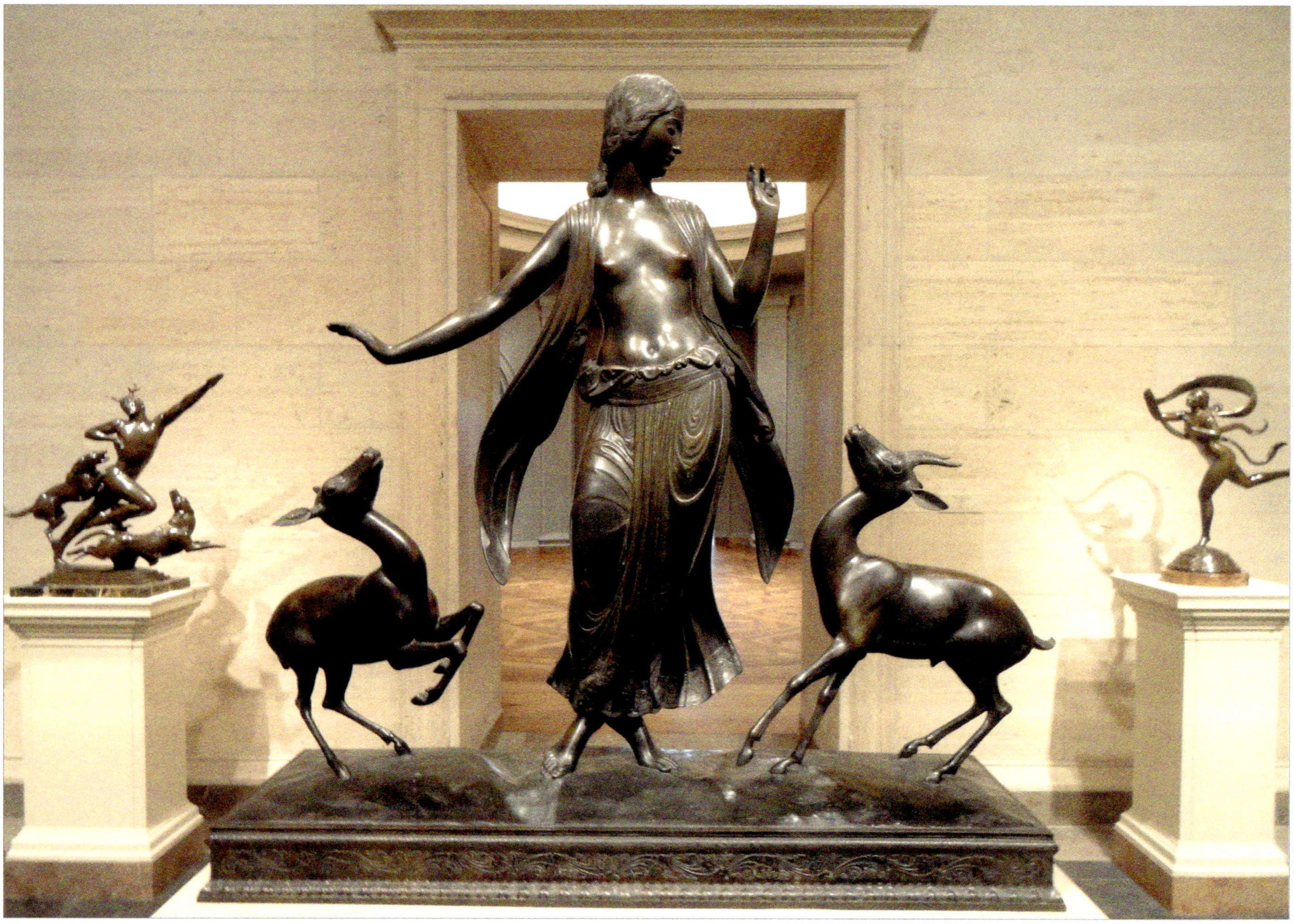

Paul Manship, *Dancer and Gazelles*, 1916. USA, bronze, 32.5" × 34.75" × 11.25".
Photo: Daderot at the National Gallery of Art, Gift of Mrs. Houghton P. Metcalf / Wikimedia Commons / public domain.

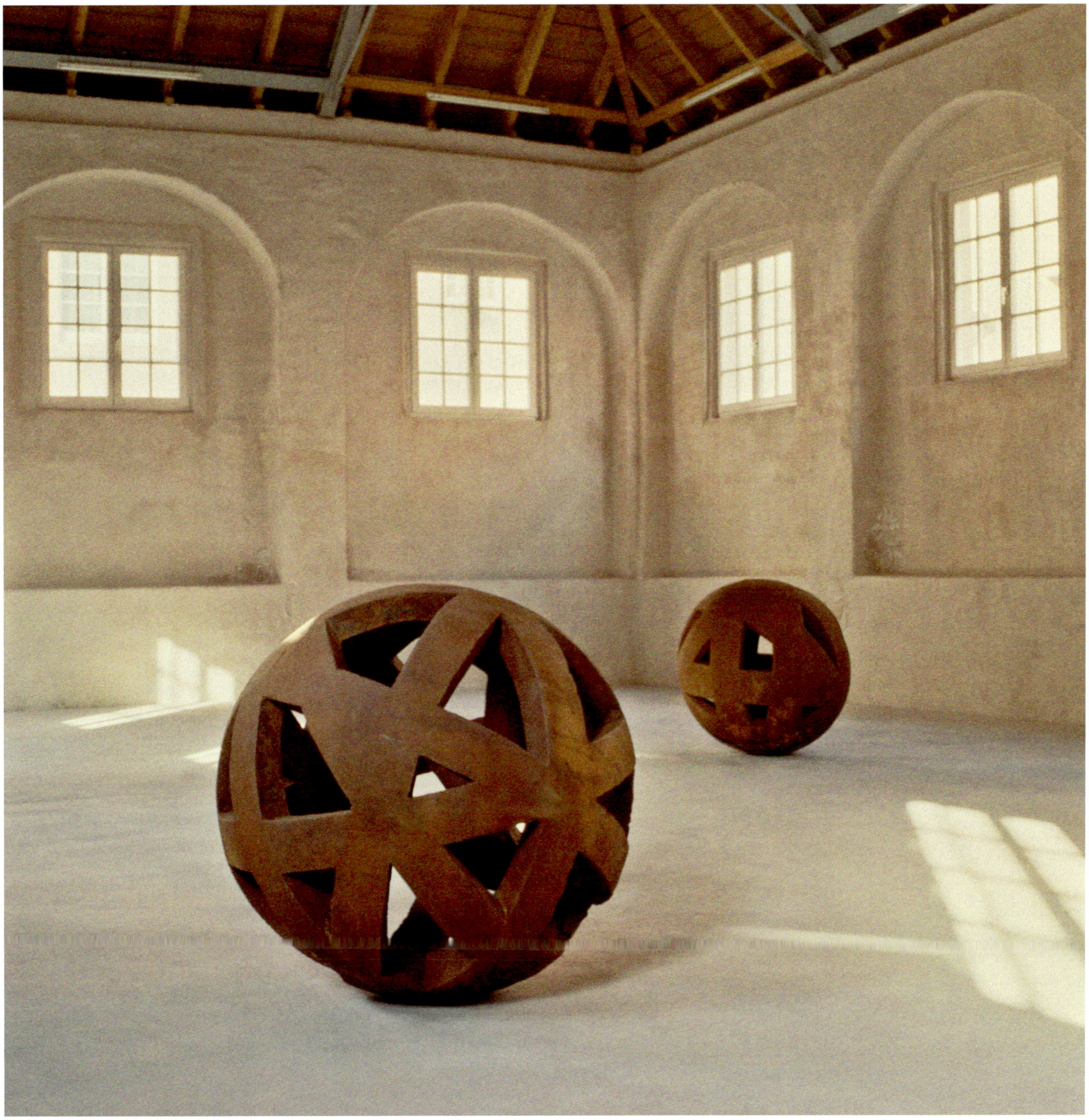

Ingo Ronkholz, *Two Spheres (one)*, 1994. Germany, cast iron, 3'4.5" diameter (each).
Courtesy of Atelier Ingo Ronkholz, City of Iserlohn, Germany.

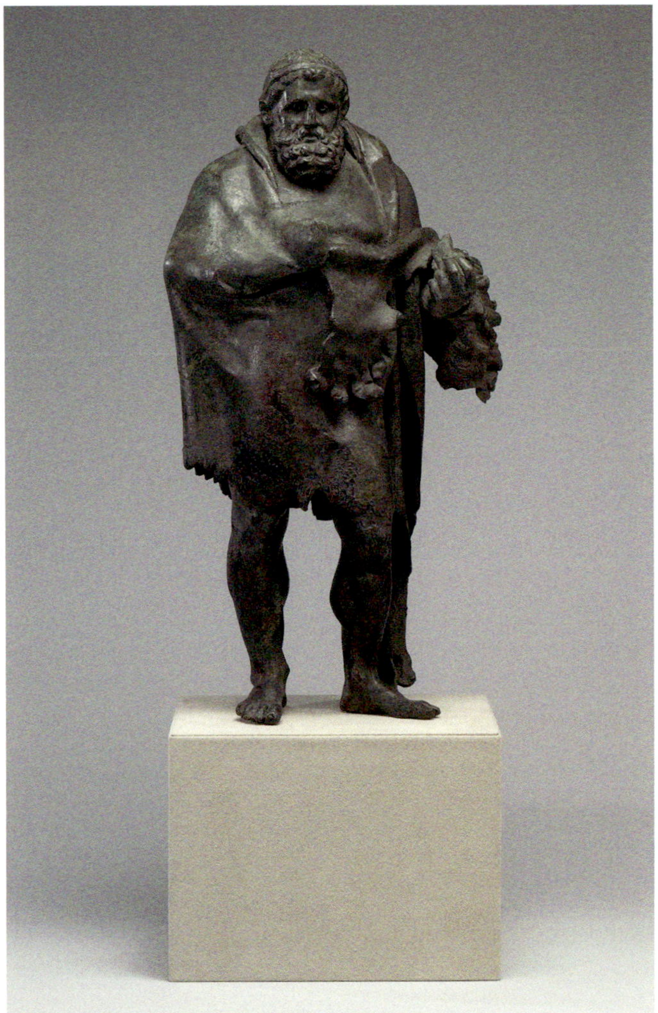

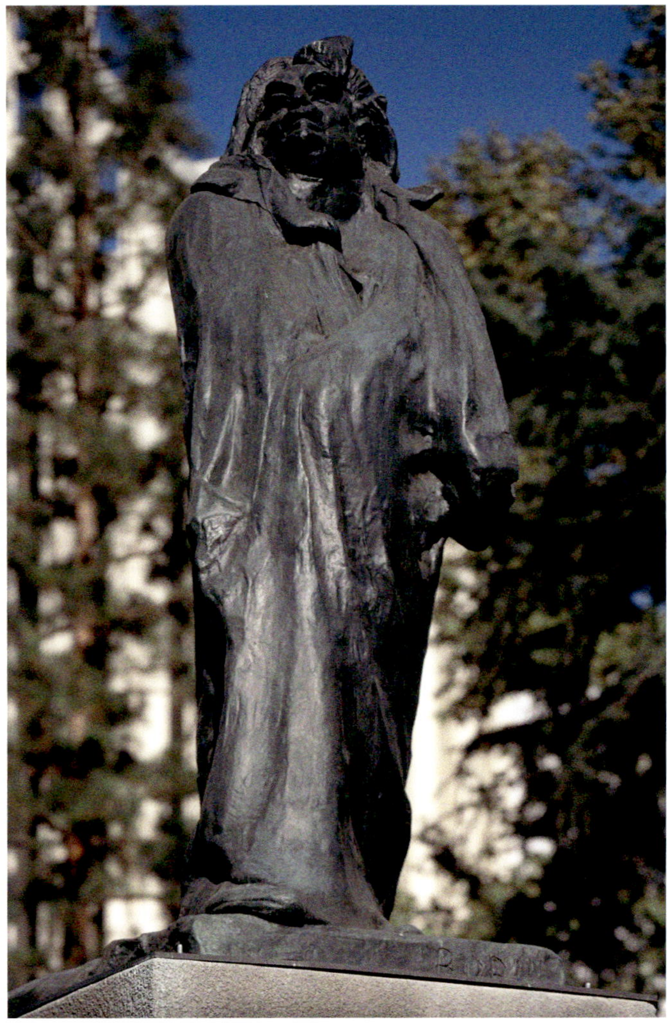

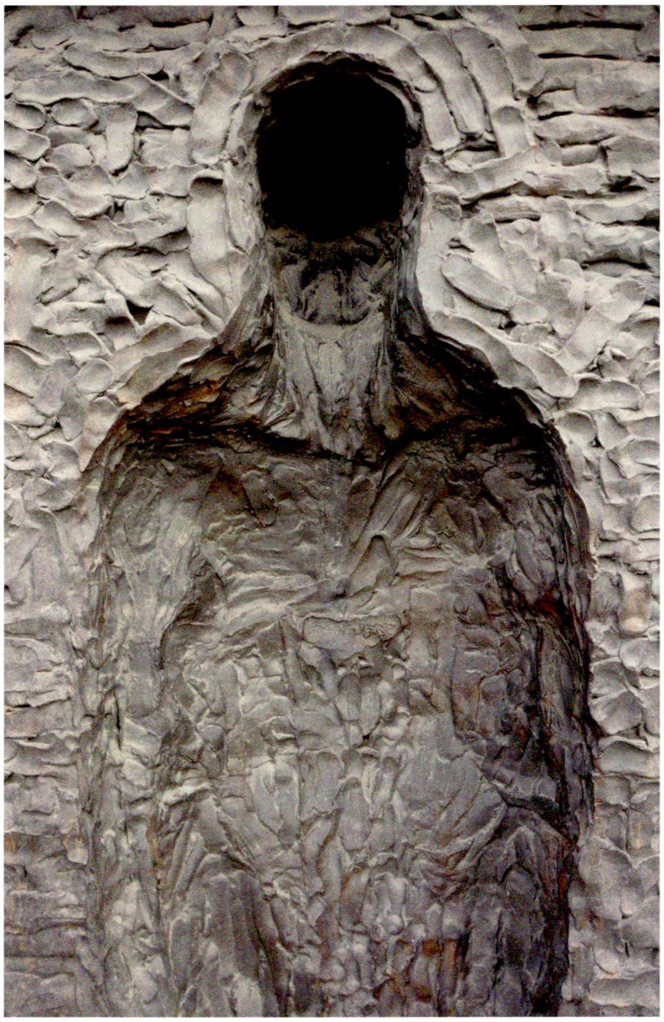

Opposite page, left Aged Herakles, Roman copy, first century BCE to first century CE; Greek original, third century BCE (Hellenistic). Discovered in Caesarea Cappadociae (present-day Kayseri, Turkey), cast bronze with silver and gilding, 6.93" × 9.37" × 5.5".
Courtesy of The Walters Art Museum, Acquired by Henry Walters, CC0, GNU FDL.

Opposite page, right Auguste Rodin, *Monument to Honore de Balzac*, first modeled 1897, this cast 1967 (Musée Rodin 9/12). France, bronze, 117" × 47.25" × 47.25".
Courtesy of LACMA, Gift of the B. Gerald Cantor Art Foundation, www.lacma.org, LACMA public domain.

Left Zenos Frudakis, *Freedom* (detail), 2000. Bronze, 20' × 2' × 8'. Photo: Zenos Frudakis, Collection of 16th and Vine Streets, Philadelphia, Pennsylvania.

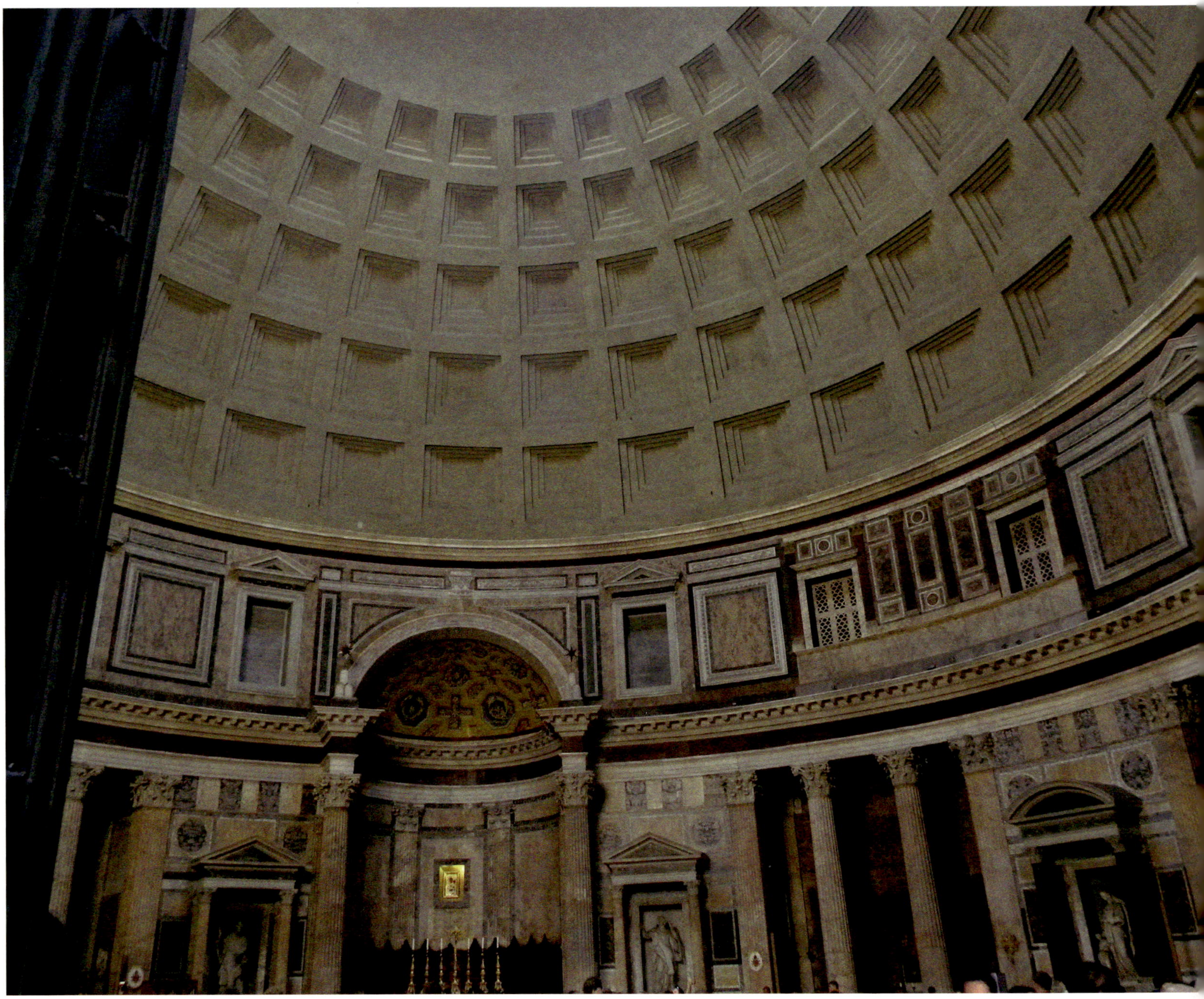

Above The Pantheon, 120 to 124 CE. Rome, concrete. The dome is 141' tall and wide, making it the largest dome in the world until modern times.
Photo: © Camélia78140 / Wikimedia Commons / CC-by-4.0 International.

Opposite page Louis Kahn, *Phillips Exeter Library Atrium*, 1969–1971. Exeter brick, stone, teak, white oak, cast concrete.
Photo: Carol M Highsmith at the Phillips Exeter Academy / Wikimedia Commons / public domain.

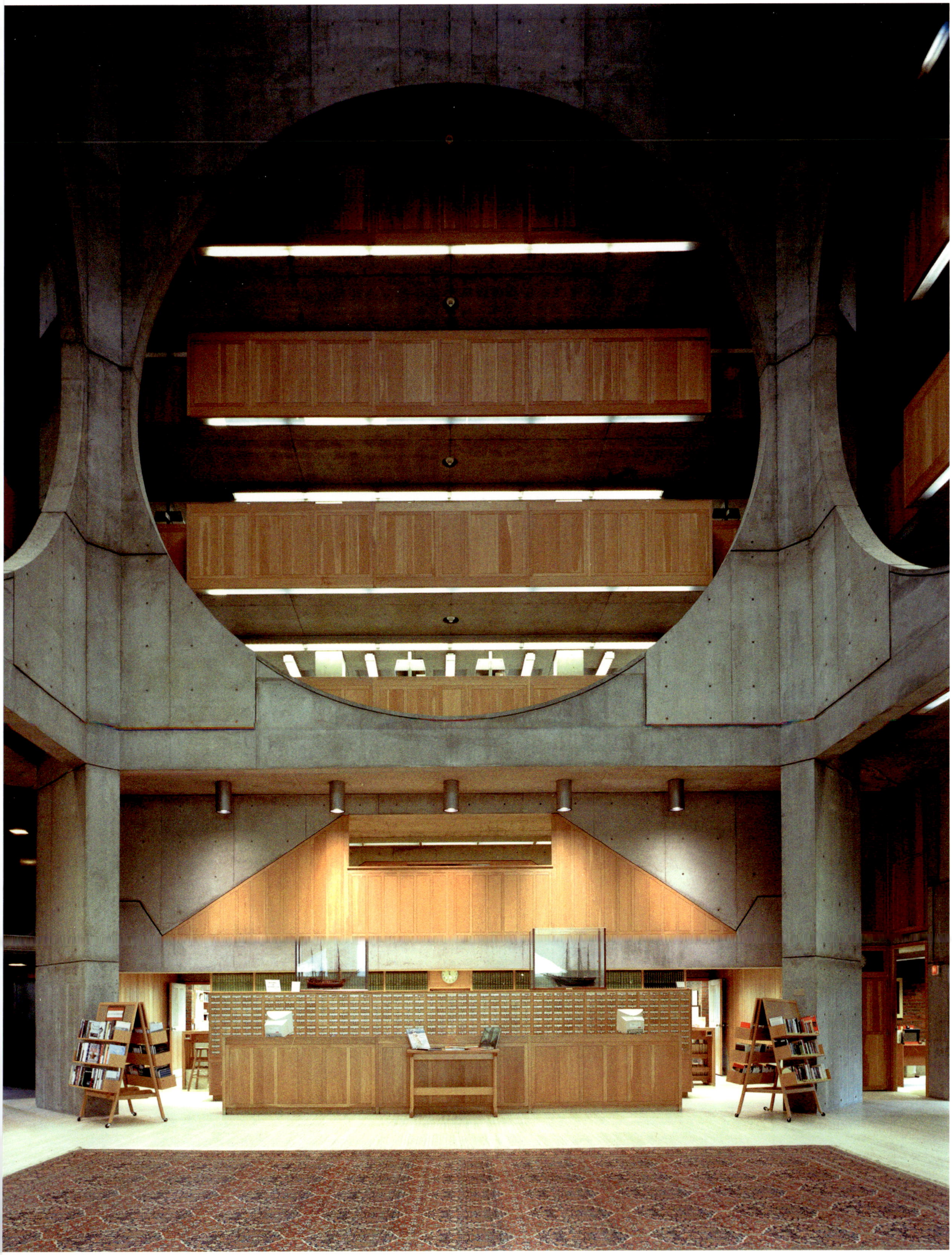

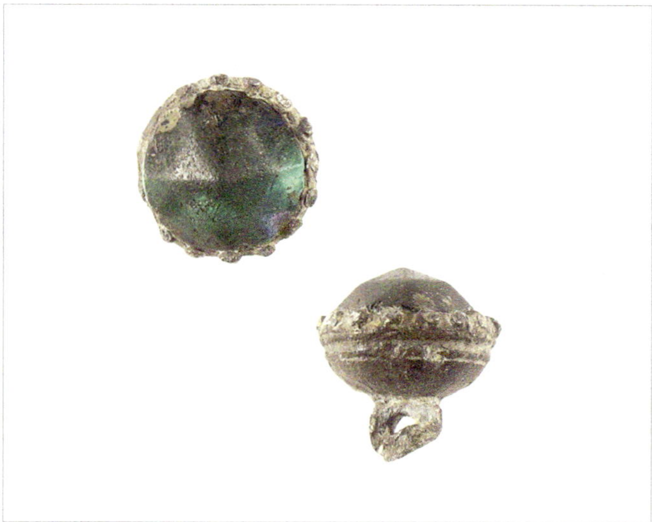

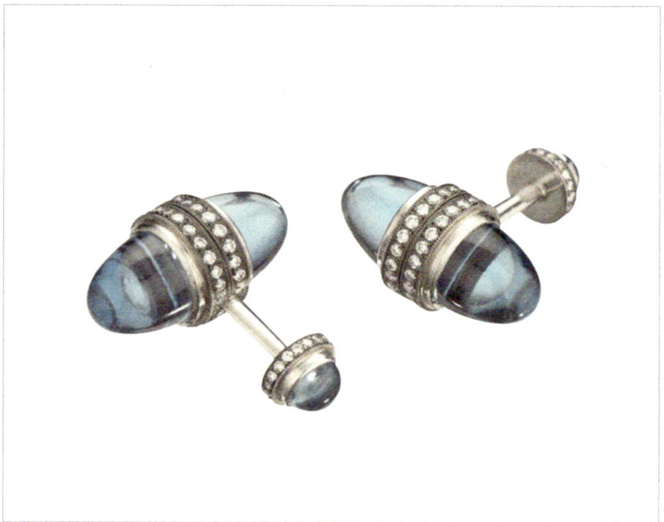

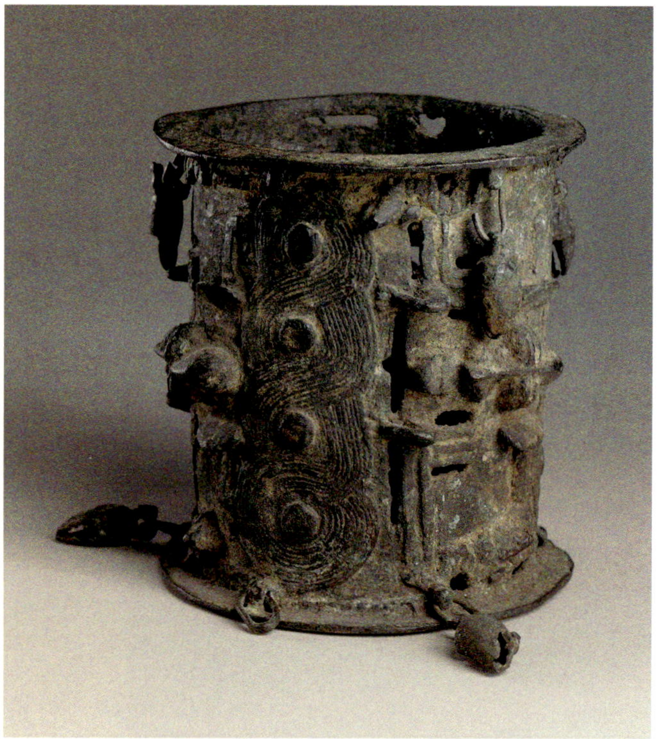

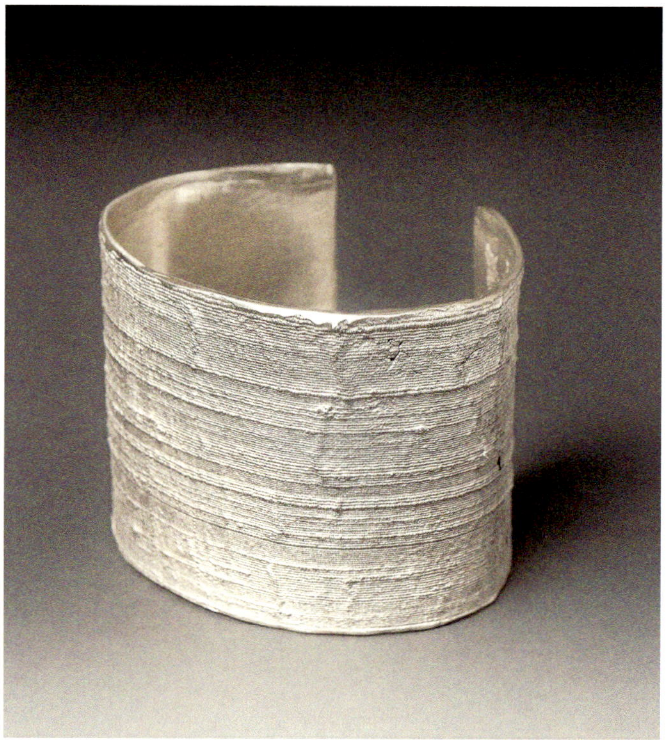

Top left Cufflink, late seventeenth to mid-eighteenth century. Lead alloy and glass with some gilding, cast, 0.51" × 0.48". Courtesy of The Portable Antiquities Scheme, CC-by-2.0.

Top right Ashleigh K. Branstetter, *Gavel Cufflinks*, 2014. Platinum, London blue topaz, diamonds, 1" × 0.25" × 0.25". Courtesy of the American Gem Trade Association.

Bottom left Bracelet of Ogoni society. Nigeria, Africa (Yoruba peoples), bronze, 4.31" × 4.6".
Courtesy of LACMA, Gift of Mr. and Mrs. Herbert Baker, www.lacma.org, LACMA public domain.

Bottom right Maya Kini, *Silk Cuff*, 2014. Silver, 3" × 2.5" × 1.75". Photo: Rik Keller, Collection of Rada Sahney.

Right Fluorescent uranium Depression glass, mid-nineteenth to early twentieth century. Glass, 0.1% to 0. 2% uranium dioxide.
Photo: © JJ Harrison / Wikimedia Commons / CC-by-2.5.

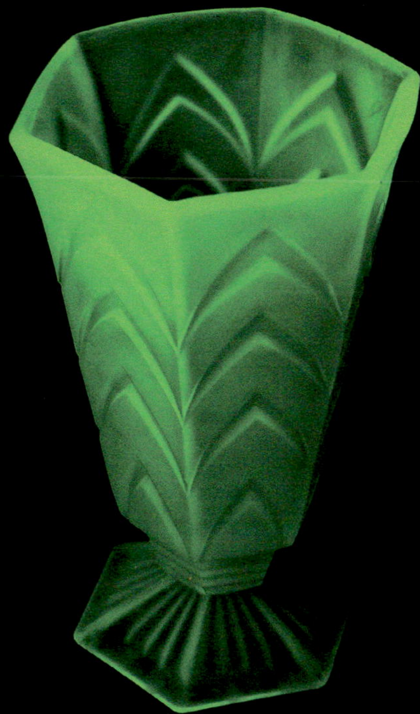

Below This piece is shown in incandescent light and glowing in the dark.

Jean M. Fernandes, *Trapped Luminescence*, 2015.
Glass, phosphorescent powder,
16" × 3.5" × 3.5".
Photo: Jean M. Fernandes.

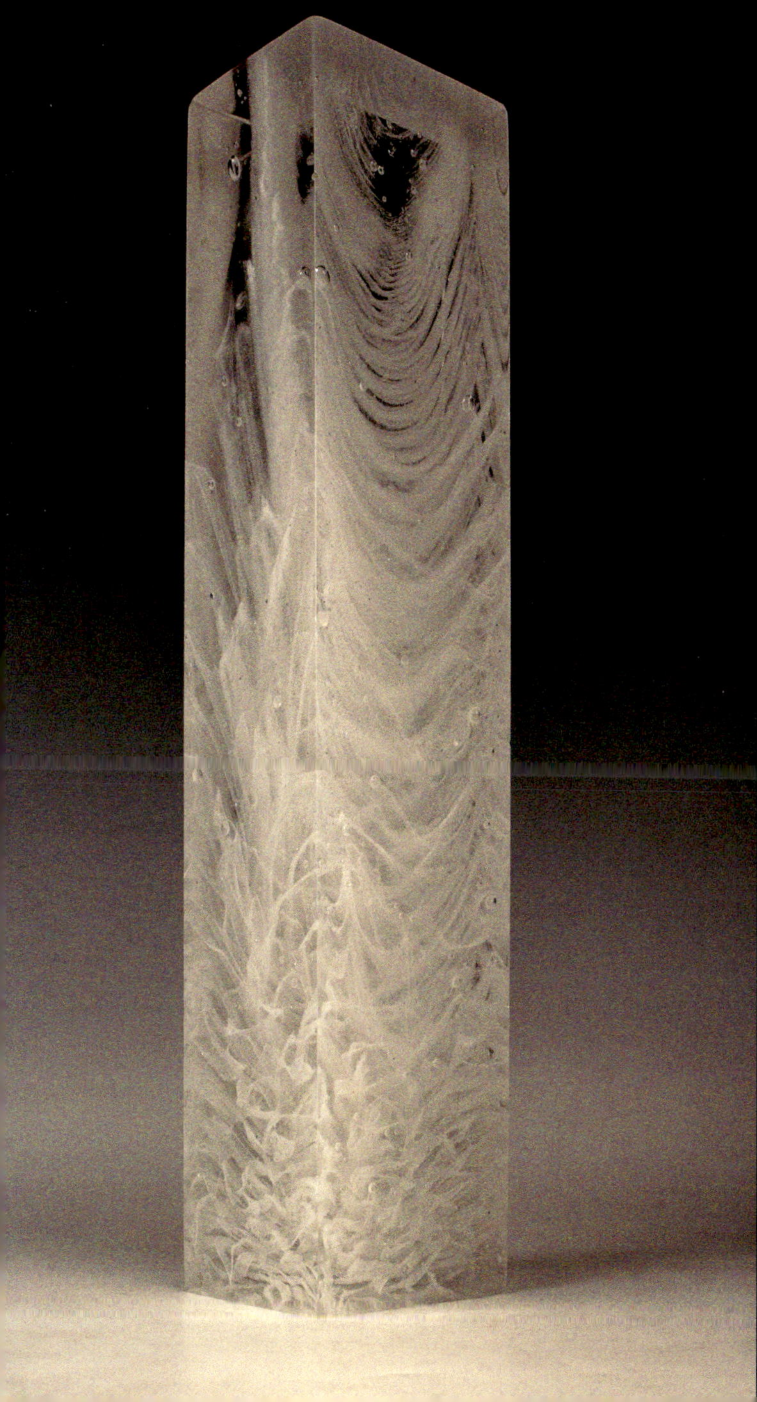

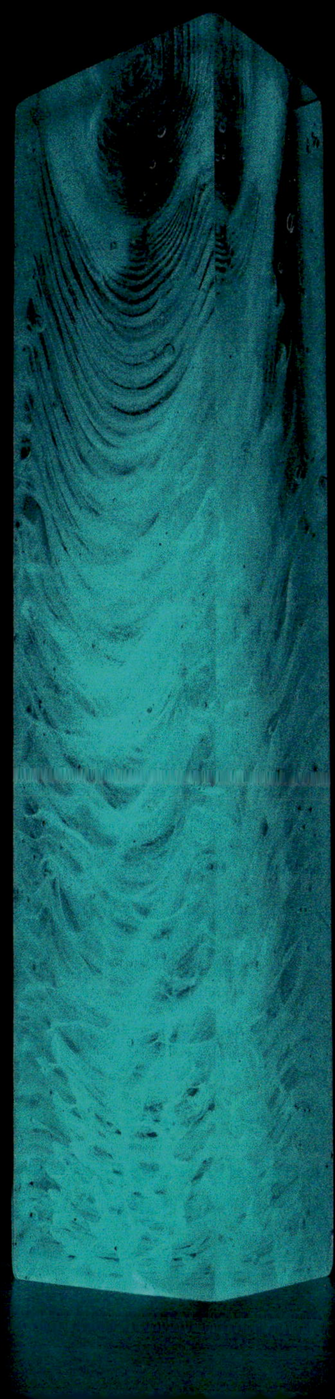

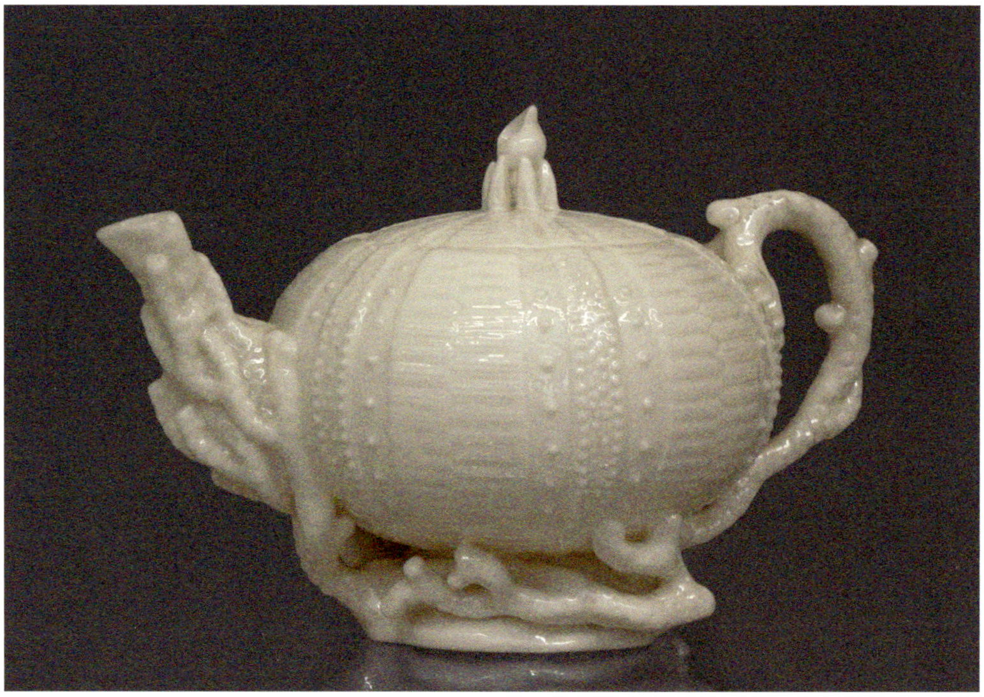

Belleek Pottery, *Echinus' Teapot*, 1869–1882. Ireland, County Fermanagh, glazed porcelain, slipcast. Photo: © David Jackson at the Victoria and Albert Museum, Gift of Lieutenant-Colonel Kenneth Dingwall D.S.O. through The Art Fund / Wikimedia Commons / CC-by-2.0 UK.

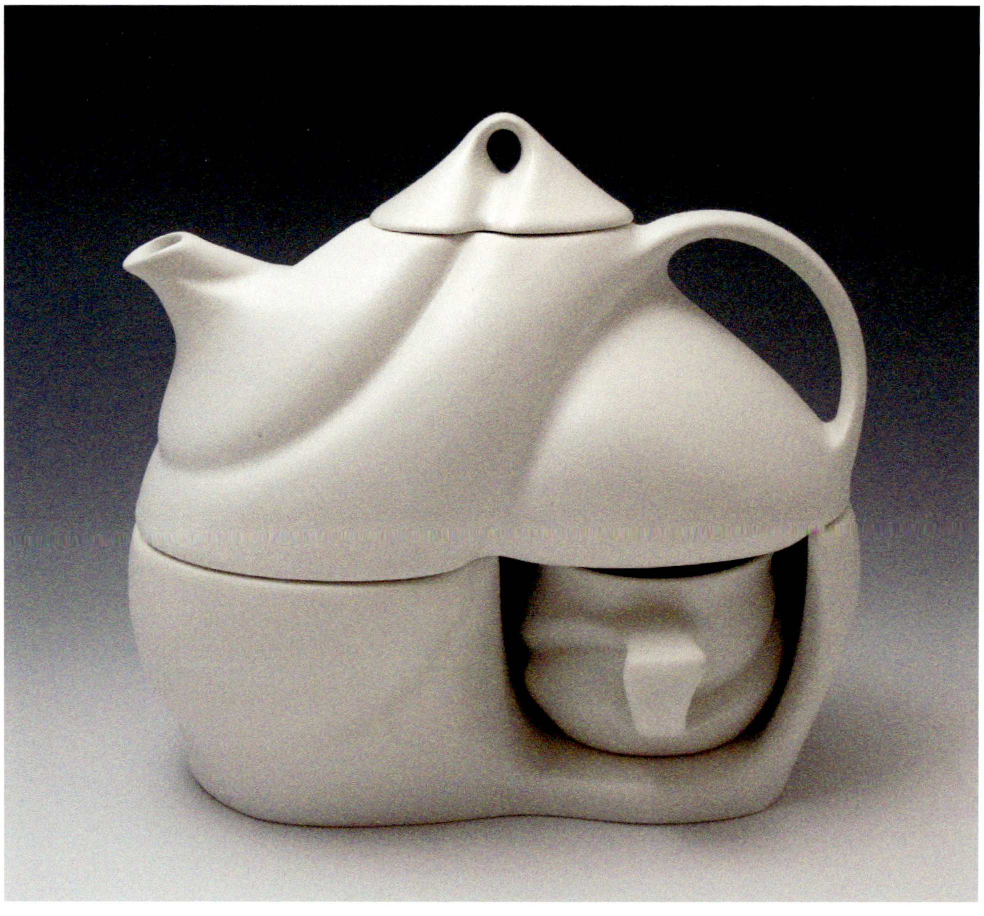

Peter Saenger, *Tea For Two*, 2014. Porcelain, 8.5" × 9" × 4.75". Photo: Peter Saenger.

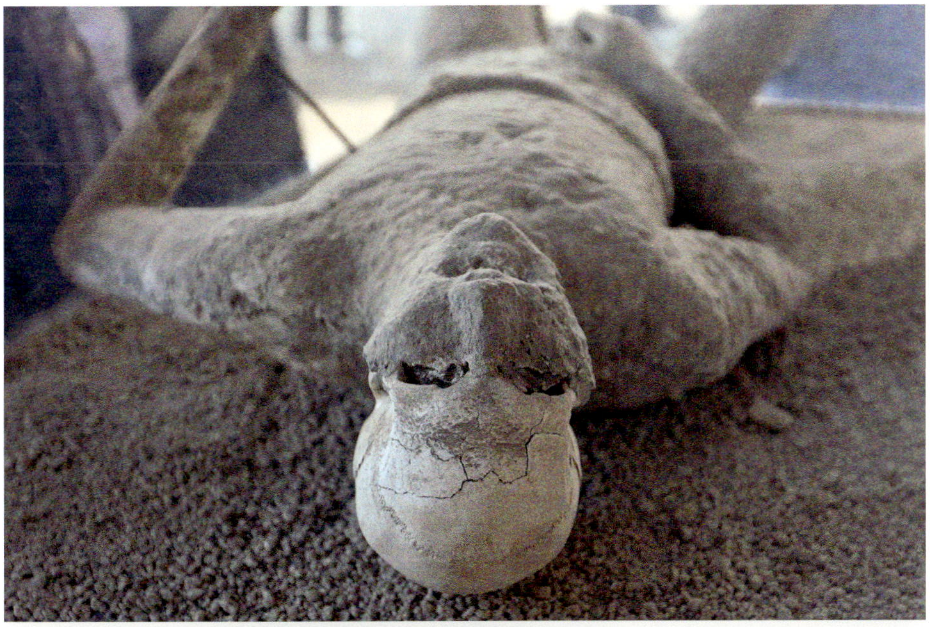

Casting of a victim from Pompeii, eruption of Mount Vesuvius, 79 CE; discovery of the site late sixteenth century, excavation began in 1748. Pompeii, Italy, cement. Photo: © Porojnicu Stelian, Depositphotos.

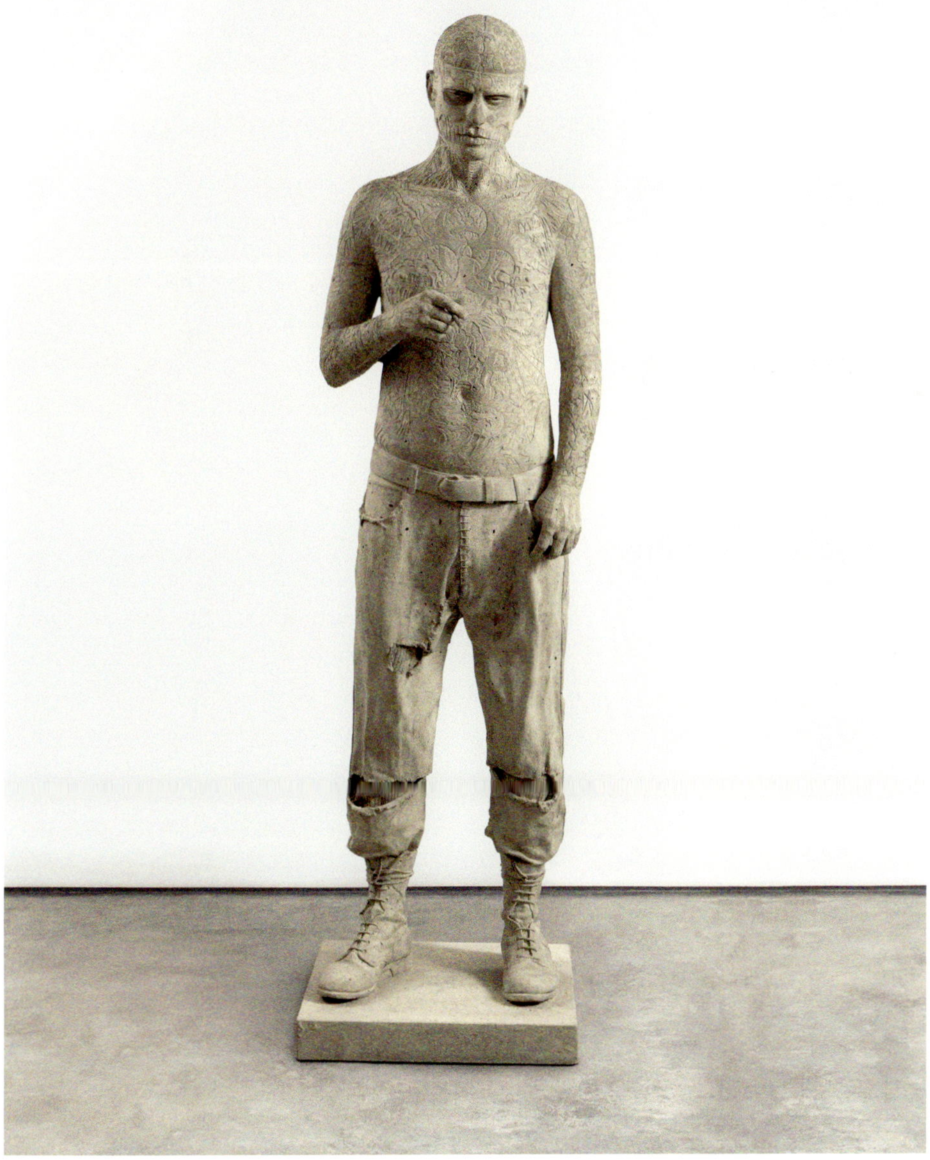

Marc Quinn, *Zombie Boy*, 2011. Concrete, 70" × 22" × 13.75". Photo: Todd-White Art Photography, Courtesy of the Marc Quinn Studio.

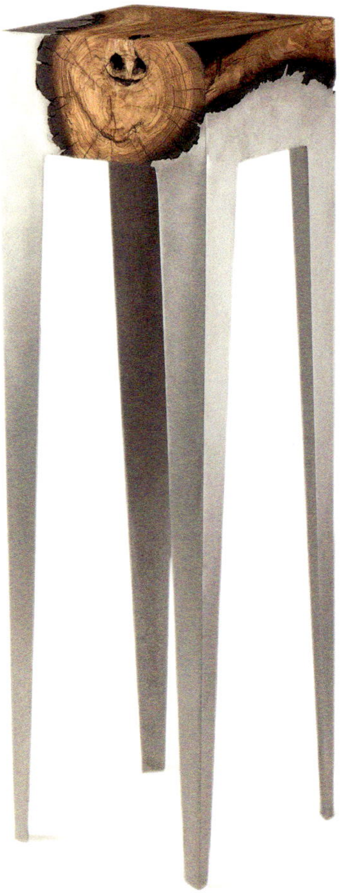

Hilla Shamia Design Studio, *Wood Casting™ Limited Edition Olive Short Console*, 2015. Wood Casting™, aluminum, olive wood, 11.75" × 11.75" × 39.4". Photos: Ido Adan.

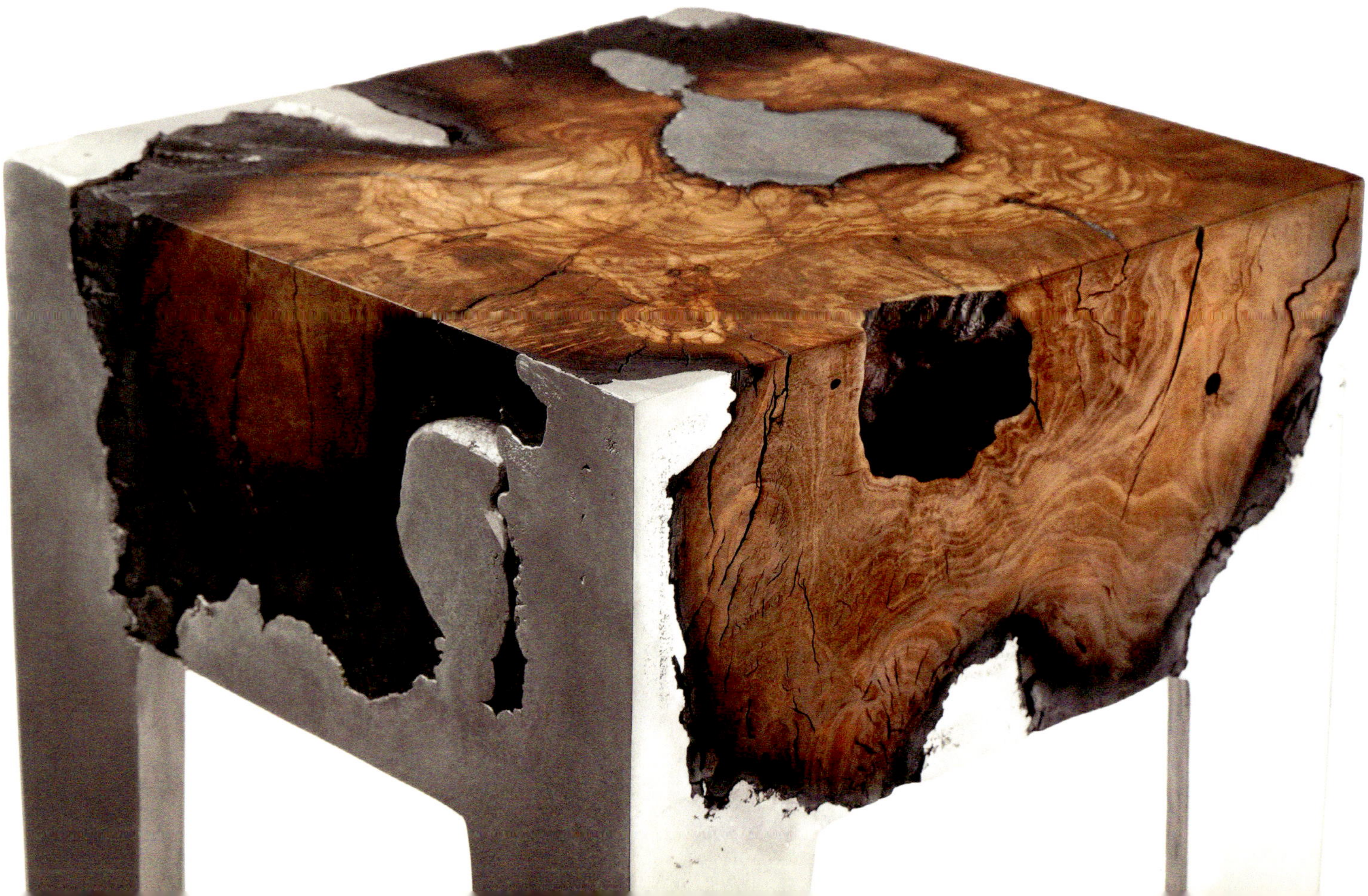

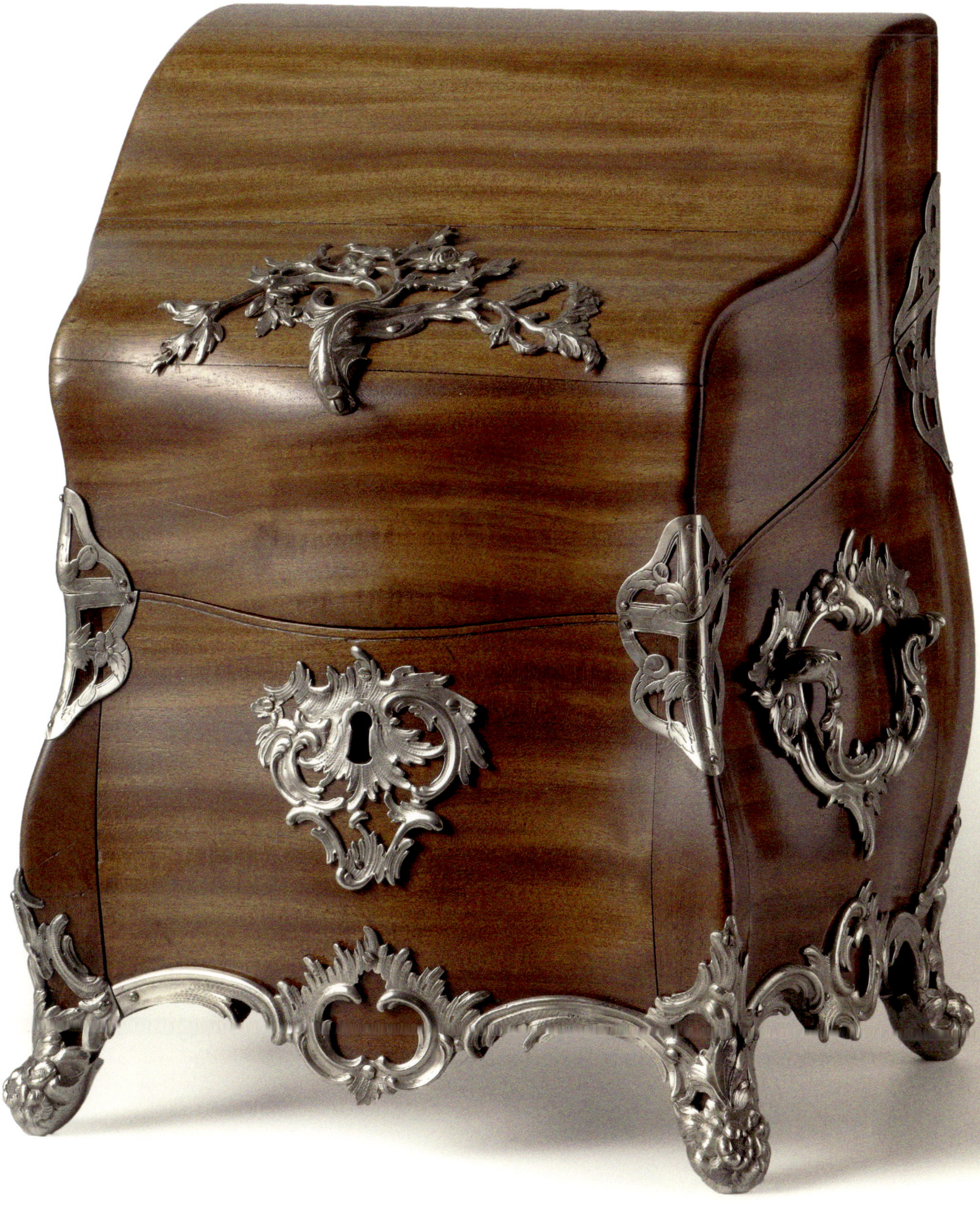

Johannes Schiotling, cutlery canteen, 1772. Amsterdam, the Netherlands, mahogany, oak, silver, velvet, silk lanyards, 14" × 11.6" × 8.7". Courtesy of the Rijksmuseum, purchased with the support of the Stichting tot Bevordering van de Belangen van het Rijksmuseum.

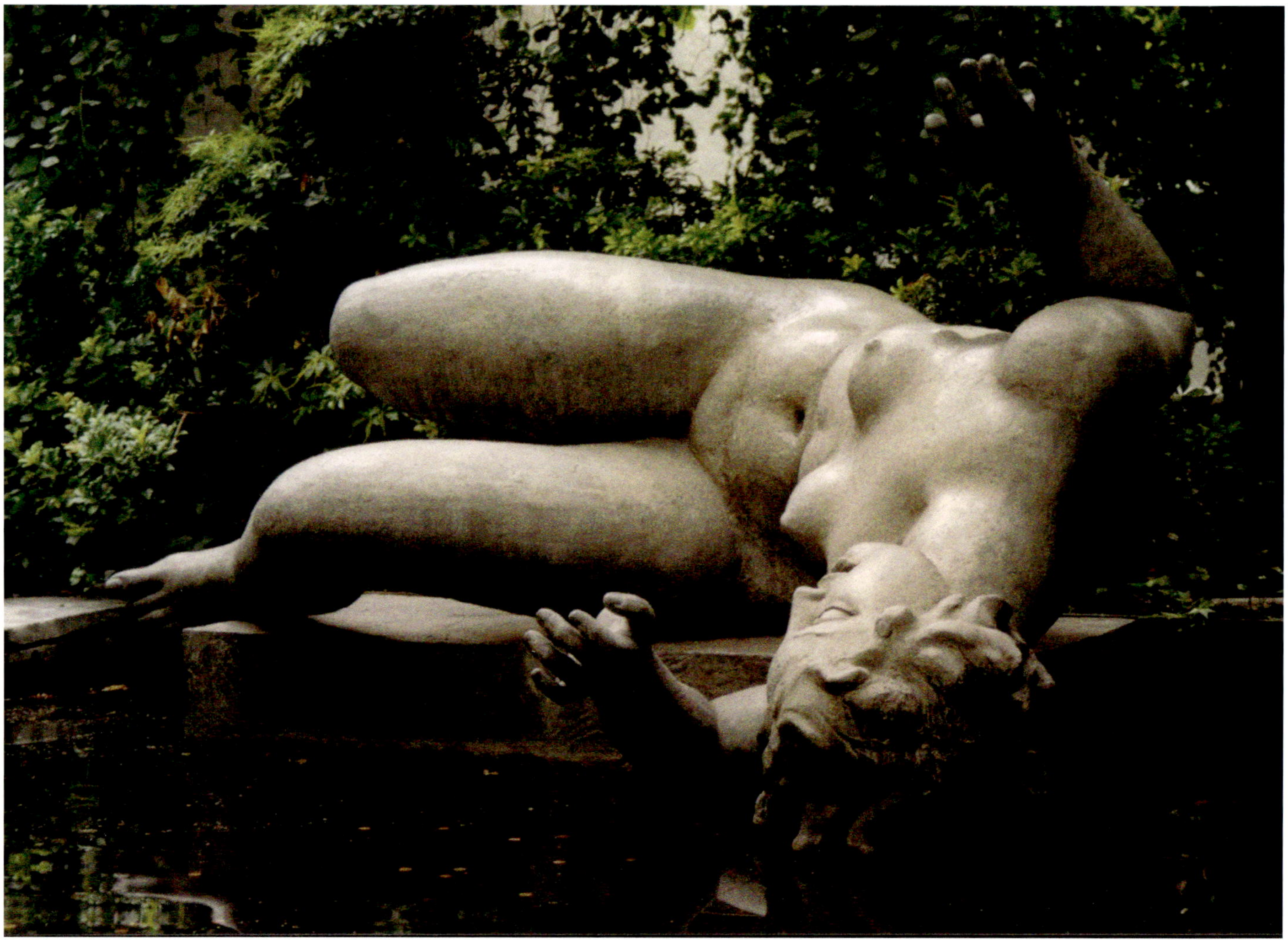

Aristide Maillol, *The River*, begun 1938–1939; completed 1943 (cast 1948). France, lead, 4'5.75" × 7'6" × 5'6" (on base measuring 9.75" × 5'7" × 2'3.75").
Mrs. Simon Guggenheim Fund, © 2015 Artists Rights Society (ARS), New York / Photo: © Janine Curry, PhD at MOMA / CC-by-2.0.

Opposite page Cameron Stalheim, *and then I saw Colby on the street and my fantasy died.*, 2014. Plastic, foam, steel, acrylic, 20' × 3' × 4'. Model: Colby Keller. Photo: Cameron Stalheim.

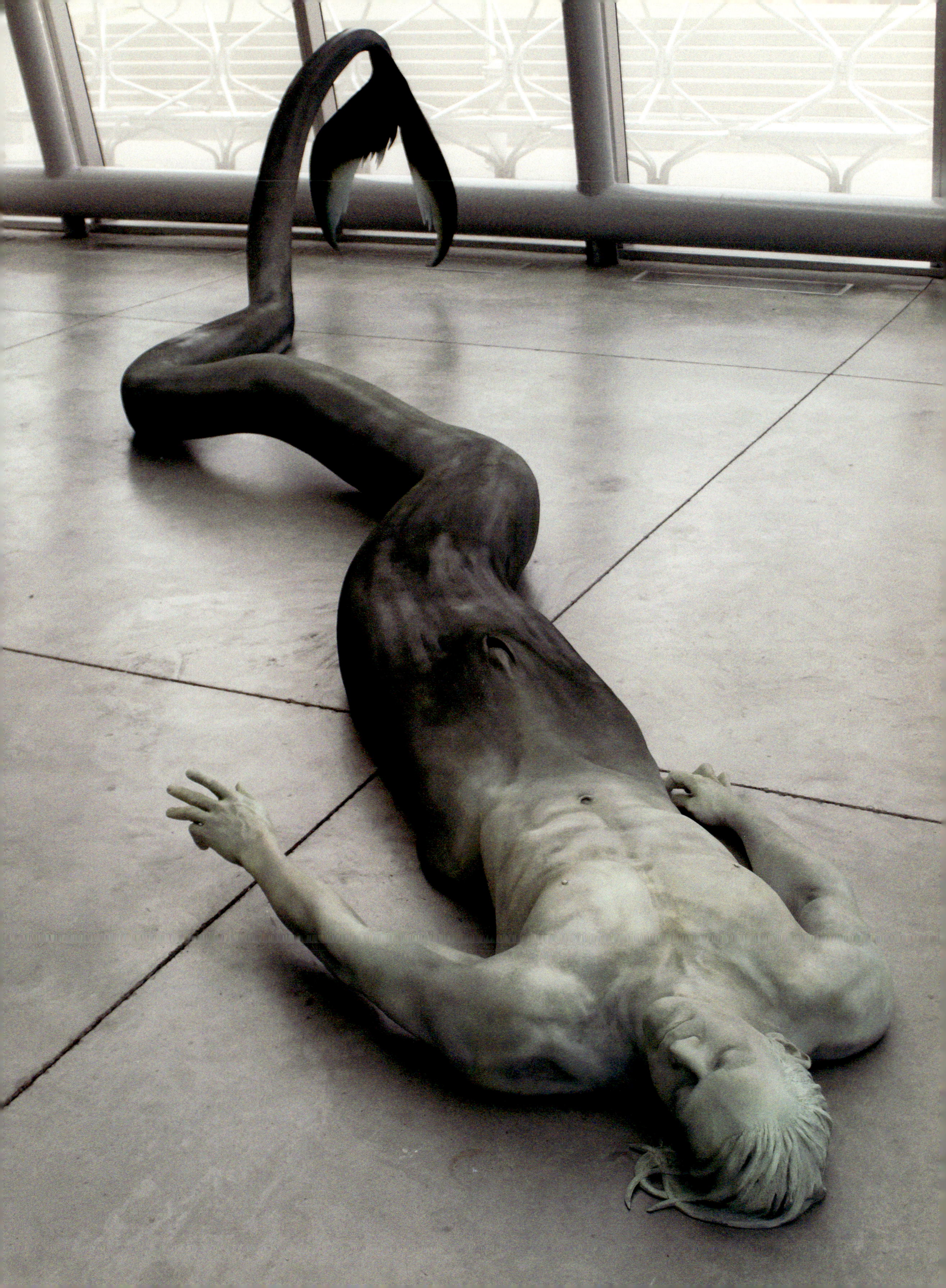

CREATION

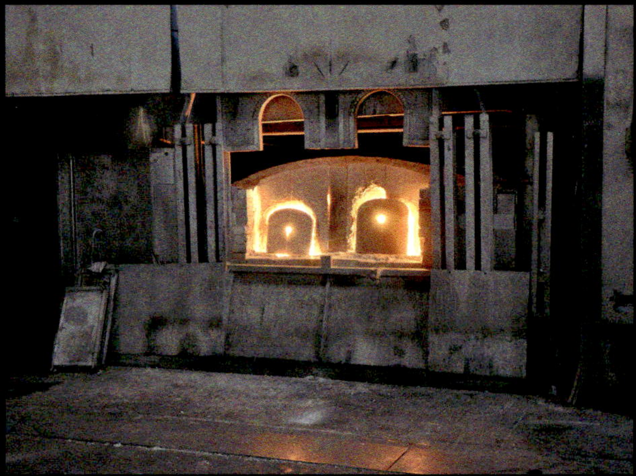

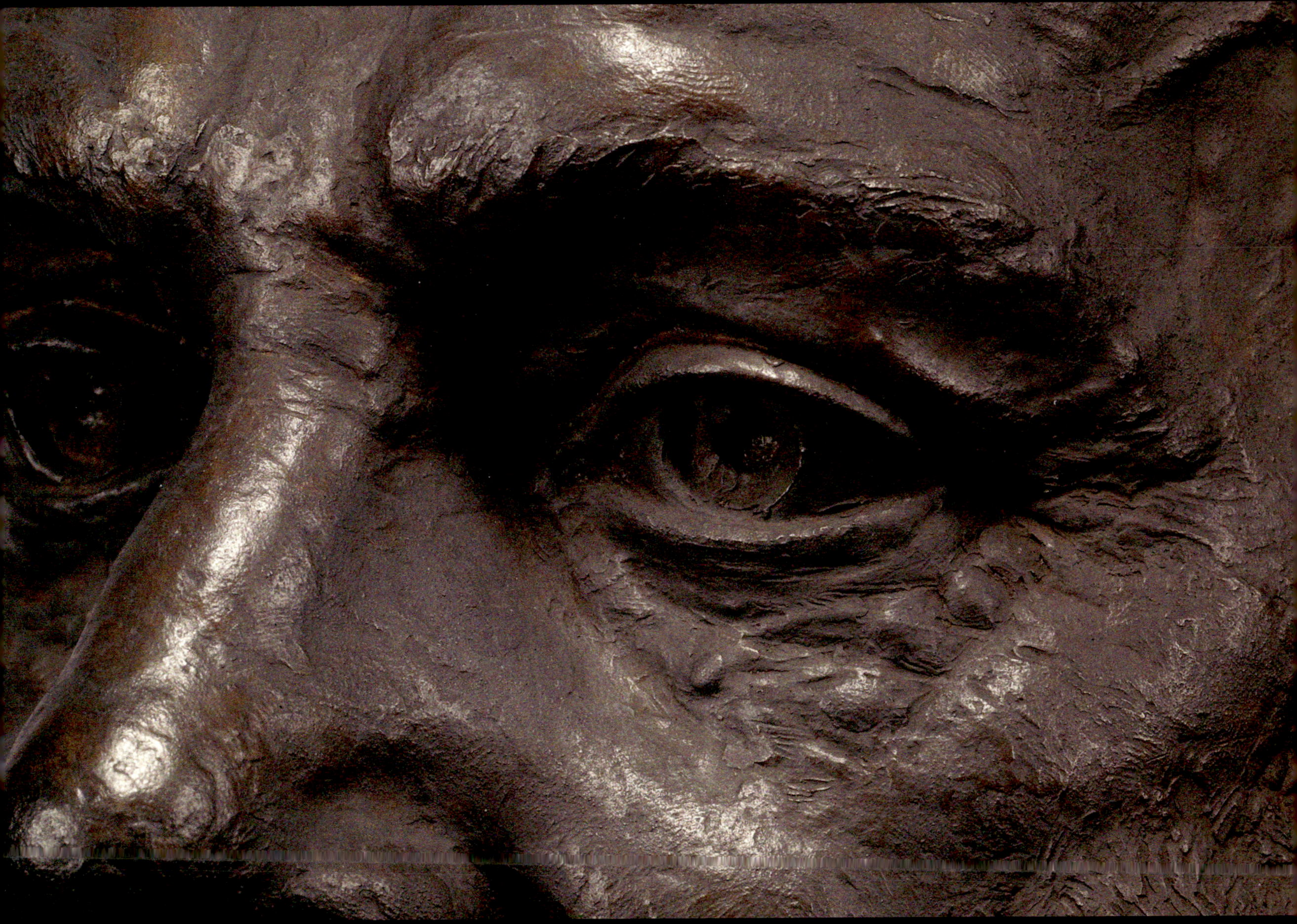

Ernest Dagonet, *Portrait Bust of George A. Lucas*, 1899. France, bronze, 19" tall.
Courtesy of The Walters Art Museum, Acquired by William T. or Henry Walters, CC0, GNU FDL.

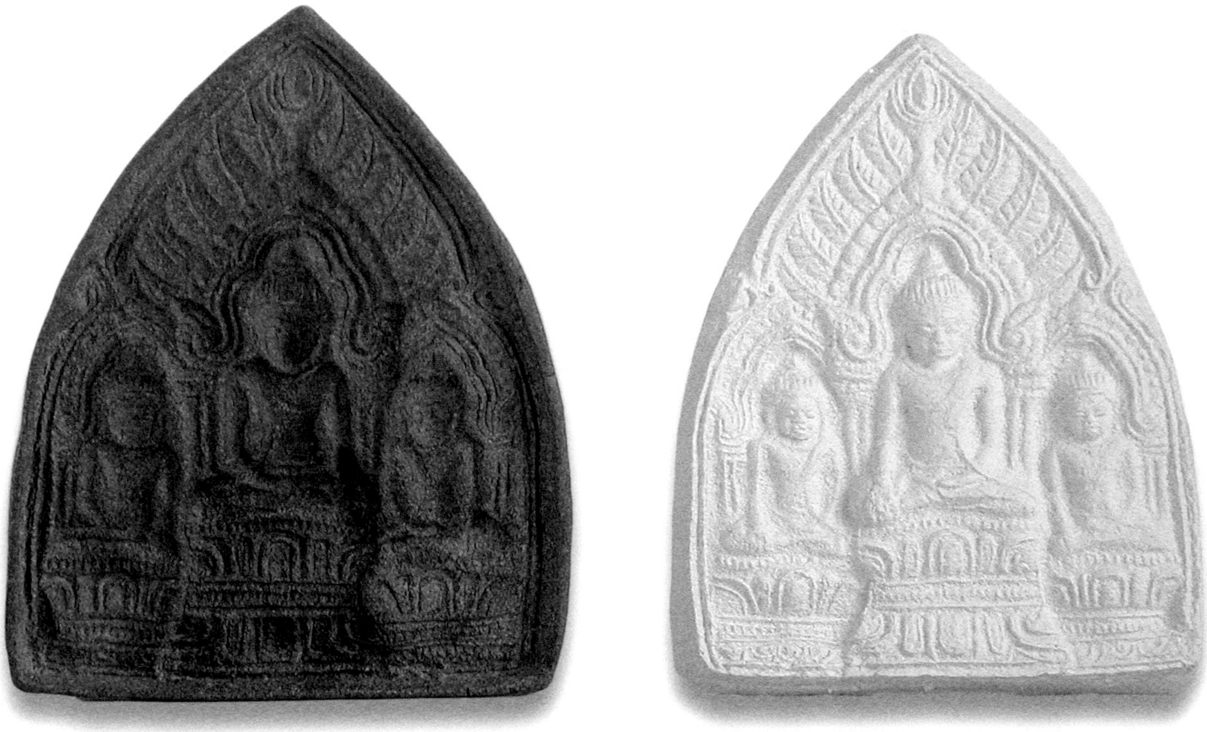

Mold for a Buddhist votive tablet (left) and cast tablet (right), twelfth to thirteenth century. Thailand, mold of copper alloy, casting is most likely plaster, 4.5" × 3.75" × 1.37".
Courtesy of LACMA, Gift of Marilyn Walter Grounds, www.lacma.org, LACMA public domain.

People have been casting for more than 10,000 years. As technology has evolved and materials have been refined, the variety of approaches, media, and applications are endless. This chapter investigates the thinking behind the making and highlights seventeen contemporary casters working in a variety of materials, each with their own methodology.

Casting has commonalities regardless of the materials employed. There are four parts to the process: finding or making a model, creating a mold, the cast itself, and whatever cleanup is required.

Models can be a human face or hand, a carefully carved wax ring, a monumental clay figure, an architectural element made of wood, a dried flower, and on and on. There are no limits.

Casting always involves a mold. Often, molds are single-use, as in most traditional lost wax casting processes; investment, ceramic shell, sand, and alginate all lead to molds that get destroyed in the removal of the casting (this is called a waste mold). Other molds are intended for producing multiples; these are made of more durable materials like rubber, plaster, graphite, metal, and wood.

The mold provides a cavity into which the casting medium is poured, pressed, injected, or flung. Molds can be simple open-faced devices that produce bas-relief forms, or they can be multi-piece and complex-to-create forms that would be impossible to achieve any other way. Molds can also require multiple materials to function properly, as in the case of lifecasting. A skin-safe, brushable rubber is painted onto a face, hand, or other part of the body creating a thin, flexible, rubber shell. After the rubber sets, it requires what's called a "mother mold" to support the floppy rubber shell during casting. Mother molds are typically made of plaster or plastic. Whether simple or complex, made for a single use or mass-production, a good mold is key for a quality cast.

Materials can be cast hot, like metal and glass, or cold, like ceramic slip, plaster, or resin. Hot or cold, they almost always start as liquids and solidify to form the final object. The transformation from liquid to solid has an alchemical feeling. Many artists, upon seeing casting for the first time, fall in love with it. This transformative dynamism is especially strong in mediums like glass and metal that are cast hot; the fiery sparks of an iron pour or the honey-like glowing hot glass can be mesmerizing, sensual, thrilling, and—at times—confounding… all the ingredients for a super-charged love affair.

The MINDSET of CASTERS

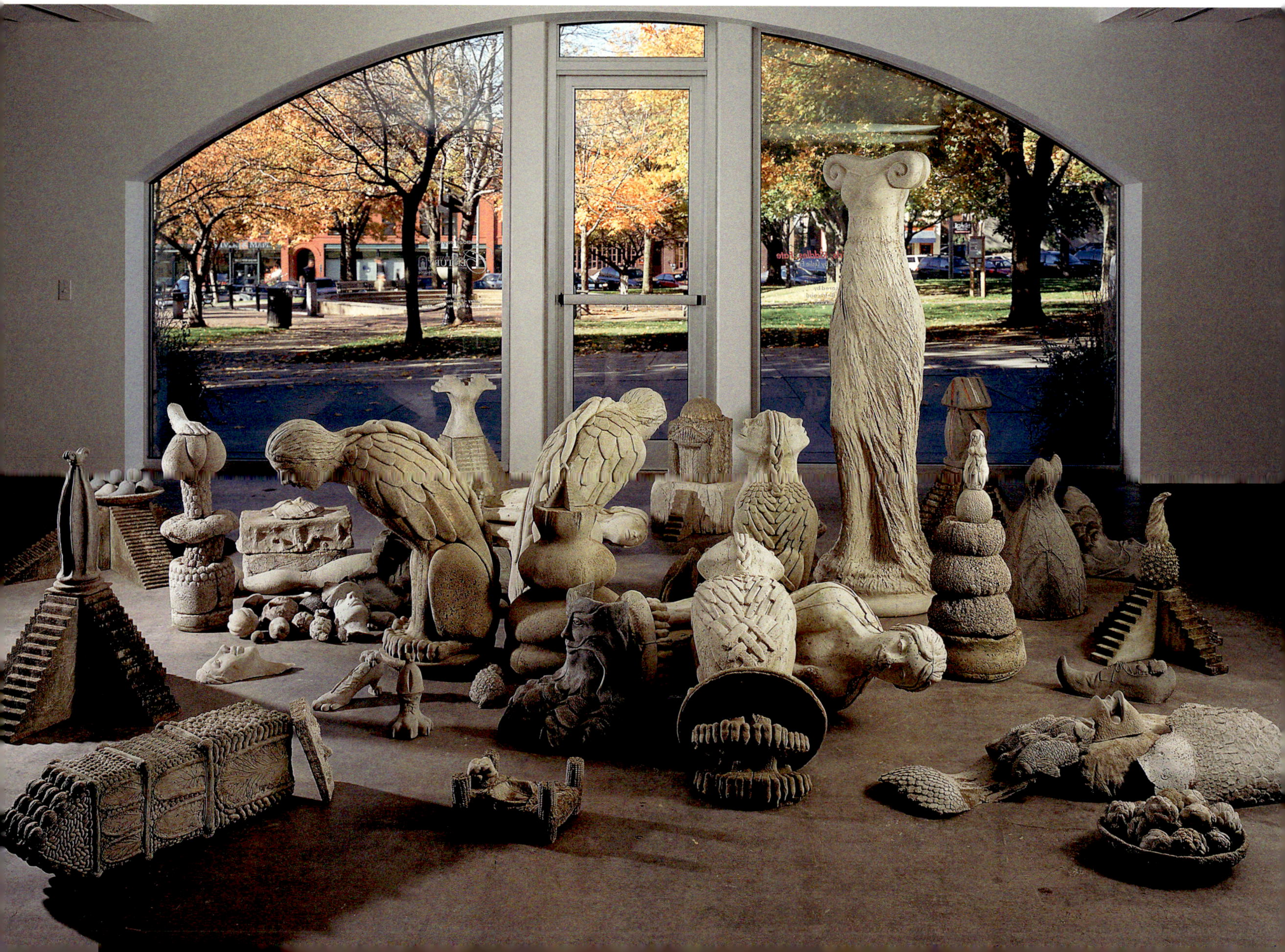

Casters working in different mediums share certain ways of thinking. Whether it's creating a model that will lend itself to the casting process or making a successful mold, ensuring a high-quality result is a common goal. A focus on achieving good surface fidelity free of bubbles, flaws, and voids, or on planning the piece for effective finishing, are commonalities that span not just contemporary makers of all stripes, but makers throughout the centuries.

These technical concerns are just one part of the shared mindset, though. Casters think in the round. One of the strengths in this way of working is the ability to create forms that would be impractical, and sometimes impossible, to create by other means.

Casters often build up a library of molds in order to produce multiples or iterations of their forms. This collection of molds can become an incredibly valuable resource, allowing makers to reuse, recombine, and recontextualize components or entire forms. In the case of Leslie Fry's installation, *Cast-offs: Girls, Riddles, Fate*, the molds are an actual part of the work. Fry has composed a collection of cast concrete forms, while in the corner of the gallery, the rubber molds used to create them are "cast off" in a heap.

Each of these forms could be shown as a sculpture in its own right, but en masse, they have a different kind of presence and reference past civilizations, myth, the ravages of time, and with the addition of the molds heaped up in the corner, casting itself.

Casting gets into artists' blood, though it's hard to pinpoint one thing that makes the process so captivating. There are many aspects to love, and different parts of the process appeal to different personalities. For some, it's carving a wax model, and for others it's the ability to create multiples; some people are passionate about the intricacies of moldmaking, and others are drawn to the fire and drama of the cast itself.

Opposite page & right Leslie Fry, *Cast-Offs* (from the solo exhibition *Cast-Offs: Girls, Riddles, Fates* at Burlington City Arts, Burlington, Vermont, USA), 2004. Concrete, rubber, variable dimensions. Photos: Burk Uzzle.

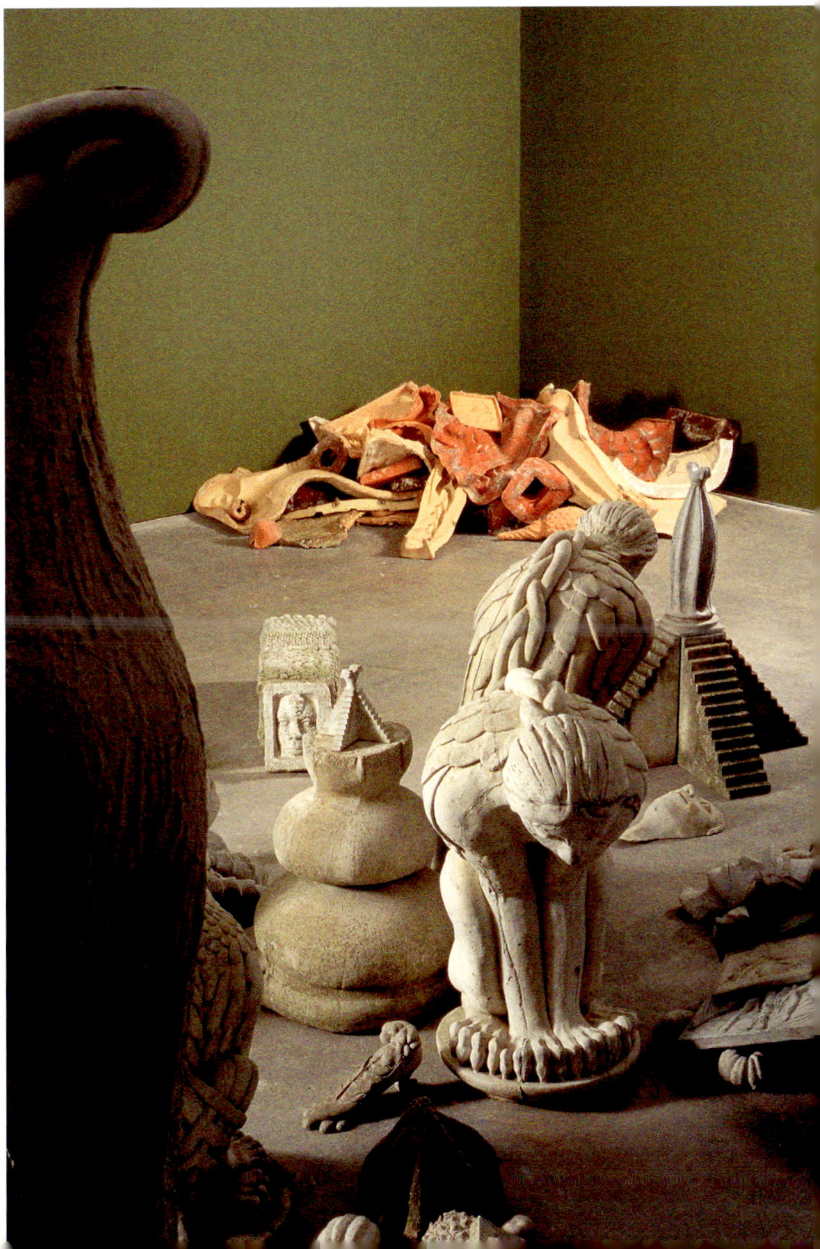

Casters on Casting

This fascination leads many artists to make work *about* casting. Yoko Sekino-Bové made a series of cast cups during her residency at the John Michael Kohler Arts Center that chronicles what she saw and did in the center of one of the most intense casting facilities in the world. There is something that speaks to her, and many artists', fascination with the process—the cups themselves are cast and the imagery on them is that of casting. Kohler, known worldwide for its beautiful and high-quality kitchen and bathroom fixtures cast in both porcelain and iron, established a highly competitive residency for artists in 1974. This coveted residency is a chance for artists to collaborate with the highly skilled Kohler casters in their state-of-the-art facilities in Wisconsin. It is a beneficial arrangement for everyone involved—exciting and energizing for the Kohler casters to work next to creative artists of all disciplines, and a rare opportunity for the artists to tap into the technical know-how and facilities of Kohler. The work that comes out of this residency is of the highest quality. Many artists featured in this book held a Kohler residency.

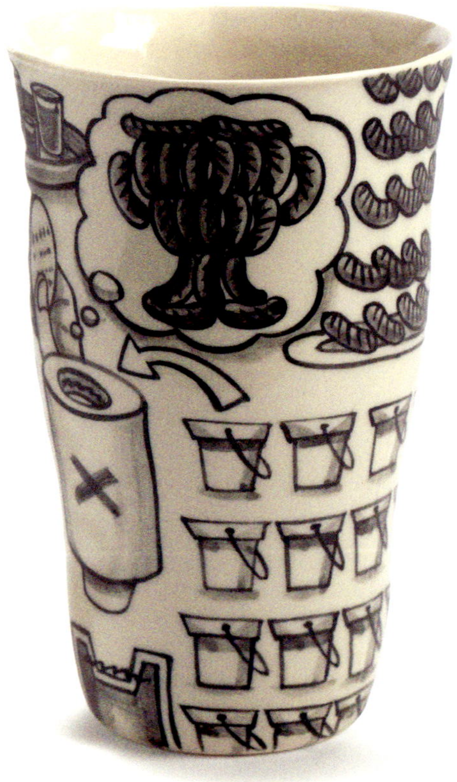

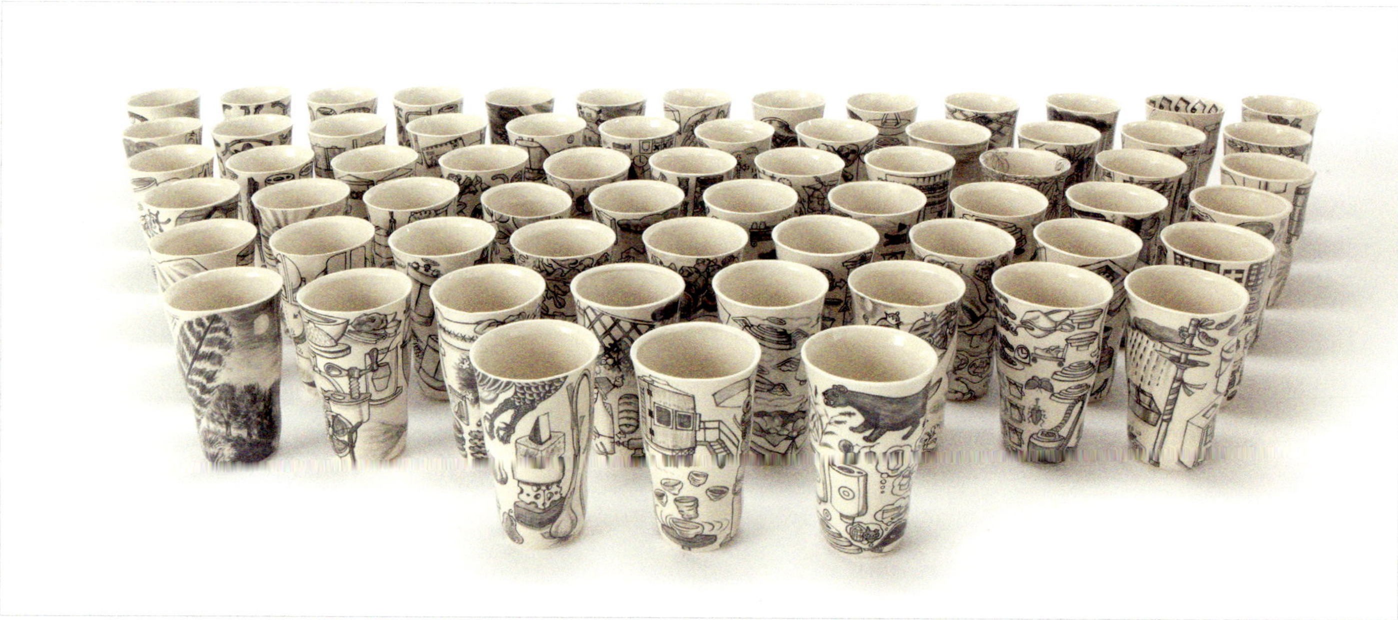

Yoko Sekino-Bové, *Diary Cup Project*, 2014. High-fire stoneware, 7" × 3" × 3" (each cup). Created in Arts/Industry, a long-term residency program of the John Michael Kohler Arts Center in Sheboygan, Wisconsin, USA. Arts/Industry takes place at Kohler Co.
Photos: Courtesy of Rich Maciejewski.

Casters Are Fastidious

Any casting is dependent on the quality of its model. Experienced casters know all too well how aggravating it is to fix the same mistake over and over again, and so painstaking effort goes into perfecting the original model before it is molded, so that problems aren't duplicated in its reproduction. Models can be made from a wide range of materials; traditional options are wax, clay, plasticene, and plaster, but anything that can be molded can work. Each material requires a different approach—some are carved and some are modeled with fingers; others, like organic materials, need to be very carefully sealed with lacquer or polyurethane to ensure that an accurate mold can be taken. Regardless, the level of care given to the model creation is directly proportionate to the quality of the final casting.

Cleanup, also known as "finishing," "de-spruing," "sanding," "grinding," and "polishing" in various fields, can be just as labor-intensive as the model creation itself. These processes all boil down to the same goal: addressing any unwanted aspects of the raw casting. Typically, this is anything in the casting that looks different than the original model. Sometimes, leaving evidence of the casting process is part of the concept and intention behind the piece, in which case less finishing would be done; but this is relatively rare. Parting lines left from the mold, sprues, bubbles, and porosity are common flaws that require correction.

Corners are often cut when it comes to the finishing done on low-quality mass-produced objects, because it requires patience and finesse. Poor cleanup is, in part, what contributes to the misguided notion that casting is an inherently substandard process. Attention to cleanup requires a surprising amount of time and skill; camouflaging a scar where a sprue attached to a form, blending a surface over a flaw, or hiding a pit in a polished area is the mark of a skilled caster.

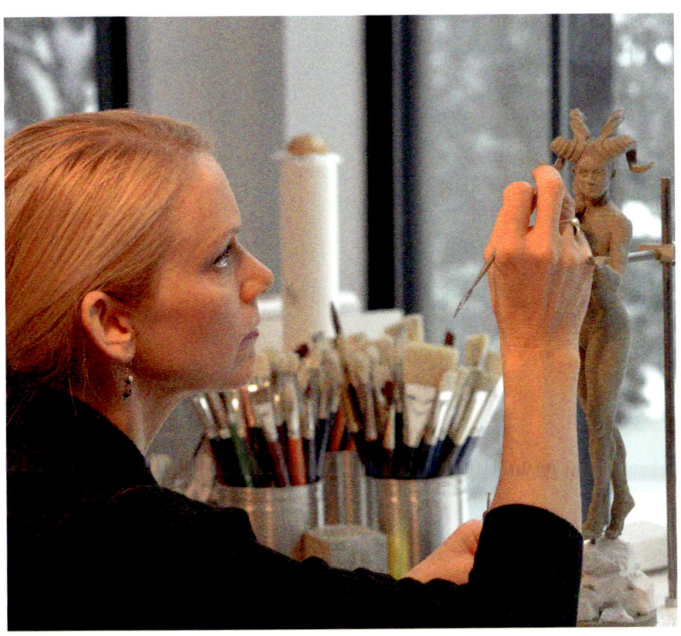

Kristine Poole refining details on the clay model before the molding process. *Spectrum Muse* original clay model, 2015. Clay, 16" × 6" × 5".
Photo: Colin Poole.

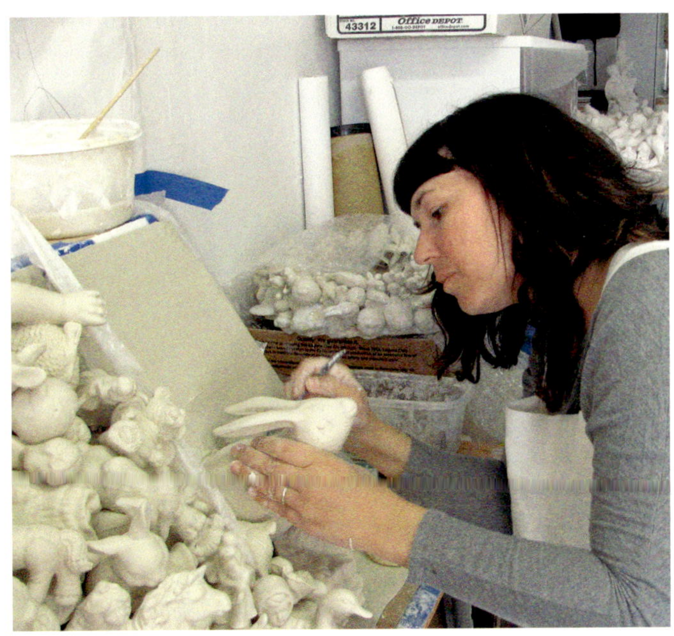

Shenny Cruces working in her studio, on her piece *Accumulations*, attaching a rabbit figure to a pile of slipcast porcelain figurines made from collected objects, 2015.
Photo: Lucy Phillips.

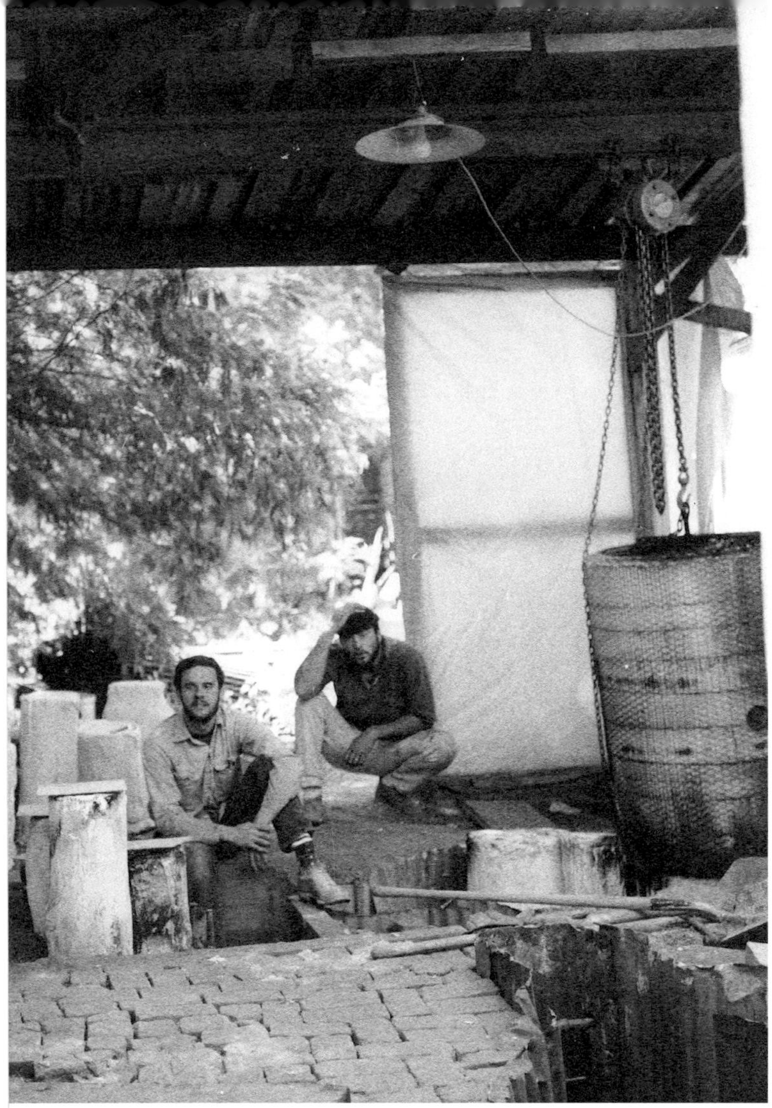

Kindred

Casting is often a group effort, and in fact, some castings are impossible to carry out alone. Generally speaking, any casting of significant scale in any medium requires a group effort.

To achieve a successful casting in certain mediums, the process resembles a choreographed performance. Bronze and iron pours are the most obvious examples. Iron is rarely used industrially anymore, so iron pours have developed a cult-like following for blacksmiths, sculptors, and artists of all varieties. Iron pours are often an annual event at universities and art centers and people will drive from several states away to attend. An iron pour requires time, labor, and fuel, so they are special occasions. There is a primal and seductive quality to iron pours; the sparks, smoke, and drama are captivating to watch, and many people attend to do simply that. The pours are sometimes done at night, which creates a pyrotechnic display and, as is true with fireworks, there is a breathtaking "wow factor." Big casting events are unparalleled by any other process in art-making as far as dynamism and drama are concerned.

The end results of iron pours are, in some ways, less important than the events themselves. The gathering of like-minded makers and enthusiasts bonds communities together and the sheer joy of the voyeurism is infectious. Some artists even create iron pour performance pieces. Allen Peterson's *Colony* is one example of this. Each cast iron bee is fused to the end of a short steel rod, which then slides into a telescoping piece of pipe that acts as a long handle. Participants move the hot glowing bees through the air like kabuki puppeteers disappearing in the blackness of night. The bees are bright against the night sky and their movement creates a dramatic display.

Whether made into an intentional performance or just by their very nature, casting events can connect people. Iron pours are the most extreme example, but many casting processes draw a crowd of excited onlookers and often turn casual observers into practitioners.

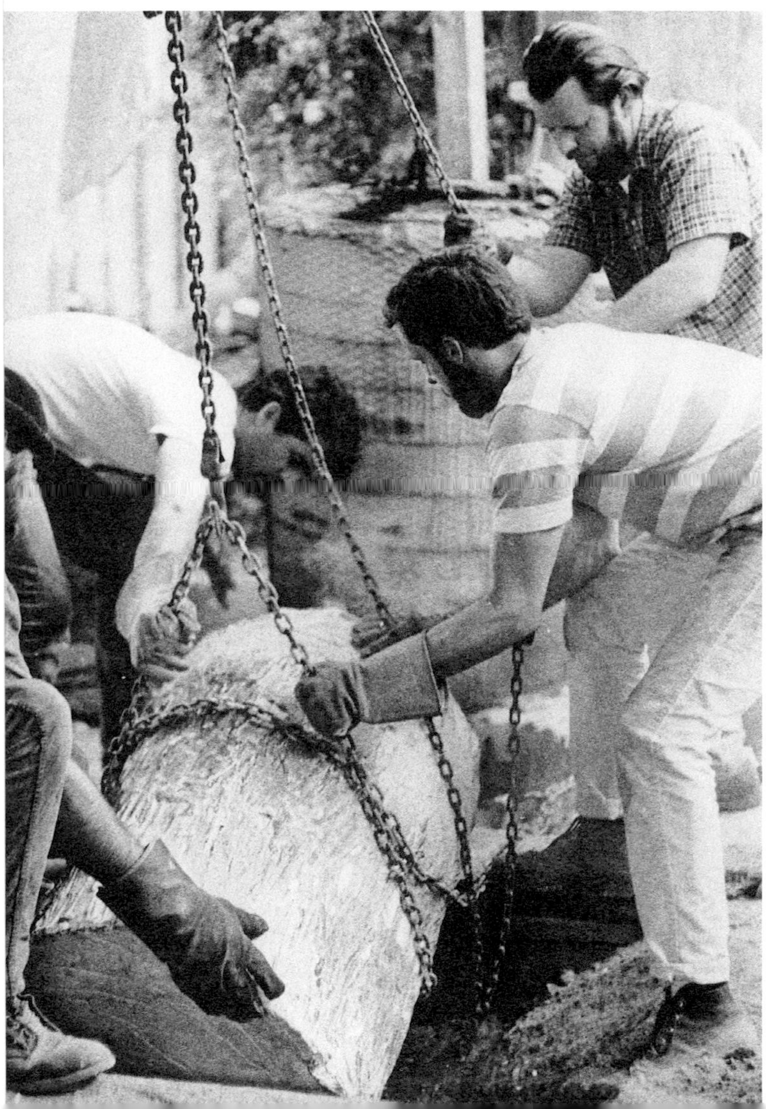

Top Lining up standard investment molds at the Sirenworks Foundry. Stephen Daly *(left)* and Noah Osteen *(right)*, 1963. San Jose, California, USA.
Courtesy of Margaret and Richard Mills.

Left Lowering large standard investment mold into casting pit prior to the pour at Sirenworks Foundry. *Left to right*: Noah Osteen, Richard Mills, Stephen Daly, and Daryle Webb, 1964. San California, USA.
Courtesy of Margaret and Richard Mills.

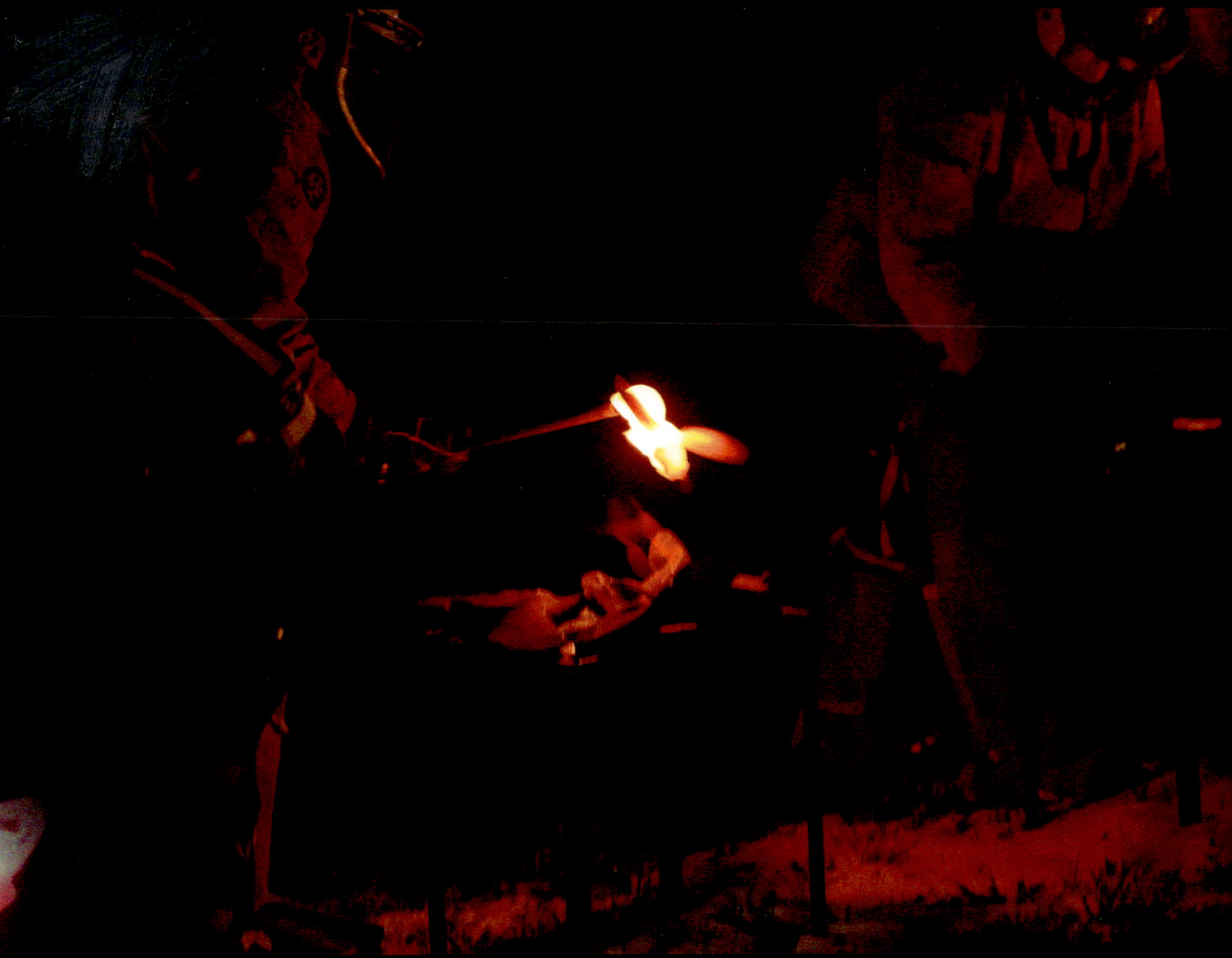

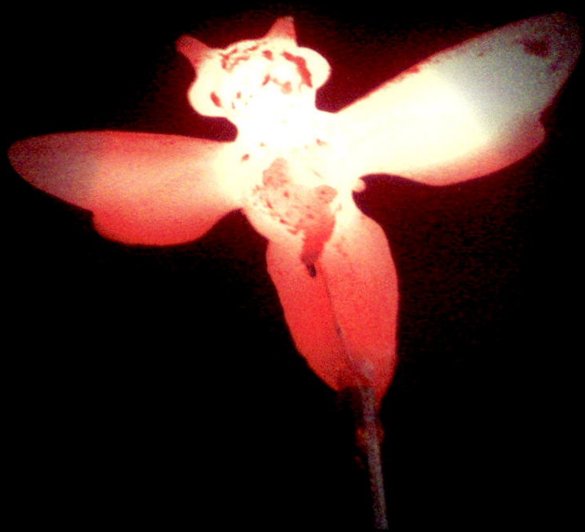

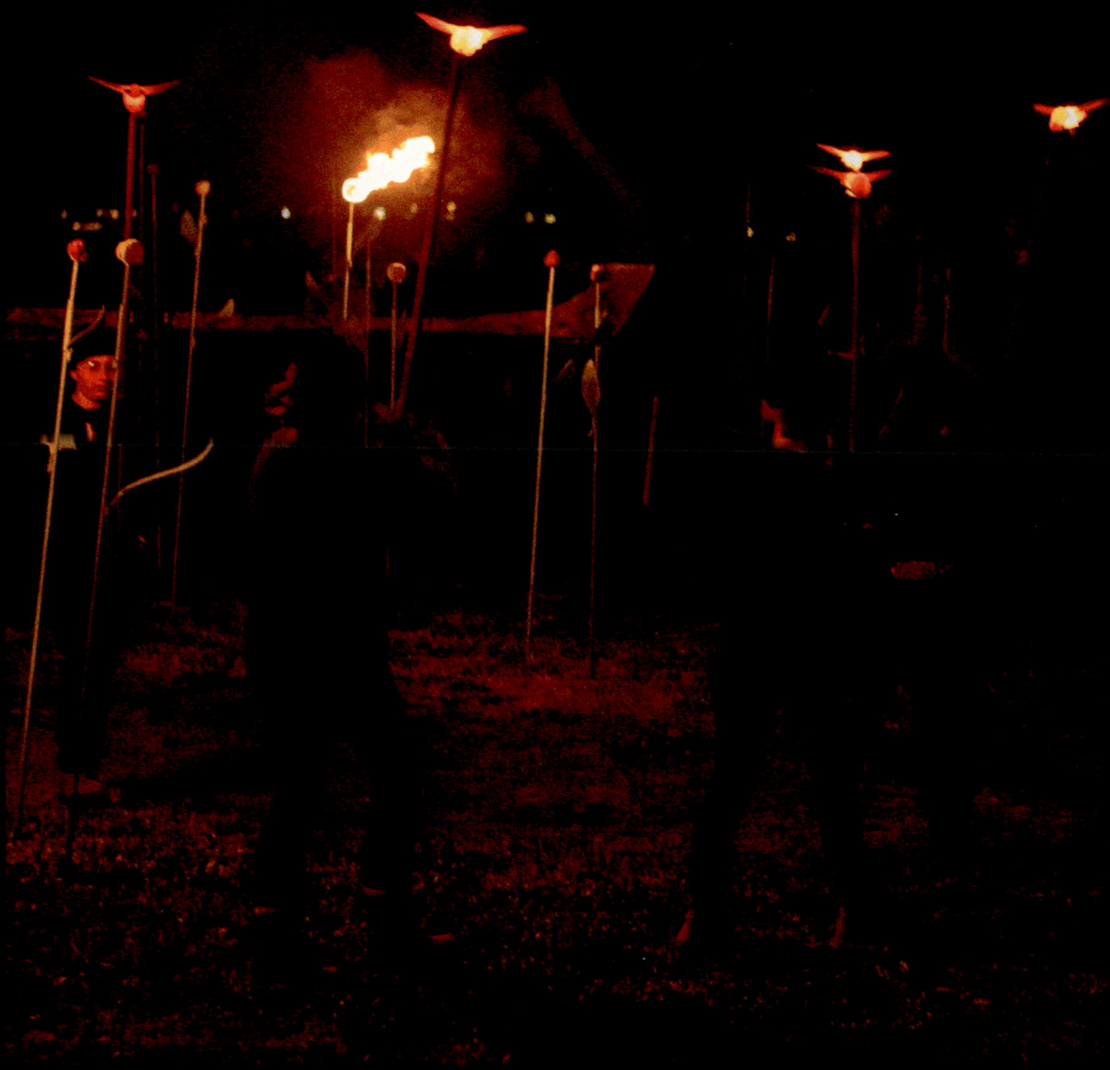

Above Allen Peterson, *Colony*, 2003. Cast iron, 6" × 11.5" × 6.5" (11.5" is the wingspan). Photo: Allen Peterson.

Top right Allen Peterson, *Colony*, 2003. Cast iron, 24" × 18" × 36". Photo: Allen Peterson.

Bottom right Allen Peterson, *Colony*, 2003. Cast iron, 24" × 18" × 36". Performers: Keelin Cavanaugh, Danelle Murphy, Paige Smith, Marisa Mitchell, Anna Weible, Jason Brown. Photo: Allen Peterson.

Cast Your Mind

In conclusion, casters share a mindset regardless of their particular disciplines or mediums. People who are predisposed to this way of thinking can often be turned into casters after witnessing a single event. For many, coming to casting is like answering the siren's call. Casters are passionate and fastidious about process; they think dimensionally and consider objects in the round, and they understand both positive and negative space in unique ways; they are experimental, they exhibit a fervor and zeal, and they are not interested in instant gratification, but instead exhibit patience and a willingness to work toward long-term outcomes; and finally, they are drawn to each other, united by their passion for this bewitching process.

Below Allen Peterson, *Colony*, 2003. Cast iron, 24" × 18" × 36".
Photo: Allen Peterson.

OPPOSITE PAGE

Top Marlene Rose pouring molten glass (2,000°F), 2004.
Photo: Gay Ribisi.

Bottom Bruce Beasley pouring a large cast iron sculpture, Krefeld, Germany, 1987.
Photo: Unknown.

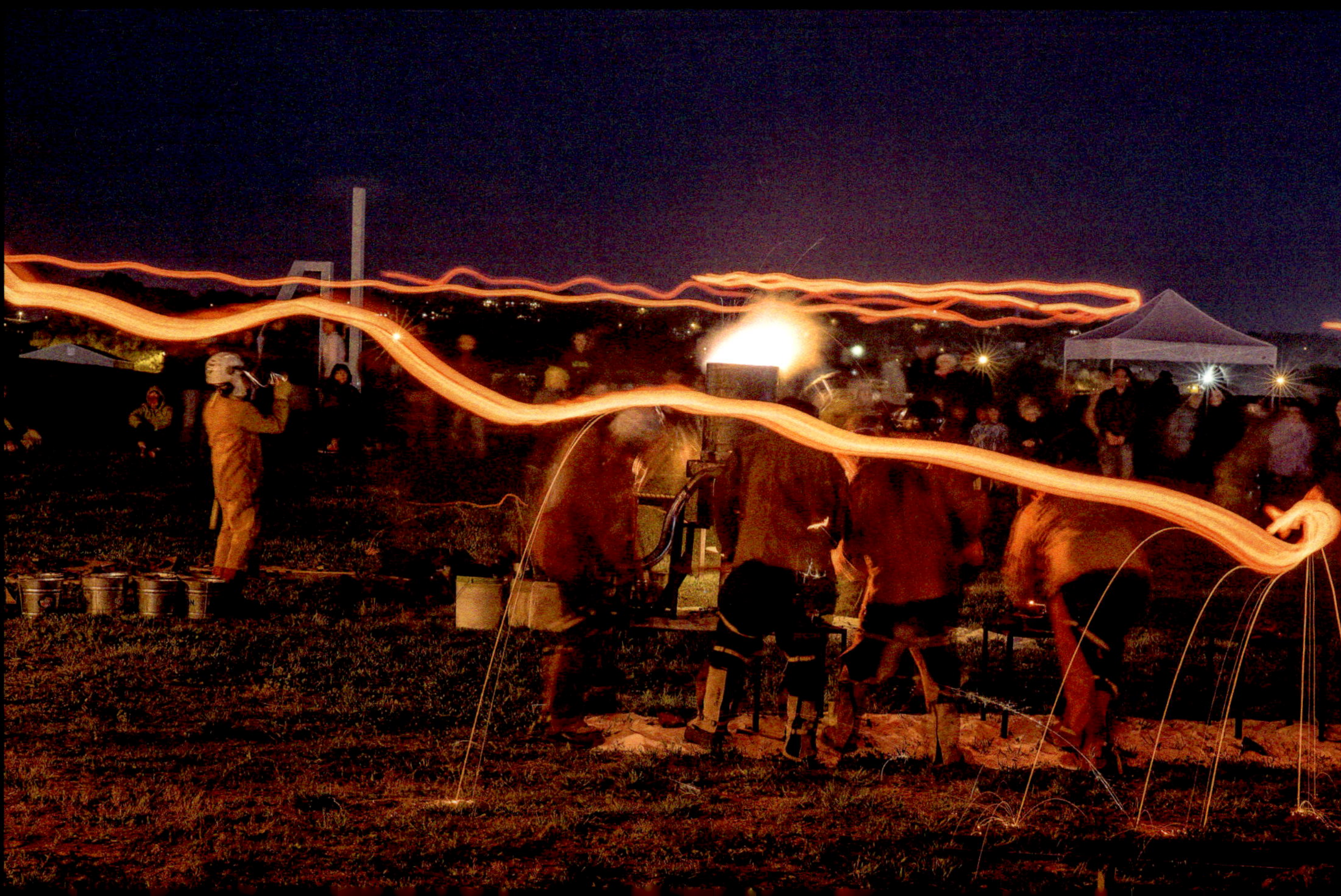

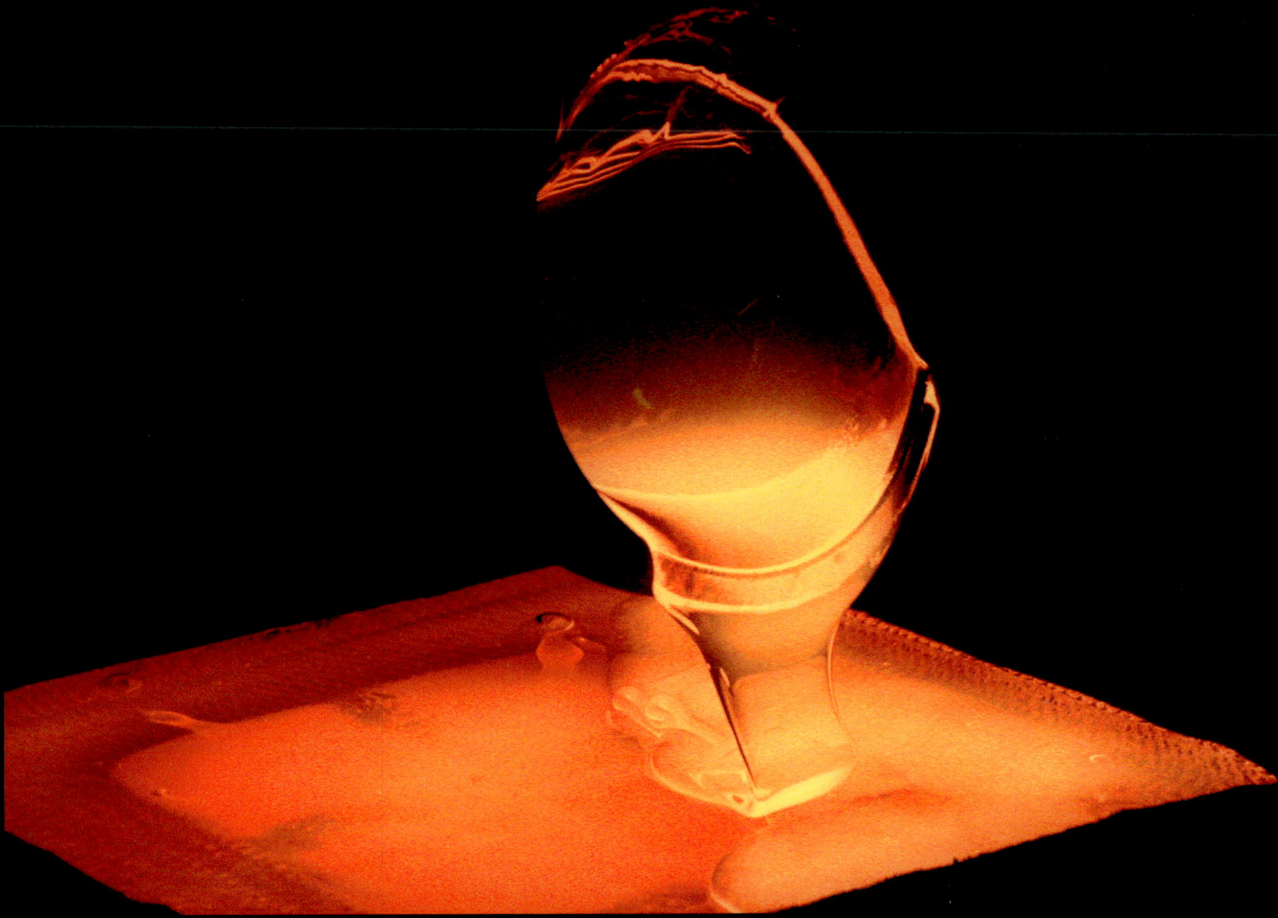

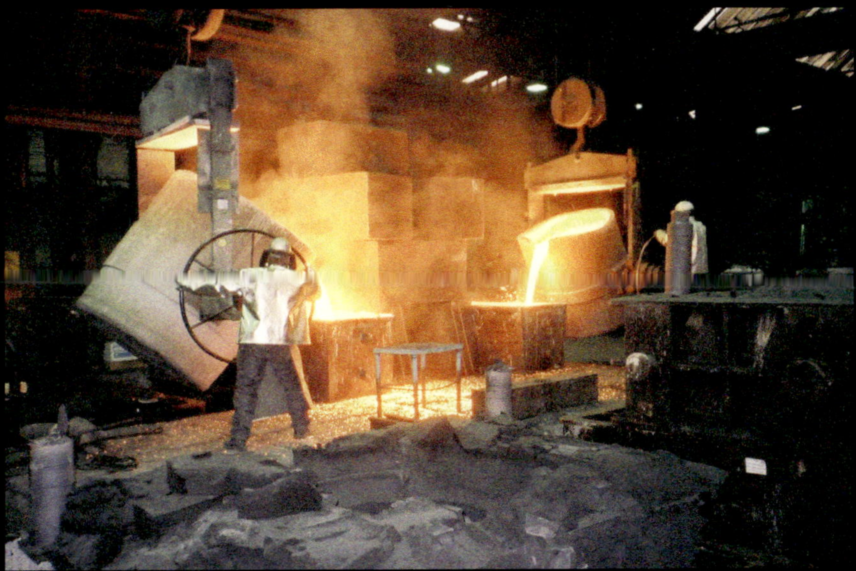

MATERIAL CHOICE

Material choice is one of the most interesting aspects of casting because of the sheer number of materials that can be used. Each one has its own characteristic properties, implications, and spirit. The sparkling transparency of glass, the chalky opacity of plaster, and the flesh-like translucency of silicone would make the same form feel profoundly changed when cast in any of these materials.

Some artists explore this by casting familiar forms in unexpected materials. A glass gun, stage curtains in eggshell-thin plaster, and a ceramic oilcan are a few examples of this idea featured in later chapters. This idea is woven through the world of casting. Artists transform familiar forms and found objects by casting them in new materials for a variety of reasons, both conceptual and practical.

Casting materials can also be chosen for their physical properties rather than their appearances. For example, the art department or special effects team for a superhero movie might cast "boulders" in foam and paint them in a trompe l'oeil style so that they look like the real thing, but can be easily lifted and thrown, giving the impression of a character possessing super strength. A maker may use bronze not just for its durability, but also for its resistance to corrosion by water; a jeweler may choose gold not just for its yellow color, but also because it's nonreactive with human skin; or industrial manufacturers may need a particular kind of ceramic for its ability to insulate electrical current.

Once a model has been made, an artist can proceed in a number of directions. Some mold materials are compatible with many casting mediums, and experimentation is simple once an artist has made a mold. In other cases, the master mold of the model is used to create multiple wax models that will each get its own single-use waste-mold.

Some artists achieve stunning results exploring the same form cast in different materials. Every medium has its own essence and temperament though. A single material can take a lifetime to master, so makers who can adroitly pivot from one material to another are exceptionally rare. Karen LaMonte, who has savant-like abilities in every medium she touches, often explores these possibilities in her work. *Odoriko*, cast in iron, glass, bronze, and ceramic, is a beautiful example of how different the same form can feel when cast in different materials.[2]

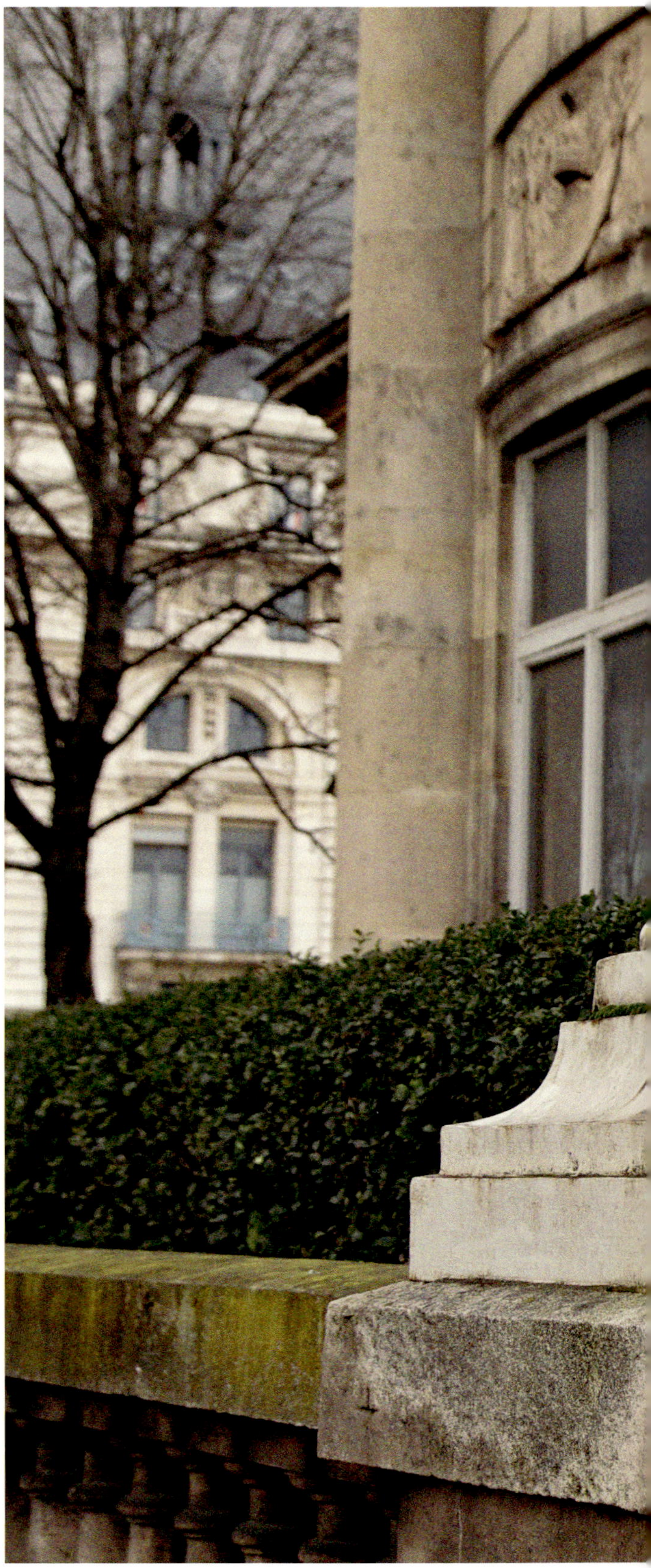

Sphinx, facade of the Hotel de Salm, Paris, France. Cast iron, JJ DUCEL & FILS (foundry).
Photo: © Jebulon / Wikimedia Commons / CC-by-1.0.

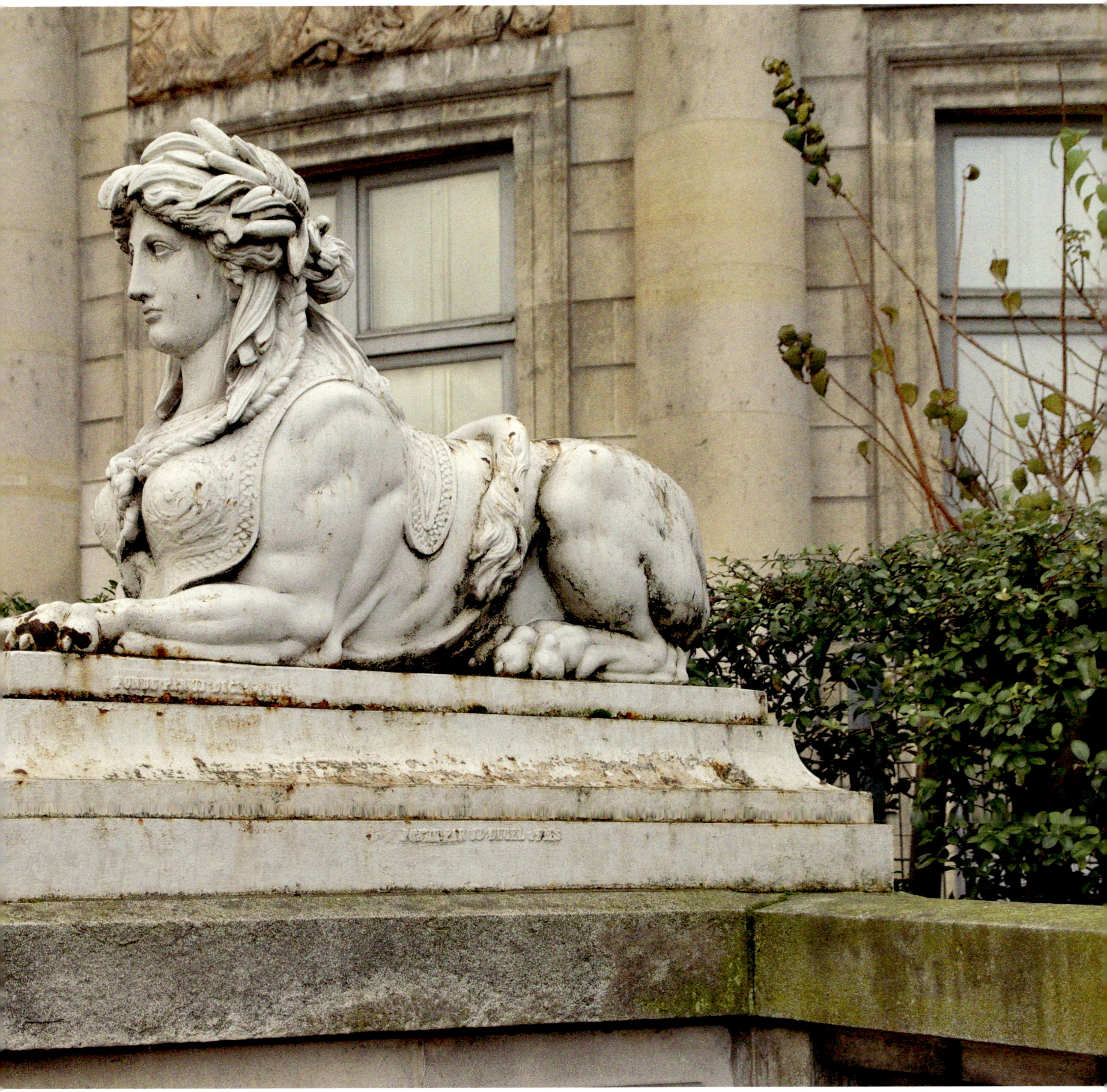

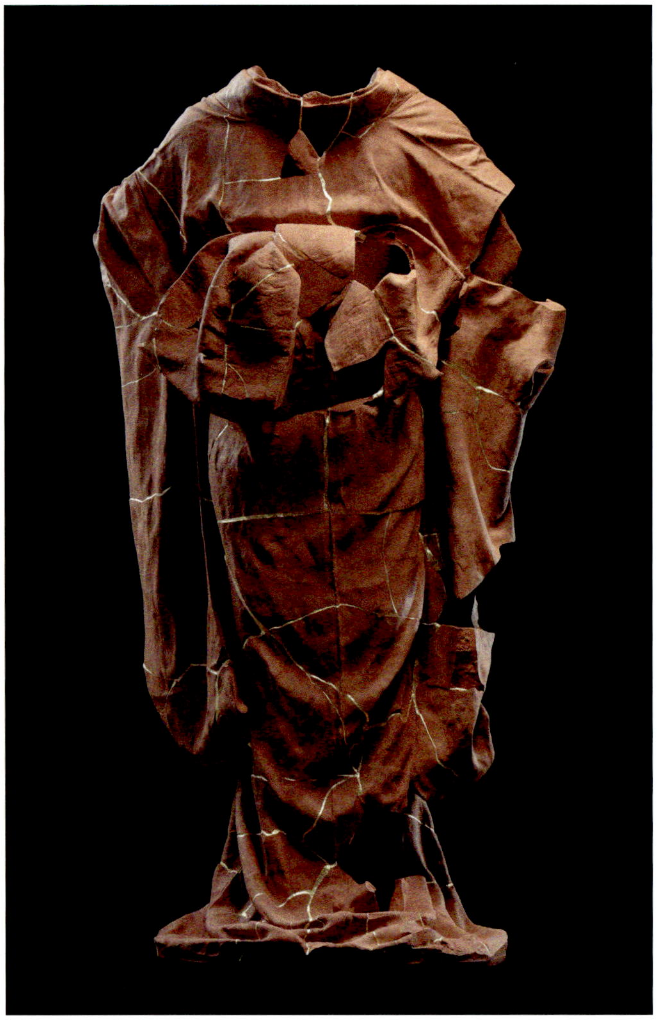

Karen LaMonte. *Odoriko*, 2011. Ceramic, 49.5" × 26.75" × 17.25". Photo: Martin Polak, © 2011 Karen LaMonte, Collection of The Chazen Museum of Art.

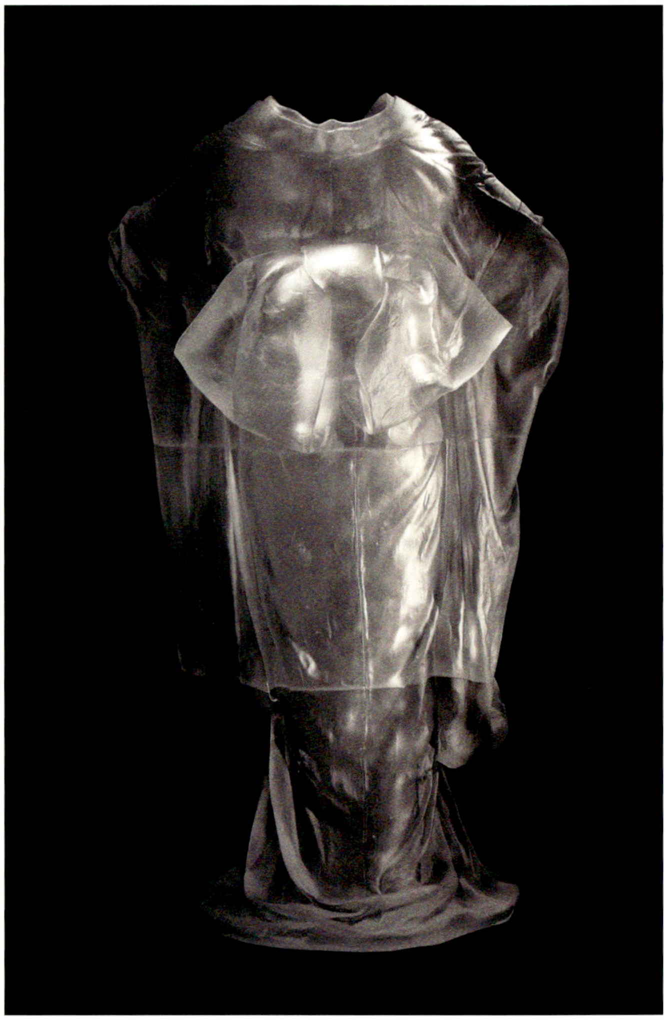

Karen LaMonte, *Odoriko*, 2012. Glass, 54" × 30" × 17.25". Photo: Martin Polak, © 2012 Karen LaMonte, Collection of The Speed Museum of Art.

Ceramic

Perfect and Imperfect Beauty

"I regard clay as the most humble and human of materials. Unpretentious, it is both the body and earth, a universal corpus. Clay parallels the simplicity of Japanese aesthetics. A kimono in clay, the idea of wrapping oneself in earth, struck me as a powerful metaphor for human impermanence."

Glass

Visible and Invisible Beauty

"Glass weaves together the visible and invisible. It is the material in my palette that has the lightest touch. Despite its enormous solidity, it is intangible and fragile.

"The Japanese vision of beauty is enigmatic and elusive. Whereas I had previously used glass for its translucency in order to reveal an absent figure, this time I used it to embody absence, using its immateriality as a reference to the nonphysical, spiritual realm."

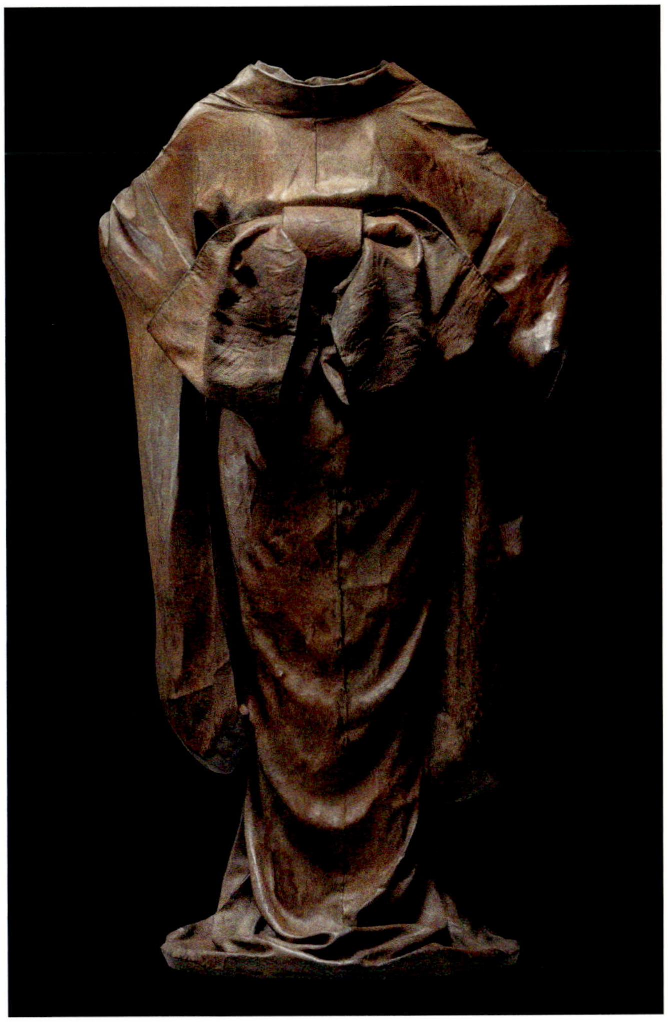

Karen LaMonte, *Odoriko*, 2012. Bronze, 53.75" × 30" × 17". Photo: Martin Polak, © 2012 Karen LaMonte.

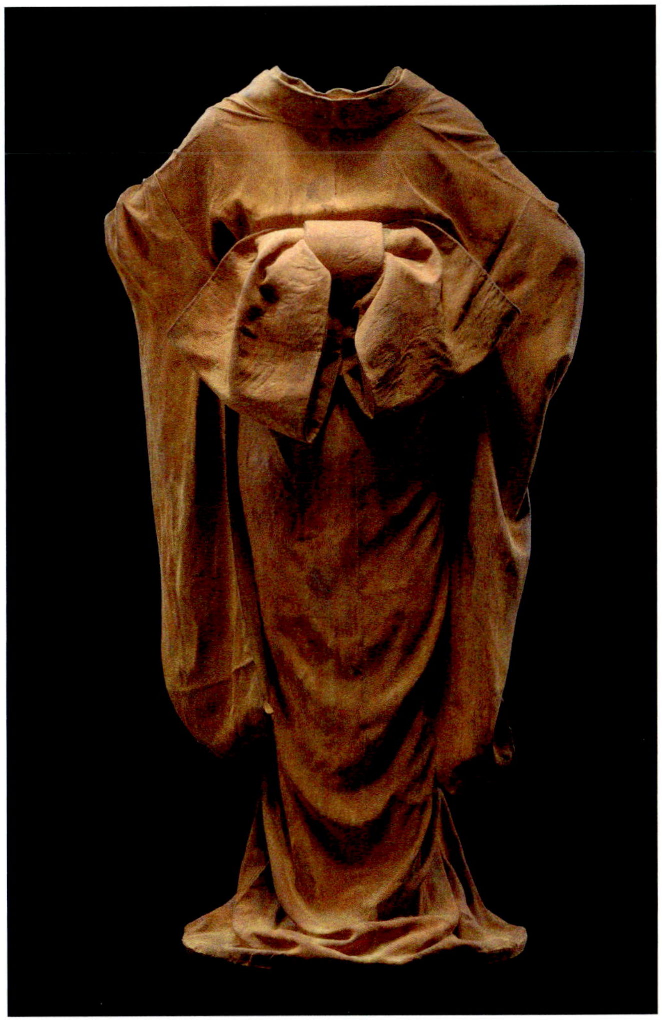

Karen LaMonte, *Odoriko*, 2012. Iron, 53.75" × 30" × 17". Photo: Martin Polak, © 2012 Karen LaMonte.

Metal

Permanent and Impermanent Beauty

"I chose metal to represent the realm of human intervention.

"Bronze is a timeless but manmade material that we use to record victories and failures. Our history is written in the monuments that catalogue our efforts and achievements. Bronze is ancient and speaks of relics, civic monuments, religious images and archeological finds.

"In contrast, rusted iron is temporality and ephemerality. Rust is a self-perpetuating process, each reaction producing the molecules needed for the next stage. Constantly changing and unstoppable, it is transience embodied. I use it as a metaphor for our physical vulnerability, the temporal nature of one lifetime. It is a reminder that even those things we may consider to be the greatest achievements of industry or progress are fleeting."

PEEK into the CASTER'S STUDIO

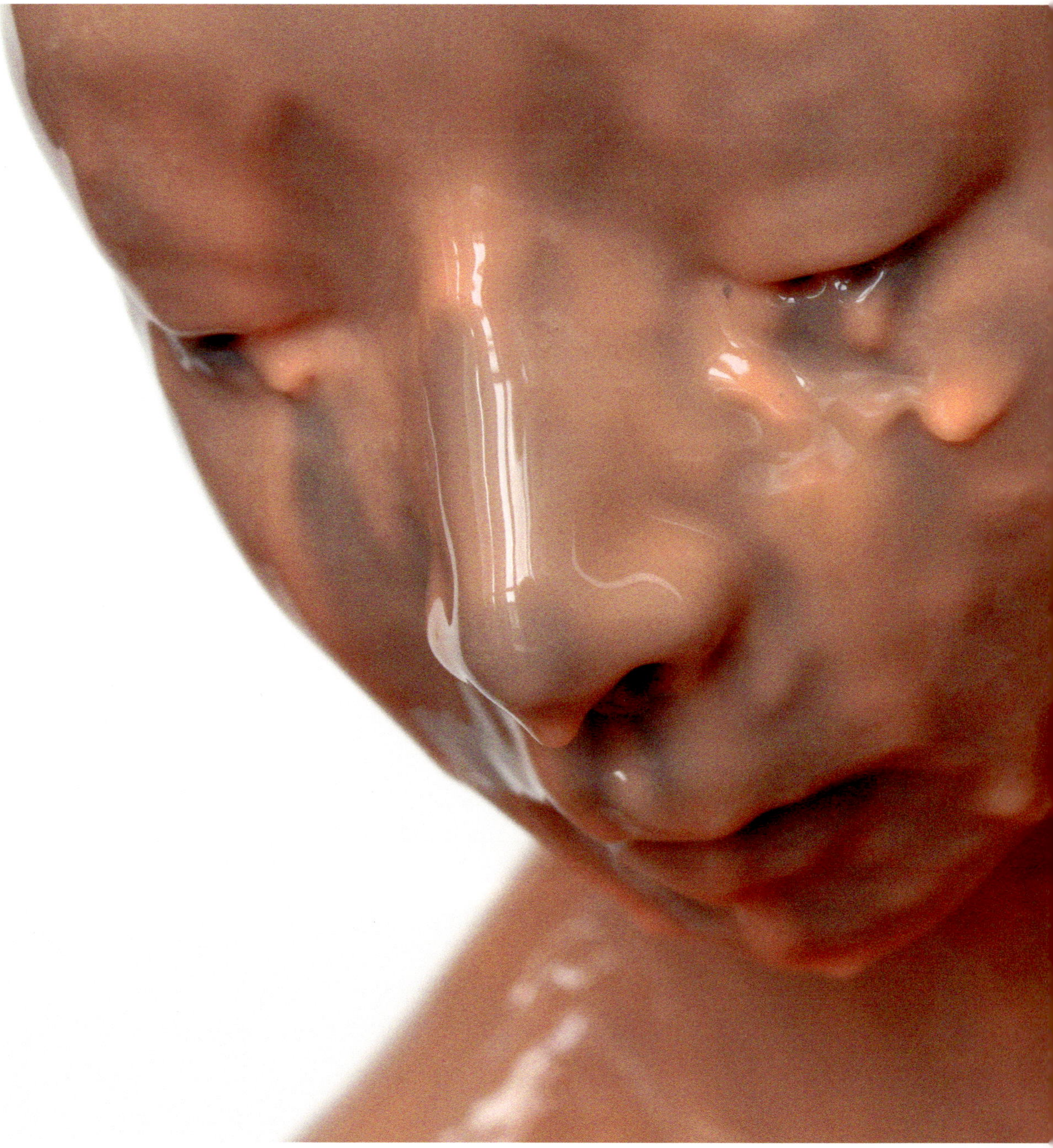

Gosia's model covered in silicone, the first stage in the moldmaking process, 2013.
Photo: Gosia, gosia.ca.

Seventeen casters open their studios to share parts of their processes.

Model creation, mold creation, casting, and finishing are the four parts of the casting process. There are countless variations even within a single medium, so this is by no means an exhaustive exploration, but rather a sampling of approaches.

This chapter gives a taste of how various makers work, highlighting the strengths in their processes. Some makers take a diverse and creative approach to model making: Rebecca Rose collages plastic toys to make models that become richly narrative rings. Others create exacting and beautiful molds: Donald Friedlich's 3-D printed graphite mold is a exquisite standalone object, function aside. Some makers use age-old casting techniques that still render superb and timeless results: Rachel Dein uses a very straightforward casting technique that results in nuanced and sublime bas relief plaster castings with a magnificent fidelity to the plants she originally collected. Other makers do astonishing finishing work on their castings: Masters & Munn create lifecastings that look so real that they could be passed by without the viewer realizing they're not human beings. This diverse process can have many faces; these are a few that exemplify the multifaceted world that is casting.

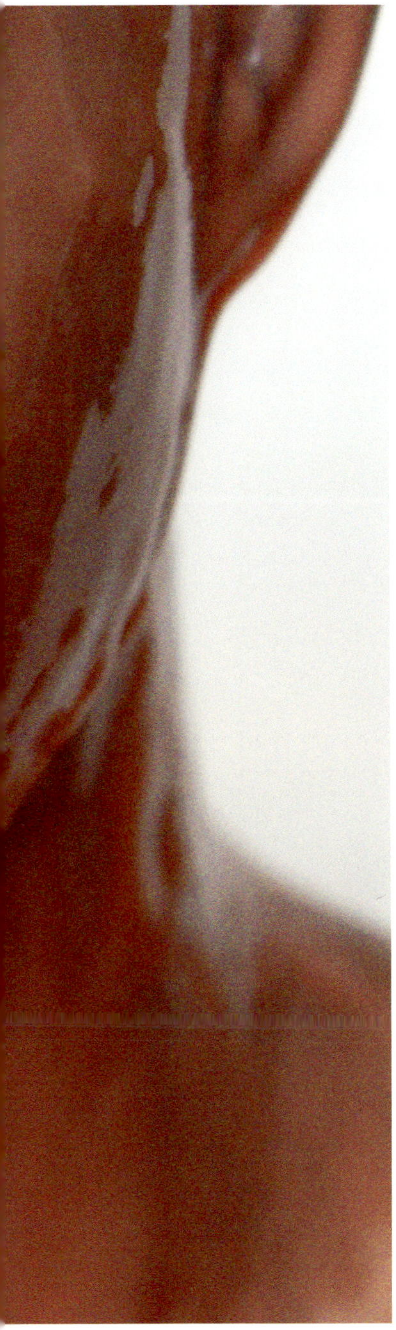

Authors' Note: While this is a book about a process, we did not want to write a how-to book in the age of the Internet. We did want to touch on technique, since it is so important to all casters, and after all, process is what ties every object in this book together. We realize that it is impossible to cover every medium in a meaningful way, so this is our nod to process. Enjoy a glimpse at how different artists approach casting.

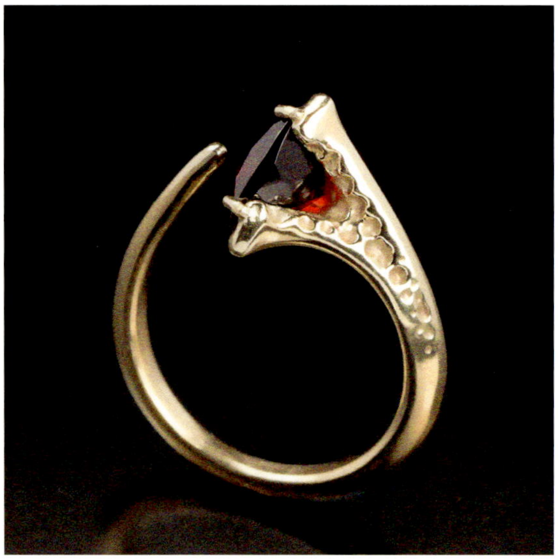

Aleksandra Vali, *Infinity Hope Ring*, 2011. 18K gold, red trillion cut 8-mm garnet, 0.75" × 0.70" × 0.50". Photo: Aleksandra Vali.

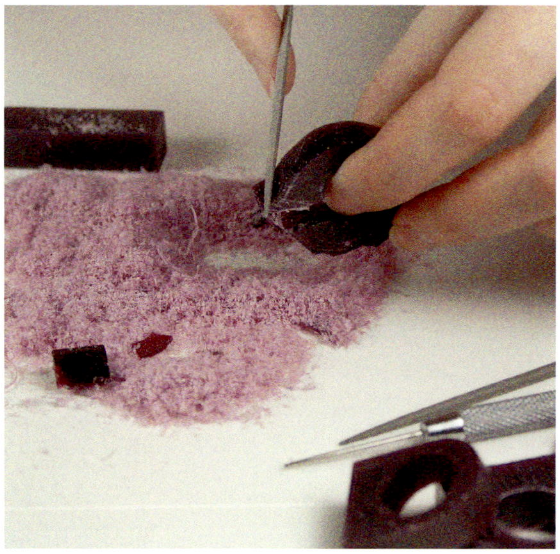

Carving wax and initial filing, 2011. Photo: Aleksandra Vali.

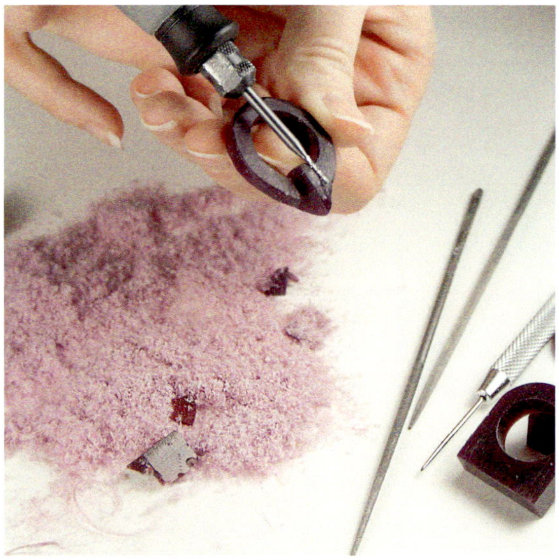

Carving wax using the flexshaft, 2011. Photo: Aleksandra Vali.

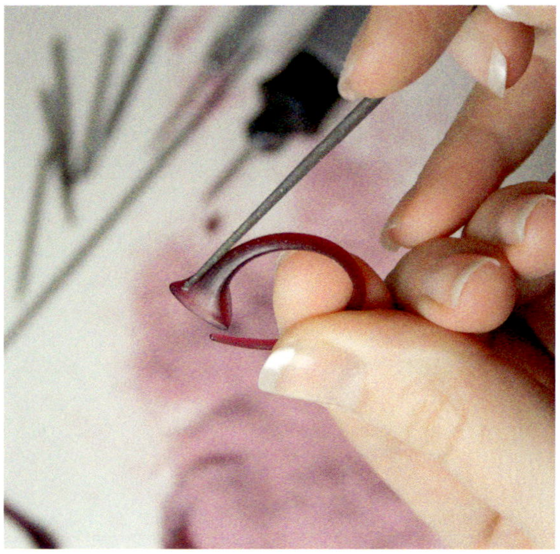

Carefully refining the wax form before casting in 18K gold, 2011. Photo: Aleksandra Vali.

"To carve a ring from wax, the first step is cutting a section from a wax tube with a predrilled hole for the finger. Thinking about this ring's design, I start by cutting out the side profile of the ring, because the side view is very important in this particular design. Next, I begin sculpting the wax to resemble my sketches. After that, I focus on cutting out the top view, first with a file, and then with a flex shaft for finer detailing. It is very important to use the flex shaft lightly with minimal pressure to avoid deep marks and unwanted errors. Finally, I polish the surface of the wax ring with an ultra-fine file, and the wax carving is complete."[3]

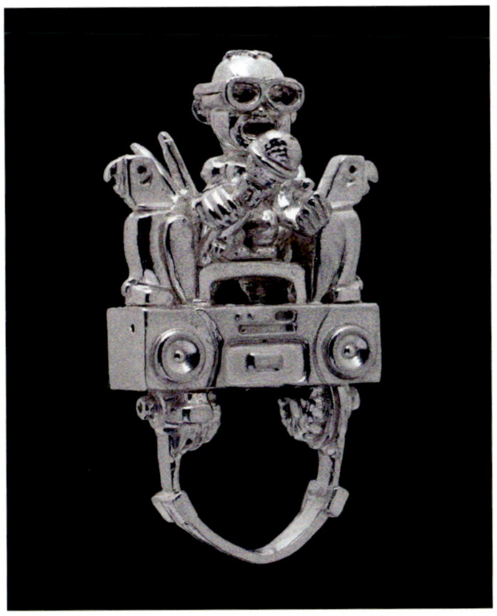

Rebecca Rose, *Soaring*, 2013. Sterling silver, 2.5" × 1.5" × 0.75".
Photo: Marcus Adkins.

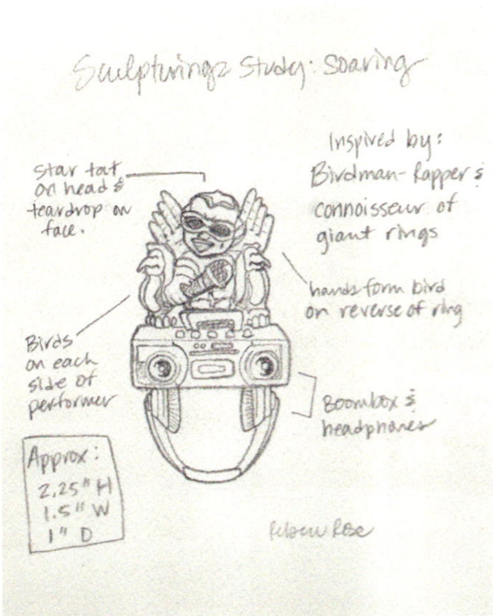

Concept sketch for *Soaring*.
Photo: Marcus Adkins.

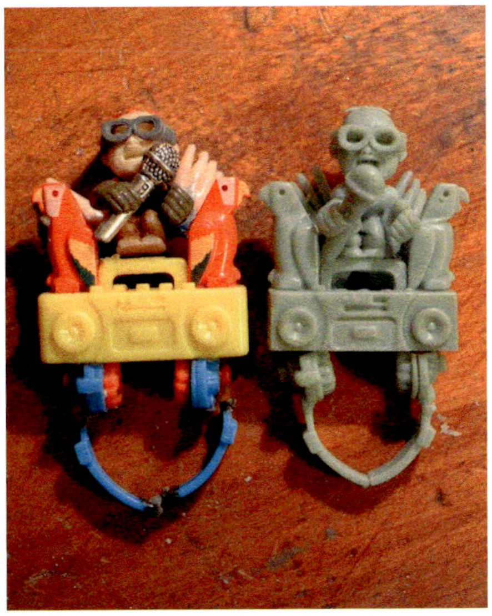

Original model, made with found plastic objects and a wax pulled from a mold of the original model.
Photo: Rebecca Rose.

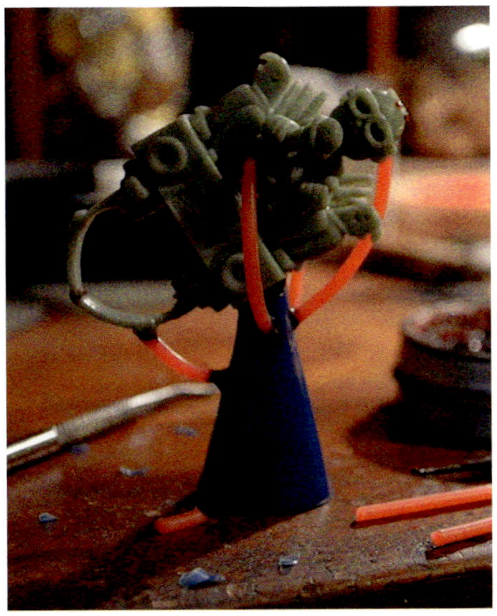

Sprued wax model ready to go into a flask and be invested.
Photo: Rebecca Rose.

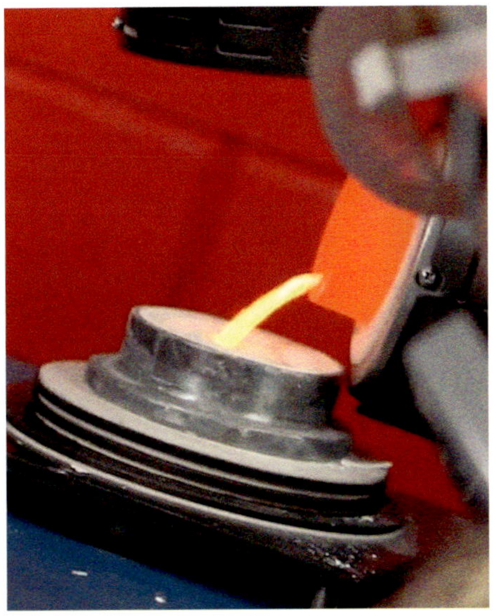

Sterling silver vacuum casting.
Photo: Rebecca Rose.

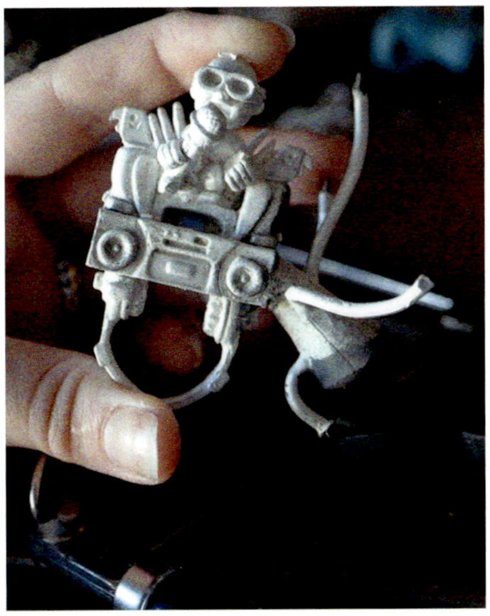

Removing sprues from the raw casting with wire cutters and a jeweler's saw.
Photo: Rebecca Rose.

"I illustrate my design, assemble the prototype, make a mold of the prototype, inject wax into the mold, prepare the wax for casting, pour the metal, cut the sprues off, and finish the piece through intense polishing to obtain a brilliant shine. I cast because the piece is created in one single form.

"The ancient transformative process of casting molten metal, as it goes from solid to liquid to solid form again, is fascinating to me, and harkens back to Ancient Mesopotamia. The challenge of a perfect casting in one shot is a stressful and exciting endeavor, prone to so many variables that can lead to error, so when the piece is cast perfectly on the first try, a huge wave of relief and success washes over me."[4]

PETER KOVACSY

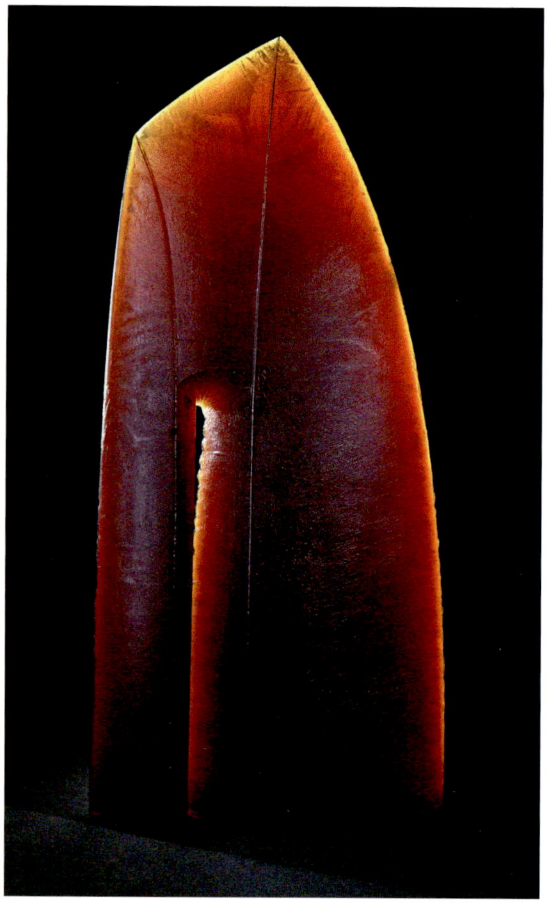

"The process I employ for glass casting and moldmaking was pioneered in the Czech Republic in the early 1950s. To make a sculpture, I start by carving a full-scale pattern block in softwood. Then, I set up a framework surrounding the pattern block and pour a refractory mold into it. After removing the pattern, the "green" mold is placed into a kiln and pre-dried. Next, I pack Blackwood Casting Crystal (supplied as unannealed chunks) into the mold and start the firing schedule. When the firing process is complete, I remove the mold material from the cast glass form very carefully, and I carry out detailing, cleaning and polishing as required. A sculpture of this scale is a three-month project."[5]

Peter Kovacsy, *Karri Sunrise 2*, 2005. Mead color, lead crystal, 37.8" × 17.3" × 3.9".
Photo: Peter Kovacsy.

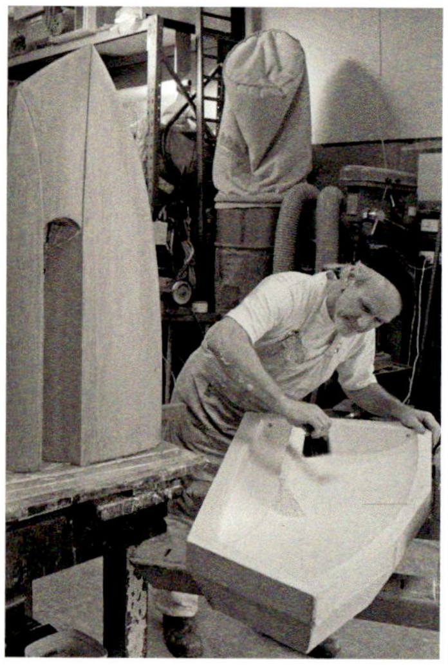

Cleaning dust from the inner face of the mold (made of foundry sand, white silica, and general purpose casting plaster).
Photo: Peter Kovacsy.

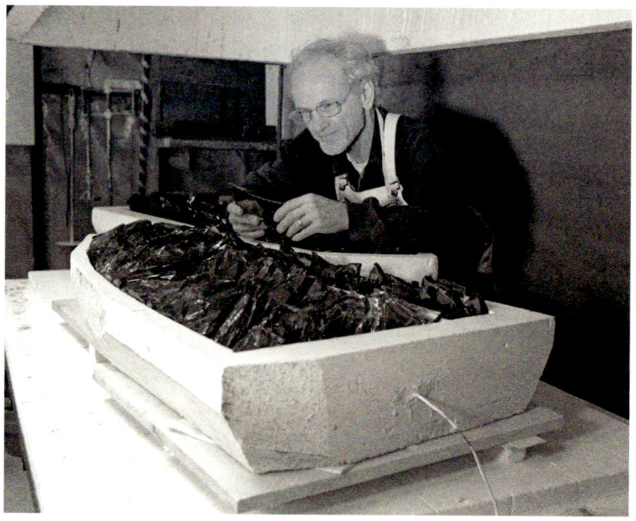

Packing unannealed chunks of casting crystal into the mold. The wire is an embedded thermocouple for monitoring and adjusting annealing temperatures.
Photo: Peter Kovacsy.

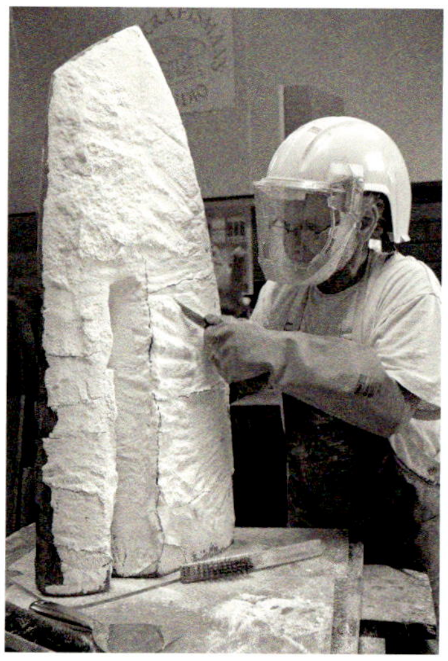

Carefully removing the mold material from the glass casting, which can take two full days. Cleaning and finishing the casting can take an additional two days.
Photo: Peter Kovacsy.

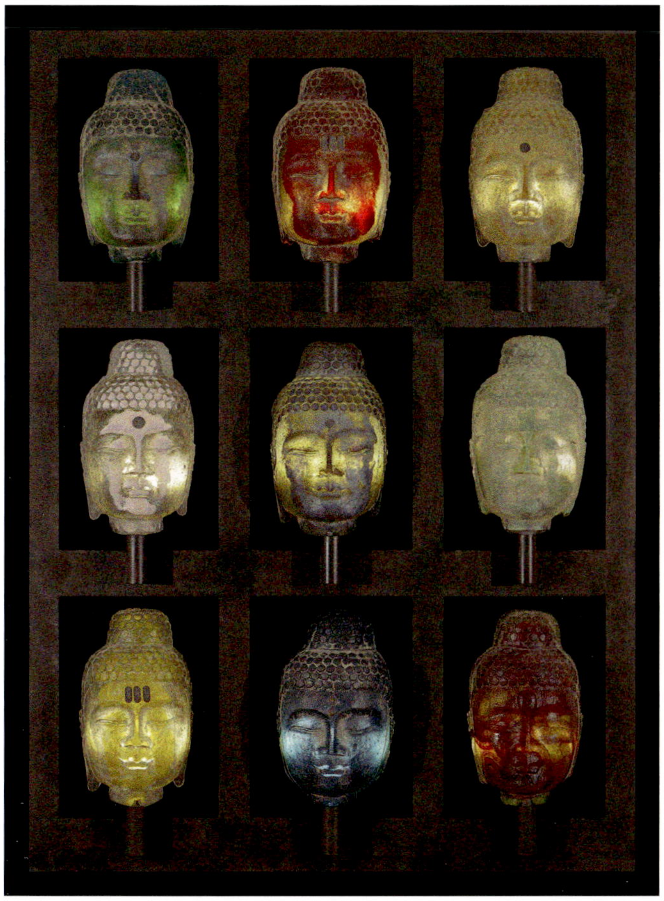

"First I carve a model. I press this carving into the sand, work the shape, and sprinkle it with fine layers of powdered colored glass. This work in the sand is the most private, intimate, and painterly time for me, a time of introspection and inspiration as I manipulate the shapes and colors.

"Then the action starts. My heavy-lifting team and I pour liquid molten glass into the molds. The heat of the glass melts the powdered layer permanently coloring the surface of the glass as it takes on the shapes I made in the sand. Pouring the glass is a team sport, a dangerous dance of heat and light with sweat, sound, and intense concentration.

"Then the glass gestates for a week in a cooling oven to relieve the stresses of its fiery making. I crack off the remains of the sand to see what has been birthed.... The glass retains the memory of what it has been poured into and the sandy texture makes it seem as if the work has just recently been unearthed after ages sleeping."[6]

Marlene Rose, *Buddha Wall*, 2011. Glass, steel, and copper, 4'6" × 3'2" × 9".
Photo: David A. Monroe.

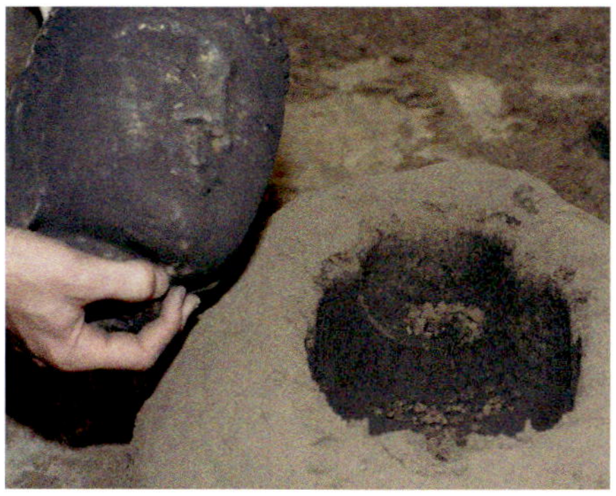

Pressing the Buddha head model into the sand mold.
Photo: Thomas Coates.

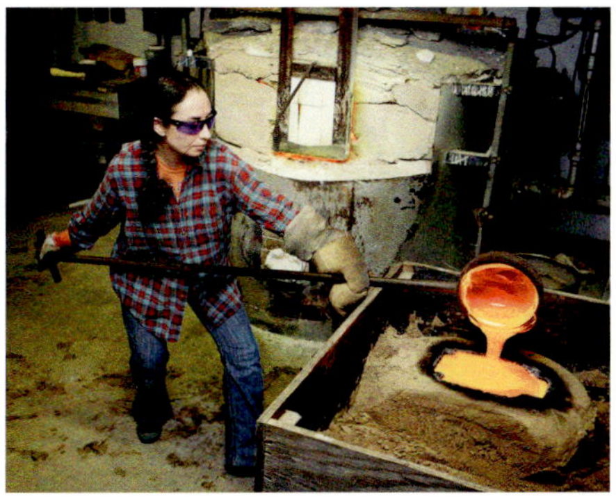

Ladling liquid molten glass into the sand mold, 2013. At this stage, the glass is about 2,000°F.
Photo: Felix Kuntz.

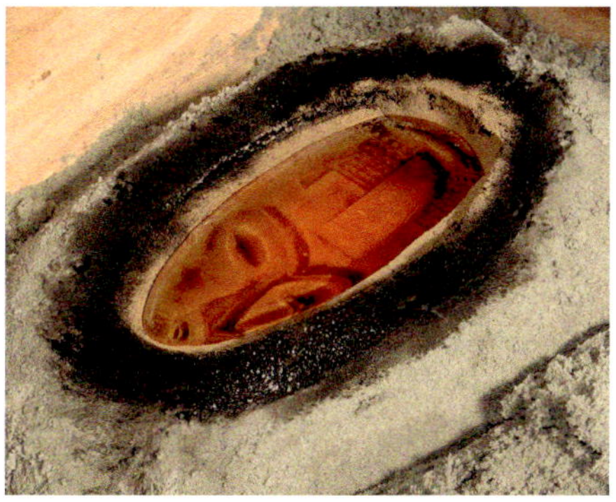

Hot glass in sand mold, 2013.
Photo: Richard Collins.

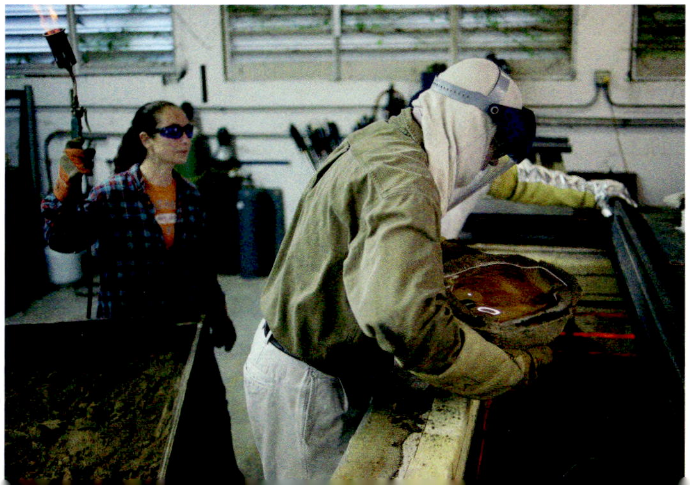

Placing the hot glass casting of a Buddha head into the annealer to cool slowly, 2013. The cooling process brings the glass Buddha head from 900°F to room temperature over six days. *Left to right*: Marlene Rose (holding torch) and Thomas Coates.
Photo: Felix Kuntz.

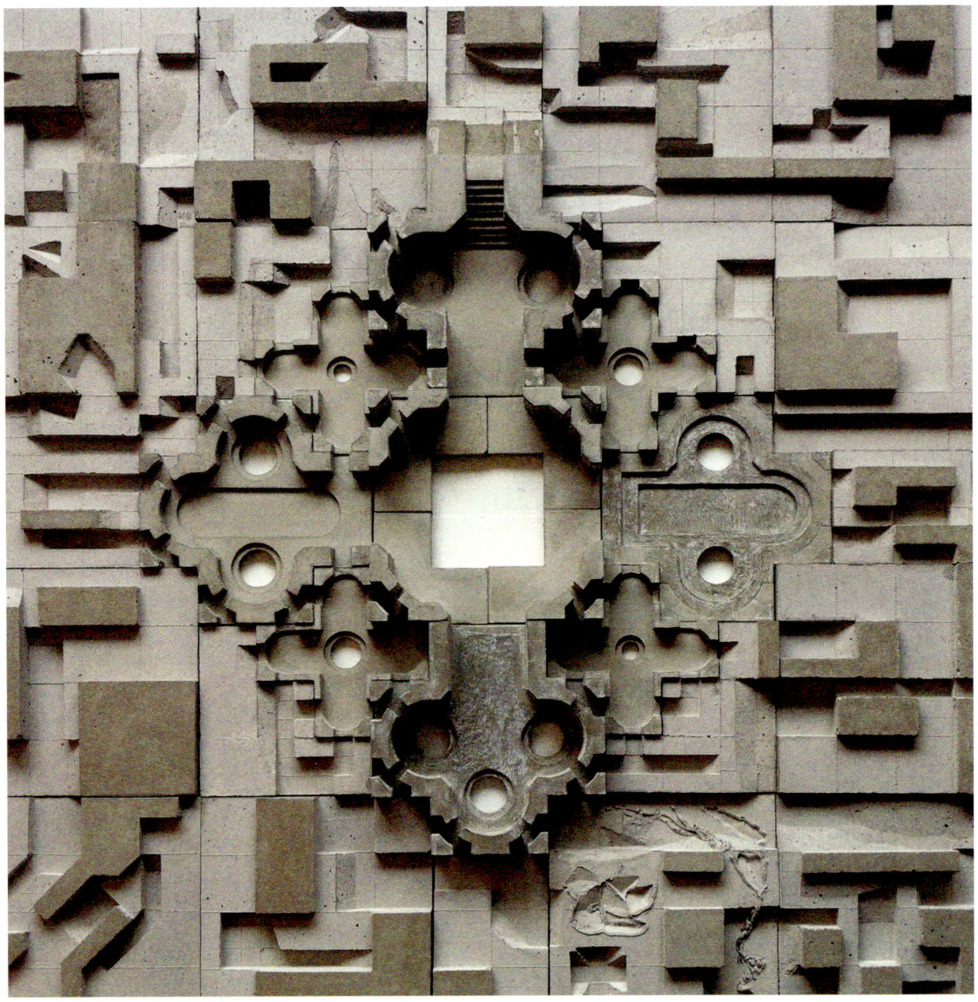

Left David Umemoto, *Foundation*, 2015. Concrete, 47.25" × 47.25" × 11.81". Photo: David Umemoto.

Bottom left Creating foamcore and polystyrene molds for *Foundation*, 2015. Photo: David Umemoto.

Bottom right Casting concrete in single-use molds for *Foundation*, 2015. The molds are destroyed when de-molding the concrete forms. Photo: David Umemoto.

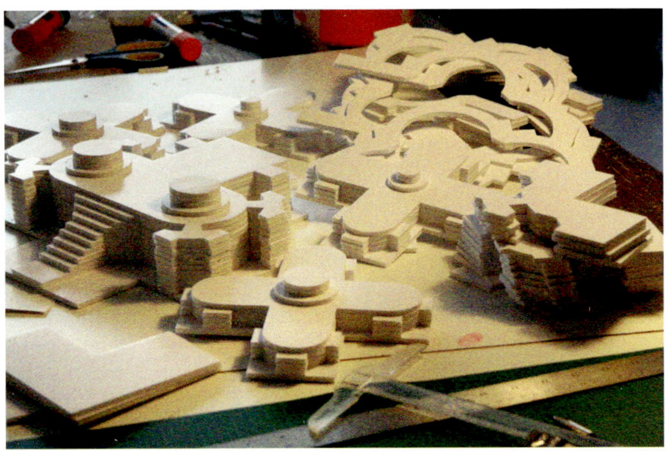

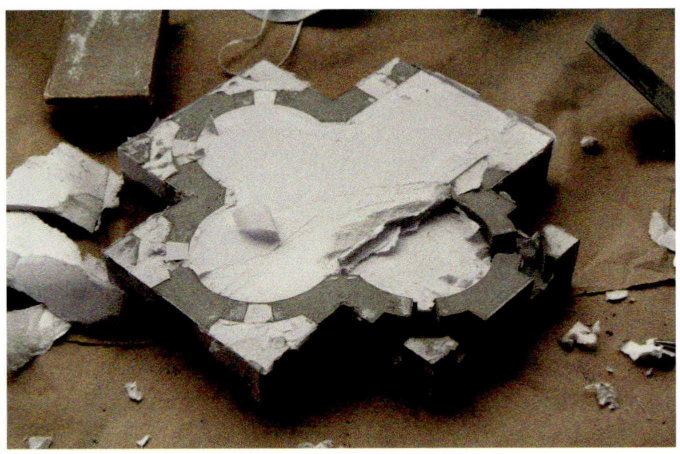

"The process used here is similar to building a foundation on a real construction site. Given the scale of the pieces, foamcore, which is cost-effective and easy to manipulate, is used rather than plywood and steel. These single-use molds rely heavily on the accuracy of the technical drawings, given that they are built directly as negatives, but they also grant more artistic freedom by allowing undercuts. Layering the foamcore (which is not strong enough in single-ply to hold the concrete) insures the integrity of the shapes and also the sturdiness of the molds. In addition, it makes it easier to create curved walls and produces a texture that is in line with the stratification concept of this artwork, *Foundation*."[7]

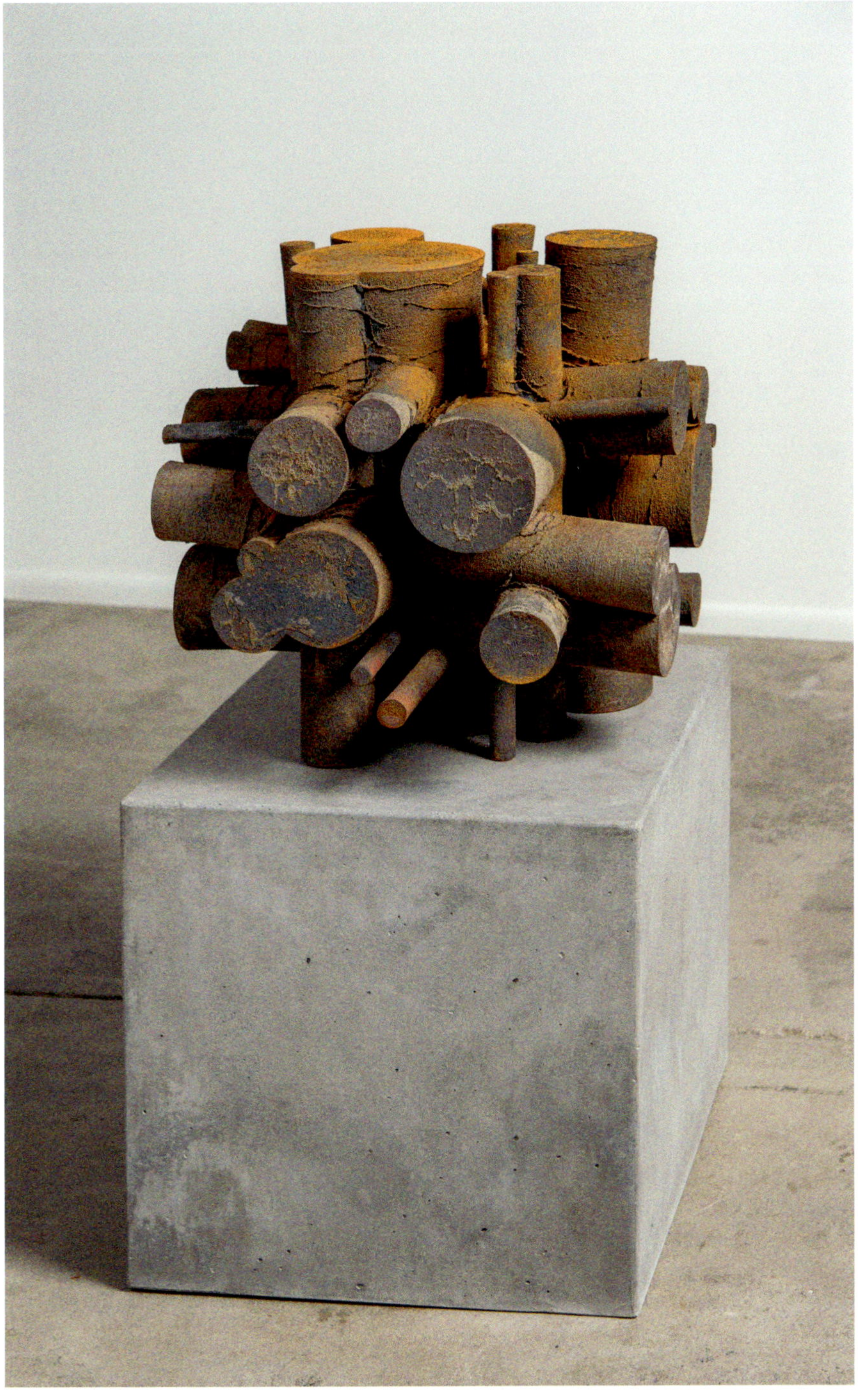

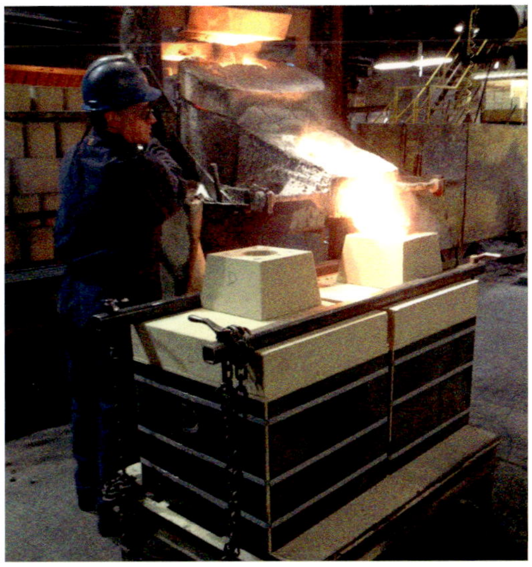

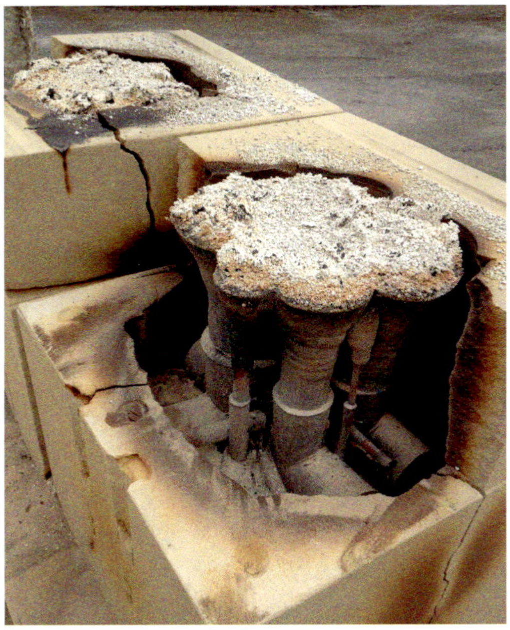

Above Tom Joyce, *Core X*, 2013. Cast iron alloy made from iron remnants retrieved from projects 1977 to present, concrete, 20" × 20" × 40".
Photo: © 2015 Daniel Barsotti, Collection of Alicia & William Miller.

Top Kohler Co. associate casting *Core X* and *Core XI* during a 2013 John Michael Kohler Art/Industry residency.
Photo: © 2013 Tom Joyce Studio.

Bottom right Removing *Core X* and *Core XI* from their resin bonded sand molds during a 2013 John Michael Kohler Art/Industry residency.
Photo: © 2013 Tom Joyce Studio.

"Core X results from molten iron moving through carved passages as freely as water seeking its own level to fill the mold. Three process-related concepts are at play: making an object all but impossible to reconstruct employing my usual forging methodology; injecting ample iron into the mold to break down the resin-bonded sand before the iron chilled, allowing liquid to migrate into a delicate web-like matrix where cracks had formed; and investing the sculpture with a 'DNA-rich' alloy concocted from iron remnants saved over thirty-nine years of studio practice. The reservoir and gating that emerged from the top, seen here as the mold sand drained away, was saved for use in other artworks."[8]

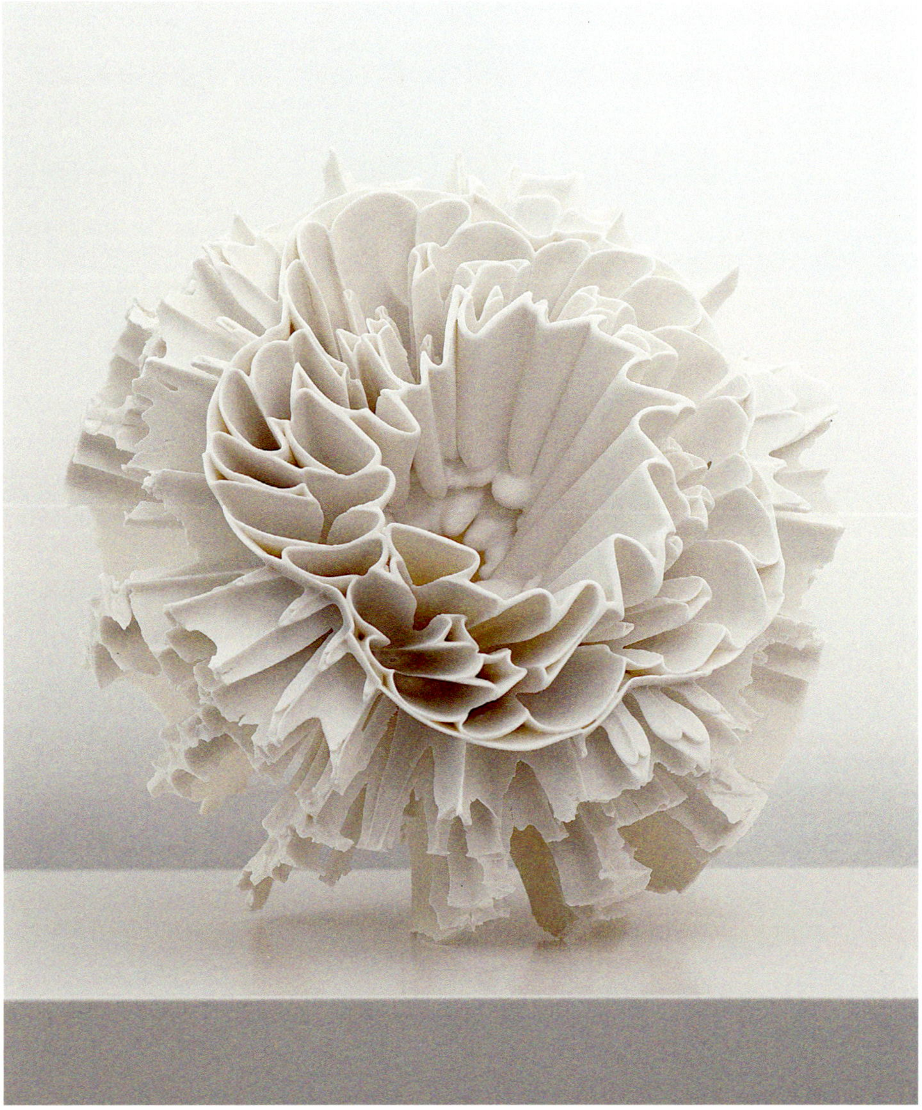

"To make my sculptures, first I pour a big cylindrical block of plaster. When the plaster is dry, I draw on it so I know where to cut. I use only a handsaw to cut the block into 100 to 150 pieces. Then I cut, carve, and shape them one by one. Next, the inside of the block is shaped with forms that will work for the final porcelain.

"Then, I place small, leather-hard porcelain shims in between the mold parts to leave space so the porcelain slip can fill those gaps and create the wings (which are both aesthetic and functional) on the sculptures.

"After I've poured the porcelain slip into the mold, I have to wait for the perfect moment. When the porcelain is stable enough (but still a bit flexible) I de-mold the piece, completing the process."[9]

Zsolt József Simon, *Untitled*, 2012. Porcelain, no glaze, 13.75" × 13.75" × 9.5". Photo: Stephen Papandropoulos.

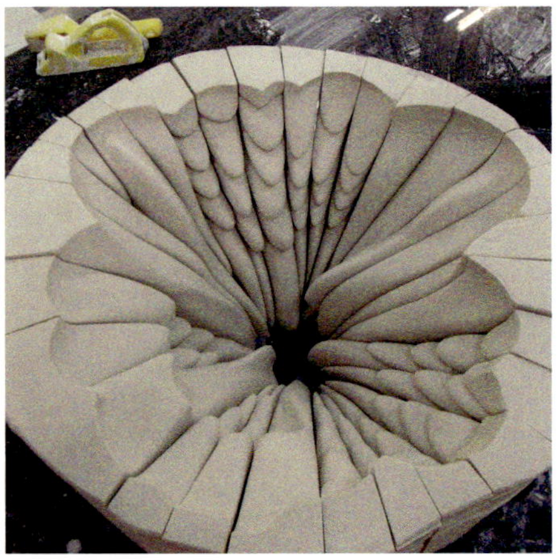

Interior view of plaster mold, 2011. Photo: Zsolt József Simon.

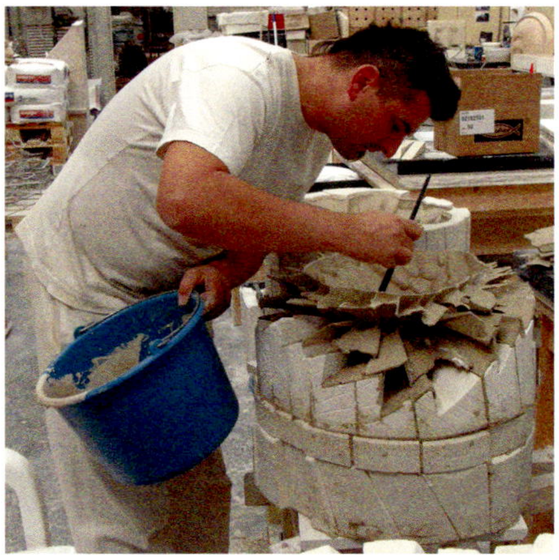

Refining *Crazy Mold 1*, 2011. Stoneware, pigment, no glaze. Photo: Zsolt József Simon.

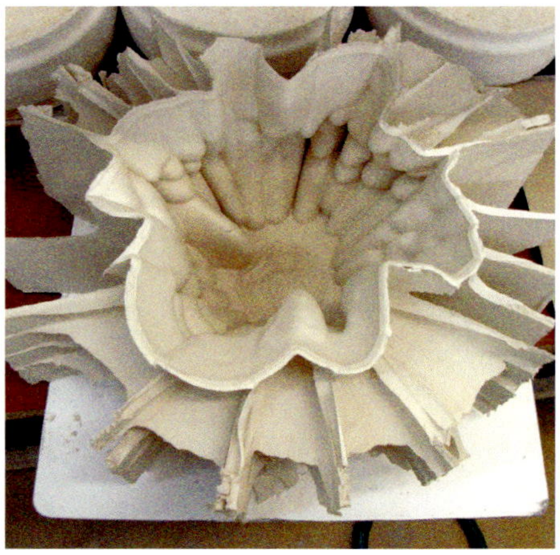

Raw porcelain casting, 2011. Photo: Zsolt József Simon.

PETER HOOGEBOOM

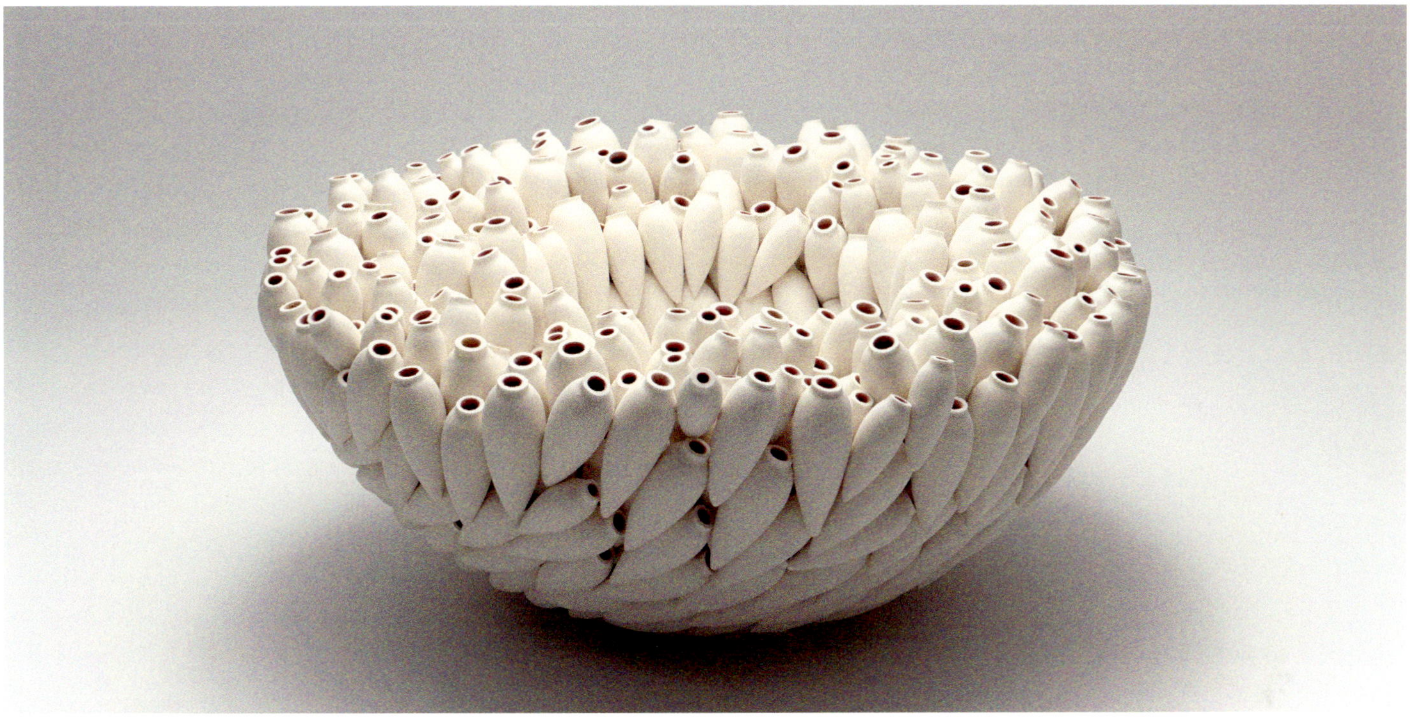

"Casting is executed in this work in two ways. First, small hollow vessels are separately slipcast and glaze-fired. Then they are fused into in a custom-made, heat-resistant mold. In this work, the glass technique of kiln casting has been introduced into ceramics. Casting clay slip gives relatively light, hollow, but remarkably strong elements."[10]

Peter Hoogeboom, *Polyamphora cruentata*, 2013. Fused ceramics, 11.4" × 11.4" × 4.9". Photo: Rob Bohle.

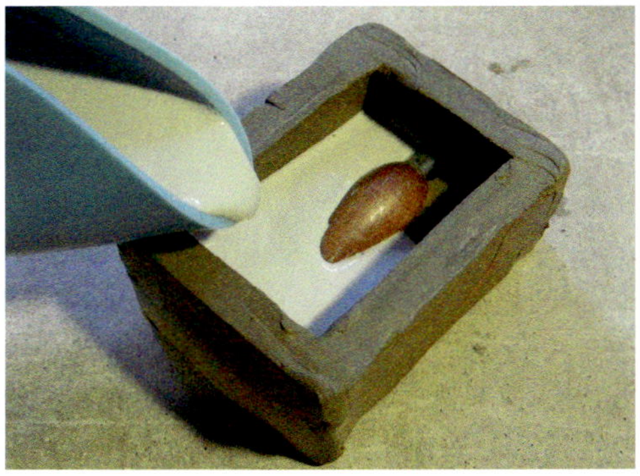

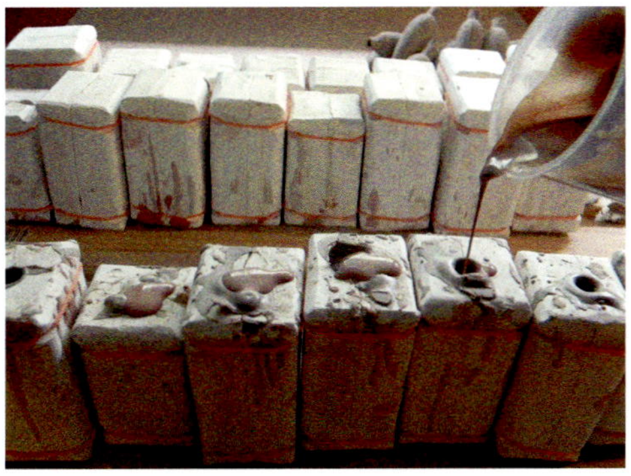

Left Pouring a plaster mold, 2013. Photo: Peter Hoogeboom.

Right Slipcasting ceramic slurry into plaster molds, 2013. Photo: Peter Hoogeboom.

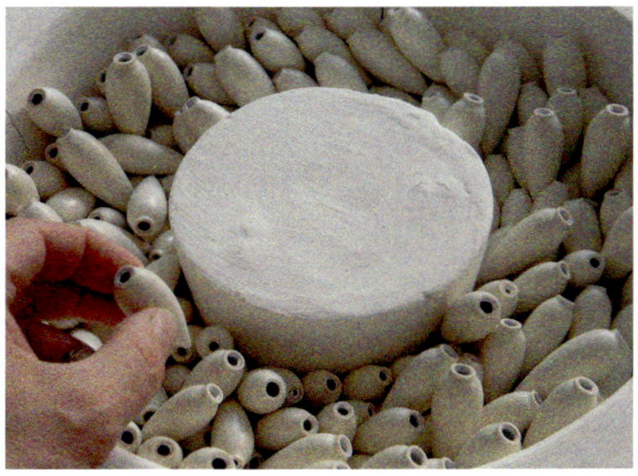

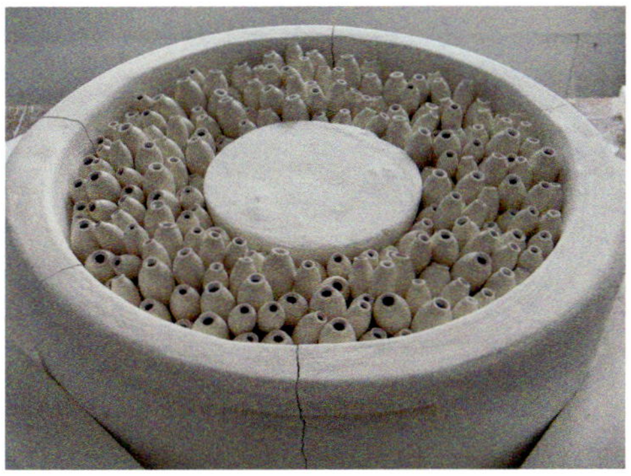

Left Placing slipcast and glazed ceramic components into a mold for fusing, 2013. Photo: Peter Hoogeboom.

Right Placing slipcast and glazed ceramic components into a mold for fusing, 2013. Photo: Peter Hoogeboom.

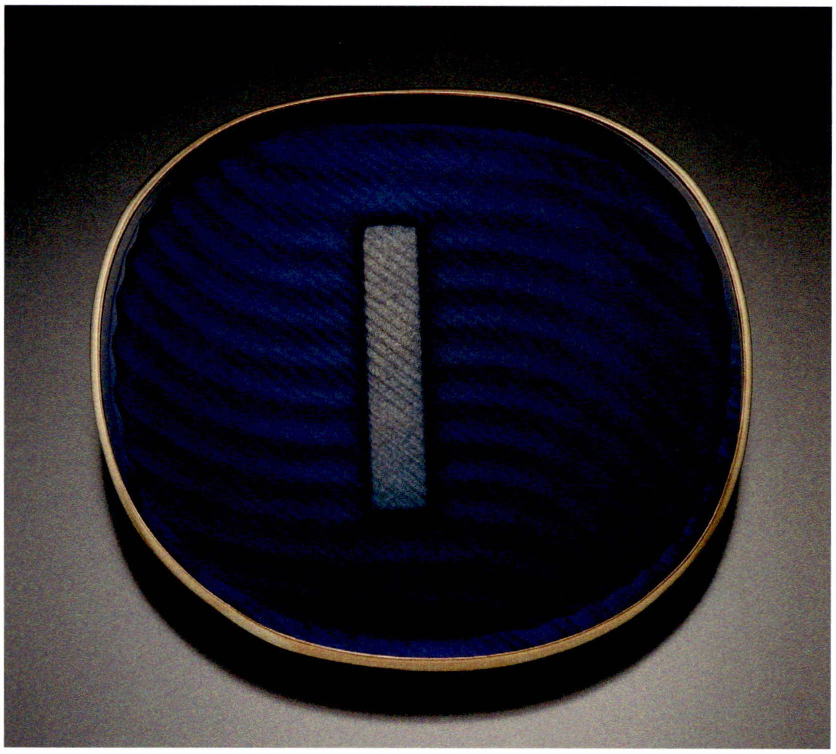

"My forms are designed in computer-aided design (CAD) software and then computer numeric controlled (CNC) machining is used to carve a graphite mold. Hot glass is gathered from the furnace and pressed into the open-face mold with a steel paddle. After annealing overnight, I do extensive grinding with diamond abrasives to refine the glass castings. Working in this process has greatly expanded the form and surface options at my disposal.

"I've intentionally made the molds oversized and this allows me to cut a wider variety of shapes from the castings taken from any given mold. There is a lot of time and expense in machining the graphite molds, so having the option of a variety of outcomes for each mold is advantageous."[13]

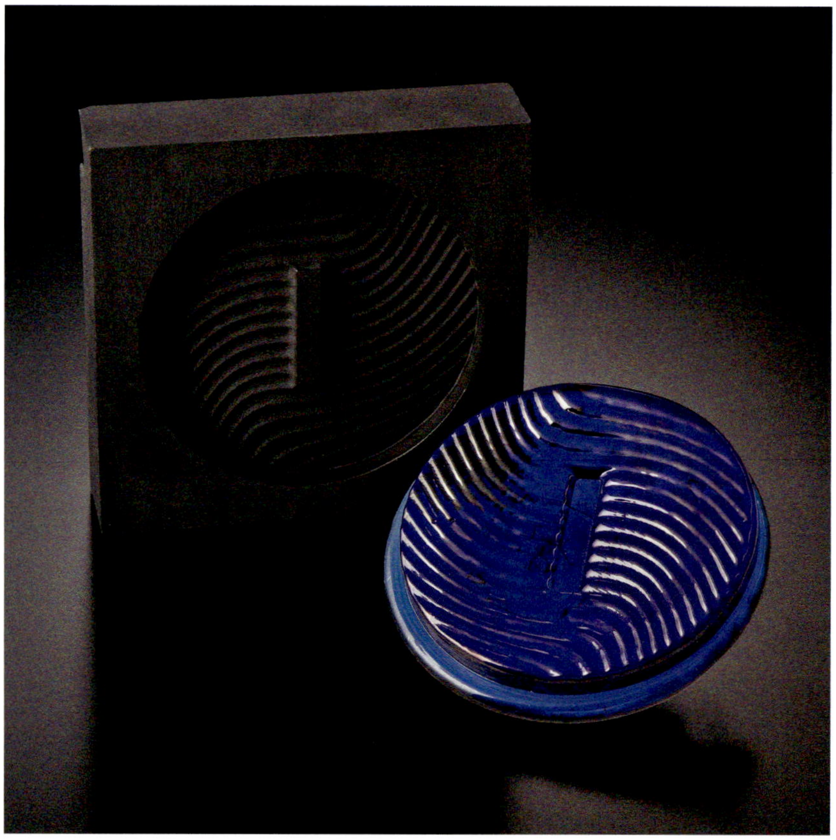

Top Donald Friedlich, *Aqua Series Brooch*, 2013. Glass, 22K gold, niobium, sterling, 18K gold, 14K gold, CNC machining of graphite mold, press molding, coldworking, fabrication, bezel setting. 2.4" × 2.6" × 0.4". Photo: Larry Sanders.

Left Graphite mold and casting, Donald Friedlich, 2013. Mold is 4" × 4" × 1". Photo: Larry Sanders.

"Casting glass using traditional methods can be very time consuming. I became interested in the visual power of multiples and simultaneously disenchanted with the amount of effort and materials necessary to create a plaster/silica mold. I came up with the idea of casting an open-faced negative impression of a small manhole cover that I could ladle-cast glass into. I created a wax of the area surrounding one of these pieces, then had it cast in bronze by a local foundry. The result was a tool I have used over a hundred times. I pour molten glass into the bronze mold, allow it to cool for a few moments, then pop the glass out of the mold, load it into an oven for annealing, then immediately pour another."[14]

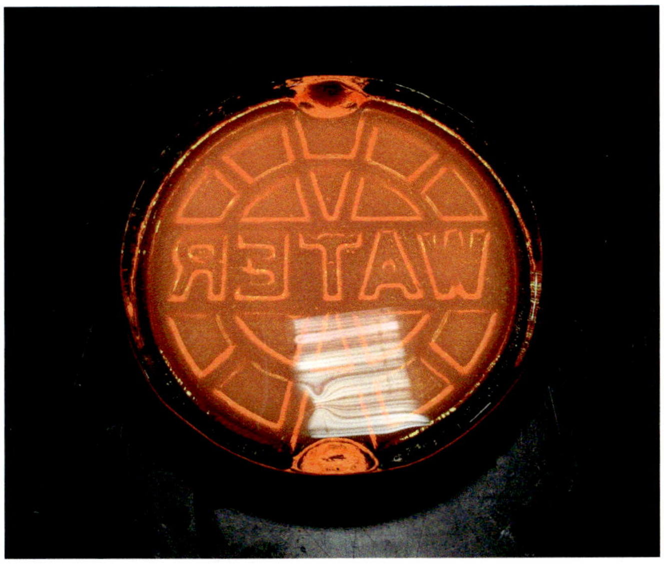

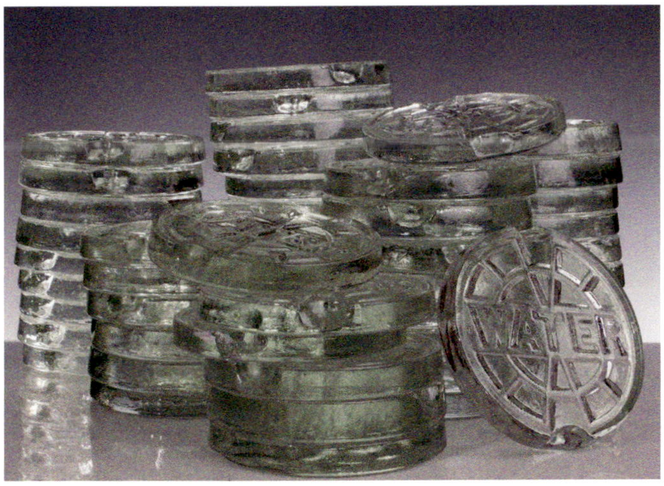

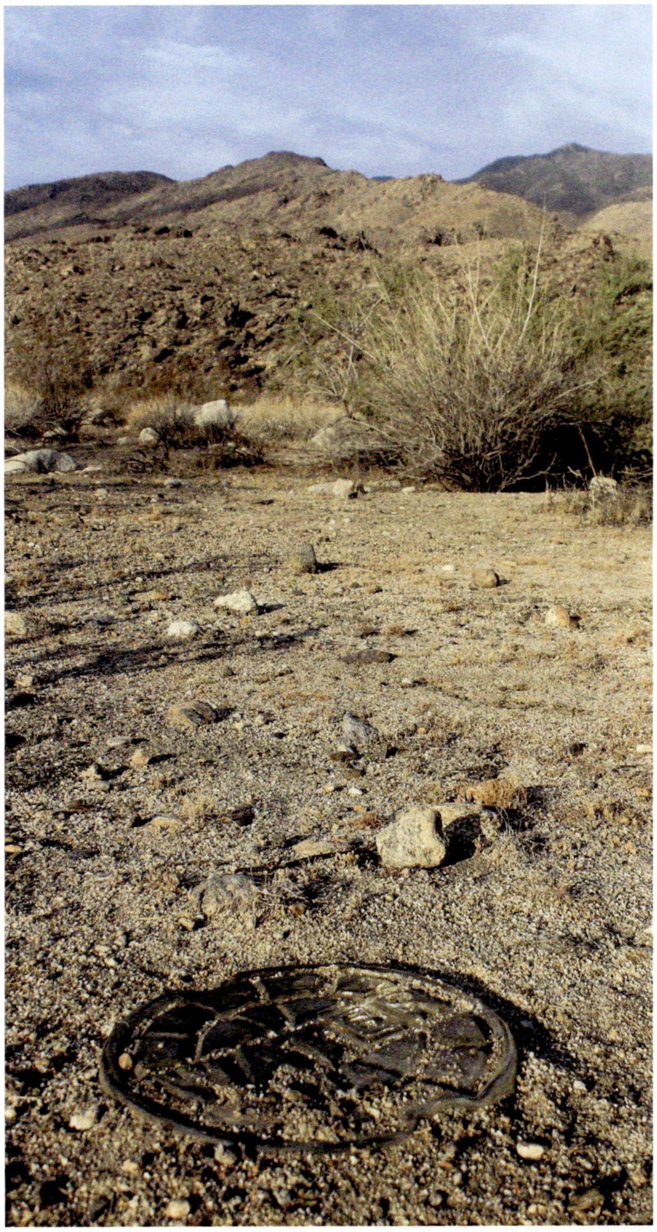

Above Neal Paustian, *Oasis* (*Water Meter* casting installed in the desert), 2013. Glass ladle-cast into a bronze mold, 1" × 6" × 6". Photo: Neal Paustian.

Top left Ladle-cast glass in its molten state, *Water Meter* casting, Neal Paustian, 2015. Photo: Neal Paustian.

Left Stockpile of *Water Meter* castings, Neal Paustian, 2015. Glass. Photo: Neal Paustian.

"I arrange flowers and foliage onto a rolled-out slab of clay and press them into the clay to transfer all their detail. I carefully remove them, place a wooden frame over the plant impressions, and then pour plaster onto the clay. Once it's set, I peel away the clay to reveal the 'plants' in relief.

"My compositions can be as simple as a single stem or as complex as a field of wild flowers, leaves, and grasses. There is a memorial as well as a celebratory quality to the casts, as they preserve a fleeting moment of glory long after the plants have faded and died. Pendulous bleeding hearts, curly fiddleheads of ferns, and wispy poppies are some of my favorite flowers to cast."[15]

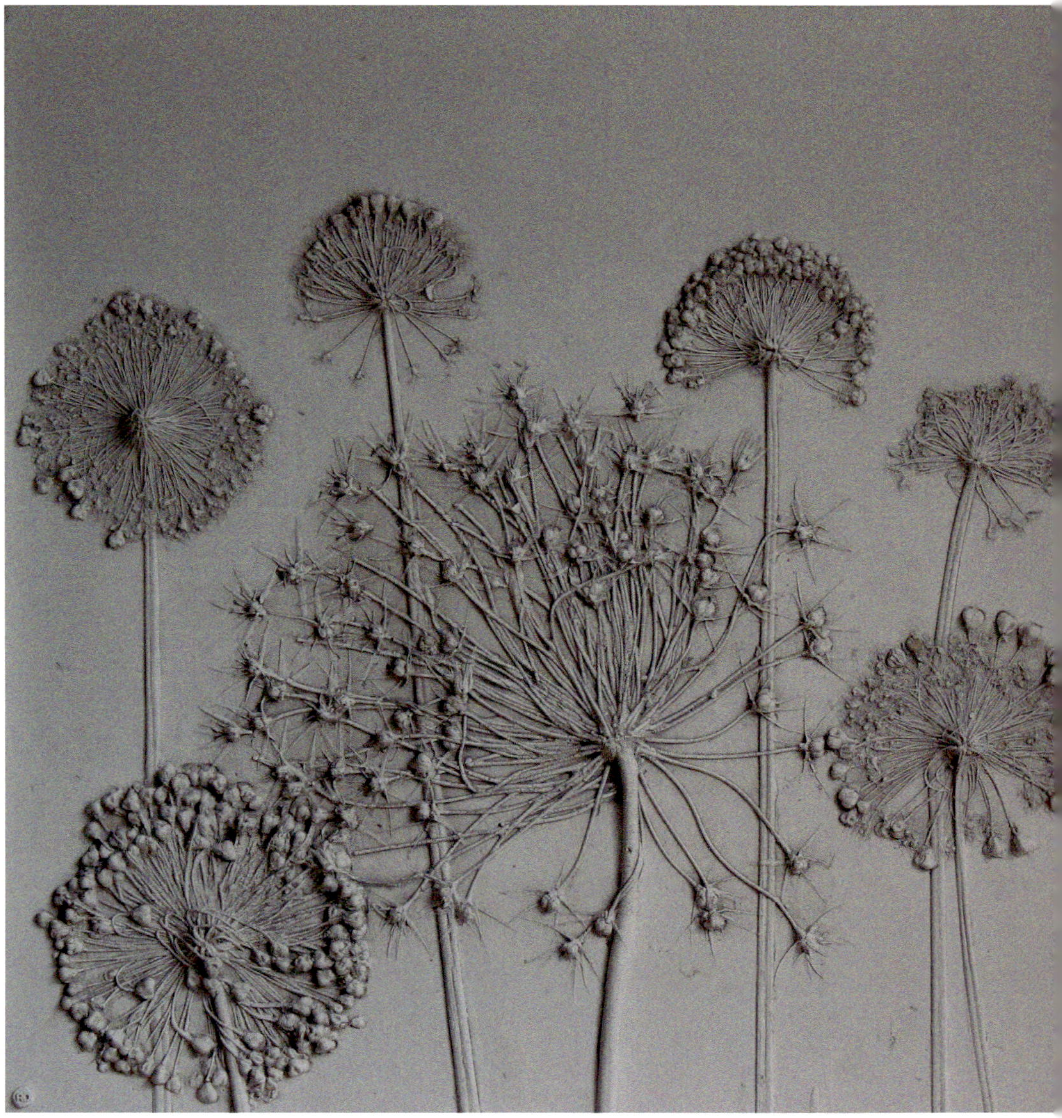

Right Rachel Dein, *Alliums*, 2014. Plaster, 15.75" × 15.75" × 0.6". Photo: Rachel Dein.

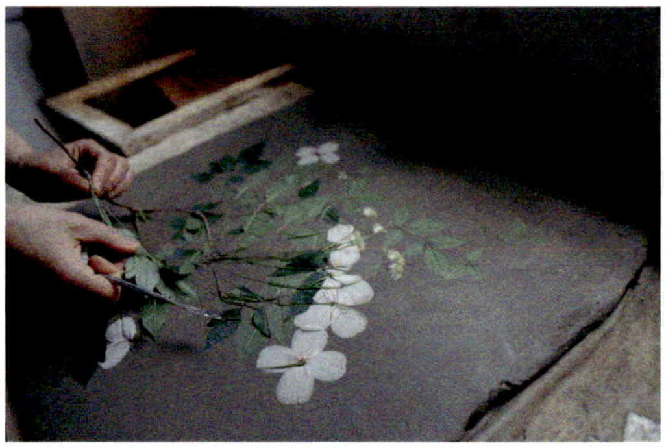

Center left Removing flowers that were pressed into a bed of clay, the clay becomes a single-use waste mold for the poured plaster. 2012. Photo: Gerard Wiseman.

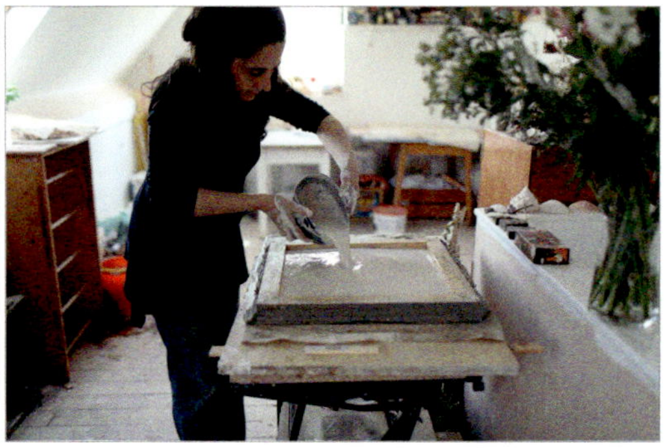

Bottom left Pouring plaster over the clay with the flower impressions. 2012. Photo: Gerard Wiseman.

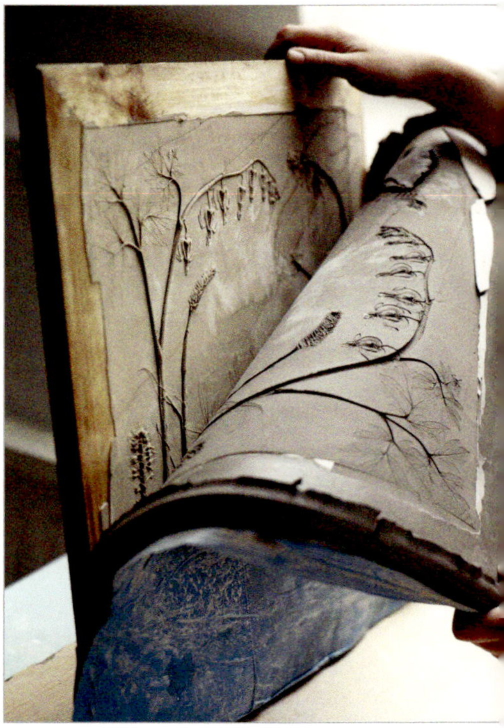

Far right Peeling the clay mold off the plaster casting. 2013. Photo: Andrew Montgomery.

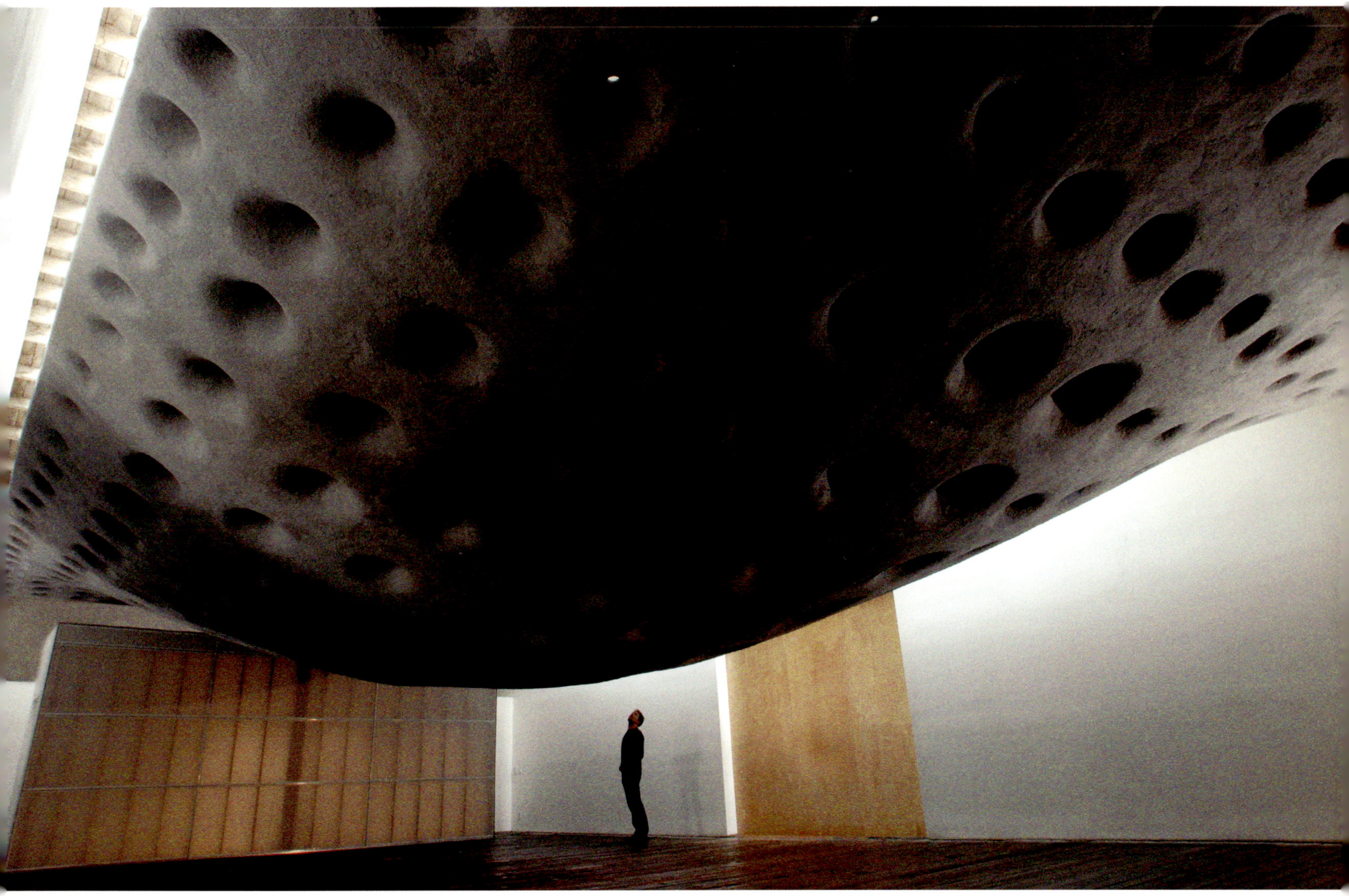

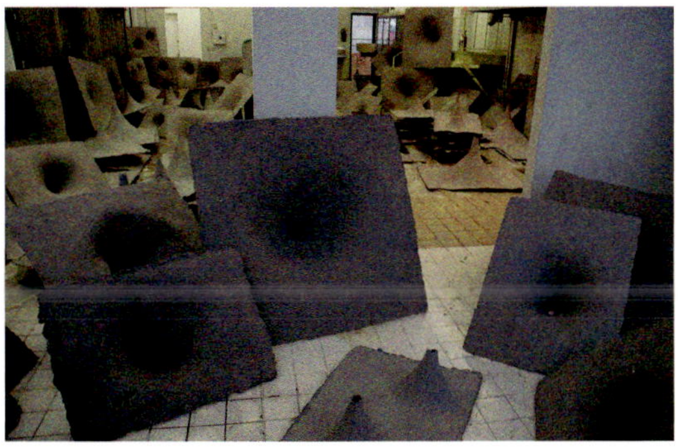

Above John Grade, *Seeps of Winter*, 2008. Installed at Suyama Space, Seattle, Washington, USA, newspaper pulp, methylcellulose, wax, fumed silica, plywood frame molds made with salvaged rigid foam, 10' × 30' × 50'.
Photo: John Grade.

"With the *Seeps of Winter* installation, I wanted to create an enveloping sense of the weight and mass of a bog as experienced from below. I imagined narrow pores through the peat conducting shafts of light from the surface above.

"I chose to use newspaper pulp for its economy and lightness. The pulp was mixed with a binder and applied to carved Styrofoam forms using thin netting as a release. Individual sections were bonded to a curving plywood structure then more pulp was applied to the seams between sections. The whole suspended mass of the sculpture swung above the heads of viewers in Suyama Space in Seattle, Washington."[16]

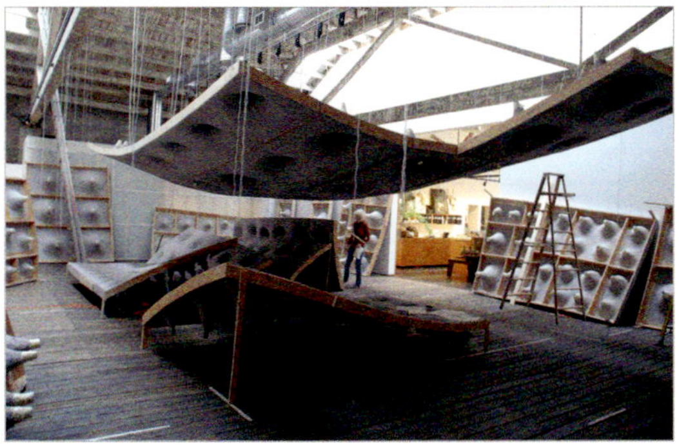

Center left Cast paper pulp forms before assembly.
Photo: John Grade.

Bottom left Assembling cast paper pulp forms.
Photo: John Grade.

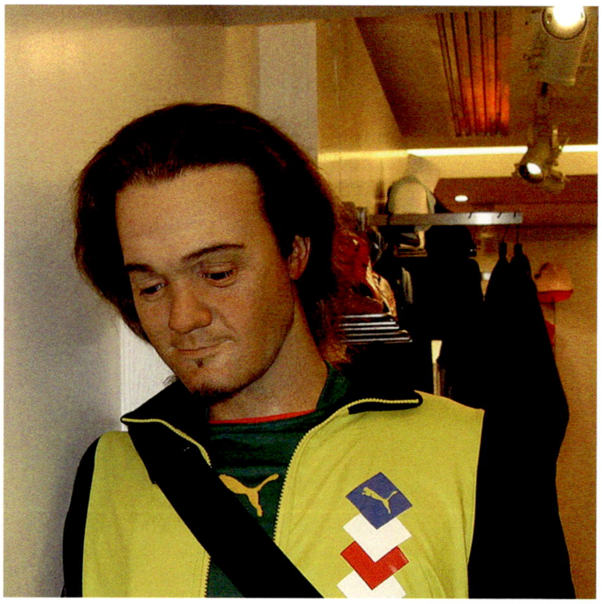

Lifelike male sportswear mannequin created for the Paris Puma International store, 2004. Silicone with a fiberglass core/armature, hand-painted, hair adapted lifecast sculptures, remolded in fiberglass, and cast out in silicone with a fiberglass core/armature, hand-painted, hair punched and individualized to suit their intended country of display, 5'10". André Masters of Masters & Munn.
Photo: André Masters, Client: Puma.

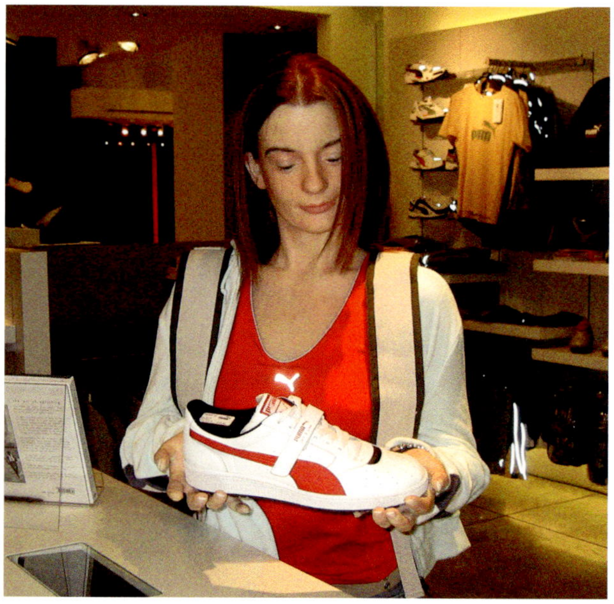

Lifelike female sportswear mannequin created for the London, Carnaby Street Puma International store, 2004. Silicone with a fiberglass core/armature, hand-painted, hair. Adapted lifecast sculptures, remolded in fiberglass, and cast out in silicone with a fiberglass core/armature, hand-painted, hair punched, 5'5". André Masters of Masters & Munn.
Photo: André Masters, Client: Puma.

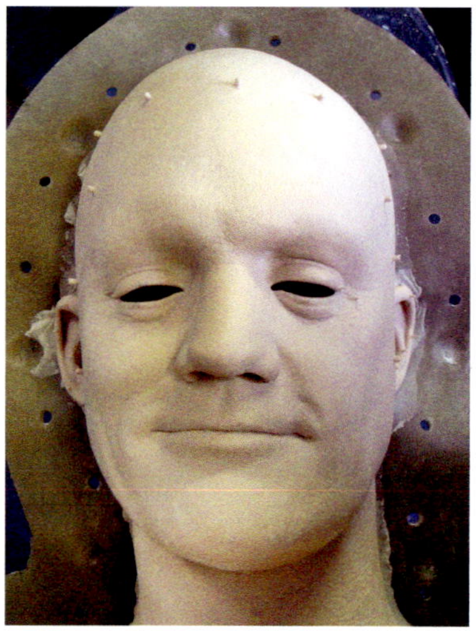

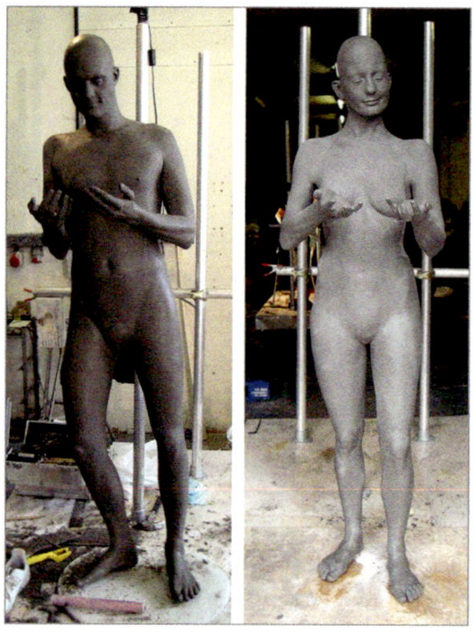

Left Silicone full head half-embedded in the fiberglass mold (front half of the fiberglass mold has just been removed), 2003. André Masters, Masters & Munn.
Photo: André Masters, Client: Puma.

Right The lifecasts are initially cast in plastiline (meltable plasticene-like clay) to enable further sculpting, refinement, and replacement of any missing skin texture. 2003. André Masters from Masters & Munn.
Photo: André Masters, Client: Puma.

"These ultra-realistic mannequins were created as a PR stunt for international sportswear company Puma. Using techniques developed in the film special effects industry, André Masters led a team of twenty artisans to produce fifteen uniquely styled humanoids to be stationed in Puma stores around the world. Cast in silicone with a fiberglass core, the sculptures were hand painted with individual hairs punched into the silicone to create different hairstyles, brows, lashes, leg and chest hair, and even nostril hairs! The resulting casts were so realistic that shoppers regularly stopped to ask them for the time or for directions before realizing they were talking to a model."[17]

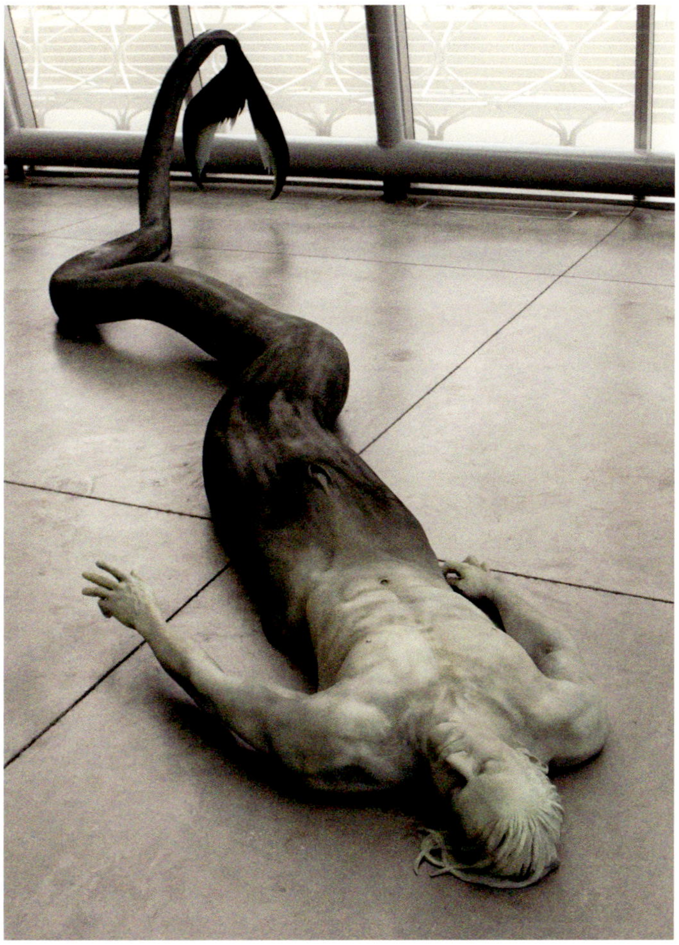

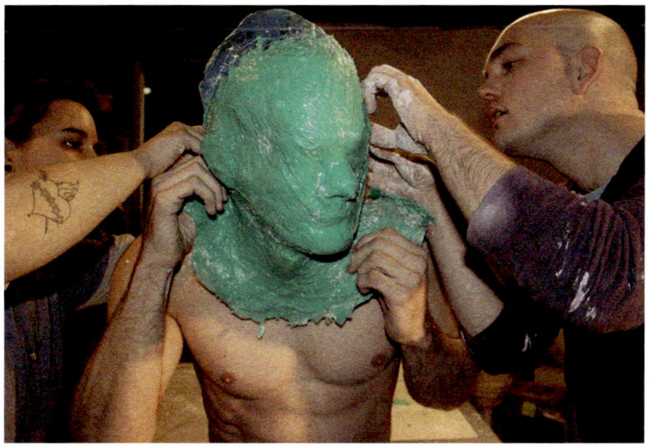

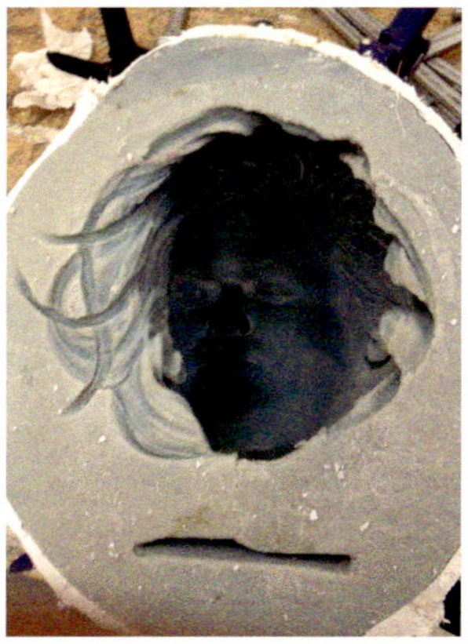

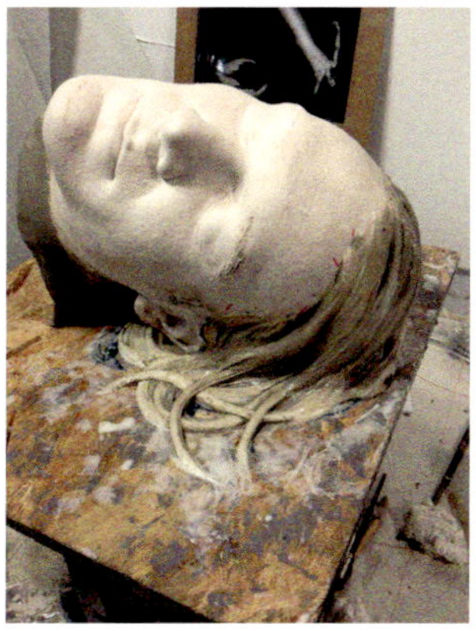

Above Cameron Stalheim, *and then I saw Colby on the street and my fantasy died.*, 2014. Plastic, foam, steel, acrylic, 20' × 3' × 4'.
Photo: Cameron Stalheim, Model: Colby Keller.

Top right Removing the silicone mold from the model's head, 2014. *Left to right:* Magali Hébert-Huot, model Colby Keller, and Cameron Stalheim.
Photo: Cameron Stalheim.

Center, left Production mold of finished head, 2014.
Photo: Cameron Stalheim.

Center, right Plaster lifecast of model Colby Keller's head with sculpted hair, 2014.
Photo: Cameron Stalheim.

"Skin-safe silicone was used to capture the model's facial details. Once that had cured, plaster was cast into the mold. Clay was added to the plaster cast at the neck to ensure a seamless transition into the body. The general movement of the hair was created with additional clay. Synthetic hair was dipped into urethane plastic and individually placed on top of the clay to create a realistic look. The use of urethane plastic kept the hair bonded to the clay without the hair lifting. The production mold was then made from layers of tinted urethane rubber. Tinting every other layer of rubber provides a visual aid to ensure proper layer thickness. The rubber is allowed to cure overnight, at which time the mold is complete and castings can be made."[18]

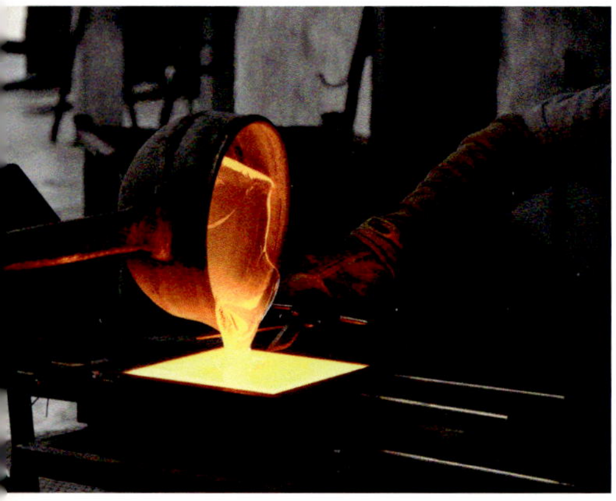

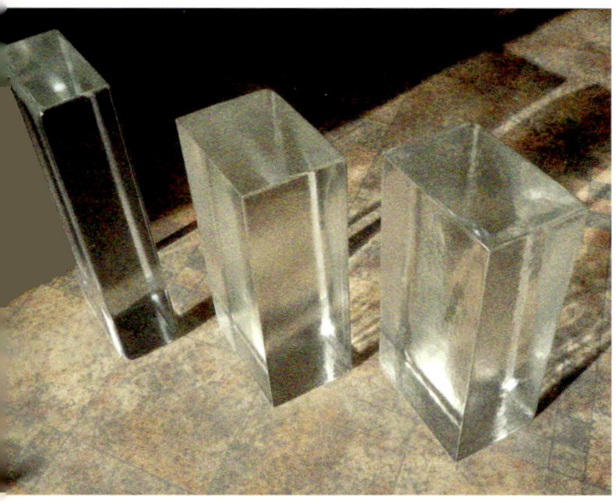

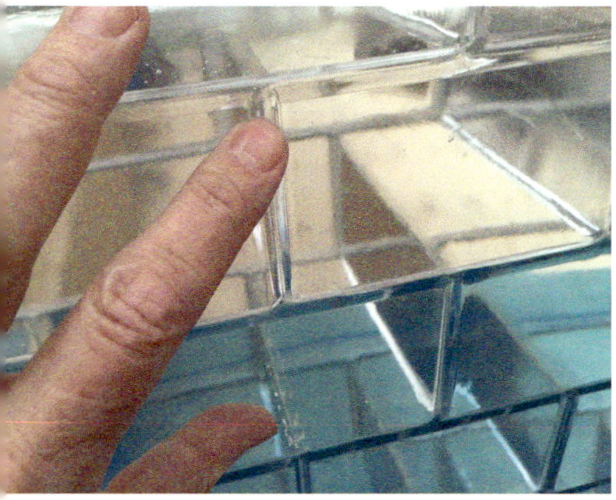

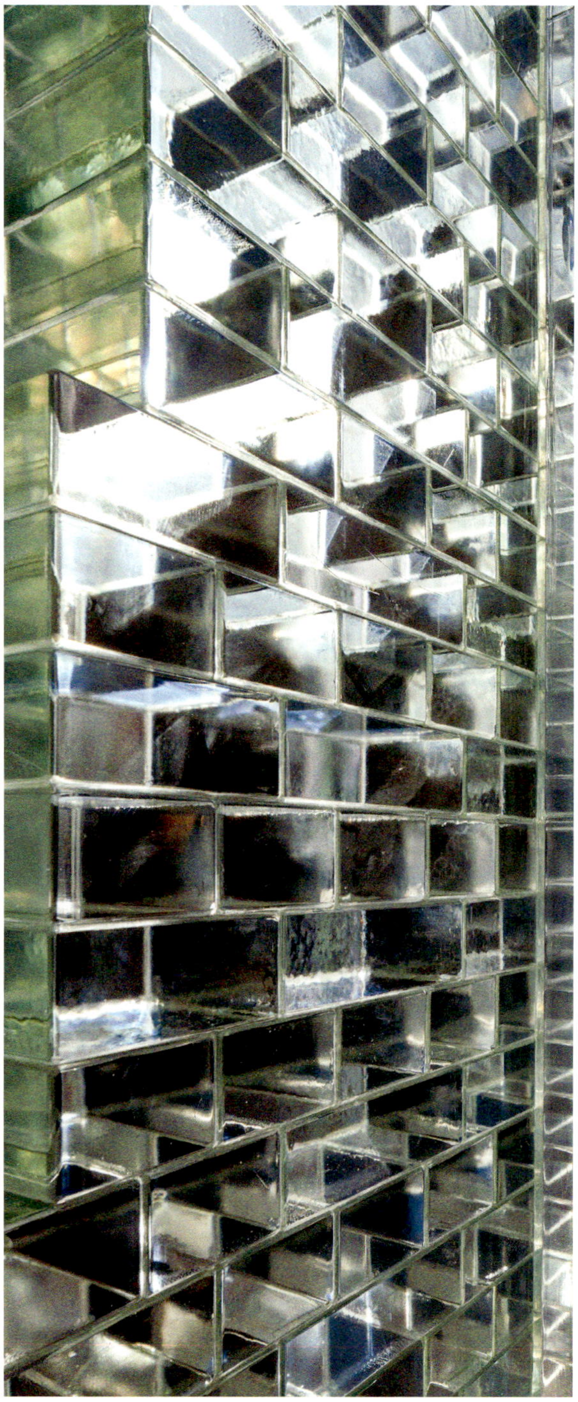

Top, far left Pouring glass bricks, 2016. Glass, steel mold, cast by Poesia in Venice, Italy, for MVRDV. Courtesy of MVRDV.

Center left Glass bricks, 2016. Cast by Poesia in Venice, Italy, for MVRDV.
Courtesy of MVRDV.

Bottom left Assembling glass bricks, 2016. Cast by Poesia in Venice, Italy, for MVRDV. Courtesy of MVRDV.

Left Over 6,000 glass bricks bound with transparent UV-setting adhesive. *Crystal Houses* for © Warenar, designed by MVRDV, Amsterdam, the Netherlands, 2016. Glass building components including window frames, etc.
Photo: Daria Scagliola and Stijn Brakkee.

Opposite page *Crystal Houses* for © Warenar, designed by MVRDV, Amsterdam, the Netherlands, 2016. Glass building components including window frames, etc.
Photo: Daria Scagliola and Stijn Brakkee

"*The Crystal Houses* took on an extremely collaborative and intense research, design, and construction process. Working with unknowns meant that the materials and building technologies had to be invented and refined throughout the process with constant collaboration between the architects, researchers, construction experts, and the glassmakers themselves.

"The development of the glass bricks was integral in the design process itself. From the hot furnaces in Italy to the laboratory-like conditions of the building site in Amsterdam, the bricks went through rigorous testing and experimentation in order to get the desired effect, aesthetically with as much transparency as possible, and also structurally. The end result was not only a stunning facade, but the team made groundbreaking advances in glass construction technologies."[19]

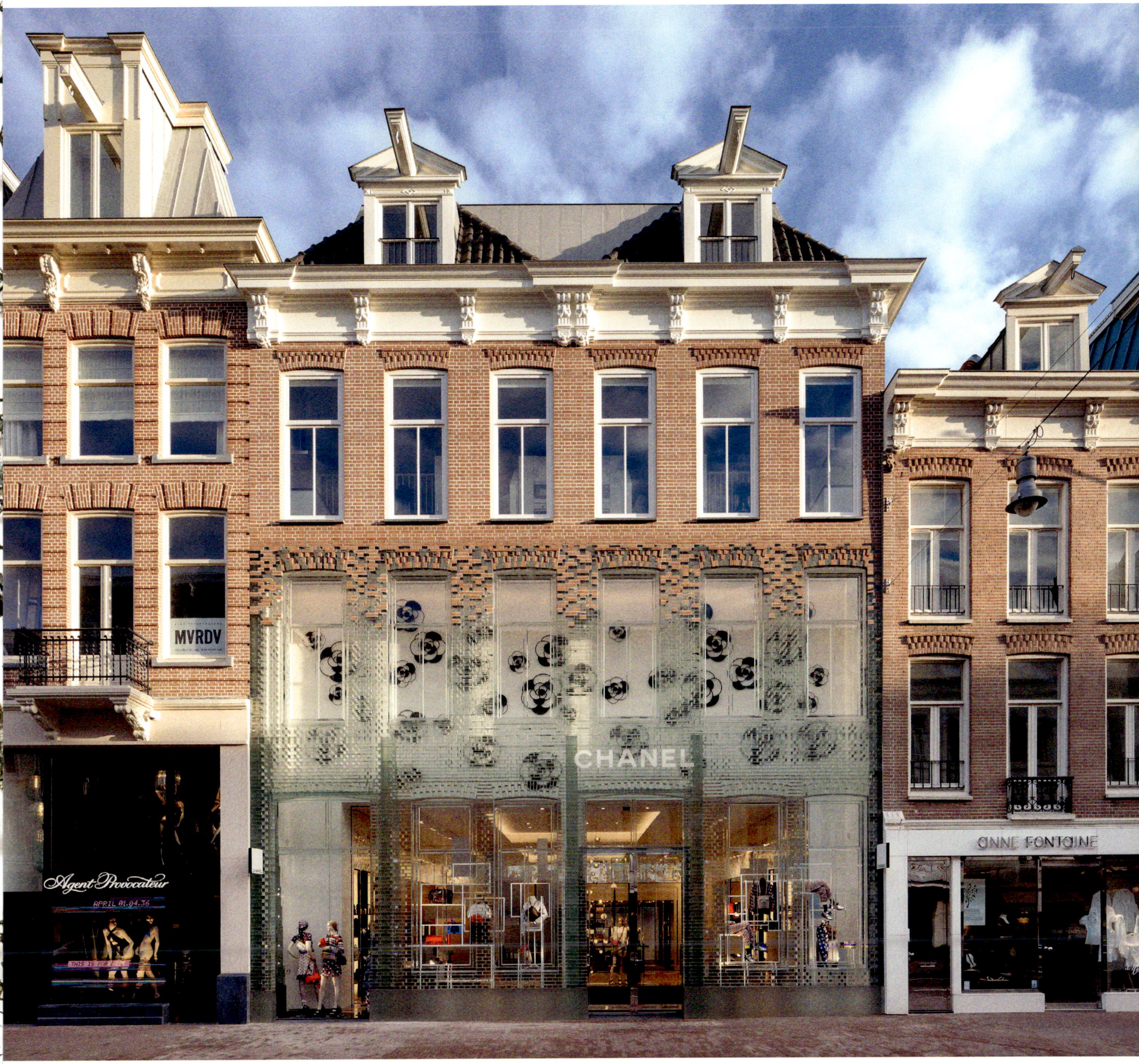

Notes

1. Email from James Thurman, February 7, 2016.
2. Email from the Karen LaMonte, June 13, 2016.
3. Email from Aleksandra Vali, April 30, 2016.
4. Email from Rebecca Rose, June 27, 2016.
5. Email from Peter Kovacsy, October 20, 2015.
6. Email from Marlene Rose, June 28, 2016.
7. Email from David Umemoto, May 2, 2016.
8. Email from Tom Joyce, June 28, 2016.
9. Email from Zsolt József Simon, May 2, 2016.
10. Email from Peter Hoogeboom, April 23, 2016.
11. Email from Rebecca Strzelec, May 6, 2016.
12. Email from Dukno Yoon, May 9, 2016.
13. Email from Donald Friedlich, April 29, 2016.
14. Email from Neal Paustain, May 9, 2016.
15. Email from Rachel Dein, April 25, 2016.
16. Email from John Grade, June 14, 2016.
17. Email from Masters & Munn, June 18, 2016.
18. Email from Cameron Stalheim, June 24, 2016.
19. Email from Alex Davidson, Public Relations Assistant for MVRDV, May 4, 2016.

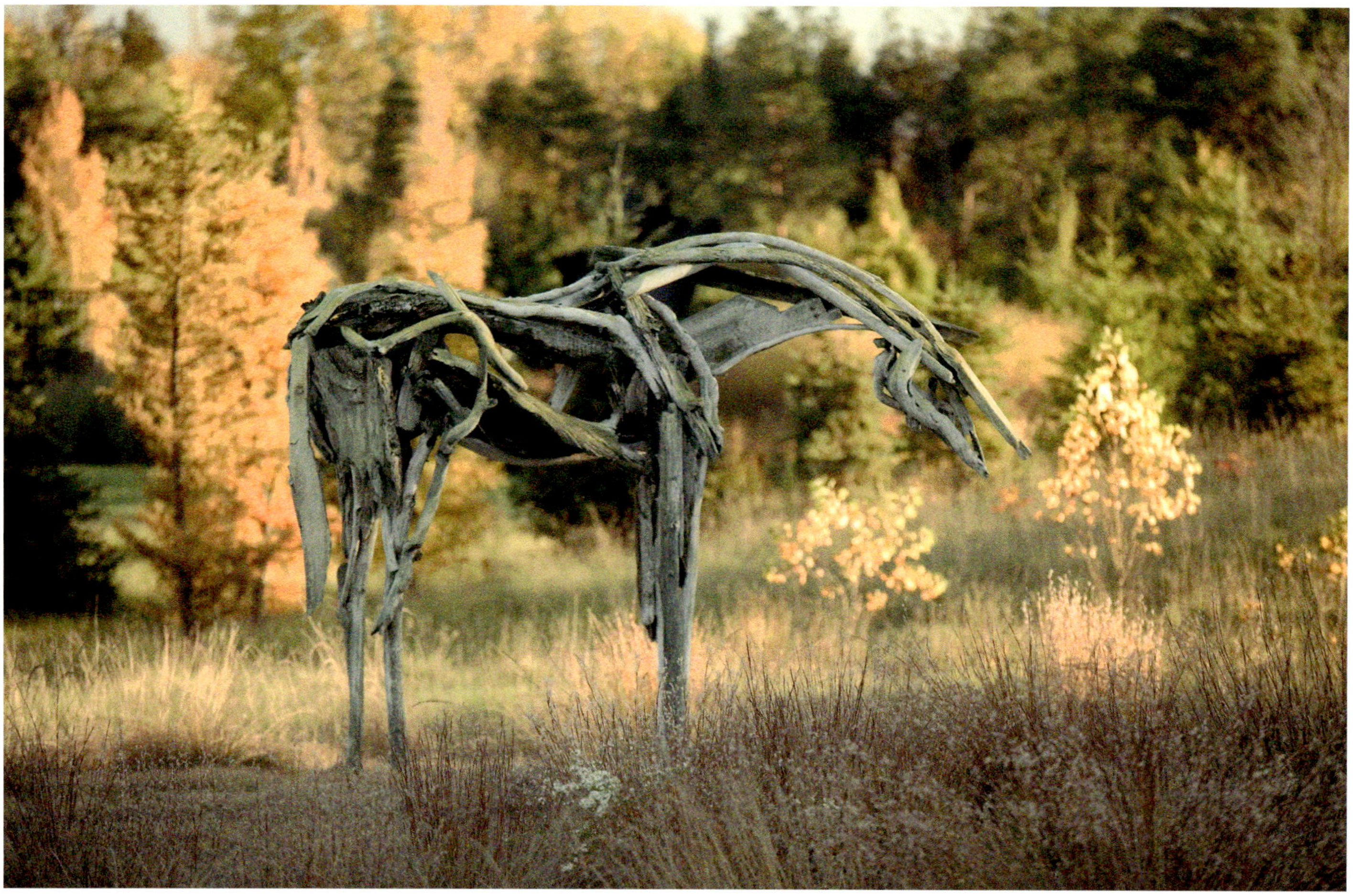

Deborah Butterfield, *Cabin Creek*, 1999. Bronze, 7'4" × 10'2.5" × 2'6.5".
Photo: William J. Hebert, Courtesy of the Frederik Meijer Gardens and Sculpture Park, Grand Rapids, Michigan, Licensed by VAGA.

Butterfield's early scavenged and constructed sculptures were limited to interior display. However owing to the encouragement of California artist Manuel Neri and the open and willing nature of the Walla Walla Foundry, Butterfield began to cast her work in bronze in the 1980s. Specifically, artist and foundry painstakingly translated her wooden originals into unique bronze masterworks that maintained the detailed textural surfaces and coloration characteristic of the original scavenged branches and boards. Iconic works like *Cabin Creek*, 1999, astound and delight even the most casual audiences, but more importantly, they represent how sculptors like Butterfield embraced both the current trends of using experimental materials in the original and those more distant historical longings for illusion and permanence through a final translation into bronze; Renaissance demands and nineteenth-century innovation are very much at play. Through all avenues of the artist's determination and the foundry's dedication, Butterfield's goal for the mimetic is met, if not exceeded.

At first encounter, it may seem as if the expressive repertoire of the German sculptor Dietrich Klinge (b. 1954) is parallel to that of Butterfield both in terms of transformation and mimesis. Since capturing initial critical acclaim in the late 1980s, the sculptor and graphic artist has perfected a technique of carving and casting, however his demands for and definitions of the finished sculpture are unique. After selecting and intensely studying large pieces of wood, Klinge boldly carves his figures, as a whole or in parts, with chainsaws. When satisfied with the initial composition, the artist works directly with the foundry throughout the moldmaking and casting processes; the application of patina is solely his domain. For Klinge, the wooden original is but a means to an end and the completed bronze is the definitive work of sculpture. The images of his 2006 *Gordian I* offer a sample comparison of the original model and the finished work; the former to remain silent and subservient, while the latter is available to public and private collectors in a limited edition.

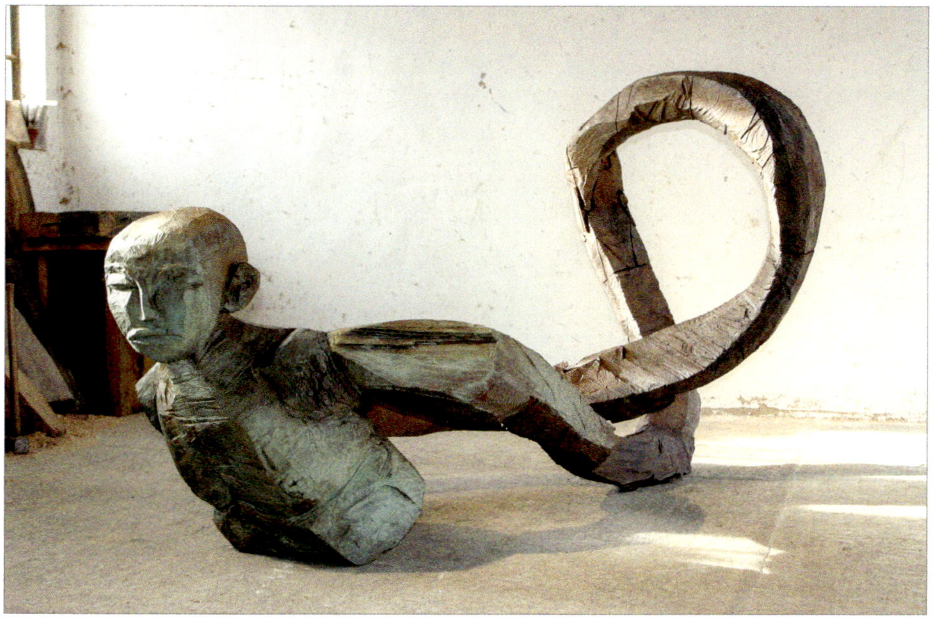

Left Dietrich Klinge, *Gordian I*, 2006. Wood model, 4'2" × 3'6" × 6'6". Photo: Dietrich Klinge.

Below Dietrich Klinge, *Gordian I*, 2006. Bronze, 4'2" × 3'6" × 6'6". Photo: Dietrich Klinge.

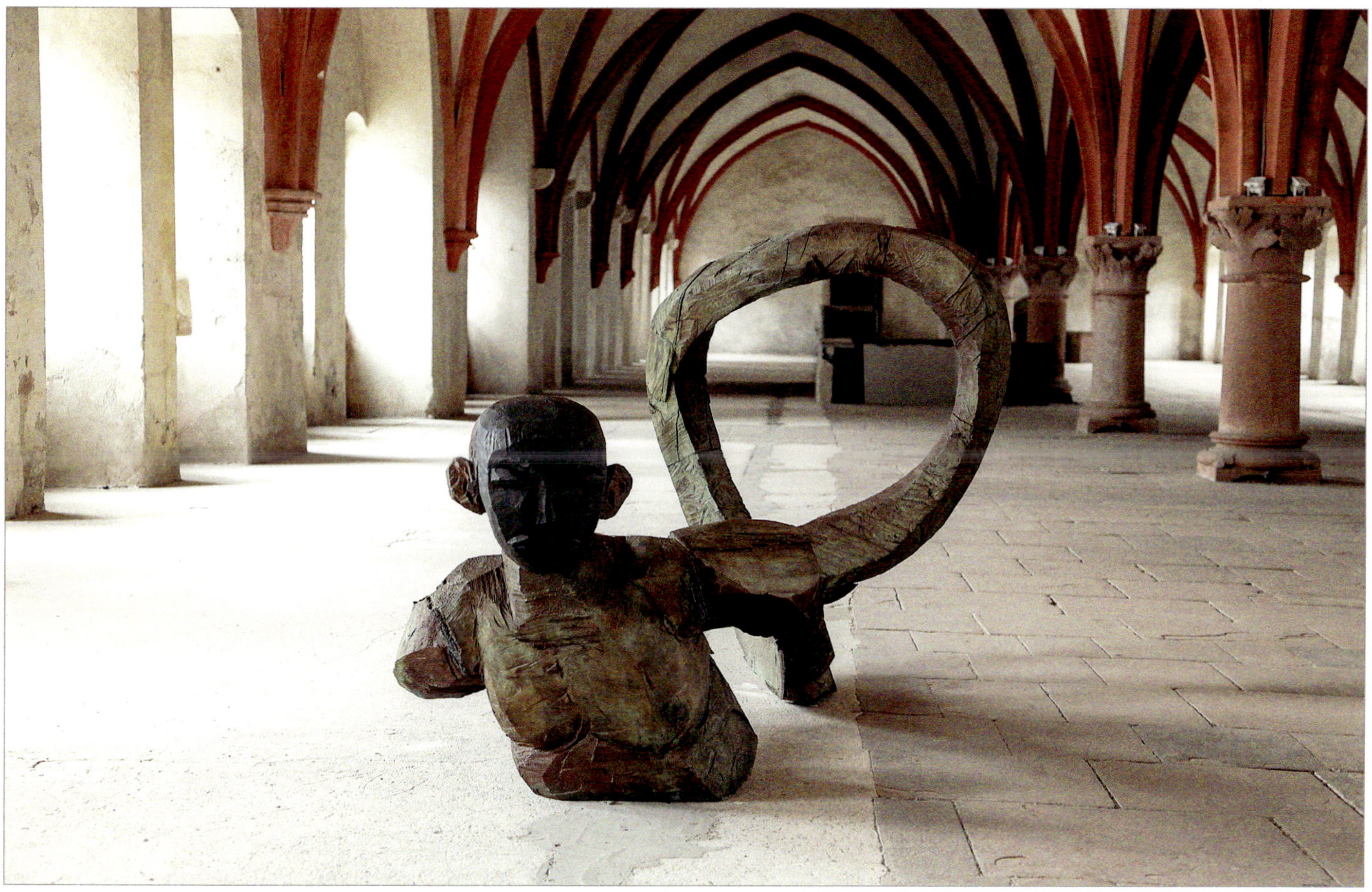

Any of today's major art fairs that stimulate and guide the global world of Contemporary art will likely feature the work of Abakanowicz, Butterfield, and Klinge as there is once again a market for large-scale metal casting. Most of the definitive sculptors of our age from Antony Gormley to Tony Cragg, Kiki Smith to Thomas Houseago, and Jaume Plensa to Ai Weiwei are working creatively and successfully with bronze, aluminum, and iron. In part, one must acknowledge the demands of the art market and the role of both artists and galleries to sustain it. There is now near equality between the acquisition of painting and sculpture, whereas nearly every historical period witnessed the supremacy of the painted image. There has been a widespread development of commissions and acquisitions, both public and private, for large interior spaces, gardens and parks that can easily accommodate large-scale metal casts. Against such powerful currents, many foundries in Europe and America have closed owing to both the costs of labor and materials. Many sculptors have chosen instead to follow the long lead of industry and are having their work cast in Asia.

On multiple fronts, China in particular has taken a leadership role in the history of large metal casting. There are a seemingly infinite number of foundries available for sculptors. At the largest and best foundry, it is not unusual to experience myriad projects created by artists and destined for sites across the globe. Within are a significant number of Chinese masters of international stature like the conceptual artist Ai Weiwei (b. 1957). *Iron Tree*, 2013, is the artist's largest outdoor sculpture to date. Composed of ninety-nine individually cast tree elements bolted together, the colossus reads as a singular element from a distance, but upon approach the illusion breaks down as each element comes into focus.

Symbolically, the sculpture speaks to issues of individuality and uniformity, as well as the fragility of the environment, and it pays homage to Chinese building and cultural traditions. The work also speaks to the ingenuity and vibrancy of large metal casting today. For example, iron has become a potent choice of material for many Chinese sculptors and solid cast objects are frequently encountered. Although much of the present discussion looks at large metal casting in the Western tradition, it is honorable in that there is cohesion from antiquity to Renaissance to the nineteenth century, to rebellion and rediscovery in the twentieth century.

Although China has played a major economic role on the world stage for decades, leadership in the visual arts has been slow to develop. In considering the former, technology, material, and labor costs have long since been inexpensive; however, the critical impact that many of China's artists now exert on the world stage may greatly change our definition of the latter.

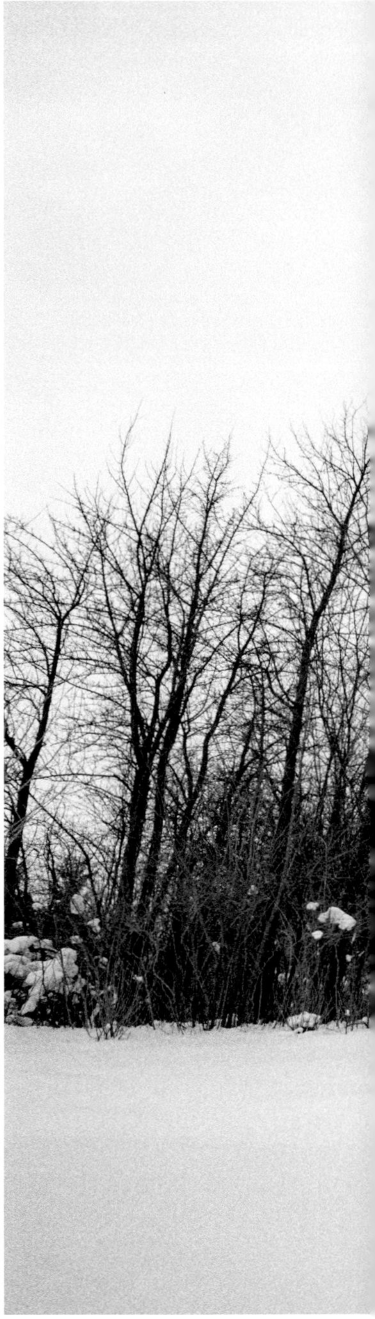

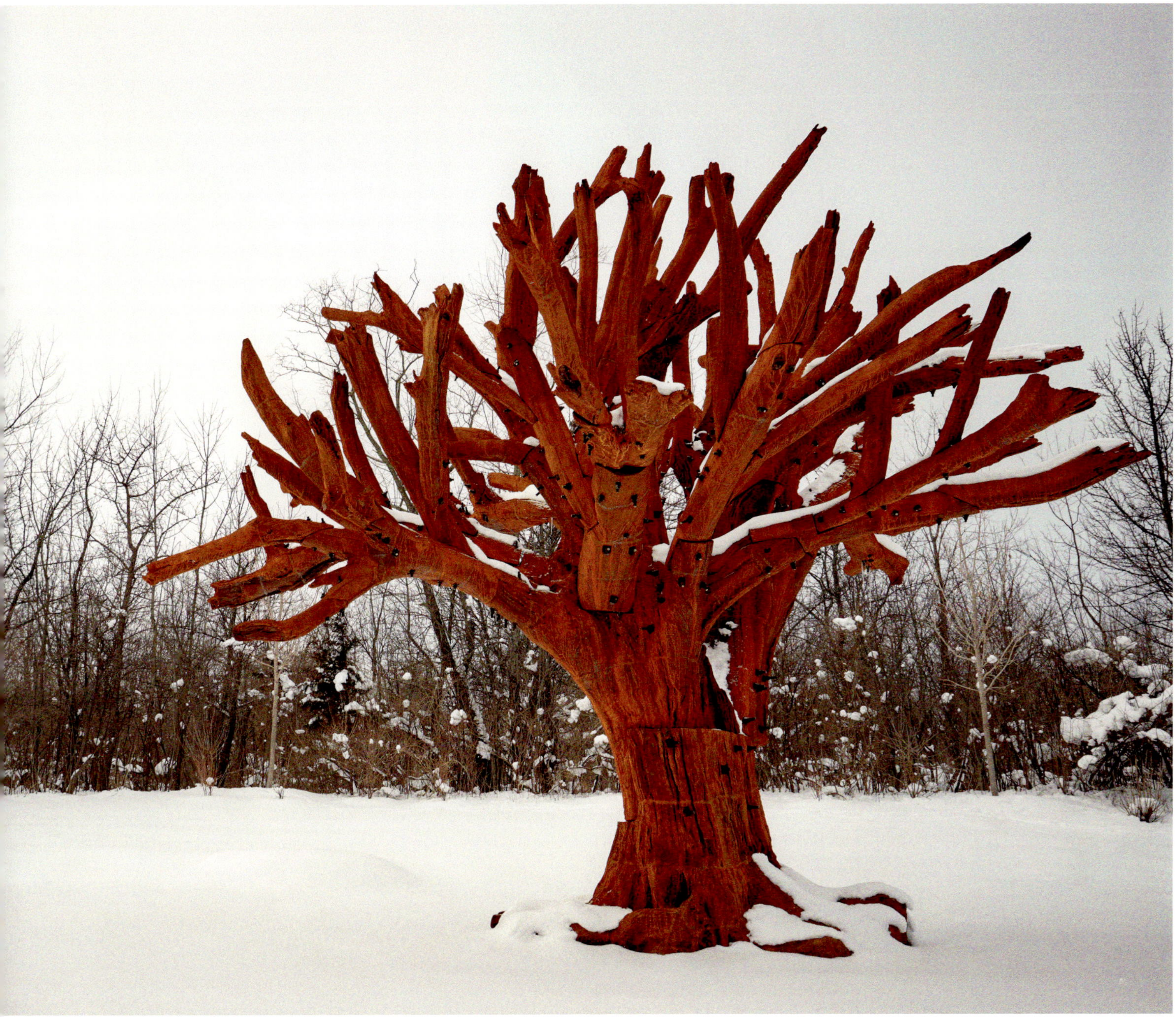

Ai Weiwei, *Iron Tree*, 2013. Iron and stainless steel, 22' × 22' × 22'.
Photo: Dean Van Dis, © Ai Weiwei, Courtesy of the Frederik Meijer Gardens and Sculpture Park, Grand Rapids, Michigan.

Joseph Becherer is the founding Director and Curator of the Sculpture Program at Frederik Meijer Gardens and Sculpture Park in Grand Rapids, Michigan. He currently serves as Chief Curator and Vice President of Sculpture and Horticulture, Collections and Exhibitions. In addition, he is the Lena Meijer Professor in the History of Art at Aquinas College.

FUNCTION

Function can be considered in two ways: the purpose of objects themselves and the characteristics of materials. Objects have certain material requirements in order for them to perform, so it follows that an object's function can dictate choices in process, material, and design. Metal is one of the most commonly cast materials and has had a monumental impact on human existence. Metal is malleable, durable, and has good tensile strength; it can be machined and it can hold a sharp edge. Cast components can be soldered, welded, and cold-connected. These qualities explain why cast metal is so useful and prevalent.

When it comes to the characteristics of individual casting materials, metal has the greatest range of possibilities as it is actually a family of materials, each with different properties and attributes. Bronze is a surprisingly durable metal almost immune to corrosion, even when exposed to water. Aluminum is lightweight, lead has a low melting point, precious metals are nonreactive with the human body, and iron and steel are the darlings of the Industrial Revolution. There are a multitude of choices when it comes to metal casting. The function of an object helps to determine what metal is best suited for a casting based on the material's attributes (for example, lead would be a poor choice for cast bridge construction as it is both soft and heavy). Metal's physical properties make it a natural choice for tools, machine components, architectural elements, hardware, weapons, sculpture, and jewelry.

Metal casting revolutionized domestic life early on; seemingly endless household applications have been at the center of human existence since the (aptly named) Bronze Age. Stamps and seals helped people indicate and authenticate ownership and authorship; while the public at large no longer uses them today, they are largely the province of notaries public. Stamps and seals were mainstays in cultures across Europe, Africa, and Asia for thousands of years and still play a role in some cultures.

Locks are often cast and machined, as were many early keys. Other door hardware like hinges, decorative escutcheon plates, door knobs, and door knockers are also cast. This everyday hardware provided previously unknown ease in securing property, whether small valuables or entire households.

Another part of domestic life affected by casting was—and is—in the kitchen. Metal cooking and drinking vessels have a durability unlike any other material. While something like a decorative rhyton may be antiquated, cast iron pots and pans are still valued today by cooks everywhere.

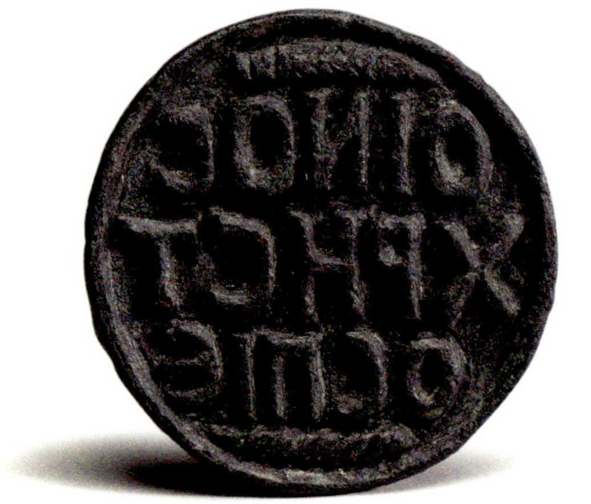

Commercial stamp, sixth to eighth century (Late Antique). Bronze, 2.4". Courtesy of The Walters Art Museum, Museum purchase in 1993, CC0, GNU FDL.

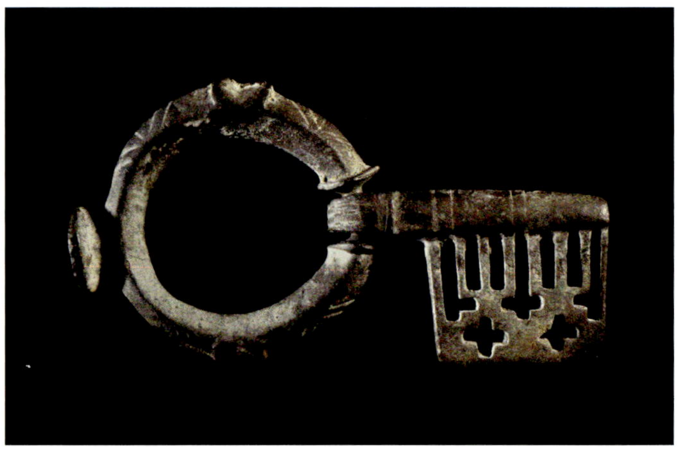

Byzantine key, fifth century. Egypt (?), bronze, 3.7" × 2.1" × 0.6". Commonly used during the Byzantine period, this type of key was easily carried, an important feature at a time when clothes had no pockets. Courtesy of The Walters Art Museum, Acquired by Henry Walters, CC0, GNU FDL.

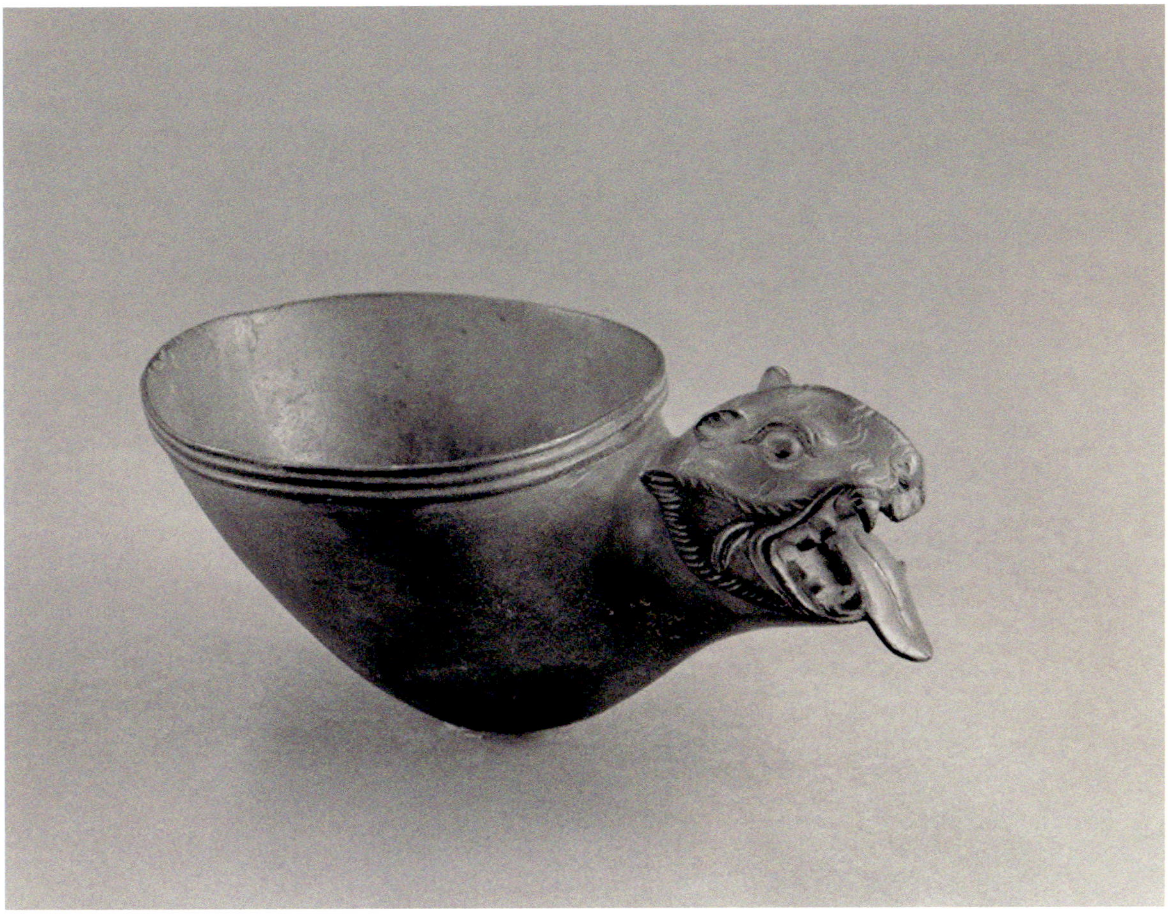

Rhyton, The Six Dynasties, 265–589 BCE. China, bronze, 2.5" × 5.25".
Courtesy of LACMA, Gift of the Ahmanson Foundation, www.lacma.org, LACMA public domain.

Architecture has also been indelibly changed by metal casting. The Baptistery doors in Florence, Italy, considered to be masterpieces of the Early Renaissance, are one of the most famous examples. Andrea Pisano sculpted the South doors to the Baptistery and Lorenzo Ghiberti sculpted doors for the North and East entrances. In a pre-printing press world when many people could not read, institutions such as governments and churches used imagery to disseminate information. These doors are covered in scenes from the Bible from the story of Adam and Eve to David and Goliath. This mix of narrative sculpture and architecture has been reinterpreted ever since. Lee Lawrie made doors in a similar format for the Library of Congress in Washington, DC. Working in this way memorializes and elevates the narrative in a durable format.

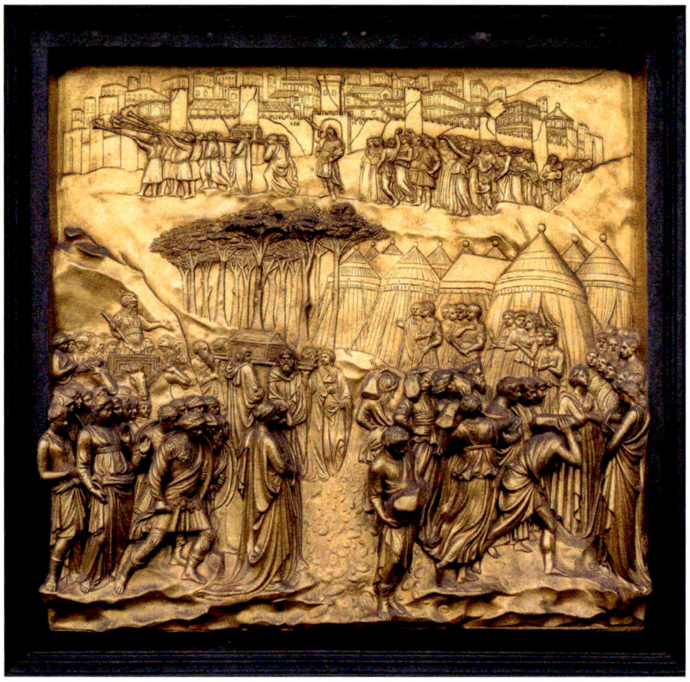

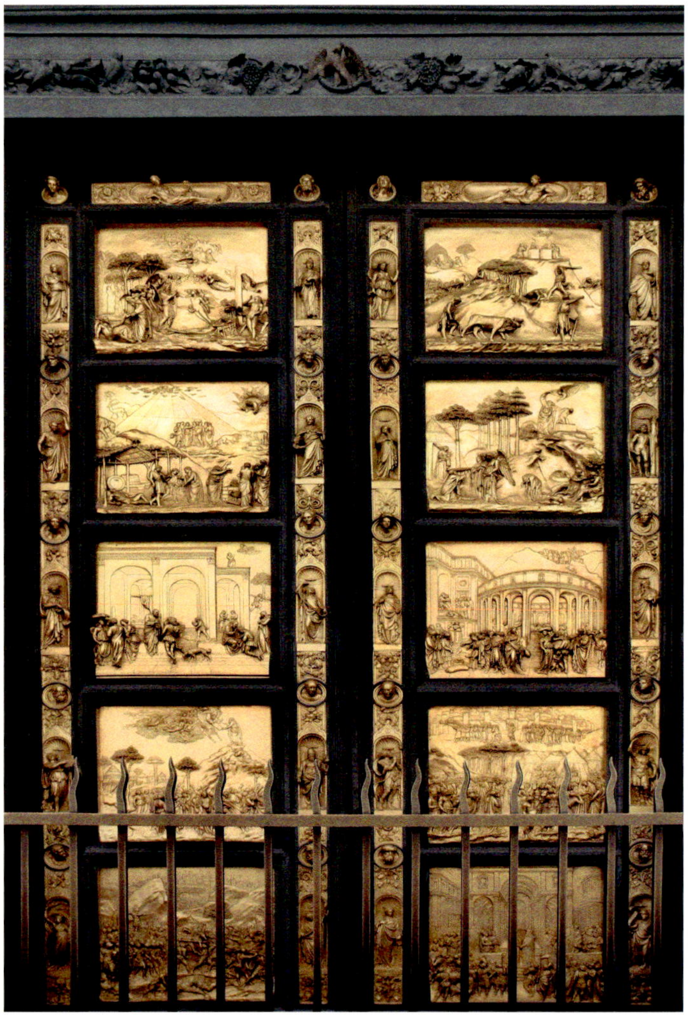

Top Lorenzo Ghiberti, *The Fall of Jericho* panel from the Baptistery Doors, *The Gates of Paradise* (detail), 1422. Florence, Italy, gilded bronze, 20' tall, 3 tons. Photo: © Mihai Andritoiu, Depositphotos.

Right Lorenzo Ghiberti, *Gates of Paradise* / Baptistery Doors, 1422. Florence, Italy, gilded bronze, 20' tall, 3 tons. Photo: © Yu Lan, Depositphotos.

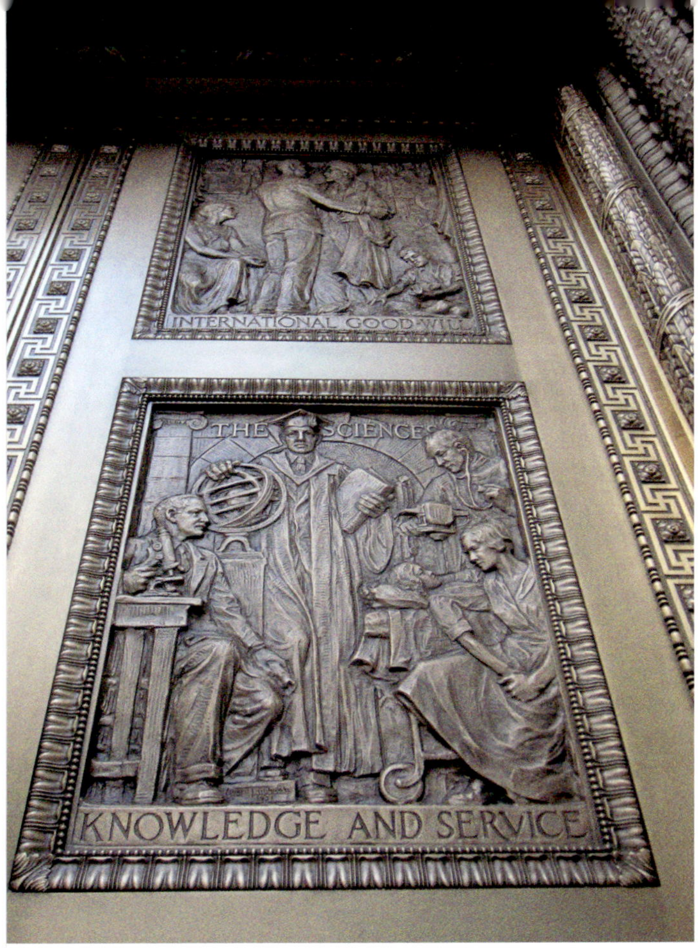

Left Arvard Fairbanks, central entrance doors from the National Bank Building, Portland, Oregon, USA, 1931. Bronze.
Photo: © Another Believer / Wikimedia Commons / CC-by-3.0.

Below In the parlance of American history, bronze doors have been used to record historic events.

Lee Lawrie, reliefs on bronze doors featuring *Ogma and Sequoyah* (door detail, East entrance), Library of Congress, John Adams Building, Washington, DC, USA, 1939. Bronze, 11' tall.
Photo: Carol Highsmith.

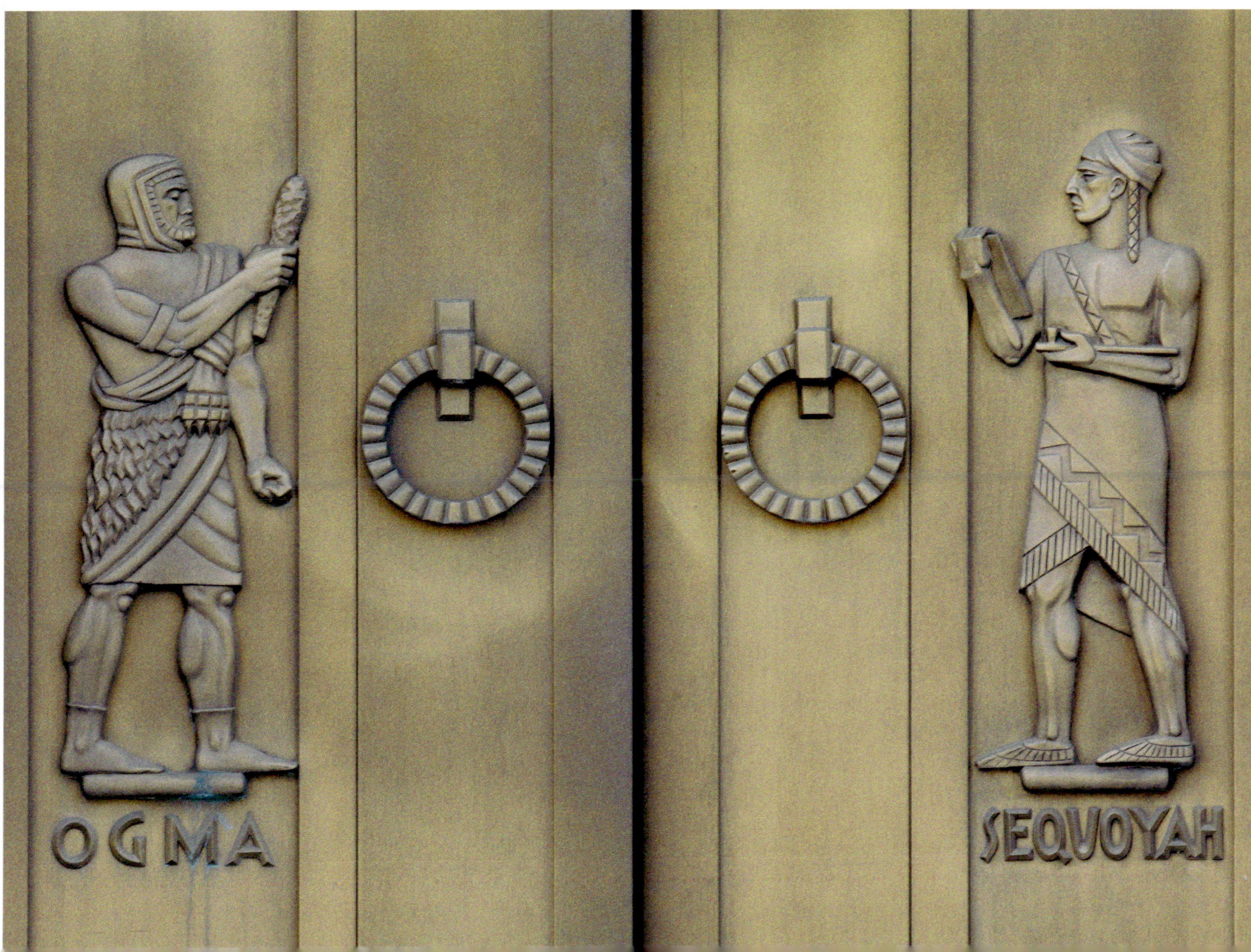

previously unknown devastation and could strike from greater distances than were ever possible before. Casting made cannons possible; they needed to be massive in order to function properly and propel something as heavy as a cannonball accurately. Beyond being functional, cannons were sometimes symbolic, and even if they were never fired, indicated authority, as with the Tsar's cannon. Its ornate nature belies its real purpose, which was a show of power.

Cast iron has also been used extensively in cemeteries in the form of mausoleums, headstones, and statuary, commonly found as a way to house remains or memorialize someone after death. New Orleans has a rich tradition of cast iron railings, fences and gates, balconies, and window grills, but some of the most spectacular cast iron is in their cemeteries. Since New Orleans is below sea level, burial below ground is forbidden by law, so above-ground mausoleums are very common. They can be used for an individual, but often house the remains of entire families.

Cemeteries are not the only religious context to heavily feature cast metal; in fact, the sacred spaces of many religious traditions are often graced with likenesses of gods, goddesses, angels, and other supernatural figures. This kind of statuary can also be used to exalt and commemorate church leaders.

We frequently encounter cast metal in public spaces in the form of sculpture and fountains as well. Most public gardens and parks have, at the very least, a simple cast bronze plaque, but many are robust sculpture gardens full of cast works.

Figurative bronze sculpture is a genre all its own that spans cultures and millennia but here, we are focusing on its simple worldwide prevalence. Bronze is a perfect choice for sculpture because of its resistance to corrosion. Bronze sculpture and memorials have been used in many ways and occupy public spaces around the world. Leaders, intellectuals, and public figures of all kinds have been glorified through bronze sculpture.

Whether a sacred space devoted to prayer or a quiet courtyard sculpture garden, these places offer us a quiet spot to reflect on our own humanity, mortality and/or the divine. Cast sculpture helps define these spaces and captures our imaginations.

Public and religious works are only a part of the world of sculpture, however. The art gallery is perhaps one of the most natural sites where we find cast metal. Before the twentieth century, sculpture was represented almost exclusively by figurative bronze. Sculpture has vastly expanded in the last hundred years though, and artists are using cast metal in new and innovative ways that leave the figurative bronze as just a small part of the world of sculpture.

These are by no means every place we encounter cast metal, but they begin to express how omnipresent it is.

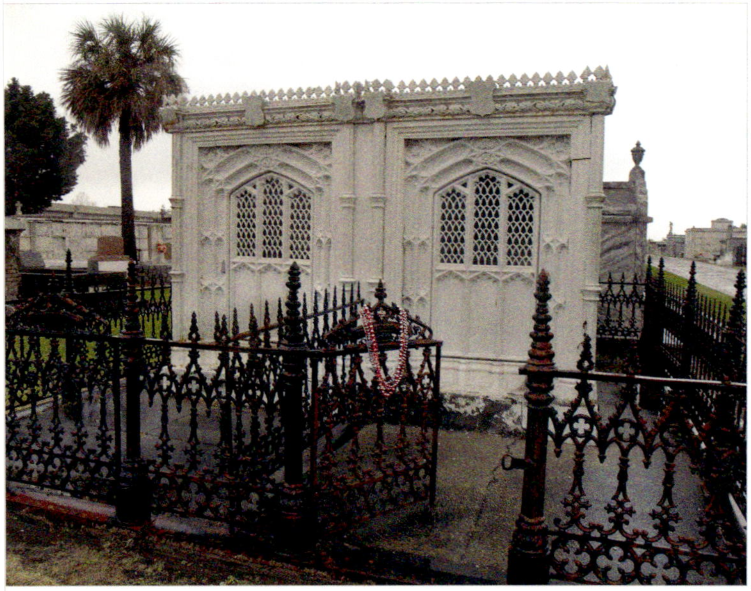

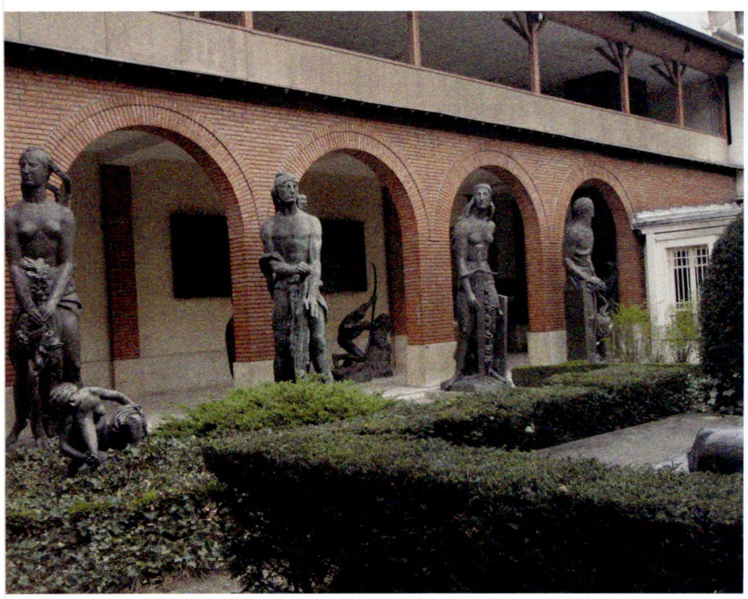

Top Not only is this crypt cast iron, but the fence and the plastic Mardi Gras beads are cast as well.

Cast iron crypt, Green-Wood Cemetery, New Orleans, Louisiana, USA, iron.
Photo: © Michael Homan / Wikimedia Commons / CC-by-2.0.

Bottom Garden in the Museé Bourdelle, France, bronze statuary.
Photo: © Langladure / Wikimedia Commons / CC-by-3.0, GNU FDL.

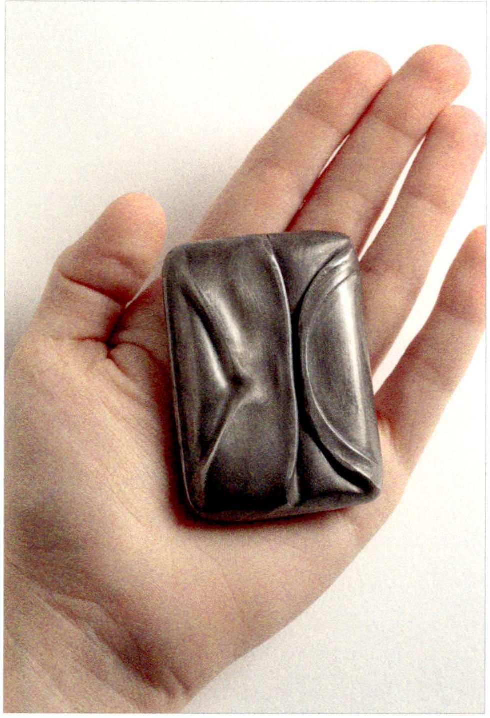

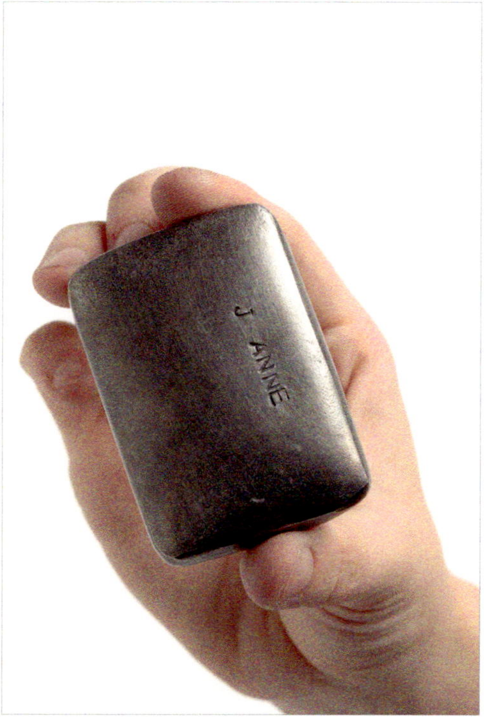

While the front side (left) of each of these hand-held vessels is fabricated from old pewter serviceware, the back side (right) is cast from the remaining scrap. Derickson made one vessel for each of her Facebook friends with the intention of eventually giving them their designated piece. Photos: David Hunter Hale.

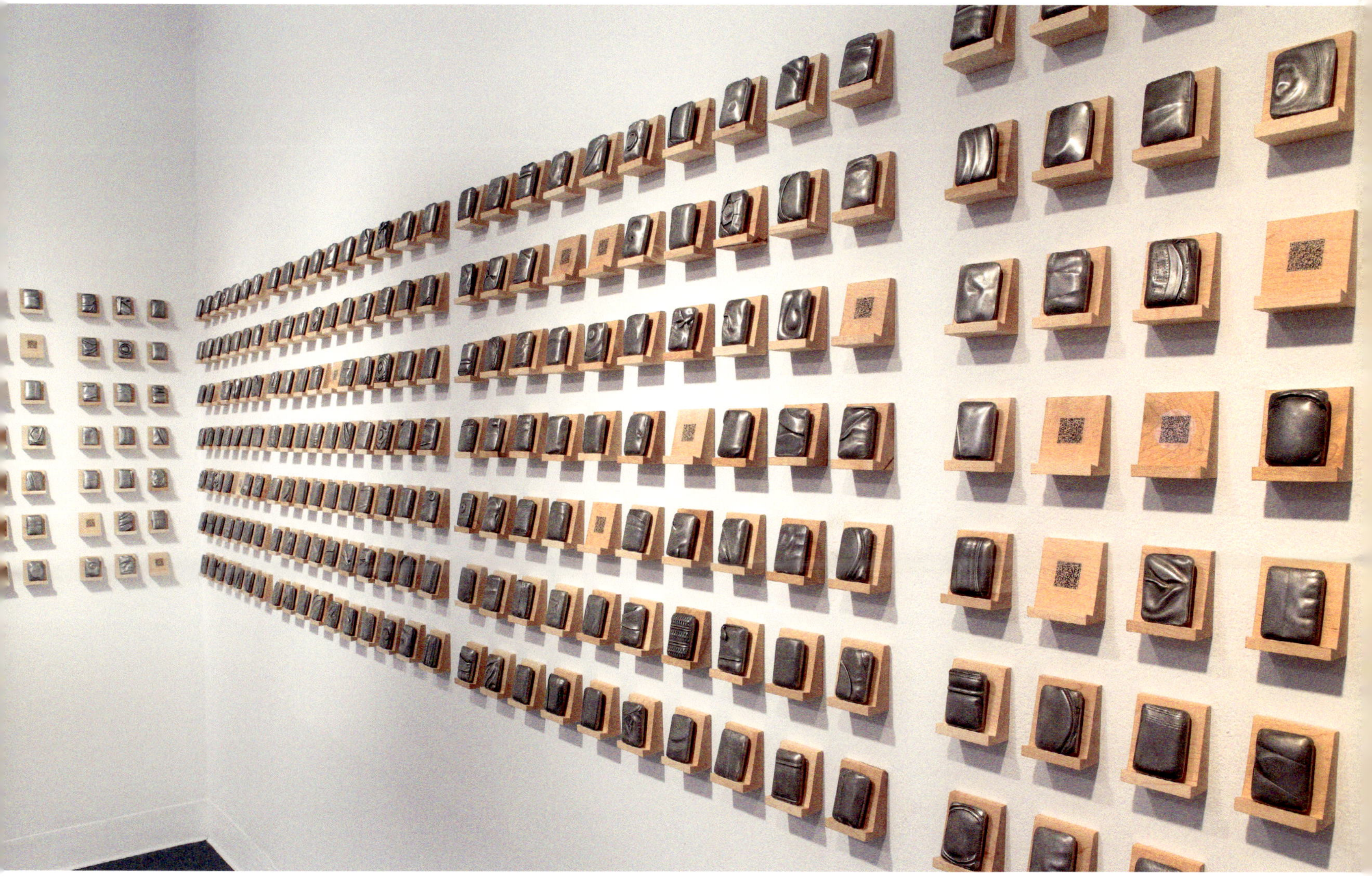

Lucy Louise Derickson, *I look forward to seeing you again, or for the first time*, 2015. Repurposed pewter serviceware, sealed message, mixed media, 3" × 2" × 1.5".
Photo: David Hunter Hale.

FOCUS on POSTPRODUCTION

Castings are often standalone items—straightforward, one-part objects like a slipcast ceramic tile, a glass vase, or a plastic toy, that make it easy to visualize the process used to produce them. Many materials require this simplicity as they cannot be retooled or altered after casting. Certainly, ceramic can be drilled, glass can be sandblasted and carved, and plastic can be used as a component in very complicated assemblages, but these materials have nothing on the multiple post-casting alterations possible in metal.

Metal can be manipulated post-casting more than any other material. It can be combined with other metal parts, creating objects so complex that they defy easy understanding and explanation. Metal can be soldered, fused, and welded to other metal parts; it can be cold-connected in a variety of ways, from rivets to tabs-in-slots; it can be forged and manipulated with hammers, and it can be easily combined with fabricated metal parts. In this way, a metal casting can either be a standalone work unto itself or it can be just small a component within a complex whole. Tadashi Koizumi's work is a perfect example; while the main body of this playful figurative vessel is fabricated, the arms, watering can, and tiny figures below are cast. The organic nature of the cast parts is what gives a human quality to the piece and sparks a feeling of connection. The gestural nature of these components could only be achieved through Koizumi's skillful modeling of wax parts and when the castings are combined with the fabricated body and the parts are soldered together, they create a whimsical and relatable work that would be much trickier to produce by other means.

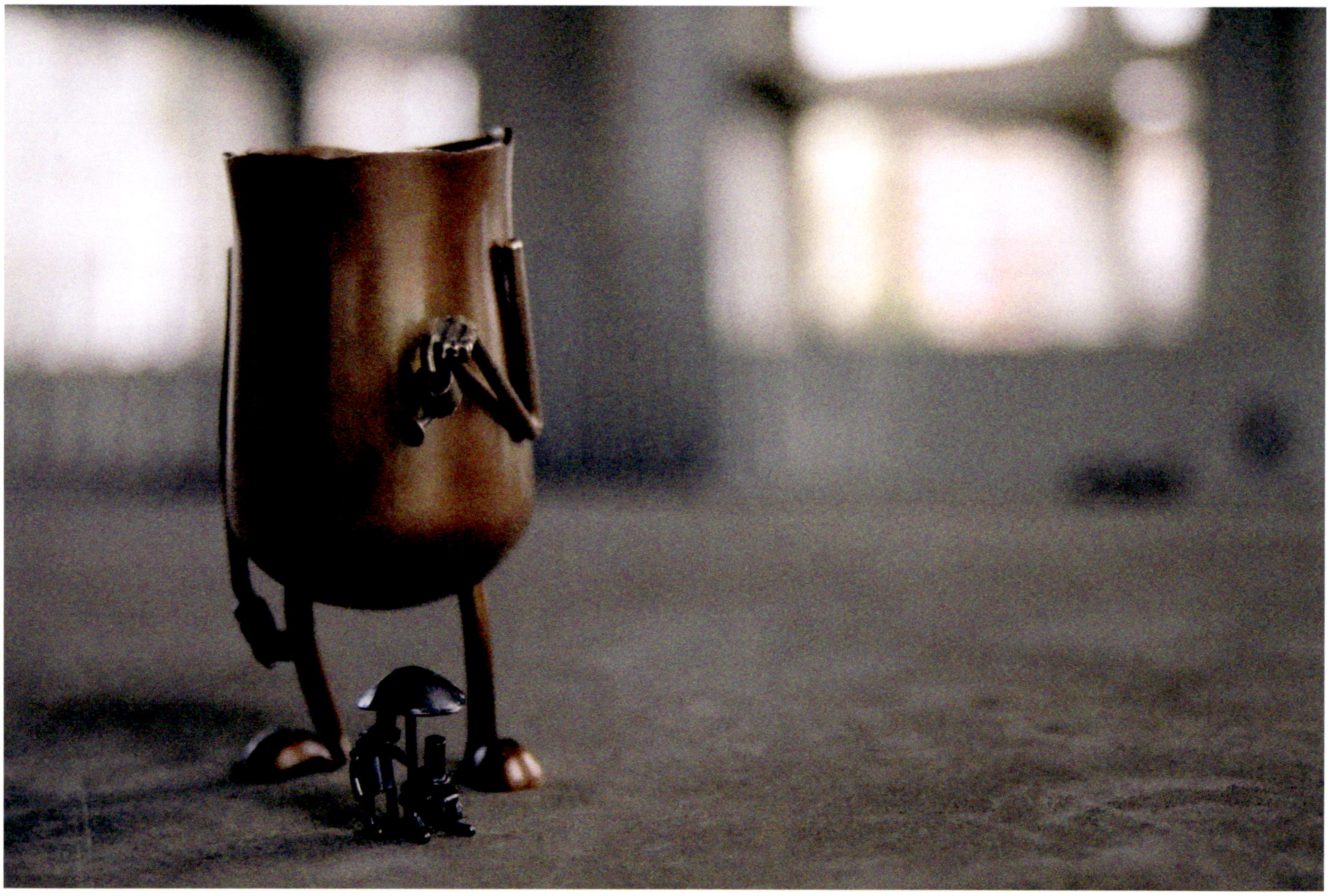

Tadashi Koizumi, *Rain Maker*, 2010. Bronze, copper, cast and raised, 3" × 3" × 4". Photo: Tadashi Koizumi.

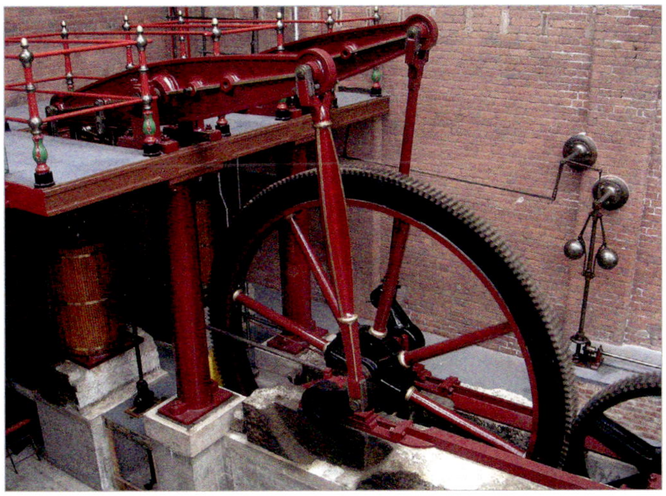

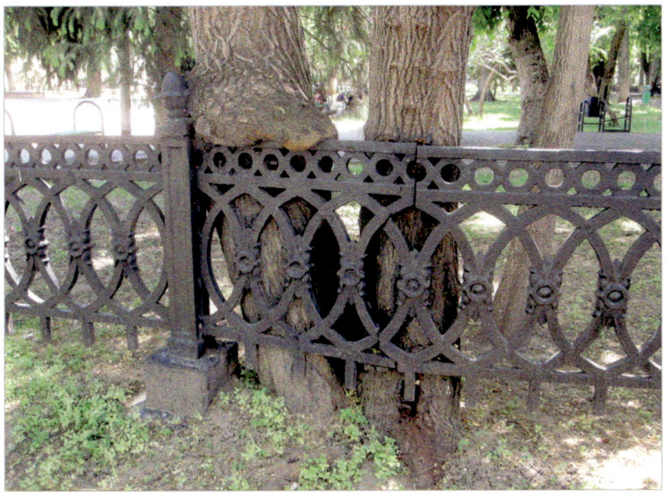

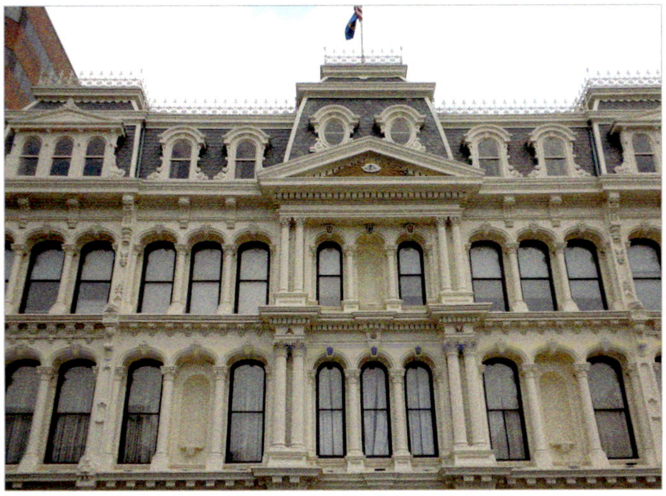

These traits are indispensable for artists, and at the same time, metal is unmatched in industrial applications because of these innate qualities. Many machines are made up of complex forms, milled and lathe-turned gears and mechanical components, and fabricated parts made from sheet metal, pipe, and rod. The ability to cast heavy-duty complex forms that can be combined with precision parts for mechanisms, and thin, strong, lightweight, fabricated elements make metal the best possible material for machines and mechanisms of all kinds. The Industrial Revolution marked an explosion of machines being produced, many of which relied heavily on cast iron components, like this twin beam engine.

The mechanization of factories and foundries that occurred during the Industrial Revolution resulted in a previously unimaginable number of cast iron fences, gates, railings, and other architectural elements. Even daunting projects like buildings with cast iron facades were suddenly within reach and entire neighborhoods flaunted buildings clad in cast iron. The SoHo neighborhood in Manhattan is a magnificent example. Architects recognized the ornamental possibilities of cast iron and capitalized on them, designing elaborate neoclassical buildings complete with Corinthian columns and other flourishes.

The Industrial Revolution wasn't the first time cast iron buildings were created, however. Cast iron pagodas were made as early as the ninth century in China during the Song Dynasty as told by the Buddhist monk, Ennin, who traveled China from 838 to 847. These pagodas, towering at seventy plus feet tall and cast in octagonal sections, were triumphs for Chinese metalsmiths.

Whether casting in eleventh-century China, eighteenth-century Europe, twentieth-century America, or today, monumental works must be approached and created in bite-sized pieces. Molds allow the reproduction of elements that can be combined to create these colossal structures. Metal is uniquely suited to this approach as these elements can be welded, riveted, or keyed together. The resulting structures are durable, complex, ornamental, and unlike anything else.

Above This Second-Empire-style building was originally constructed for the Grand Lodge of the Masons. The cast iron façade makes it one of Wilmington, Delaware's most recognizable landmarks.
Thomas Dixon, *The Grand Opera House*, 1871. Wilmington, Delaware, USA, cast iron facade.
Photo: © Smallbones / Wikimedia Commons / public domain.

Top left Crossfield twin beam engine, 1840. Cast and wrought iron, cylinders 12" and 20" diameter, stroke 3'6".
Photo: © Terry Whalebone / Wikimedia Commons / CC-by-2.0.

Center left Cast iron "orgrazhdenie" (landscaping fence). Almaty, Kazakhstan, iron.
Photo: Dennis Keen.

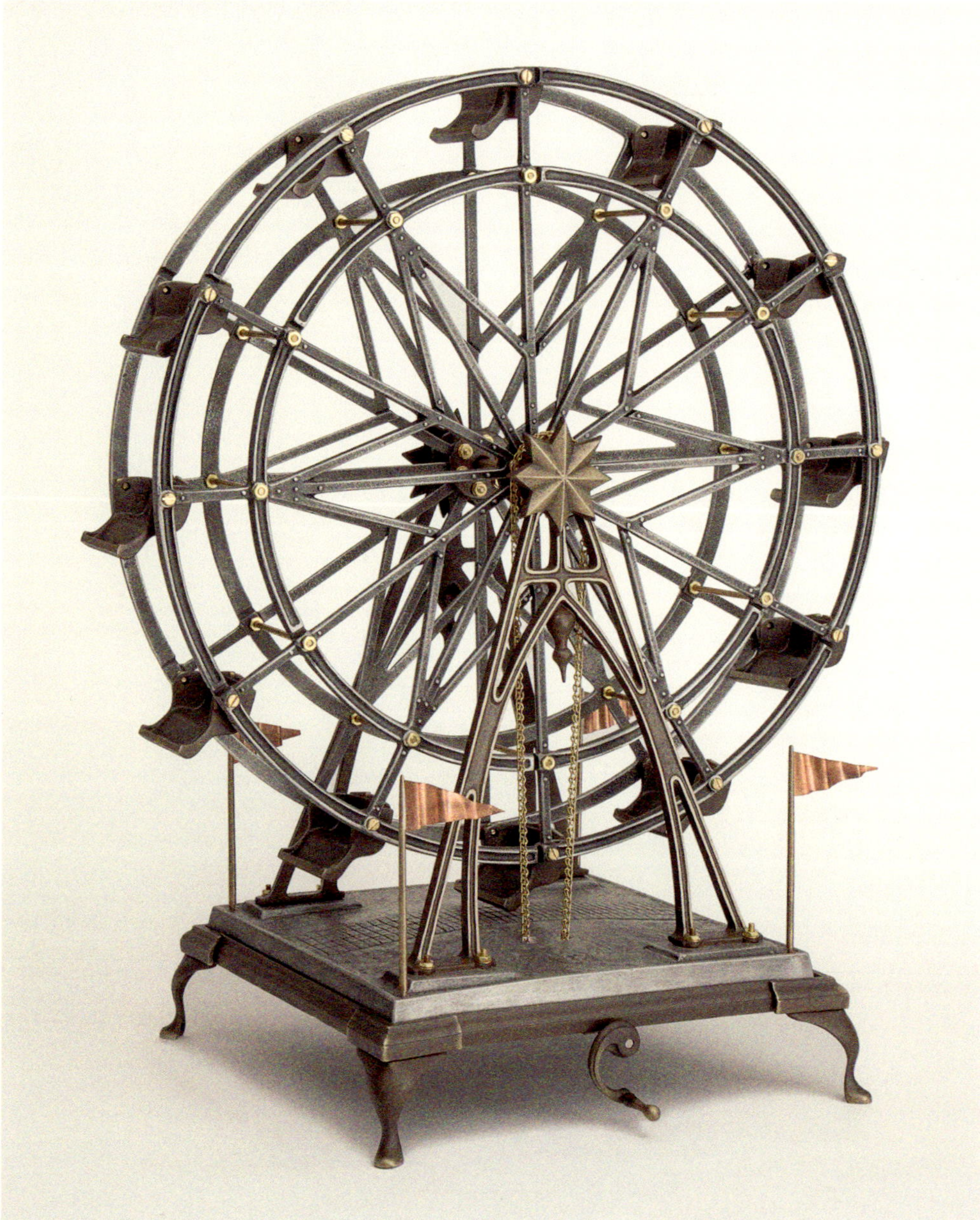

Scott Nelles, *Ferris Wheel*, 2013. Bronze, aluminum, 29" × 13" × 13".
Photo: Don Rutt.

Casting components is a logistical necessity when it comes to anything approaching the scale of buildings, but even in smaller sizes, it can be a savvy and powerful way to work. Scott Nelles's piece, *Ferris Wheel*, is a perfect example. Nelles sandcast forty-eight elements in bronze and aluminum to create this miniature functional ferris wheel, complete with a functioning crank that engages a "ladder chain" attached to brass sprockets.[3]

Karen Westland's *Infinity Oval Bowl* is another example of casting multiple small elements to be assembled after casting. Using a clever tab-and-slot approach, Westland found her inspiration in tools built for space exploration, recreating the construction methods of telescopes and satellites.[4]

These last two examples demonstrate using components and repeated elements to achieve a visual richness. Another way of working with components is in the service of a larger form. In monumental architectural structures, this is a matter of practicality—it would be impossible to cast a form so large in a single piece. First of all, melting enough metal all at once would be problematic; moving it would be next to impossible; and if the casting failed, it would be better to lose only a small piece rather than the whole form and have to start over on a brand new model.

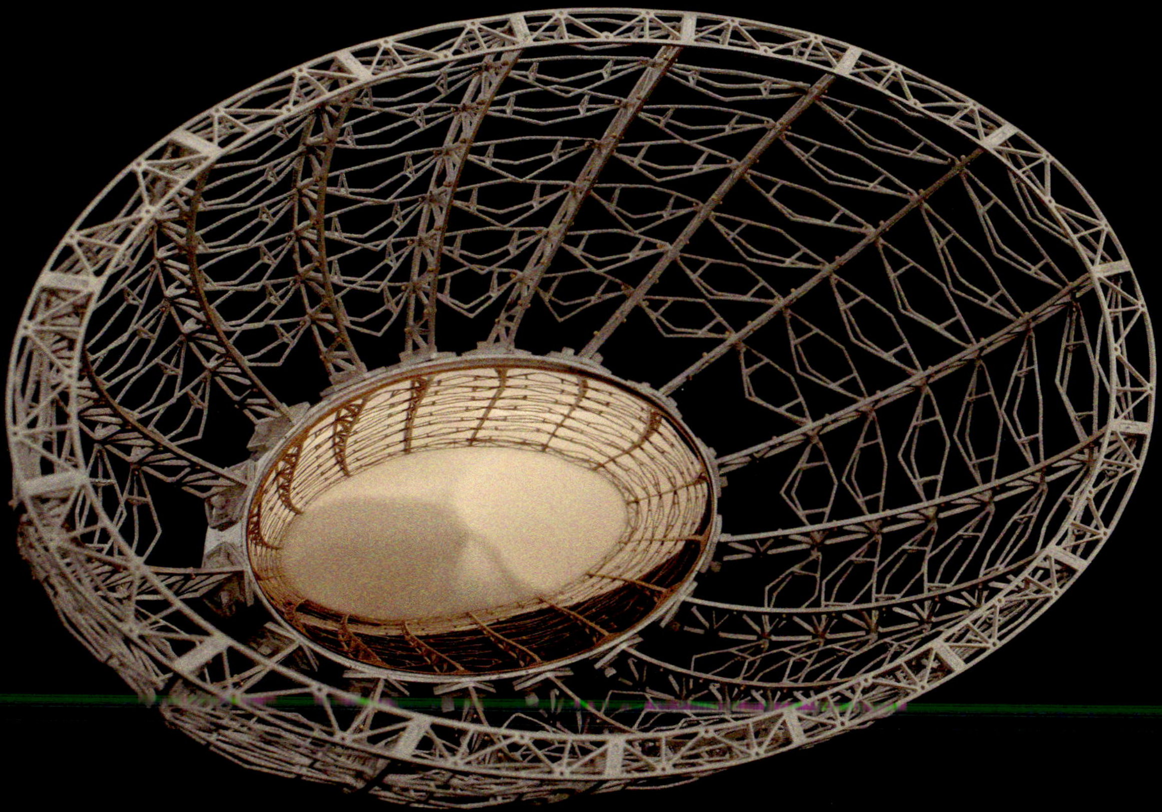

Karen Westland, *Infinity Oval Bowl*, 2015. Recycled sterling silver, recycled 18K gold, silver gilt, laser cut silver, SLS nylon, lost wax casting, PUK welding, hand fabricated, 10.25" × 6.3" × 5.5". Photo: Graham Clark Photography.

The Daibutsu (literally, "big Buddhas") of Japan are a great example of monumental sculptures cast in many pieces and assembled afterward. There are Daibutsu carved in stone and cast in concrete, but we will focus on the bronze examples here. The casting of the *Great Buddha of Kamakura* that began in 1252 is a perfect example. Each cast element is unique unto itself, but combined, the pieces add up to a breathtaking giant Buddha almost forty-four feet tall. There are many bronze Daibutsu in Japan; some are older or larger than the one at Kamakura, but this one remains in its original condition, never having been recast.

The *Great Buddha of Kamakura* is so large that visitors can walk inside it to revere it as a sacred monument, but it is also of great interest to makers of all kinds. Inside, evidence of its assembly process is clearly visible. The interior, which looks like a three-dimensional quilt, reveals countless square components, each unique, fitted, and assembled together to create an almost-seamless exterior. It looks like small square sections were welded together in great horizontal bands and then these completed slices were put in place and the final assembly was completed.

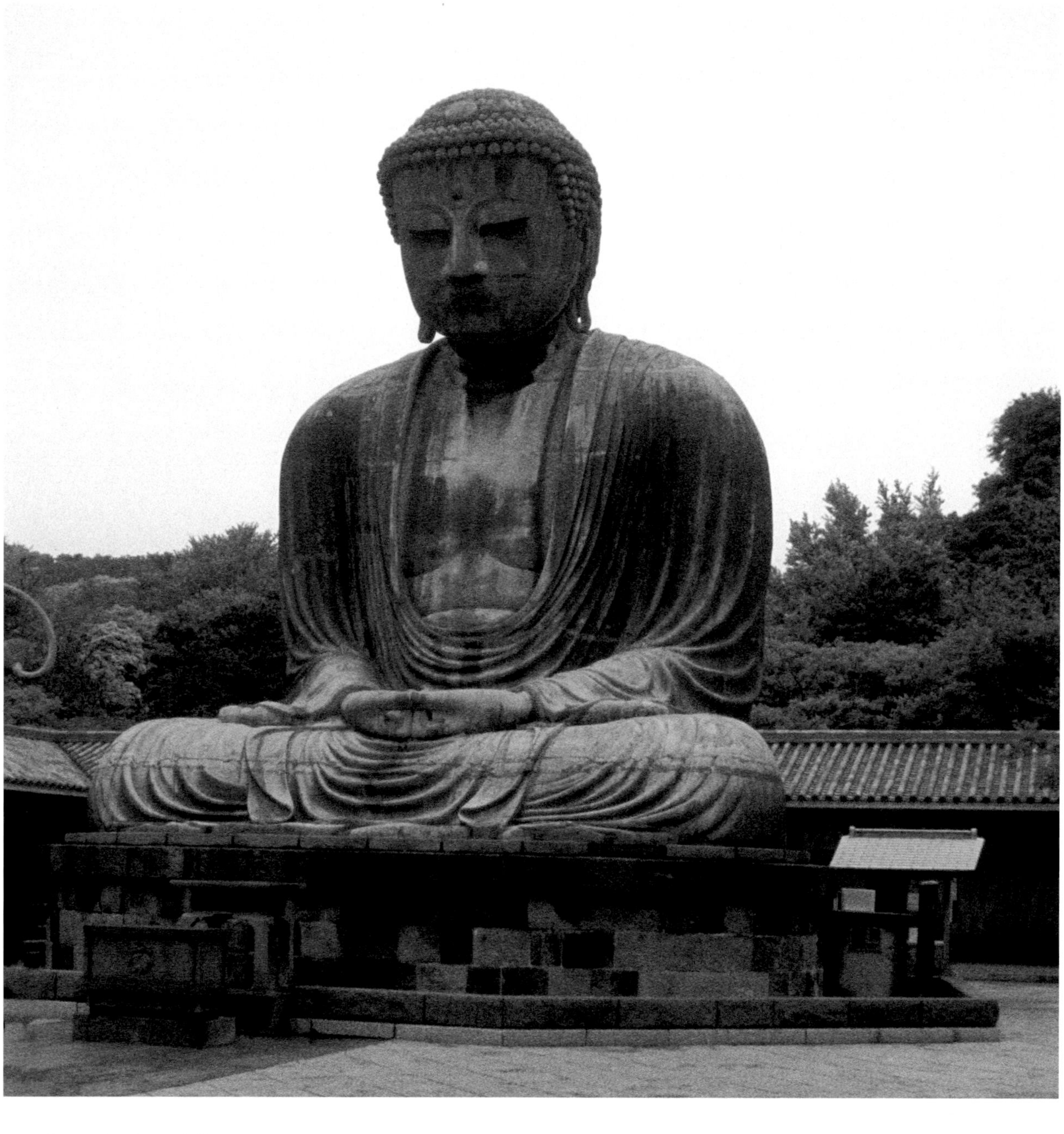

The Great Buddha of Kamakura, 1252 (the Kamakura Period). Kōtoku-in Temple in Kamakura, Kanagawa Prefecture, Japan, bronze, 43'10", 93 tons.
Photo: Myriam Thyes / Wikimedia Commons / public domain.

These monumental examples are helpful as a learning tool and source of inspiration for casters of all kinds, but scale is actually immaterial when thinking about component casting. If the *Great Buddha of Kamakura* is the macro, the *Cradled Gem Earrings* by Sasha Newkirk are the micro. Casting and assembling even the tiniest parts can be a great help in making a piece of jewelry. Cleanup is more easily and often more thoroughly performed on small simple components, rather than on complicated forms with nooks and crannies too small to access. Small cast pieces often only make up part of a piece of jewelry, as in the case of these earrings; tiny fabricated tube-settings for the diamonds, as well as small jump rings, were soldered in place after the stone setting and cleanup were completed, resulting in crisp, elegant little forms.

There can be great value in casting components whether the work is large or small. Makers can benefit from looking beyond casting as a means to simply make a standalone object and think in terms of parts. Casting can be an effective way to make grand components or the smallest of elements. The plasticity of form, the capacity for reproduction, the ability to replicate organic and found objects, as well as being able to manipulate cast objects, make casting a valuable process for a multitude of applications.

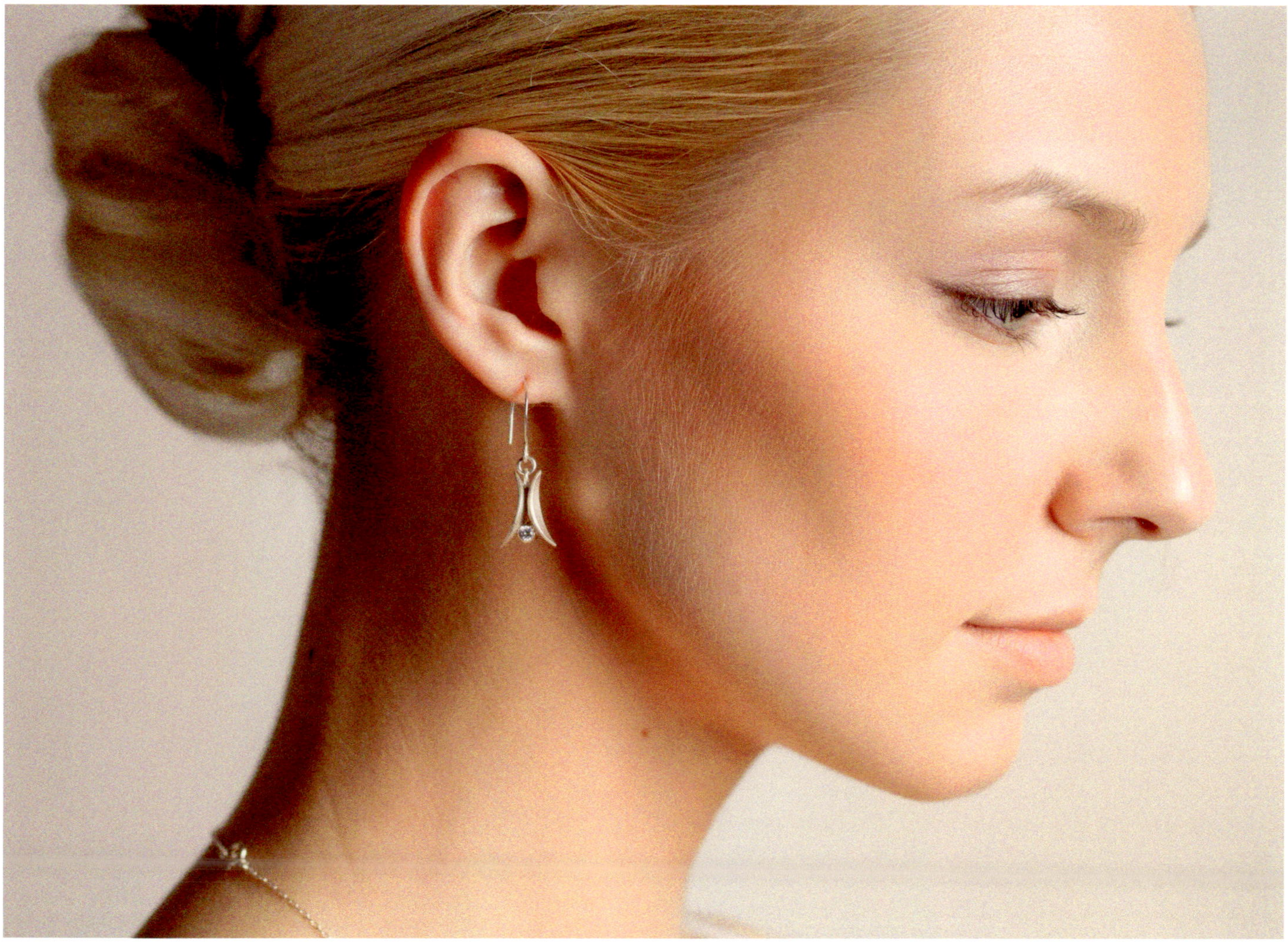

Sasha D. Newkirk, *Cradled Gem Earrings*, 2000. Sterling silver, white sapphire, 1" × 0.3" × 0.3".
Photo: Lisa Walker, Model: Kari Shaevitz.

Notes

1. McNeil, Ian, *An Encyclopedia of the History of Technology* (Routledge Companion Encyclopedias) (London: Routledge, 1990), 50.
2. Goddard, Jolyon, ed., *National Geographic Concise History of Science and Invention: An Illustrated Time Line* (Washington, DC: National Geographic, October 27, 2009), 92.
3. Email from Scott Nelles, April 28, 2016.
4. Email from Karen Westland, May 22, 2016.

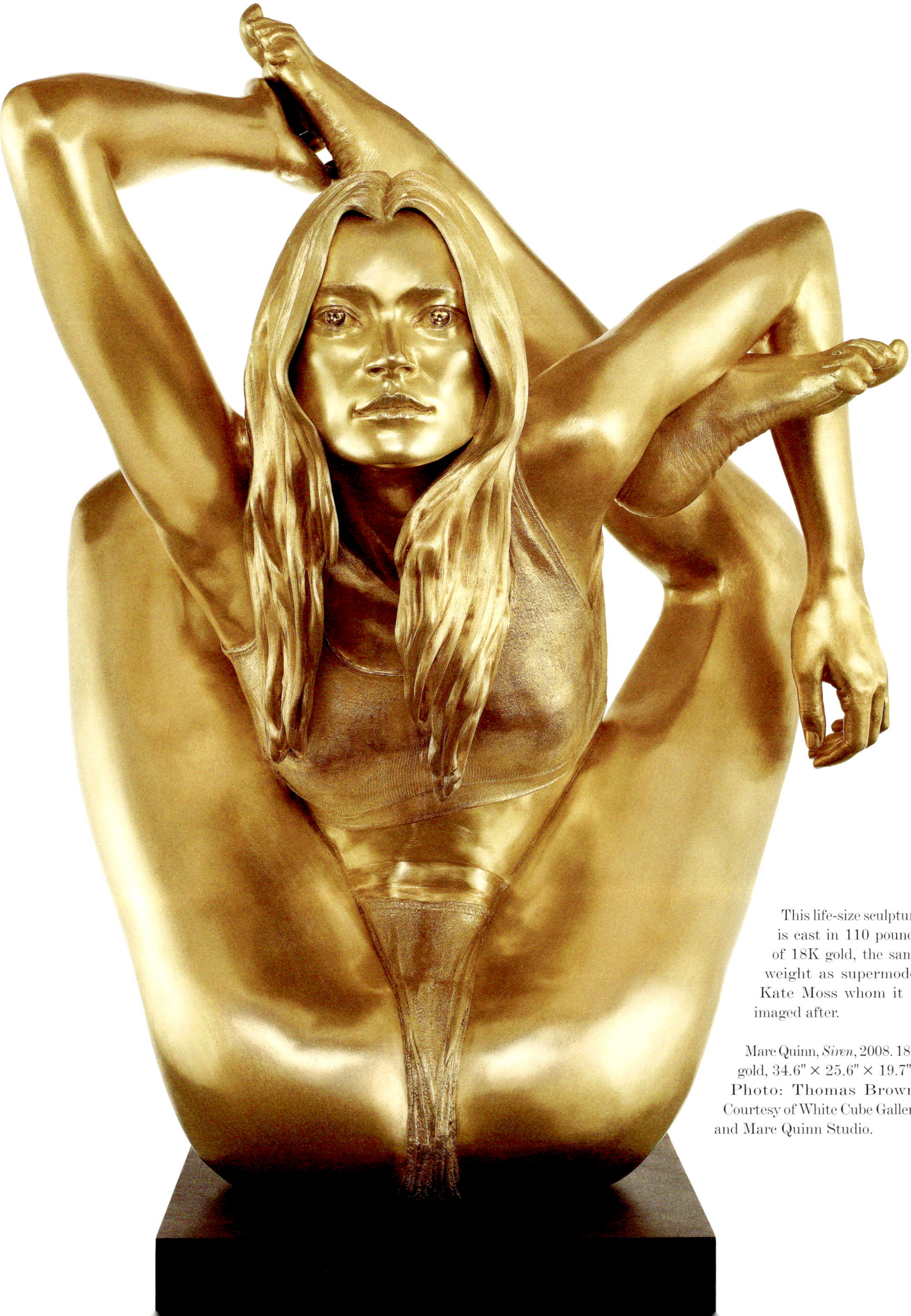

This life-size sculpture is cast in 110 pounds of 18K gold, the same weight as supermodel Kate Moss whom it is imaged after.

Marc Quinn, *Siren*, 2008. 18K gold, 34.6" × 25.6" × 19.7". Photo: Thomas Brown, Courtesy of White Cube Gallery and Marc Quinn Studio.

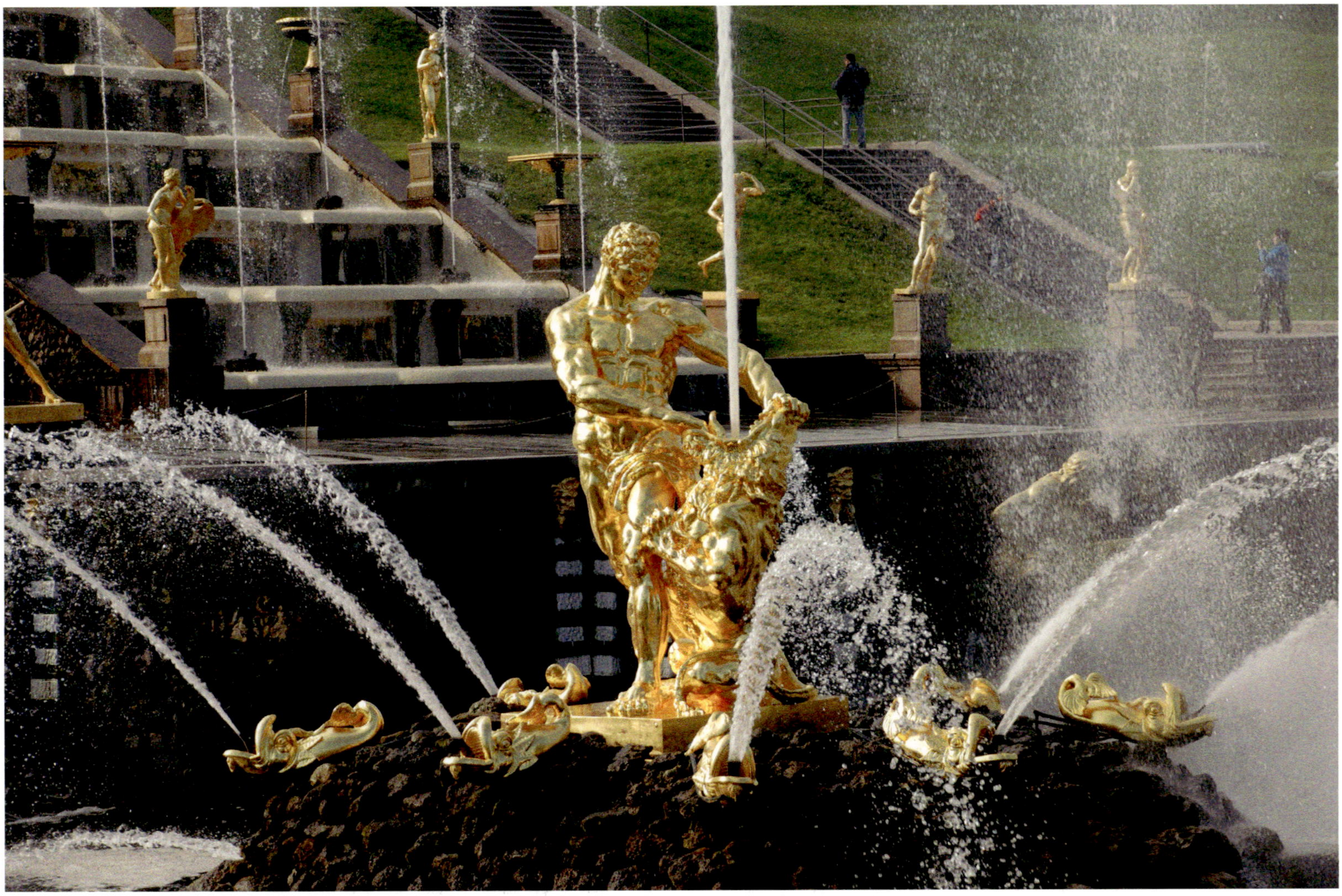

Mikhail Kozlovsky, *Samson Fountain*, 1735, restored in 1802, remade in 1947. Russia, bronze, sculptural group is 11.5' tall (with pedestal it is 19'8").
Photo: © Mykhailo Gorbachov / Wikimedia Commons / CC-by-3.0.

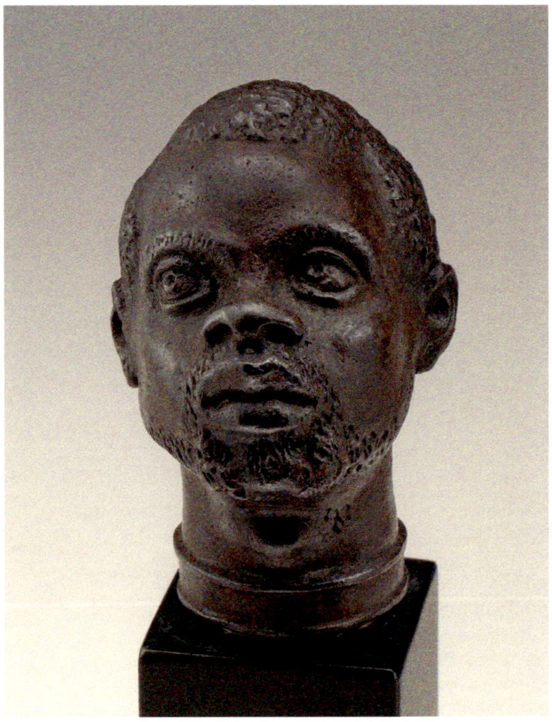

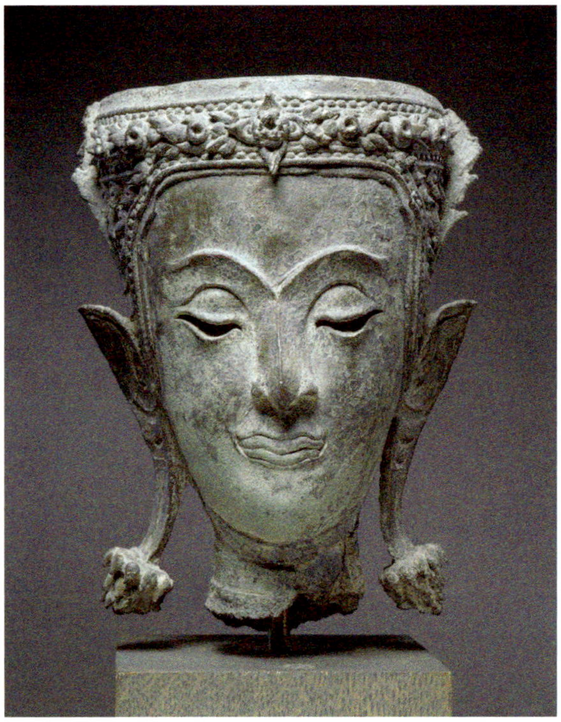

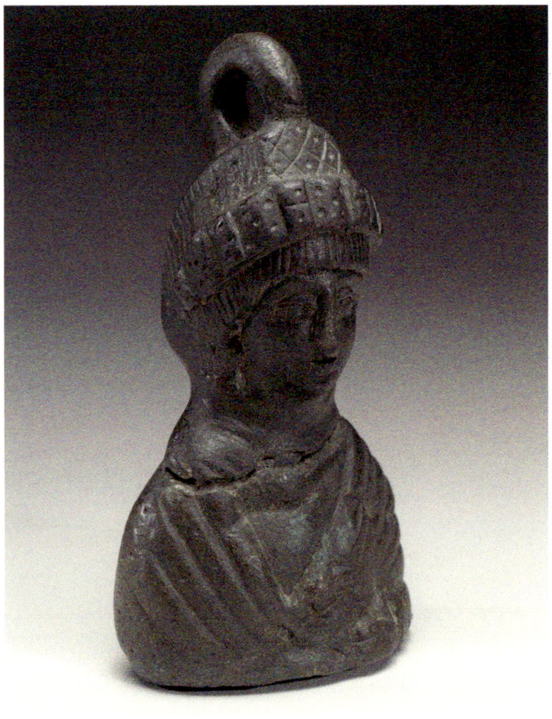

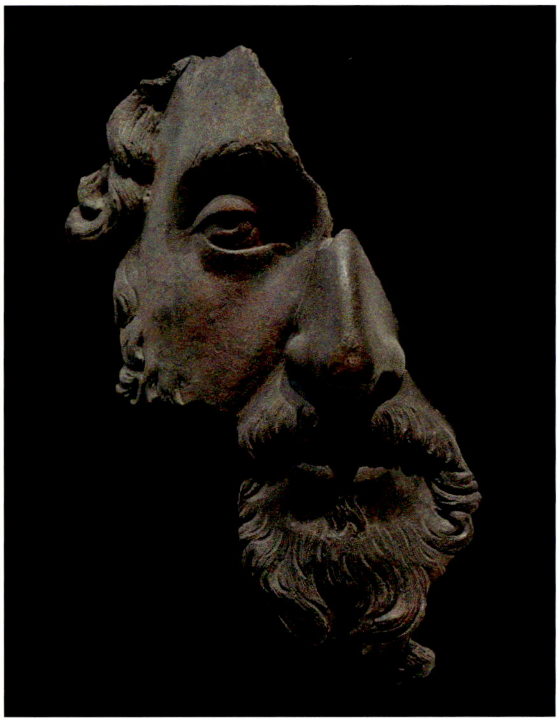

Top left *Head of a Moor*, second half of sixteenth century. Venetian, cast iron, 2.25" × 1.4" × 0.8".
Courtesy of The National Gallery, Samuel H Kress Collection.

Top right *Head of the Crowned Buddha*, second half of seventeenth century. Ayutthaya, Thailand, leaded tin brass, 5.9" tall.
Courtesy of The Walters Art Museum, CC0, GNU FDL.

Bottom left *Byzantine Bust of an Empress*, fifth century (Late Antique). Bronze, 4.62" × 2.37" × 1.75".
Courtesy of The Walters Art Museum, Museum purchase in 1945, CC0, GNU FDL.

Bottom right *Marcus Aurelius Antoninus Augustus*, after 170 CE. Roman, bronze, 9" tall.
Photo: © 2007 Marie-Lan Nguyen at the Louvre / Wikimedia Commons / CC-BY 2.5.

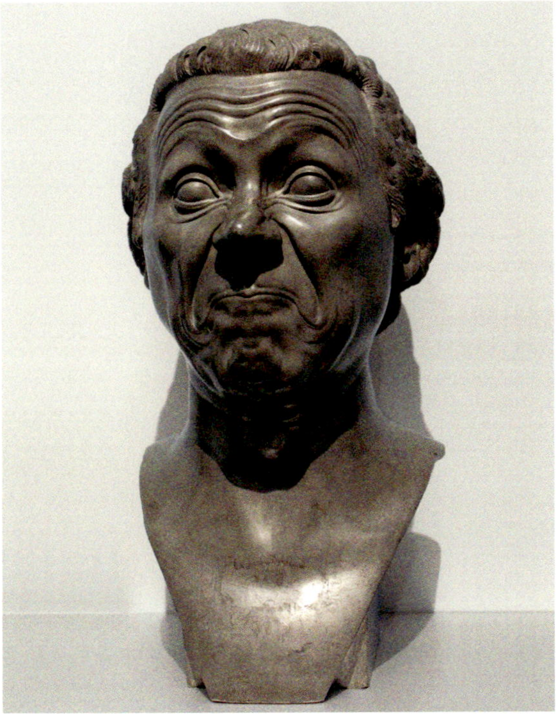

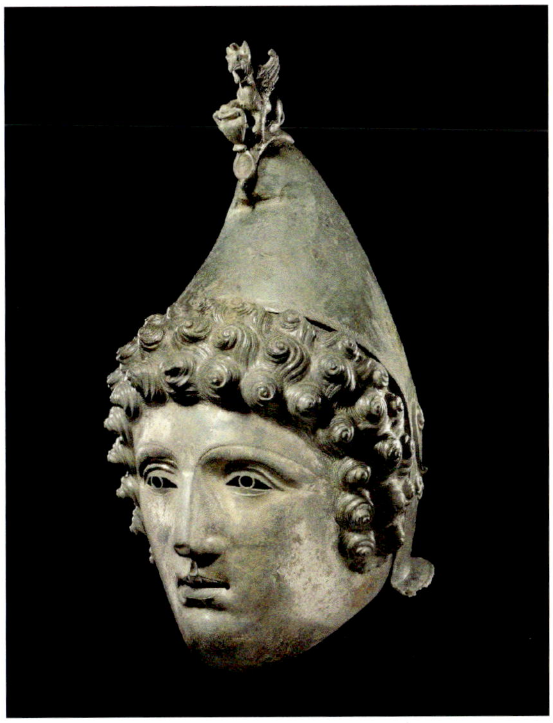

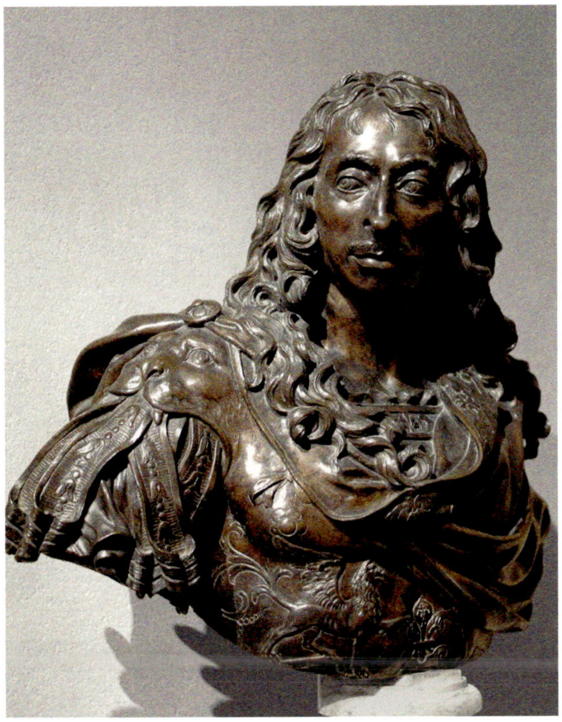

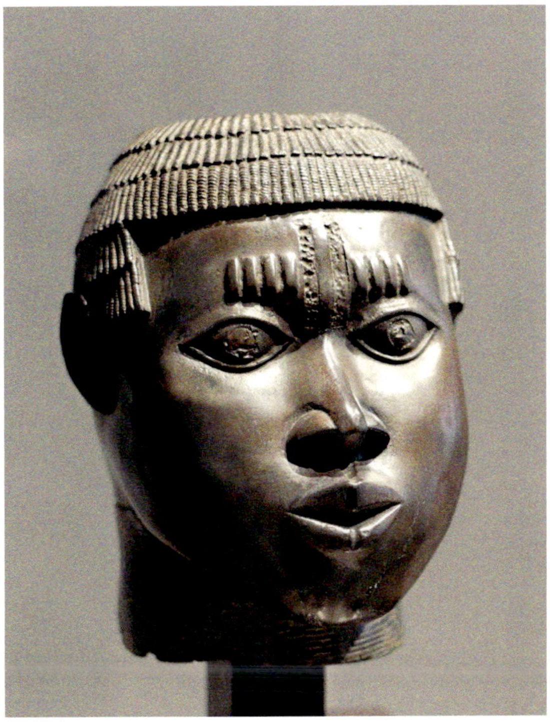

Top left Franz Xaver Messerschmidt, *No. 26 Character Head, The Satirist*, 1770–1780. Vienna, Austria, lead, hollow cast (with old patina), 17.7" H × 8.25" × 9.6". Photo: © Anagoria at the Germanisches National Museum / Wikimedia Commons / CC by 3.0.

Bottom left Antoine Coysevox, *Bust of Louis de Bourbon, Prince of Condé*, also known as *le Grand Condé*, 1688. Bronze, 23.6" × 26.8" × 13.4". Photo: © Marie-Lan Nguyen at the Louvre / Wikimedia Commons / public domain.

Top right *The Crosby Garrett Roman Cavalry Helmet*, first to third century CE. Roman, copper alloy, 16" tall. Courtesy of Portable Antiquities Scheme, CC by 2.0.

Bottom right *Benin Bronze*, late fifteenth to mid-sixteenth century. Benin Kingdom (Nigeria), bronze, 12" tall. Photo: © Marie-Lan Nguyen at the Louvre / Wikimedia Commons / public domain.

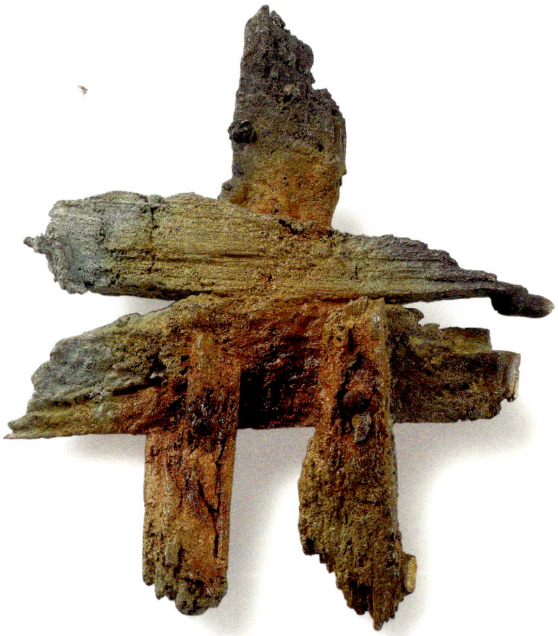

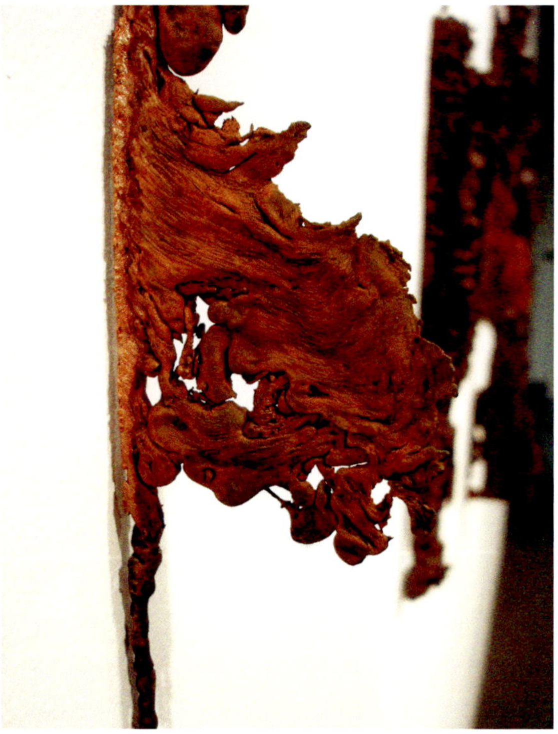

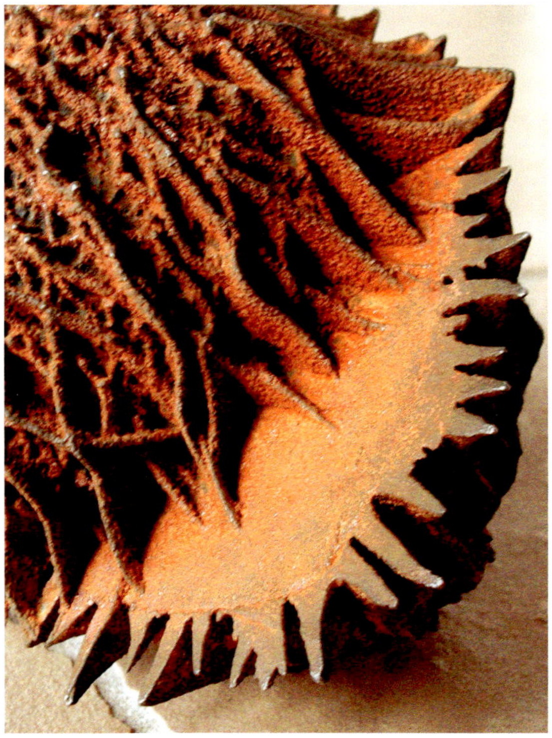

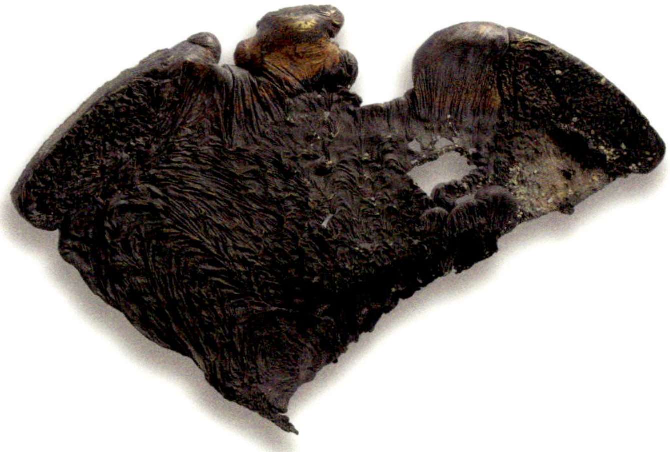

Top left Yasmin Vinograd, *Memory Figurine #2*, 2015. Silver, iron, 2.2" × 1.9" × 0.7". Photo: Alex Kucherenko.

Bottom left Tom Joyce, *Seed I* (detail), 2013. Cast ductile iron, 6" × 6" × 6". Created in Arts/Industry, a long-term residency program of the John Michael Kohler Arts Center in Sheboygan, Wisconsin. Arts/Industry takes place at Kohler Co. Photo: © 2002 Tom Joyce Studio, Collection of Diane and Sandy Besser.

Top right Casey Westbrook, *the blind make the blind*, 2011. Cast iron, cottonwood, 12" × 12" × 48". Photo: Teresa Johnson, Courtesy of Carbon Arts, Collection of Sentinel Iron Works, Ellinwood Kansas.

Bottom right Nikolai Balabin, *A Glide*, 2013. Bronze, 3.13" × 5.13" × 0.25". Photo: Ilja Balabin.

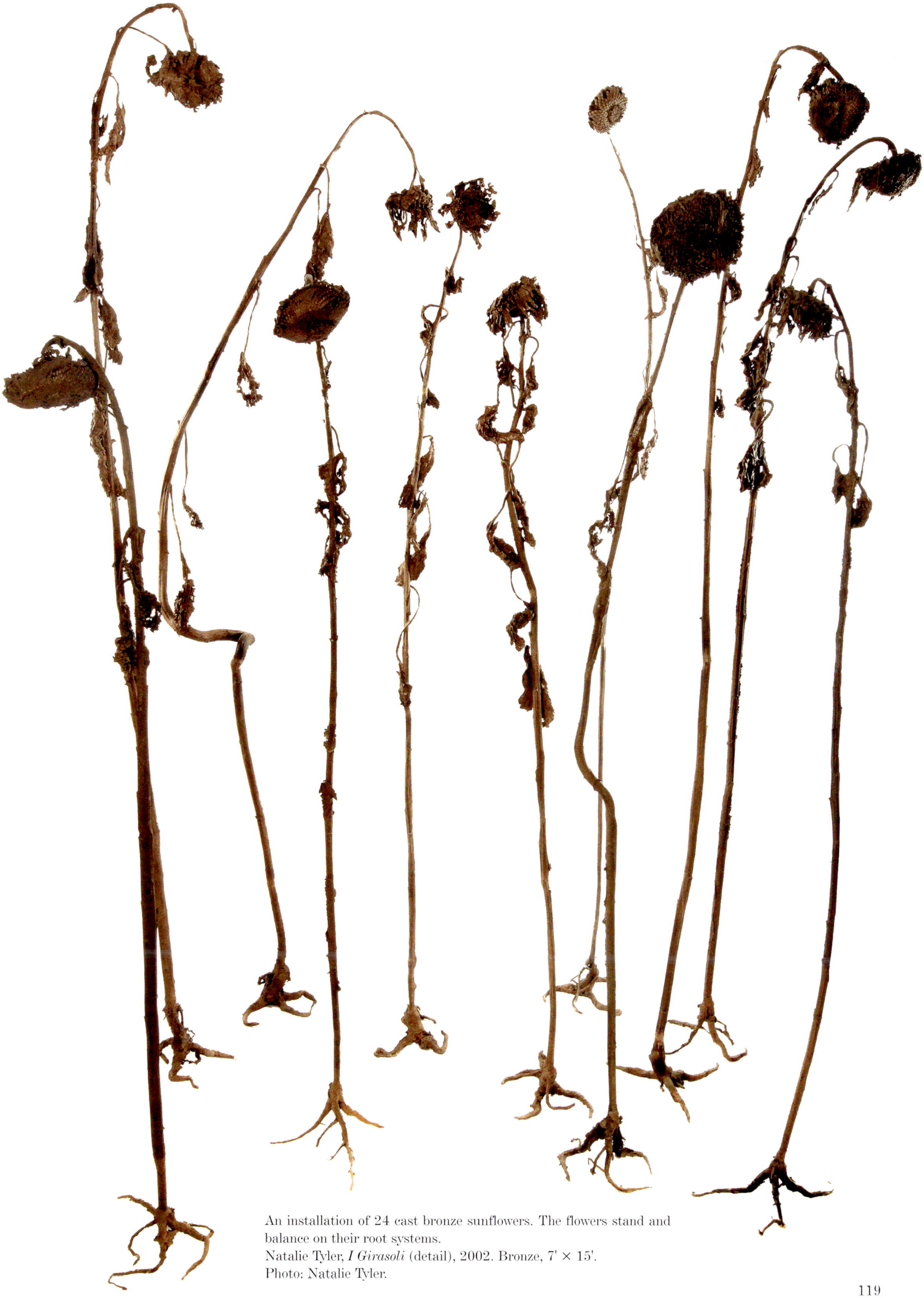

An installation of 24 cast bronze sunflowers. The flowers stand and balance on their root systems.
Natalie Tyler, *I Girasoli* (detail), 2002. Bronze, 7' × 15'.
Photo: Natalie Tyler.

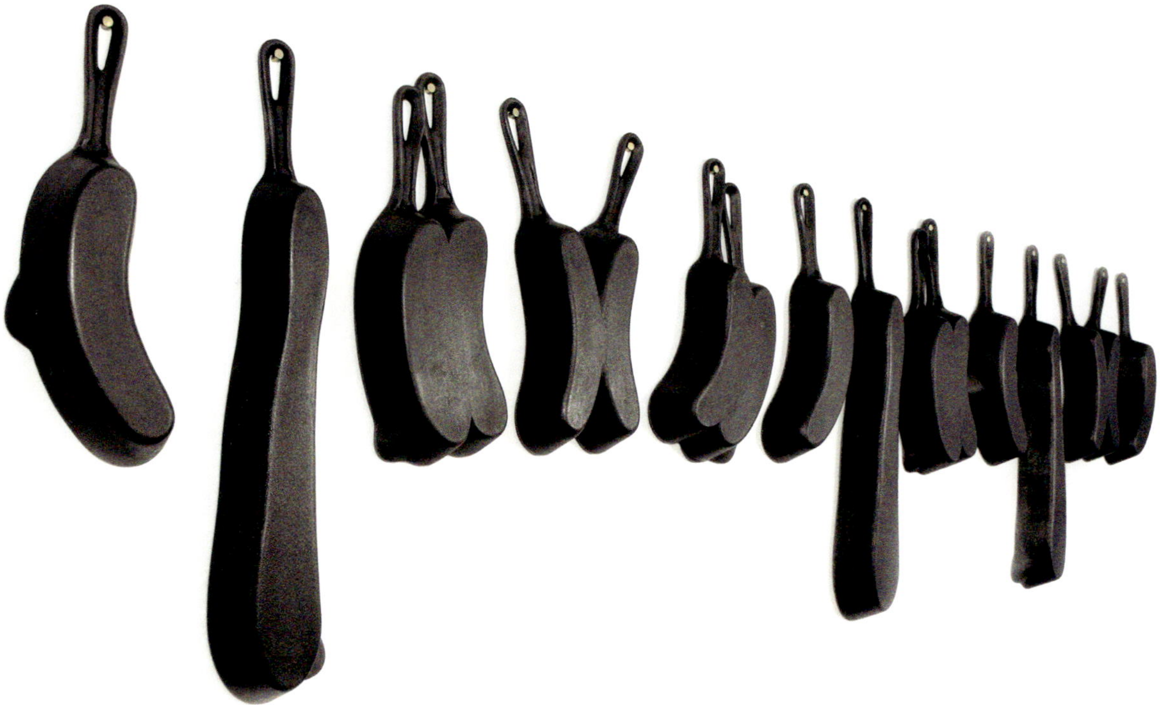

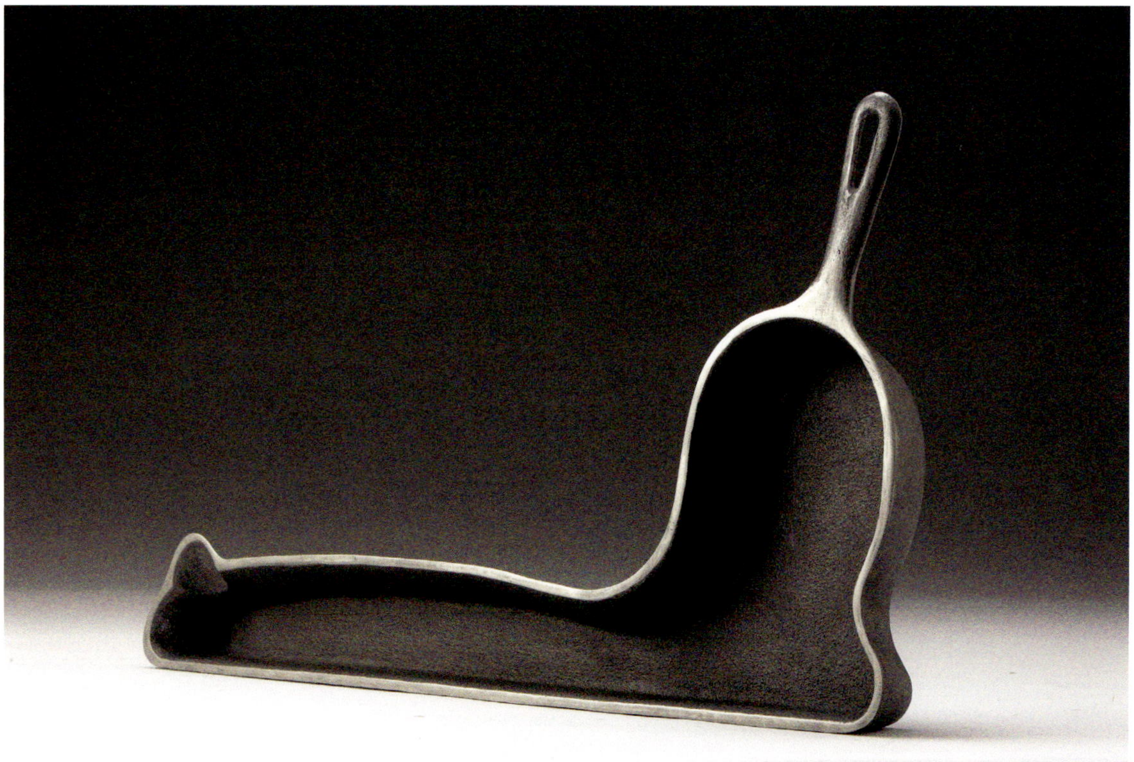

Top Myra Mimlitsch-Gray, *Brat Pans*, 2007. Cast ductile iron, dimensions varied, displayed at 8' across. Created in Arts/Industry, a long-term residency program of the John Michael Kohler Arts Center in Sheboygan, Wisconsin. Arts/Industry takes place at Kohler Co.
Photo: Myra Mimlitsch-Gray.

Bottom Myra Mimlitsch-Gray, *Freestanding Skillet*, 2007, Cast ductile iron, 13" × 23" × 2". Created in Arts/Industry, a long-term residency program of the John Michael Kohler Arts Center in Sheboygan, Wisconsin. Arts/Industry takes place at Kohler Co.
Photo: Rich Maciejewski, Collection of the Kohler Co.

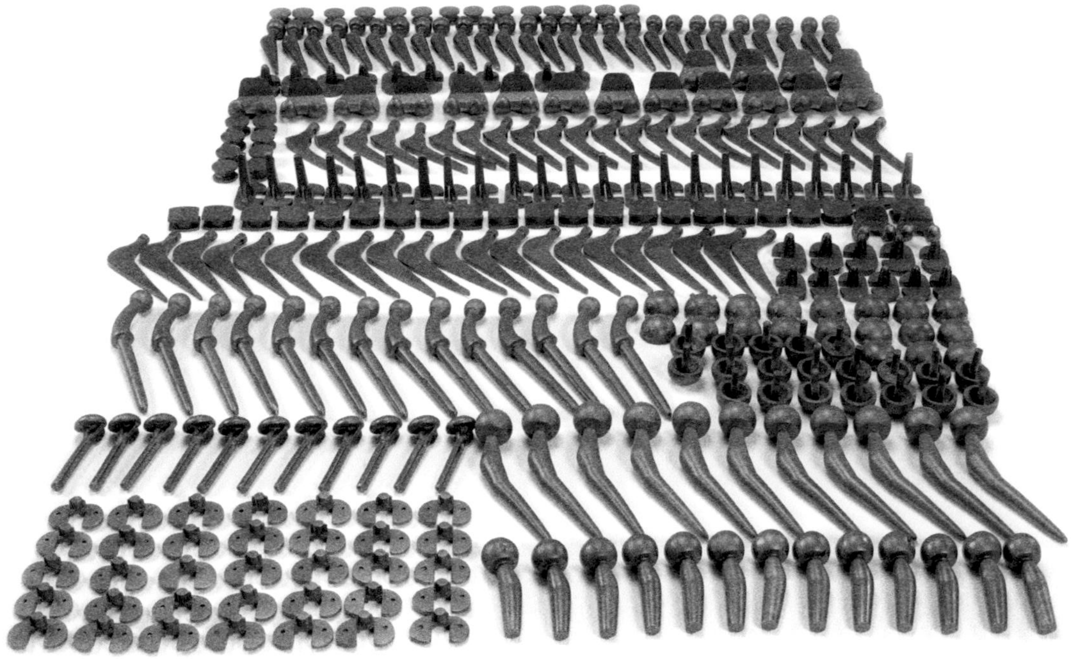

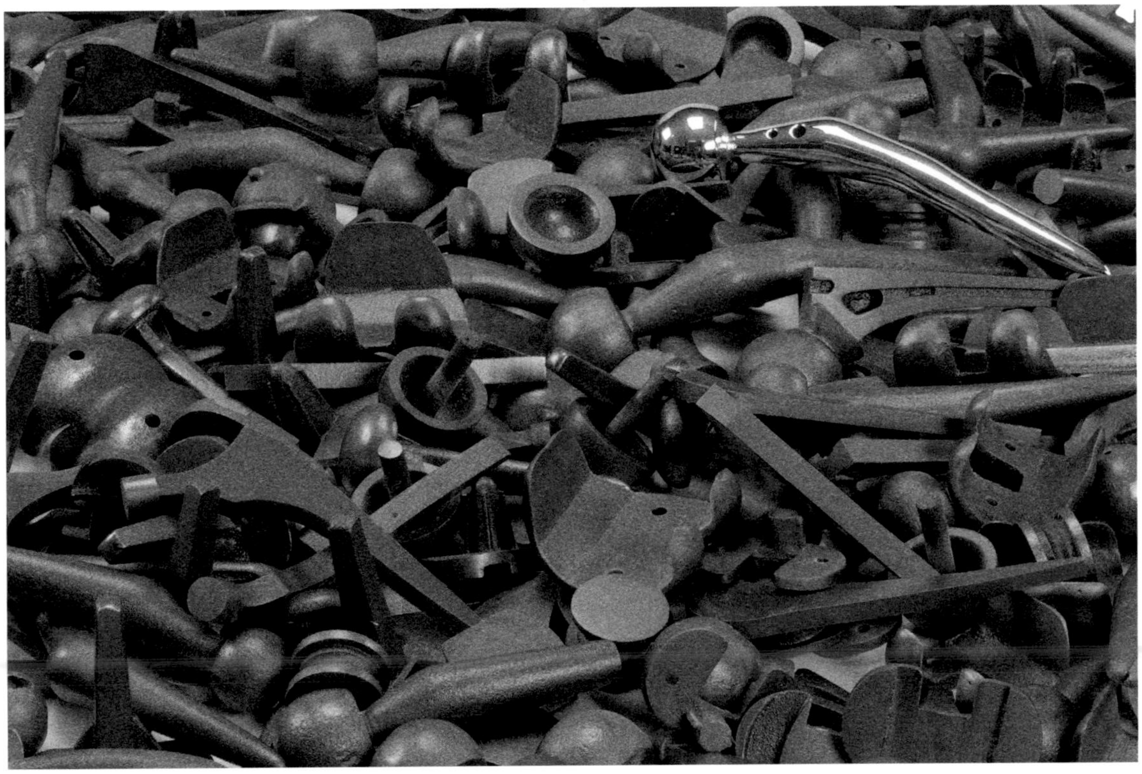

Houdek has produced medical implant-inspired forms en masse in materials the human body would reject were the devices actually implanted.

Holland Houdek, *Mechanization (Implants Series 5)*, 2015. Iron, dimensions variable. Created in Arts/Industry, a long-term residency program of the John Michael Kohler Arts Center in Sheboygan, Wisconsin.
Photo: Jeff Machtig.

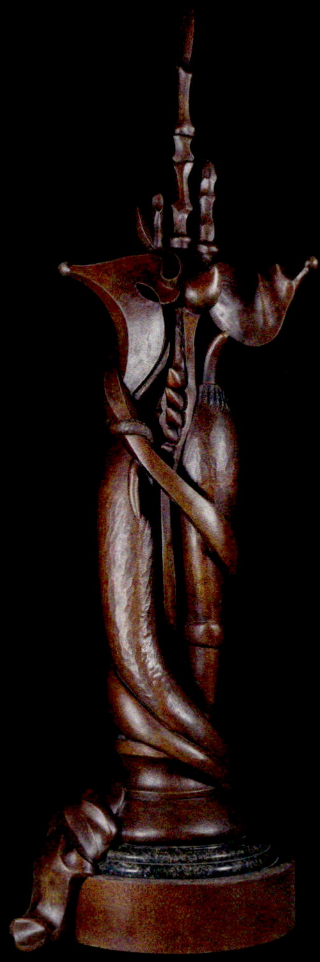

Left Michael Boles. *Daphne*, 2013. Cast bronze; edition of 6, 7'8" × 2'2" × 1'10".
Photo: Christopher White, Collection of Mr. Frederic G. Levin, Pensacola, FL.

Below These cast bronze awards were created for *Spectrum: The Best in Contemporary Fantastic Art*.

Colin and Kristine Poole from KC Poole Studios, *The Spectrum Muse Award*, 2015. Bronze, sterling, silver, 24K gold, marble, 16" × 6" × 5".
Photo: Colin Poole.

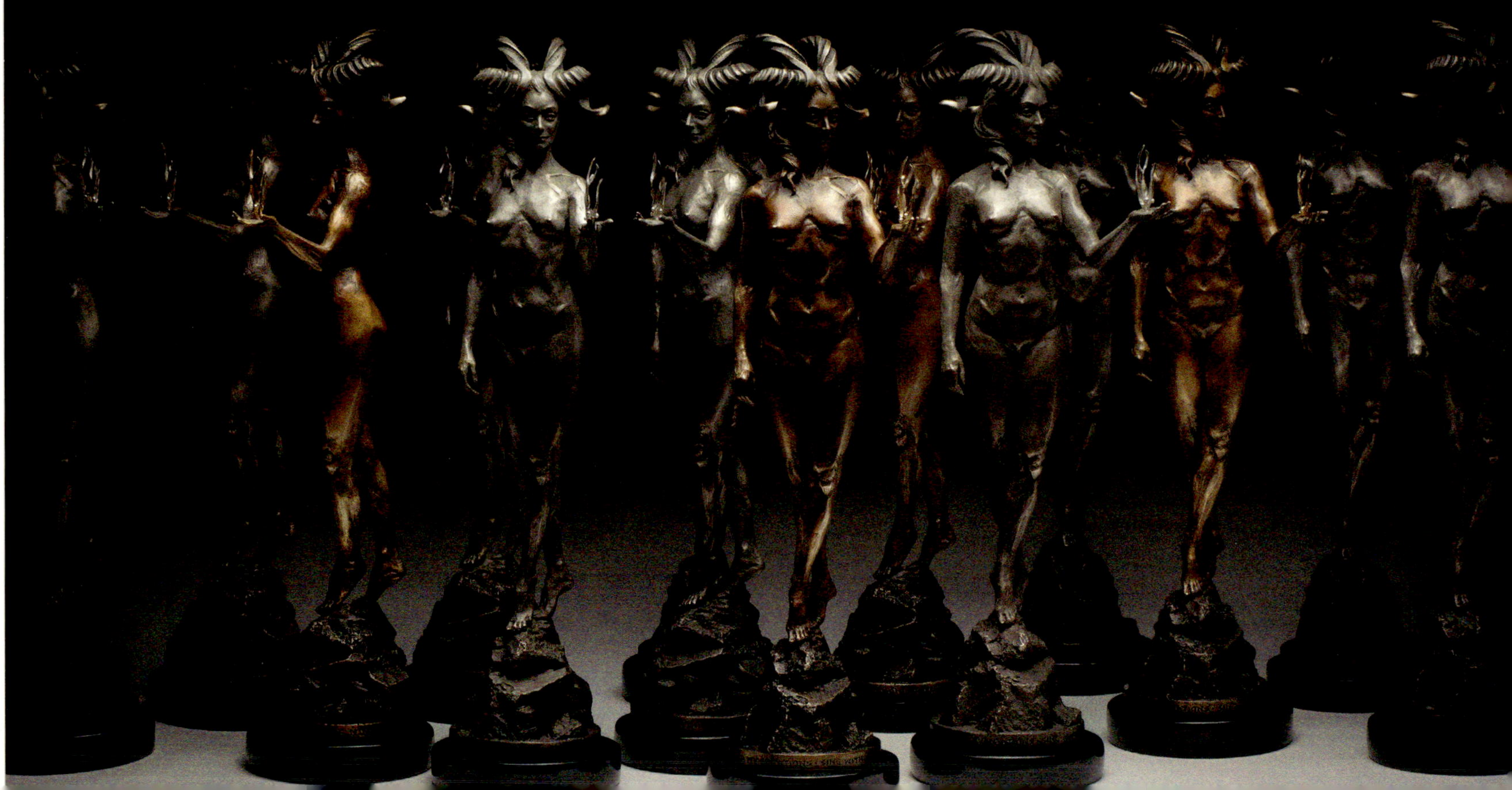

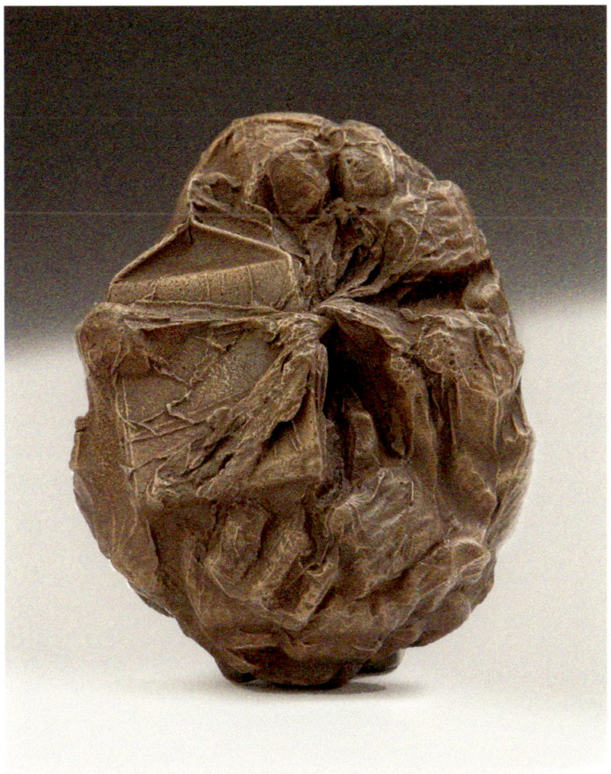

Right Penny Harris, *The Petrified Poem* (detail brush), 2013. Bronze, lost form casting technique, 3.5" × 2.75" × 11". Photo: Rowan Conroy.

Left This hyper-realistic Lilliputian bronze trash bag can fit in the palm of a hand.

John Rais, *Refuse with Milk*, 2006. Cast iron, patina, 6.50" × 4" × 5".
Created in Arts/Industry, a long-term residency program of the John Michael Kohler Arts Center in Sheboygan, Wisconsin. Photo: Kohler Co.

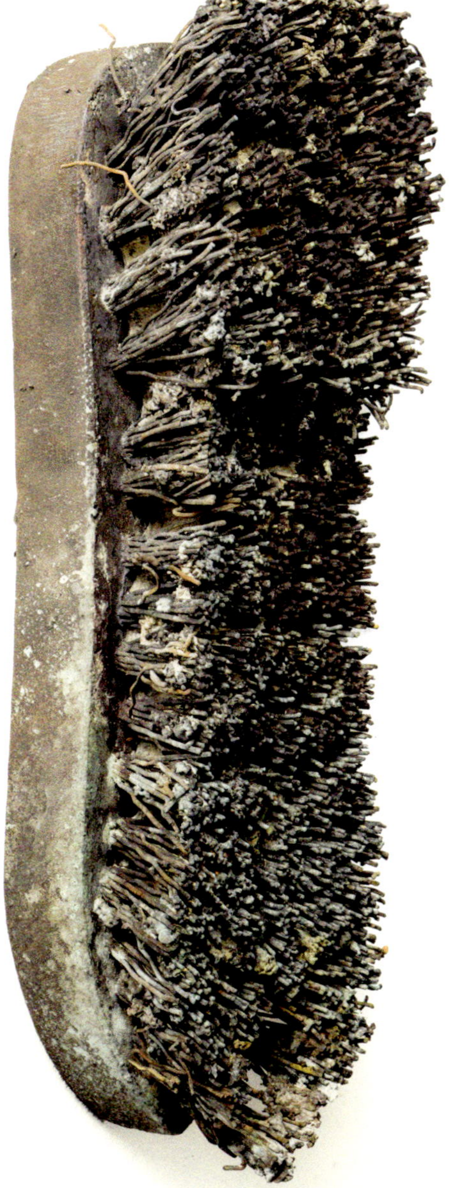

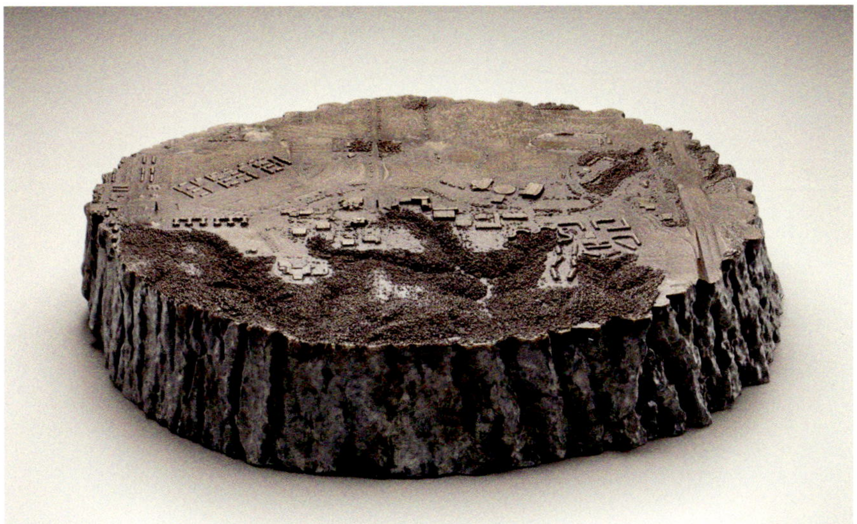

Above The model for this work was rapid prototyped using material and process initially designed for the automotive industry. It explores land use and industry in Michigan.

Norwood Viviano. *Making Way* from the *Recasting Michigan* series, 2010. Bronze, 3-D printed pattern and lost wax bronze casting, 20" × 20" × 3.5". Photos: Tim Thayer.

Right Norwood Viviano, *Making Way* from the *Recasting Michigan* series (detail), 2010.

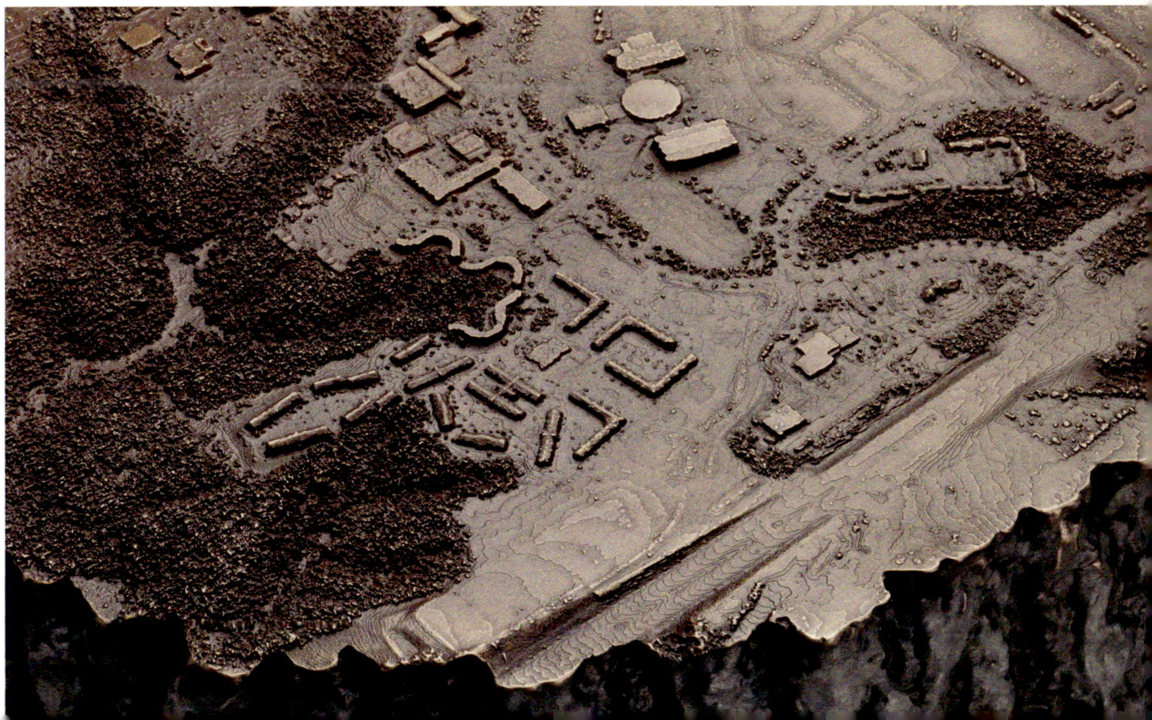

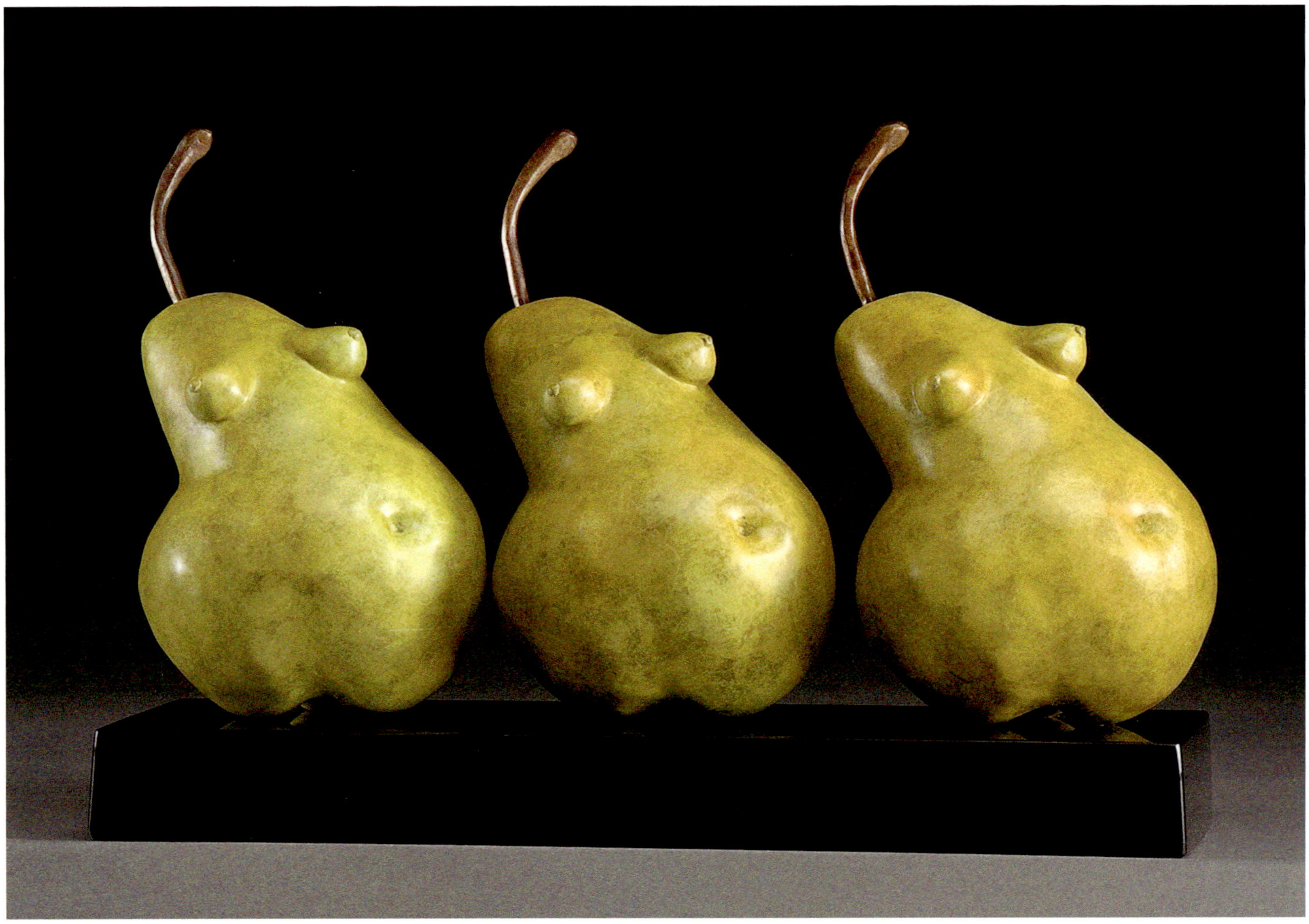

Above Jack Hill, *The Bartlett Sisters*, 2004. Cast bronze with heat applied patina, 6" × 15" × 5". Photo: Mel Schockner.

Opposite page Stephen Daly, *Mentoring*, 2006. Bronze, patina, 19'7" × 8'11" × 4'3". Photo: Stephen Daly, Collection of The Polytechnic University, Valencia, Spain.

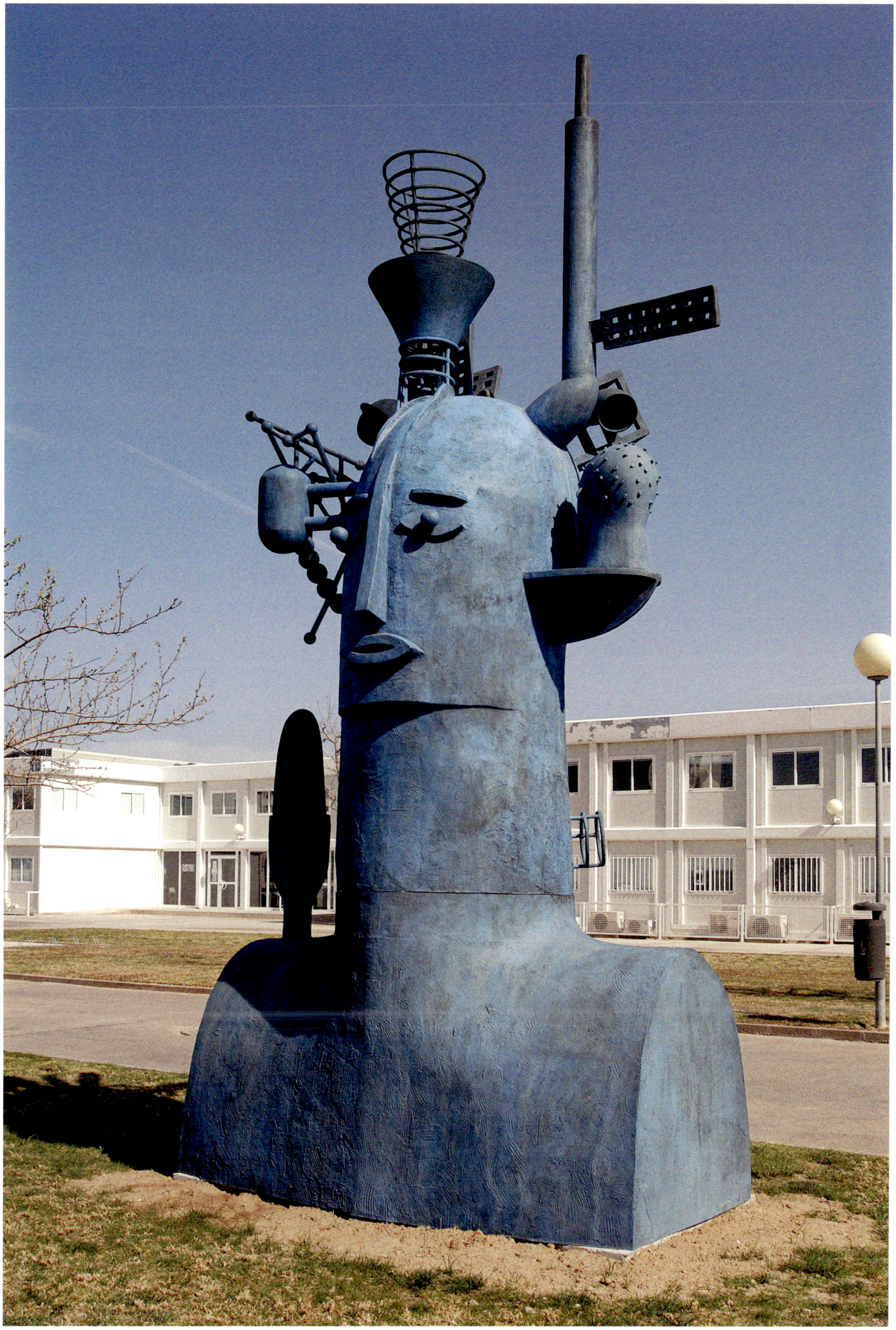

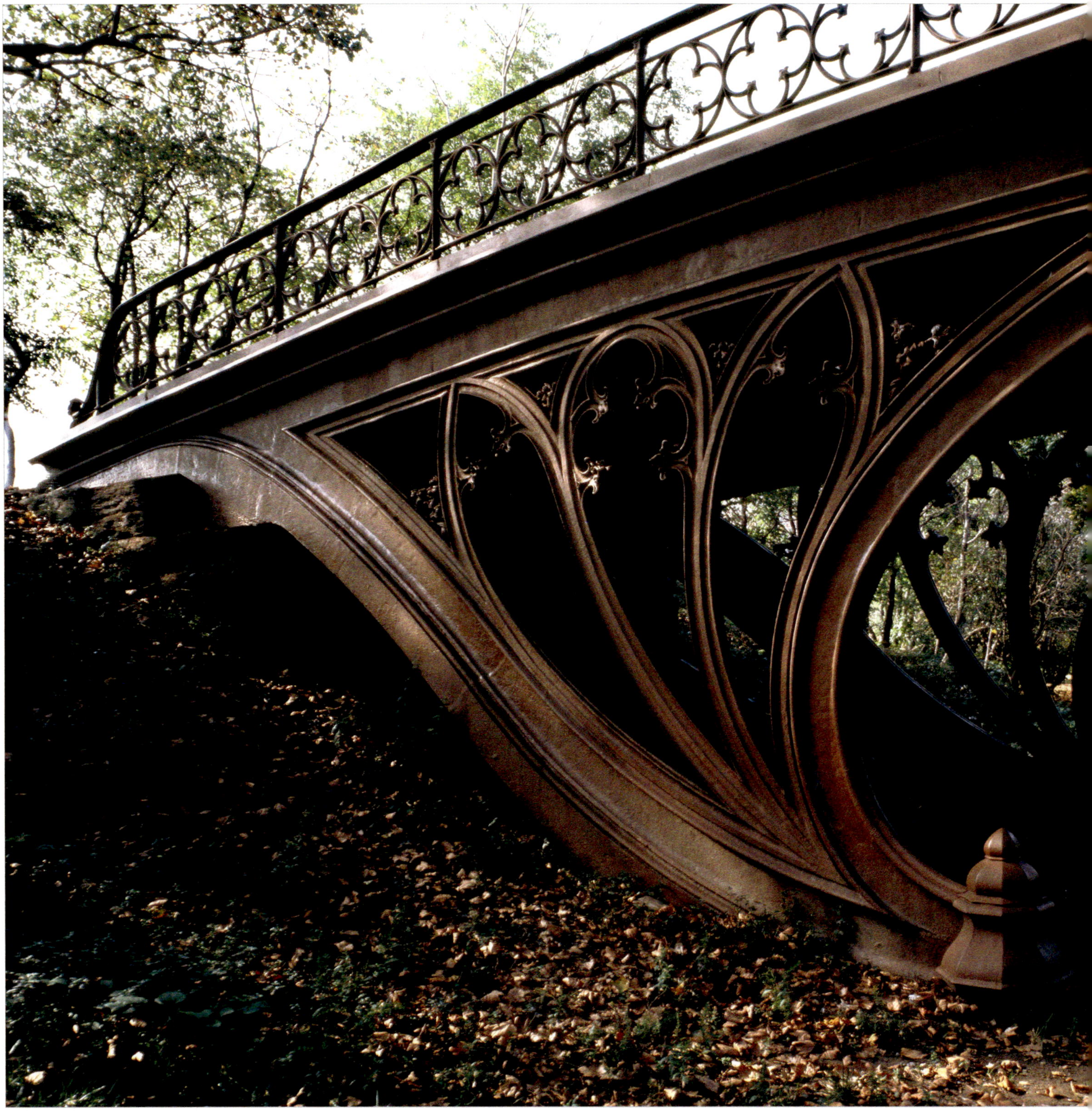

Calvert Vaux, *The Gothic Bridge* (*Bridge No. 28*), 1864. Central Park, New York, New York, USA, cast iron, 11'7" wide, 93' railing. Photo: Jet Lowe, Library of Congress, Prints and Photos Division, Washington, D.C., public domain.

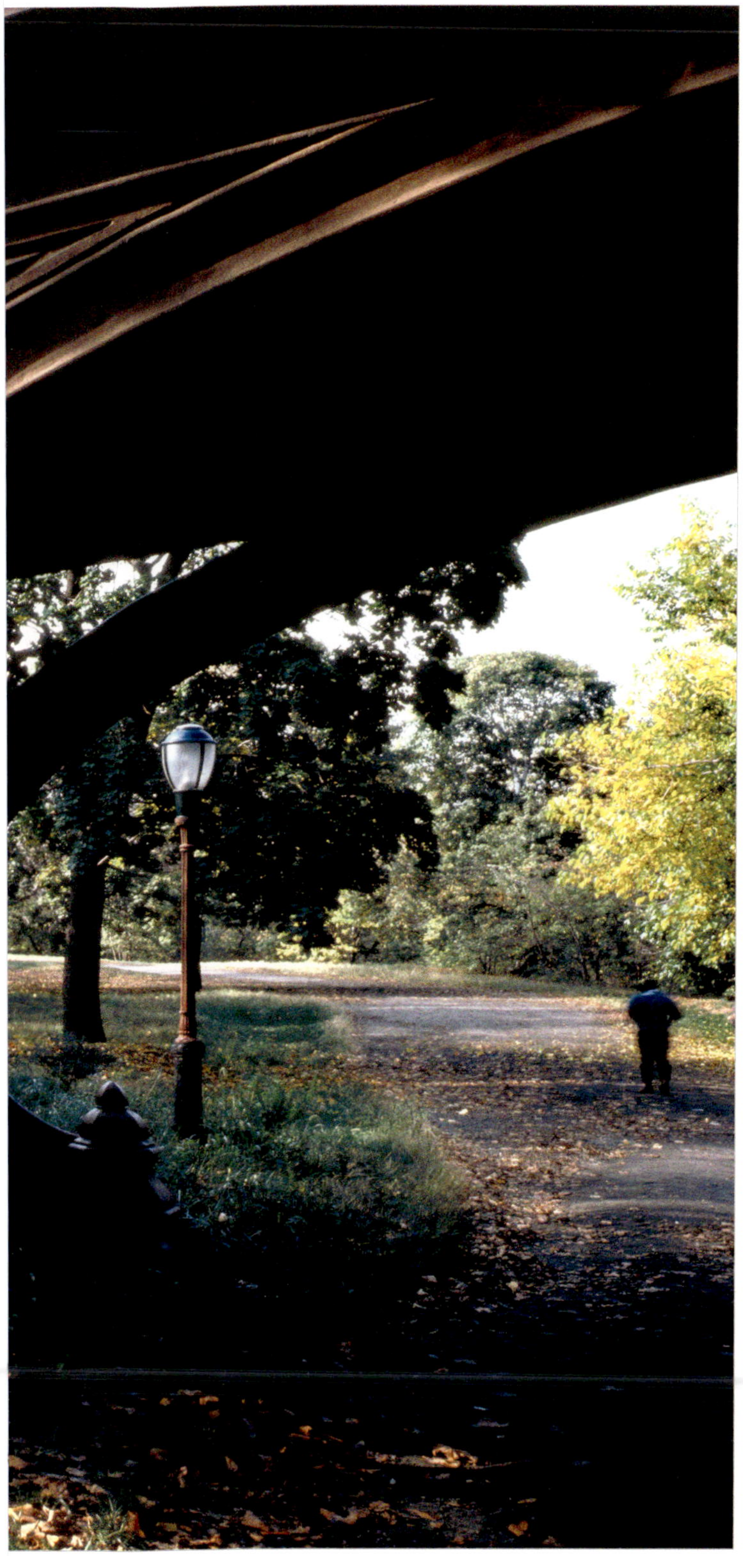

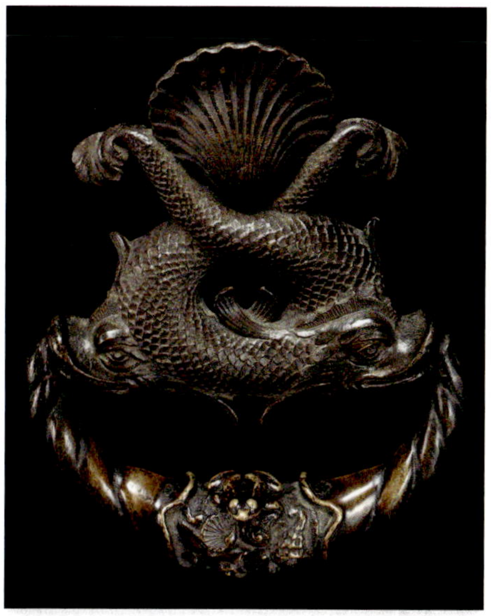

Doorknocker, fifteenth or sixteenth century. Paduan (probably), bronze, medium brown patina, 8" × 6.3". Courtesy of The National Gallery, Samuel H. Kress Collection.

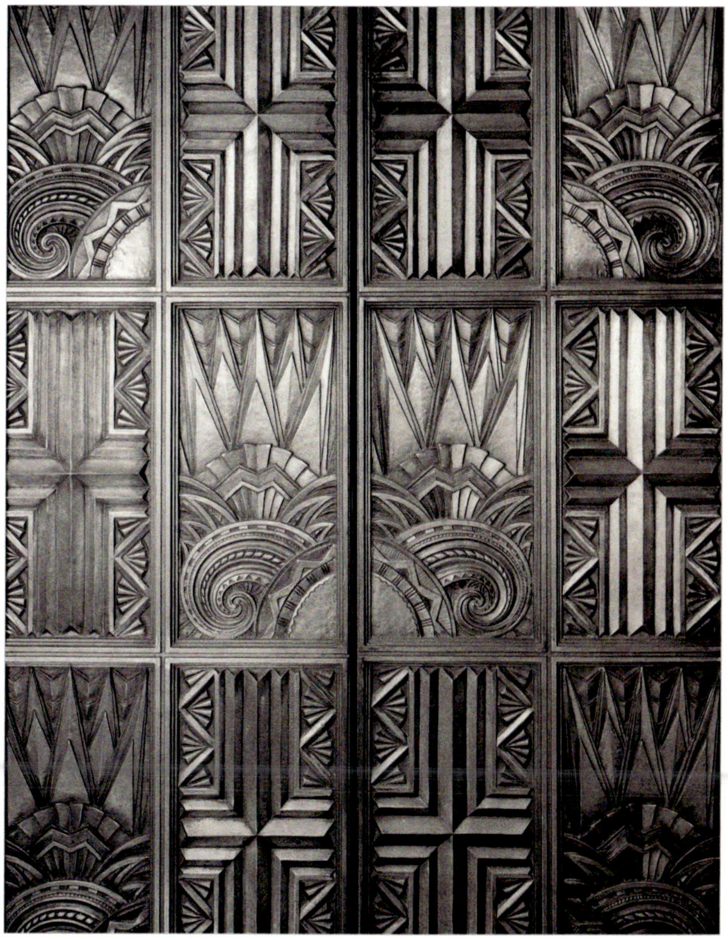

Stiles O. Clements of partners Morgan, Walls, & Clements, elevator doors, Richfield Oil Building (detail), demolished 1969, Los Angeles, CA, USA, bronze.
Courtesy of the National Park Service, provided by the Historic American Buildings Survey (HABS).

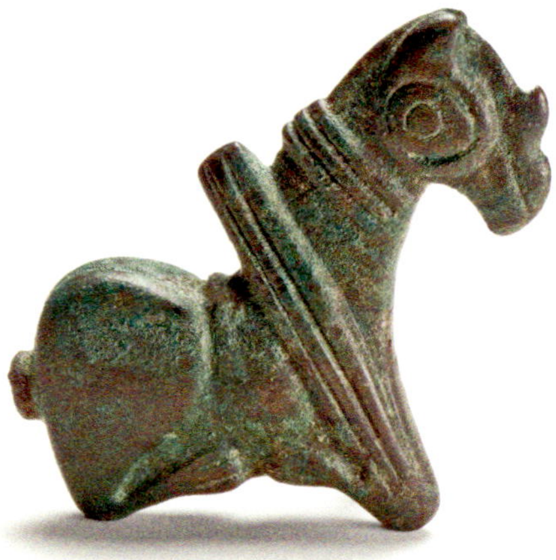

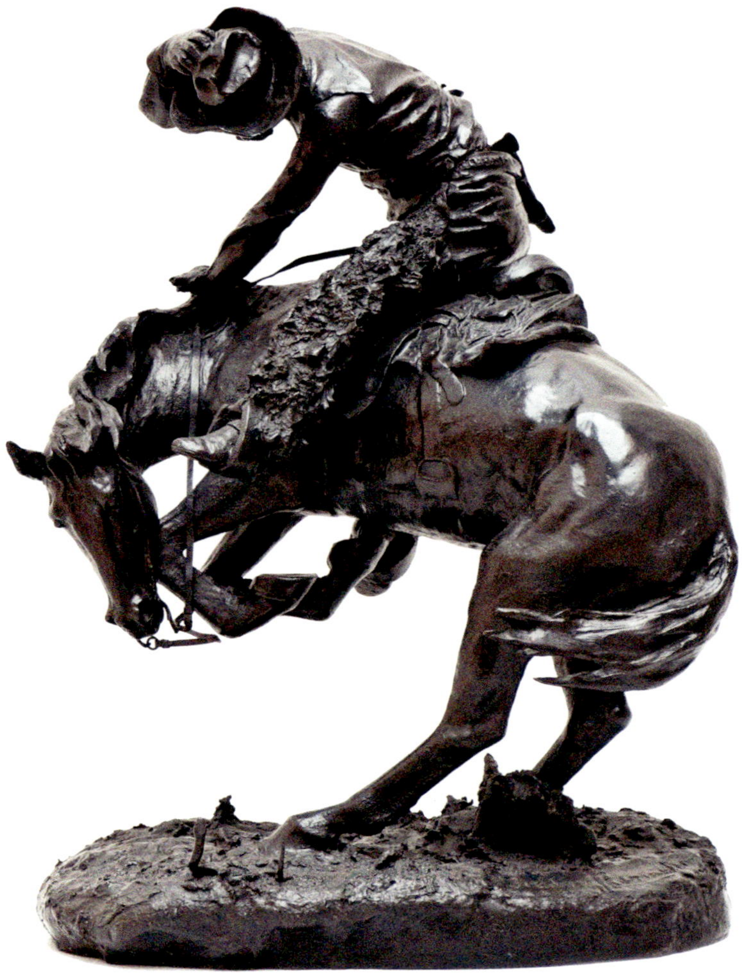

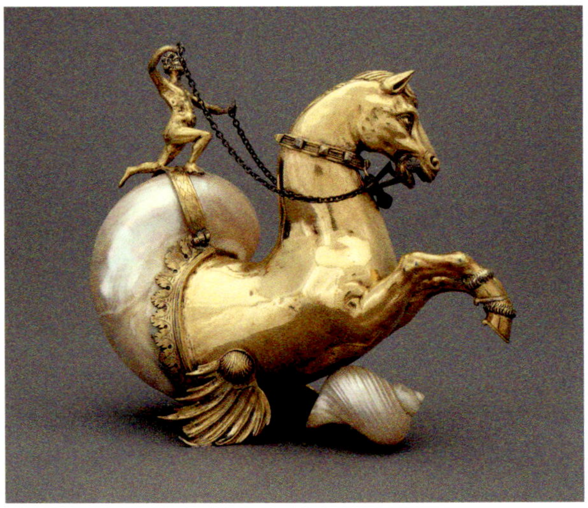

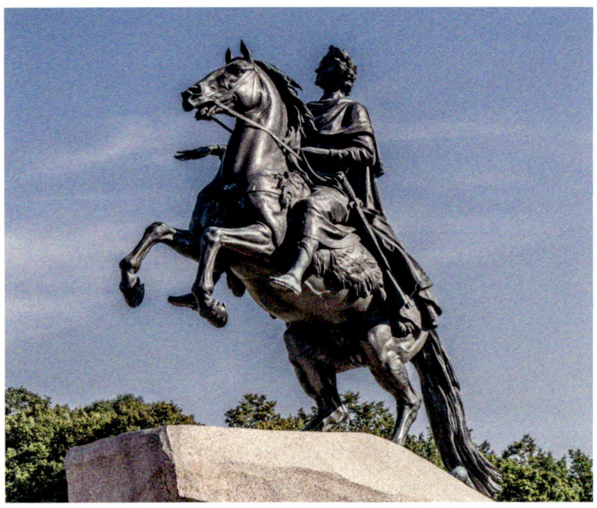

Top left Pinhead, 1350 to 800 BCE. Iran, Luristan Province, bronze, 3.12".
Courtesy of LACMA, Gift of The Ahmanson Foundation, www.lacma.org, LACMA public domain.

Top right Frederic Remington, *The Rattlesnake*, 1903. Bronze, 23.5" × 18" × 12".
Courtesy of the Grand Rapids Art Museum and The Wege Foundation.

Bottom left *Seahorse*, 1590 to 1600. Germany, silver gilt, nautilus shell, 7" × 7.87".
Courtesy of LACMA, Gift of Varya and Hans Cohn, www.lacma.org, LACMA public domain.

Bottom right Étienne Maurice Falconet, *The Bronze Horseman*, 1768–1782. Saint Petersburg, Russia, bronze.
Photo: © Alex (Florstein) Fedorov / Wikimedia Commons / CC-by 3.0.

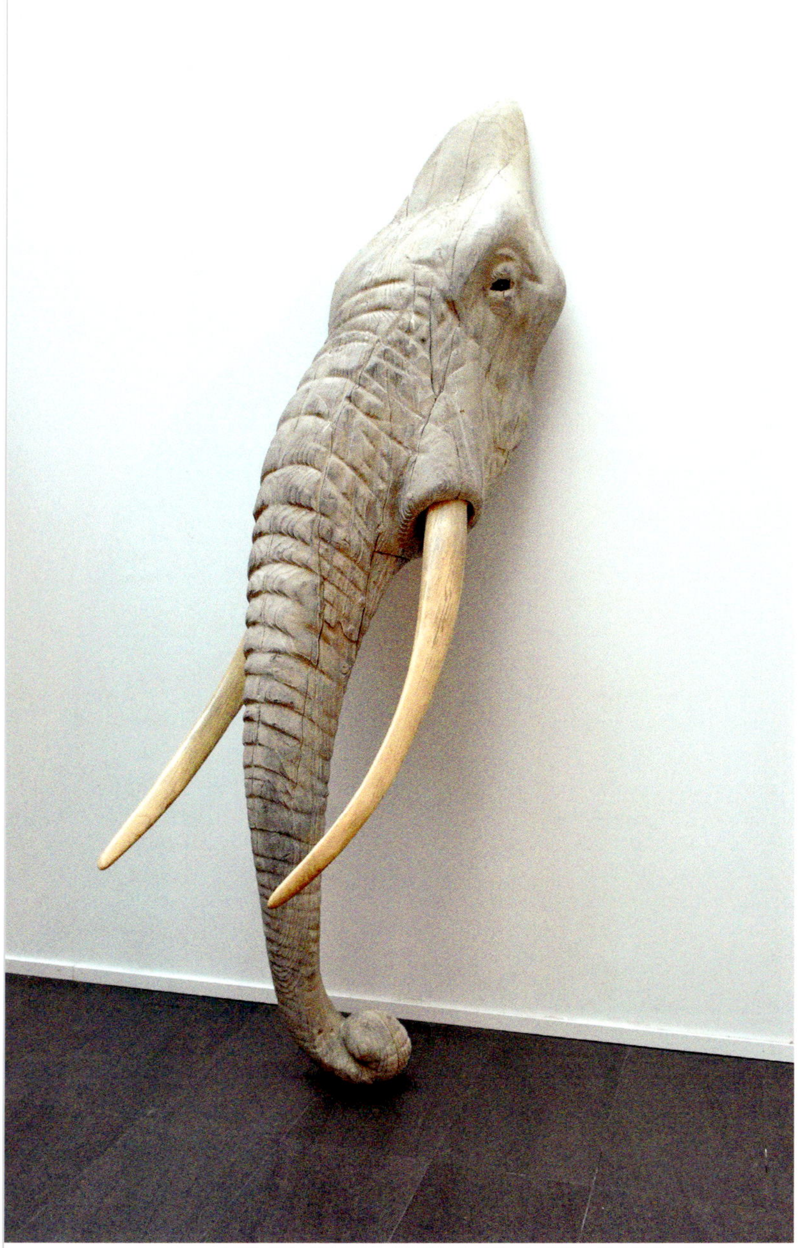

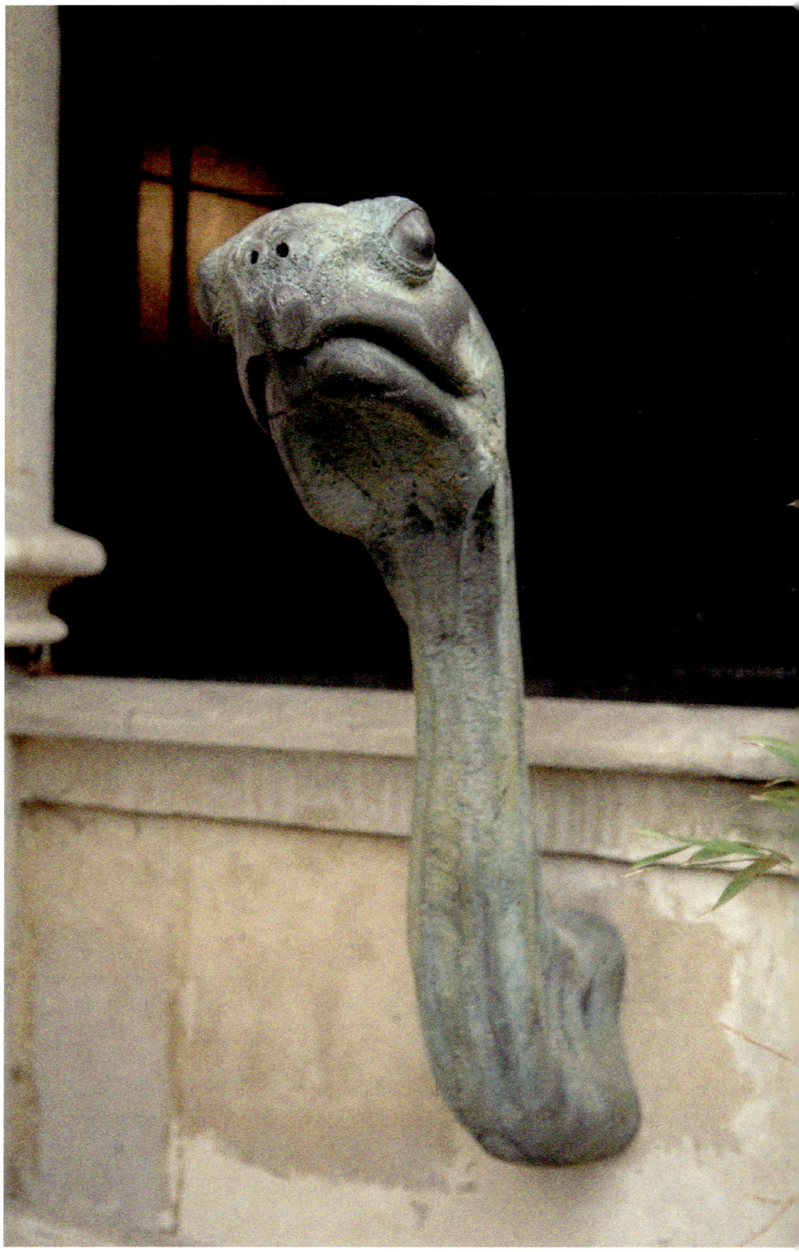

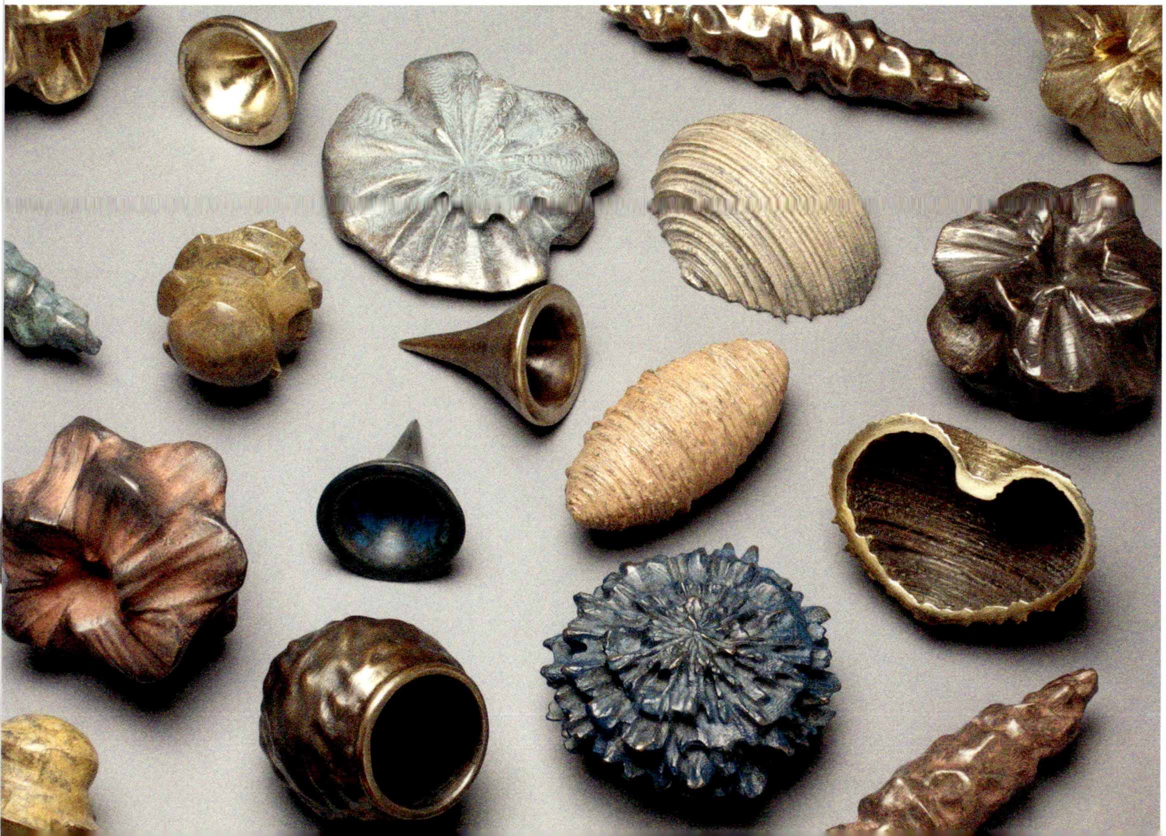

Top left Quentin Garel, *Eléphant II*, 2014. Bronze, 12'8.5" × 2'2" × 2'2". Photo: Quentin Garel, Courtesy of Bertrand Delacroix Gallery.

Top right Quentin Garel, *Tortue*, 2008. Bronze, 3'.5" × 2'6.6" × 9". Photo: Quentin Garel, Courtesy of Bertrand Delacroix Gallery.

Bottom Cast bronze data visualizations of geophysical phenomena made from ABS prototypes.

Mary Neubauer, *Data Visualizations*, 2011. Bronze, 4" × 4" × 3.5". Photo: Damian Johnson.

Bruce Beasley, *Icarus*, 1963 (won the Purchase Prize at the Paris Biennale,1963). Aluminum. Photo: Unknown.

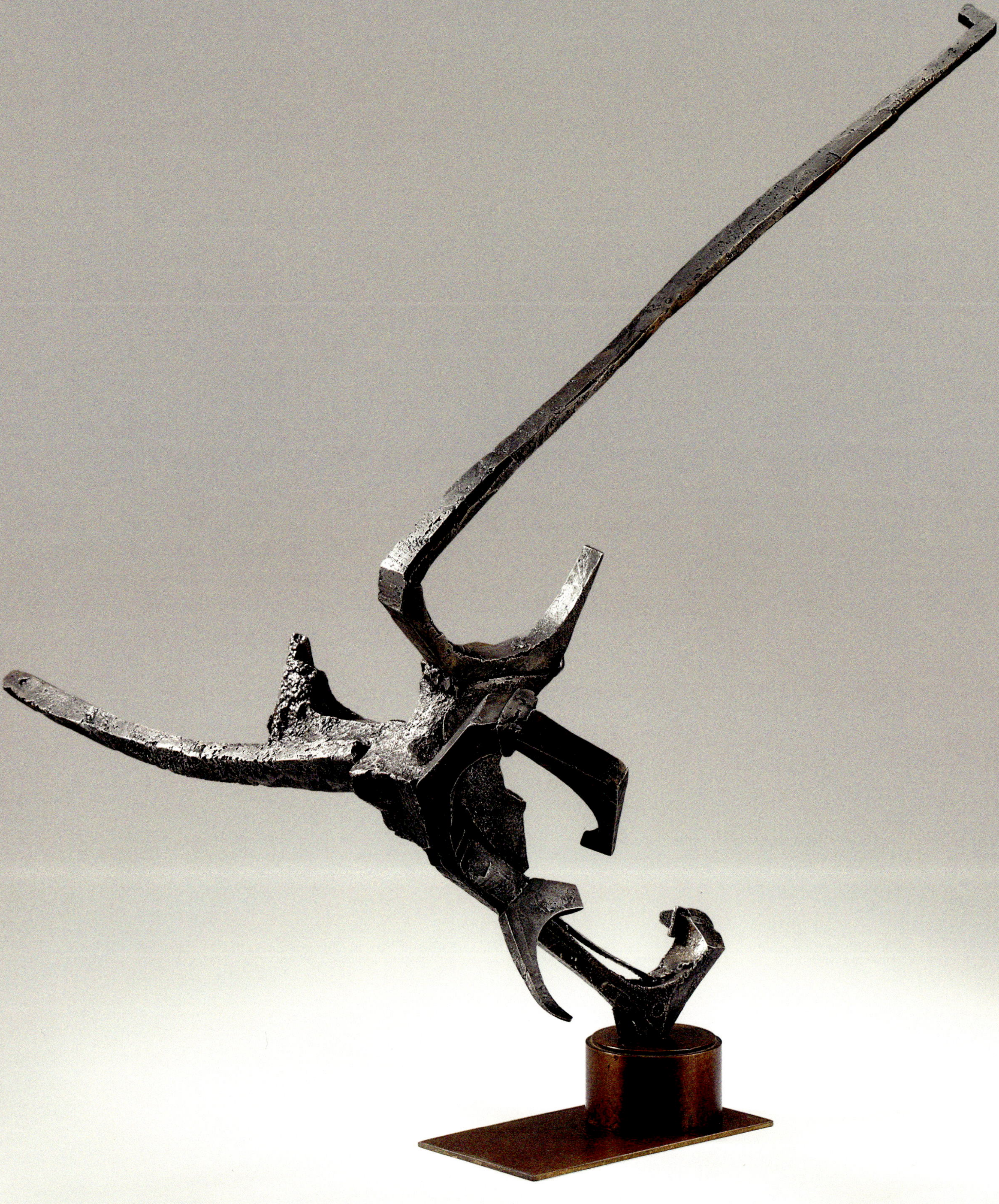

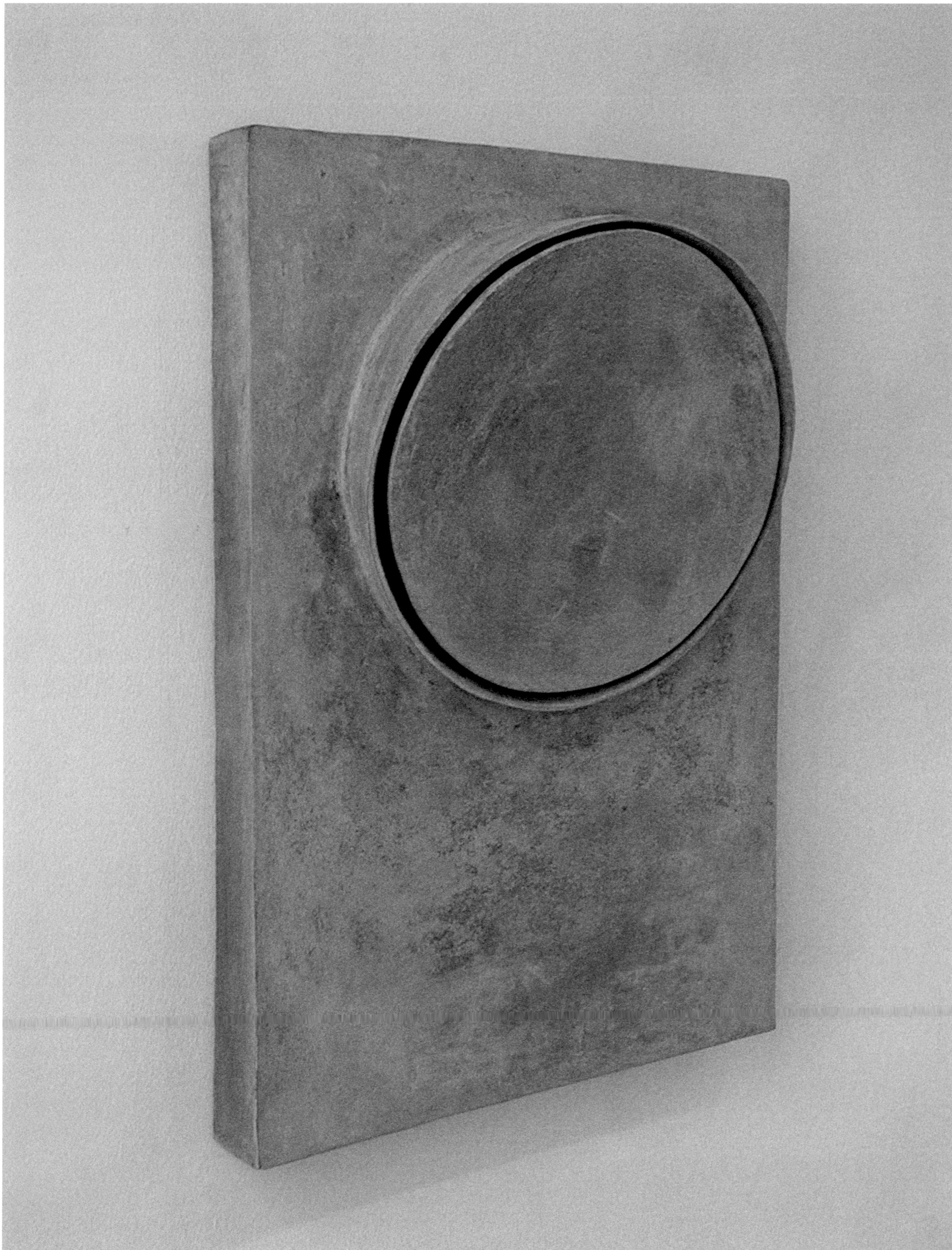

Ingo Ronkholz, *Wandskulptur 2002 No. 14*, 2002. Bronze, 19.75" × 27.12" × 6.75".
Photo: Atelier Ingo Ronkholz.

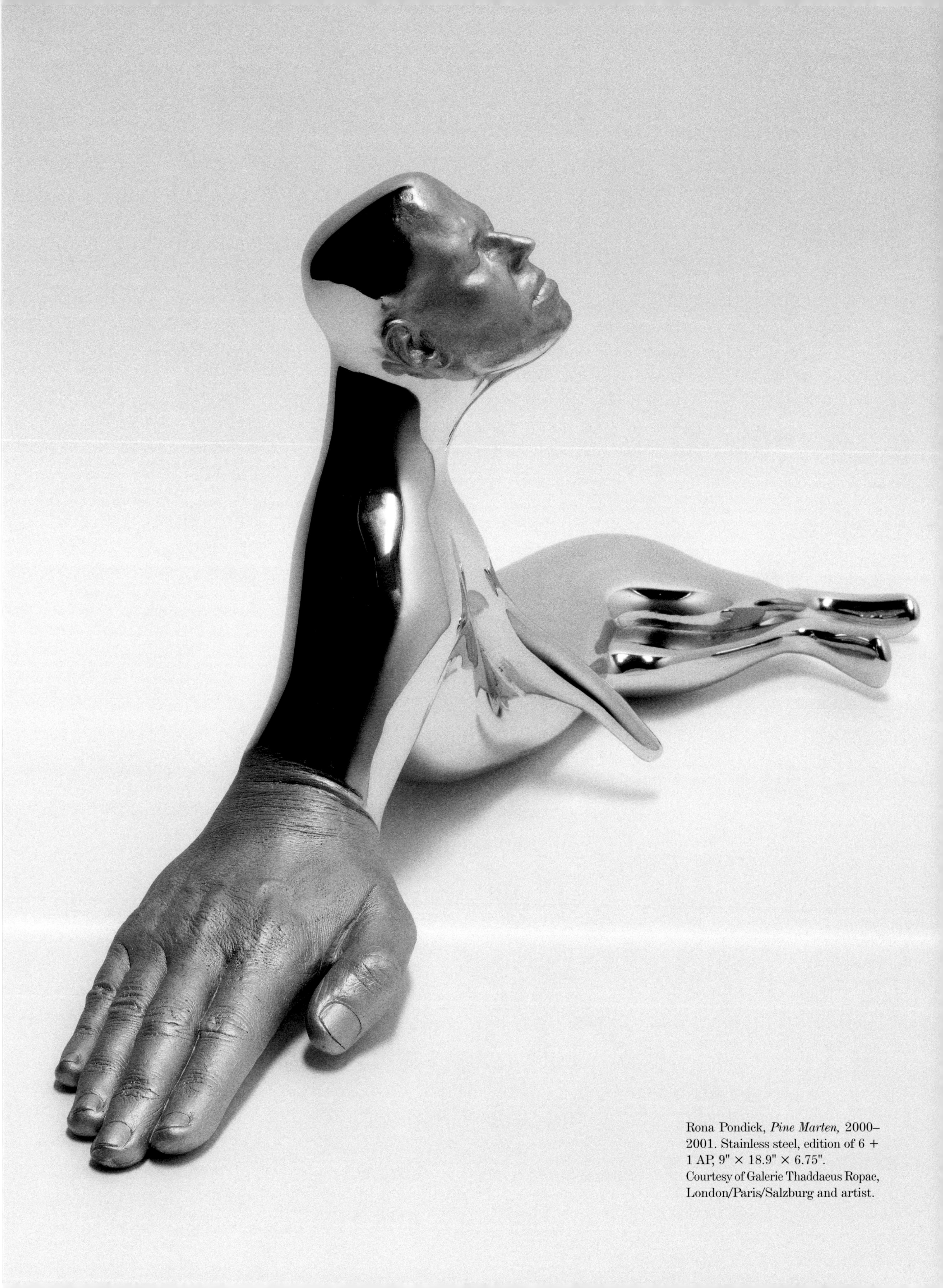

Rona Pondick, *Pine Marten,* 2000–2001. Stainless steel, edition of 6 + 1 AP, 9" × 18.9" × 6.75". Courtesy of Galerie Thaddaeus Ropac, London/Paris/Salzburg and artist.

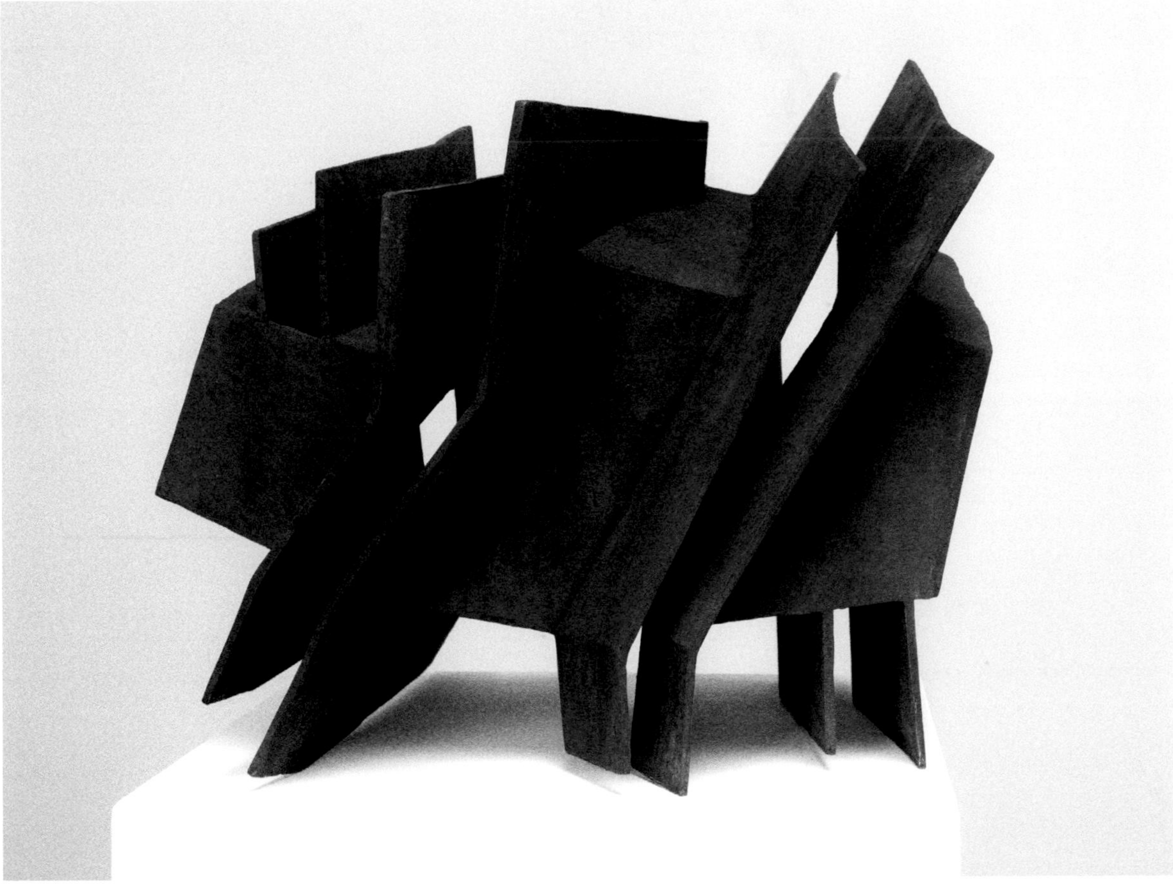

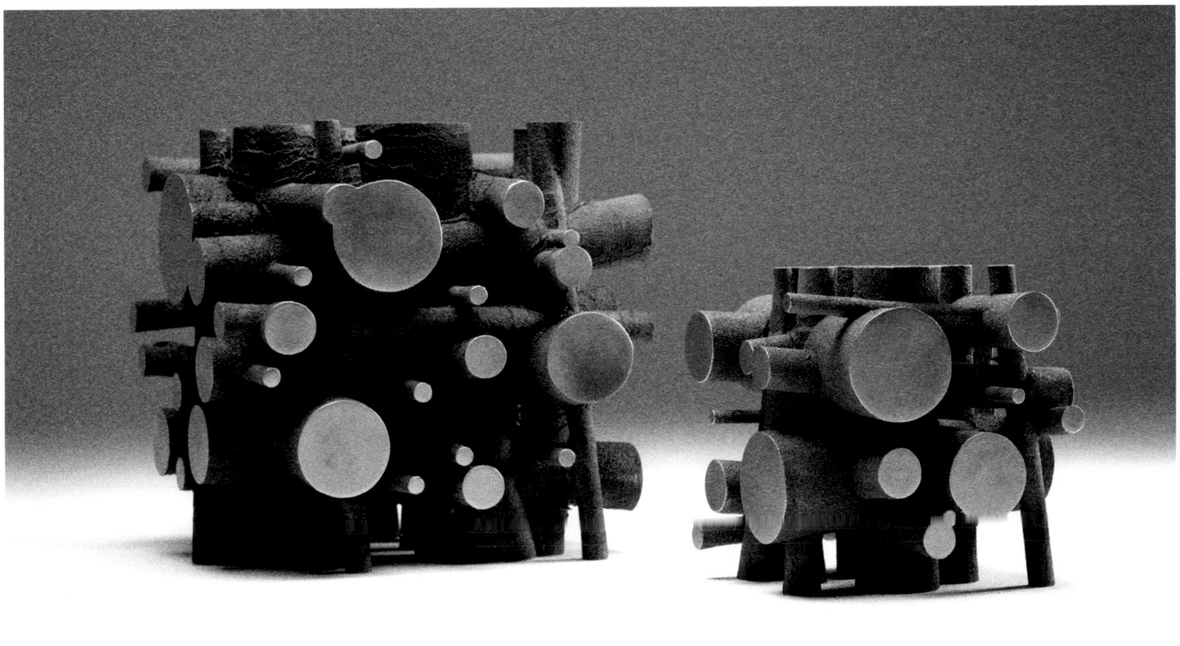

Top Ingo Ronkholz, *Sculpture 2014 No. 15,* 2014. Bronze, 19.7" × 20.9" × 21.25".
Photo: Atelier Ingo Ronkholz.

Bottom Tom Joyce, *Core V* and *VI*, 2013. Cast ductile iron alloy made from steel filings from projects 1977–present, 10" × 10" × 10" (right), 15" × 15" × 15" (left).
Photo: 2002 © John Michael Kohler Art/Industry, Created in Arts/Industry, a long-term residency program of the John Michael Kohler Arts Center in Sheboygan, Wisconsin. Arts/Industry takes place at Kohler Co. *Core VI*–Tia Collection.

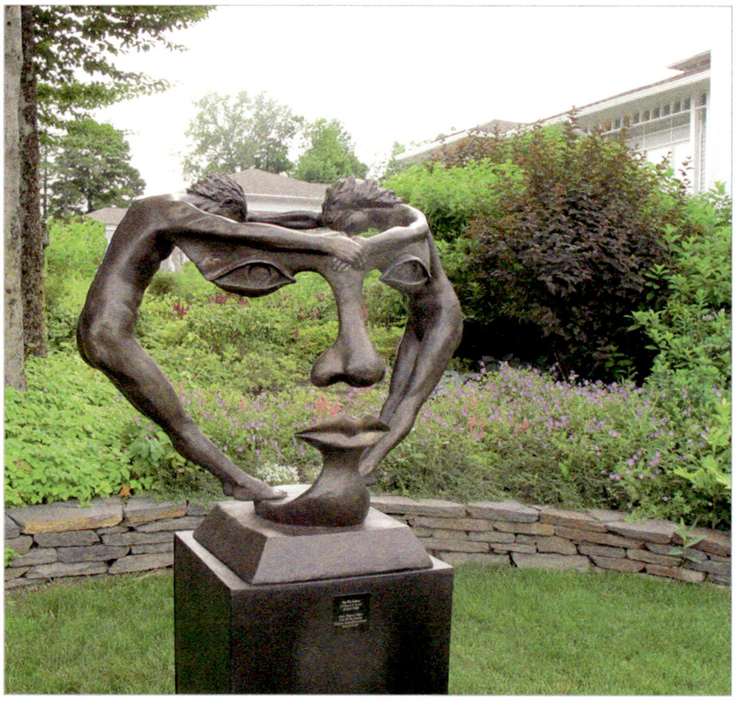

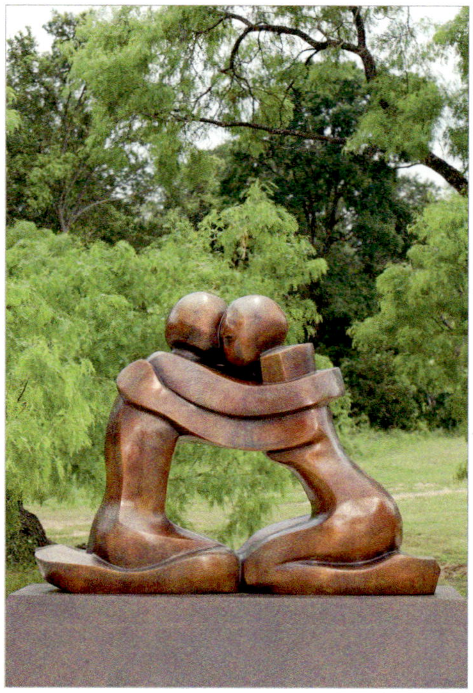

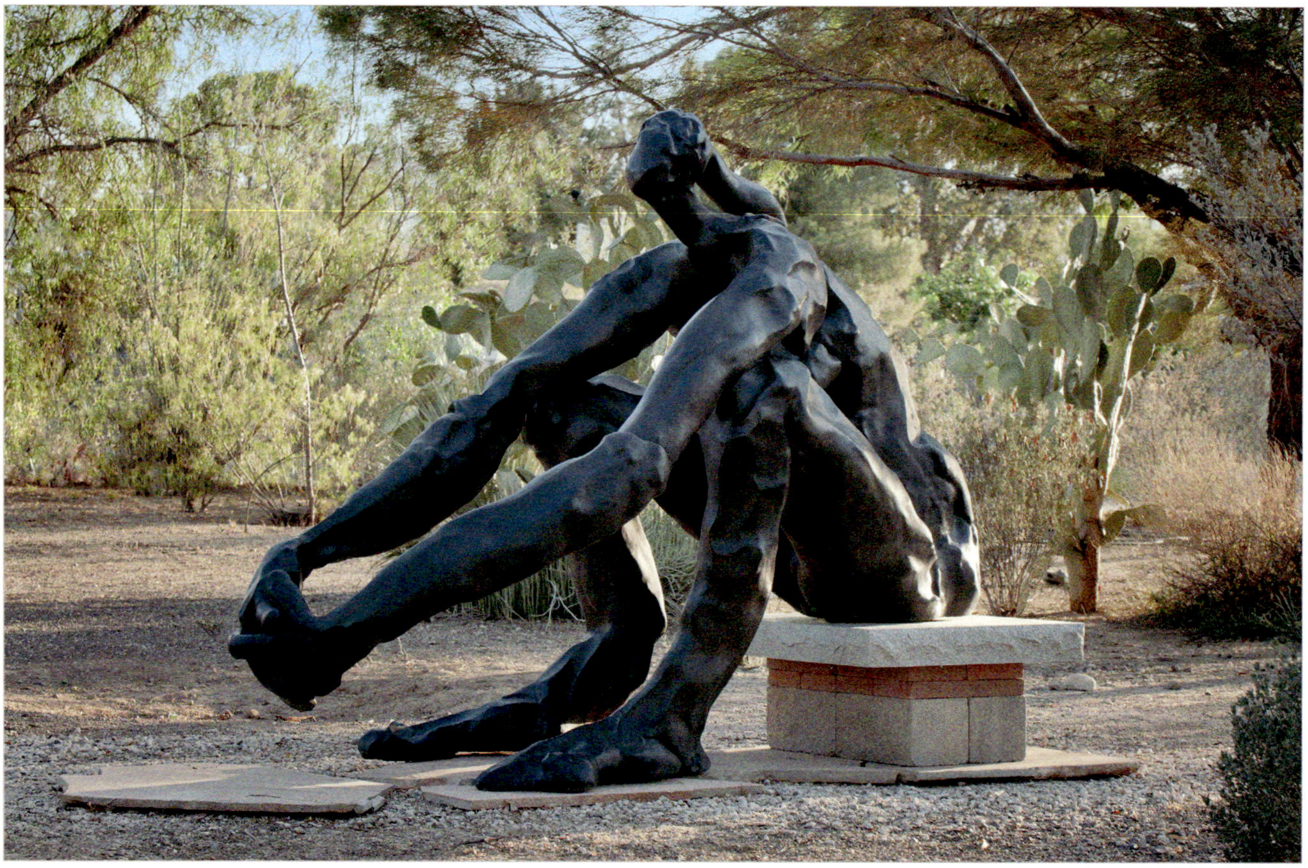

Top left Michael Alfano, *We Two Together*, 2014. Bronze, 2'9" × 3'7" × 3'7". Photo: Michael Alfano.

Top right Mark Yale Harris, *Crush*, 2007. Bronze, 5'1" × 6'8" × 1'8.5". Photo: Jim Lincoln, Courtesy of ARTWORKinternational INC.

Bottom Curt Brill, *Diana*, 2012. Bronze, granite, 6'3" × 4'9" × 8'2". Photo: Troy Tuttle.

Top to bottom, right

Akan Gold Weight, Bird and Its Chicks, nineteenth century. Akan Culture, Africa, cast brass (0.14 to 0.19 g). Former collection of French botanist and ethnologist Jean Baur, donated to the Museum of Toulouse in 2010.
Photo: © Didier Descouens at the Museum of Toulouse / Wikimedia Commons / CC by 3.0.

Akan Gold Weight, Scorpion, nineteenth century. Akan Culture, Africa, cast brass, (0.14 to 0.19 g). Former collection of French botanist and ethnologist Jean Baur, donated to the Museum of Toulouse in 2010.
Photo: © Didier Descouens at the Museum of Toulouse / Wikimedia Commons / CC by 3.0.

Akan Gold Weight, Bat, nineteenth century. Akan Culture, Africa, cast brass (0.14 to 0.19 g). Former collection of French botanist and ethnologist Jean Baur, donated to the Museum of Toulouse in 2010.
Photo: © Didier Descouens at the Museum of Toulouse / Wikimedia Commons / CC by 3.0.

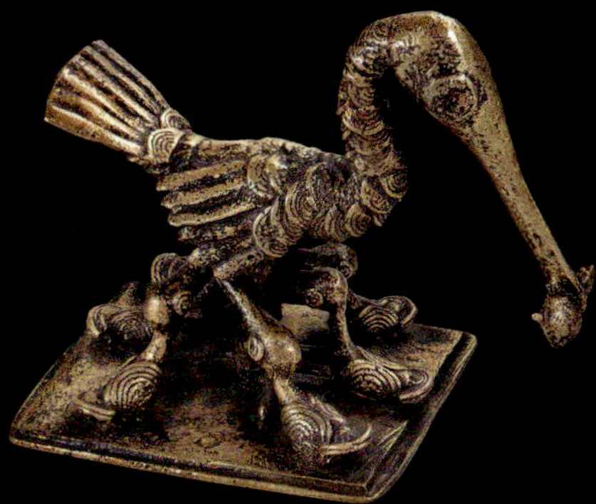

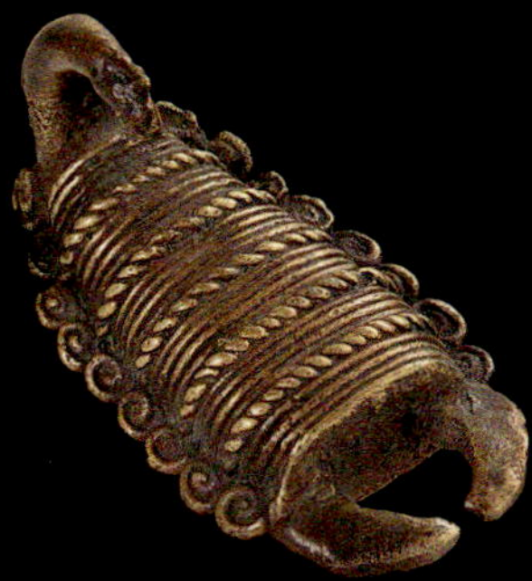

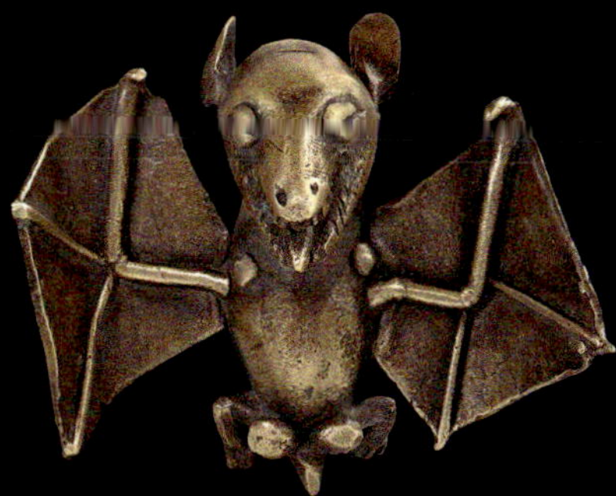

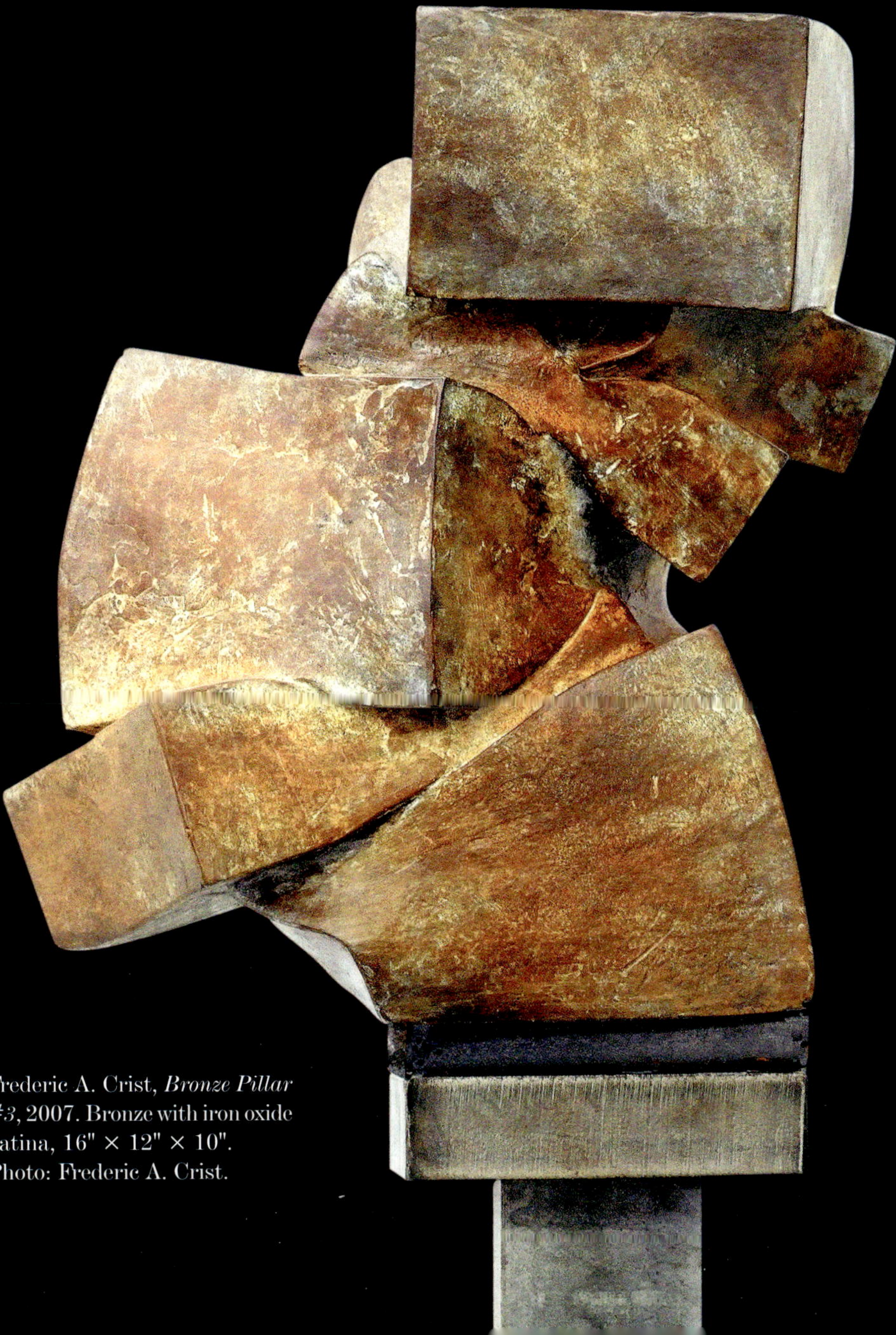

Frederic A. Crist, *Bronze Pillar #3*, 2007. Bronze with iron oxide patina, 16" × 12" × 10".
Photo: Frederic A. Crist.

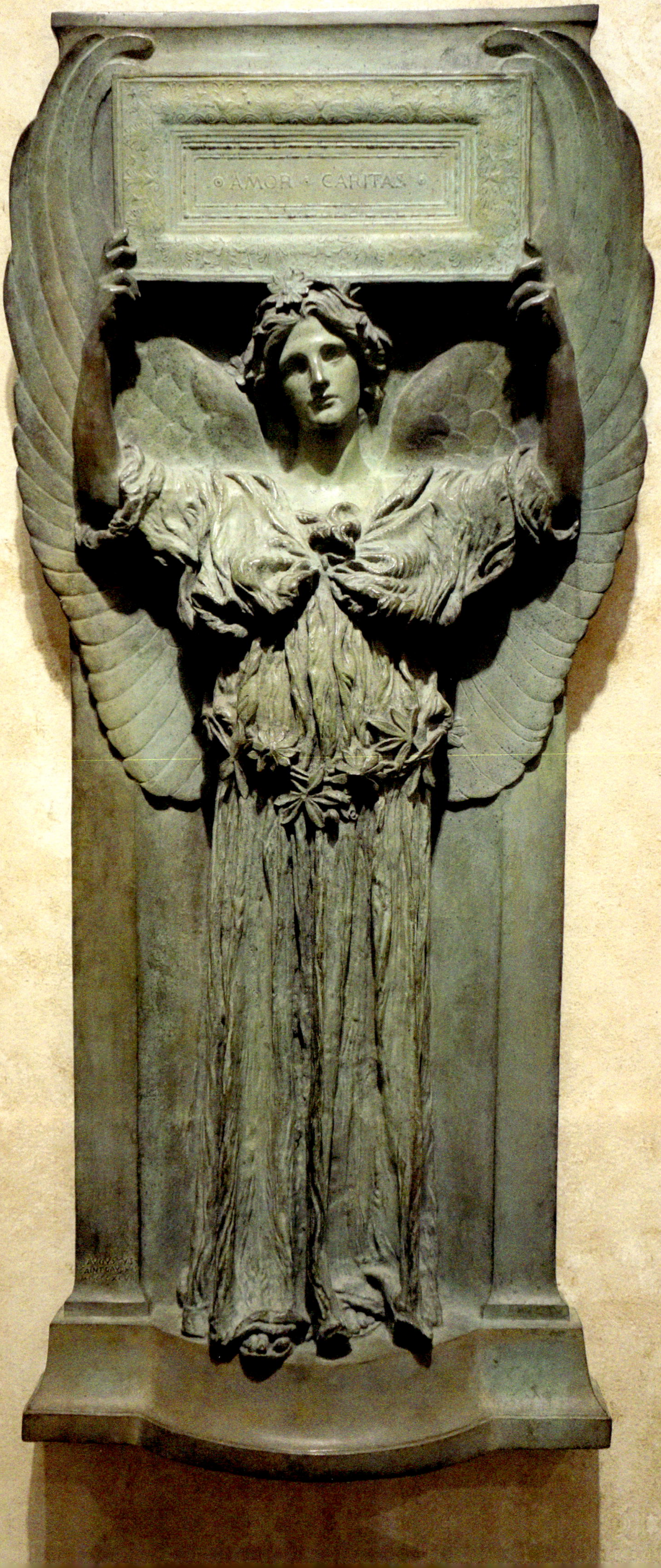

Augustus Saint-Gaudens, *Amor Caritas*, 1898. Bronze, 3'4" × 1'5.75" × 5". Courtesy of Brooklyn Museum, Gift of Mr. and Mrs. Harris Klein, CC-BY.

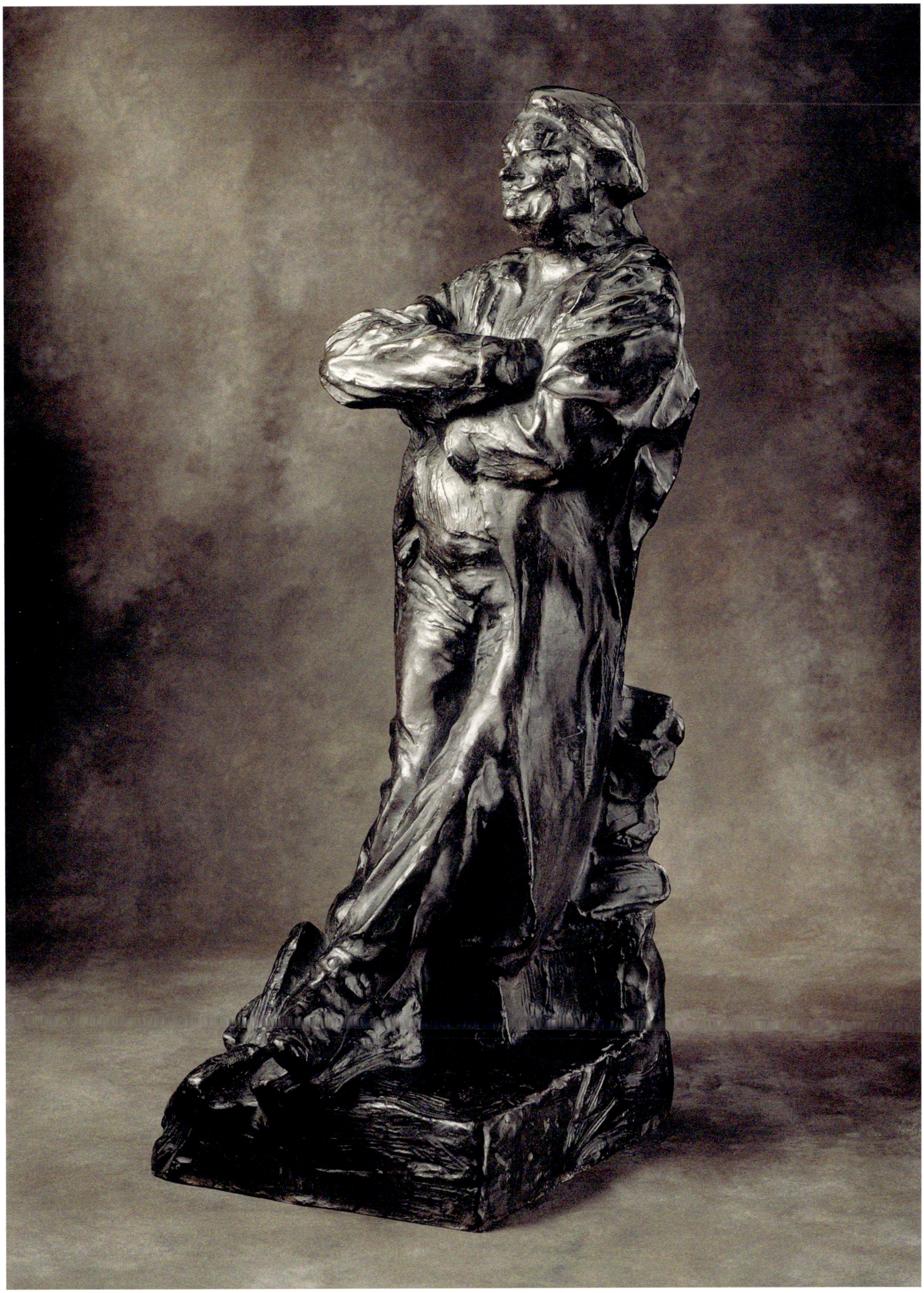

Auguste Rodin, *Balzac in a Frockcoat,* first modeled ca. 1891–1892 (Museé Rodin, 11 from an edition of 12, 1980). France, bronze, 23.5" × 8" × 10.25".
Courtesy of LACMA, Gift of B. Gerald Cantor Art Foundation, www.lacma.org, LACMA public domain.

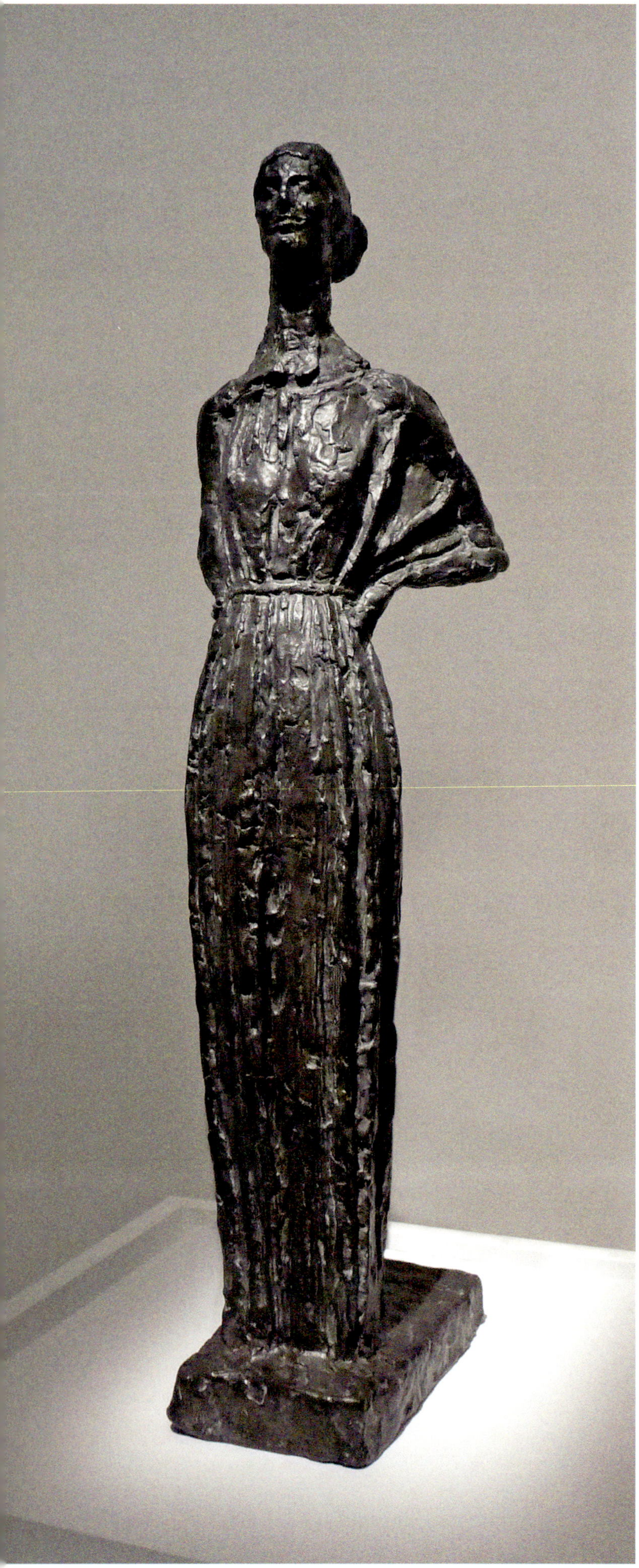

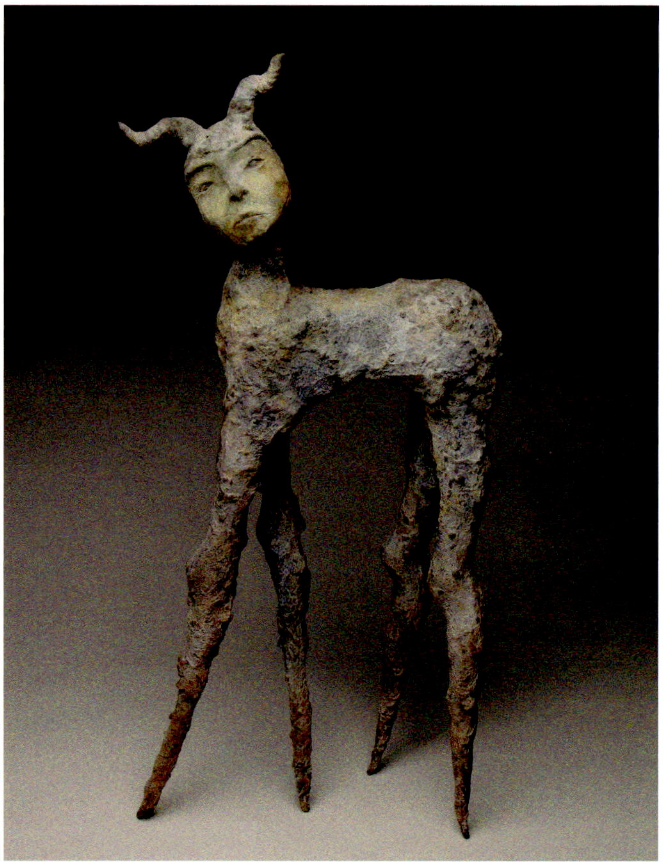

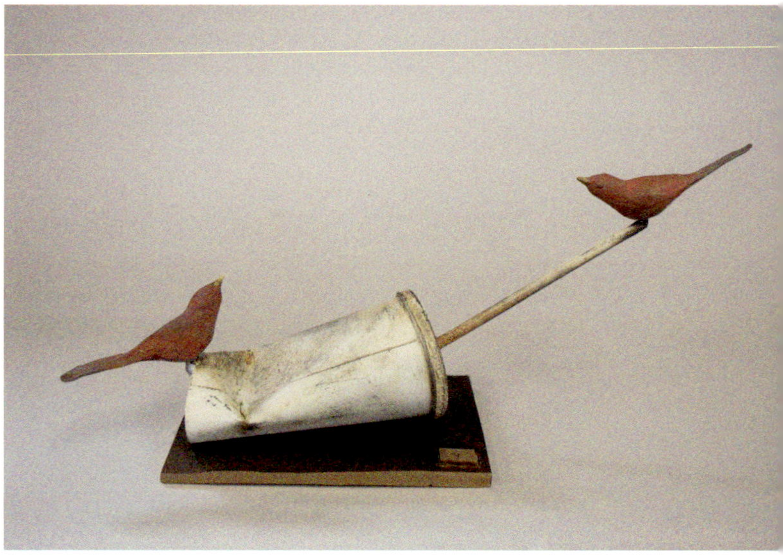

Left Antoine Bourdelle, *Madeleine Charnaux*, 1917. France, bronze, 24.25" × 6.75" × 7.25".
Photo: Ji-Elle, Collection of Museé Bourdelle / Wikimedia Commons / public domain.

Top right Casey Stueber, *A Spectral Appearance*, 2012. Cast iron, patina, 23" × 11" × 14".
Photo: Casey Stueber.

Bottom right Bill Heiderich. *Cup With Birds*, 2014. Bronze with patina and paint, 6" × 16" × 5".
Photo: Bill Heiderich.

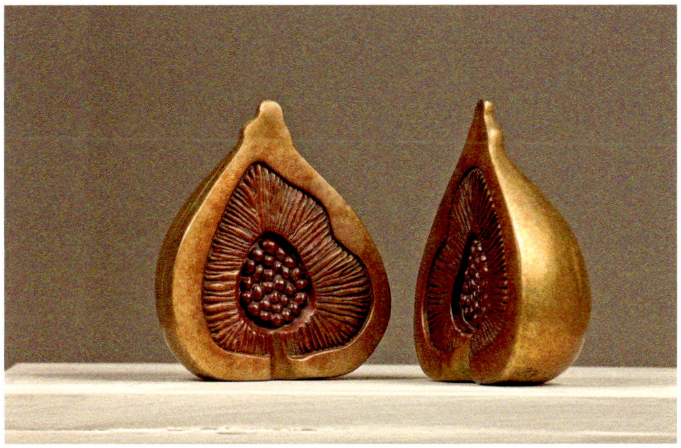

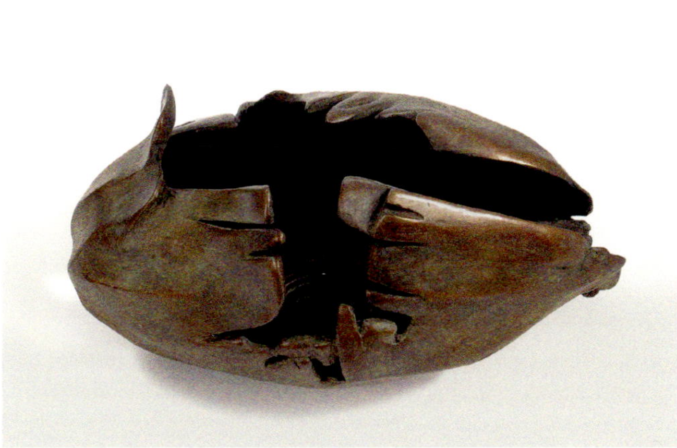

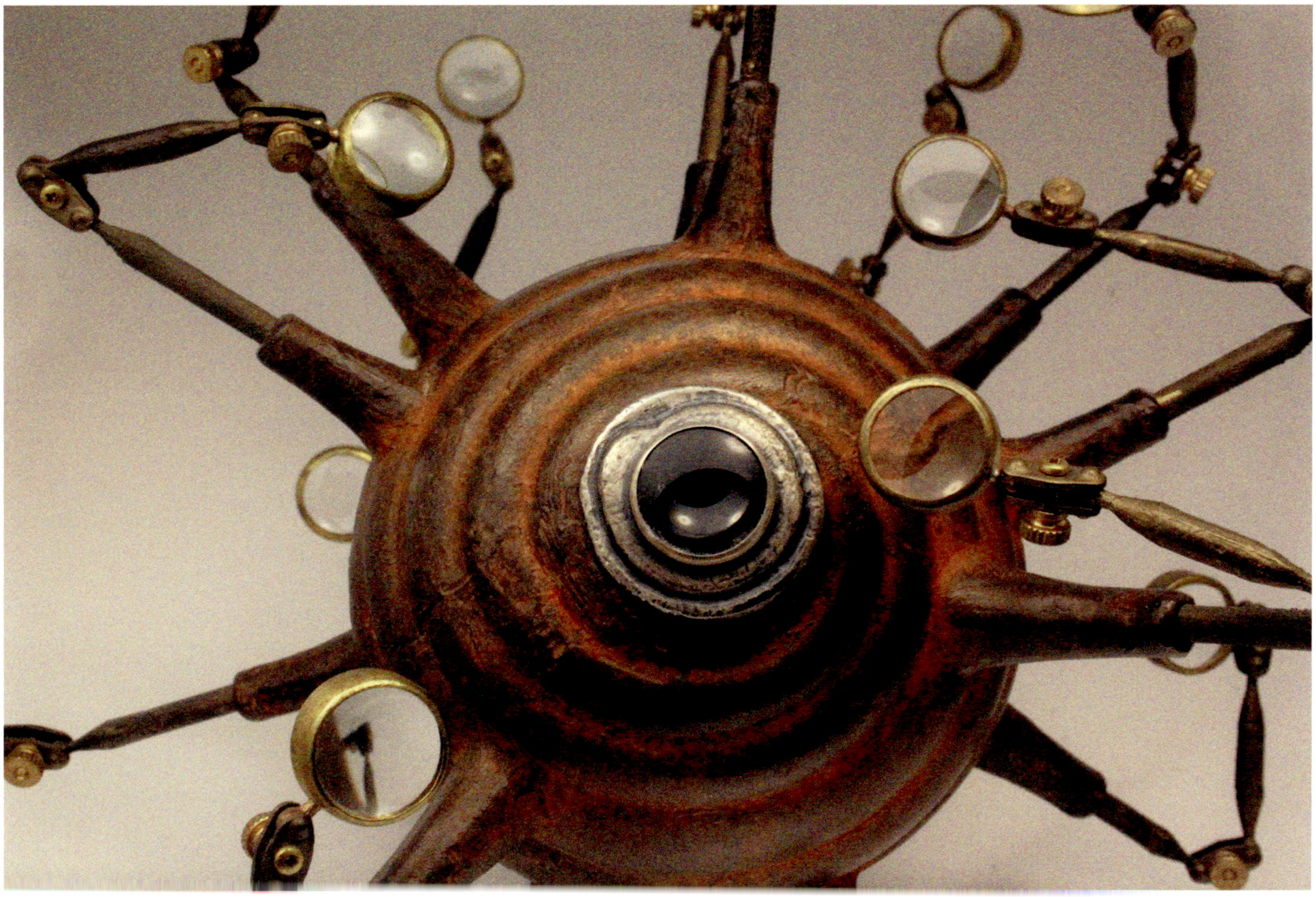

Top left Jan Kirsh, *Fig*, 2009. Bronze, 6" × 5" × 3".
Photo: Stephen Cherry.

Top right Marianne Weil, *Loughcrew*, 2007. Cast bronze, 12" × 8" × 7".
Photo: Matt Flynn, Collection of Evan Schwartz, NYC.

Bottom David Barnhill, *This Idea Shall Too Evolve*, 2008. Bronze, brass, magnifying glass, sterling silver, 18" × 14" × 10".
Photo: David Barnhill.

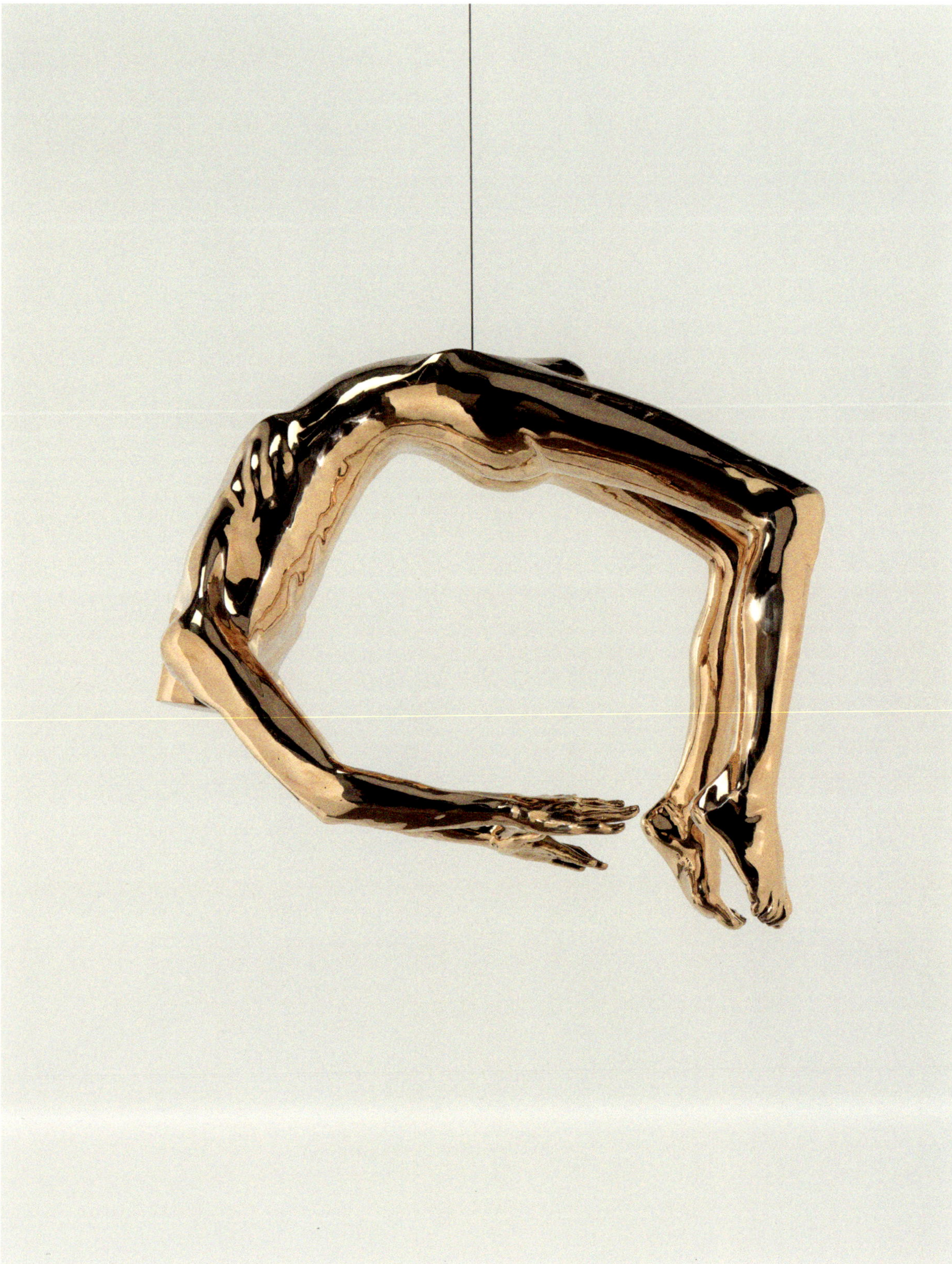

Louise Bourgeois. *Arch of Hysteria*, 1993. Bronze, polished patina, 33" × 40" × 23".
Photo: Christopher Burke, Art © The Easton Foundation / Licensed by VAGA.

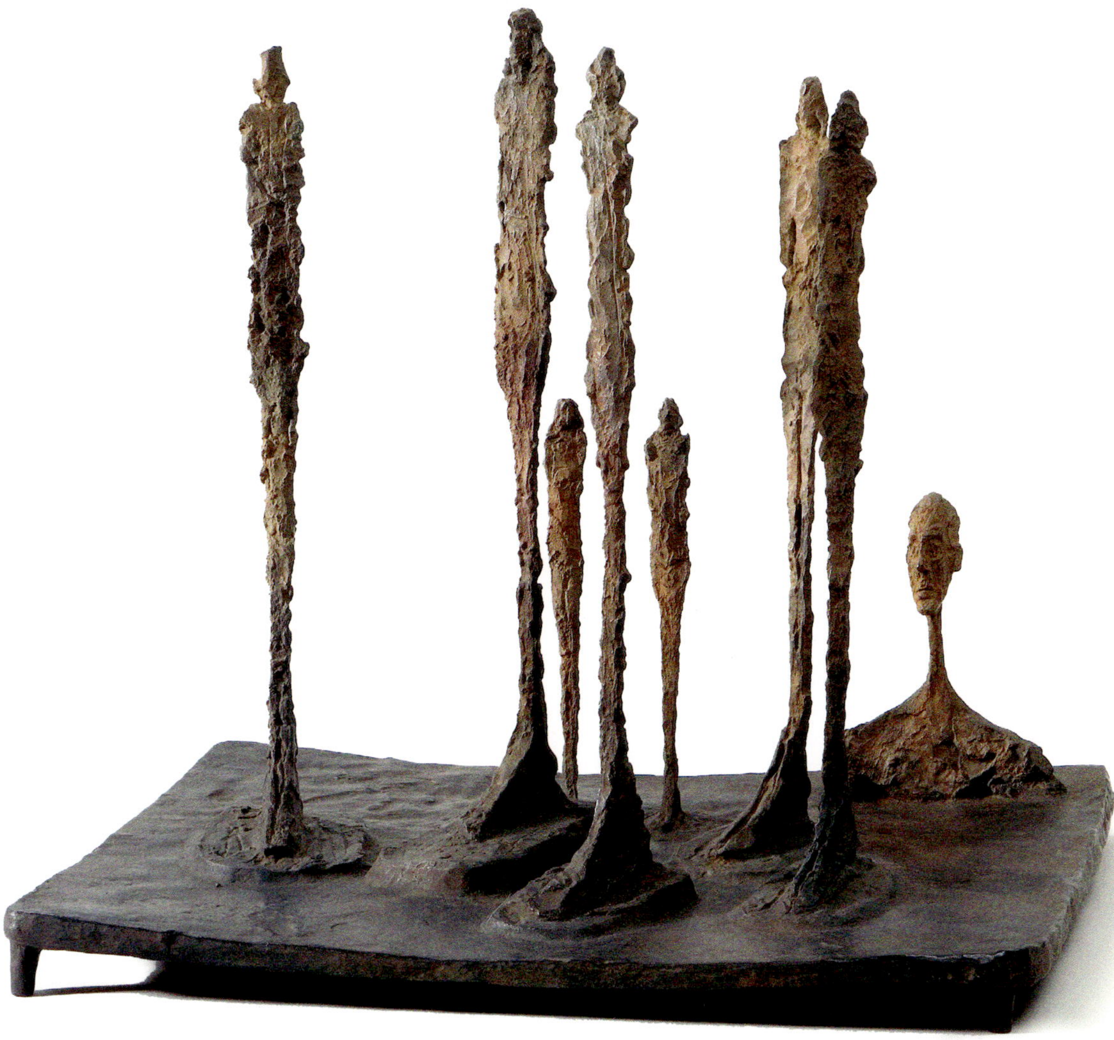

Alberto Giacometti, *The Forest*, 1950. Bronze, 23" × 23.62" × 19".
Photo: Unknown, Art © 2015 Alberto Giacometti Estate/Licensed by VAGA and ARS, New York, NY.

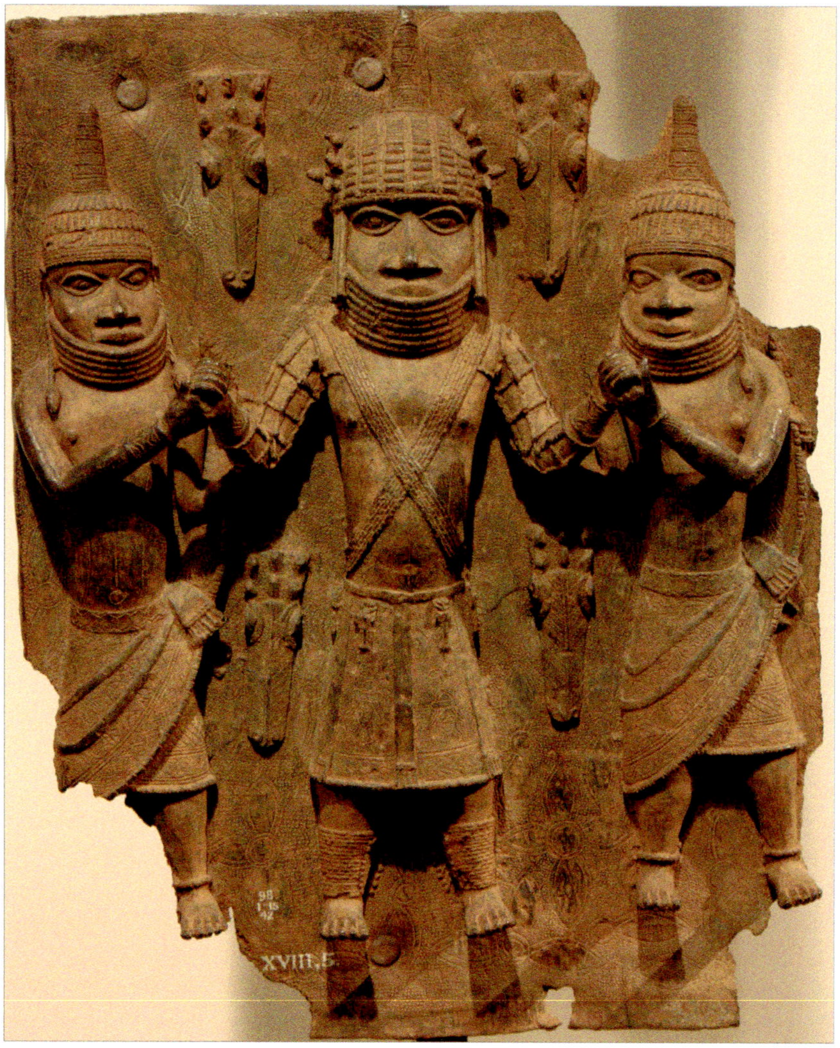

Right Benin Bronze Plaque, sixteenth century to seventeenth century. Benin kingdom (Nigeria), brass, 19" × 15" × 3.1". Donated by the Secretary of State for Foreign Affairs to The British Museum. Photo: © Machel Wal / Wikimedia Commons / CC by 3.0.

Below Karen LaMonte, *Kimono* Installation, 2012. Iron, dimensions variable. Photo: Martin Polak, © 2012 LaMonte's studio.

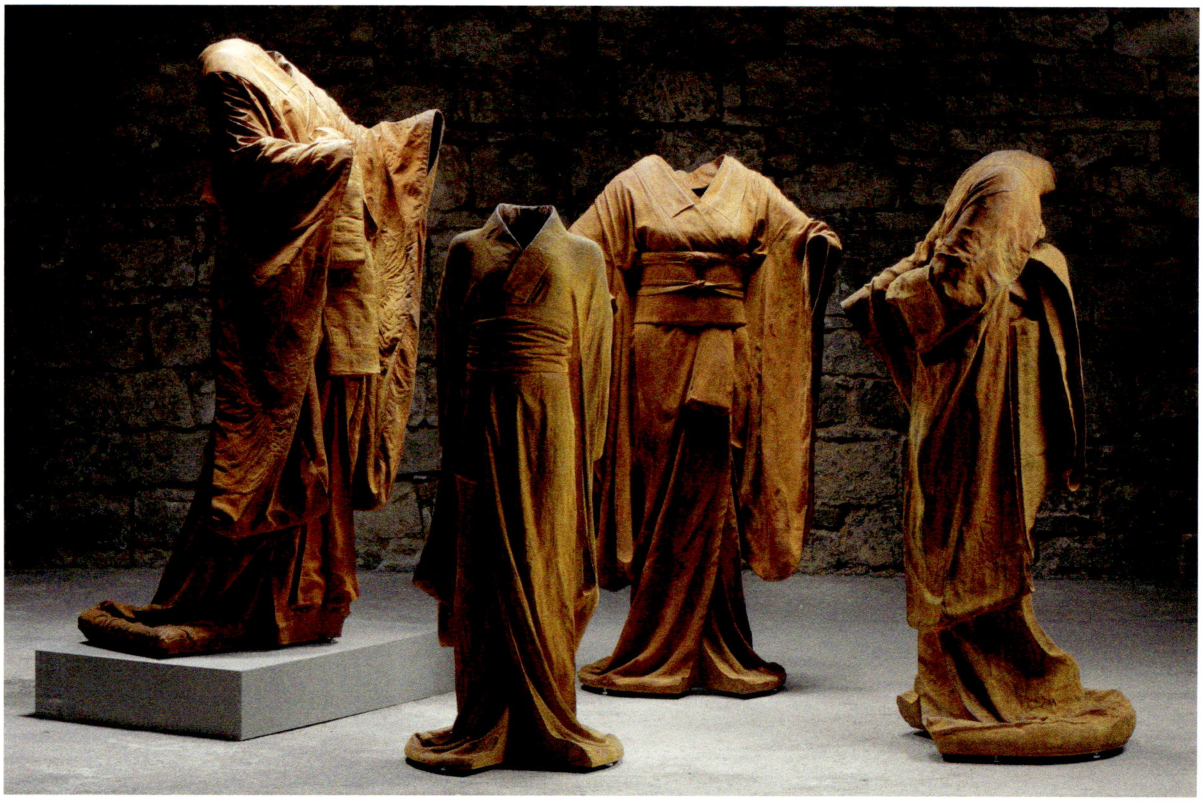

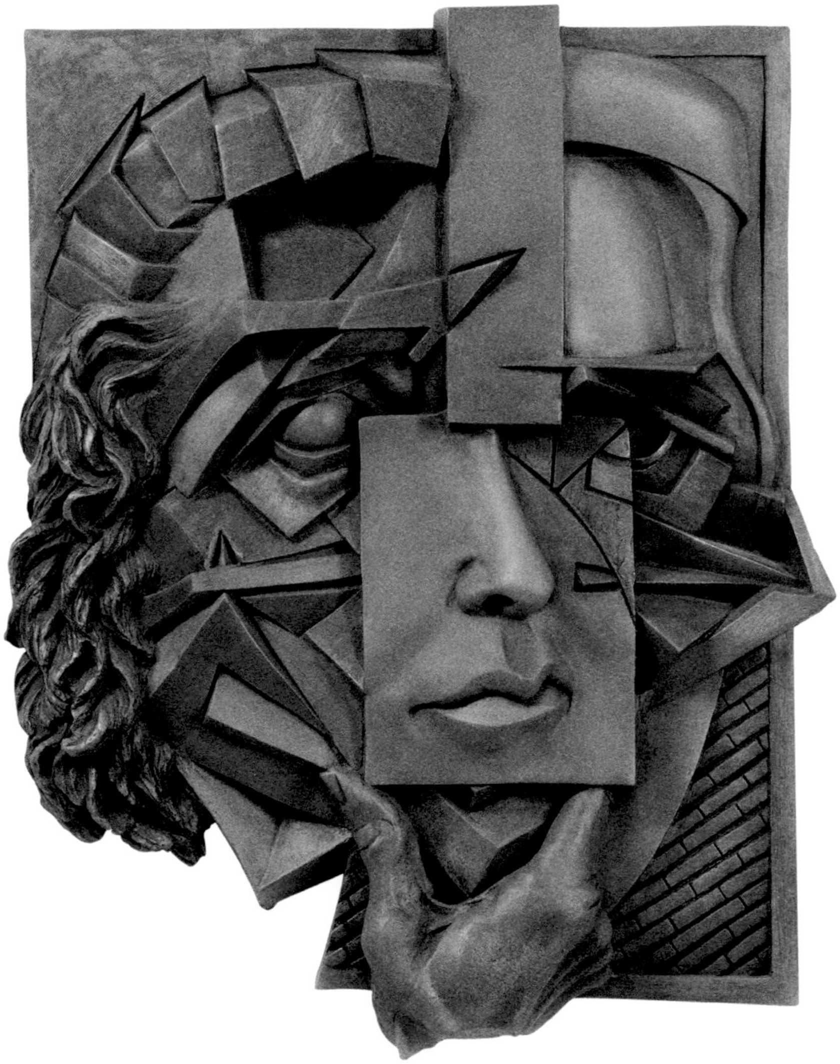

Left Suzanne Horwitz, *Façade*, 2011. Aluminum, 26" × 21" × 3". Photo: Tom Van Endye.

Below Bronze Relief of Artemis, 225 to 200 BCE. Greece, bronze. Photo: © Zde / Archaeological Museum of Delos / Wikimedia Commons / CC by 3.0.

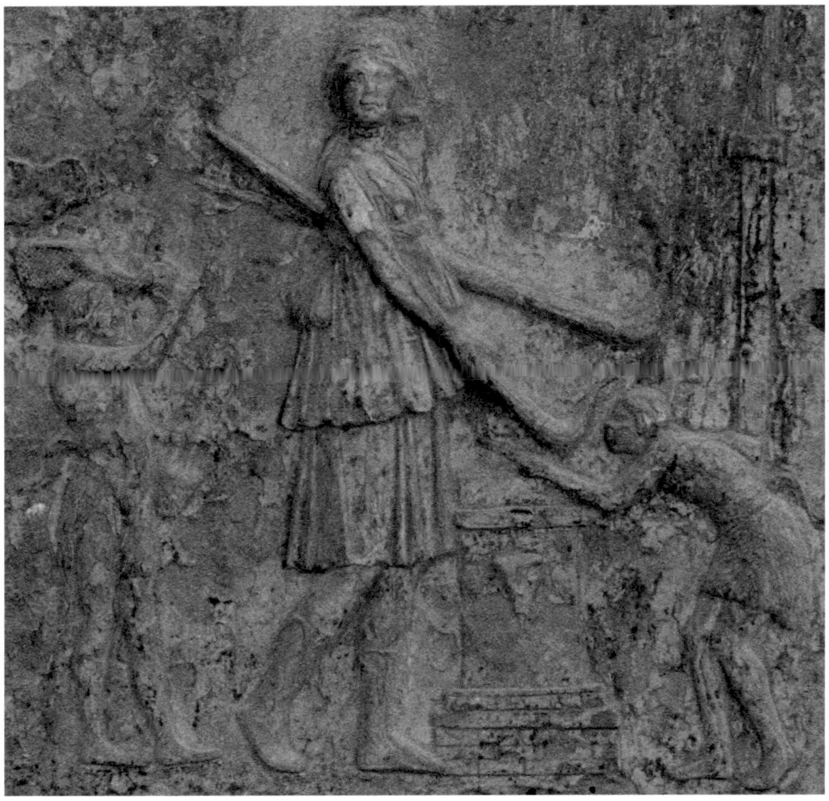

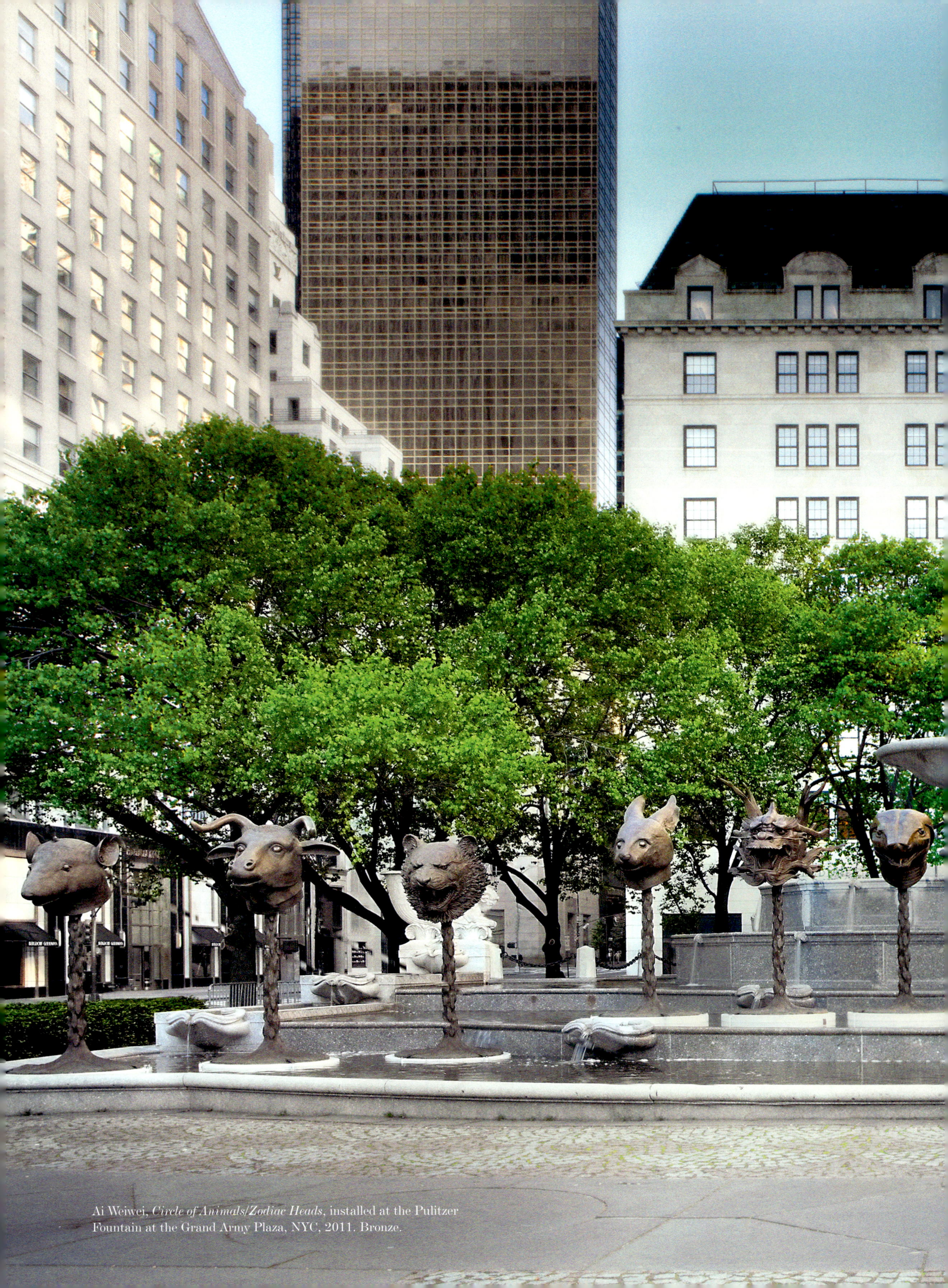

Ai Weiwei, *Circle of Animals/Zodiac Heads*, installed at the Pulitzer Fountain at the Grand Army Plaza, NYC, 2011. Bronze.

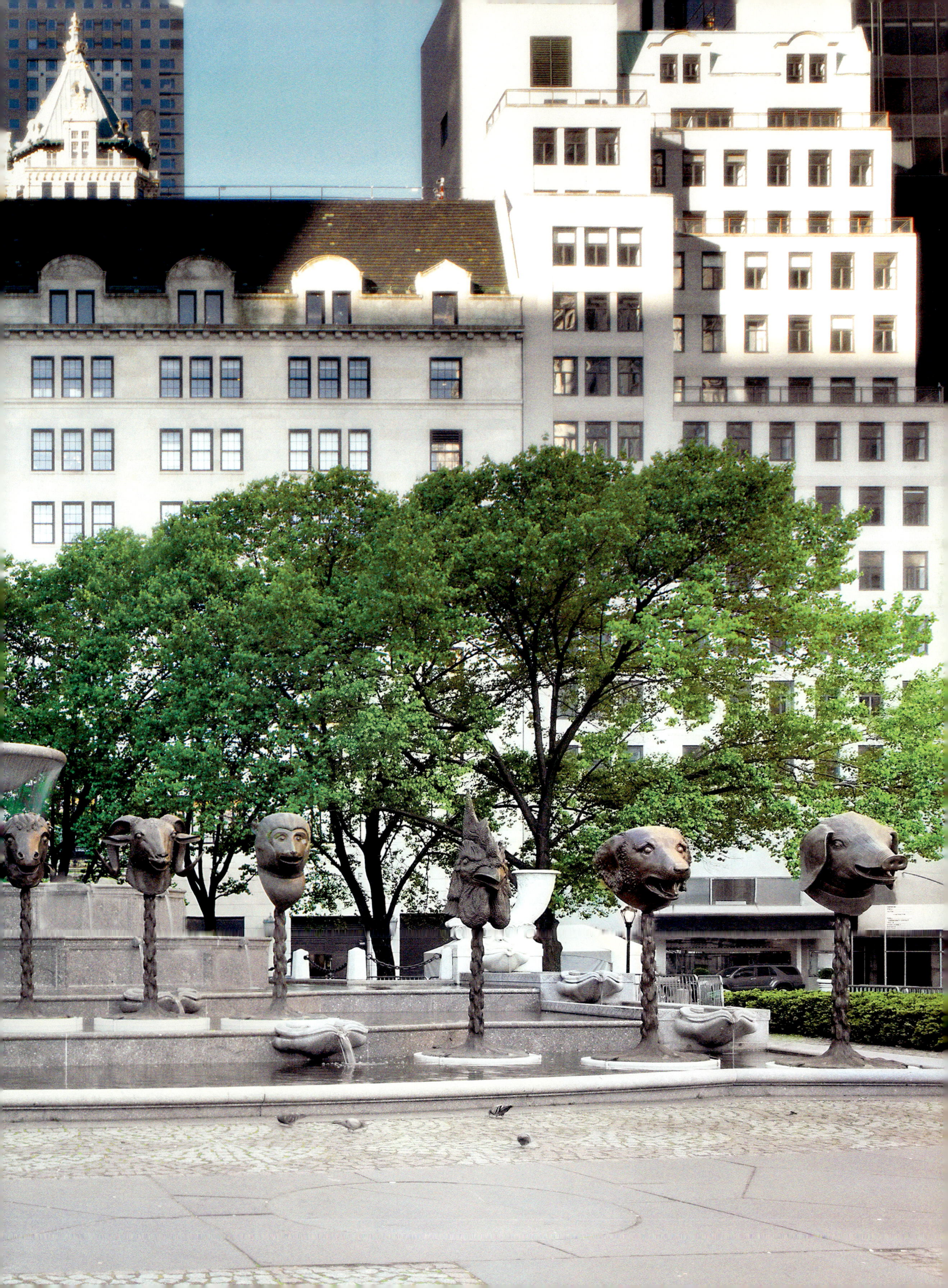

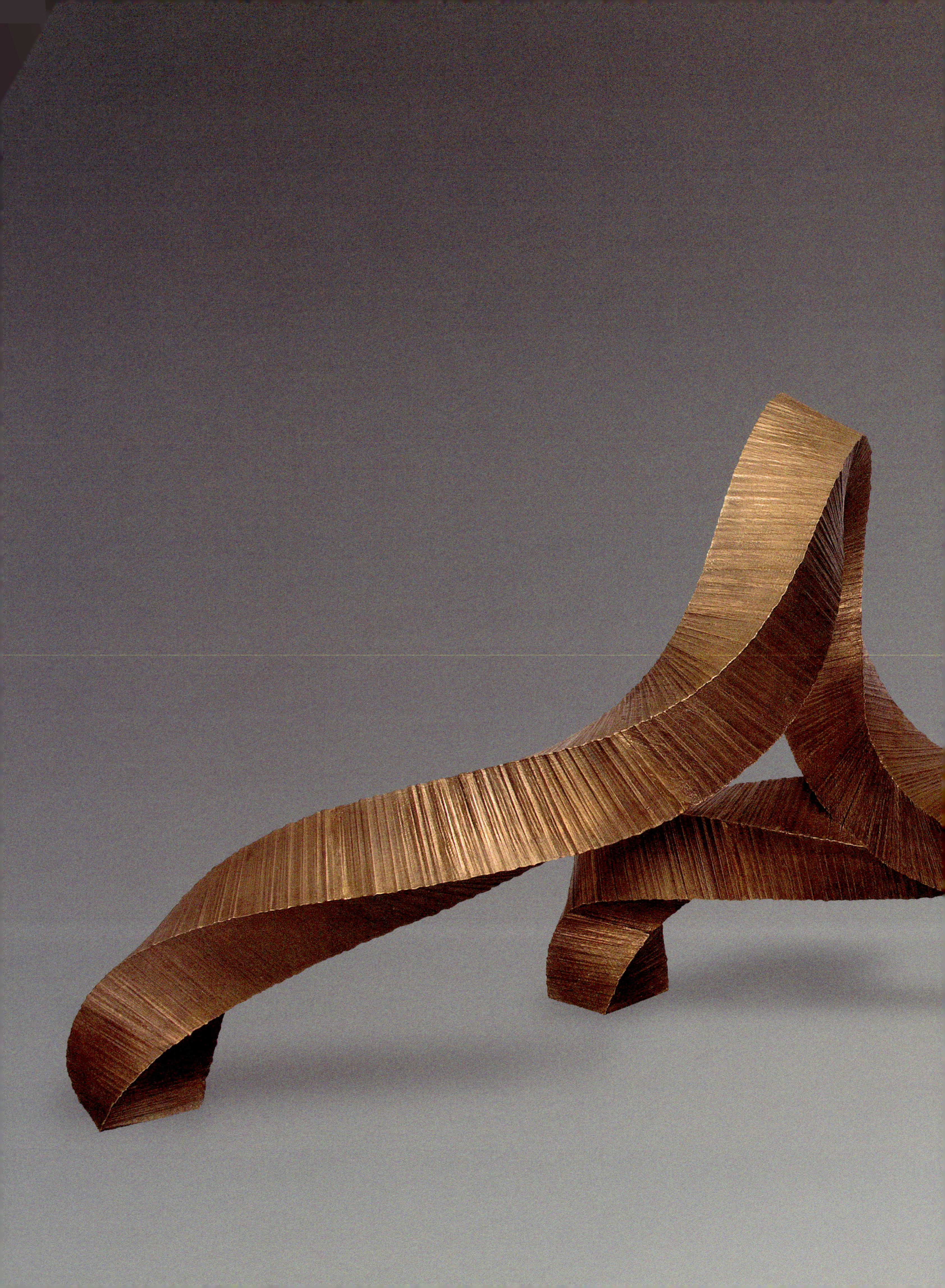

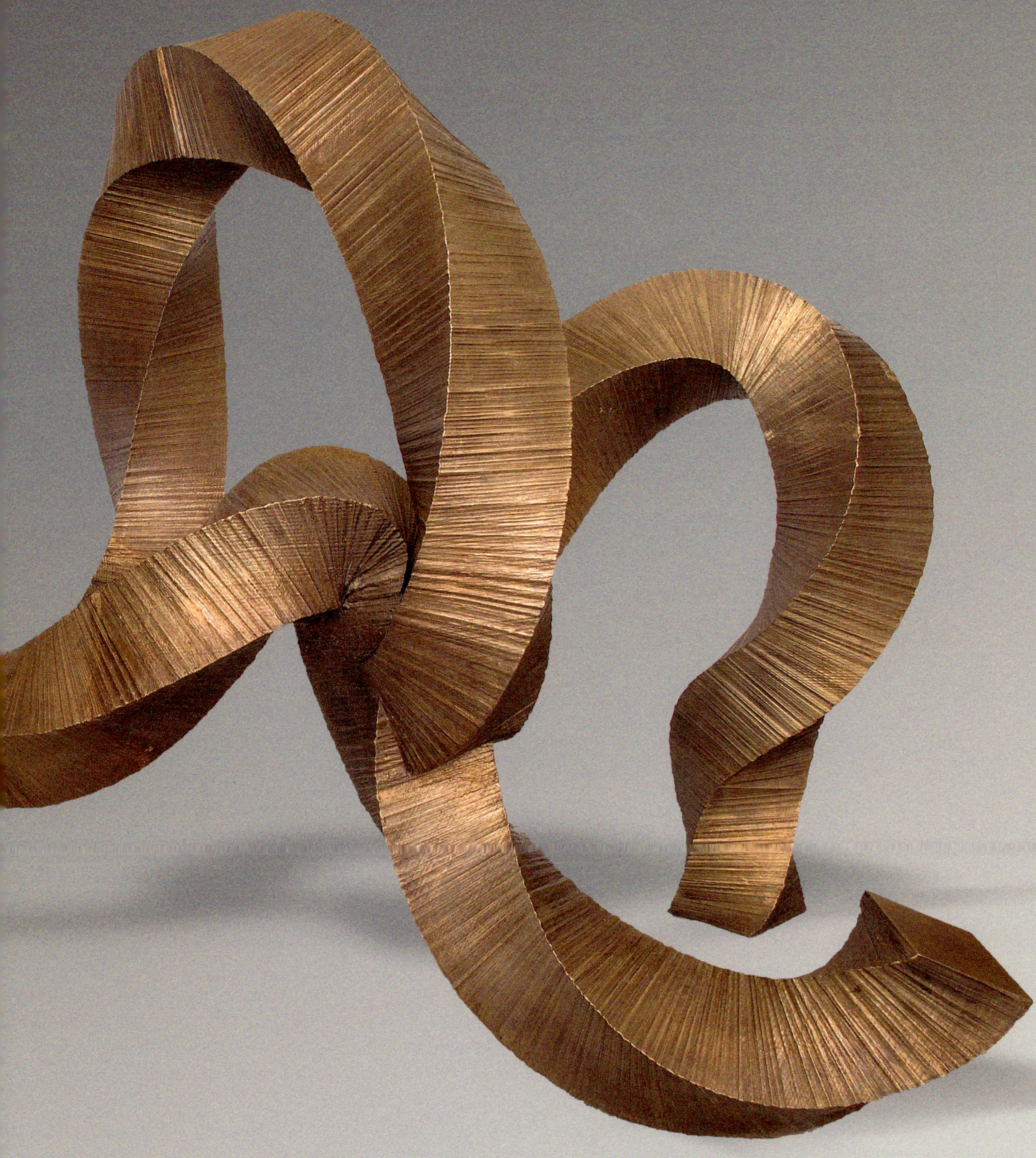

Bruce Beasley, *Torqueri V*, 2015. Cast bronze, 39" × 78" × 31".
Photo: Bruce Beasley.

CERAMICS

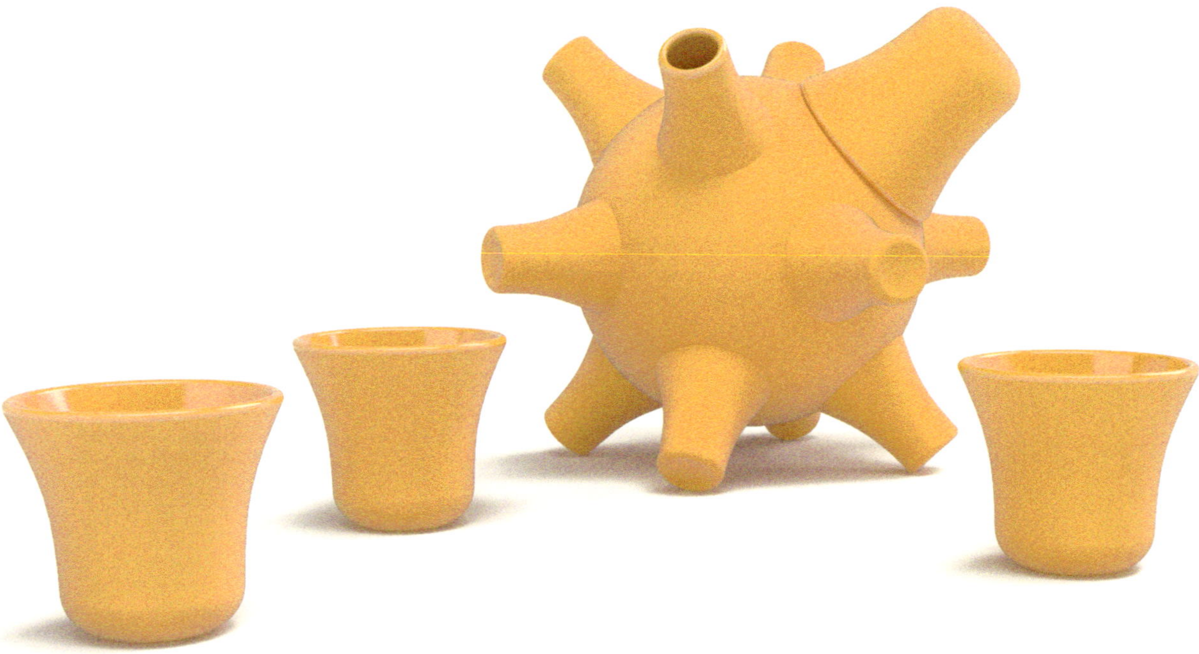

Alexander Purcell Rodrigues, *Sake Bomb*, 2009. Ceramic (earthenware body), mold designed in 3-D CAD software and printed using a 3-D FDM printer, 6" × 6" × 6".
Courtesy of © 2011 Alexander Purcell Rodrigues Design Studio, © Alex Purcell / Wikimedia Commons / CC by 3.0.

Opposite page CJ Jilek, *Flounce*, 2010. Porcelain, underglaze, monofilament, flocking, 10" × 8" × 8". Photo: CJ Jilek.

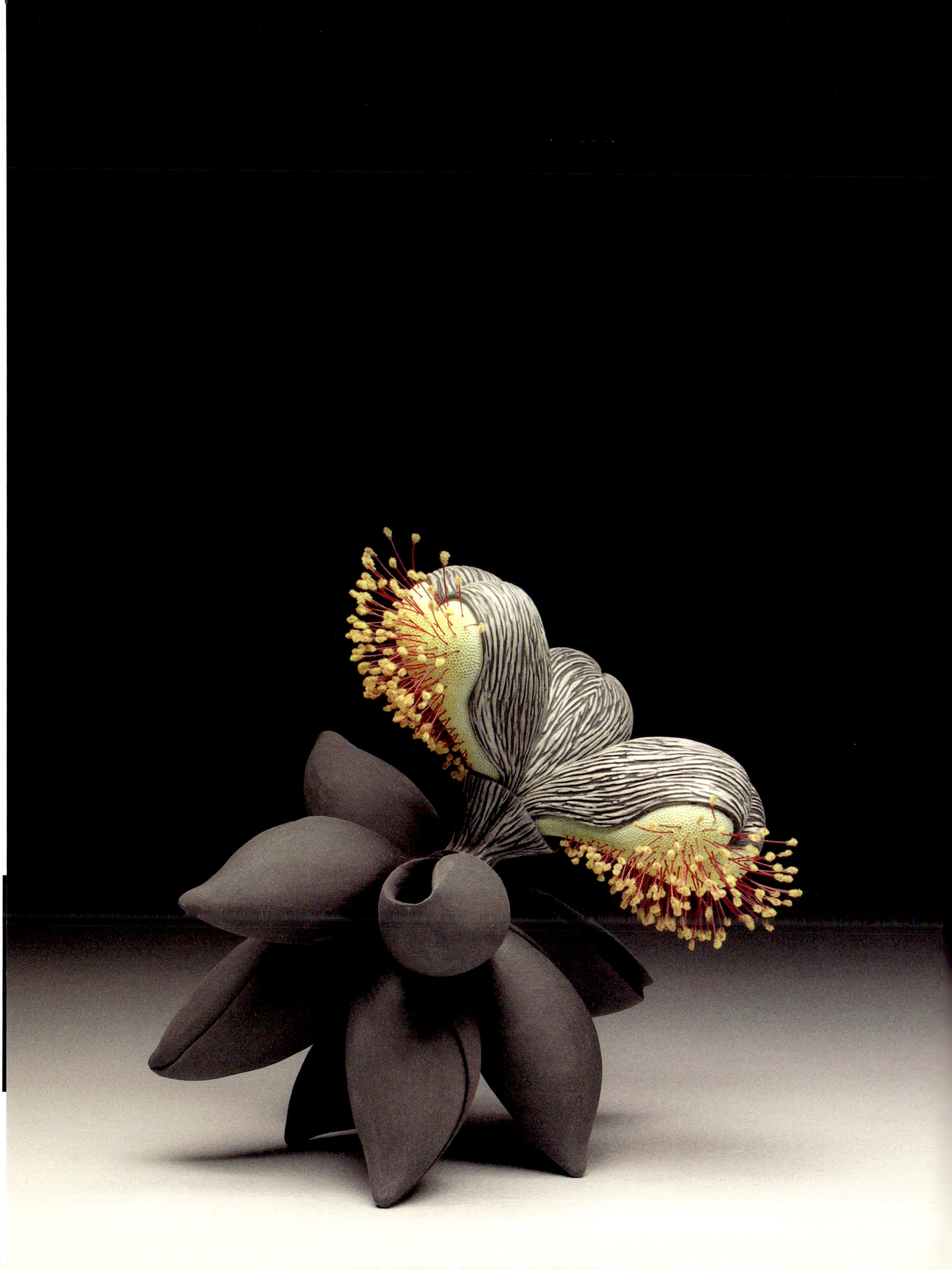

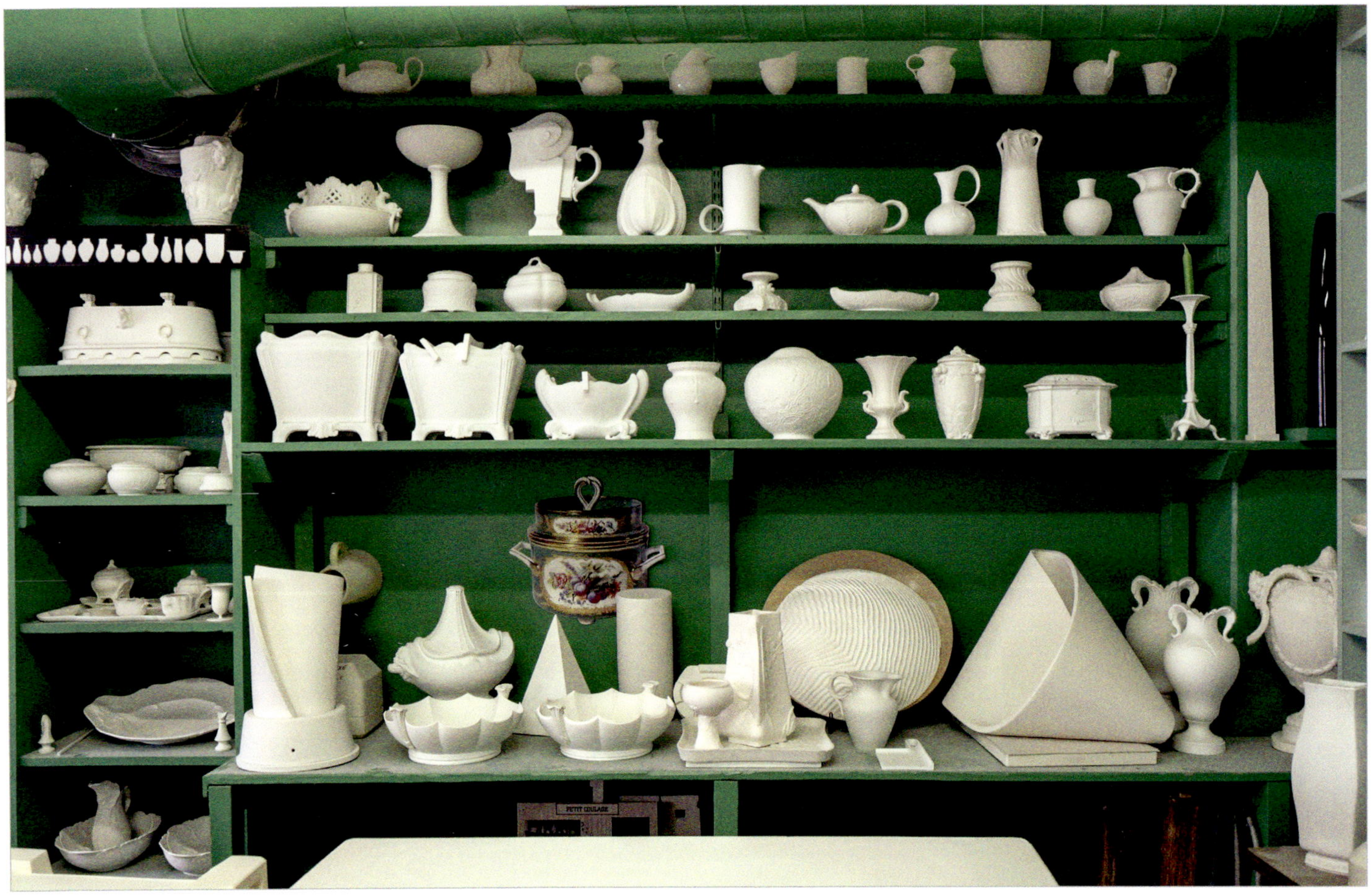

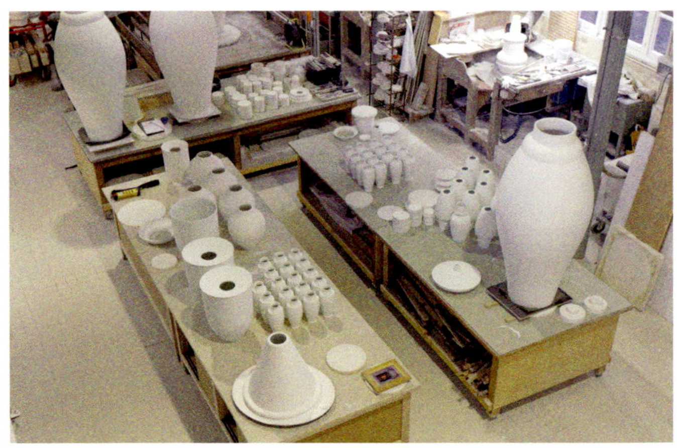

This painstaking approach to creating perfect models (followed by perfect molds) is an absolute necessity for ceramicists whether in a small studio or in an industrial setting. Complex pieces with handles, spouts, and lids are often cast as separate parts for ease of production. Each element has its own mold, is individually cast, and the parts are attached before firing. Unlike metal, ceramic cannot be altered, manipulated, or assembled after firing, so parts must be joined and all finishing must be done beforehand.

The vast majority of ceramic objects in the world are mass-produced. Ceramic factories (often called "potteries") produce far more than just dinnerware, though; vases, pots, urns, and statuary of all kinds are commonly made on a mass scale. Casting objects at any size allows for lids that fit perfectly, dishes that stack flawlessly, and assures that the perfect slope of a plate or the curve of a spout is meticulously reproduced. No one wants sauces sliding off the edge of their plate or a teapot that drips. While a poorly designed, drippy gravy boat is annoying, the stakes get a lot higher with products like toilets. Toilets, sinks, bricks, tiles, and ceramic insulators are some of the most ubiquitous cast ceramic objects, each of which has very real consequences if designed or executed imperfectly.

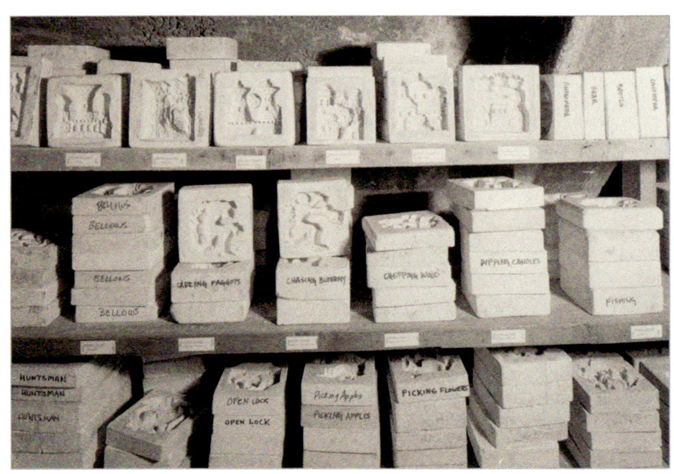

As mentioned in earlier chapters, modern bathrooms are defined by cast ceramic and the importance of ceramic sinks and toilets is paramount. Hygiene and sanitation have been drastically improved by these affordable fixtures and the indoor plumbing that makes them function.

Tiles are another defining part of modern bathrooms. The impermeable nature of tile means that bathrooms don't develop mold and can be thoroughly cleaned. Most tiles are produced by press molding. They can be made by hand as is shown in this image from the Moravian Pottery Works, or, when industrially produced, they can be made by a process known as "dust pressing." Dust pressing involves drying a ceramic slip until it is a very fine powder that is hydraulically pressed into a mold using enormous pressure, and is then fired. Dust pressing produces strong and perfectly uniform tiles by the hundreds of thousands.

OPPOSITE PAGE, top to bottom

Sèvres production samples (small), 2013. France, slipcast porcelain.
Photo: © Myrabella / Wikimedia Commons / CC by 4.0.

Sèvres large-scale slipcasting workshop at Sèvres, 2013. France, slipcast porcelain.
Photo: © Coyau / Wikimedia Commons / CC by 3.0.

Moravian Pottery and Tile Works, production molds for tiles, 1987. Plaster.

Below Black and white bathroom, 2011.
Photo: © Gürkan Sengün / Wikimedia Commons / CC by 3.0. Photo: Jet Lowe / Wikimedia Commons / public domain.

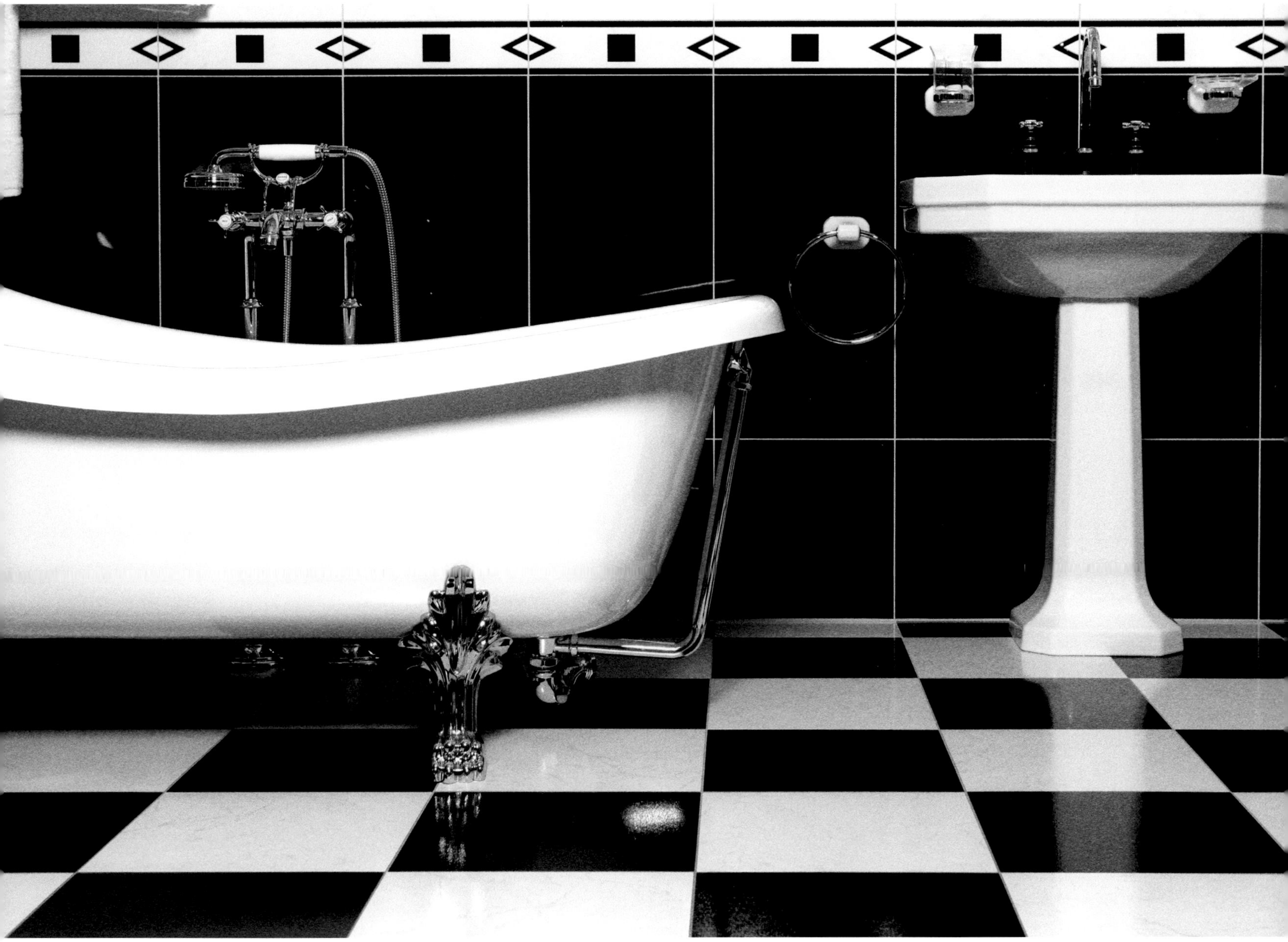

Bricks, which are also produced in mind-boggling quantities, are often press-molded. Many inexpensive bricks today are made using extrusion, but fine bricks are still made using molds and are cast and fired individually. Brick and tile have shaped our cities, our homes, and our lives for thousands of years. The durability of brick means that buildings can last for centuries. Its distinctive color defines cities all over the world.

Bricks were first made approximately five thousand years ago in modern day Iraq, but they were sunbaked, and weren't suitable for exterior use. The Babylonians developed fired bricks that could be used for building construction, the finest example being the Ishtar Gate. Built by Nebuchadnezzar II in 575 BCE, the Ishtar Gate was thirty-eight feet tall and built entirely from cast brick and tile. Bas relief tile animals grace a plane of cobalt-blue glazed tiles that is breathtaking even today. The Ishtar Gate was the main entrance to the ancient city of Babylon, in modern day Iraq. Through the gate was a brick corridor known as "The Processional Way," a half-mile-long passageway of fifty-foot-tall brick walls adorned in more tile and tile animals. The Ishtar Gate was excavated in the early twentieth century and pieces of it went to many different museums, but the main gate is installed at the Pergamon Museum in Berlin.

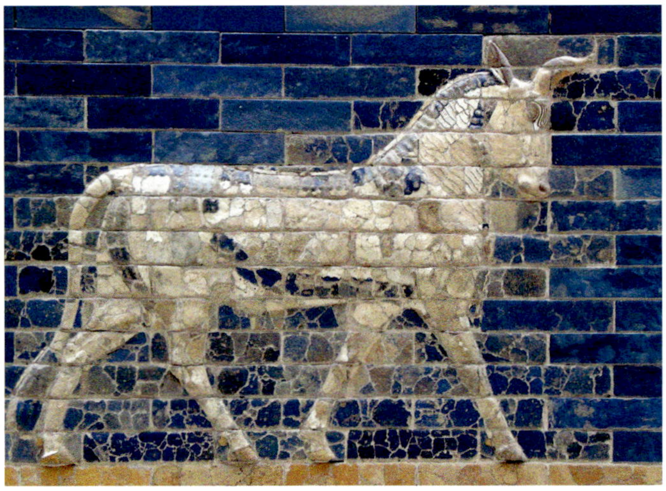

Top Detail: Bas relief bull from the Ishtar Gate.
Photo: © Josep Renalias, Pergamon Museum, Berlin / Wikimedia Commons / CC by 3.0.

Bottom The Ishtar Gate, 575 BCE. Babylon, glazed brick, 46' × 100'.
Photo: © Radomir Vrbovsky at the Pergamon Museum, Berlin / Wikimedia Commons / CC by 4.0.

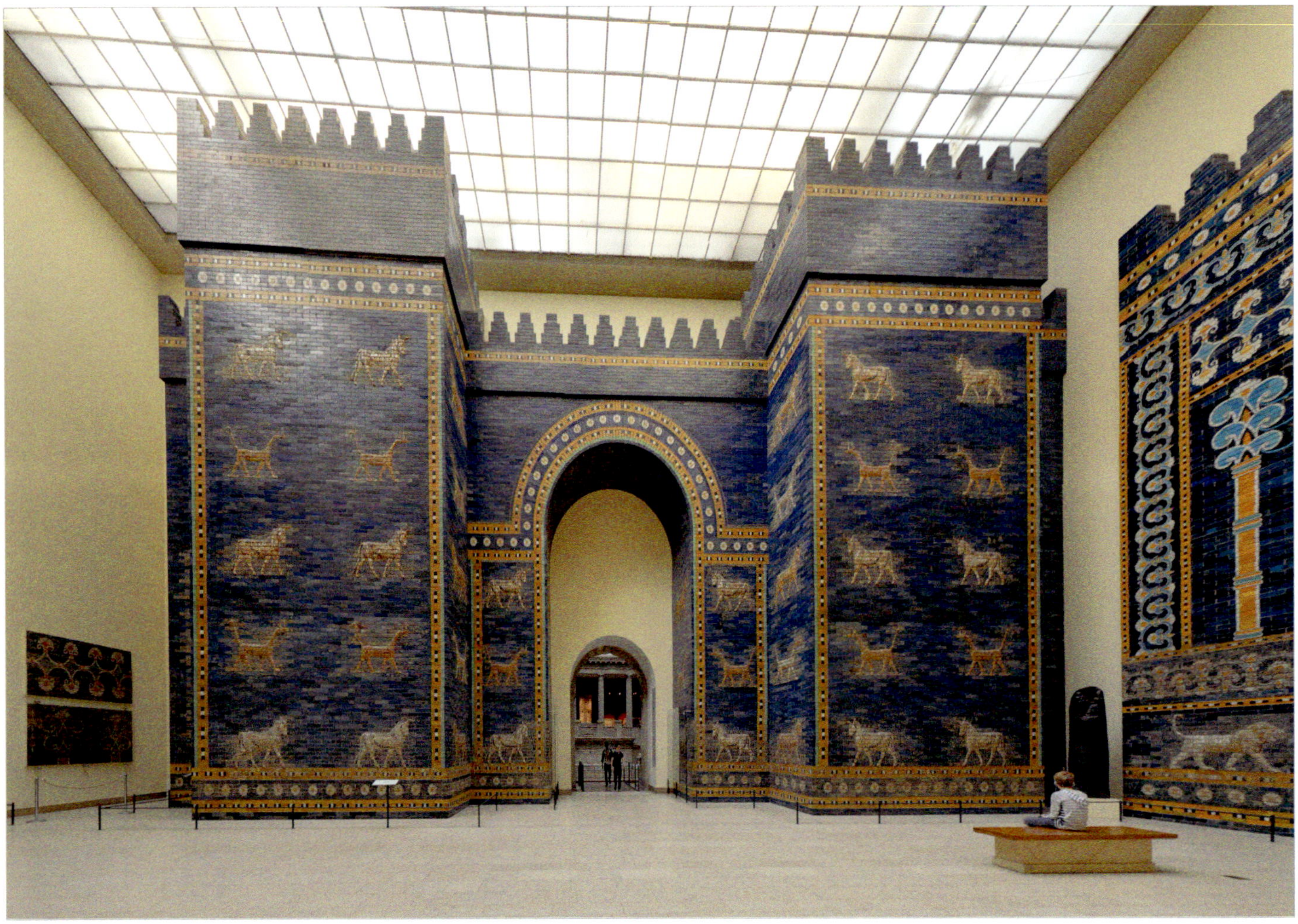

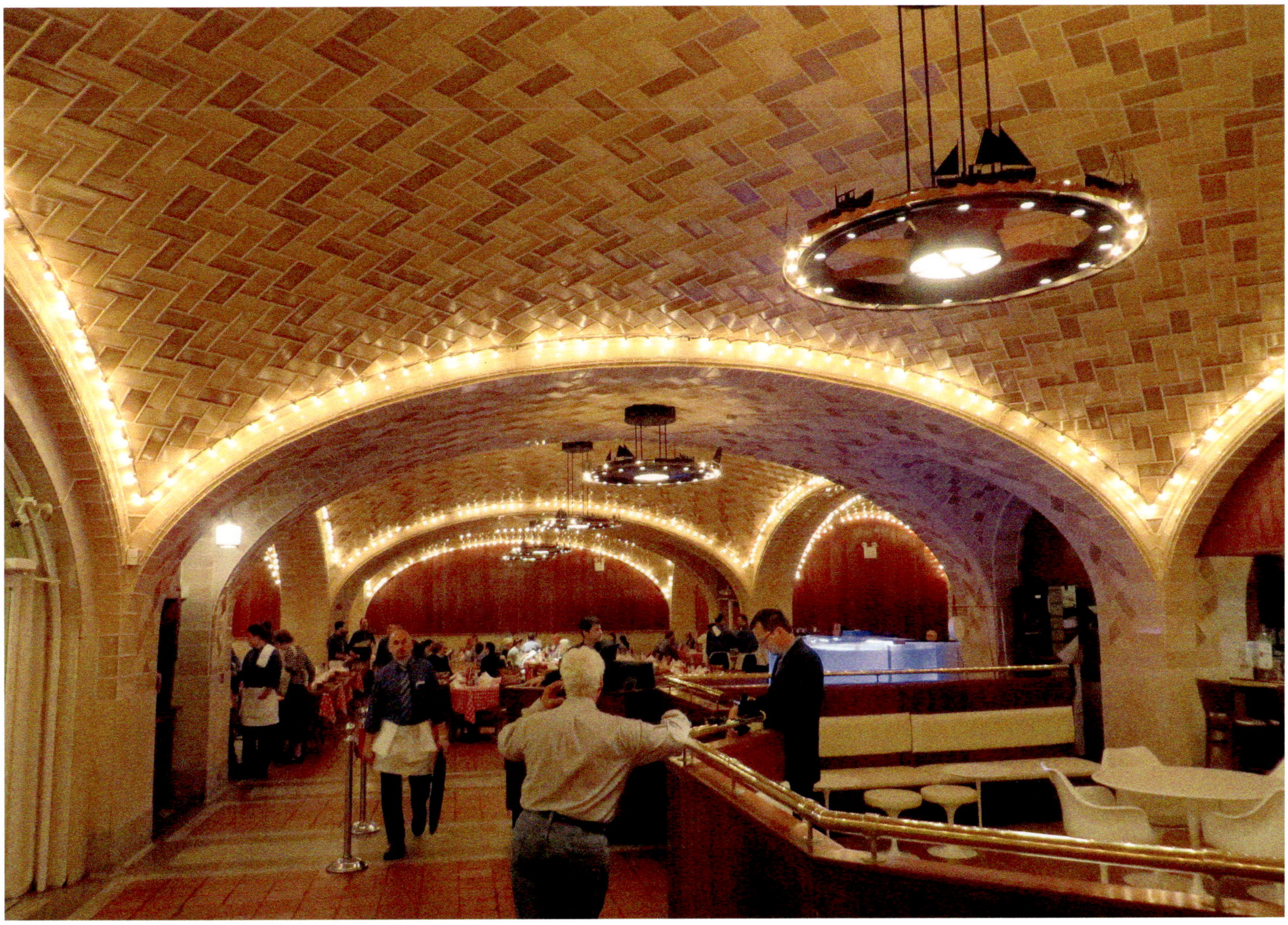

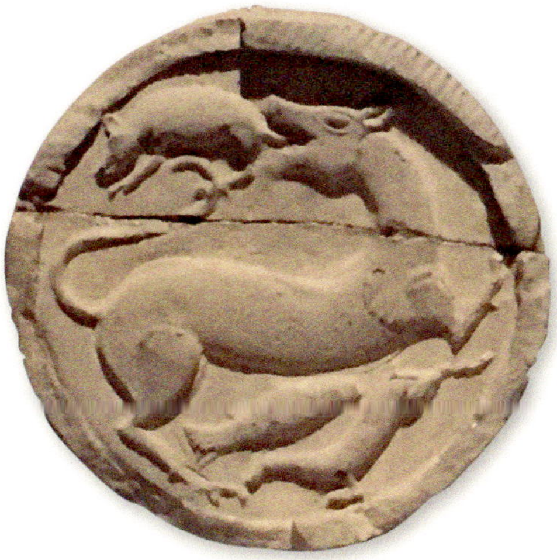

Today, there are many spaces that use tile in eye-catching ways. The Grand Central Oyster Bar is a subterranean gem hidden in the lower level of Grand Central Station in New York City. This fabulous vaulted space is defined by its (now) iconic tile ceiling

Tile work can be sumptuous and exquisite or it can be so understated that it's barely noticed. Tile fills our kitchens, bathrooms, and covers buildings both inside and out. Tile roofing dates back to 10,000 BCE and was independently developed in both China, and later, the Middle East.[1] This exquisite example of a Chinese roof tile mold would have been used to make ornamental end tiles to decorate the eaves of buildings with symbols of luck and protection.

Top Raphael Guastavino (designer), *The Grand Central Oyster Bar*, 1913. Guastavino tile (developed for vaulted spaces).
Photo: © Leonard J. DeFrancisci / Wikimedia Commons / CC by 3.0.

Bottom Roof tile end mold, 770–475 BCE. China, pottery.
Photo: Ed Vates. Collection of Shaanxi Provincial Institute of Archaeology on loan to the Field Museum, Chicago.

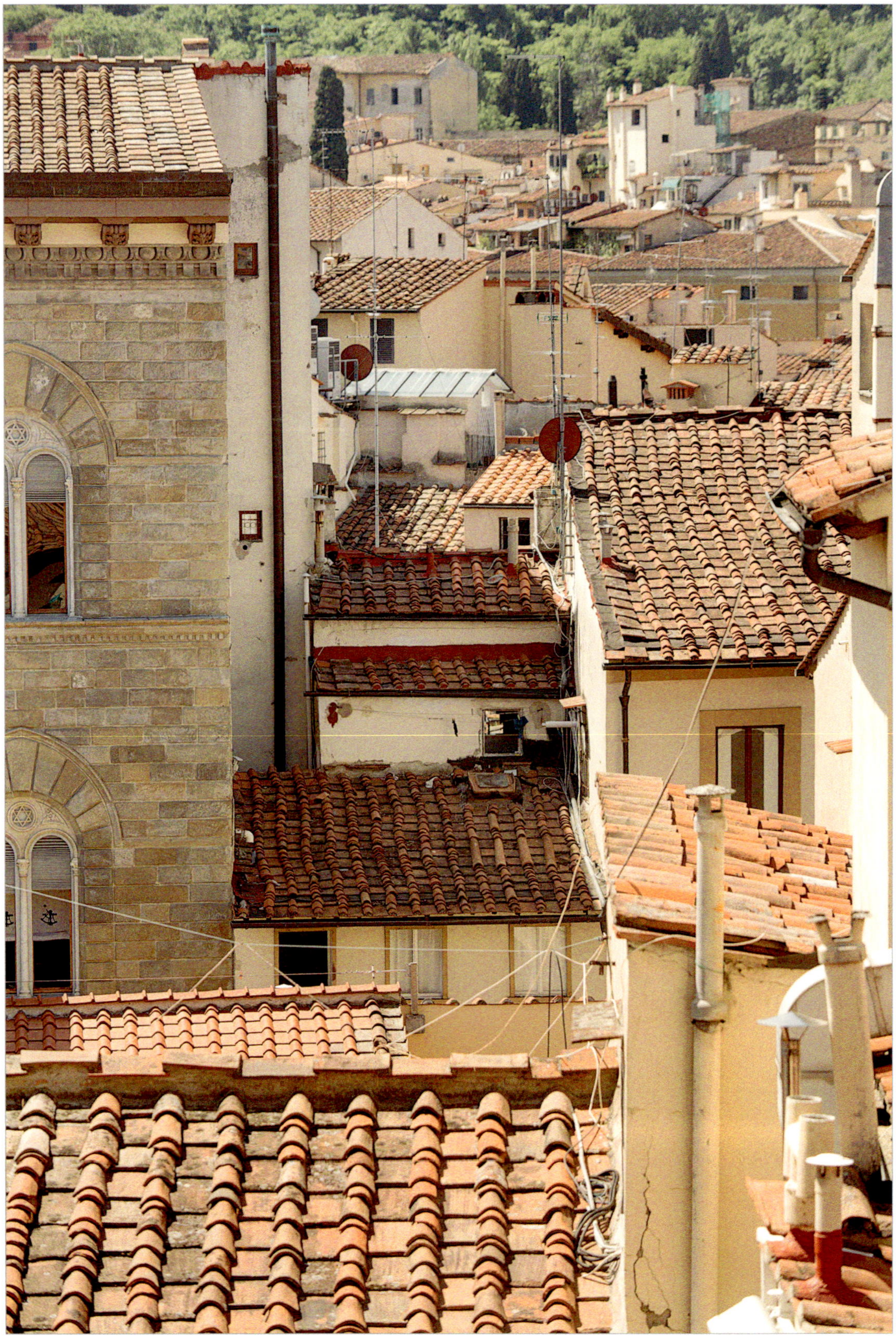

Tile rooftops. Florence, Italy, terracotta.
Photo: © Steve Lovegrove, Depositphotos.

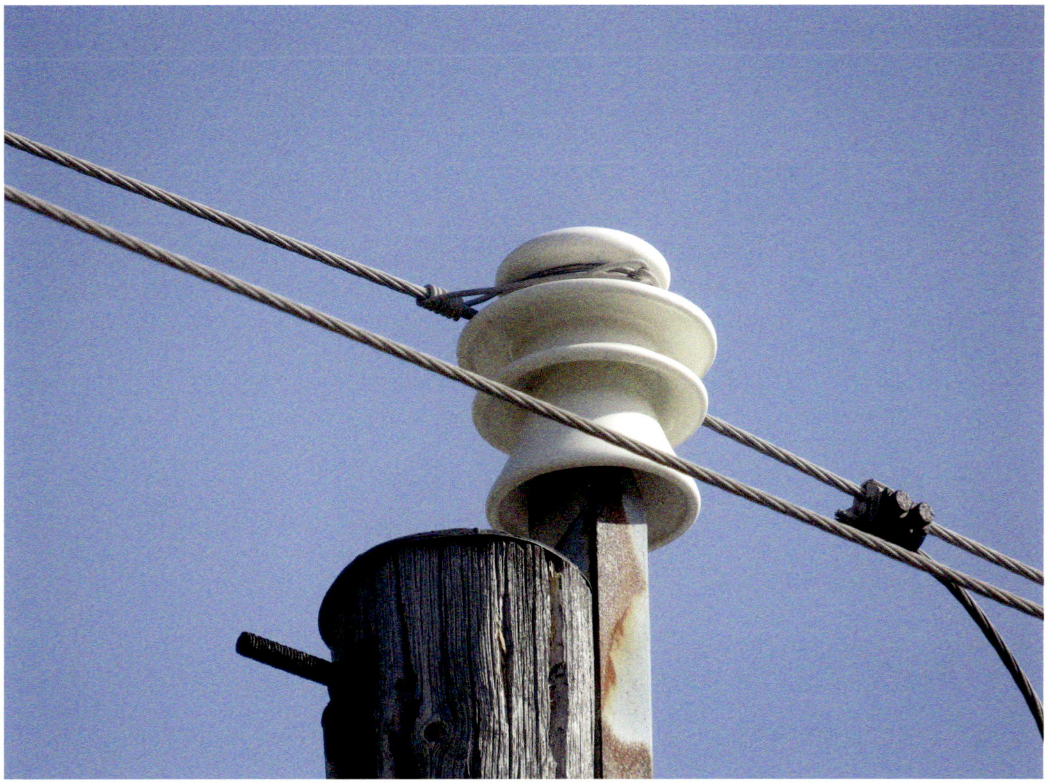

A 20 kV pin insulator, 2013. Porcelain.
Photo: © Vesa Linja-aho / Wikimedia Commons / CC by 2.0.

Terracotta roof tiles are one of the most distinctive visual elements in some regions. What would Italy be without its endless sea of terracotta rooftops? Press-molded since their inception, roof tiles are long-lasting, impermeable, fireproof, and beautiful. There are many variations in shapes and colors, but these distinctive tiles have remained virtually unchanged for 12,000 years. Like bricks, some roof tiles are now extruded, but that typically leads to a lower quality product.

Industrial ceramic production is an ever-expanding sector of manufacturing. One cast ceramic product that we rarely see or think about, but that benefits our lives immeasurably, is the insulator. Ceramic insulators are used to prevent electrical surges and arcs. There are many different materials used as insulators, almost all of which are cast. Porcelain is a favorite though, because it is strong, durable, can handle high temperatures, and doesn't collect condensation the way glass does. Porcelain insulators are used in electrical towers, on telephone poles, electric fences, and on electrified train tracks in subways. On a micro scale, cast ceramic is also used in electronics.

Cast ceramics define our lives in so many ways. From intimate activities like drinking coffee from a favorite mug or bathing in a tub, to the fundamental mechanics of safely turning on a light in a dry home, it's hard to imagine a world without cast ceramics.

SITE
The Factory

Cast ceramics are produced on a prodigious industrial scale all over the world. One of the most interesting facilities in America today is the Kohler Company in Sheboygan, Wisconsin. It is an undeniable giant in the production of high quality cast sinks, toilets, and bathtubs. Kohler contributes far more than just kitchen and bathroom fixtures though; the company hosts a highly-competitive residency for artists as well. Founded in 1974, Arts/Industry is a residency program of the nearby John Michael Kohler Arts Center, a local nonprofit arts organization dedicated to the mission of generating a creative exchange between artists and the public. If awarded one of these sought-after residencies, an artist goes to Kohler for a period of two to six months to use the company's facilities and work in their studios alongside highly skilled casters working for Kohler.

This residency benefits both the artists and Kohler employees. Artists get a chance to work in exceptionally high-tech facilities. Kohler Company's Pottery as well as its Iron and Brass Foundries and Enamel Shop provide artists with the opportunity to do large-scale casting as well as other industrial processes. They also get to work around casters who are masters of technique. Employees get a chance to see creative thinkers at work. It is a symbiotic relationship and speaks to Kohler's commitment to the arts. Above all else though, Kohler is a successful business and one of the best-known companies supplying cast kitchen and bathroom fixtures for a seemingly insatiable market.

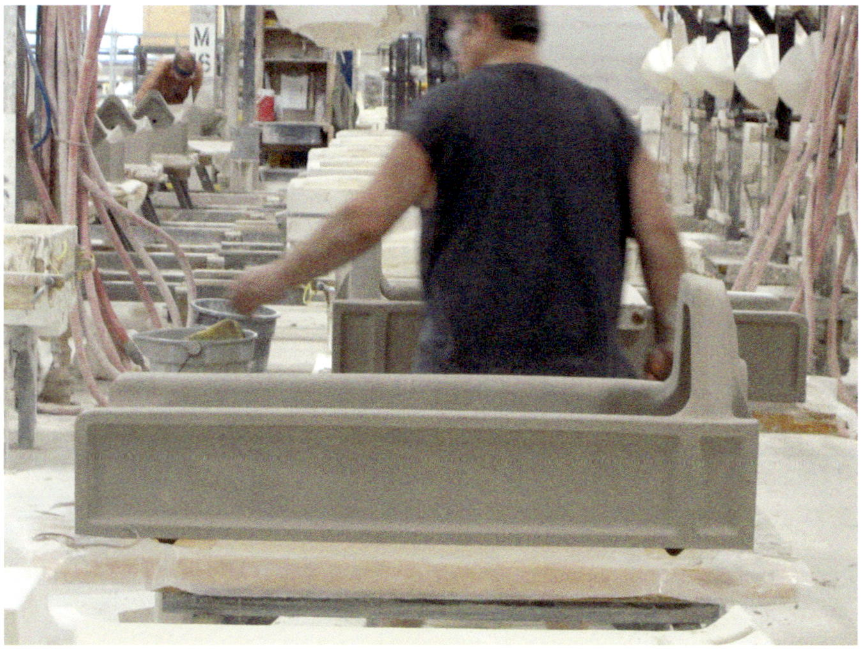

Left Kohler Co. associates in their casting facility. Sheboygan, Wisconsin. Photo: Ezra Shales.

Opposite page Yoko Sekino-Bové is known for her exquisite work in ceramics as well as her elegant and distinctive imagery. She flourished during her residency as she embraced the opportunity to work at this scale. Yoko Sekino-Bové, *Flowers in Mirror, Moon on Water*, 2014. High-fire stoneware and enamel paint, 6'2" × 1'9" × 1'9".

Created in Arts/Industry, a long-term residency program of the John Michael Kohler Arts Center in Sheboygan, Wisconsin. Photo: Rich Maciejewski.

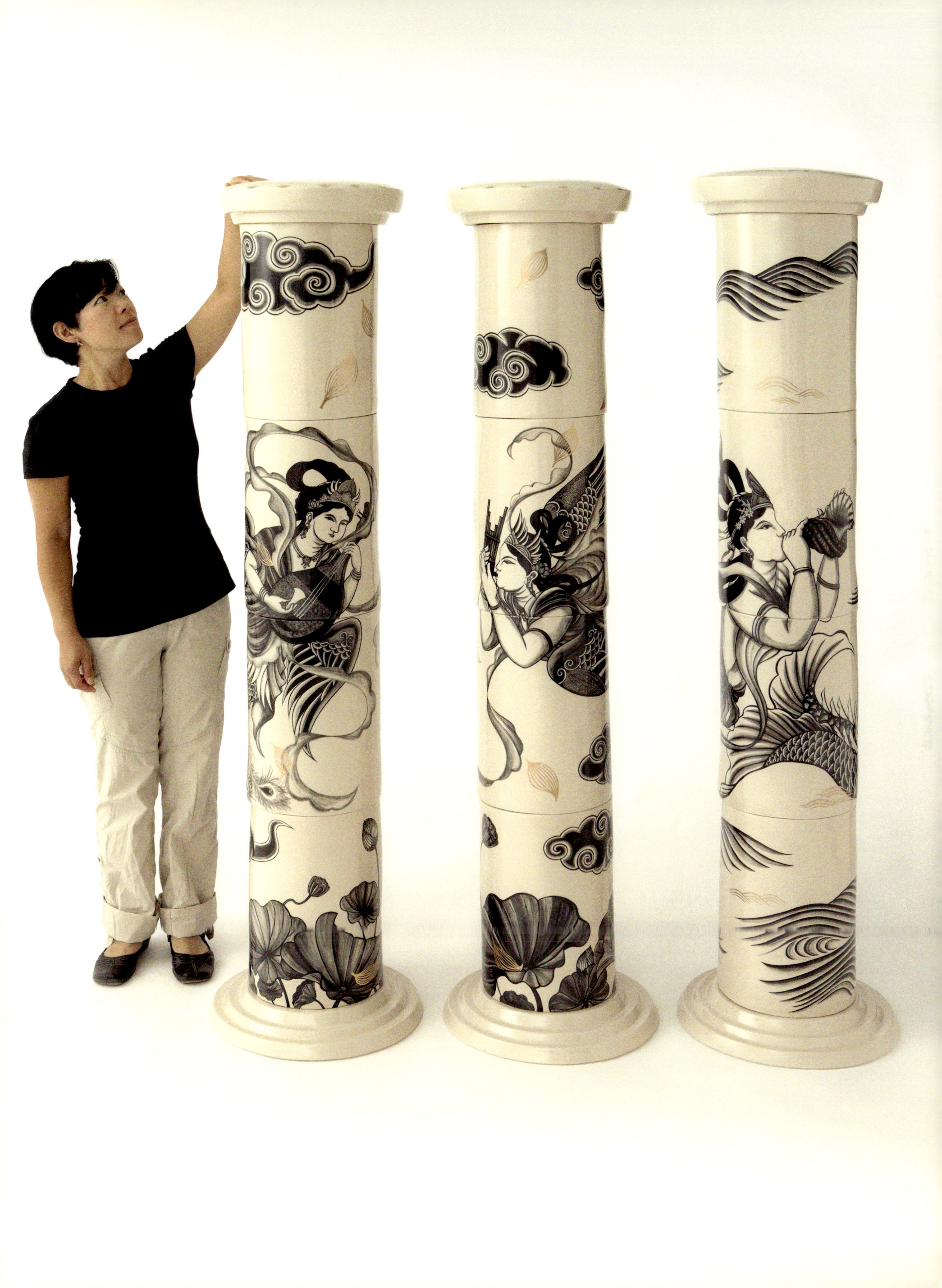

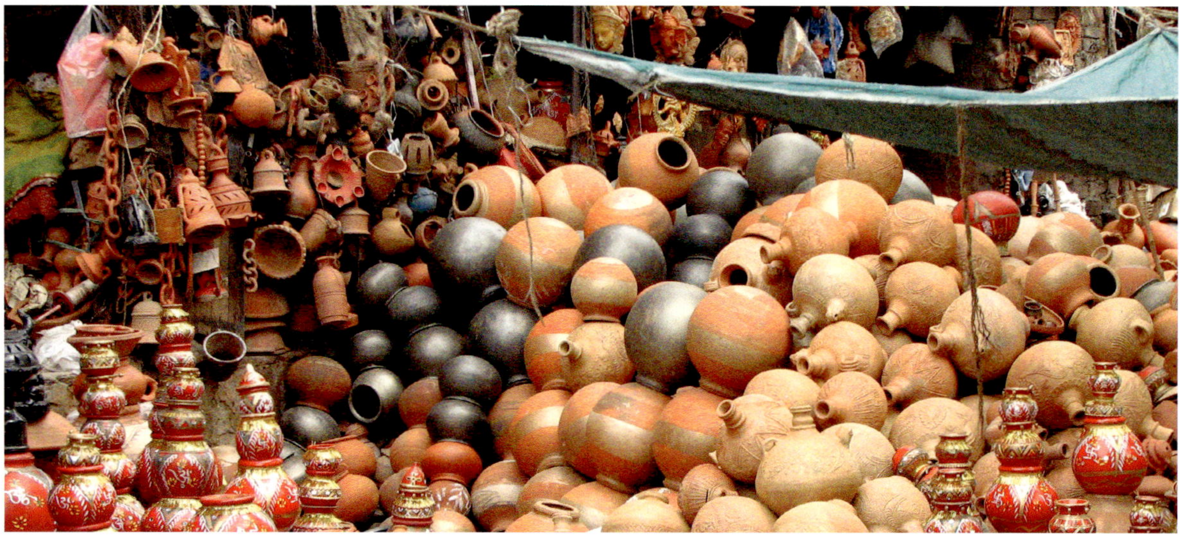

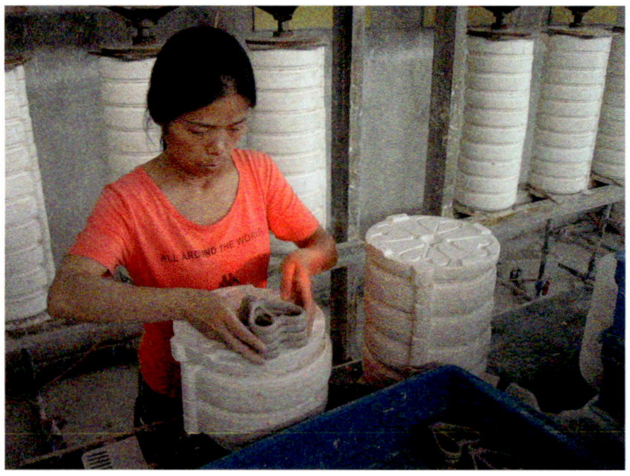

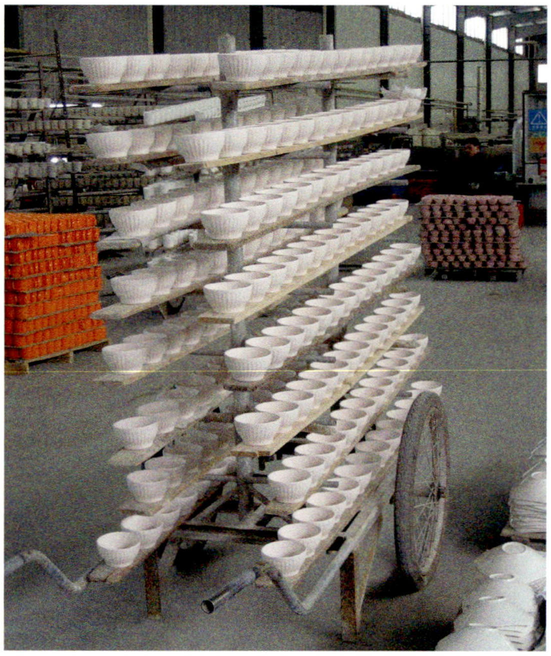

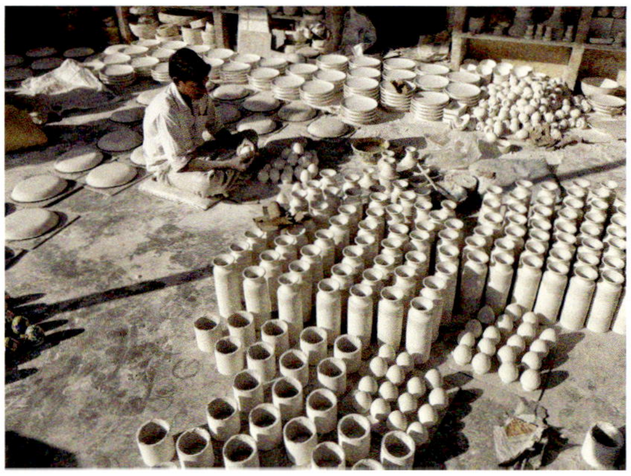

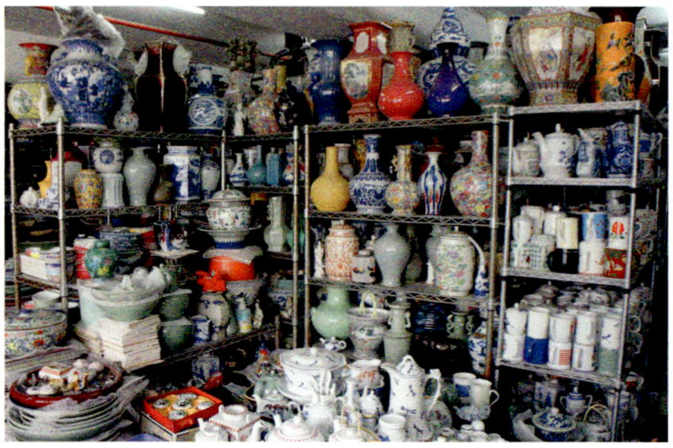

Top Pots for sale in India.
Photo: © McKay Savage / Wikimedia Commons / CC by 2.0.

Center right Transporting greenware by cart at JiaJia Factory, 2011. China.
Photo: Paul Leathers.

Top to bottom, left

Pressure casting (mug handles) at JiaJia Factory, 2011. China.
Photo: Paul Leathers.

Yuet Tung China Works. Kowloon, Hong Kong.
Photo: © Mk2010 / Wikimedia Commons / CC by 3.0.

Pots for sale in Jaipur, India, 2007. Ceramic.
Photo: © Angelo Giampiccolo / Shutterstock.com.

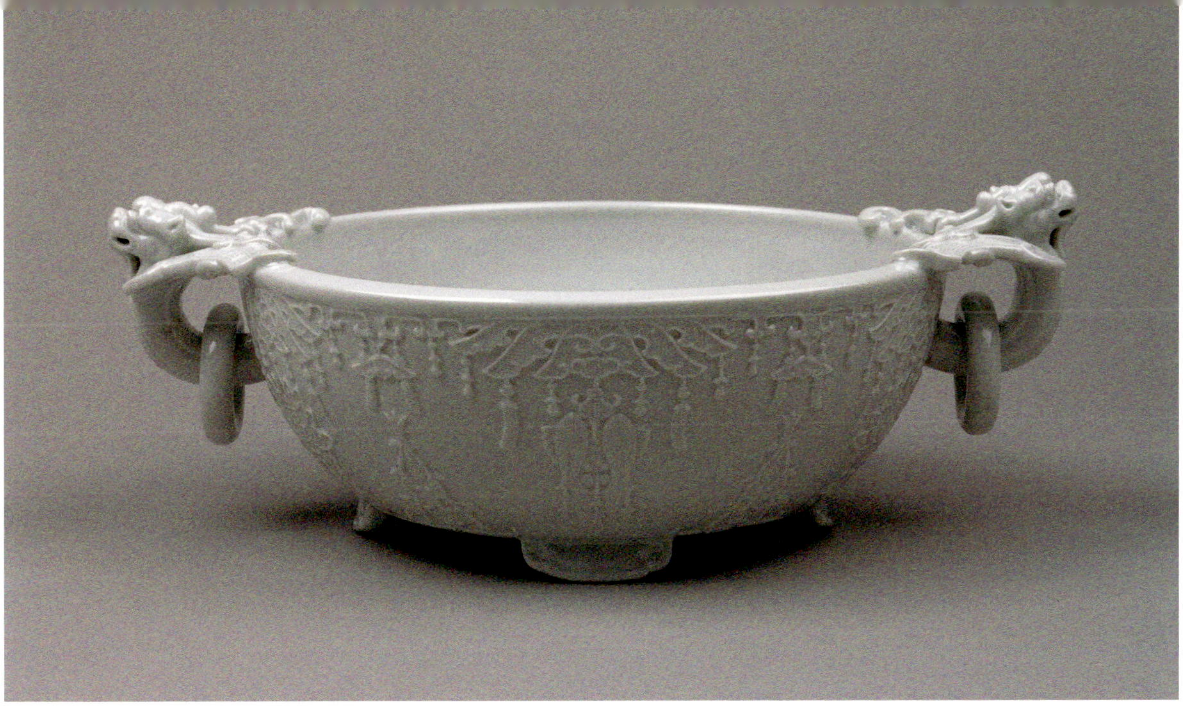

American factories are young relative to those in Asia. Each country has its own unique traditions stretching through its history, and produces distinctive ceramics.

Indian ceramic traditions were influenced by its central location. Egyptian, Turkish, Persian, and Chinese traditions melded over the centuries to become what we now know as Indian ceramics. Today, the ceramics factories in Jaipur produce a range of wares in astonishing quantities.

Chinese cities such as Jingdezhen, Guangdong, and Xiamen, have a rich history of ceramic work. Jingdezhen, which recently celebrated one thousand years as the porcelain capital of the world, produces some of the country's finest ceramics. Jingdezhen was the imperial center for ceramic production. When it was deaccessioned, it eventually transitioned through Maoist control and into today's city, notable for its massive compound of ceramic studios.

The Chinese ceramics system revolves around each individual maker being an expert in one area—one person is a model-maker, one is a moldmaker, one is a caster, etc.... all the way through the process to the kiln master. This system lends itself to mass production as well as aptitude in exacting reproduction of individual objects.

The Chinese have been both admired and maligned for hundreds of years for their almost preternatural ability to reproduce works of both Asian and Western origins. In many cases, these reproductions are cast. Reproduction has been a part of the Chinese culture of making for centuries. If a bronze funerary vessel was too expensive, a ceramic replica could be made for significantly less. This was a matter of practicality and without shame. High quality reproductions serve a purpose and are often made so skillfully that they can be easily mistaken for (or passed off as) original antiques. Even pieces clearly advertised as reproductions can fetch huge sums of money. Individual high-end reproductions plus the mass production of everyday wares makes Jingdezhen a major economic engine in the region.

Top Ornamented bowl, 1820 to 1850. China, porcelain with white glaze, 10.75" tall.
Courtesy of The Walters Art Museum, Acquired by William T. or Henry Walters, CC0, GNU FDL.

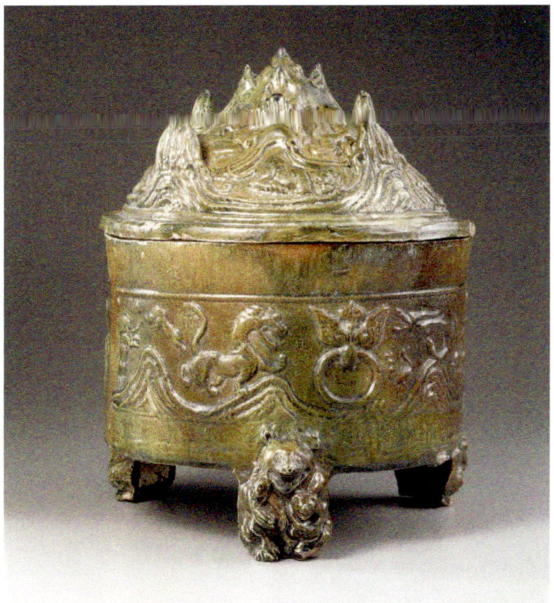

Right Hill Jar and Cover, 206 BCE to 220 CE. China, earthenware with lead glaze, 11.5" × 8.75".
Courtesy of the Brooklyn Museum, Purchase gift of Dr. Bertram H. Schaffner, Mr. and Mrs. Milton F. Rosenthal, Mr. and Mrs. Greg Fitz-Gerald, and Mr. and Mrs. Robert L. Poster, Helen Babbott Sanders Fund, Charles Stewart Smith Memorial Fund, Caroline A.L. Pratt Fund, and Caroline H. Polhemus Fund, 1995.49a-b, Creative Commons-BY.

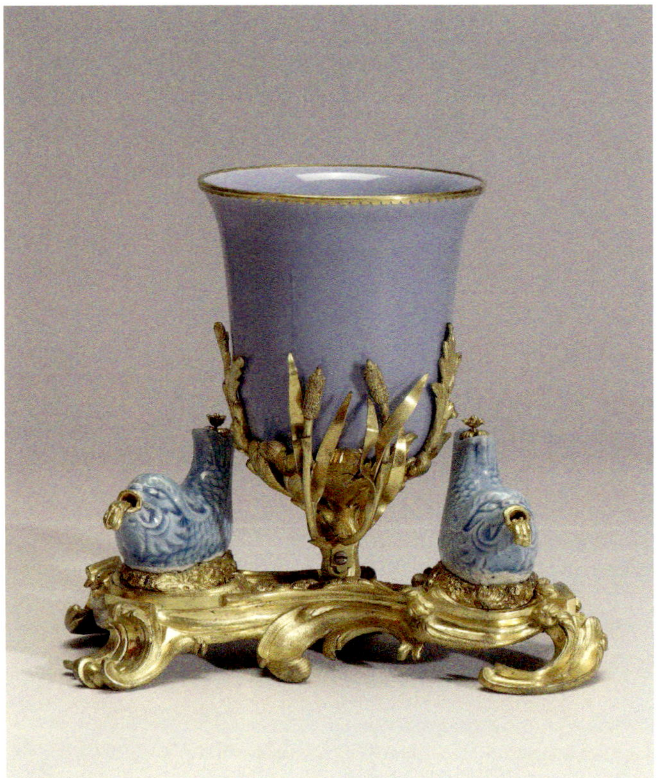

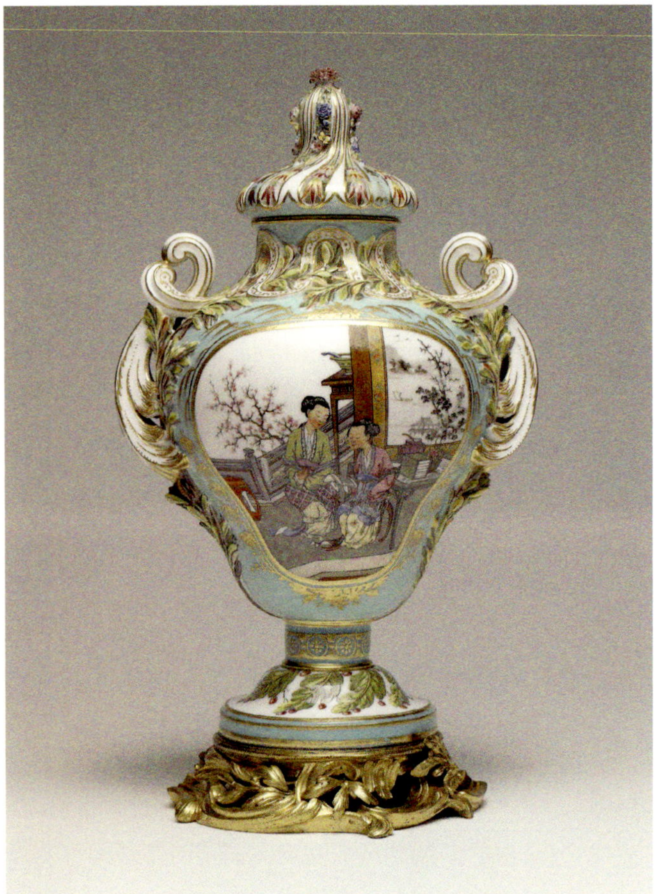

Chinese ceramics made their way into Europe via the Portuguese as early as the sixteenth century. Porcelain became highly sought after as its fine, white, translucent delicacy had no equal in the European tradition. Europeans avidly collected porcelain pieces and even created baroque metal mounts for them so they would match the fashion of the time. "Chinoiserie" became all the rage in Europe—art and objects of all varieties were covered with romanticized images of the East. Chinese makers responded to the clamoring for Chinoiserie and actually began producing works specifically for export to the West, complete with these inaccurate, nostalgic, and cliché images.

In the first half of the seventeenth century alone, three million pieces of Chinese porcelain had been hungrily snapped up by Europeans and the pressure was on for Europeans to develop their own porcelainware.[2] This remained elusive for some time, but eventually, kaolin, a naturally occurring white clay and the missing element in porcelain, was discovered in Europe. In quick succession, Meissen, Sèvres, Royal Worcester, Frankenthal, Wedgwood, Villeroy & Boch, Spode, and countless other porcelain factories popped up and began feeding the market yearning for serviceware and decorative objects of all kinds.

Recipes for clay bodies, the mixture of clays used for making objects, were closely guarded secrets, so competing companies raced to be the first to unlock the mysteries of porcelain, held in the strictest confidence by the Chinese. Because these companies were working in isolation, a wide range of recipes for new clay bodies was developed. Soft paste, hard paste, bone china, and Jasperware were all results of this highly charged period of competitive research. Remarkably, these clay bodies are still in use today. Each one has its strengths and is prized for different reasons.

Top Bowl Mounted with Two Fish, bowl: 1730–1740; fish: early eighteenth century; mounts: 1745 to 1749. China (ceramicist), France (metalsmith), porcelain with glaze, monochrome turquoise/light blue, gilding, and French ormolu mounts, 7.4" tall.
Courtesy of The Walters Art Museum, Acquired by William T. or Henry Walters before 1896, CC0, GNU FDL.

Left Sèvres Porcelain Manufactory (Manufacturer), Jean-Claude Duplessis le Père (Designer) Charles Nicolas Dodin (Decorator), One of a Pair of Potpourri Vases, 1762. France, soft-paste porcelain with enamel and gilding, overall with lid: 11.75" × 6.75" × 5.19".
Courtesy of The Walters Art Museum, Acquired by Henry Walters in 1928, CC0, GNU FDL.

One of the most significant developments in casting from this era of advancement is the process of slipcasting. The process was developed in France circa 1790 and by 1814, slipcasting was introduced at the Sèvres manufactory. The company now known as Sèvres was started at the urging of Louis XV and Madame de Pompadour in 1738. It is still one of the most prestigious factories in the world and has a history of partnering with artists from Francois Boucher to Louise Bourgeois.

Slipcasting is the process of pouring casting slip (a liquid clay) into a plaster mold. The plaster mold pulls the water out of the slip, creating a solid shell. The longer the slip remains in the mold, the thicker this wall becomes as more and more moisture is wicked out of the slurry. If the slip is left in the mold for a long time, it will solidify completely, leaving a solid cast. If a hollow casting is desired, when the shell reaches the right thickness, the remaining slip in is poured out of the mold, leaving a hollow casting. Slipcasting is a staple in the production of cast ceramic objects today, both mass-produced and one-of-a-kind.

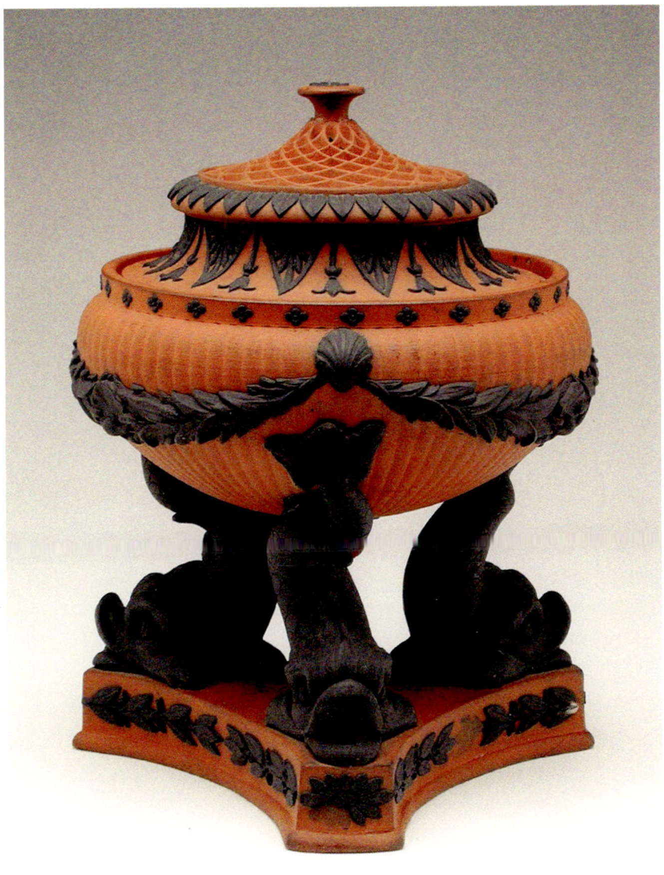

Wedgwood's proprietary clay body, Jasperware, is distinctive because its color is innate. Wedgwood's famous light blue is the most identifiable, but many other colors exist as in this example.

Wedgwood Dolphin Tripod Pastille Burner, ca. 1830–1850. England, stoneware with metallic-oxide stain, 7" × 5" × 5". Courtesy of LACMA, Gift of Alice Braunfeld, www.lacma.org, LACMA public domain.

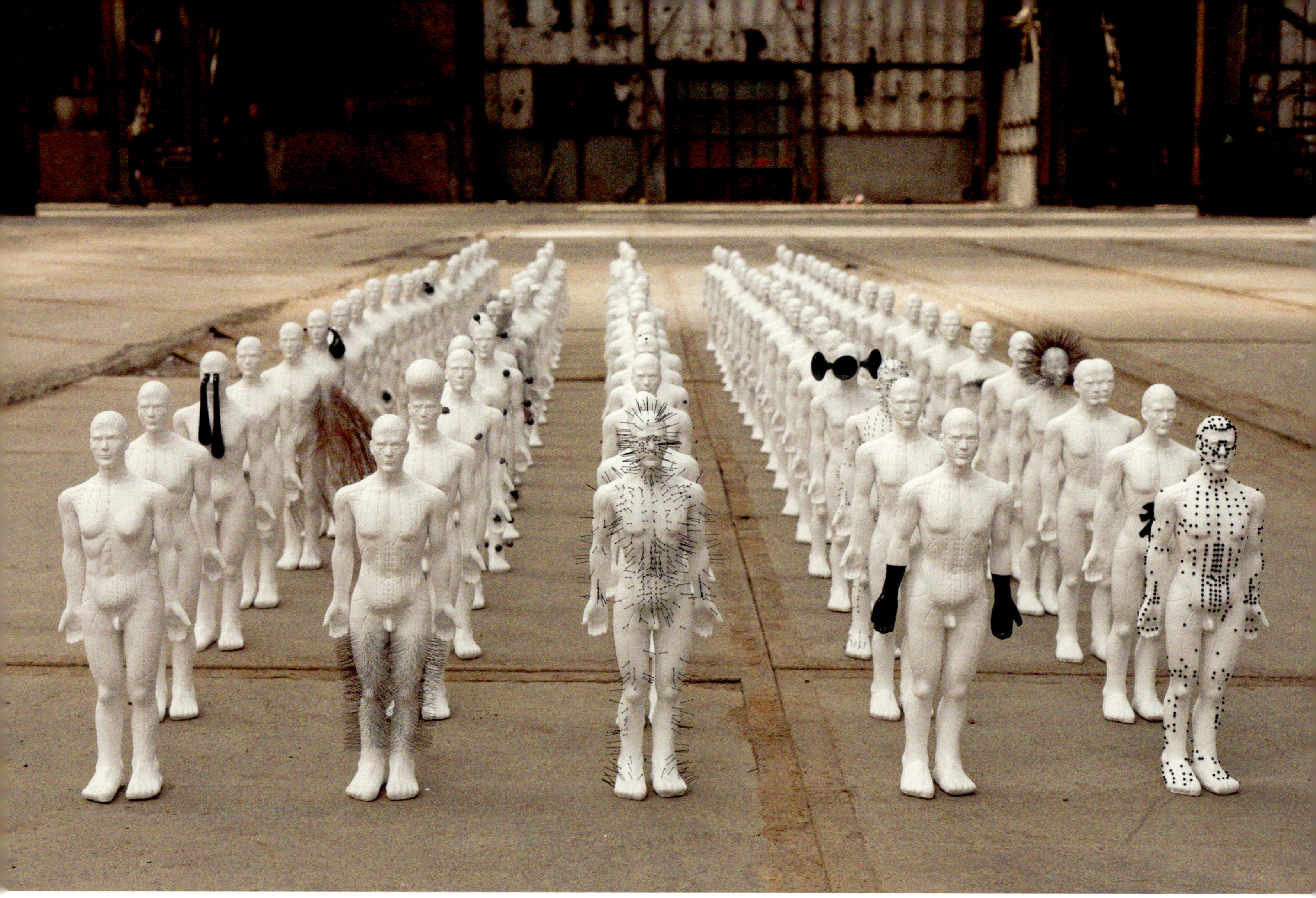

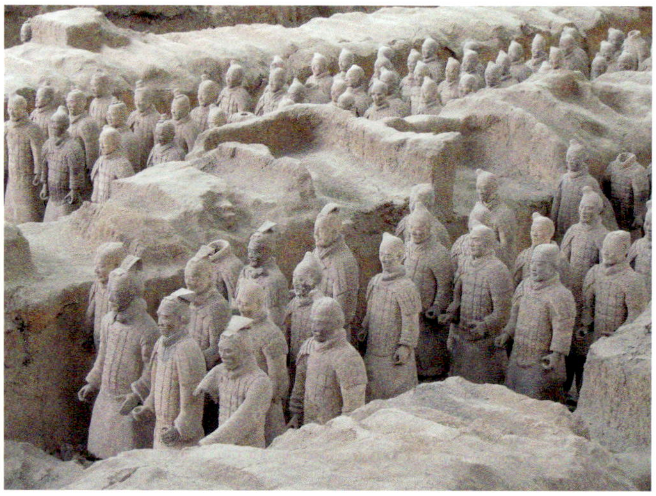

Top Ruudt Peters, *CUN ZAI* (Installation) 2015. Porcelain, silver, glass, steel, magnets, graphite, horse hair, paint, each figure is 16" × 6" × 2".
Photo: Rob Versluys.

Bottom Terracotta Army, Qin Dynasty (221 to 206 BCE). Xi'an, China, terracotta, 8,000 soldiers, 130 chariots with 520 horses and 150 cavalry horses.
Photo: © Eikenhein / Wikimedia Commons / CC-by-3.0.

Contemporary artists of all kinds are partnering with some of these production centers in the tradition of places like Sèvres and Kohler. Ai Weiwei hired 1,600 ceramicists in Jingdezhen to create his piece *Sunflower Seeds* (see page xi) made up of 100,000,000 individually cast seeds. It took two and a half years to produce.

Ruudt Peters collaborated with ceramicists in Xiamen, China to produce his work, *CUN-ZAI*. These slipcast porcelain figures were inspired by Chinese acupuncture models while Peters was traveling in China. During a residency at the Chinese European Art Center, Peters was connected with a Chinese ceramic artist who produced a mold and cast the ninety-nine porcelain figures. Peters then altered the castings after they came out of the molds. Changing the original molded figure into ninety-nine individuals is a very interesting way to work with molds; rather than clones, they each have their own personality and each one speaks to different parts of our nature. This work has a clear relationship to the *Terracotta Army*, perhaps the most famous work of molded ceramics ever made, and one that also used molds to create thousands of individual warriors rather than an army of look-alikes.

The *Terracotta Army* surrounds the tomb of Qin Shi Huangdi, the first Emperor of China, and was created to protect him in the afterlife. *The Terracotta Army* consists of approximately eight thousand individual ceramic warriors, as well as bronze horses, chariots, and real weaponry. The entourage is not limited to warriors however, there are also many other ceramic figures that were intended to do everything from entertaining the emperor to taking care of his horses in the afterlife.

Excavation has been ongoing since the site was discovered in 1974. The site is so vast that even after forty-plus years, archaeologists have barely scratched the surface. The ceramic and metal workshops have been uncovered, and with them, many interesting facts. They found that craftspeople worked in small groups or "cells" that were extremely regimented. Workers had to sign each part they produced, creating a clear record of quality, which produced a built-in system of accountability. Workers were required to report others' incompetence; they worked under a system of incentives and punishments, including death. Over 700,000 people (including women and children) worked on the Emperor's tomb and its surrounding complex.[3] This highly structured and brutal system produced "the eighth wonder of the world" but at a huge human cost.

Thankfully, factories aren't quite so brutal today, although health and safety standards are still severely lacking in many developing countries, making many ceramic factories dangerous places. Ceramic factories provide the world with everything from dinnerware to bricks and microelectronic components to fine art. The range in products also leads to a range in facilities. Ceramic factories worldwide may look different today, but when it comes to casting, certain factors are always the same: clay, molds, kilns, and the skilled hands that know how to unlock their secrets.

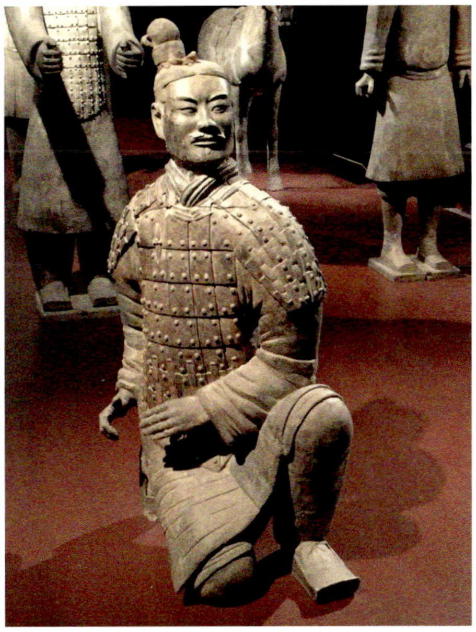

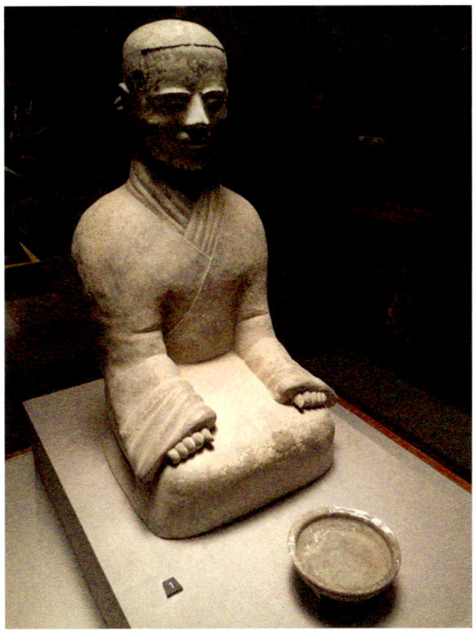

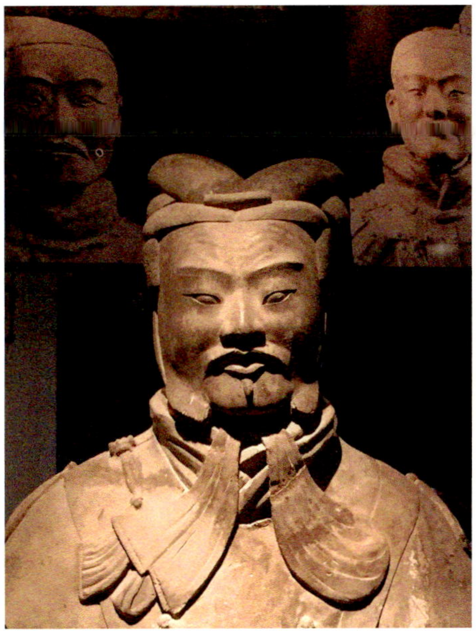

Top to bottom

Archer from the Terracotta Army, Qin Dynasty (221 to 206 BCE). Xi'an, China, terracotta.
Photo: Ed Yates. Collection of Shaanxi Provincial Institute of Archaeology on loan to the Field Museum, Chicago.

This stable boy is just one example of the variety in characters included in the Emperor's afterlife entourage, which includes personnel from scribes to acrobats.

Stable Boy from the Terracotta Army, Qin Dynasty (221 to 206 BCE). Xi'an, China, terracotta.
Photo: Ed Yates. Collection of Shaanxi Provincial Institute of Archaeology on loan to the Field Museum, Chicago.

Armored General from the Terracotta Army (detail), Qin Dynasty (221 to 206 BCE), Xi'an, China, terracotta.
Photo: Ed Yates. Collection of Shaanxi Provincial Institute of Archaeology on loan to the Field Museum, Chicago.

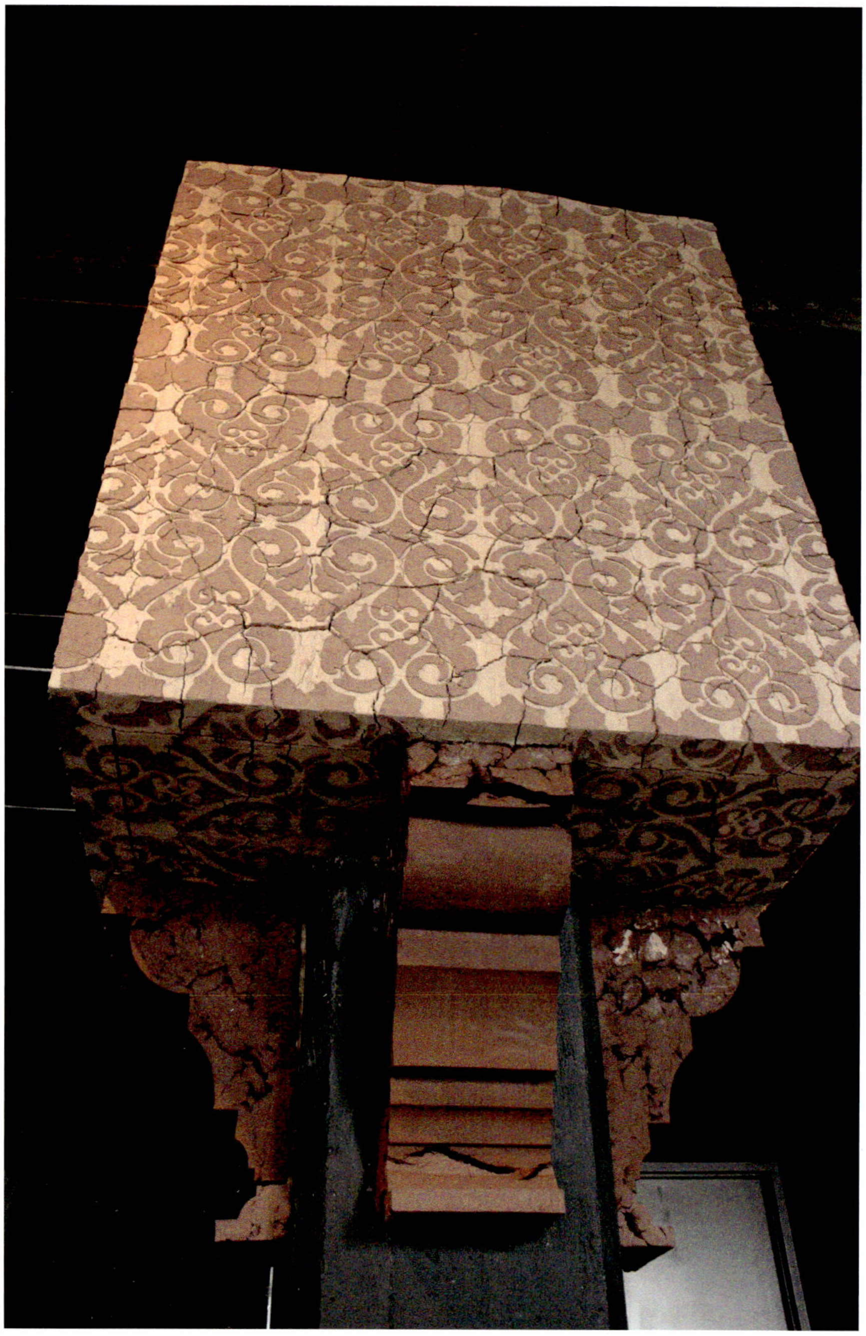

Ian McMahon achieves fascinating surfaces in all of his work, but *Transitory Architecture* is especially spellbinding in this regard. Clay is literally thrown into wooden molds/jigs then scraped to smooth the surface. Clay of a lighter color is then applied through a stencil to create the ornamental patterning. This carefully constructed and exquisite surface is then "counterintuitively" allowed to dry and crack, creating the feeling of something ancient or uncontrolled—a beauty well past her prime, but spectacular in her decay. The corbels that hold up this elaborate crumbling cube offer another surface dichotomy. The overall form of the corbels is one of perfect classical proportion and the front surfaces are clean and refined, but when viewed in profile, the process of their creation is revealed: clay was roughly thrown into the mold in chunks and left in a raw, unfinished state. If this work had pristine surfaces all around, it would be completely changed. The way McMahon employs surface ornamentation and refinement in conjunction with cracking and dilapidation creates a piece that embodies decadence in decline, every fissure adding to its appeal.

Ian McMahon, *Transitory Architecture I*, 2007. Unfired clay, wood, hardware cloth, 4' × 5' × 5'.
Photos: Ian McMahon.

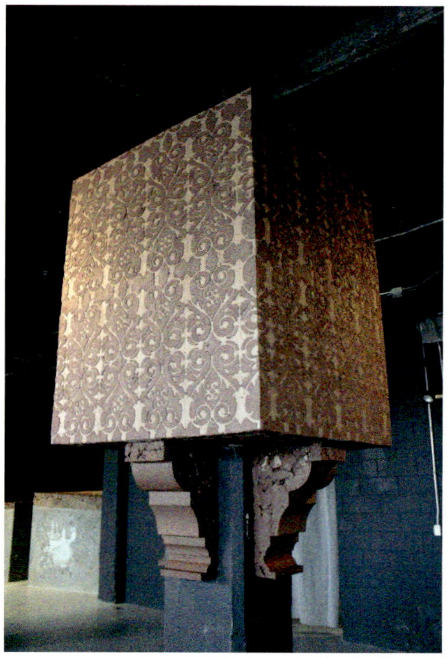

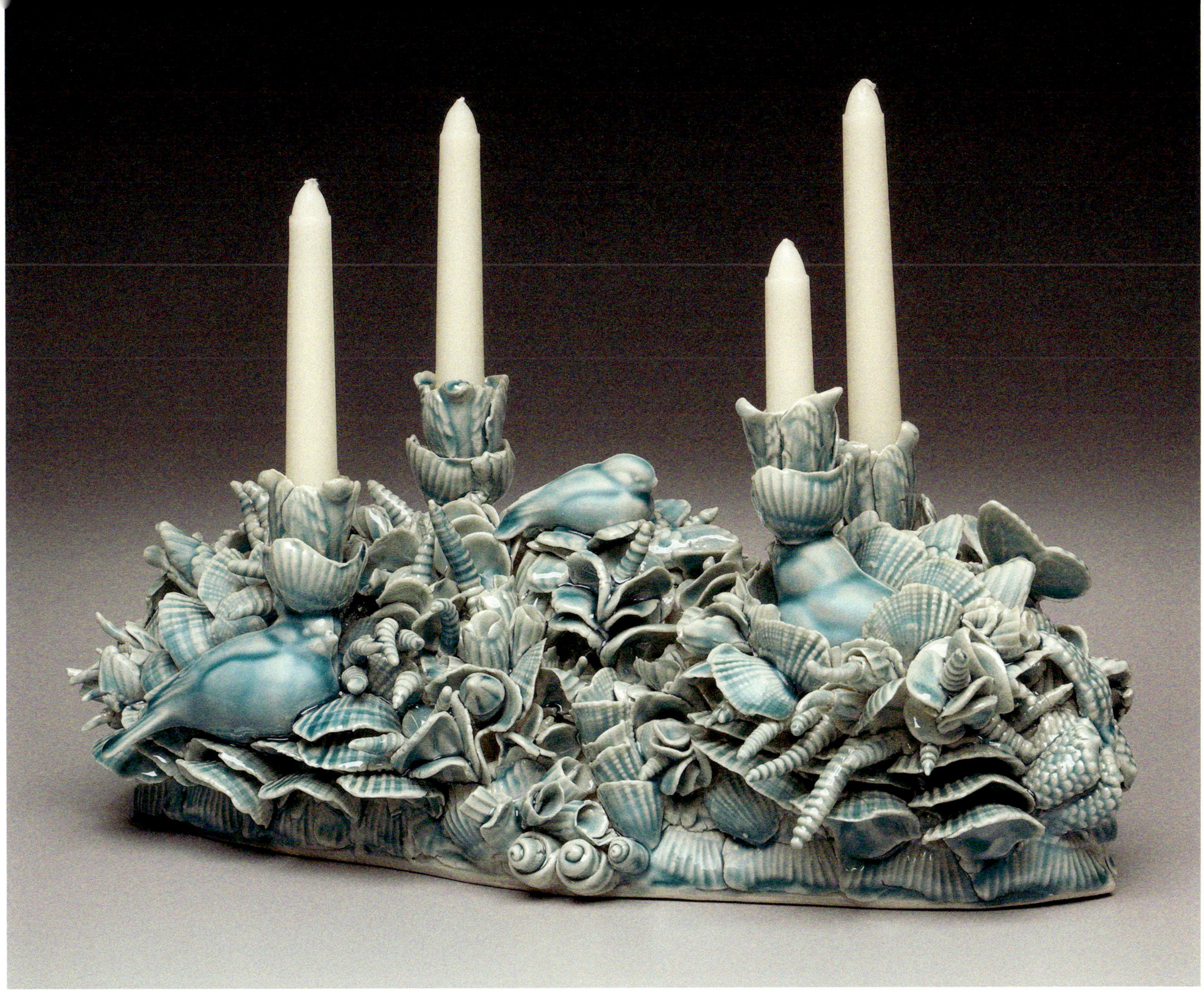

Kate Maury, *Quad Candle Holder*, 2015. Cone 6 porcelain, slab built, sprigs, cast, 14" × 7" × 9". Photo: Peter Lee.

Glaze is another, far more traditional, way to add interest to ceramic surfaces. Bright translucent colors can fill crevasses adding depth to even subtly textured surfaces, as is the case in Kate Maury's *Quad Candle Holder*. The watery aqua glaze emphasizes even the slightest indentations in her shell forms and adds a softness to the birds that would be impossible to achieve if the piece remained unglazed. This piece also has another kind of surface that is worth noting. While the textures on individual shells are surface in the "micro" sense, the "macro" surface is made up of clustered forms that undulate, protruding and receding to create a dramatic contrast against the soft smooth bodies of the glazed birds.

Glazed or unglazed, rough or smooth, ceramic surfaces offer up a wide range of delights for the eye as well as the hand.

Notes

1. Anne E. Grimmer and Paul K. Williams, "Preservation Brief 30: The Preservation and Repair of Historic Clay Tile Roofs" (National Parks Service, U.S. Department of the Interior, n.d.), Web, July 19, 2016.
2. Mary Murphy-Gnatz, "The Porcelain Trade, University of Minnesota Libraries" (N.p., n.d.), Web, July 20, 2016.
3. John Roach, "Emperor Qin's Terra Cotta Army—National Geographic" (National Geographic, n.d.), Web, accessed July 20, 2016, http://science.nationalgeographic.com/science/archaeology/emperor-qin/.

GALLERY
Ceramics

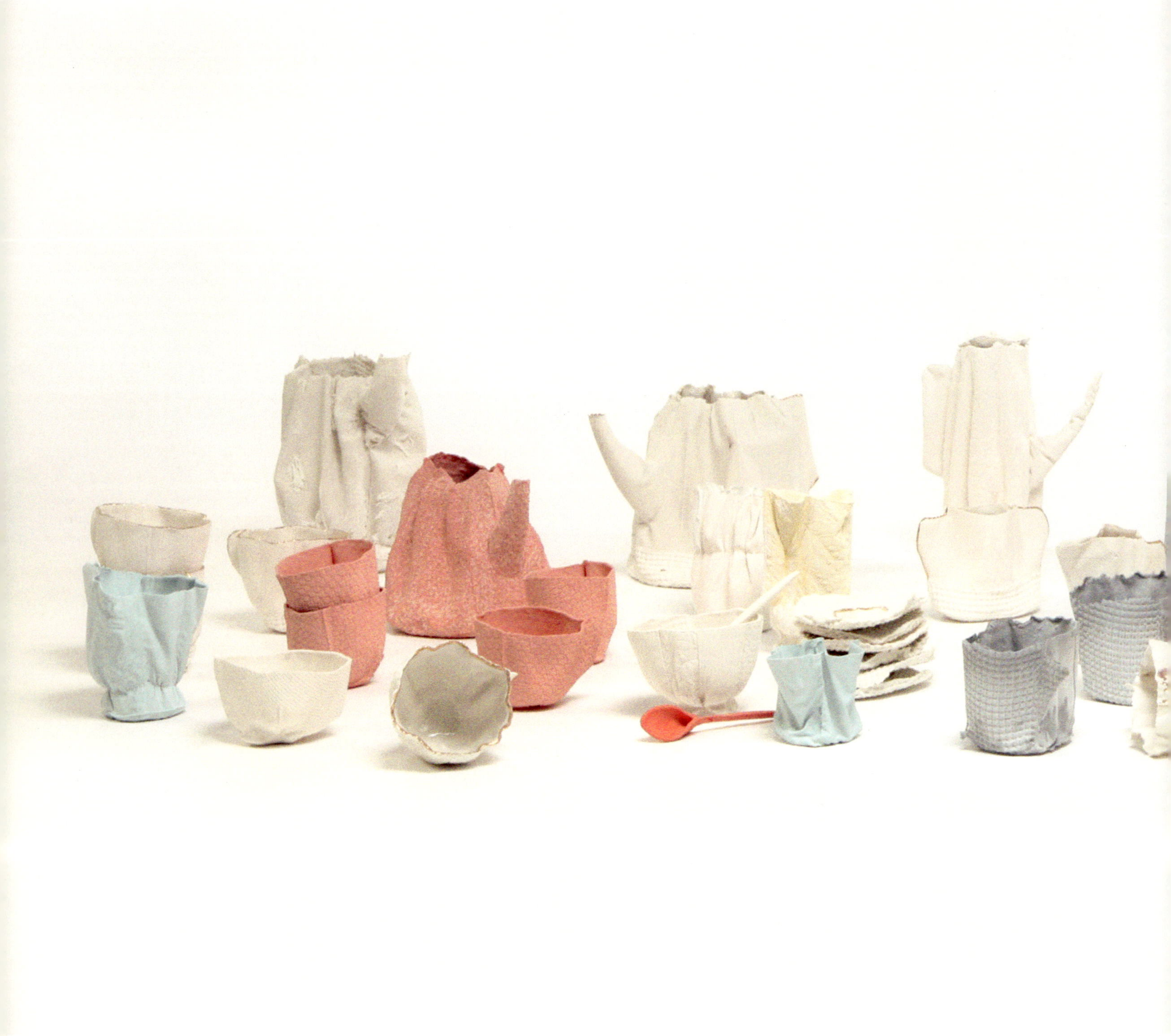

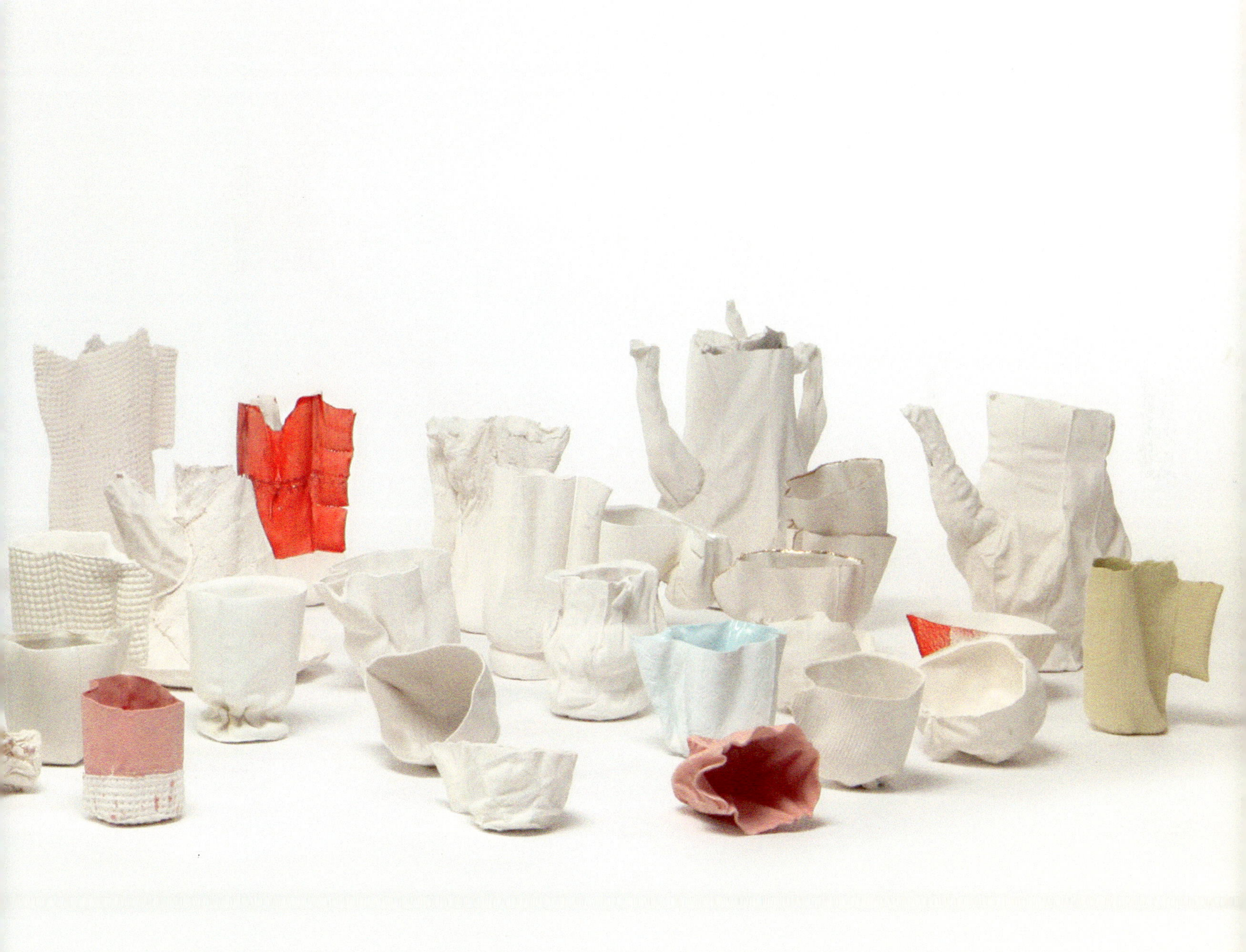

Rachel Boxnboim, *Alice Vase*, 2012. Porcelain, dimensions variable.
Photo: Oded Antman.

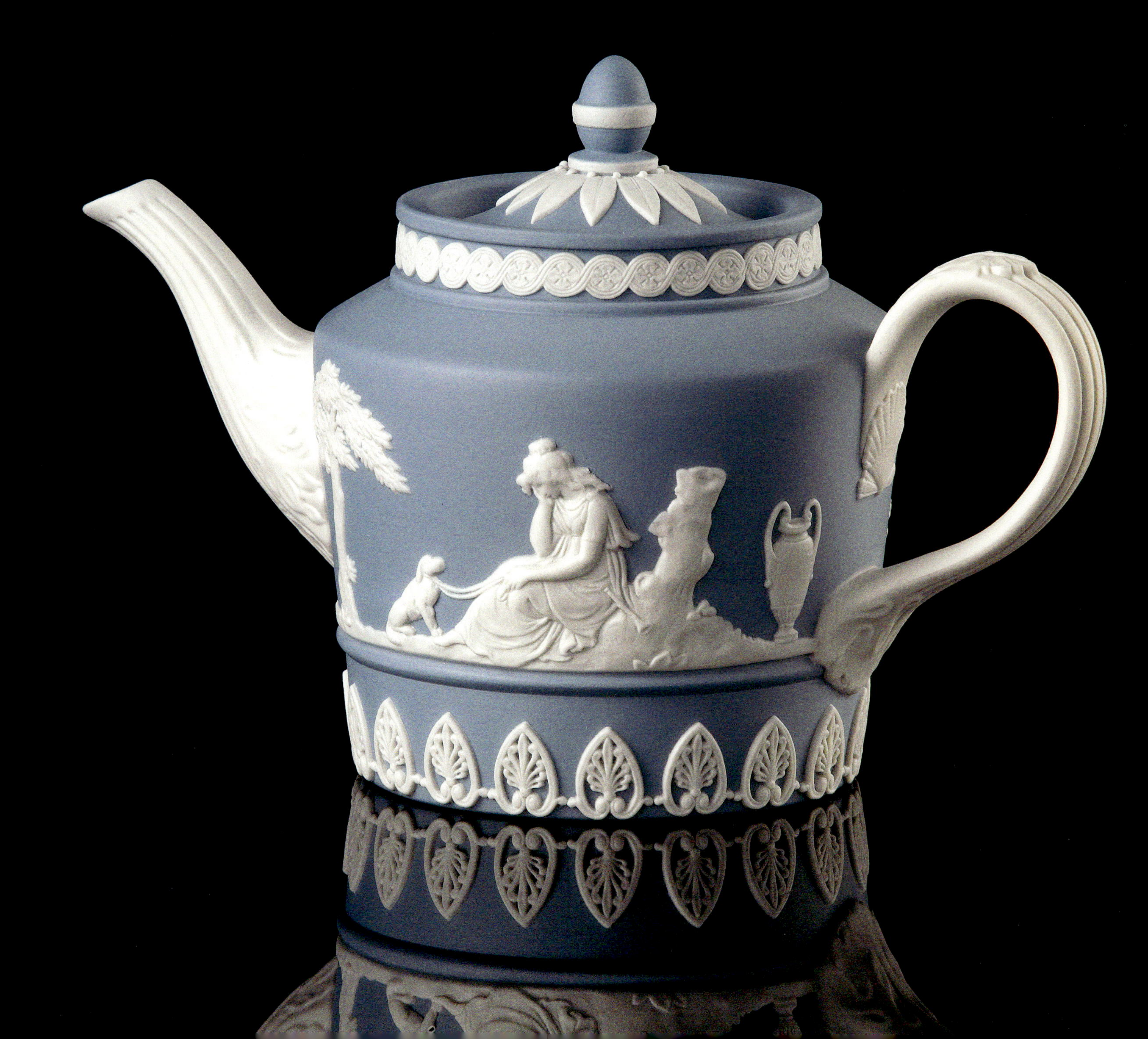

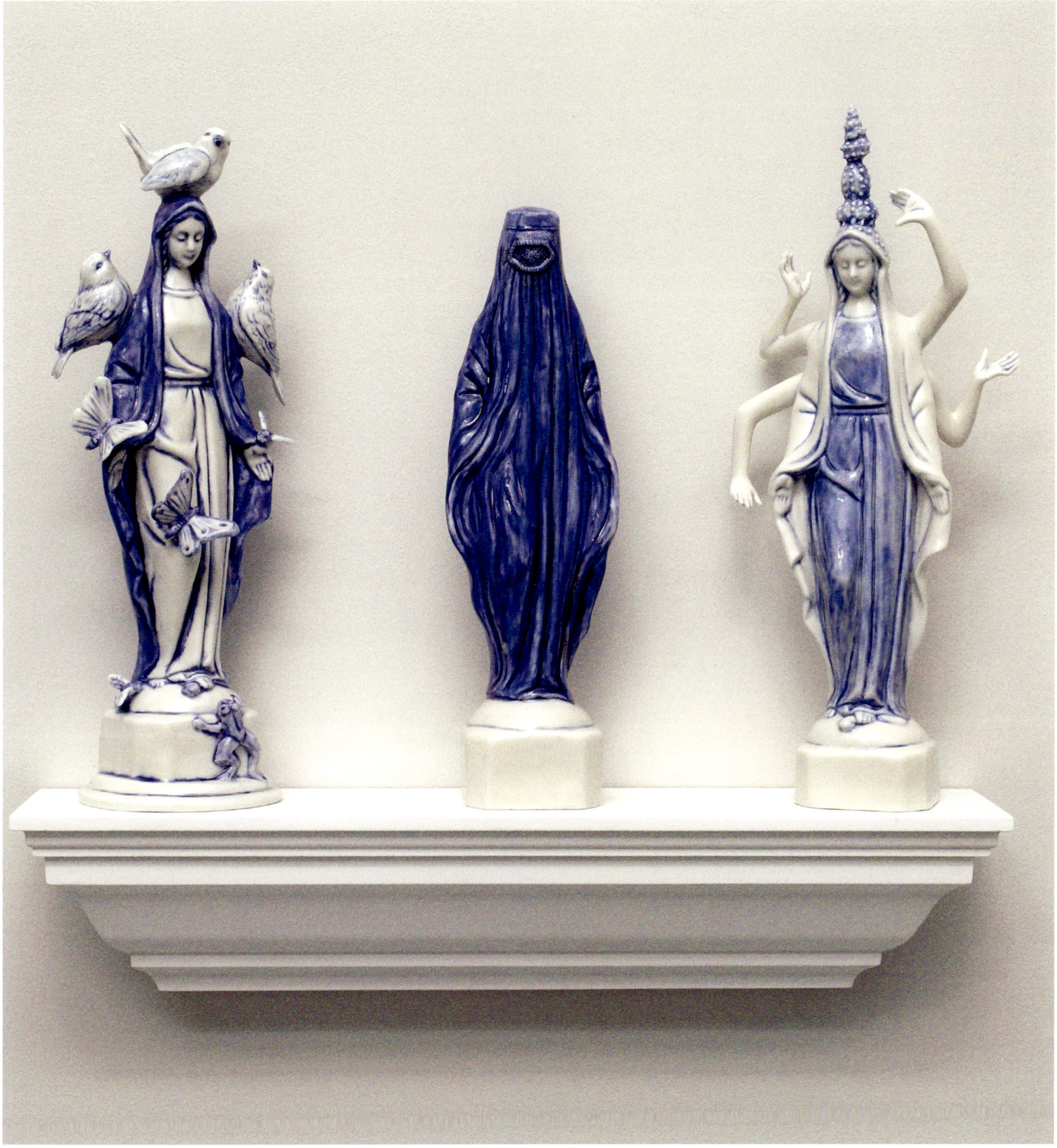

Above Cynthia Consentino, *Madonnas: Garden Mary, Burka Mary, Devi*, 2013. Slipcast porcelain, sculpted alterations, slip trailing, 18" × 18" × 4" (including shelf).
Photo: John Polak.

Opposite page *Wedgwood Prestige: The Lady Elizabeth Tea Set*, 2016, Wedgwood Jasperware.
Courtesy of Wedgwood: WWRD United Kingdom, LTD.

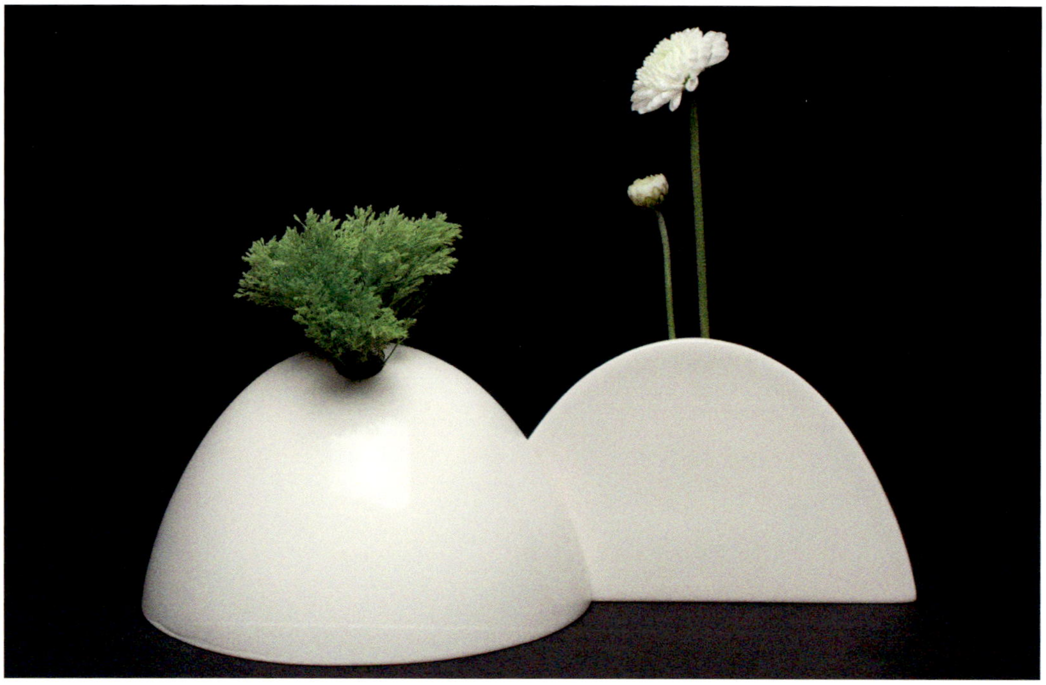

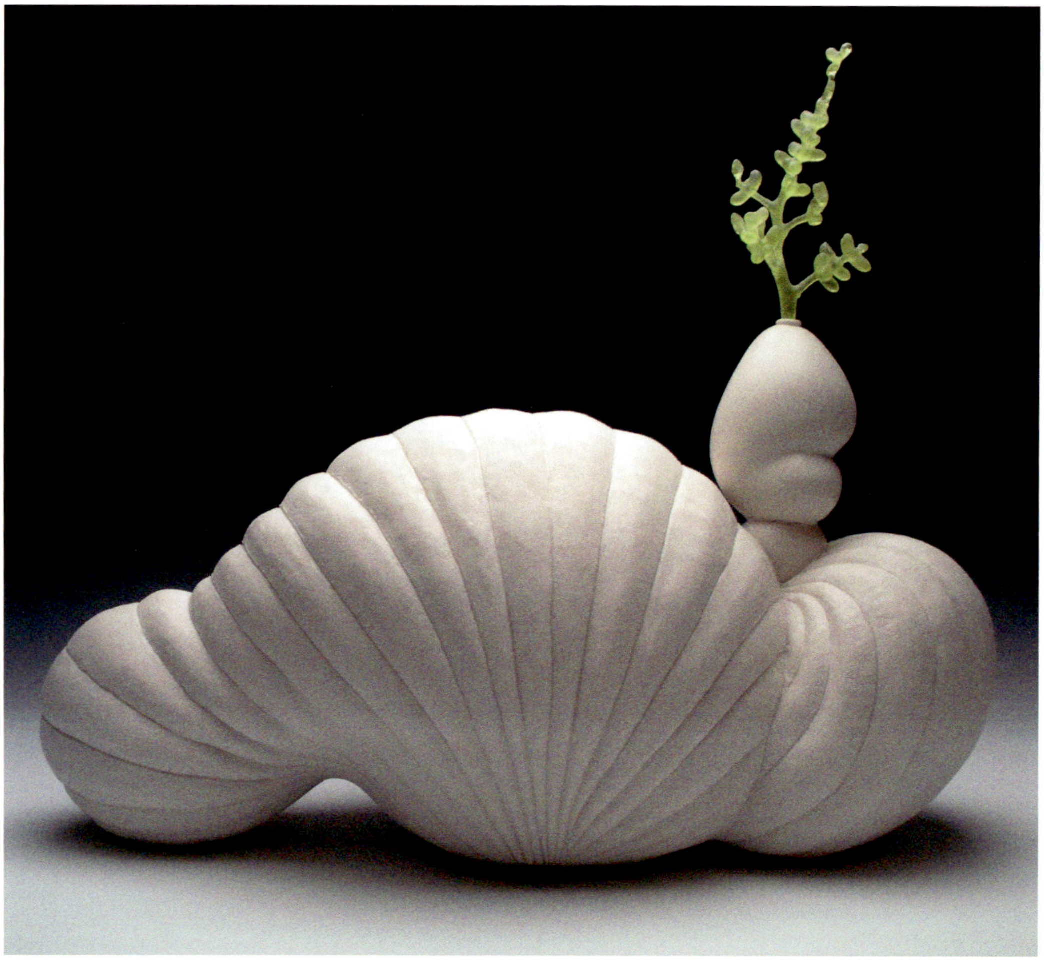

Top Molly Kite Spadone, *Flat Face Bud Vase*, 2014. Porcelain, 3.25" × 5.25" × 2.5". Photo: Jay York.

Bottom Helen Otterson, *Lepidoptera*, 2012. Porcelain, glass, 16" × 18" × 8". Photo: Helen Otterson, Collection of the Plains Art Museum.

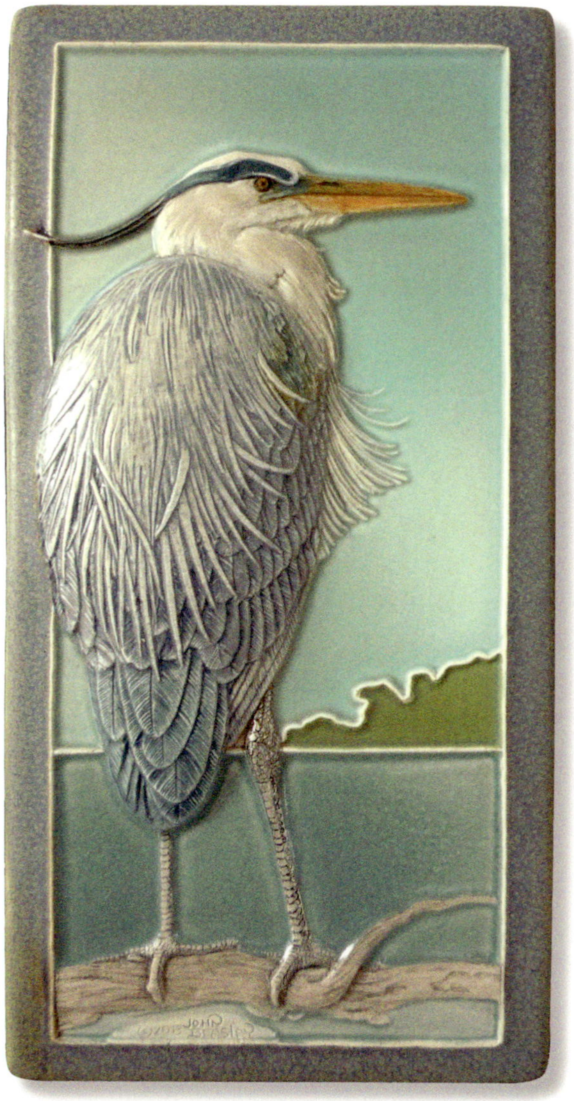

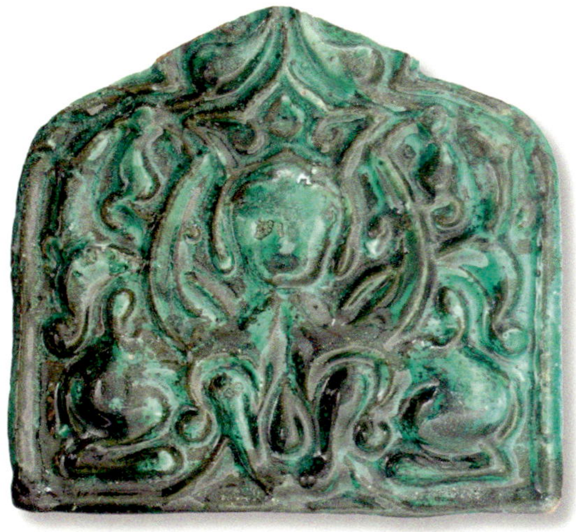

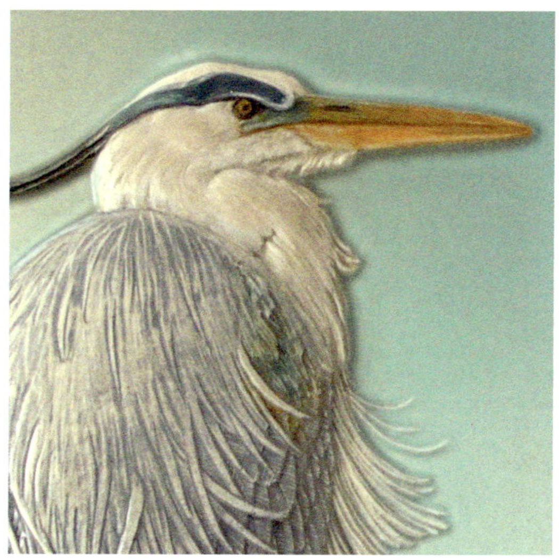

John Beasley, *Taking it Easy, Great Blue Heron*, originally designed and sculpted in 2013 and still in handcrafted production. Ceramic clay and glazes,
4" × 8" × 0.38".
Photo: John Beasley.

Top Tile, thirteenth century. Turkey, earthenware, molded and glazed, 4.25" × 4.7"
Courtesy of LACMA, Gift of Camilla Chandler Frost, www.lacma.org, LACMA public domain.

Bottom John Beasley, *Taking it Easy, Great Blue Heron* (detail).
Photo: John Beasley.

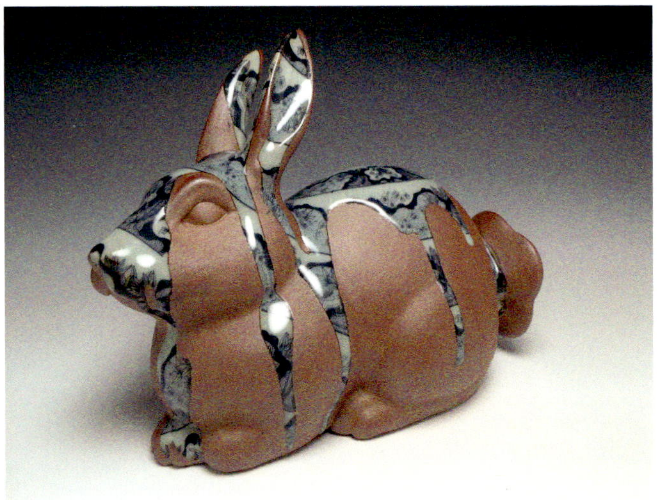

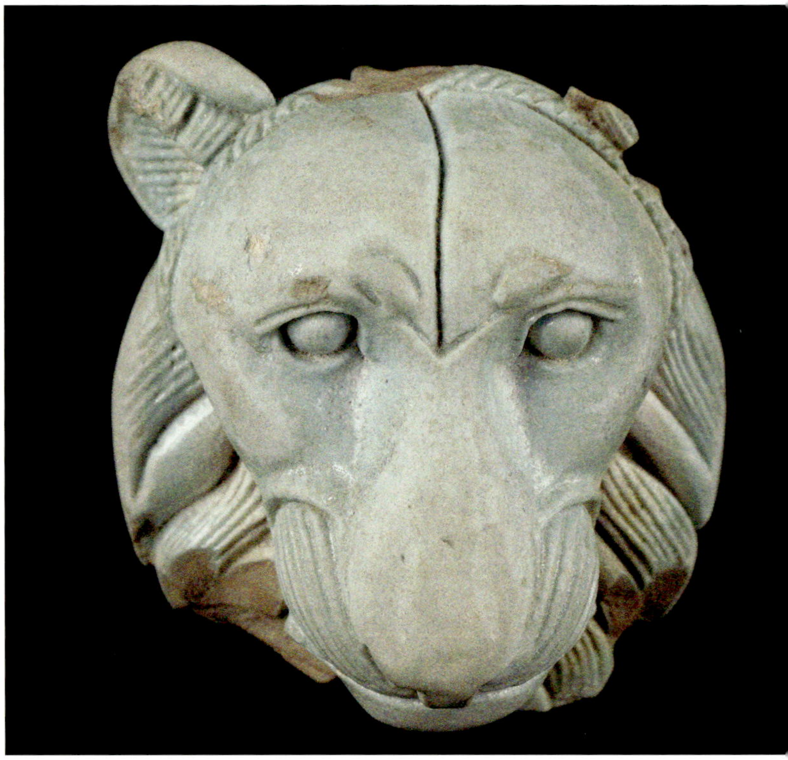

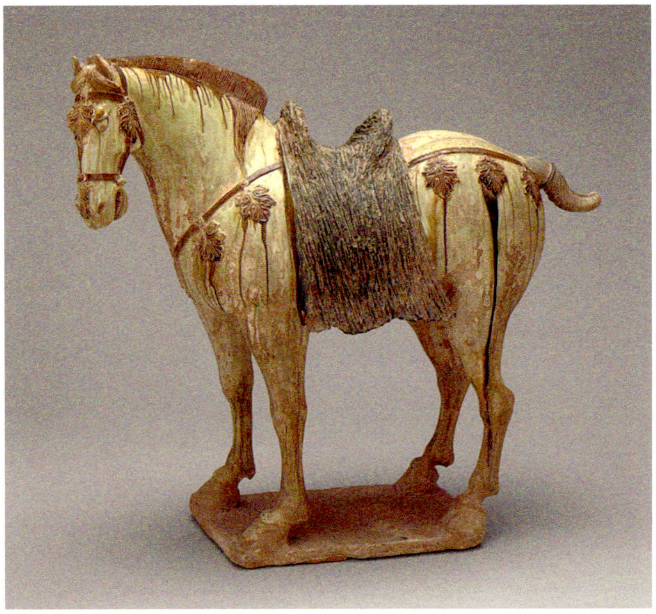

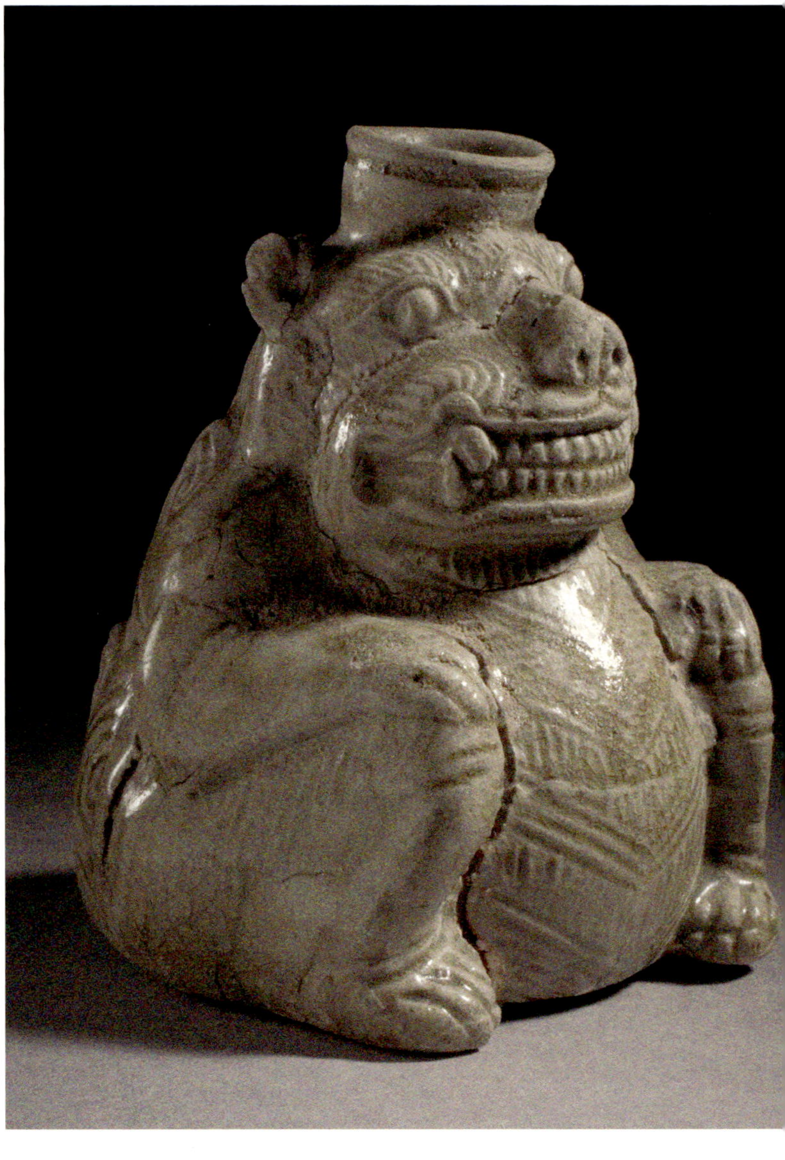

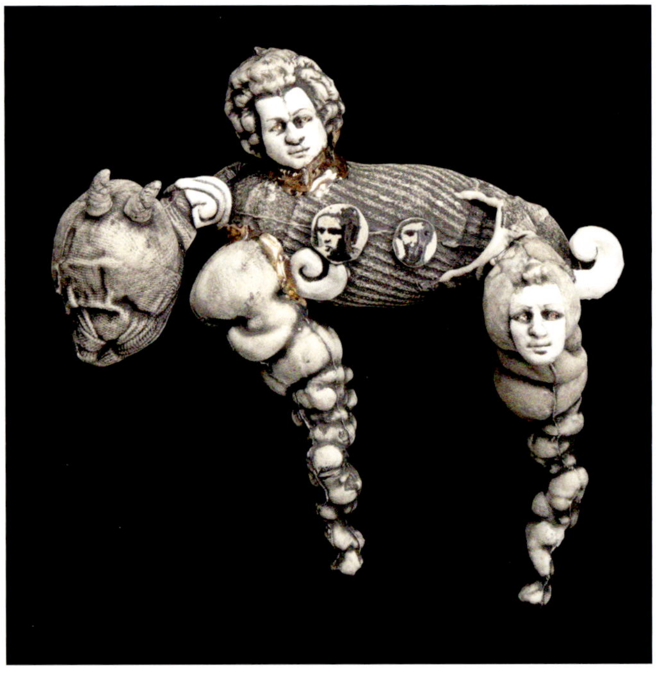

Right Shoji Satake, *On'na*, 2016. Slipcast altered porcelain and 3-D ceramic print, 7" × 6" × 6".
Photo: Shoji Satake.

OPPOSITE PAGE

Top left Lesley Baker, *Blue Willow*, 2014. Stoneware, 8" × 13" × 5".
Photo: Lesley Baker.

Top right Appliqué: Head of a Lion, ca. 380 to 342 BCE. Egypt, faience with light blue glaze. 1.81" × 1.68" × 0.87".
Courtesy of The Walters Art Museum, Acquired by Henry Walters in1923, CC-by-0, GNU FDL.

Center Funerary Sculpture of a Horse, Middle Tang dynasty, ca. 700 to 800. China, molded earthenware with molded, applied, and incised decoration and polychrome (sancai) glaze, 28" × 31".
Courtesy of LACMA, Gift of Hedwig Worch, www.lacma.org, LACMA public domain.

Bottom left Sigrid Zahner, *Dog*, 2012. Cast and assembled fabricated forms porcelain slips stains glaze gold luster, 12" × 2'6" × 4".
Photo: Wilbur Montgomery of WM Photographics.

Bottom right Vessel in the Form of a Bear, 265–316 CE. China, Western Jin Dynasty, molded stoneware with incised decoration and green glaze, 3.87" × 3" × 3.5".
Courtesy of LACMA, Gift of Mr. Robert H. Blumenfield in honor of the museum's 40th anniversary, www.lacma.org, LACMA public domain.

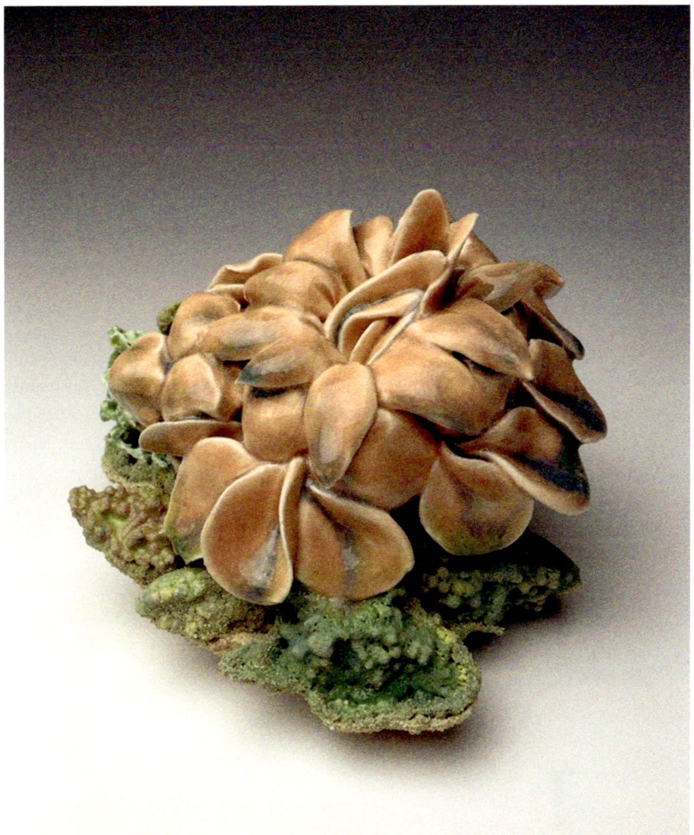

Below Rachel K. Garceau, *within the hollows*, 2014. Porcelain, 6" × 12" × 12".
Photo: Rachel Garceau.

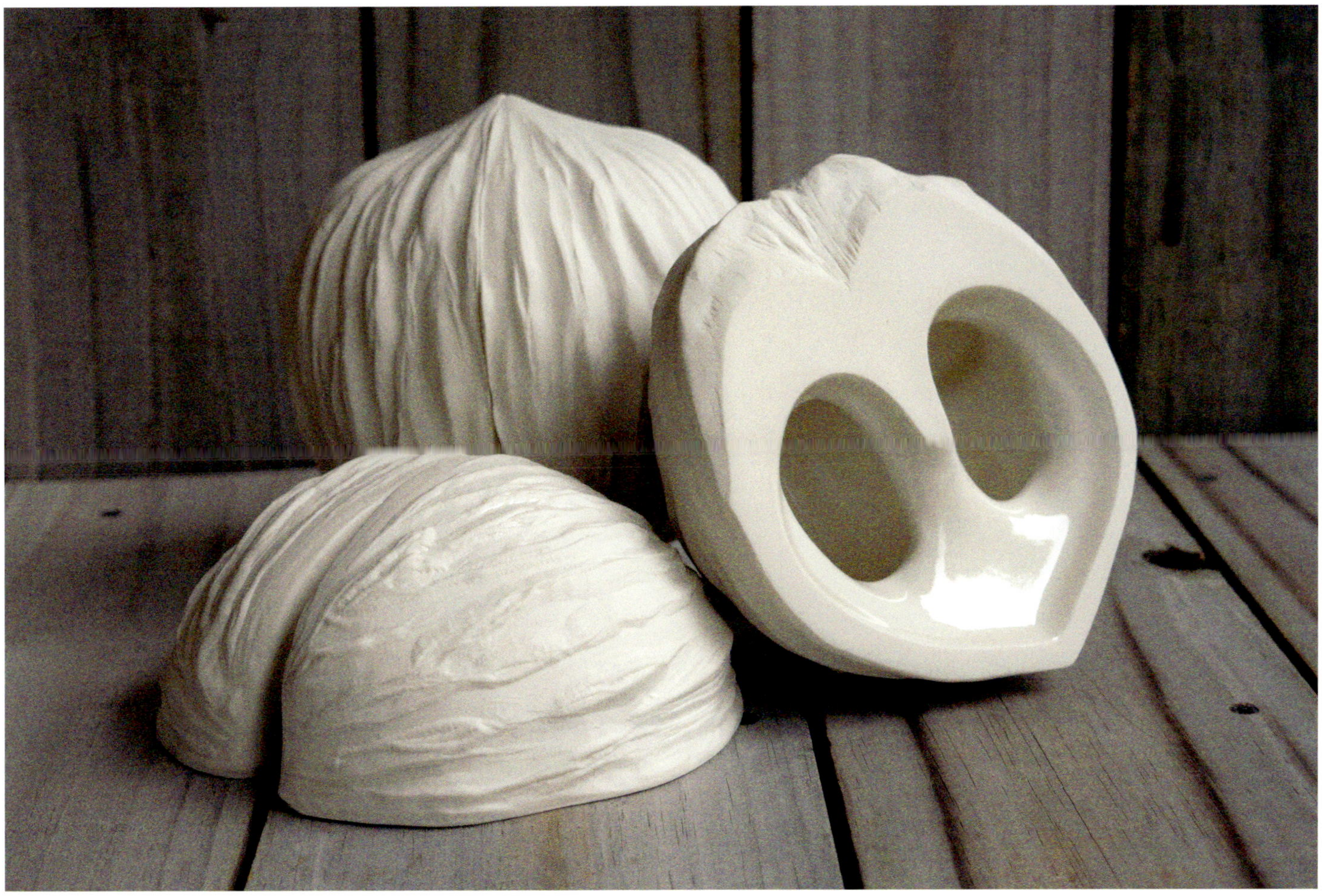

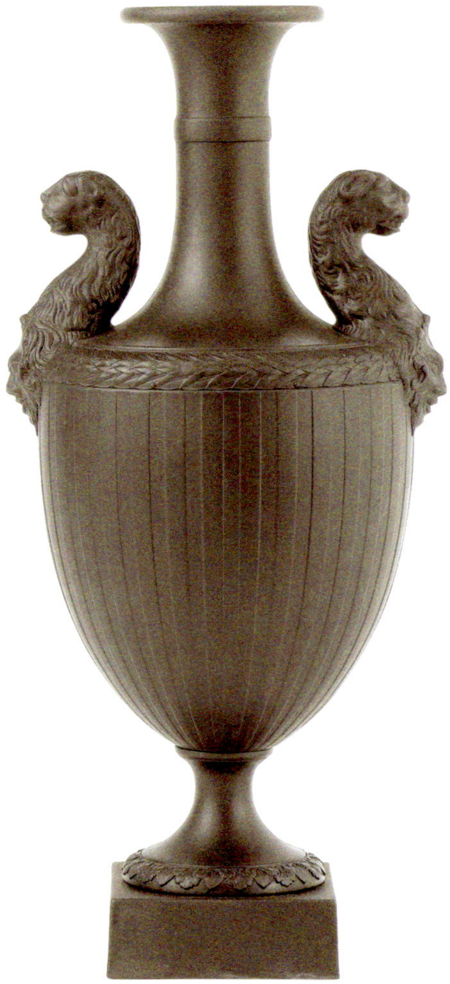

Wedgwood Prestige: Black Panther Vase, 2016. Wedgwood Jasperware, 16.9" tall.
Courtesy of Wedgwood: WWRD United Kingdom, LTD.

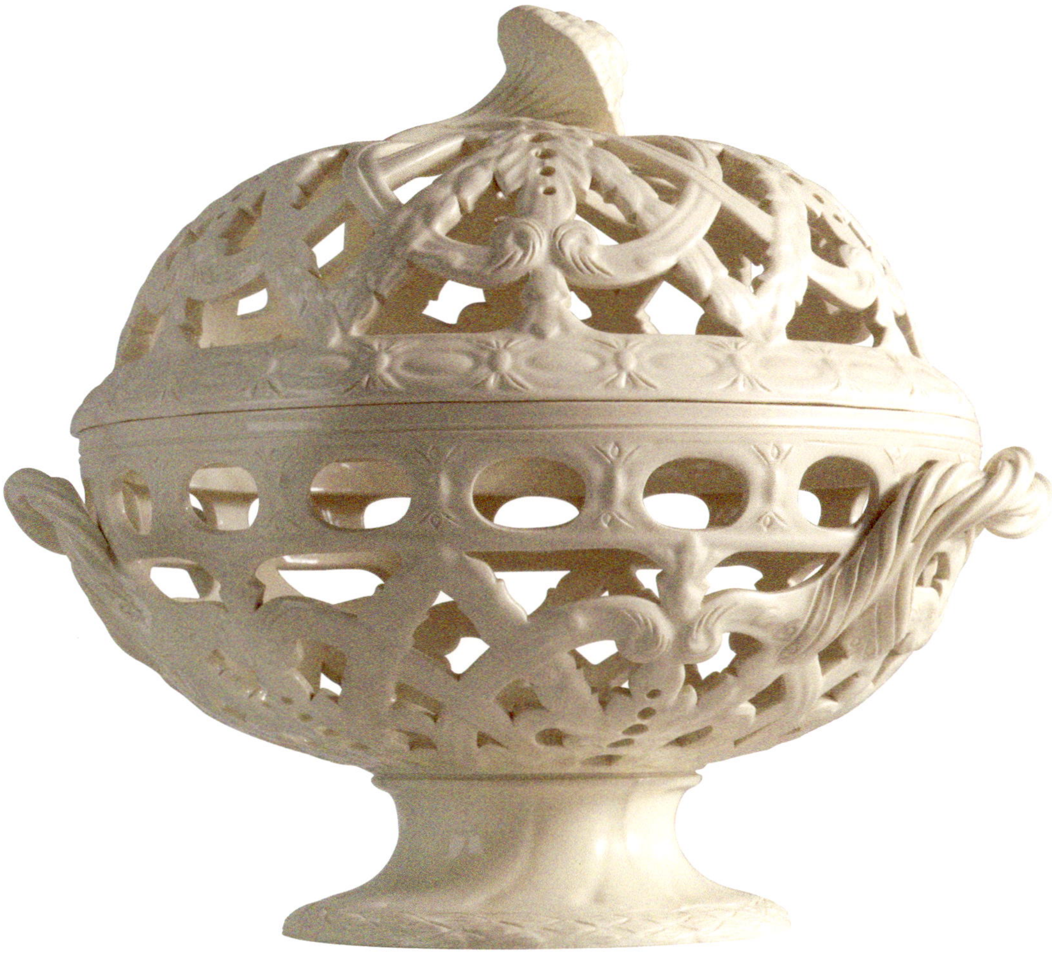

Wedgwood Prestige: Orange Bowl, 2016, Wedgwood Queen's Ware, 9" × 9.44" × 9.44".
Courtesy of Wedgwood: WWRD United Kingdom, LTD.

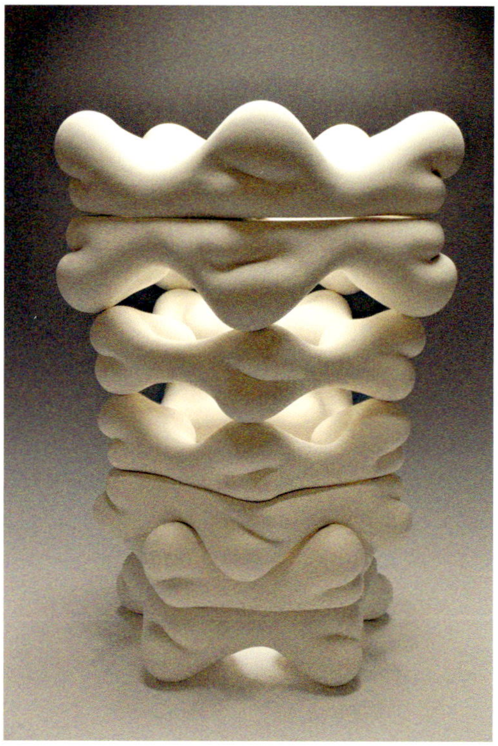

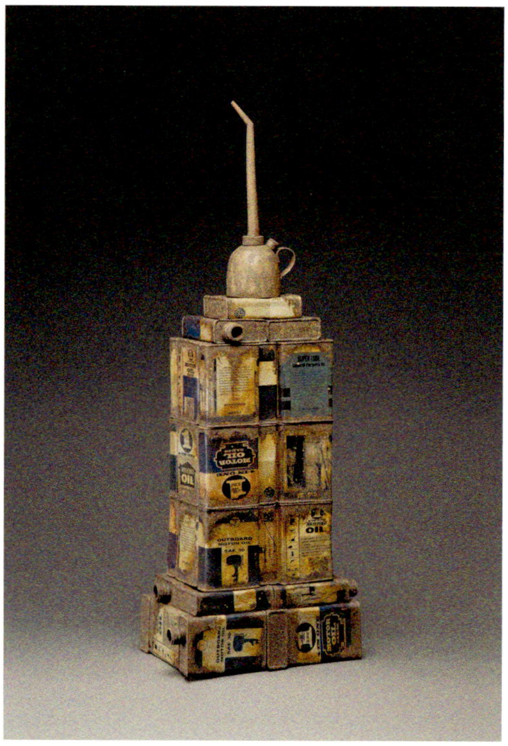

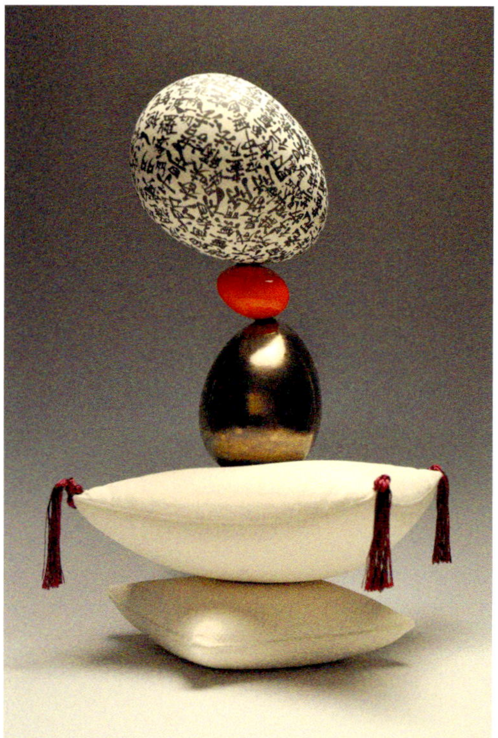

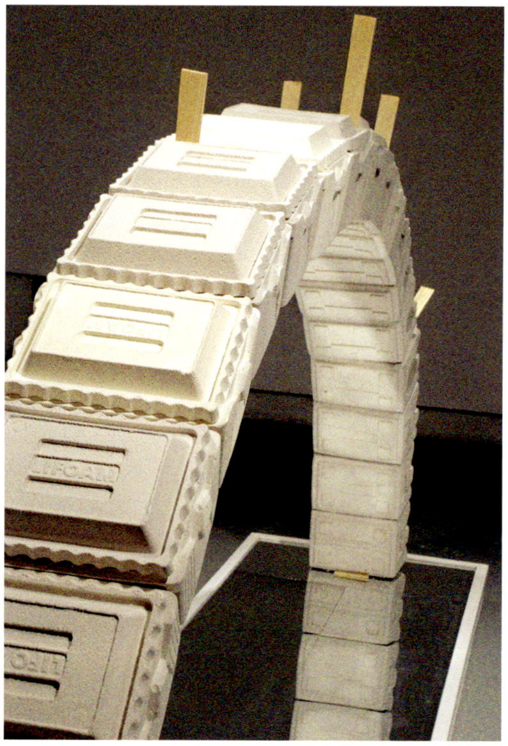

Top left Mika Negishi Laidlaw, *Jomon*, 2011. Slipcast porcelain, 27" × 17" × 17".
hoto: Mika Negishi Laidlaw.

Top right Mitchell Spain, *Consummation of Empire*, 2012. Porcelain, earthenware, ceramic decals, 2'6" × 11.25" × 8".
Photo: Mitchell Spain, Collection of Caroline B. and Daniel J. Anderson.

Bottom left Mika Negishi Laidlaw, *Correlation*, 2011. Slipcast porcelain, 19" × 11" × 11".
Photo: Mika Negishi Laidlaw.

Bottom right Danny Crump, *The Circle of Lifoam*, 2014. Porcelain, mirror, wood, 3'8" × 9' × 2'.
Photo: Danny Crump.

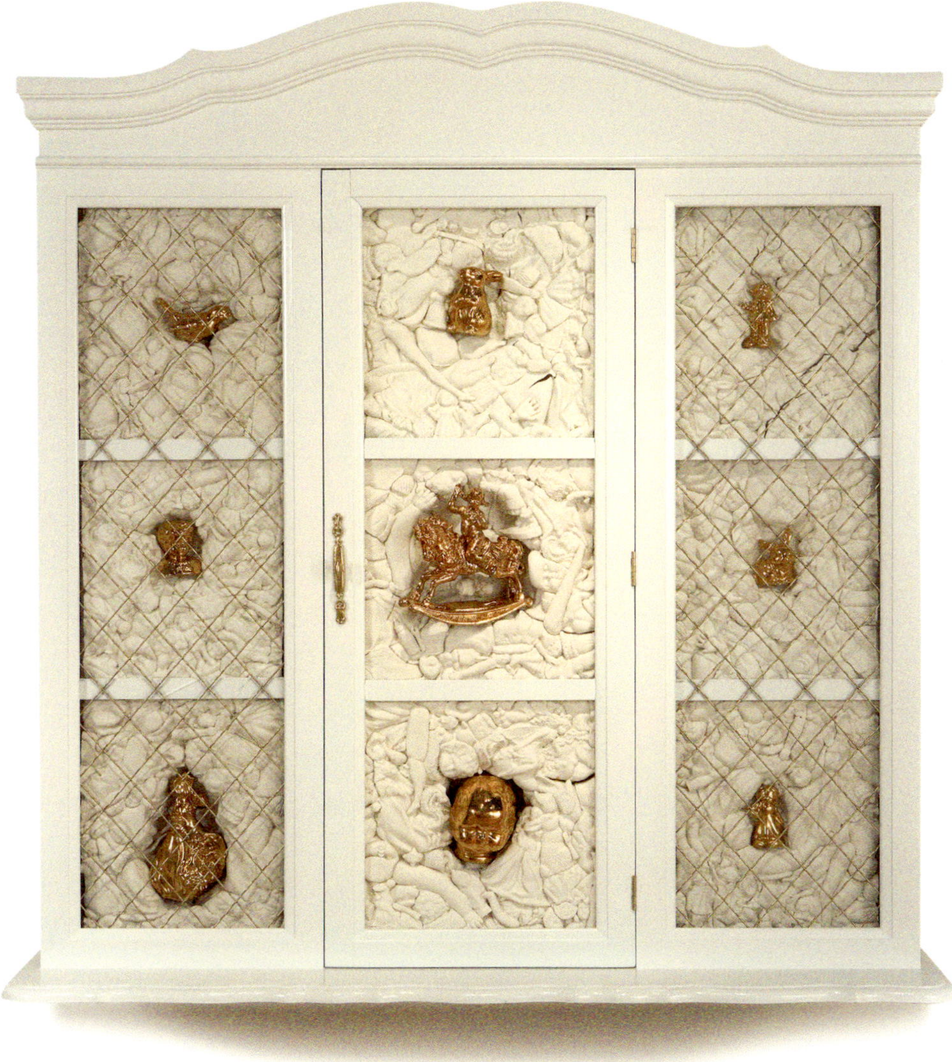

Shenny Cruces, *The Golden Cage*, 2011.
Slipcast porcelain, cone 6, mixed media,
4'3" × 4' × 1'4.5".
Photos: Sibila Savage.

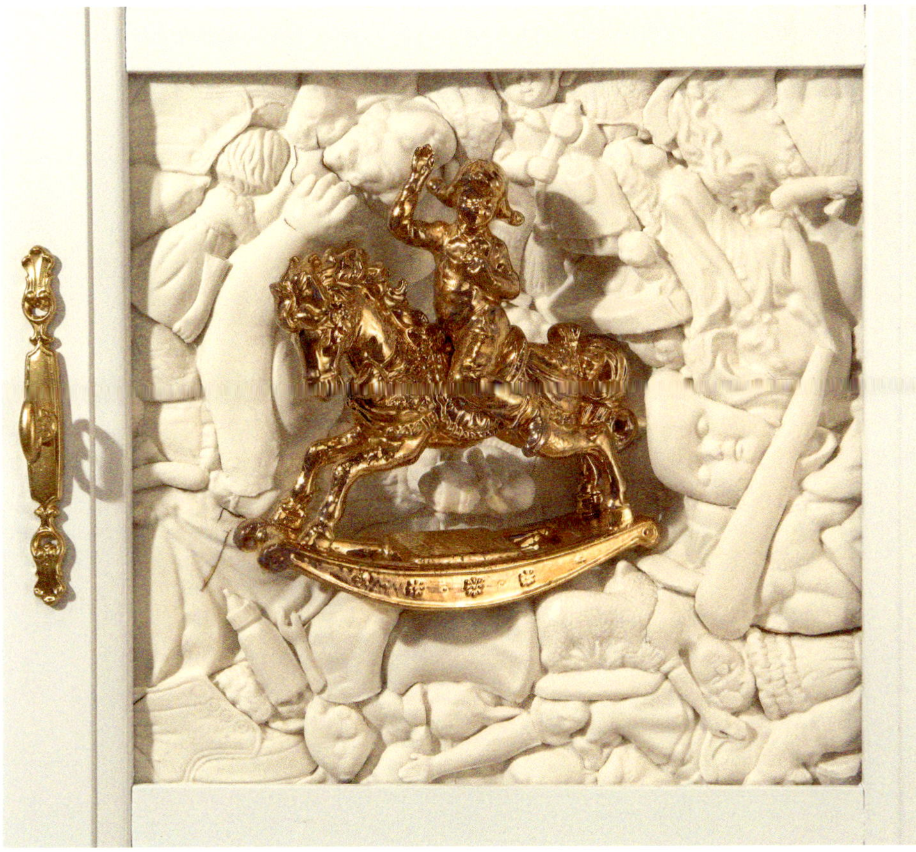

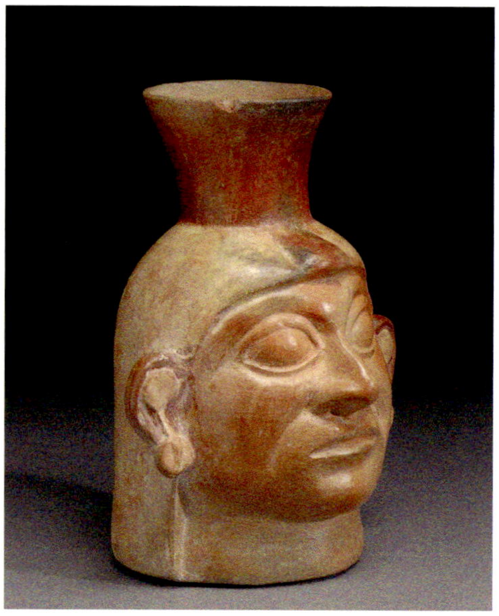

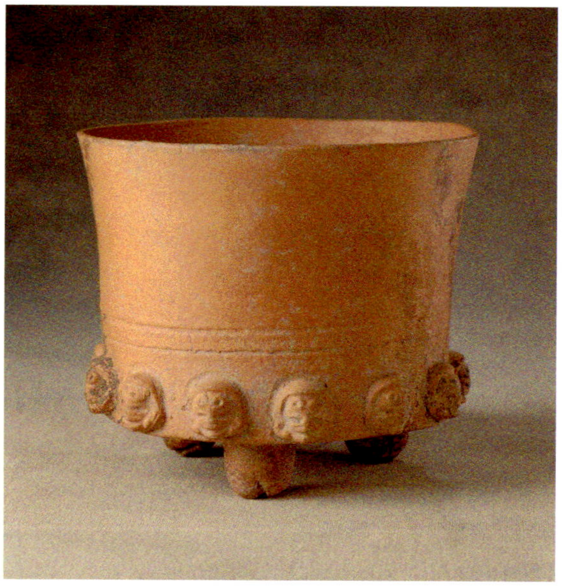

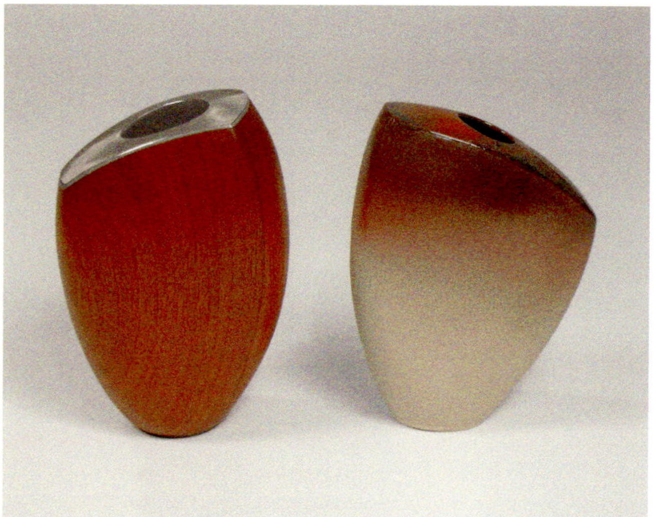

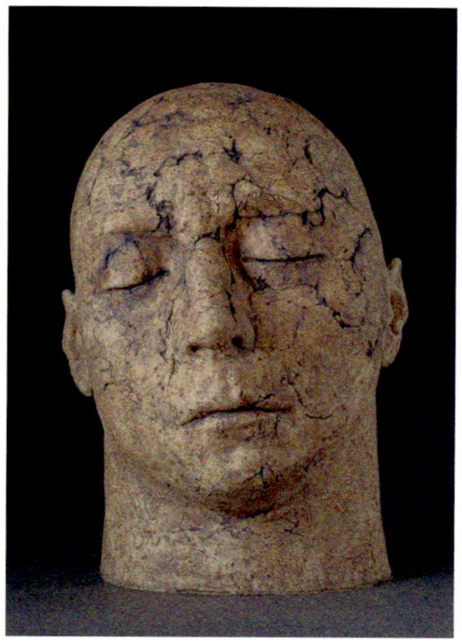

Top left Portrait Vessel, 50–800 CE (Early Intermediate-Middle Horizon). Peru, red, orange, and gray earthenware, 8.25" × 6.5" × 5.5".
Courtesy of The Walters Art Museum, Anonymous gift in 2009, Walters Art Museum, CC0, GNU FDL.

Top right Tripod vessel, 200–500. Teotihuacan, Mexico, thin orange ceramic, 4.8" tall.
Courtesy of LACMA, Gift of Constance McCormick Fearing, www.lacma.org, LACMA public domain.

Bottom left Brian Ferrell, *Two Vases*, 2014. Mahogany, pewter, rubber, stoneware, 8" × 3" × 6".
Photo: Brian Ferrell, Private Collection.

Bottom right Marlene Miller, *Earthen Head with Fissures*, 2013. Stoneware with oxide stains, 19" × 12.50" × 13".
Photo: Marlene Miller.

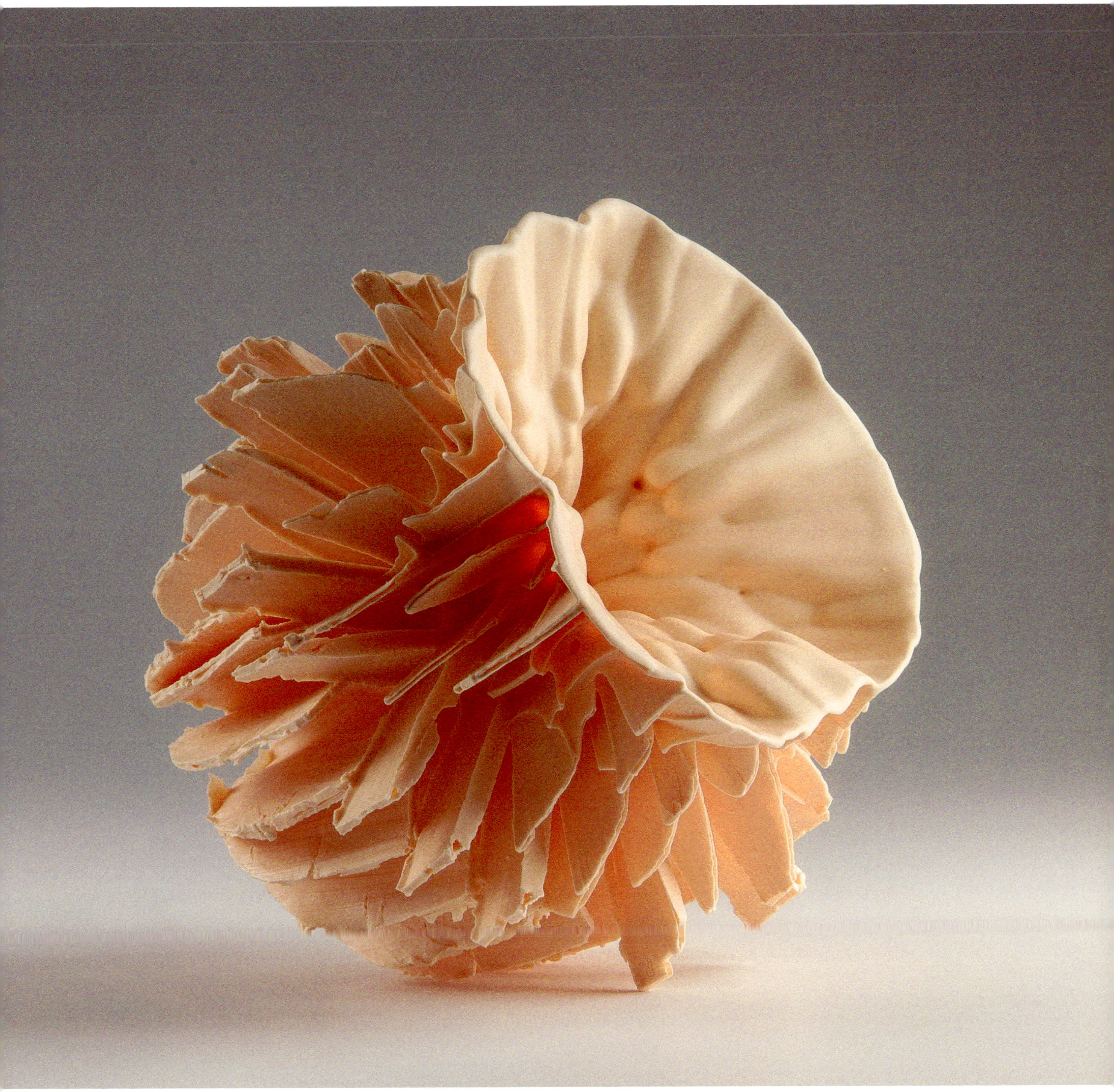

Zsolt József Simon, *Untitled*, 2014. Porcelain, pigment, no glaze, 10.25" × 10.25" × 7.9".
Photo: János Bánát.

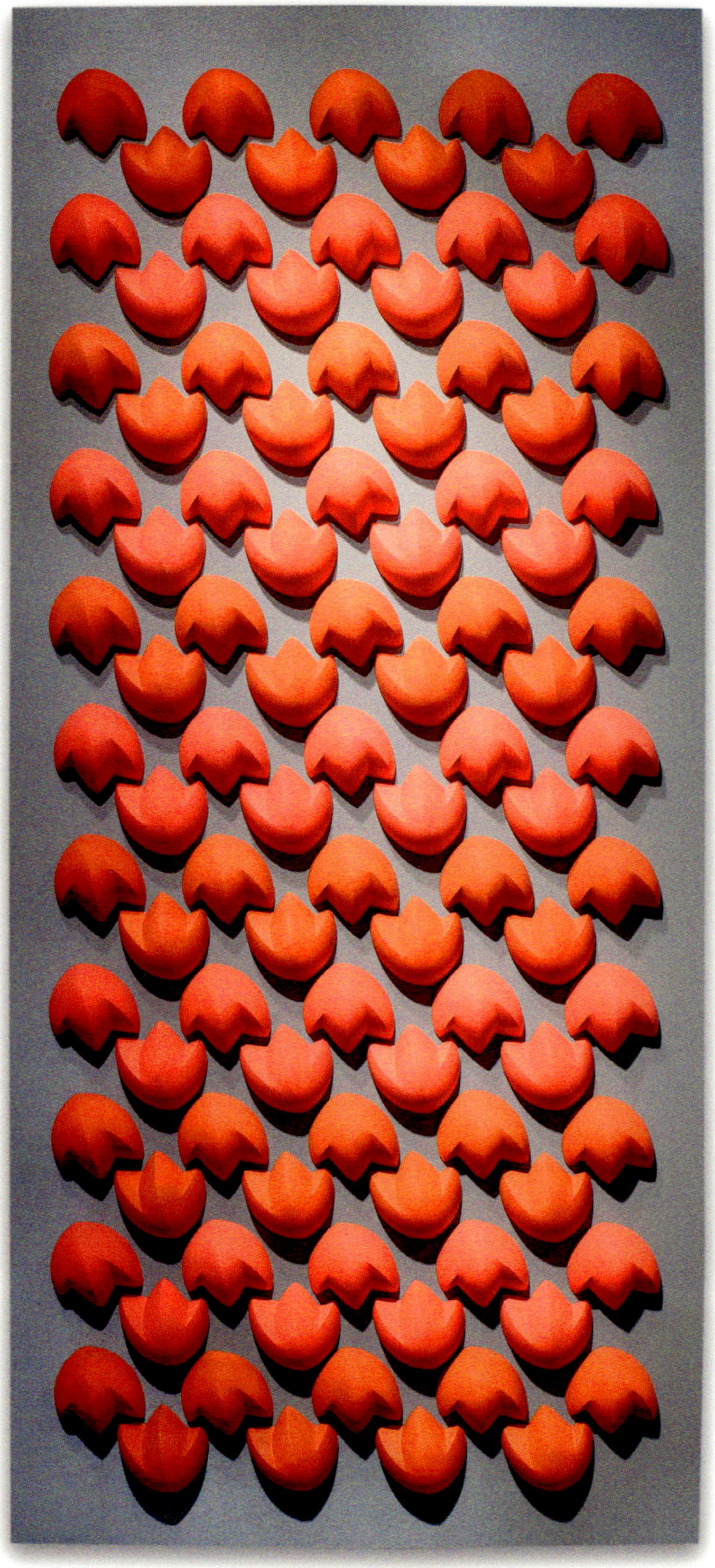

Jackie Head, *Burst*, 2014. Slipcast ceramic, spray paint, 6'4" × 2'10" × 4.5".
Photo: Anna Powell Teeter.

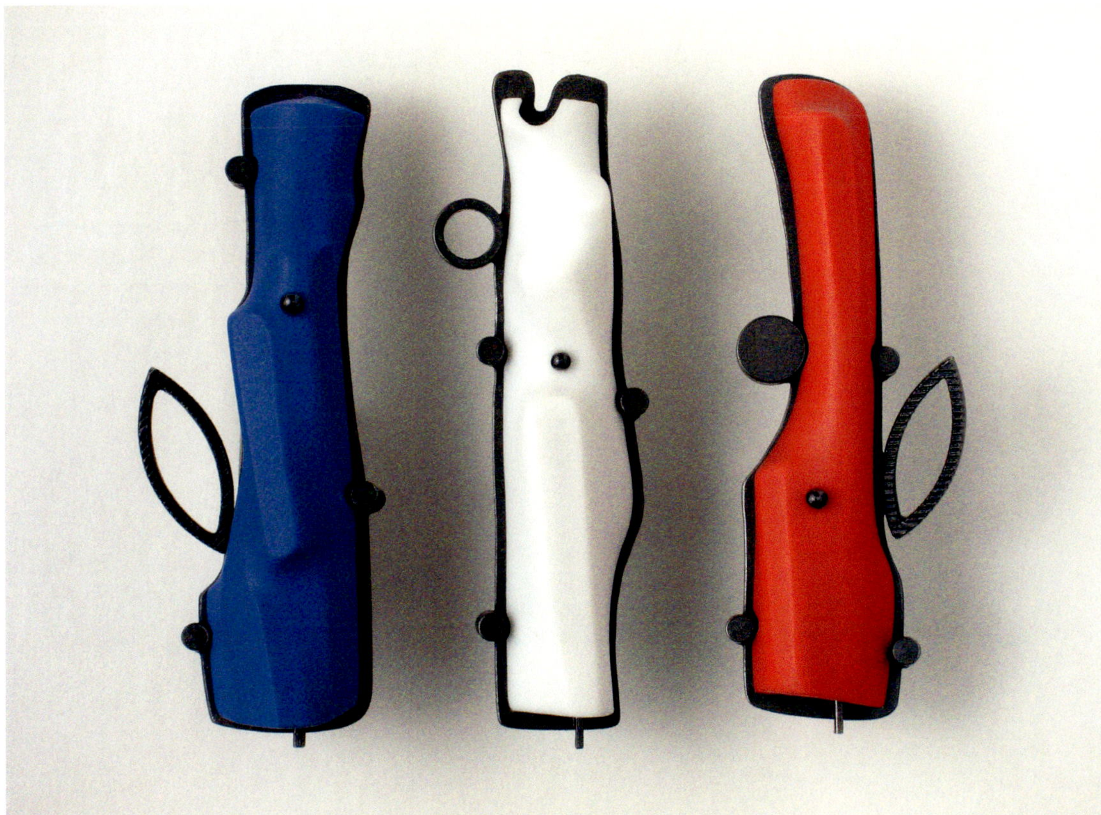

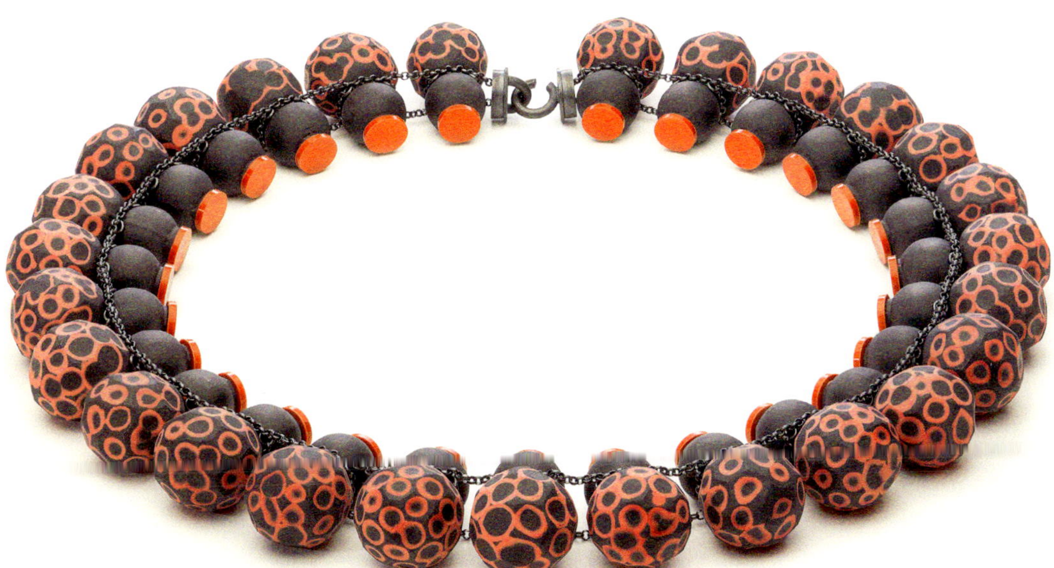

Top Paul W. Leathers, *JDZ (Blue & White; the new Red) Fragment Brooches*, 2012. Colored porcelain, sterling silver, 2.35" × 0.75" × 0.5".
Photo: Paul Leathers.

Bottom Peter Hoogeboom, *Mokume Gane Holland Hulu*, 2014. Taiwanese porcelain, silver, artificial rubber. 9.84" × 9.84" × 1.77".
Photo: Conor Vella.

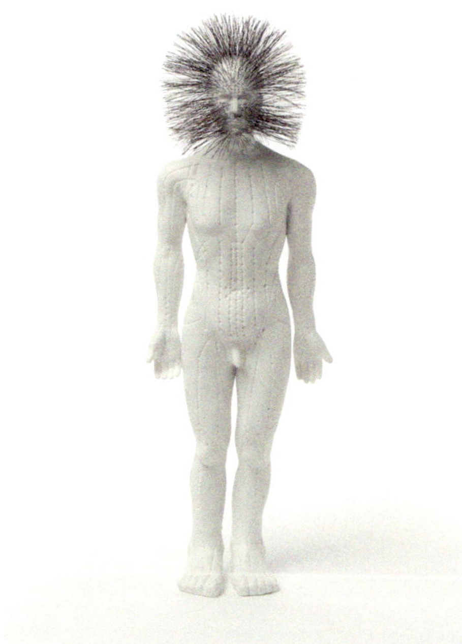

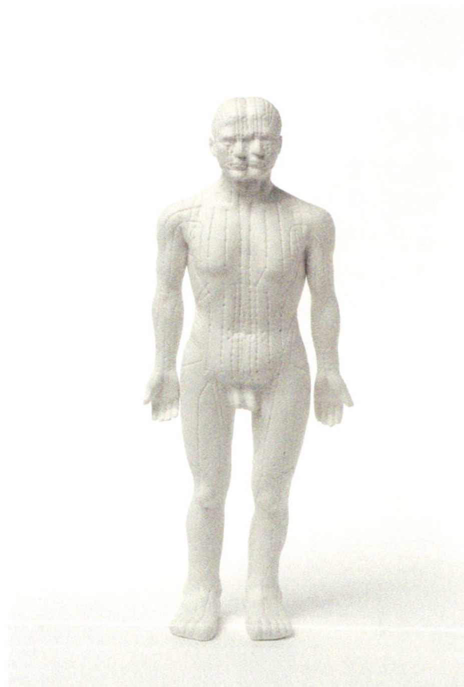

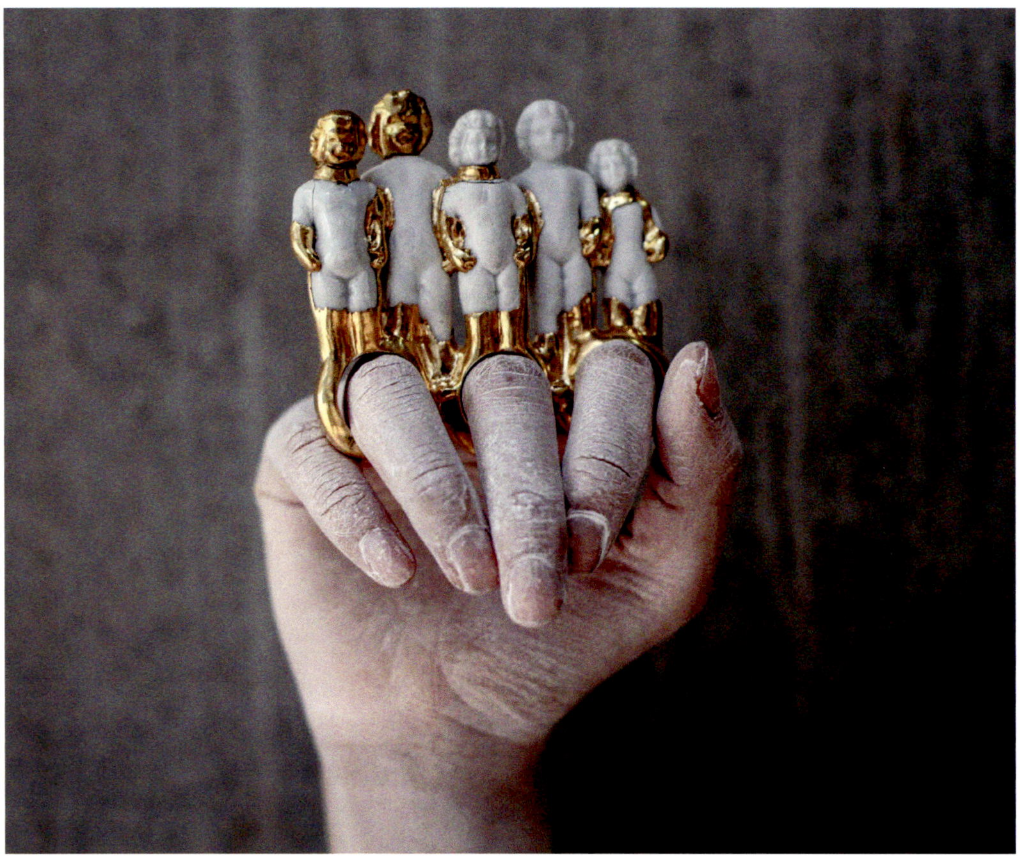

Top left Ruudt Peters, *CUN ZAI shua zi*, 2015, cast porcelain, silver, 16" × 6" × 3".
Photo: Rob Versluys, Collection of Lucie and Peter Wolfs.

Bottom Qian Yang, *Five Porcelain Dolls Ring*, 2014. Broken vintage porcelain dolls repaired with gold plated brass made into a three-finger ring.
Photo: Ksusha Zhitomirskaya & Qian Yang.

Top right Ruudt Peters, *CUN ZAI nan xing*, 2015. Cast porcelain, 16" × 6" × 3".
Photo: Rob Versluys, Collection of Lucie and Peter Wolfs.

Opposite page Maeng Wookjae, *Camouflage*, 2012. Porcelain, wood, 16" × 30" × 41".
Photo: Myoungwook Huh.

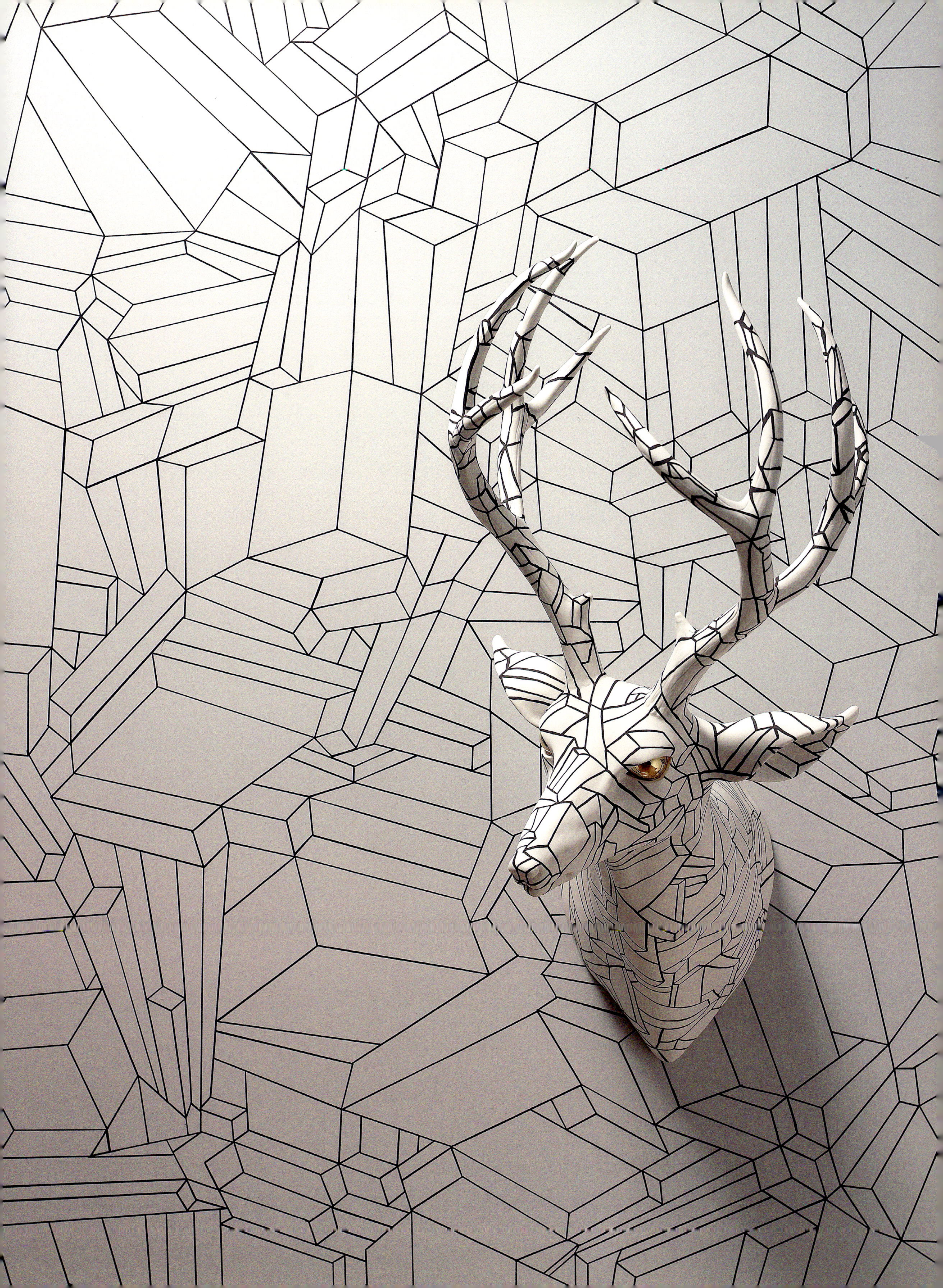

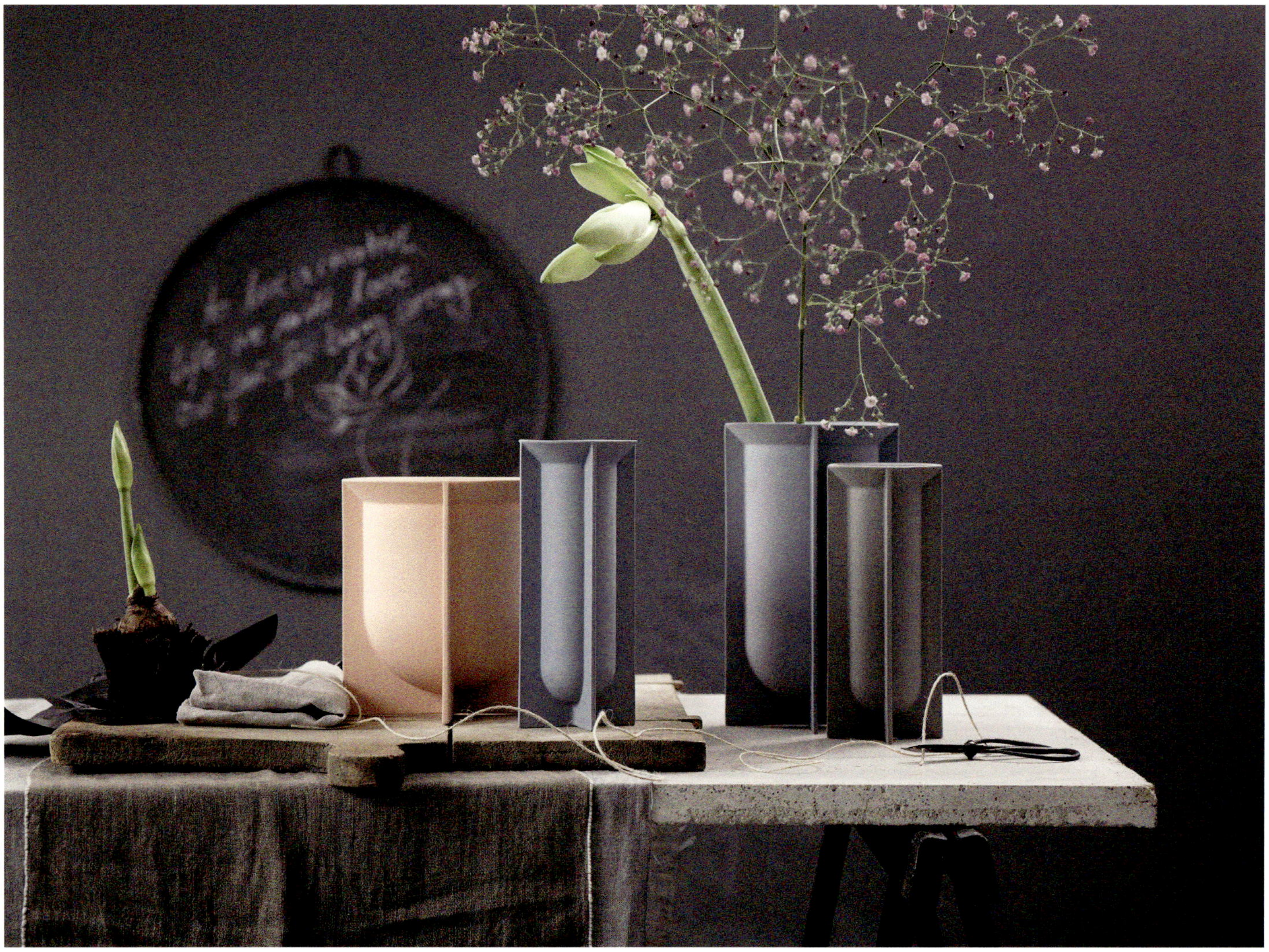

Above Sebastian Herkner for Rosenthal, *Rosenthal Domo Mood*, porcelain, dimensions variable. Photo: Andreas Hoernisch and Styling: Natascha Sanwald, Courtesy of Rosenthal GmbH.

OPPOSITE PAGE

Top Novie Trump, *Map of the Night Swallows*, 2014. Porcelain, steel, acrylic, light fixture, 14" × 14"× 8". Photos: Antonio Sirna.

Bottom Detail.

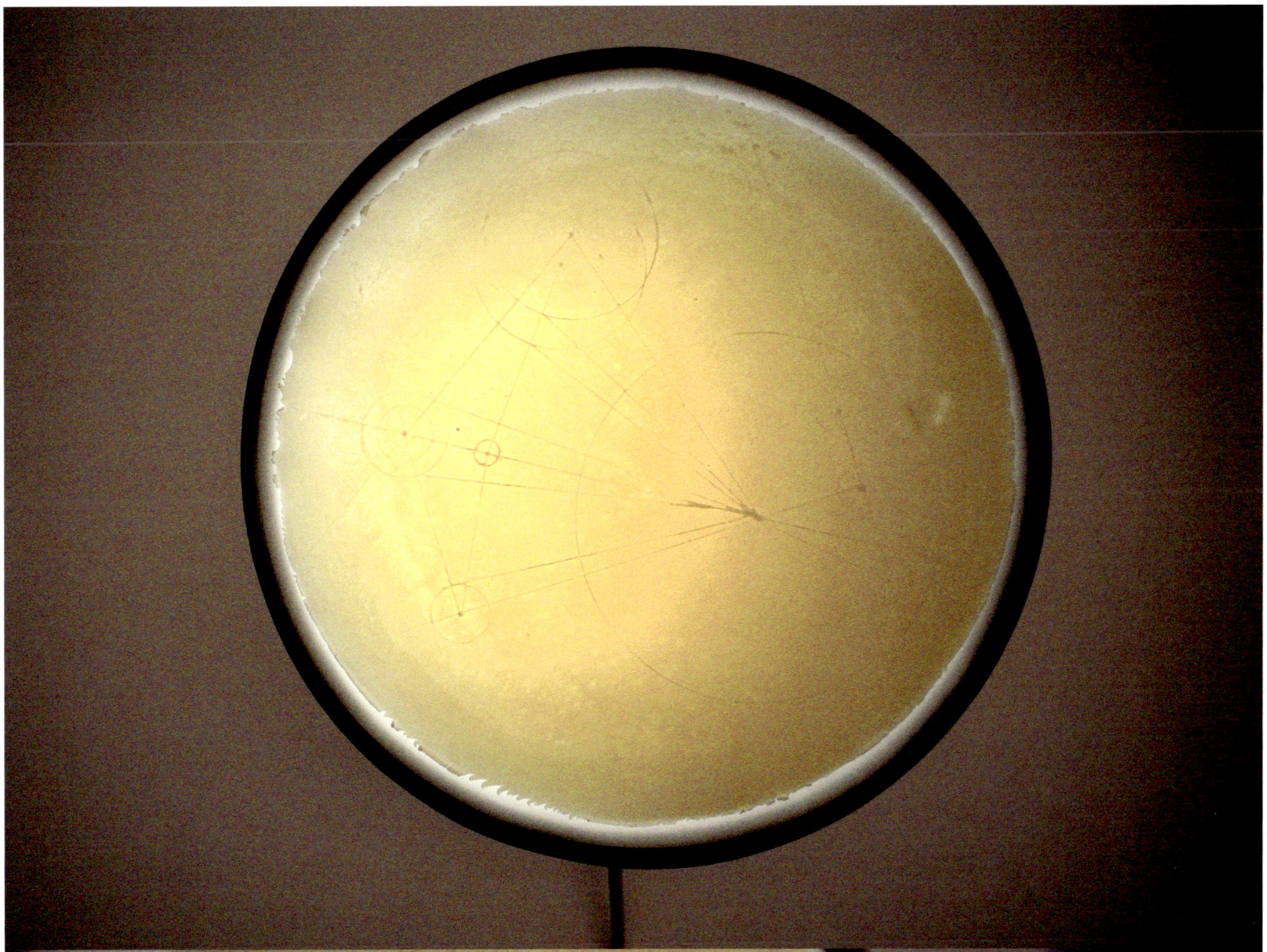

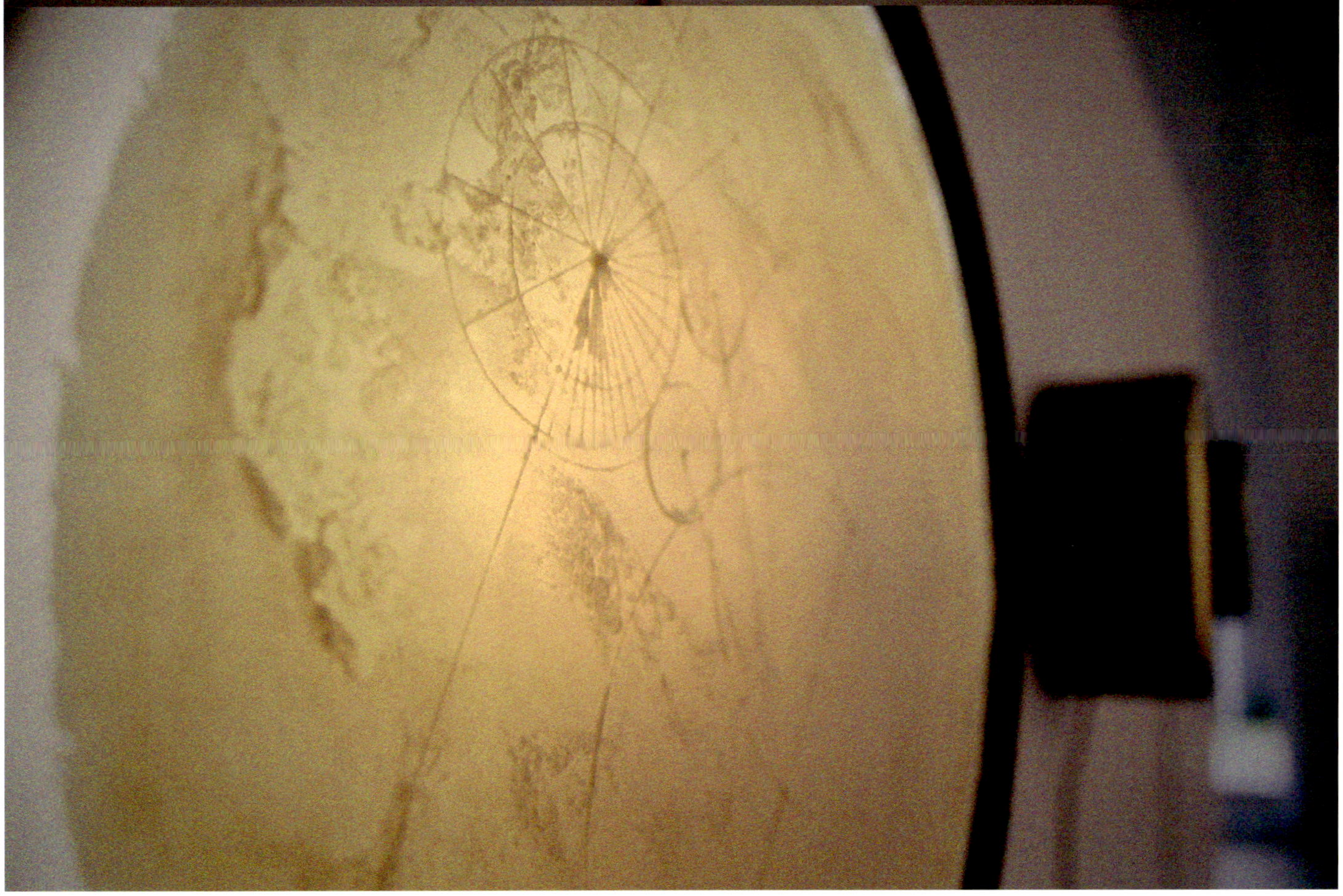

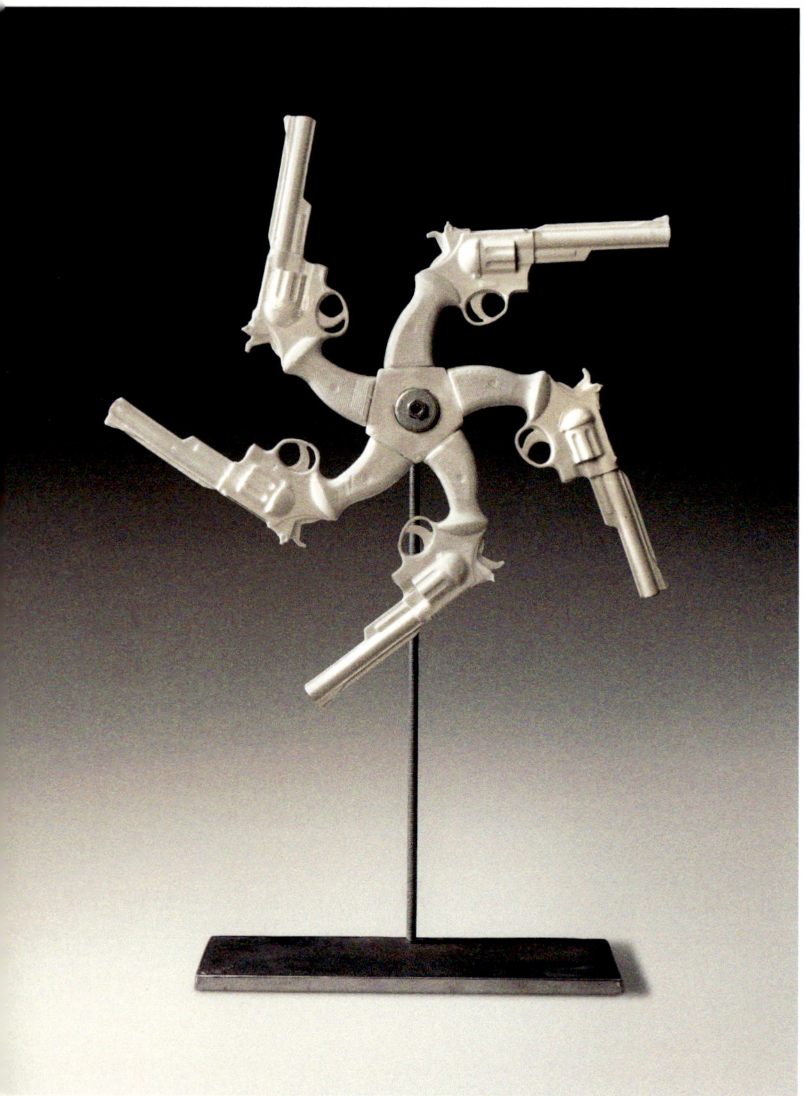

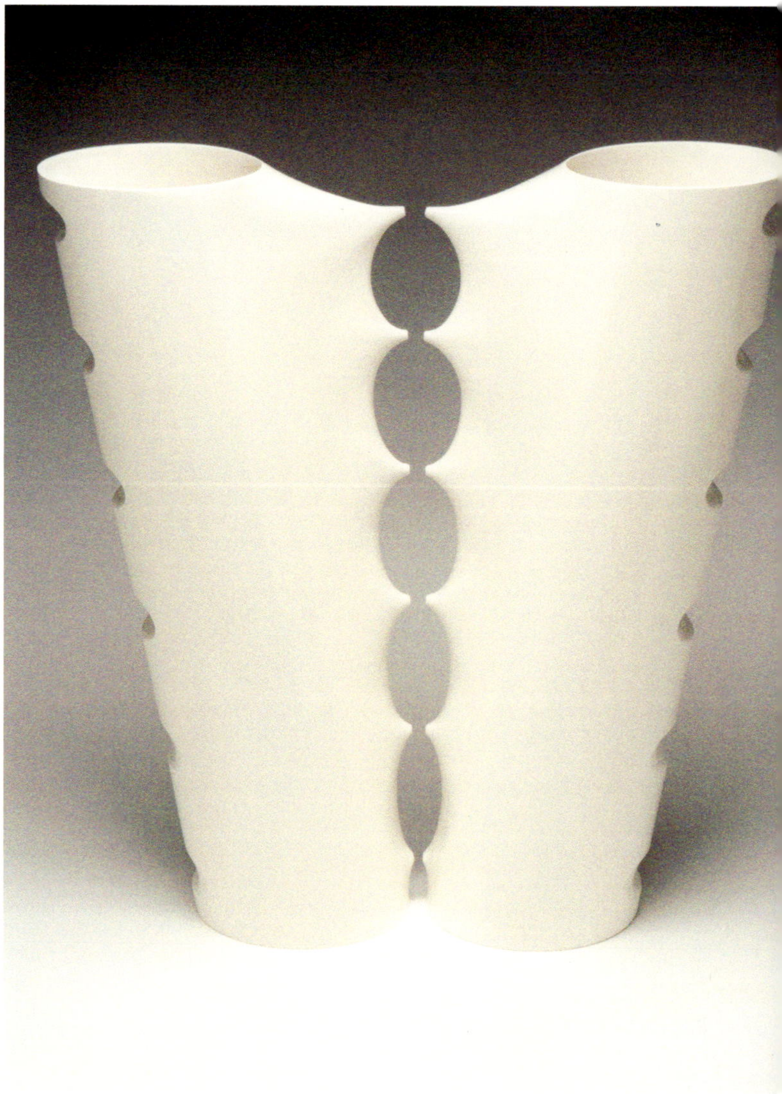

Above left Linda Lighton, Lighton Studios, *44 Magnum Mandala*, 2011. Clay, glaze, 2' × 1'11" × 2'2".
Photo: E.G. Schempf, Collection of Belger Art Center, Dick Belger and Evelyn Craft.

Above right Miles Spadone, *Mytosis*, 2014. Ceramic, tinted shellac, lacquer, 1'4.5" × 1' × 4".
Photo: Jay York.

Opposite page Trudy Golley, *Faux Banksia Vase,* 2010, Porcelain, 16.5" × 4.3" × 3.14".
Photo: Paul Leathers, Collection of Alberta Foundation for the Arts.

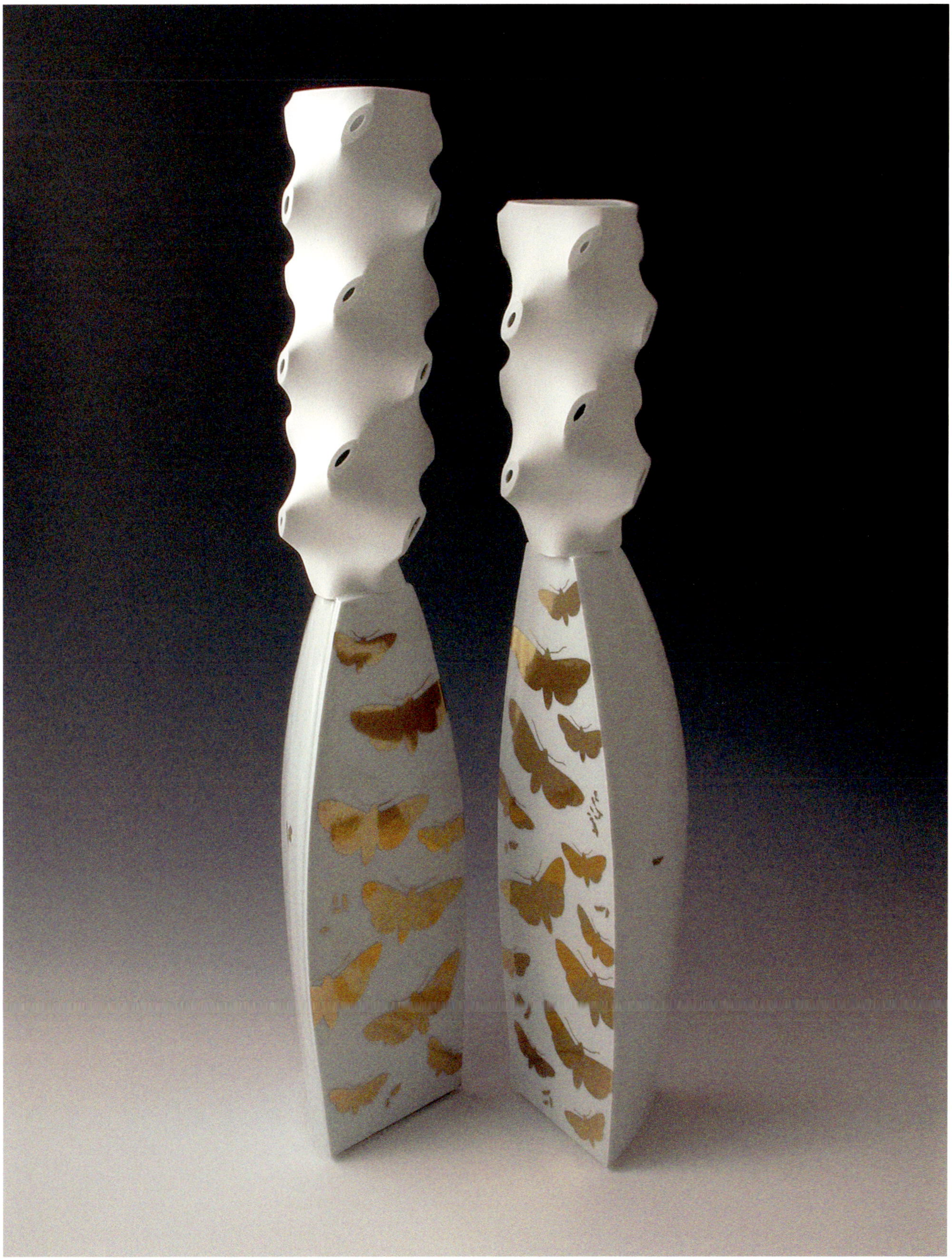

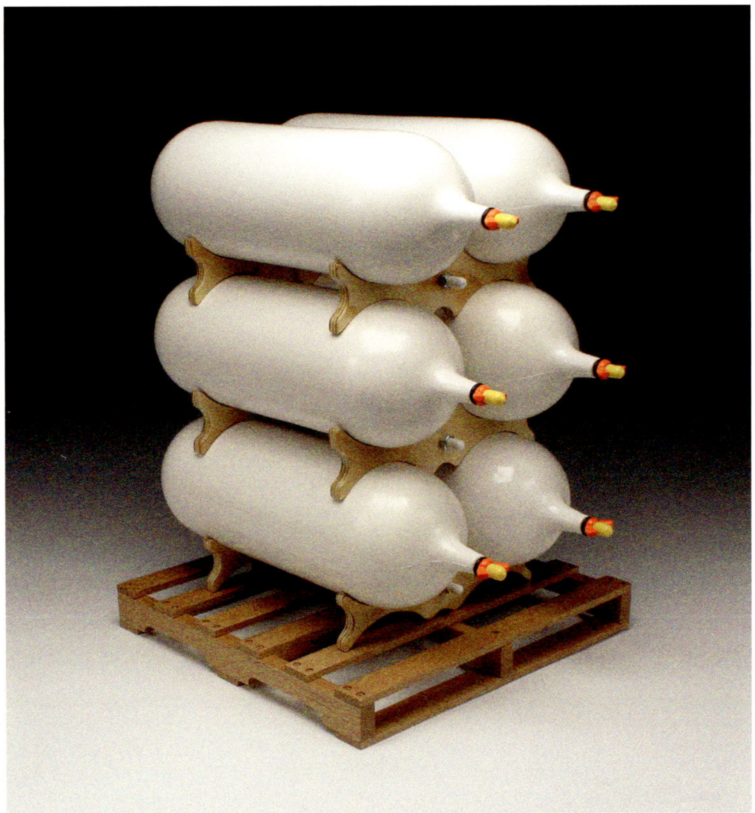

Top Robert "Boomer" Moore, *Tanks*, 2013. Whiteware, wood, 18" × 10" × 10". Photo: Trina Moore.

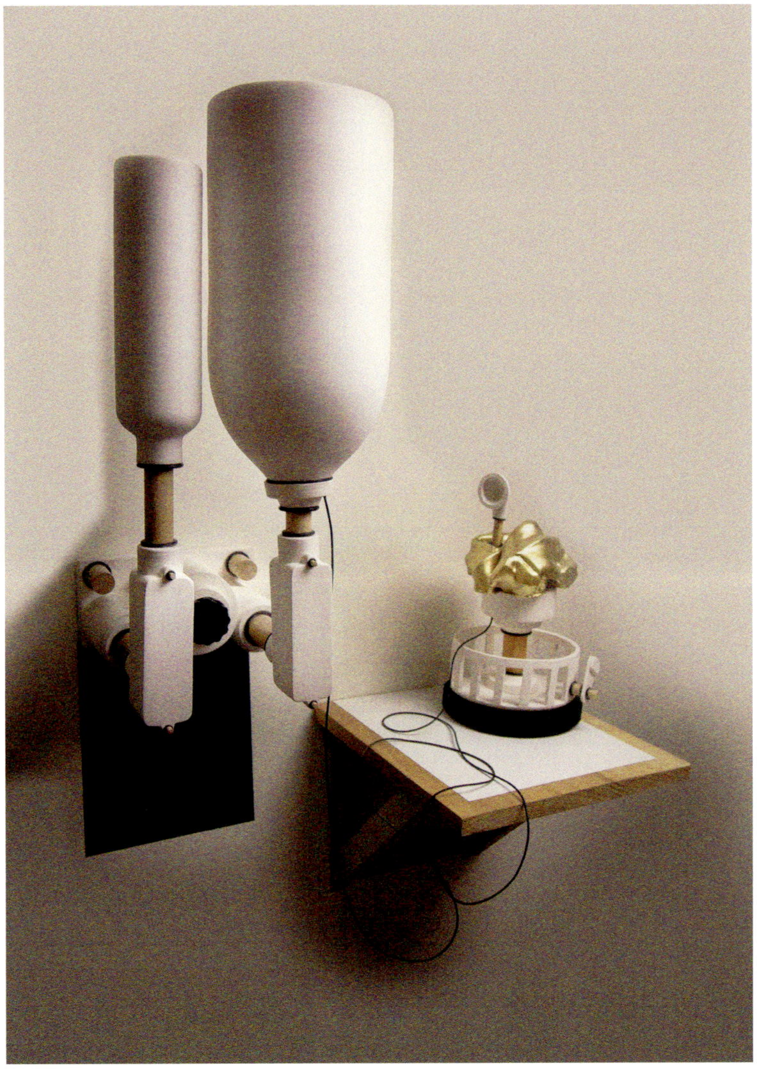

Bottom Kwok Pong "Bobby" Tso, *Moment of "Re"inventing – Ideally Speaking*, 2013. White earthenware, wood, plastic, paint, 1'5" × 1'8" × 1'. Photo: Kwok Pong "Bobby" Tso, Collection of Archie Bray Foundation for the Ceramic Arts.

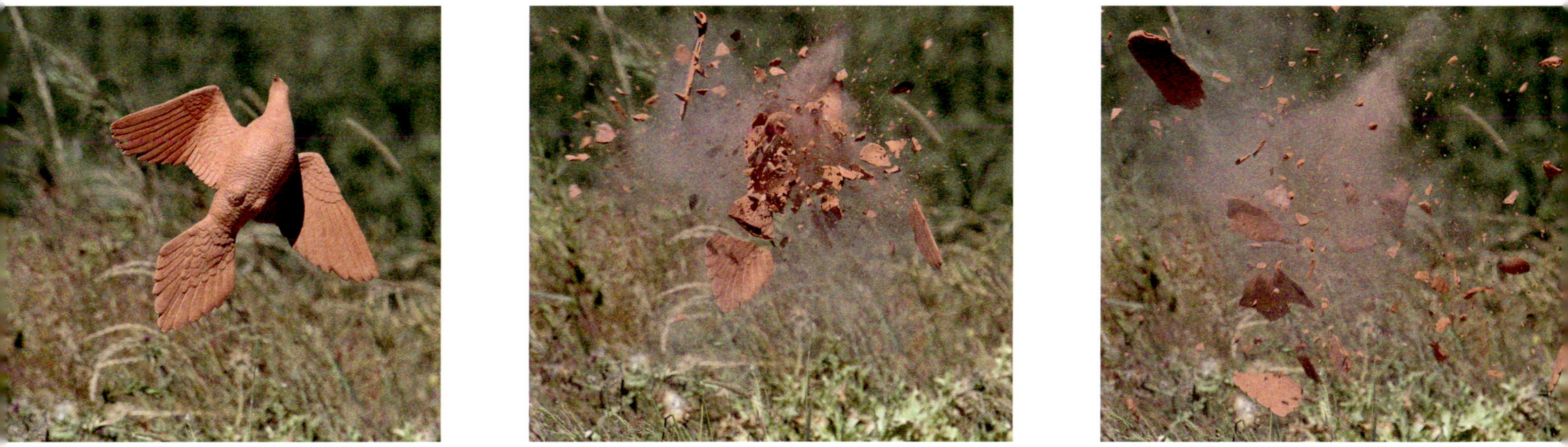

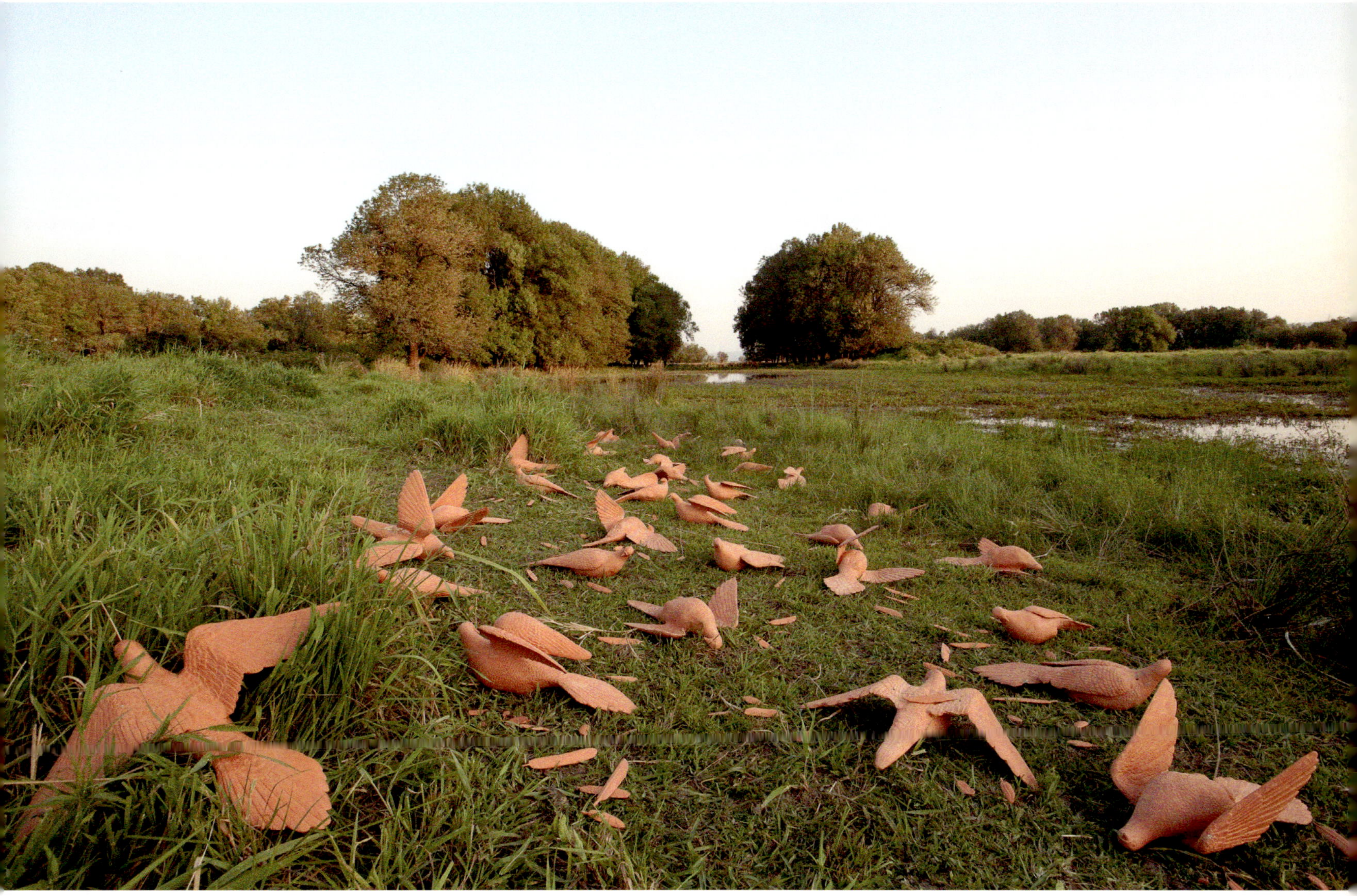

Kate MacDowell, *Clay Pigeons*, 2010. Slipcast and hand carved terracotta fired to cone 04, lead shot, dimensions variable, (each pigeon is life-sized, about 16" from beak to tail-tip).
Photos: Dan Kvitka.

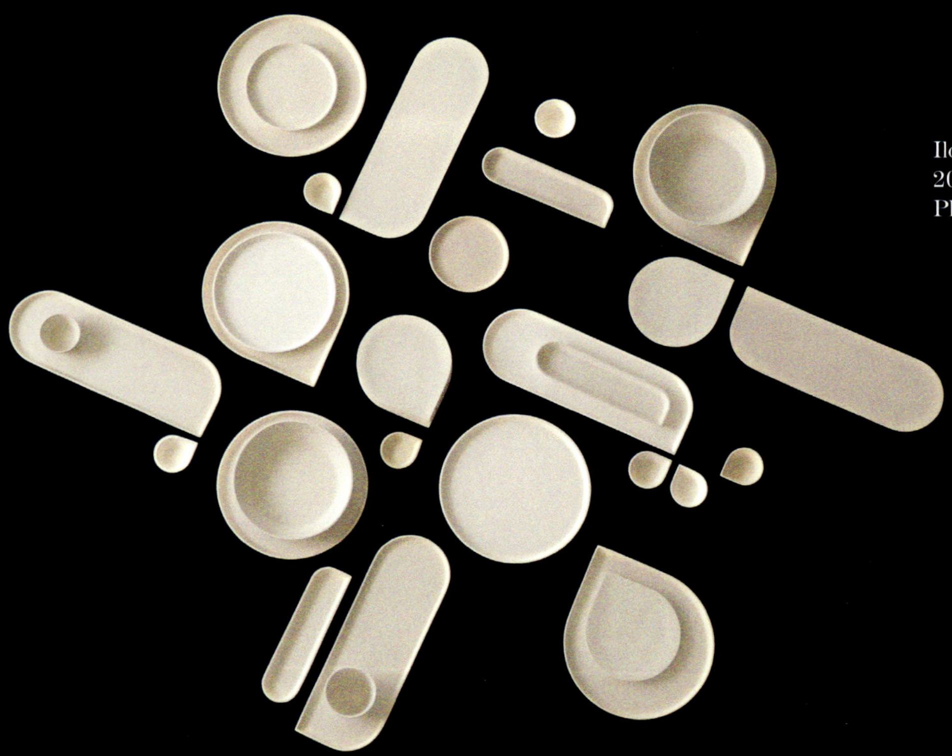

Ilona Van den Bergh, *OnA* tableware, 2012. Stoneware, porcelain. Photo: Ilona Van den Bergh.

Dryden Wells, *Stack 8*, 2010. Slipcast and assembled, found object, porcelain, fired to cone 13 reduction, 1'11" × 1' × 1'2". Photo: Dryden Wells.

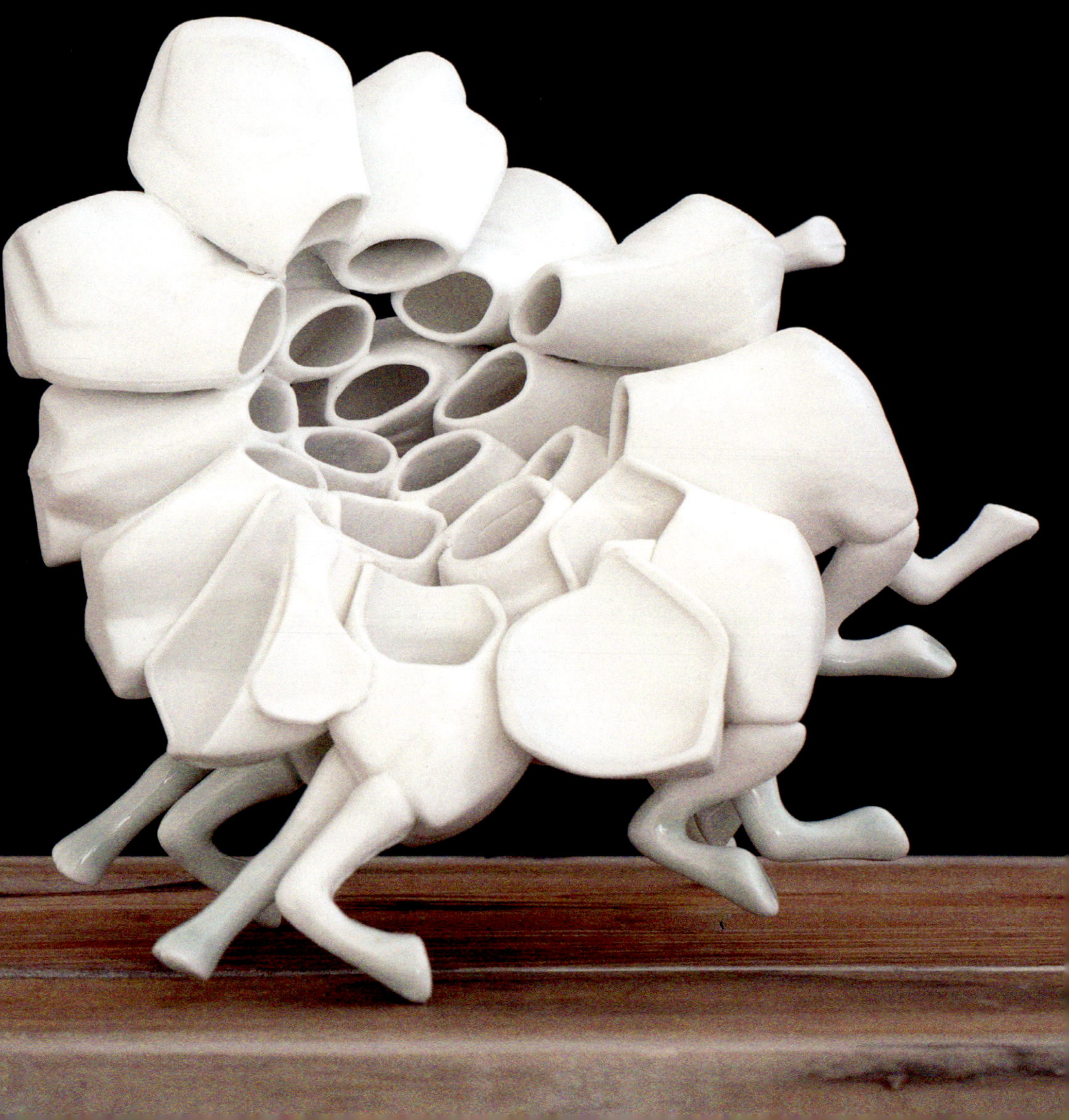

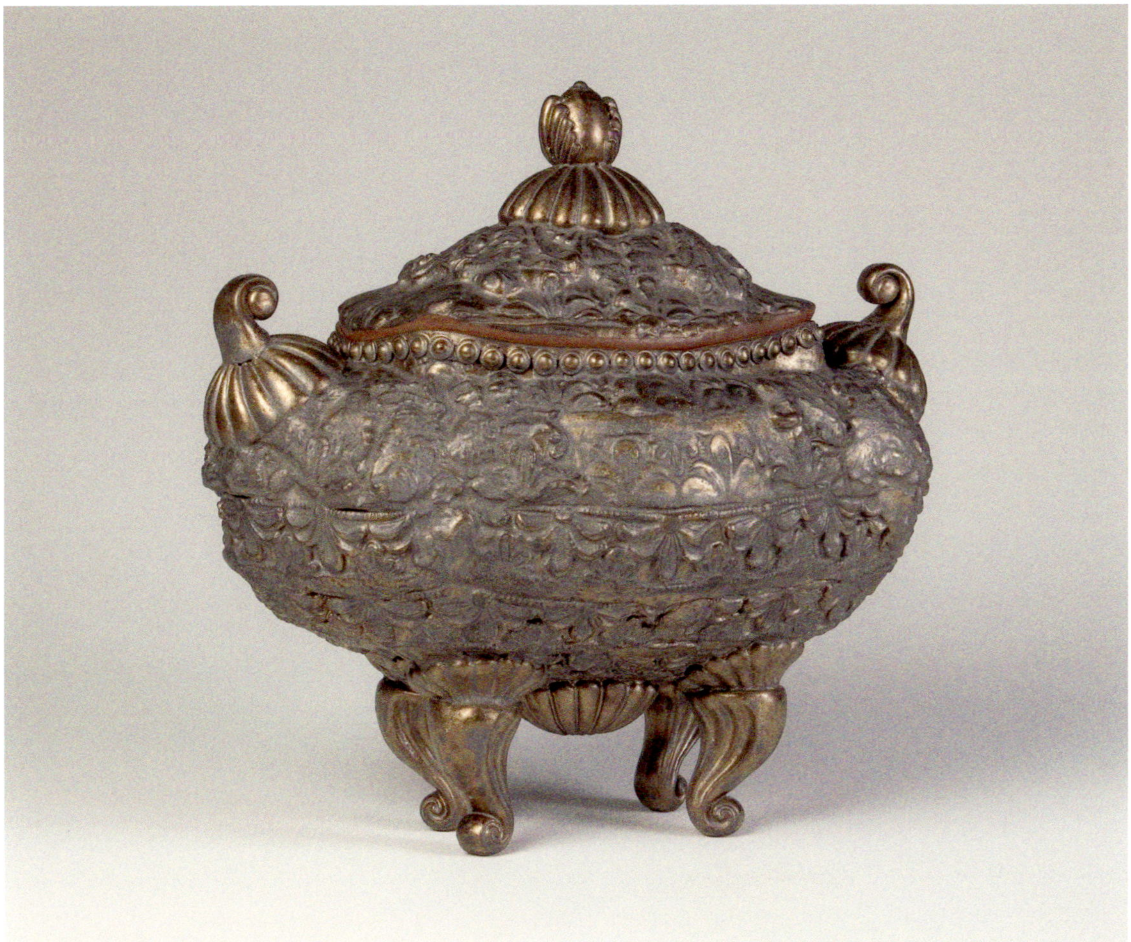

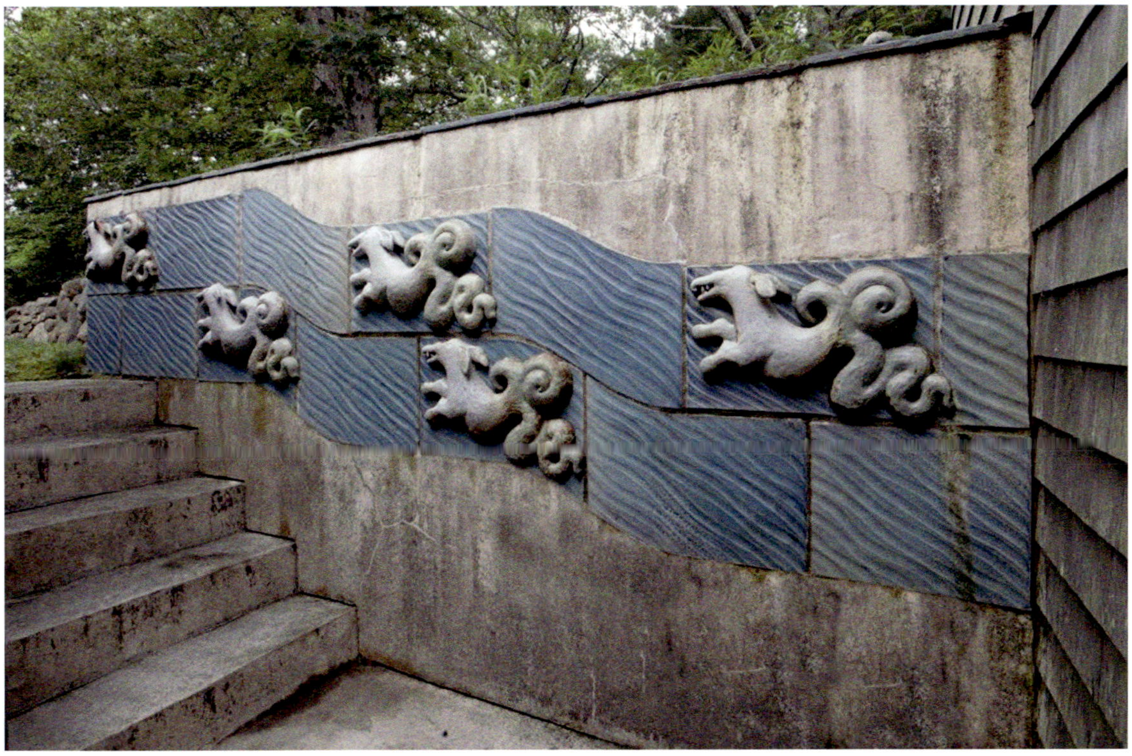

Top A. Blair Clemo, *Tureen*, 2010. Cast, press molded and handbuilt red stoneware, 17" × 10" × 13".
Photo: Mike Fleming, Collection of Schein-Joseph International Museum of Ceramic Art, Gloryhole Collection, NYSCC, Alfred, NY.

Bottom Lucy Scanlon, *Retaining wall with Bas Reliefs*, 1999. Casting slip high fired in soda and reduction kilns, 2'9" × 11'10" × 4.62".
Photo: Ken Richardson.

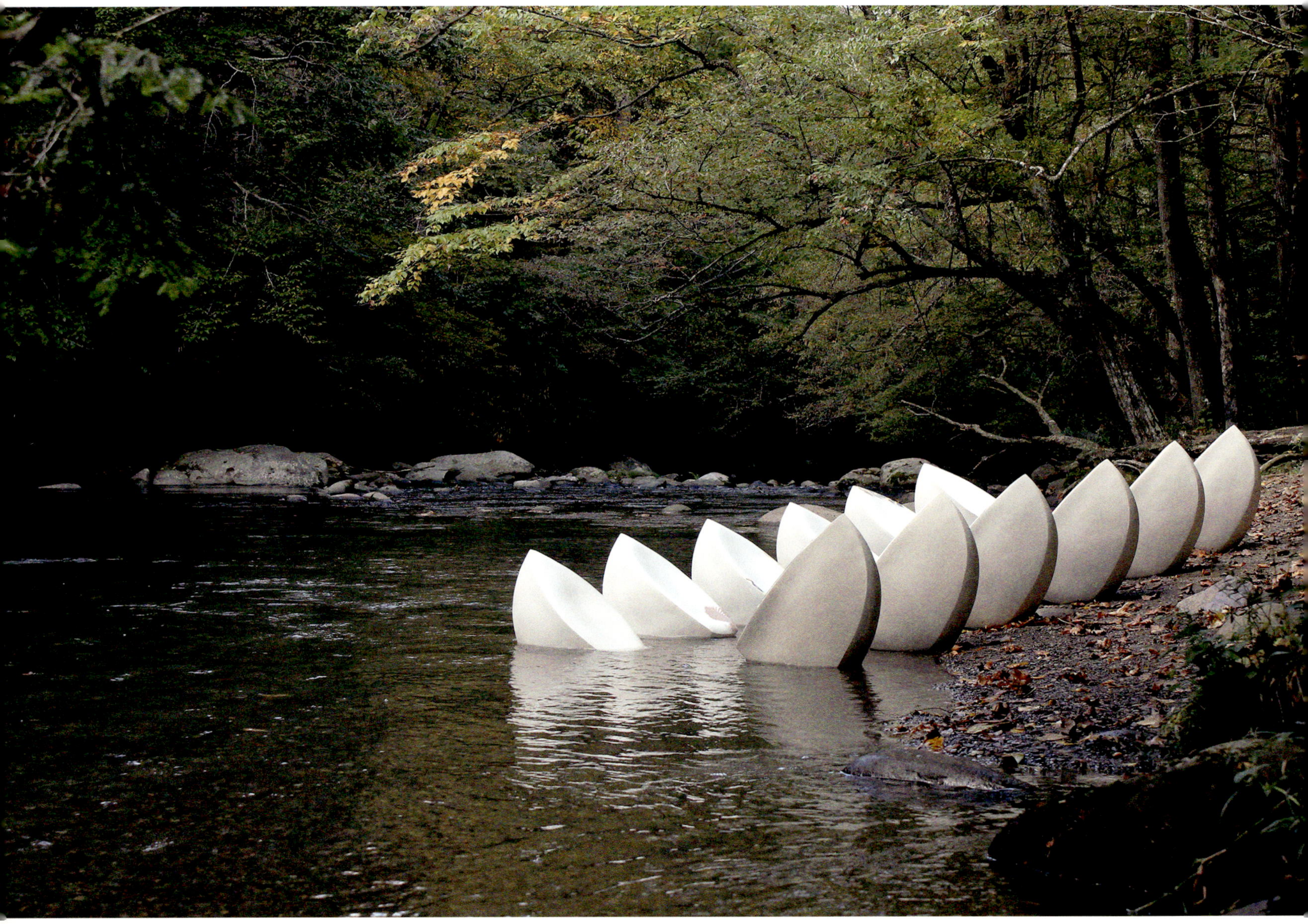

Rachel K. Garceau, *gateway*, 2013. Porcelain, each form is 24" × 14" × 22".
Photo: Rachel K. Garceau.

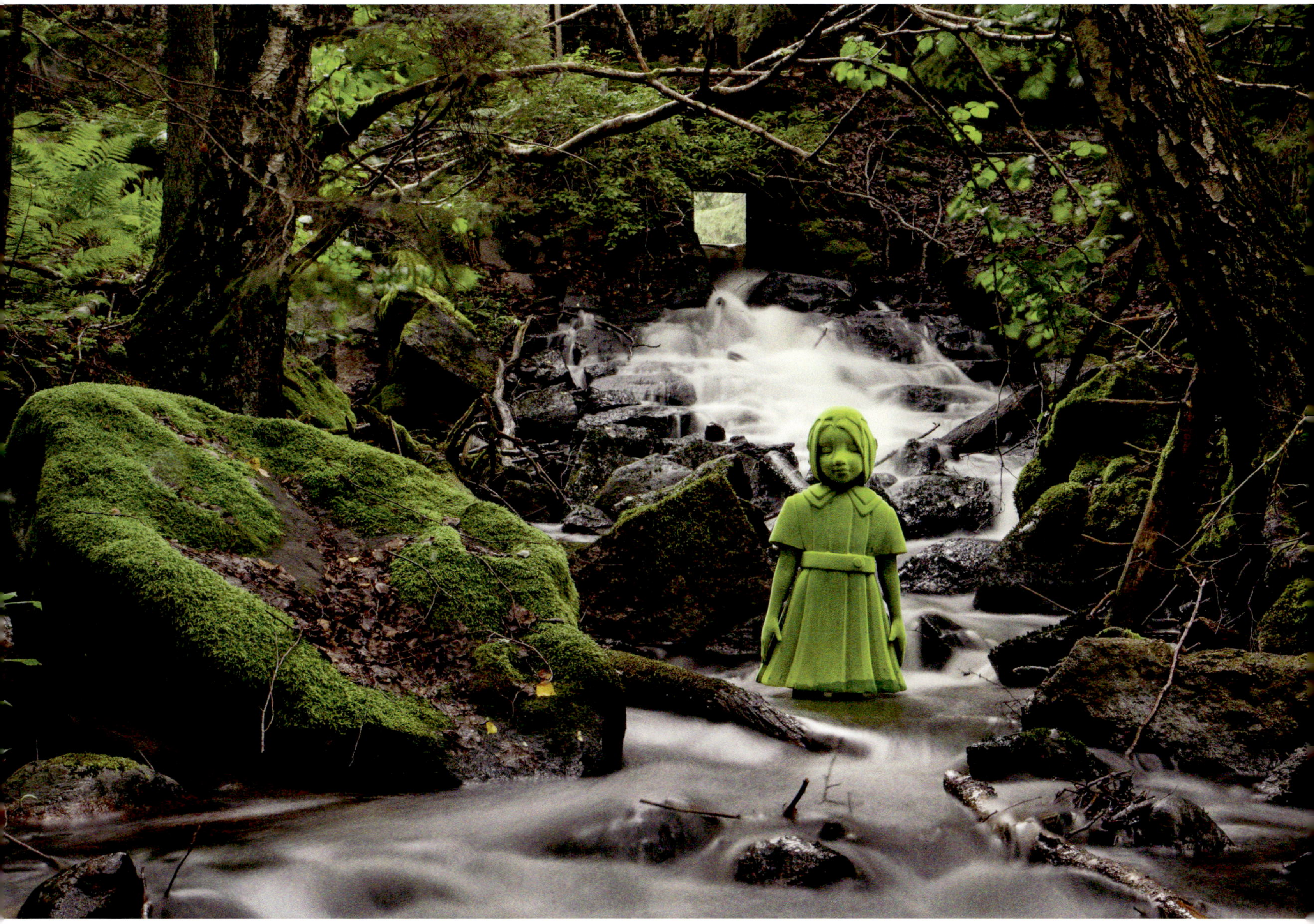

Kim Simonsson, *Moss Girl and River*, 2015. Ceramics, nylon fiber, 21.65" × 21.65" × 47.24". Photo: Jefunne Gimpel.

GLASS

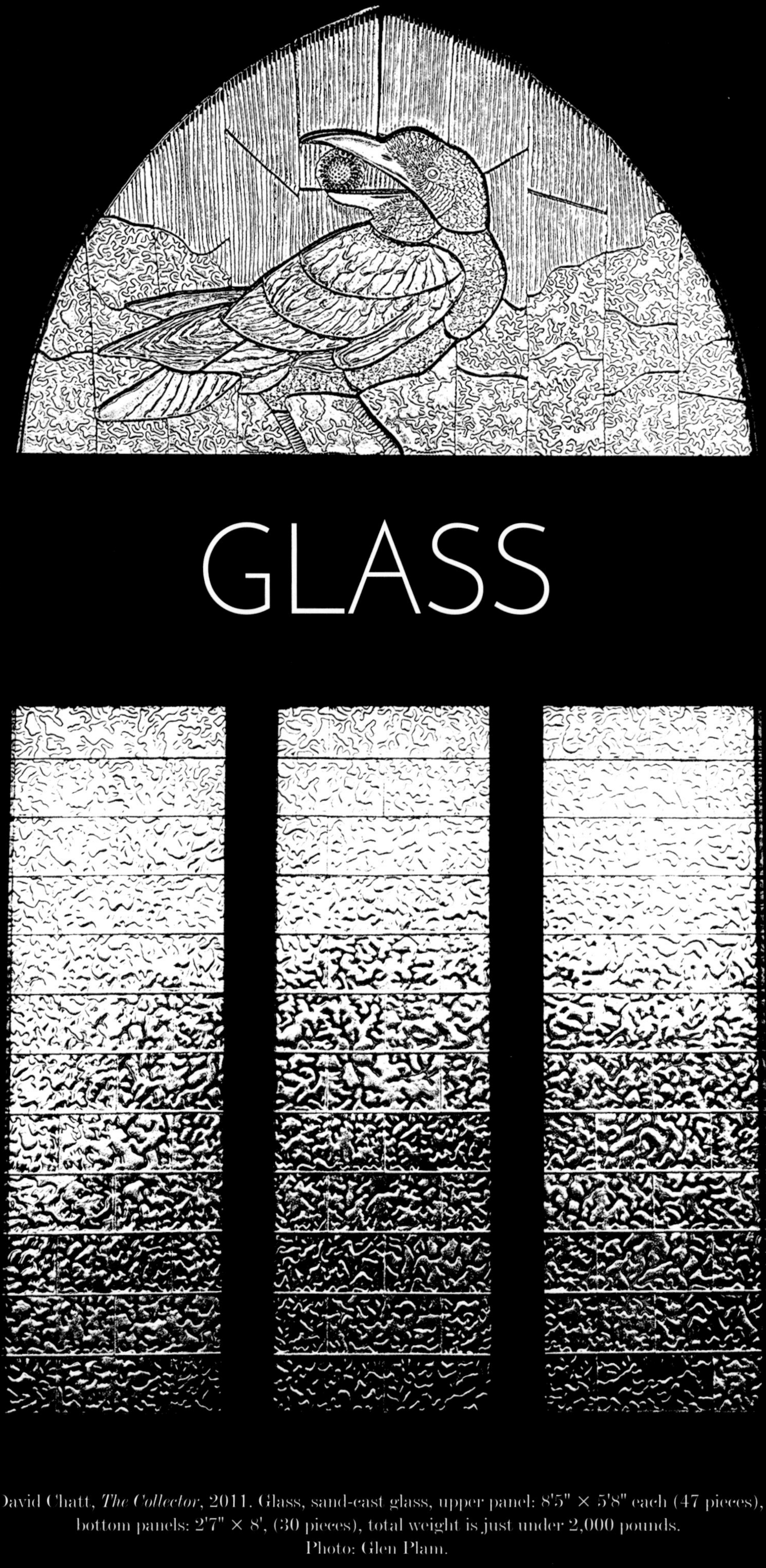

David Chatt, *The Collector*, 2011. Glass, sand-cast glass, upper panel: 8'5" × 5'8" each (47 pieces), bottom panels: 2'7" × 8', (30 pieces), total weight is just under 2,000 pounds.
Photo: Glen Plam.

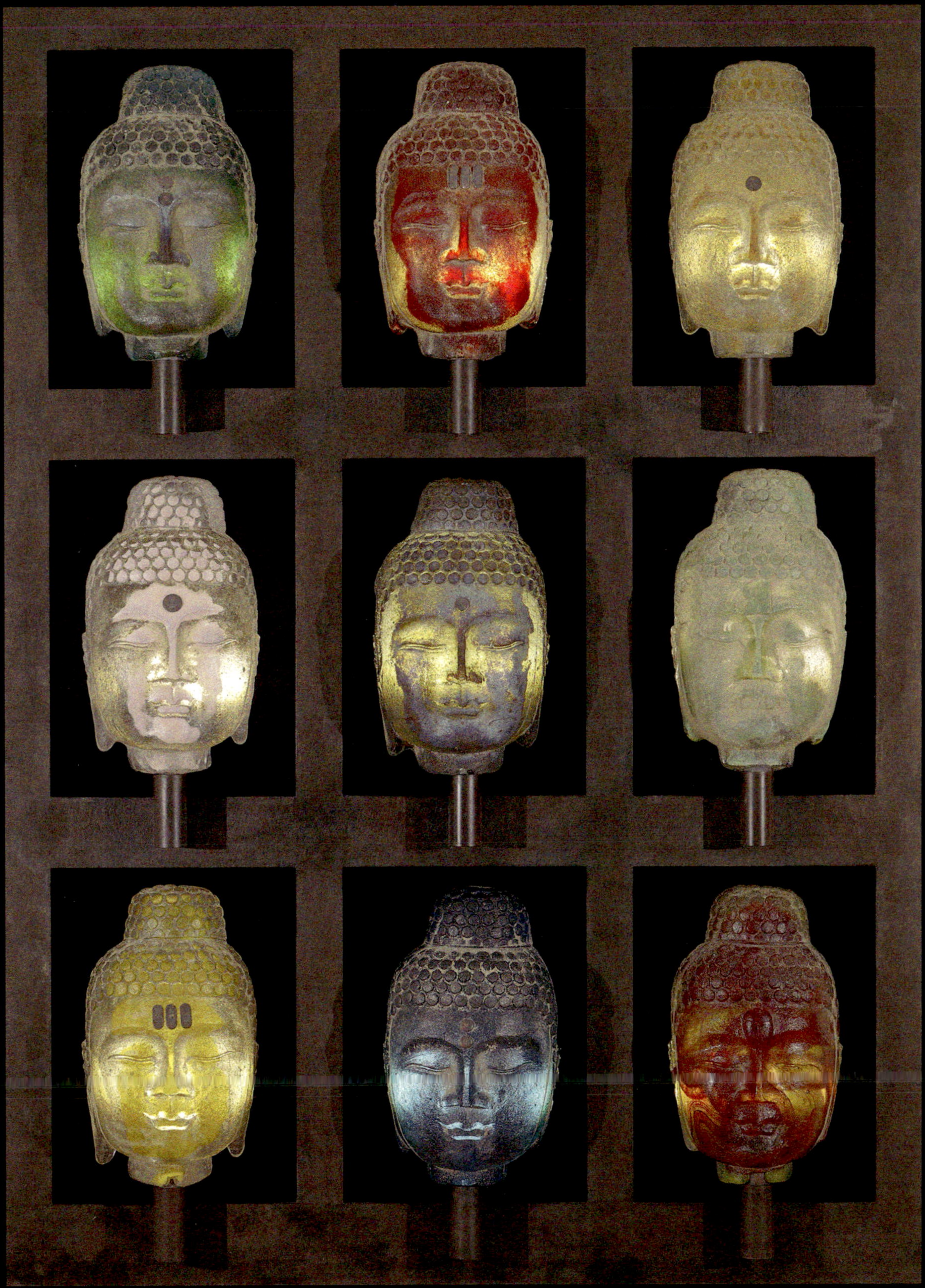

Marlene Rose, *Buddha Wall*, 2011. Glass, steel, and copper, 4'6" × 3'2" × 9". Photo: David A. Monroe.

GLASS CASTING
A History of Innovation

by Susie J. Silbert

Walking into any major glass exhibition, you are likely to encounter a variety of forms of cast glass. From haunting and monumental, like Karen LaMonte's kiln cast dresses, to mysterious and ethereal like Mark Peiser's polychromatic hot cast *Passages*, or ruminative and reflective like Norwood Viviano's *Mining Industries*, cast glass objects inhabit space, work with light, and feature the properties of glass in a way that is all their own. A collection of techniques, modern casting is primarily broken down into two categories, hot casting and kiln casting, broadly defined by the temperature used to carry out the process and the location where it occurs. Within these categories, there is significant room for exploration and innovation, making casting one of the most customizable glass forming processes in use today. Perhaps for that reason, it has been embraced by amateurs and professionals alike who seek a method of working in glass that is flexible enough to accommodate a range of ideas and aesthetics.

The adaptability of glass casting has also had key implications in the development of the material over time. From the very beginnings of glassworking nearly five thousand years ago to its contemporary moment, casting has been an entry point for new technologies and a means by which glassworkers come to understand their material and imagine new identities for it. In ancient Mesopotamia, for example, casting was one of the very first methods used by early glassworkers looking to apprehend this unprecedented material. Borrowing forming techniques and object types from ceramics and faience, a vitreous material moldable at room temperature, these early glassworkers created amulets and inlay by using molds to stamp forms into softened glass and fusing chips of glass in open faced molds made from clay or steatite.[1]

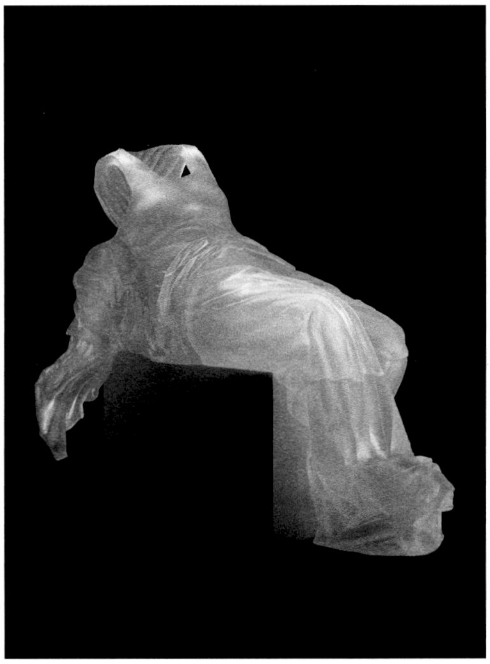

Karen LaMonte, *Semi-Reclining Dress Impression with Drapery*, 2005. Cast glass, 3'5" × 2'3" × 3'3". Photo: Martin Polak, © 2005 Karen LaMonte.

Small, opaque, and brightly colored, the objects early glass casters made not only borrowed technically from other materials, but were intended to emulate the look of other media as well. Without a sense of the range of visual possibilities of glass (for instance, transparency) and without a cultural need for those possibilities, the value of glass came from its ability to successfully mimic precious stone. This was especially true in ancient Egypt. From about 1450 BCE, glass was used in jewelry, furniture, and wall plaques in place of turquoise, carnelian, and lapis lazuli. These cast items were no mere knockoffs; instead, they were employed alongside precious stones in courtly objects including King Tutankhamen's throne.[2]

Even centuries later, when glassmakers in Assyria (eighth century BCE) and Persia (fifth to fourth century BCE) discovered how to make impressively transparent glass, they employed the property in service of emulating rock crystal. The highly accomplished vessels they produced in transparent green and colorless glass were formed as stone would be too, using lapidary tools to hollow out and refine cast glass blanks. The practice of using cast glass as a stand-in for rare stones continued throughout the Hellenistic and early Roman periods, with glassworkers producing ever more complex facsimiles as they developed new glass working techniques.

A phenomenal sea change occurred in the meaning, use, and value of glass with the invention of glassblowing and its concurrent improvement in furnace technology between 25 BCE and 25 CE. While ancient glassworkers operated like modern day kiln casters, melting premade billets in low temperature kilns, the advent of glassblowing brought widespread access to molten glass for the first time, revolutionizing the capabilities of the average workshop. Soon, the speedy technique of glassblowing transformed the material from a laboriously produced specialty item available to a relative few into an affordable and ubiquitous aspect of everyday life. Though Roman glassworkers still occasionally created luxury glass, like the incredible fourth-century-CE carved cage cups, blowing soon supplanted other glass forming processes from casting to core forming.

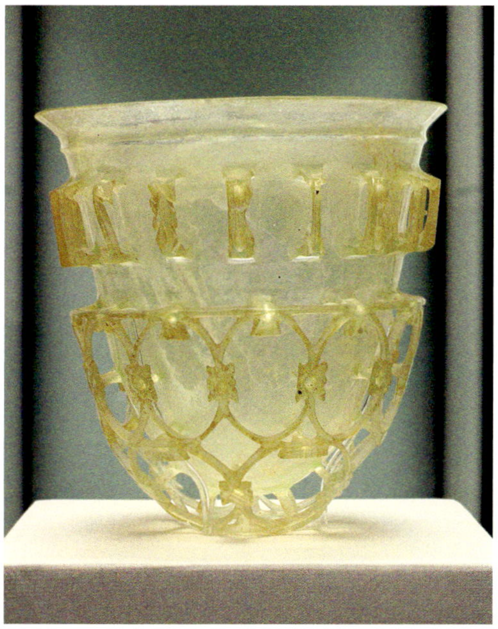

Roman Cage Cup, second half of the fourth century. Roman, glass.
Photo: Matthias Kabel, Staatliche Antikensammlung (The State Collections of Antiquities), Munich, CC-by-3.0, GNU FDL.

All expertise associated with casting was not lost, however, as molds became an essential part of glassblowing. By inflating a bubble of glass within a specially designed blow-mold, glassblowers could shape and decorate their vessels in one quick, reliable step. This innovation was widely adopted and mold-blown glassware spread throughout the Roman and, later, the Islamic world.

For the next one and a half millennia, casting was not a major part of glassmaking. When it did resurface, it was in service of the opulent courts of Louis XIV, whose *Manufacture Royale de Glaces des Miroirs* at St. Gobain used hot casting technology innovated by Italian immigrant Bernard Perrot to create mirrors of unprecedented size. Documented in Diderot's *Encyclopedia*, mirrors of up to nine feet tall and three feet wide were cast by pouring massive amounts of glass onto a metal table and spreading to size with a large roller.[3] The industrial applications of casting continued in the early nineteenth century, when American glassworkers created a hand-operated machine for glass pressing. By dropping a gather of hot glass between two bronze dies, pressers could form a vessel and its decoration in record time, making highly ornamented glass cheap, affordable, and abundant for the first time in the modern era.

A few decades later, French artists working out of the ceramic research facilities at Sèvres porcelain developed pâte-de-verre, one of the best known and most beloved of all kiln working techniques. Using finely ground glass mixed with a binder into a paste, artists like Henri Cros, Amalric Walters, Argy Rousseau, and François Décorchment, created stunningly detailed multi-colored figurines, vessels, and other small sculptures in the Art Nouveau style. However, despite the popularity of the technique in the late nineteenth and early twentieth centuries, by the 1930s pâte-de-verre had gone out of fashion, leaving little indication of the processes used in its manufacture.[4]

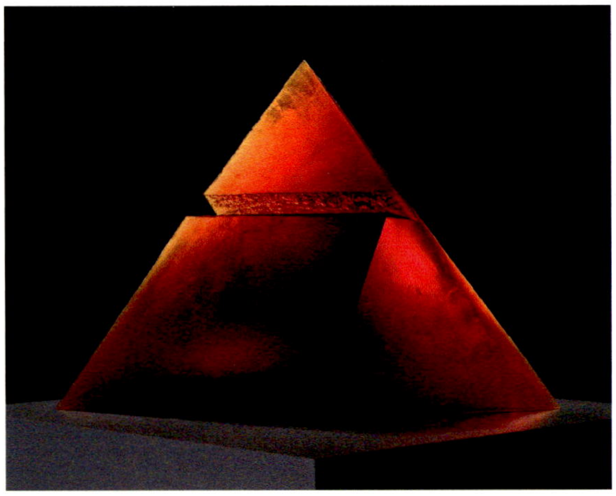

Stanislav Libensky and Jaroslava Brychtová, *Red Pyramid*, 1993. Zelezny Brod, Czech Republic, glass, 2'9" × 3'11" × 11".
Courtesy of Corning Museum of Glass, Collection of The Corning Museum of Glass, Corning, N.Y., Gift of the artists.

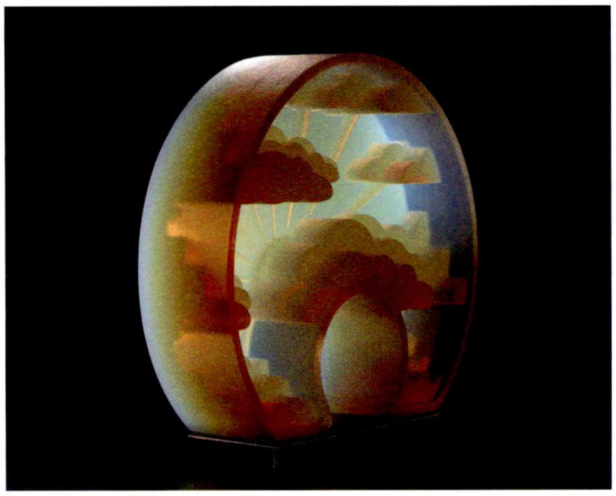

Mark Peiser, *Passage #1*, 2012, Hot cast, phase-separated glass, granite base, 23.25" × 8.625" × 25".
Photo: Steve Mann, Private Collection.

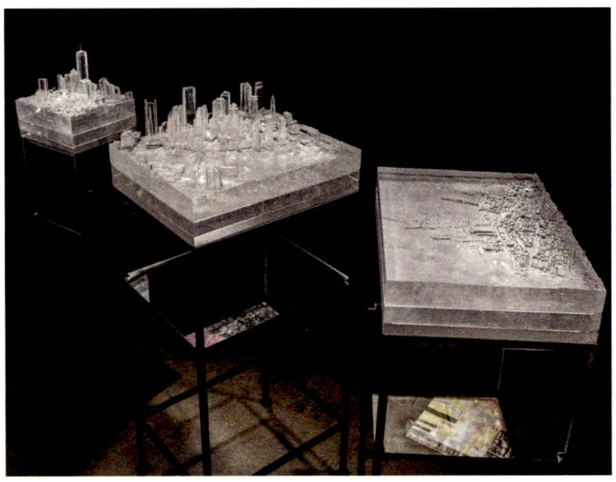

Norwood Viviano, *Mining Industries: Boston* (installation detail), 2015. Rapid prototyped pattern kiln cast glass, fabricated steel, dimensions variable.
Photo: Tim Thayer/Robert Hensleigh.

The turmoil and prosperity of the postwar period led to critical innovations in the use of both hot and kiln casting in the mid-twentieth century. Some, such as the exuberant deployment of hot casting by designers Bertil Vallien and Erik Höglund from Sweden, and Oiva Toikka from Finland, in the early 1950s and 1960s, were spurred by the fertile environment nurtured by the global studio craft movement. Others, like the monumental, geometric kiln castings developed by Stanislav Libenský and Jaroslava Brychtová in Czechoslovakia beginning in the mid-1950s were a creative response to an overly repressive government. Banned from pursuing abstract art in painting and traditional sculpture, Libenský, Brychtová, and a host of others channeled their ideas through the "minor" art of glass. The large scale castings exploring light, form, and volume they produced in concert with state-run glass foundries continue to be among the most sought after and admired examples of artist-made glass today.

In the United States, despite experiments by Edris Eckhardt in the 1950s and Henry Halem in the early 1970s, casting did not take off in a major way until the early 1980s, almost twenty years after the studio glass movement began. In part, the delay was due to early studio glass' emphasis on blowing and the overall lack of adequate glass working equipment in those first years. Nonetheless, by the late 1970s, artists drawn to the narrative and representational possibilities of casting and up for the technical challenges it represented began experimenting with casting in a concerted way. Every decade since then has a seen a proliferation of artists interested in the multifarious effects of casting and an expansion of the equipment, glasses, and technical knowledge available to them.

Today, artists cast glass for many of the same reasons that animated its historical use: the desire to replicate objects in alternate media, to apprehend the material's properties, introduce new technologies, and to explore luxury and grandeur of scale. For first- generation studio glass practitioner Mark Peiser, casting is a means to explore the material possibilities of glass and glass chemistry. Composing his own glass formulae, specialized castings furnaces, and mold-making equipment, Peiser has taken innovative approaches to hot casting since 1983.[5] For most of the last decade, he has been investigating opal glasses, and specifically, working to master the liminal moment when a striking opal transitions from perfectly clear to totally opaque. By carefully controlling the way the glass is poured as well as the design of the mold, and crucially, the rate at which the piece cools, Peiser is able to make works that are neither fully transparent nor opaque, but rather occupy all points in between and, as a result, appear as if they contain the entire spectrum of color. Only producible by hot casting, and even then, necessitating the continuous stream of a bottom pouring furnace, Peiser's works are both artistic and scientific culminations of a lifelong pursuit of knowledge through casting.

While Peiser has spent the last thirty years developing new technologies within the medium, Norwood Viviano and Daniel Cutrone use casting as a way to introduce new technologies to glass. Fluent in digital media, they each design and produce their mold positives with the assistance of computer-aided tools including CAD software, CNC (computer numerical control) milling machines, and 3-D printing.

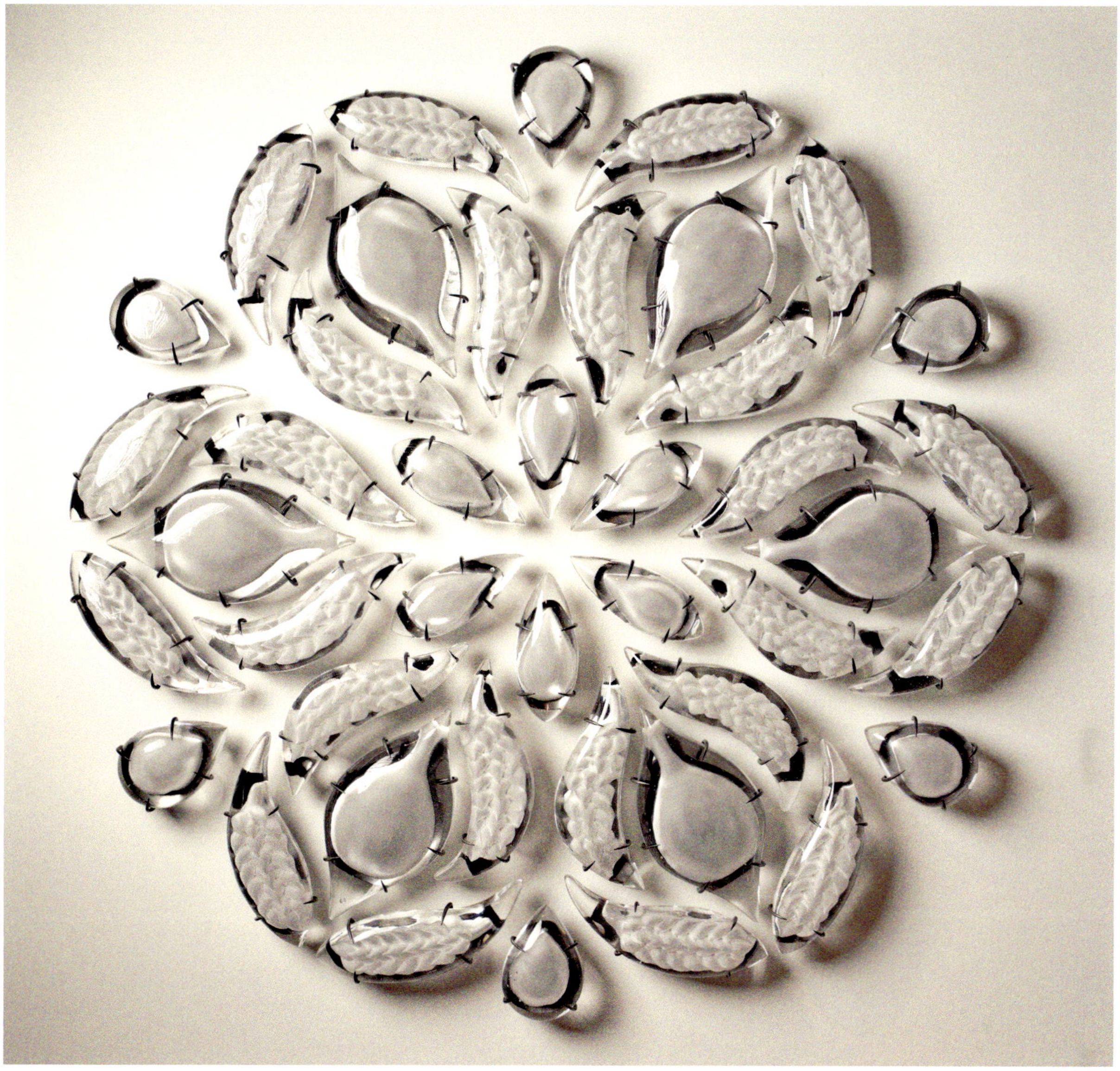

Joanna Manousis, *Demeter's Rose*, 2015. Crystal, steel, aluminum, negative core cast crystal, forged steel 4'10" × 4". Photo: Kenny Ek, Courtesy of Wexler Gallery, Private Collection, Australia.

Casting in all media has long been used to create replicas at a 1:1 scale, but digital tools enable the artists to reproduce large, complex objects in miniature with remarkable accuracy. Both use that capability to its fullest in works that reference geography. Cutrone's *Objects of Desire* series features Mount Everest within a variety of settings. Reflected into a blackened mirror, or tipped on its side within a blown glass capsule, these pieces provoke viewers to reexamine their relationship to the unattainable. In his *Mining Industries* series, Viviano reproduces full segments of cities in three-dimensions in order to examine their industrial past. Overlaid onto maps representing the same area at different points in time, Viviano's cityscapes become a series of lenses alternately occluding and giving access to the history of place.

Channeling the sumptuousness of nineteenth-century pâte-de-verre and the monumentality of Louis XIV mirrors, history and memory also animate the works of Karen LaMonte and Joanna Manousis. Harnessing the capabilities of kiln casting to create forms that embody two images at once, these artists employ the transparency and reflectivity of glass to invest found items with an otherworldly significance. Using the lost wax technique, Karen LaMonte creates sculptures of dresses that retain the shape of the body inside them. Elegant and ethereal in transparent glass, these draped dresses stand, sit, and recline like classical sculptures with the figures removed. Haunting and beautiful, LaMonte's pieces are ruminations on the impermanence of life and the endurance of beauty. Employing a similar technique, Joanna Manousis's most recent works invest the form of a cathedral's stained glass rose window with imagery of sheaths of wheat. Three-dimensional and reverse cast so that the wheat appears encased in sparkling crystal from the front, but is merely a hollow from the back, Manousis's work makes the quotidian substance of wheat precious, a status that is emphasized by her jewel-like mounting of it on the wall. For these artists, and the countless others who join them in their exploration of technique, casting is a means to attain otherwise unachievable results.

From its ancient origin as a method to successfully mimic other materials, glass casting today is being employed by ever greater numbers of artists because it offers innate effects that no other medium or technique does. From glass that hovers between transparency and opacity like a cloud on a misty morning or that simultaneously records the industrial past and present of a specific city site using the most up-to-date technologies, these artists are giving new life and direction to a millennia-old form.

Notes

1. E. M. Stern and Birgit Schlick-Nolte, *Early Glass of the Ancient World: 1600 B.C.–A.D. 50: Ernesto Wolf Collection*, English ed. (New York: Verlag Gerd Hatje, 1994), 49.
2. Andrew J. Shortland, *Lapis Lazuli from the Kiln: Glass and Glassmaking in the Late Bronze Age* (Leuven: Leuven University Press, 2012), 76.
3. Mark Pendergrast, *Mirror Mirror: A History of the Human Love Affair with Reflection* (New York: Basic Books, 2009), 152–153.
4. Keith Cummings, *Contemporary Kiln-Formed Glass*. (London and Philadelphia: A&C Black; University of Pennsylvania Press, 2009), 23.
5. Mark Peiser, *Looking Within: Mark Peiser, the Art of Glass: Works from 1967–2003* (Asheville, NC: Asheville Art Museum, 2003), 70.

Susie J. Silbert is Curator of Modern and Contemporary Glass at the Corning Museum of Glass (CMoG). Silbert holds a master's degree in decorative arts, design history, and material culture from Bard Graduate Center, and a BFA from the University of Wisconsin-Madison.

FUNCTION
Serviceware and Storage

In this book, we are including mold-blown glass and pressed glass as parts of casting since they are mold-dependent processes with results similar to casting in other mediums, specifically, slipcasting and rotocasting.

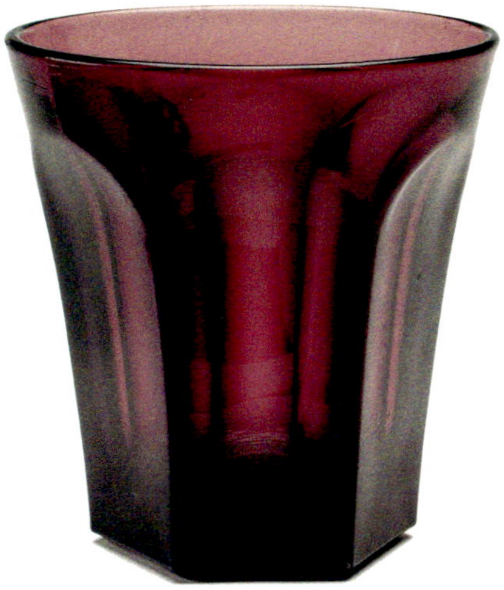

Serviceware and storage are two of the most important and common functions of mold-blown and cast glass. The earliest cast glass storage vessels were labor intensive and quite valuable, so they were primarily used for expensive oils, perfume, and other rarities. Bowls, platters, glasses, and other serviceware were made as early as the fifteenth century BCE in Mesopotamia and Egypt.[1] When glassblowing was perfected, casting fell out of favor for the most part, but one aspect that carried over was the use of metal molds. Mold-blowing has been a mainstay of glassware production ever since—simply open any cupboard or refrigerator and take a look.

The developments in molds and machinery in the nineteenth century created both decorative and functional possibilities for glass. Pressed glass, made by dropping a fixed amount of molten glass into a mold and pressing it into that mold with a plunger, created highly decorative and very affordable pieces that became mainstays of kitchens and dining rooms of the nineteenth and early twentieth centuries.

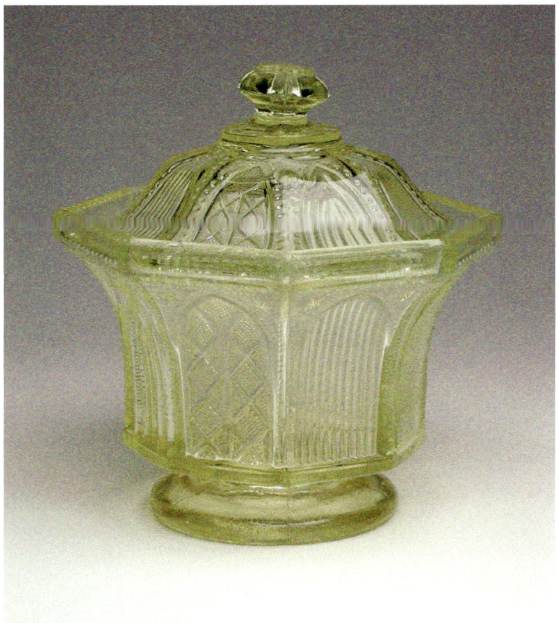

Top Bar glass, ca. 1860s. Pittsburgh, PA, Amethyst glass, mold-blown glass, 4.5" tall.
Courtesy of LACMA, Gift of the Reverend W. Daniel Quattlebaum in memory of Edith Brockett Quattlebaum, www.lacma.org, LACMA public domain.

Left Gothic Arch sugar bowl, 1840–1860. New England, USA, glass, 5.25" tall.
Courtesy of LACMA, Gift of the Reverend W. Daniel Quattlebaum in memory of Edith Brockett Quattlebaum, www.lacma.org, LACMA public domain.

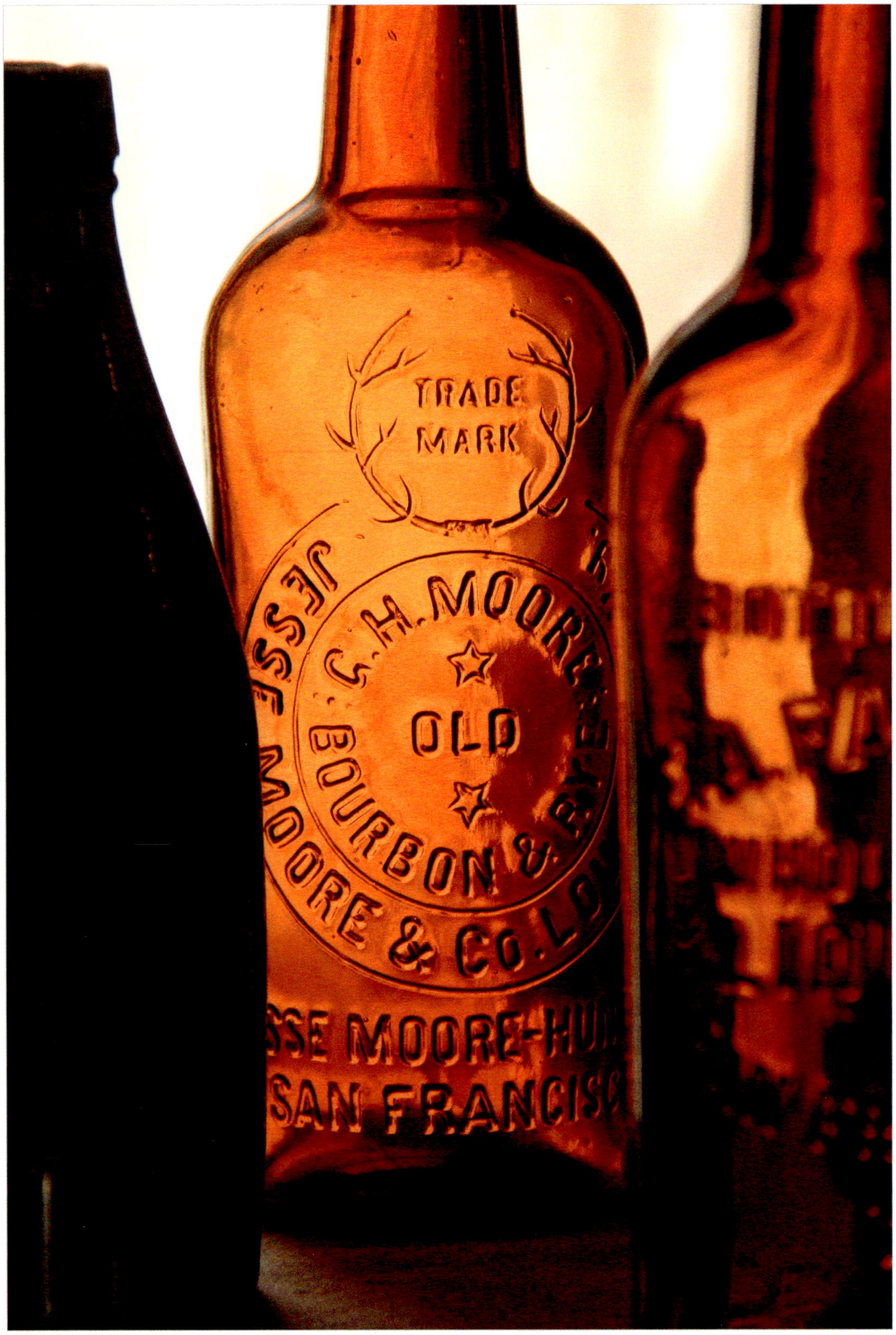

Bourbon & Rye bottle, C. H. Moore, 1800s. USA, glass, 1.75" tall × 3".
Photo: © Sivasankar / Wikimedia Commons / CC-by-3.0.

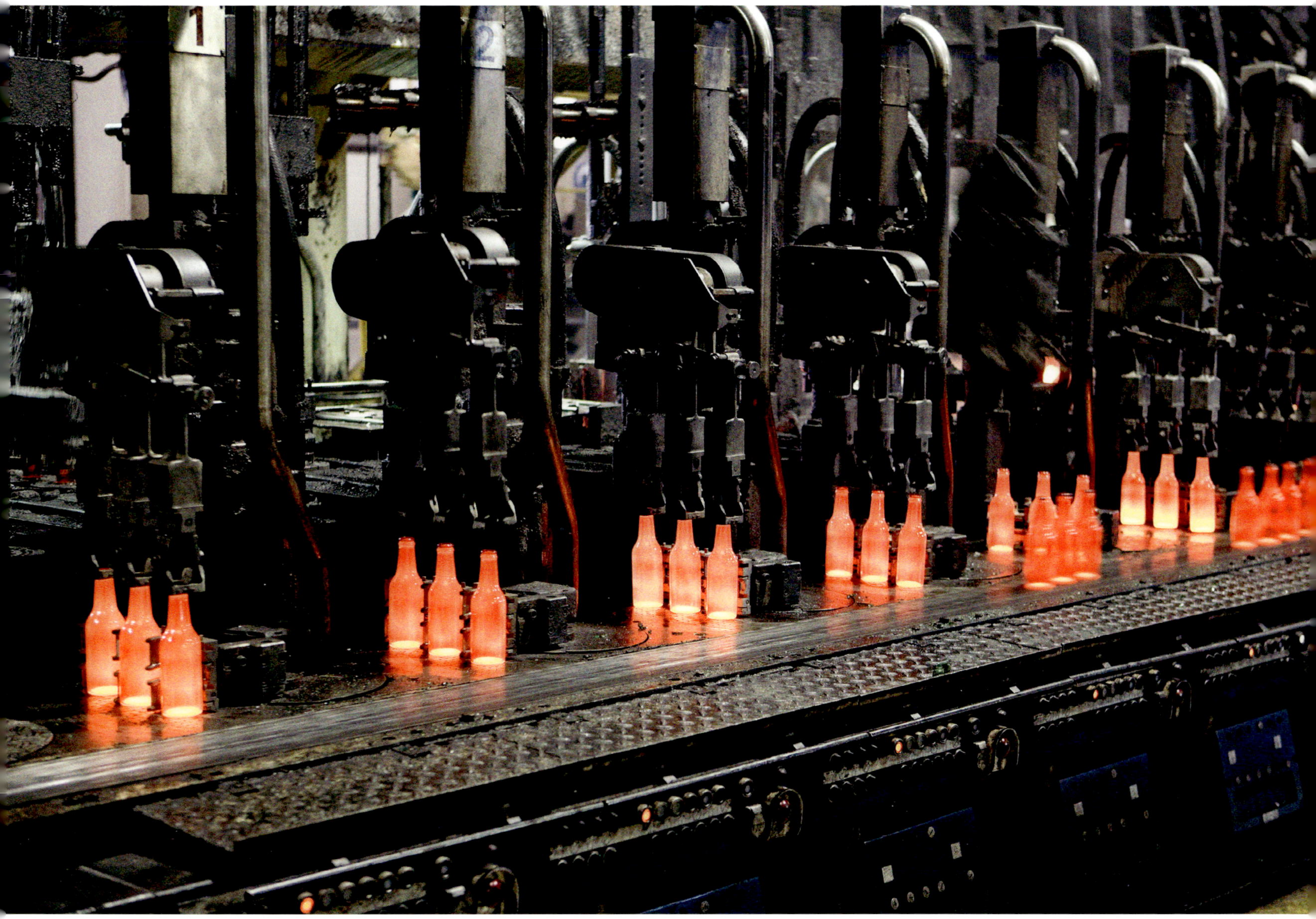

Glass became a preferred material for storage vessels with the advent of glassblowing, which made it possible to readily produce a wide range of bottles and jars. Even if slightly irregular, mouth-blown glass bottles provided storage for wine, oil, vinegar, and a variety of other liquids. Cork was the traditional method of closure—it could expand and contract to compensate for the irregularity of the openings on mouth-blown bottles, but with the mechanization of bottle production, new solutions were developed. A wide array of metal closures was developed in this period, but metal "crown caps" are a solution still used on beer and soda bottles today.

Mechanized glass container production.
Photo: © Federal Glasindustrie e.V. / Wikimedia Commons / CC-by-3.0.

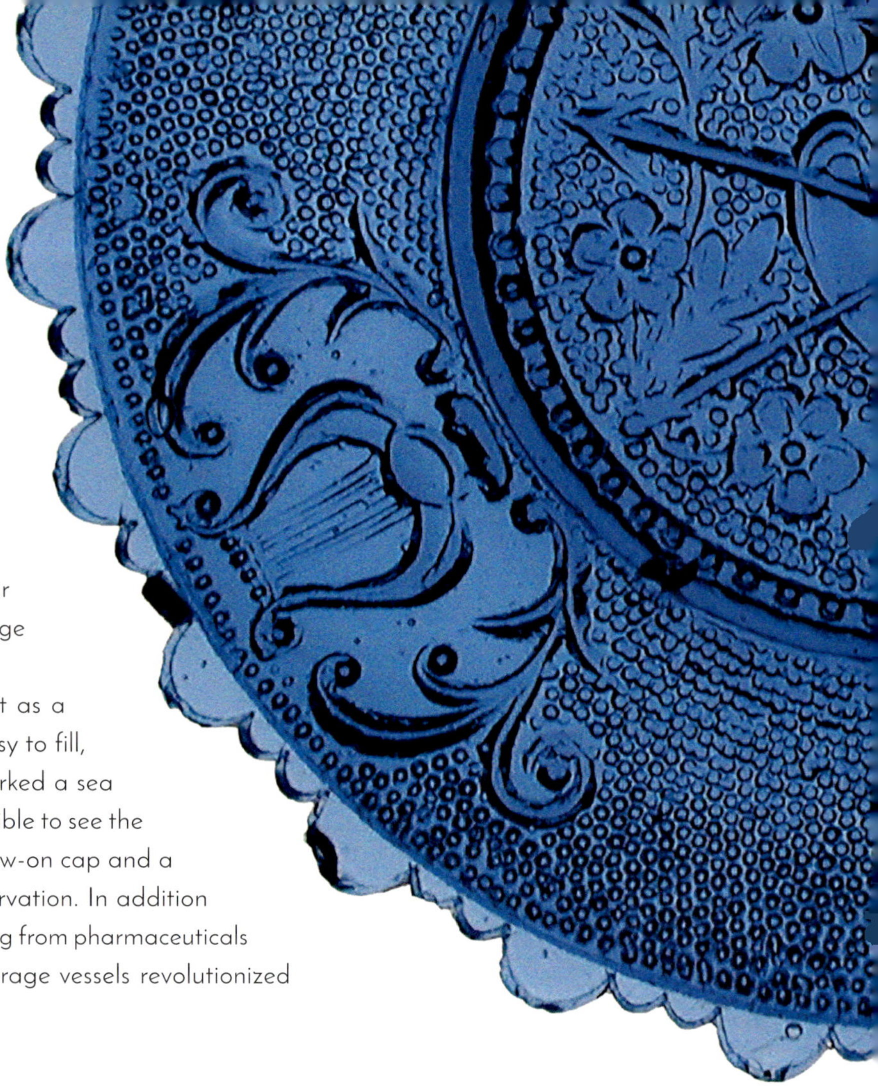

The mechanization and mass production brought about by the Industrial Revolution occurred parallel to the mass-movement of people to cities and away from farms, which created the need for more food availability where people were living and, by definition, the ability to store and ship food before it spoiled. "Heat-based canning" was discovered in 1806 and caught on when Nicholas Appert, a French cook, wanted to find a way to preserve foods during the Napoleonic wars.[2] The irregularity of vessels and their closures led to some degree of spoilage, but sterile food storage was on its way to being a reality.

Wide-mouth jars, specifically, became more important as a variety of foods needed to be stored in vessels that were easy to fill, open, and close. The invention of the Mason Jar (1858) marked a sea change in food storage. From the clear glass, making it possible to see the contents of the jar, to the threaded neck, allowing for a screw-on cap and a truly air-tight seal, the Mason Jar revolutionized food preservation. In addition to this, mold-blown glass vessels provided storage for everything from pharmaceuticals to chemicals. Automation in producing these utilitarian storage vessels revolutionized entire industries.

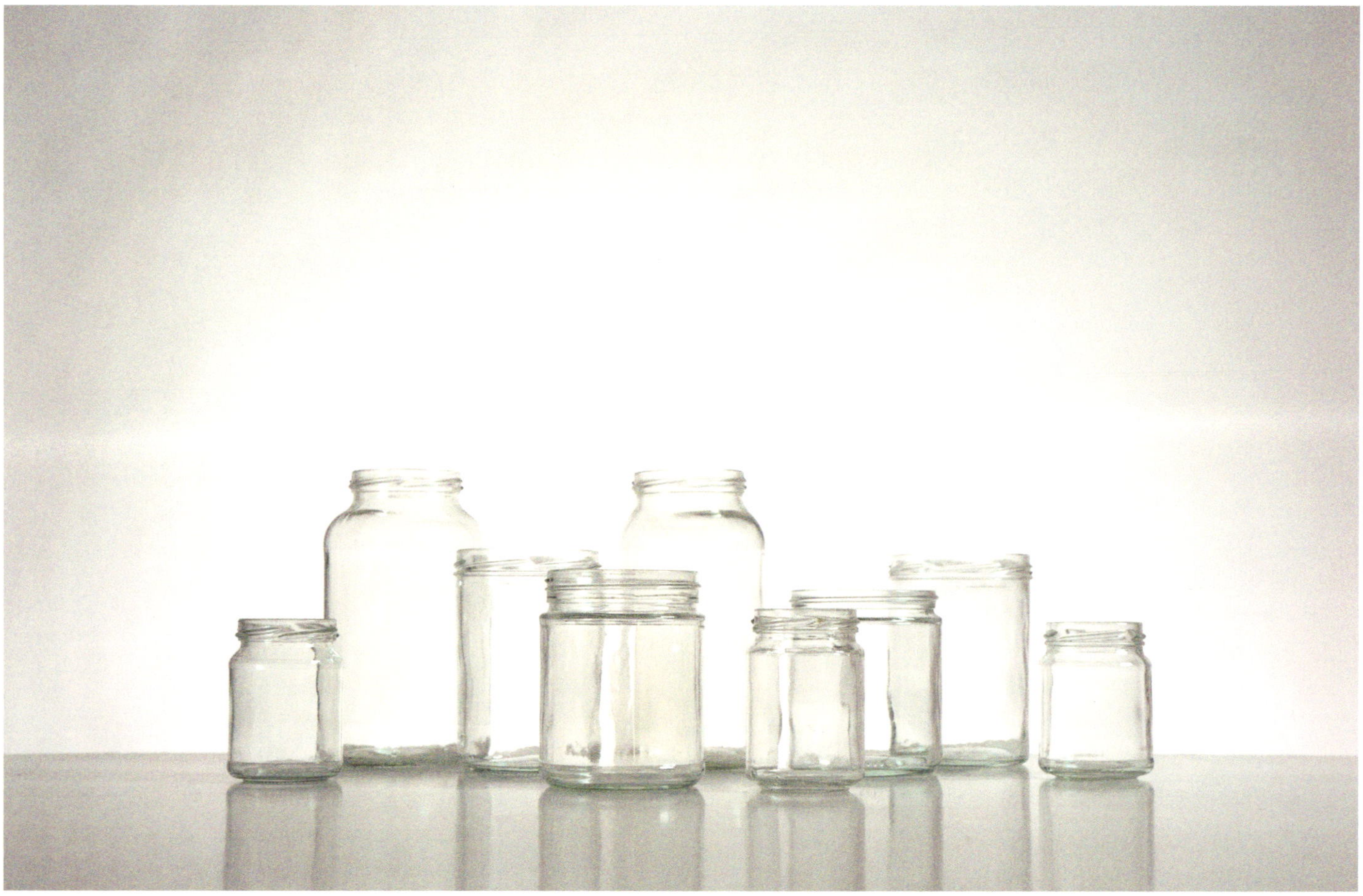

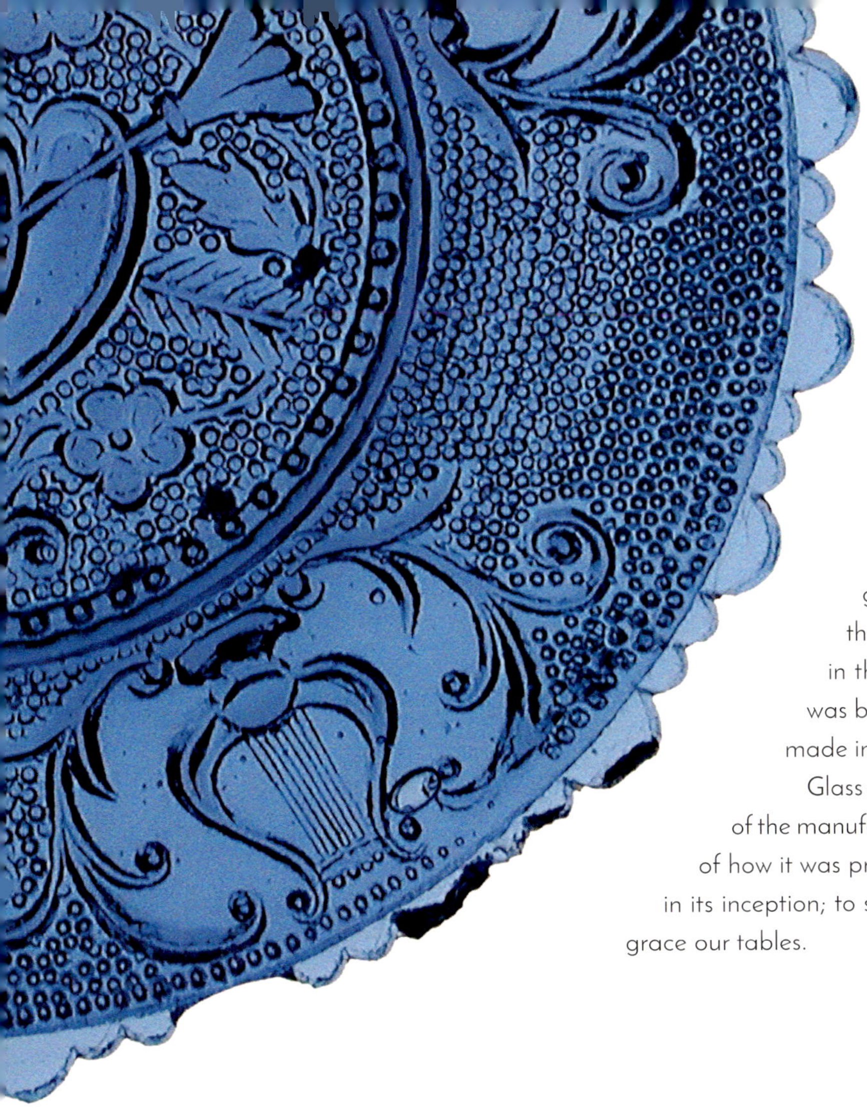

Beyond this revolution in function, the late nineteenth- and early twentieth-century developments in glass helped define the look of the times. Pressed glass was well-suited to the ornate and heavily embellished Victorian era and had its heyday in the 1880s. Highly ornamental molds were made by skilled metalworkers and a dizzying array of glasswares was produced. Most of these molds were cut up and recast into munitions for World War I and pressed glass fell out of favor for a while. It made a resurgence during the 1930s because it was domestically (and affordably) produced in the US, and was renamed "Depressionware." Depressionware was brightly colored and cheery, an antidote to the times, and was made in a variety of new patterns that fit the aesthetics of the 1930s. Glass serviceware and storage vessels fill our homes today as a result of the manufacturing developments of the Industrial Revolution. Regardless of how it was produced though, glassware serves the same function as it did in its inception; to store food and other valuable goods, fill our cupboards, and grace our tables.

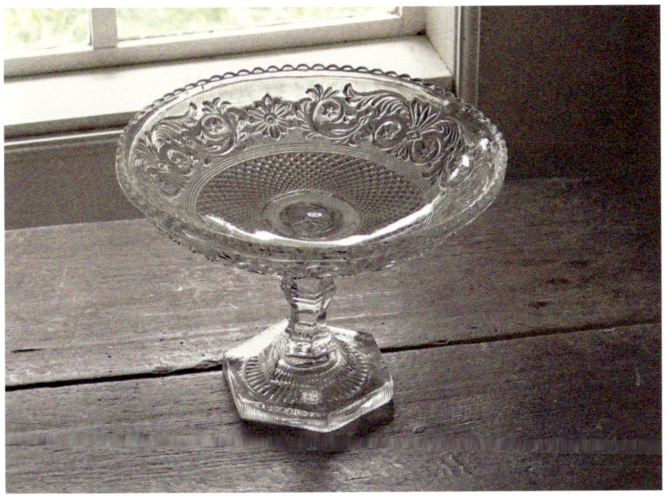

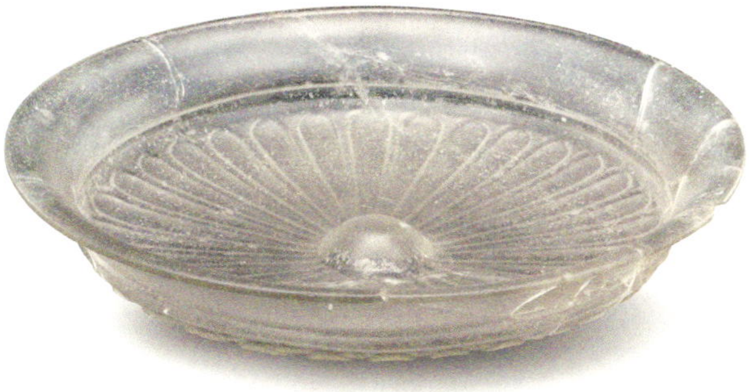

Top, center of spread *Cup Plate*, Boston and Sandwich Glass Factory, ca. 1830s. Sandwich, Massachusetts, USA, pressed sapphire glass, 3.5".
Courtesy of LACMA, Gift of the Reverend W. Daniel Quattlebaum in memory of Edith Brockett Quattlebaum, www.lacma.org, LACMA public domain.

Opposite page, bottom Threaded glass jars.
Photo: © Botho Willer, botho.cc / Wikimedia Commons / CC-by-4.0.

Above left Pressed glass bowl with foot, late nineteenth to early twentieth century. Glass.
Photo: © Wolfgang Sauber / Wikimedia Commons / CC-by-3.0.

Above right Bowl, 500 to 400 BCE. Probably Iran, overall size 1.25" × 6.9".
Courtesy of The Corning Museum of Glass, Corning, N.Y.

SITE
Unexpected Places

Cast glass can be found in obvious places like our tables and pantries, but it also has less predictable homes. As a material, cast glass has unique characteristics that lend it to unusual applications.

Two of the primary qualities of glass—its translucency and transparency—make it a perfect and obvious choice for dramatic architectural uses. It can be lit so it glows, like this cast glass bar made by Nathan Allan Glass for the Terminal City club's wine bar, *Cuvée*. The cast bar creates a focal point in the space that feels elemental, ethereal, and modern. The texture cast into the bar is on the bottom of the cast panels only (the sides and top are perfectly smooth) but the transparency of the glass creates an optical illusion that the texture covers all sides, adding visual interest. The texture also keeps the glass from revealing what is stored under the bar. For the same reason, using texture to create translucency in cast glass can be useful for objects such as countertops, sinks, and room dividers.

Another unexpected architectural application for cast glass is flooring. The historic use of glass as part of a walkable surface was to bring light to whatever lay below. Making their first appearance on ships, the aptly named "deck prisms" were cast glass, jewel-like forms, flat on top (to blend with the surface of the deck) and faceted underneath. They were set into ships' decks to light the cabins below. The faceted lower parts of the castings were designed to disperse the natural light as much as possible. Deck prisms sparked the idea for "vault lights" (also known as "pavement lights or "sidewalk prisms"), which are set into sidewalks and grates to light tunnels, basements, and other subterranean spaces. These were very popular in the late nineteenth century but mostly went out of vogue in the 1930s when electric light took hold. While most of the originals are pretty beat up today, these can still be found in New York City as well as in many other cities.

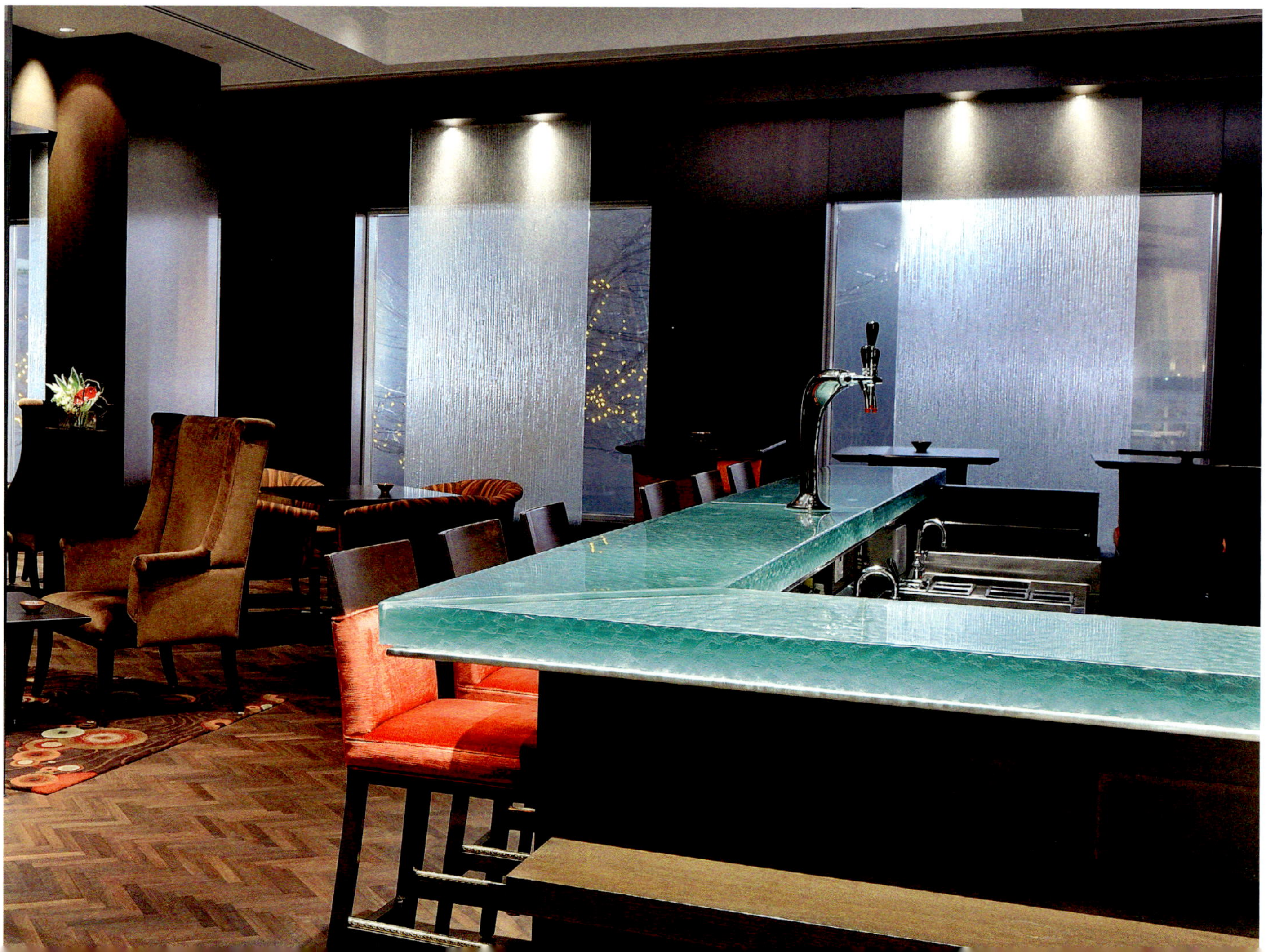

Today's glass flooring depends on its transparency rather than its light transmitting properties, although some companies still make vault lights. For the most part, contemporary glass floors are large transparent slabs used to highlight ruins or excavations below them, such as in the Acropolis Museum in Athens, Greece. There's something about walking on glass and hovering above the ground that feels unnatural, but that thrill mixed with a touch of fear is what makes glass floors compelling. They have also been used for observation decks in skyscrapers (to sweat-inducing and nauseating effect for those with a fear of heights), over water, and to showcase sculptural installations, as seen in this example made by Jockimo and installed at Grand Valley State University to house this piece of sculpture by Brad Patterson. Glass floors are a clever and efficient way to showcase sculptural installations. The work can be enjoyed and kept safely out of harm's way without sacrificing its prominent location in a busy public space.

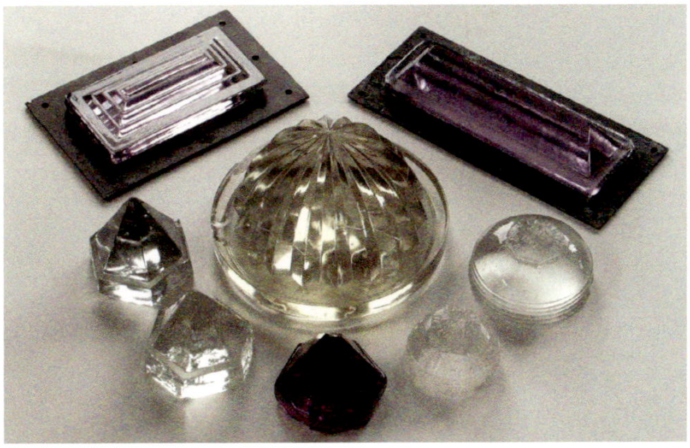

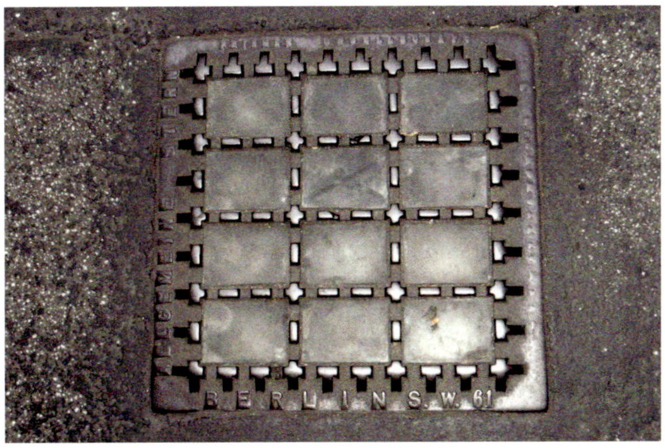

Opposite page Nathan Allan Glass Studios, *Terminal City Heavy Glasstop*, 2013. Clear glass, 3" thick.
Photo: Nick Didlick Photograph, Client: Terminal City, BC, Designer: SSDG Interior Design, BC.

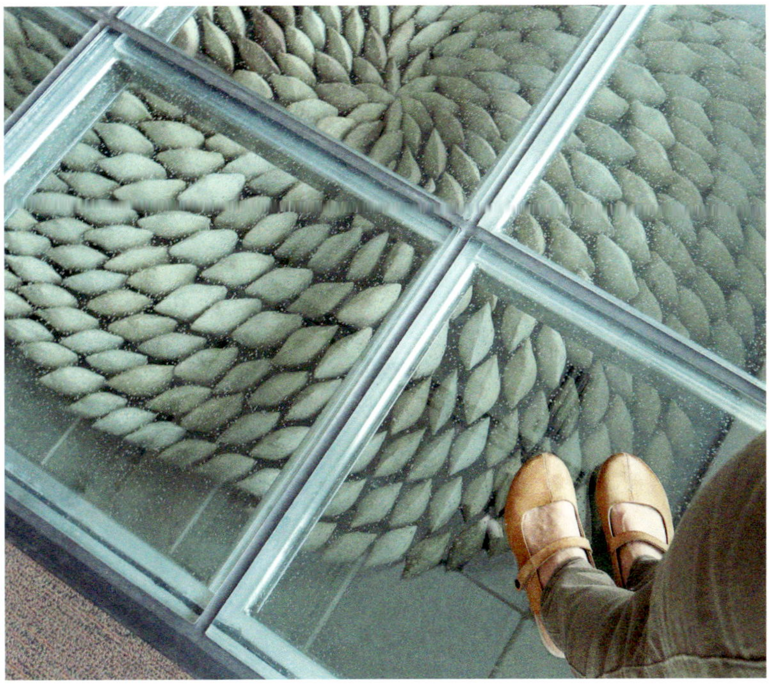

Top to bottom

This group of deck prisms features different styles of faceting. The small round Lanergan's prism on the right has a threaded top for easy replacement. Dimensions variable.
Photo: © Ian Macky / Wikimedia Commons / CC-by-3.0.

Vault lights, 2008. Berlin.
Photo: © Mark Ahsmann / Wikimedia Commons / CC-by-3.0.

Jockimo Crystal Clear™ glass flooring panels, 9 panels total, 29" × 29". Artwork below glass: Brad Patterson, *Symbiosis: Variation #1*, stoneware with colorant, 2004.
Photo: Renée Zettle-Sterling, Collection of Grand Valley State University Art Gallery Collection, Allendale, Michigan, USA.

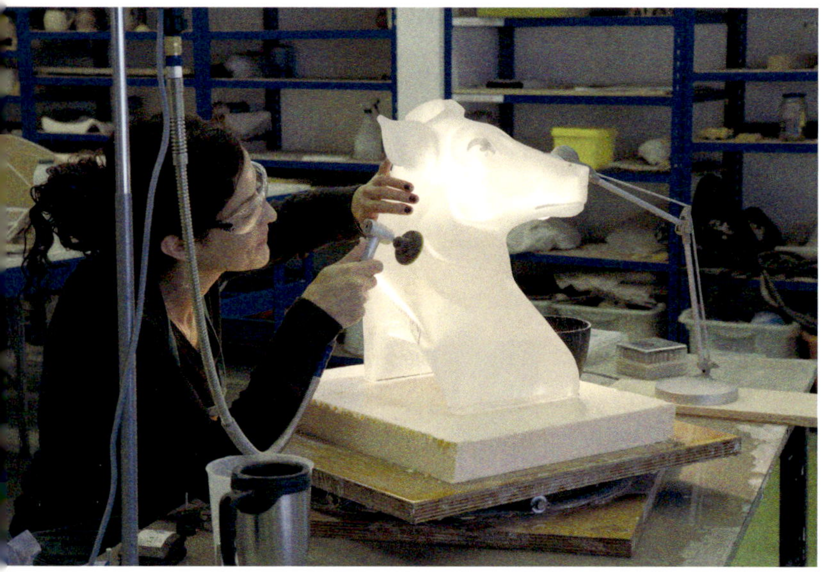

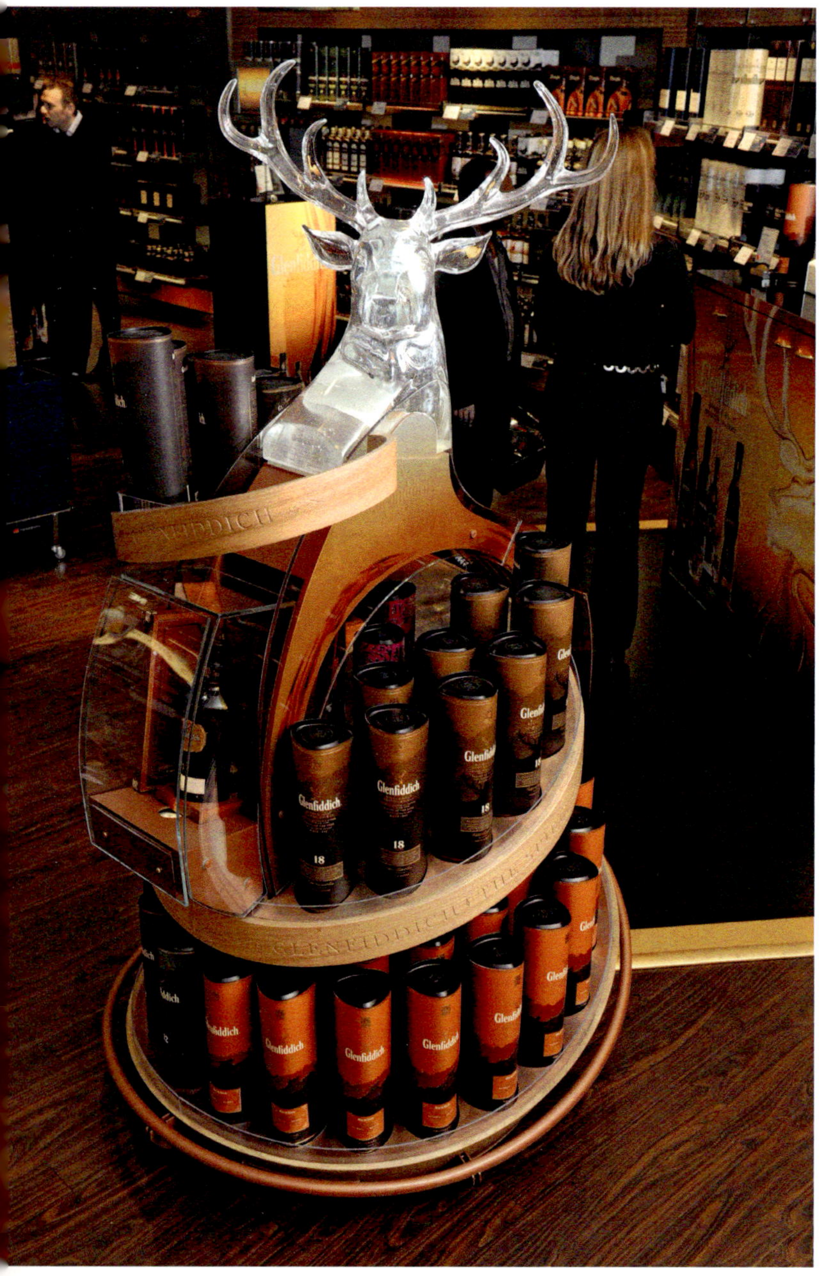

Another example of glass in a surprising location is Katya Izabel Filmus's cast crystal *Stag Heads,* made for William Grant & Sons as the centerpiece of a worldwide rebranding campaign for Glenfiddich Whiskey. At 110 pounds per head, each of these regal creatures was a significant undertaking, the annealing process alone requiring forty-one days. The antlers were cast separately from the heads and bonded after the final polish.[3]

The *Stag Heads* were installed as the central features in flagship displays in twenty-one major airports around the world. The company set a wonderful example in commissioning an artist as skilled as Filmus for their rebranding efforts. Normally, work of this caliber is kept carefully guarded (and relatively hidden) in museums or galleries; by having these works in some of the most highly trafficked places on earth, creativity is injected into everyday life. Stumbling across the majesty of these cast creatures in an airport is the best kind of surprise.

Top Once the casting is removed from the kiln and de-molded, Filmus removes the mold texture from the casting process. She uses finer and finer diamond pads on a flexible shaft until a smooth surface is achieved.
Katya Izabel Filmus, *Lead Crystal Stag Head for Glenfiddich Whiskey*, 2012–2014. Lead crystal, lost wax cast, 19.7" × 13.75" × 27.5", 110 lbs.
Project managed by Chris Blade. Photo: Chris Blade, Client: Glenfiddich Whiskey.

Left Katya Izabel Filmus, *Lead Crystal Stag Head for Glenfiddich Whiskey*, 2012–2014. Lead crystal, lost wax cast,19.7" × 13.75" × 27.5",
110 lbs. Project managed by Chris Blade.
Photo: Chris Blade, Client: Glenfiddich Whiskey.

Opposite page Greg Ondo, *Northern Lights*, 2014. Granite, steel, wood, cast glass, 9' × 10' × 1'.
Photos: Greg Ondo, Collection of St. Joseph Healthcare.

Gregory Ondo's *Northern Lights*, found in St. Joseph's Hospital in Bangor, Maine, is another example of art being in an unexpected place. Many hospitals haven't been buying art due to austerity measures, but others make art a priority; in spite of the cost, these institutions believe in its therapeutic value. Studies have shown that people who are sick or injured can benefit from art.

Hospitals, by their very nature are filled with pain, fear, and can easily feel grim and institutional; art can improve those environments by providing escape, solace, a destination, or even a simple moment of beauty. A nice example of this is Ondo's ten-thousand-pound granite slab shimmers with sixty individually cast and inlaid pieces of glass that form a glimmering veiny pattern through the stone. Some of the inlays extend all the way through the granite, leaving small glass circles on the back of the slab where they break the surface. The back of the piece is minimal but for a smattering of these tiny green circles. The celestial glow of those potent little glass dots on the back, the monumentality of the granite, and the glint of the glass veins create a sense of wonder about the piece.

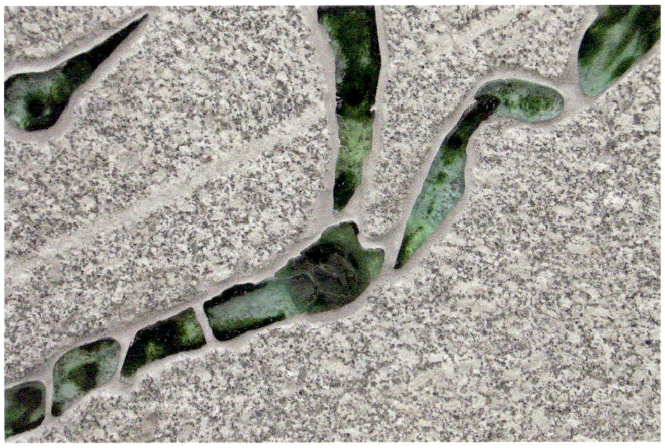

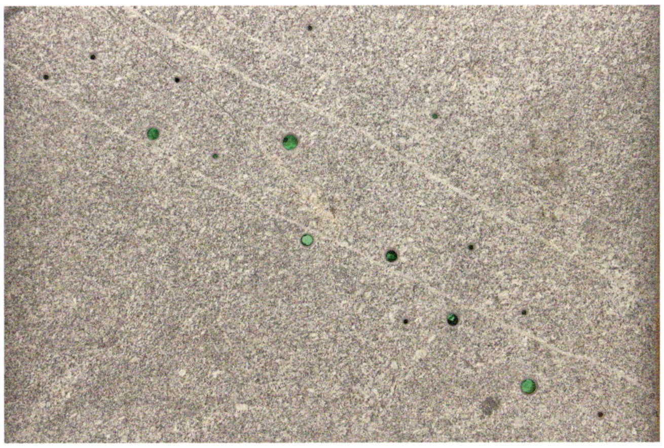

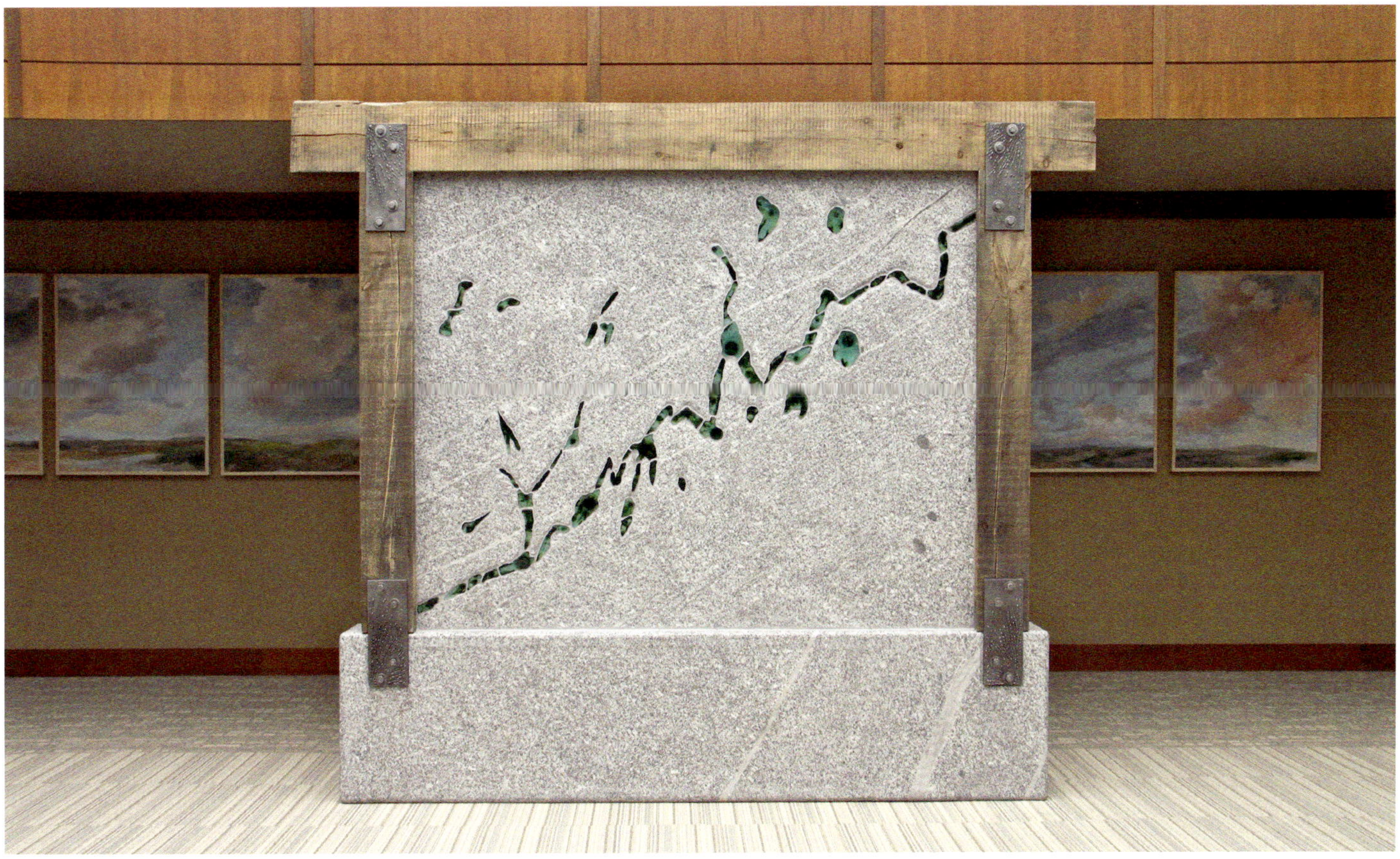

FOCUS on TRANSPARENCY, TRANSLUCENCY, and LIGHT

Glass is seductive. It's mesmerizing in its glowing, honey-like, molten state. It's clear from its prominence in the world of contemporary craft and fine art that it's also enchanting in its finished state. Glass's most emblematic characteristics are color, translucency, and its ability to transmit and refract light. Glass artists take advantage of these innate qualities in many different ways with spectacular results.

Tamara Coatsworth's *Endless Possibilities* exploits the endless spectrum of colors available for glass artists.

Tamara Coatsworth, *Endless Possibilities*, 2014. Glass, 6" × 7" × 2.5". Photo: Tamara Coatsworth.

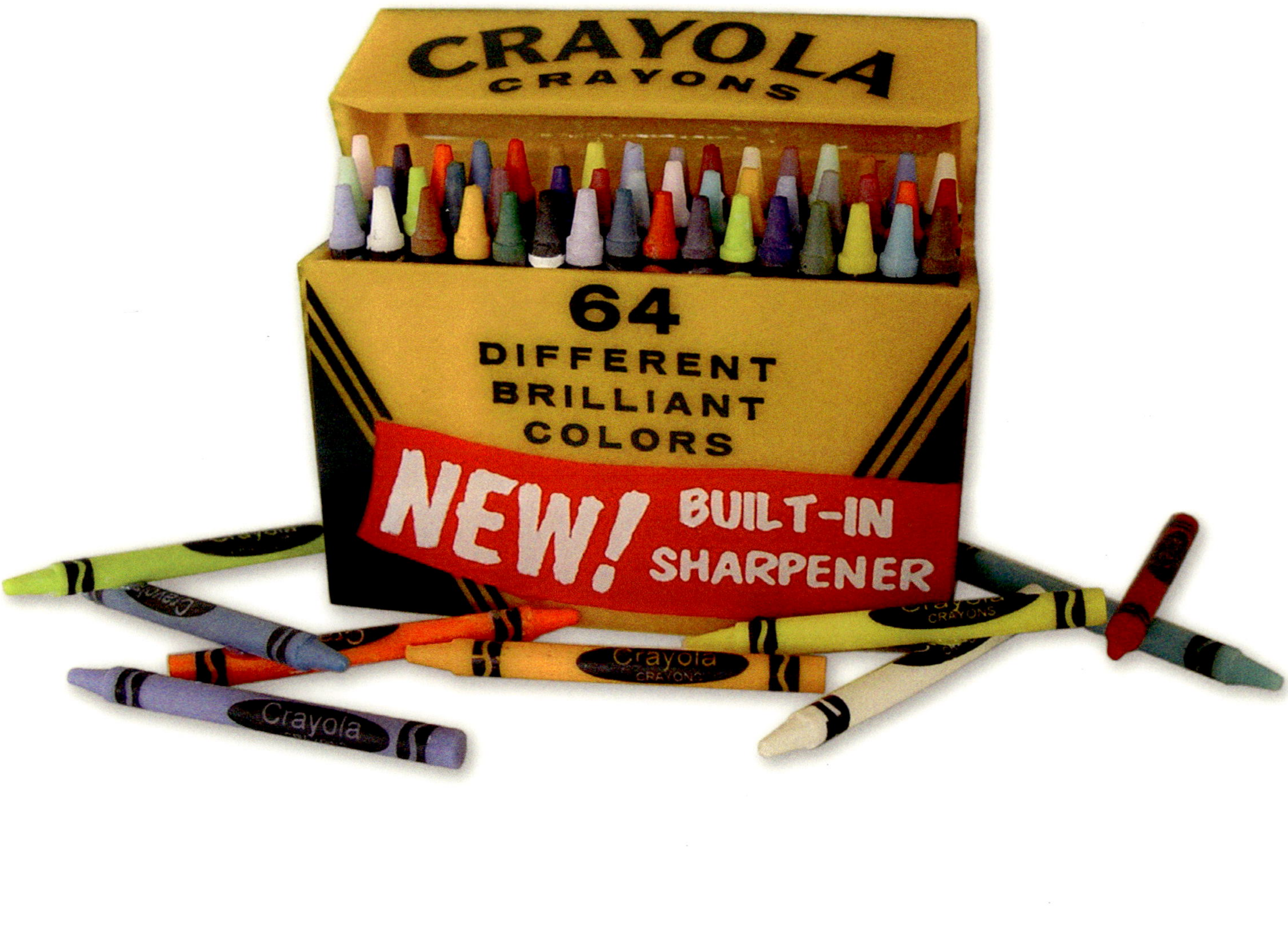

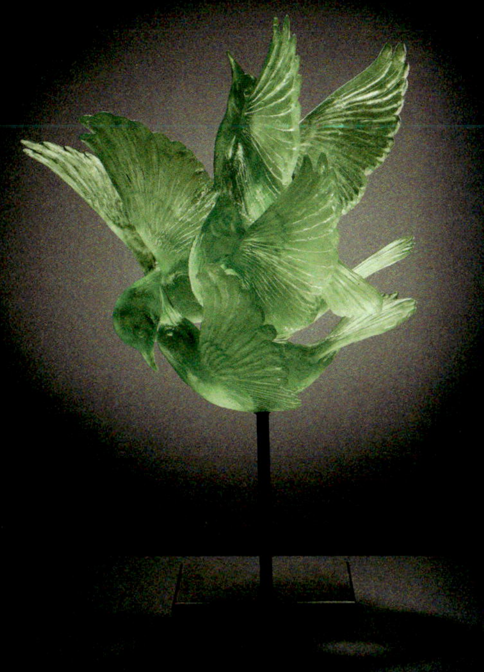

Katherine Rutecki's work utilizes the best traits of glass. Her clever use of transparent color gives her pieces a luminous quality. The aptly named *Green* is cast in a single color, but since the elegantly sculpted birds go from thick to thin, there is an illusion of light and dark, highlight and shadow. This emphasizes certain key details like eyes and outlines of feathers, but surely when viewed in the round or with shifting light, different details would emerge.

Charles Miner's *Frog Sushi* uses color in a completely different way. The pale, slightly translucent, speckled green piece emulates certain species of frogs, demanding a second look to see that the speckles don't simply sit on the surface, but rather run through the body of the glass. The brightly colored spots and the individual treatments of the eyes grant each frog its own personality, so rather than anonymous repeating forms, Miner skillfully spawned a community of camouflaged amphibians inhabiting a sushi platter.

Radiate by Carol Milne is shocking simply because it appears to be knitted glass, but the eye-popping transparent yellow paired with deep green makes the piece even more striking. Both colors are technically transparent, but the differences in both value and saturation create an important contrast between the needle and the knitting.

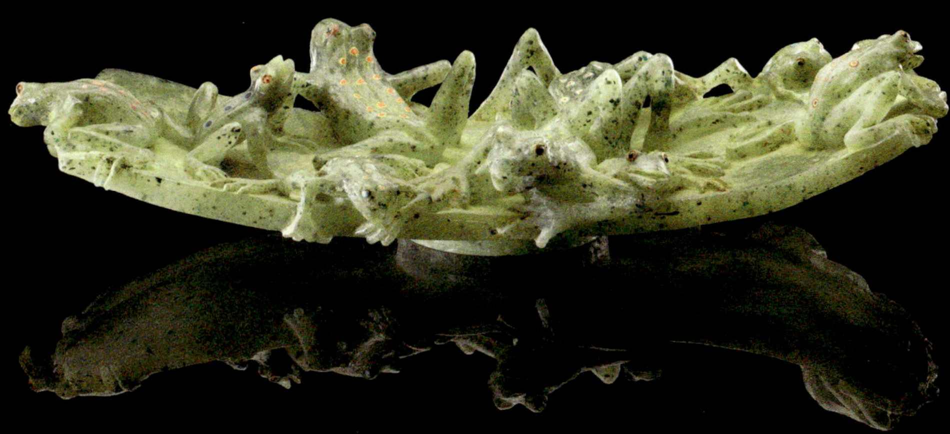

Top Katherine Rutecki, *Green*, 2014. Lost wax cast lead crystal, 14" × 14" × 27".
Photo: Jason Berge.

Center Charles V. Miner, *Frog Sushi*, 2006. Lost wax cast glass, 45% lead crystal, 21.5" × 10" × 4".
Photo: Wendy McEahern.

Right While this piece appears to be knitted glass, the actual knitting was done in the wax model. Carol Milne, *Radiate*, 2014. Cast glass, 7" × 14" × 16".
Photo: Carol Milne.

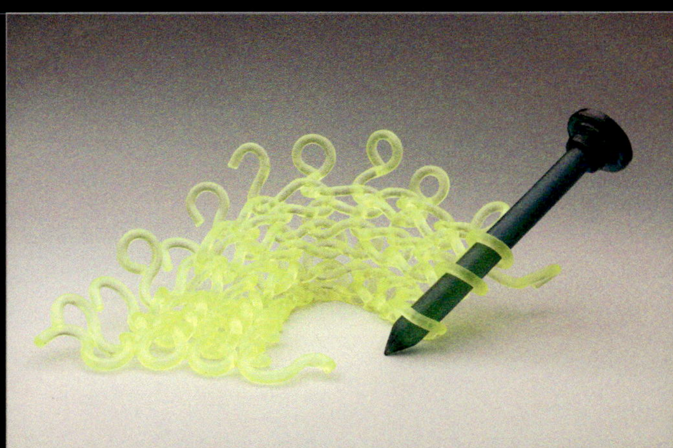

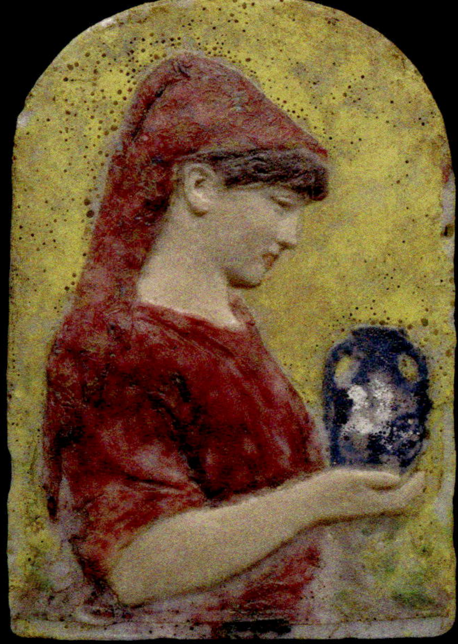

Henry Cross, *Plaque with Classical Female Figure*, 1886. Manufacture de Sèvres France, Paris, 5.3" × 3.8" × 0.66".
Courtesy of CCMG, Collection of The Corning Museum of Glass, Corning, N.Y.

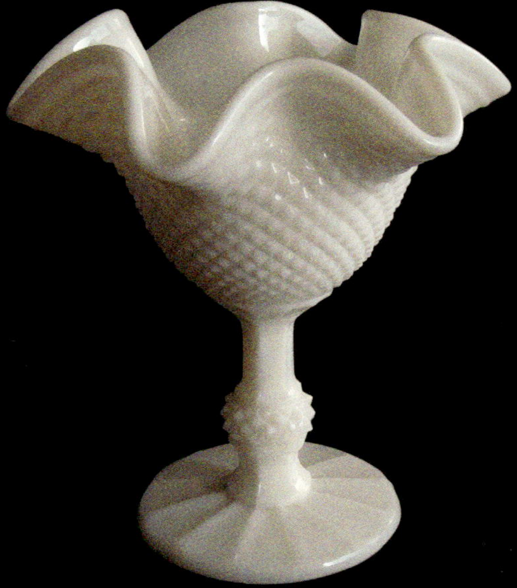

Pedestal Milk Glass Bowl, 1950–1960, Milk glass.
Photo: Manfred Heyde, CC-by-2.0.

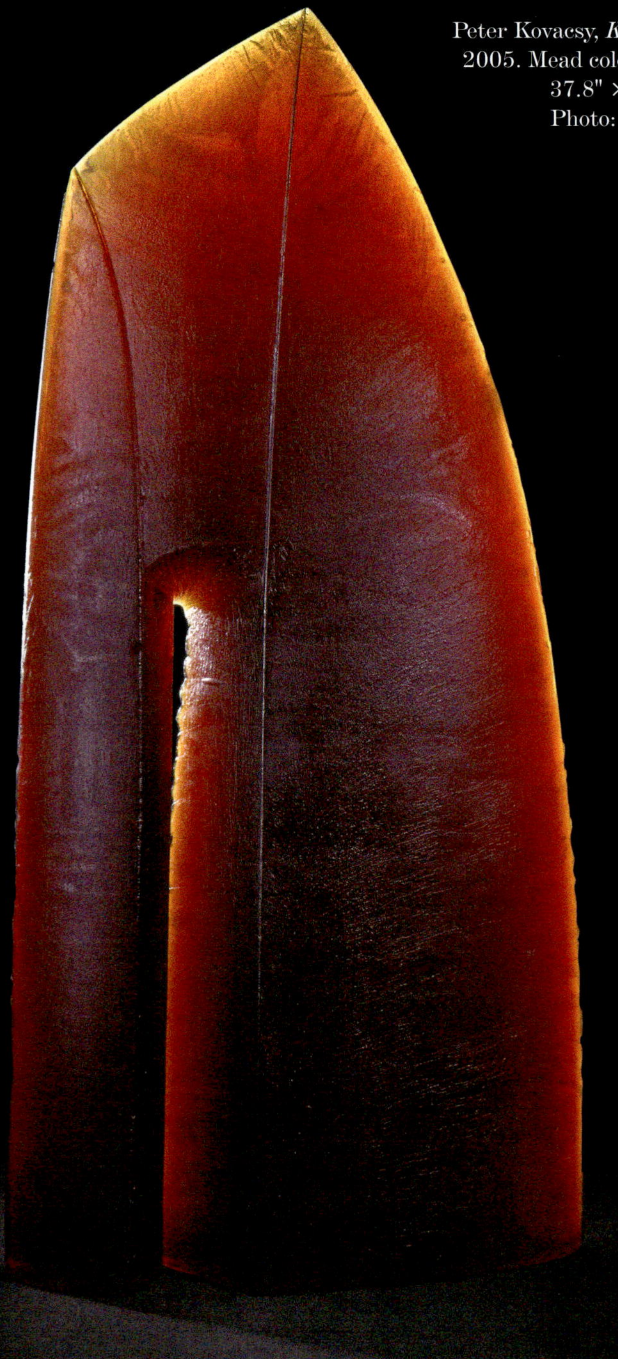

Peter Kovacsy, *Karri Sunrise 2*, 2005. Mead color, lead crystal, 37.8" × 17.3" × 3.9".
Photo: Peter Kovacsy.

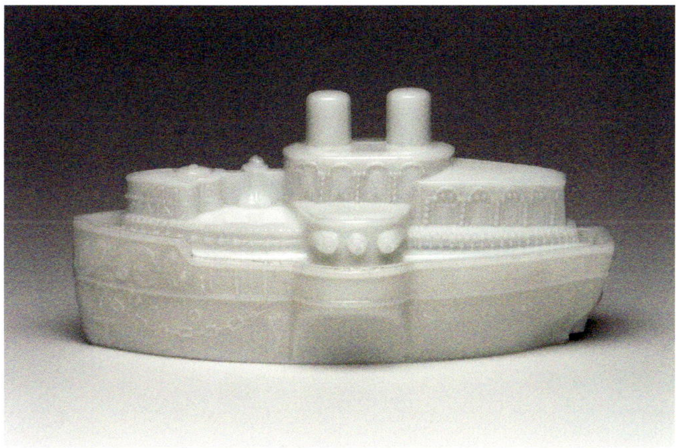

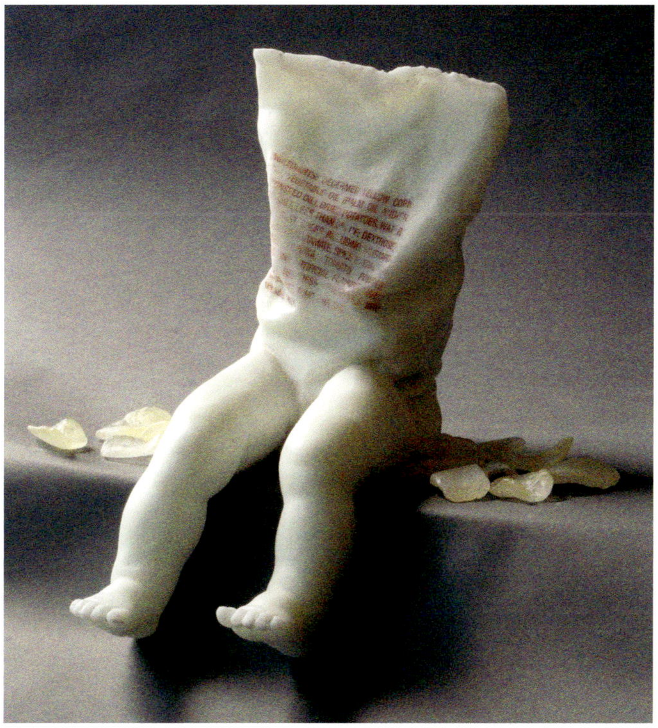

Top left Covered Dish in the Shape of the US Battleship *Maine*, ca. 1898. USA, glass, 3.75" × 7.75" × 4.12".
Courtesy of The Brooklyn Museum, Gift of Mrs. William Greig Walker. Brooklyn Museum photograph, 2004, CC-BY.

Top right Robert Bender, *Chips*, 2014. Cast glass, iron oxide transfer, 12" × 10" × 14".
Photo: Robert Bender.

Bottom left Brian Berman, *HOLOS Spiral Peace Portal 216*, 2014, Kiln cast glass, 14" × 4" × 14".
Photo: Brian Berman.

Bottom right Joanna Manousis, *Distilled Portrait I*, 2011. Crystal, mirror, stainless steel, taxidermy magpie, 16" × 6" × 6".
Photo: Joanna Manousis.

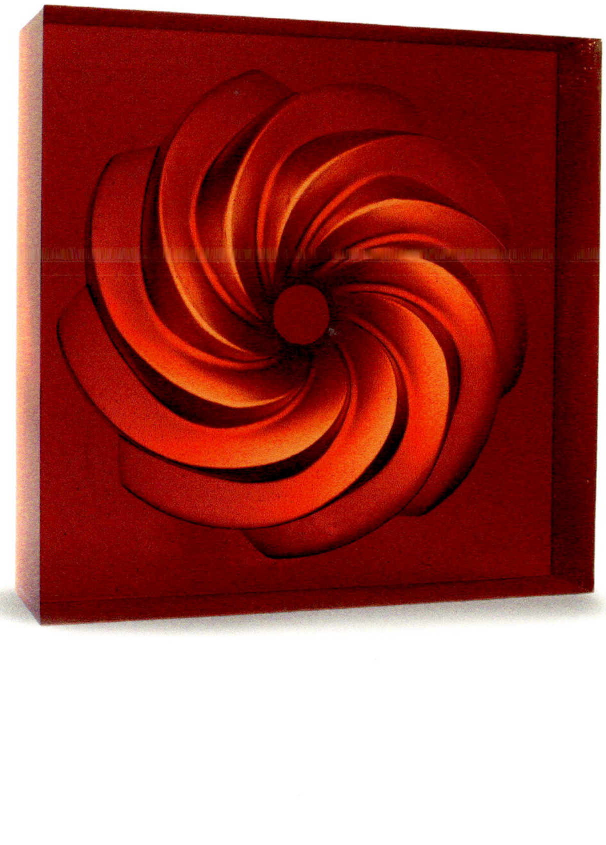

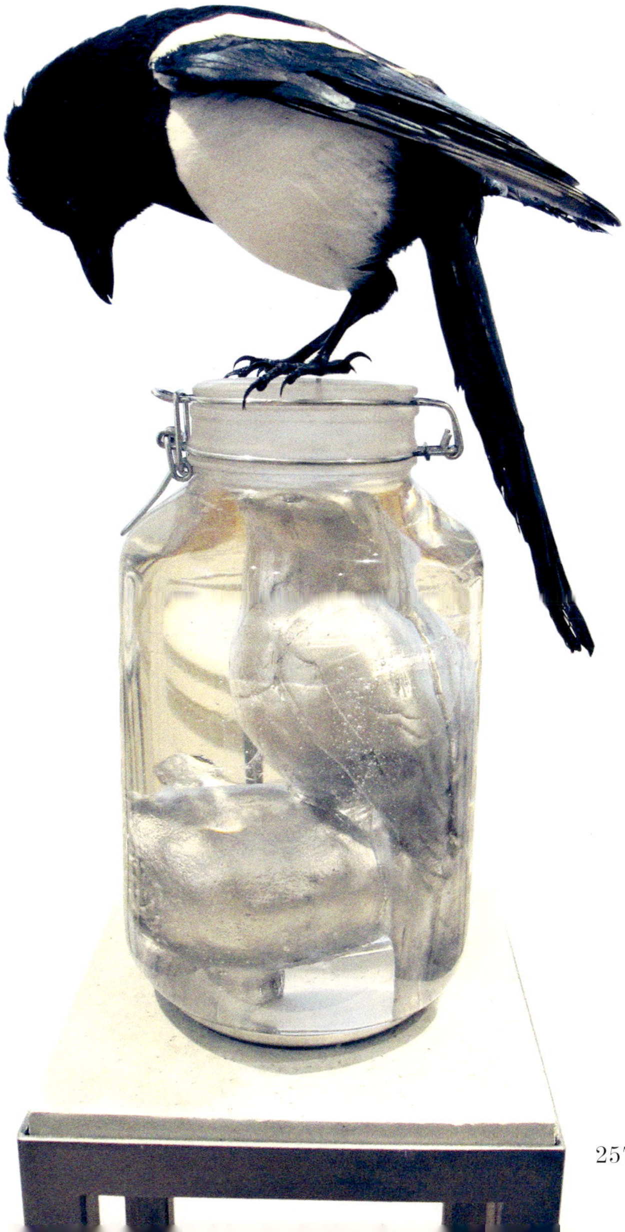

Right Christina Bothwell, *Lucky in Love*, 2015. Glass, pit fired ceramic, wood, and found objects, 4'2" × 1'8" × 1'8". Photo: Robert Bender.

OPPOSITE PAGE

Top Heike Brachlow, *Infinity*, 2015. Polychromatic cast glass, 11" × 10.6" × 9.8". Created during a residency at the Australian National University in Canberra. Photo: Heike Brachlow.

Bottom Erwin Timmers, *Rebound (Pink)*, 2011. Cast recycled glass, glass stain, 6" × 6" × 5". Photo: Anything Photographic.

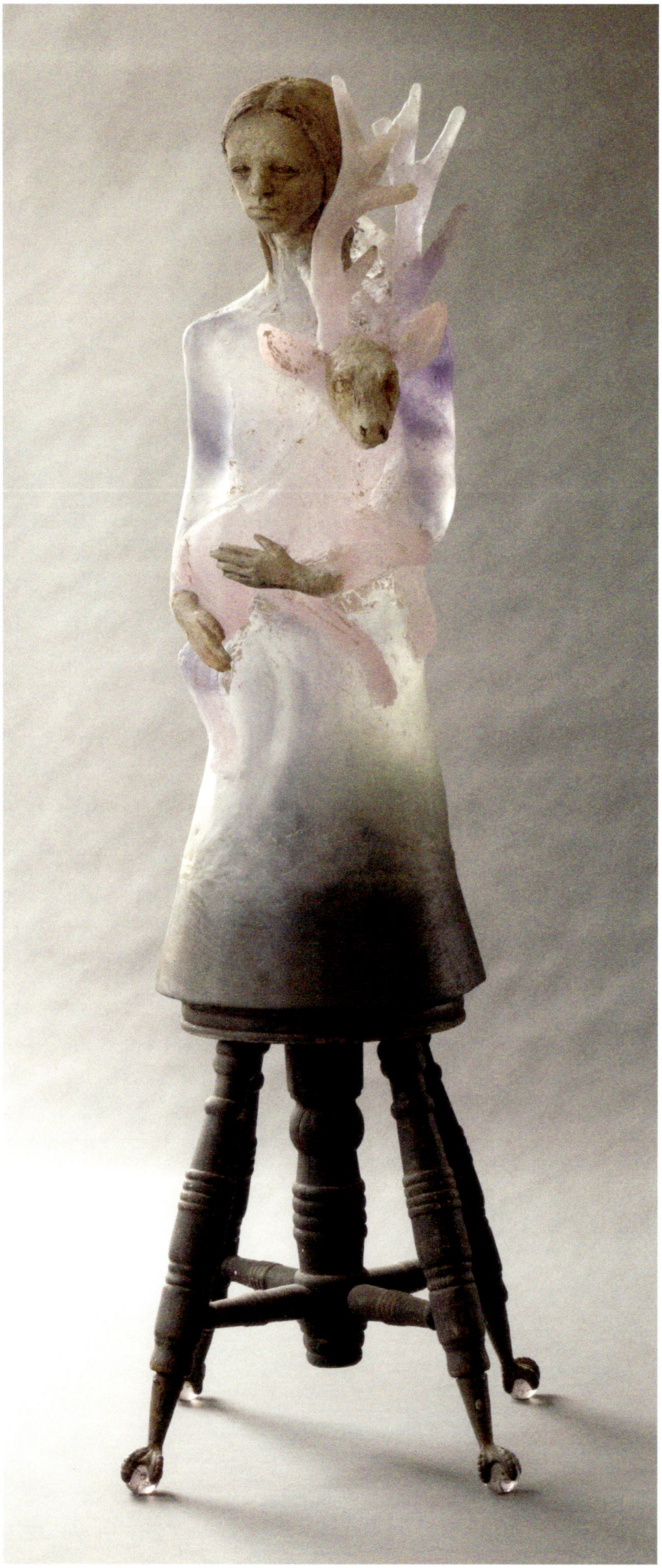

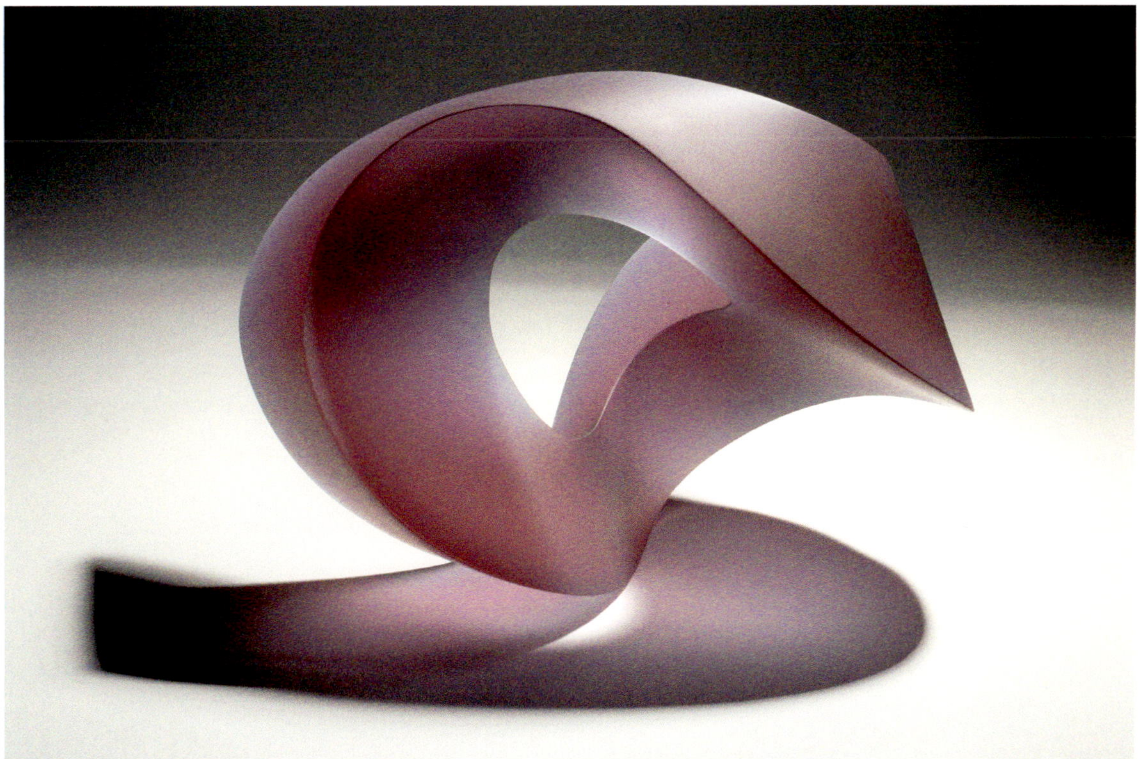

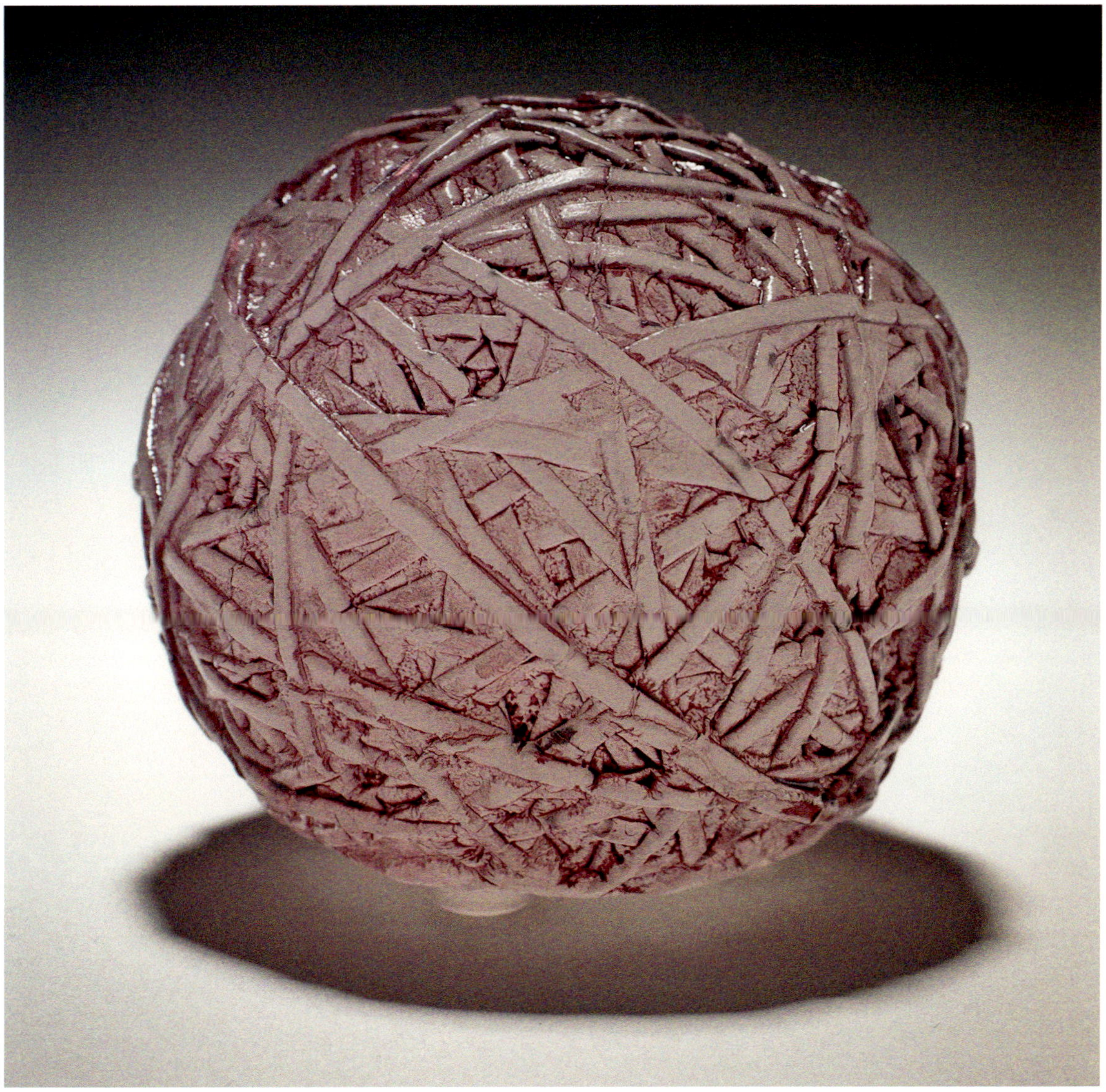

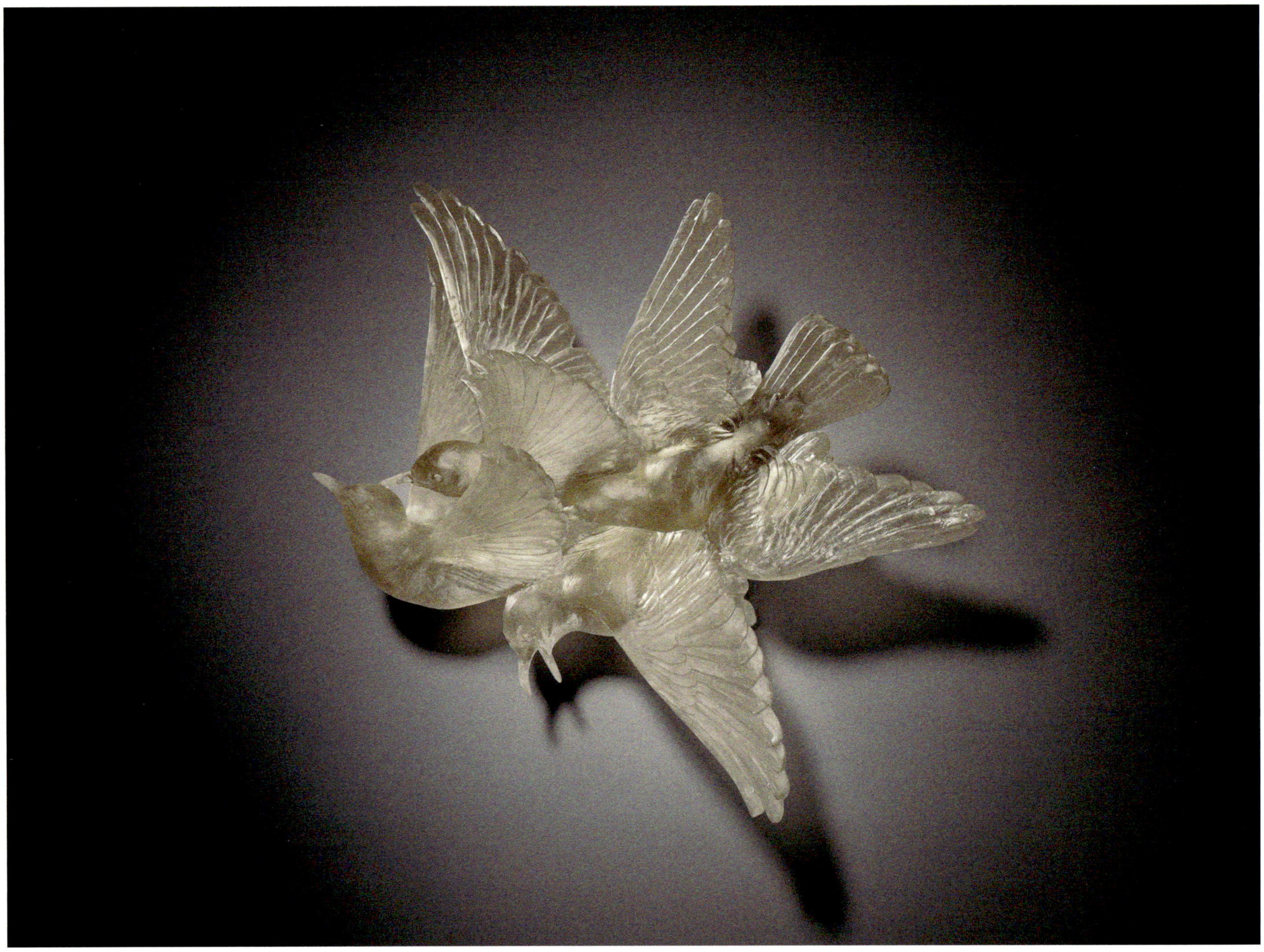

Top Katherine Rutecki, *Grey*, 2011. Lost wax cast lead crystal, steel mount, 20" × 10" × 14". Photo: Jason Berge.

Bottom Steven A. Ramsey, *Snail in Hand*, 2015. Glass, 3" × 5" × 3". Photo: Steven A. Ramsey, Courtesy of Marta Hewett Gallery.

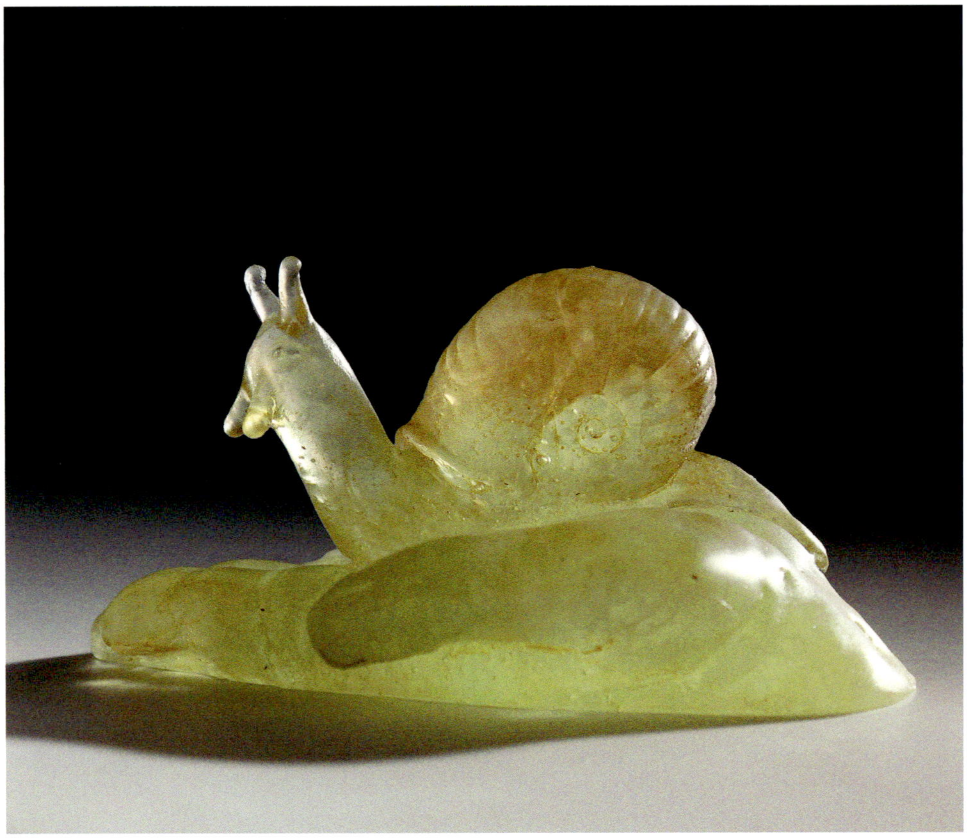

Anna Boothe, *AmForUs*, 2015. Lead crystal, 17.5" × 8" × 8".
Photos: Rick Echelmeyer.

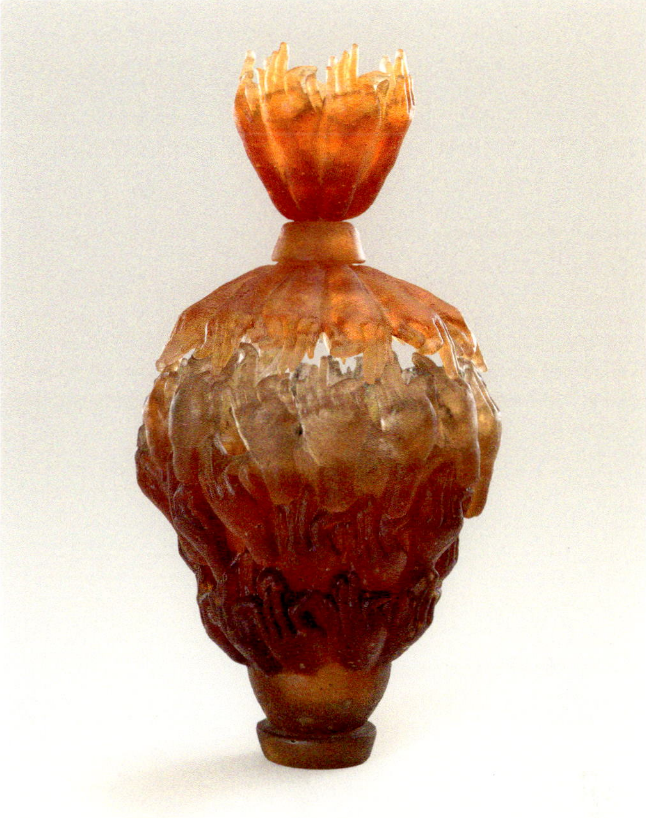

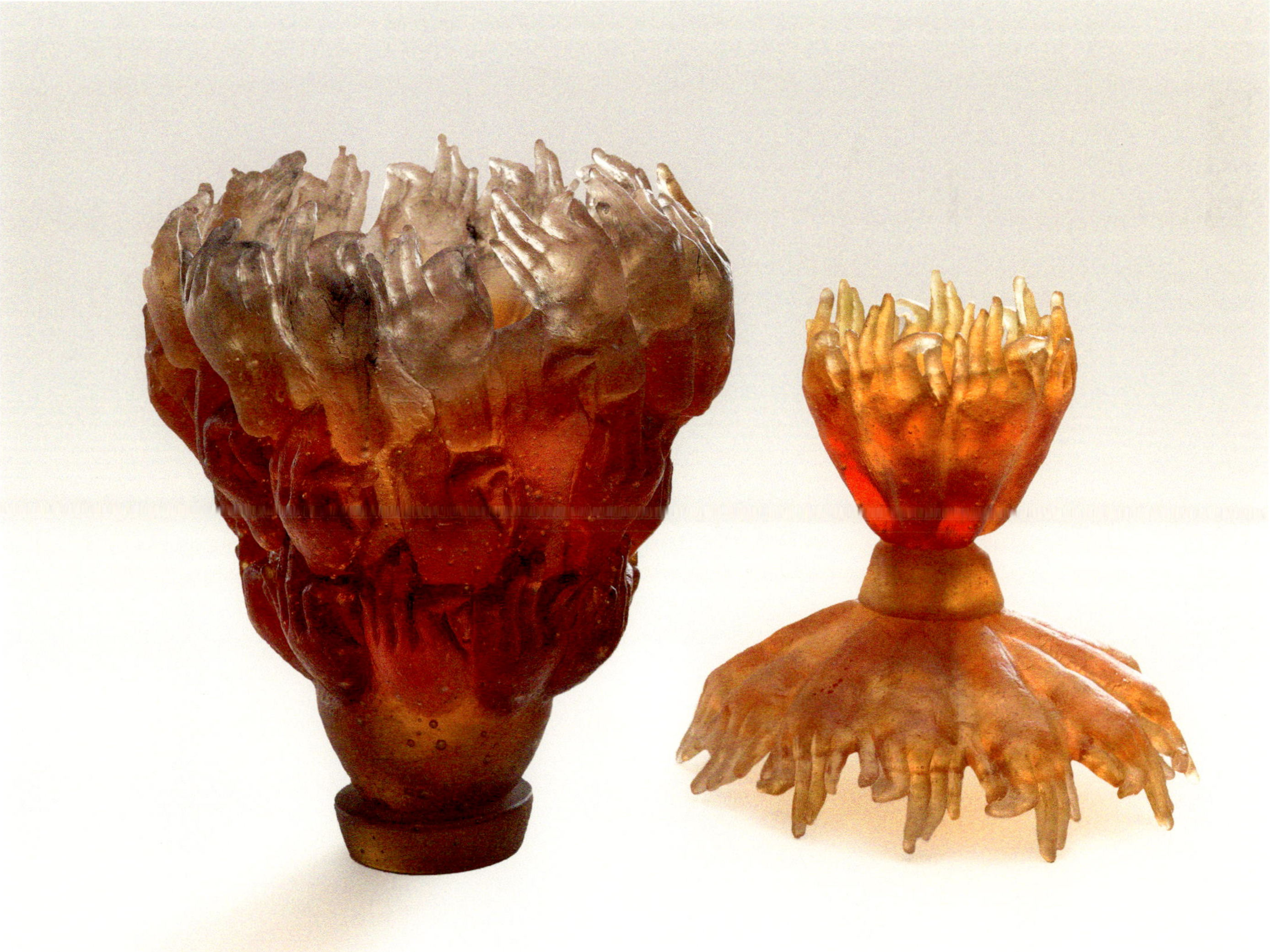

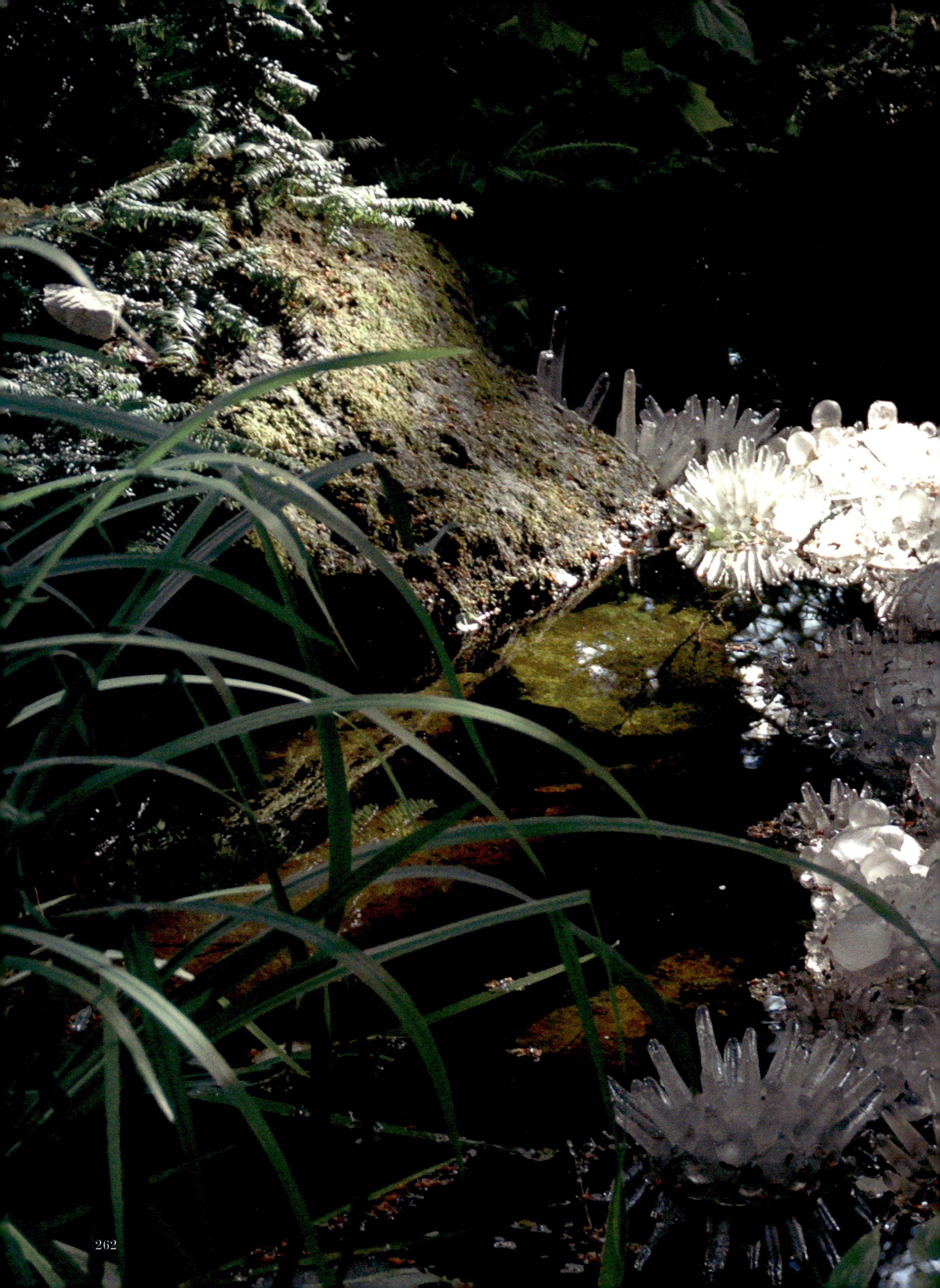

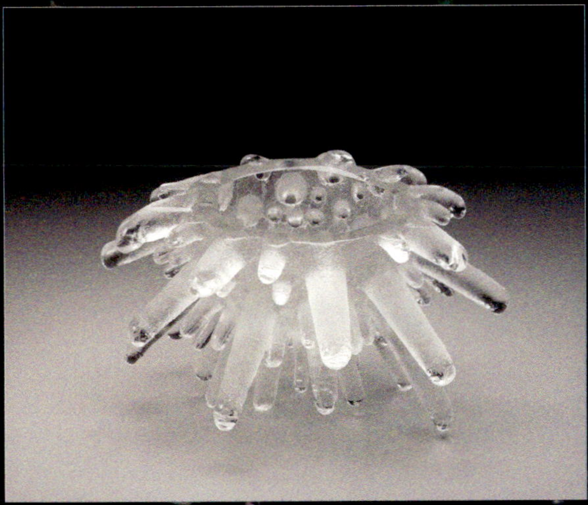

David Chatt, *Cast Bowl*, 2008. Sand-cast glass, 8" × 14".
Photo: Steve Mann Photography.

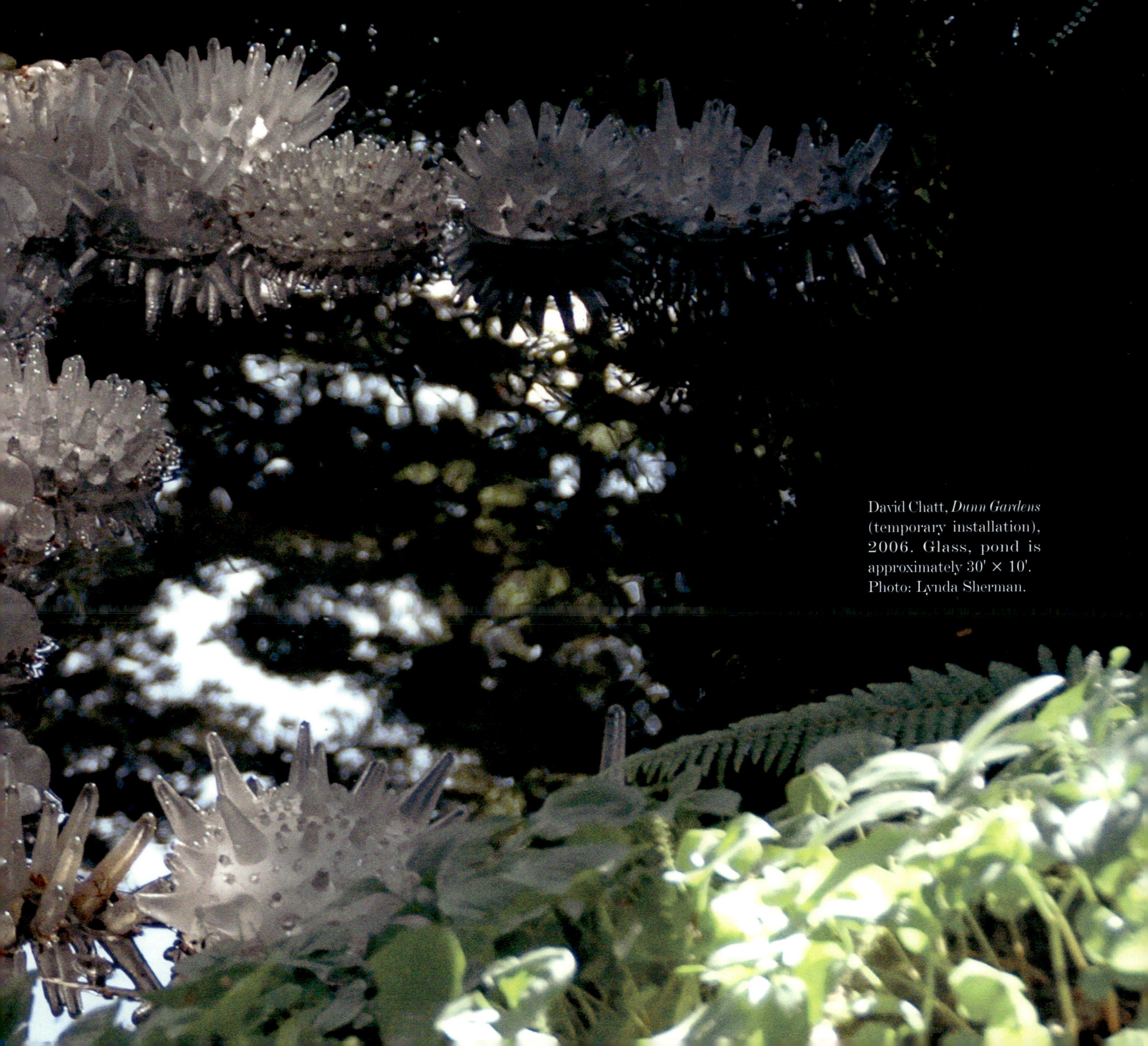

David Chatt, *Dunn Gardens* (temporary installation), 2006. Glass, pond is approximately 30' × 10'.
Photo: Lynda Sherman.

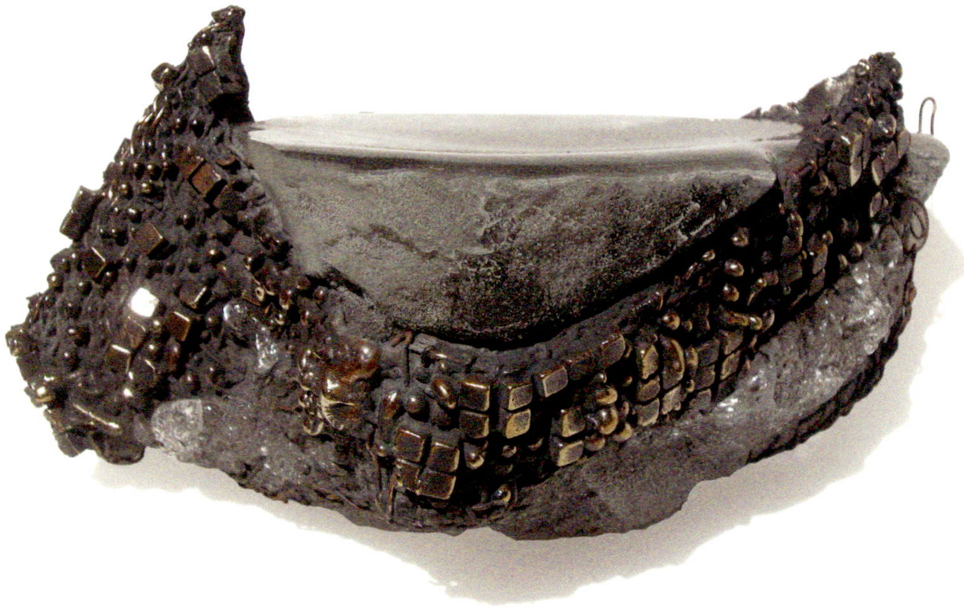

Top Marianne Weil, *Dug Out*, 2015, Cast glass, cast bronze, 9" × 7" × 5".
Photo: Marianne Weil, Courtesy of Ille Arts and the artist.

Bottom Lucy Lyon, *Wait,* 2014. Glass, steel, 21.5" × 4'2" × 14".
Photo: Addison Doty.

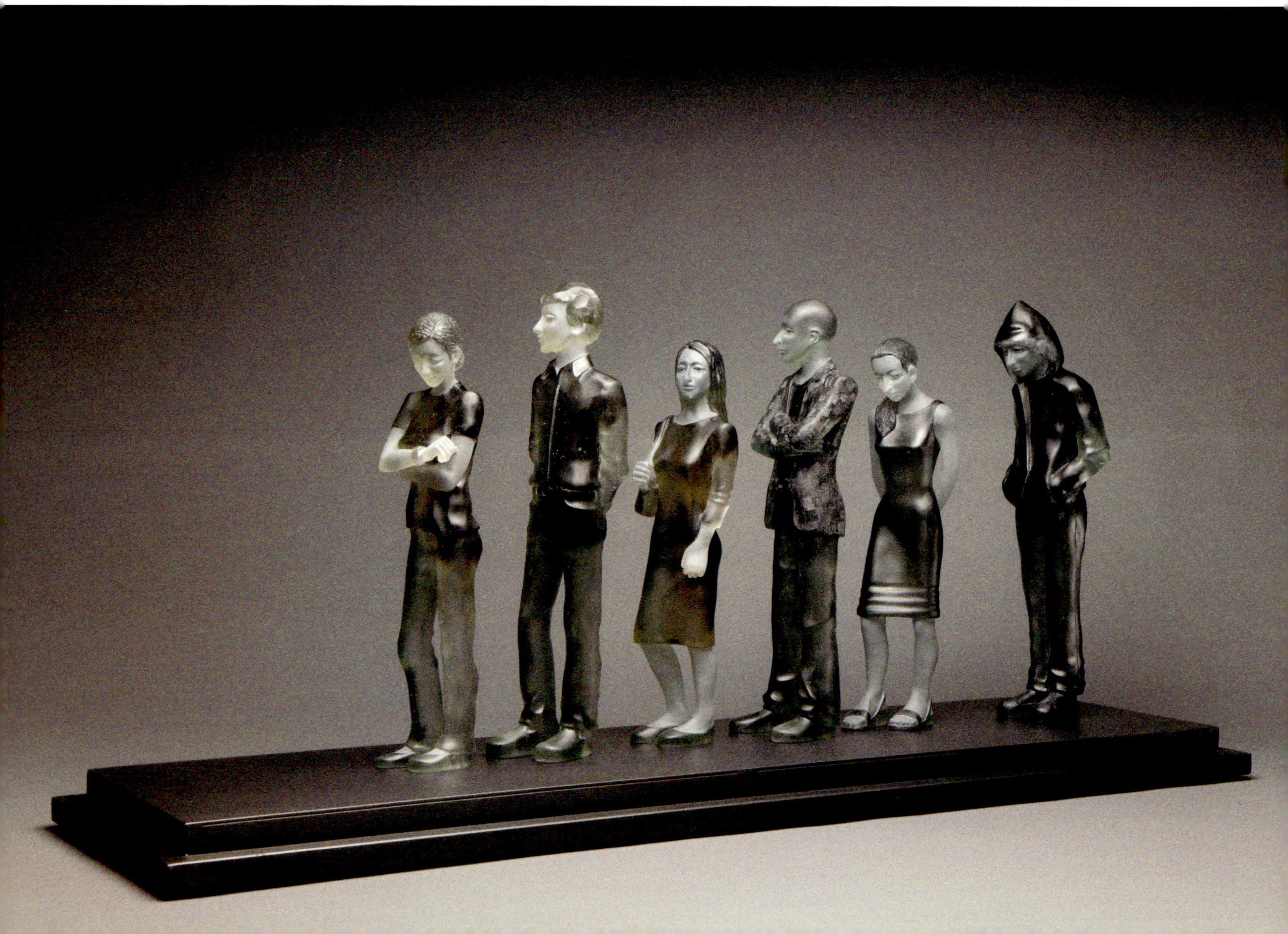

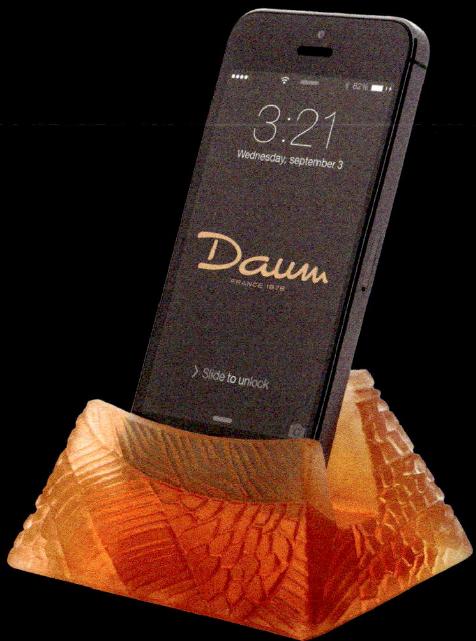

Left Daum, *Crystal Stand for Smartphone* (detail), crystal, 2.16" × 3.31" × 2.55".
Photo: Daum.

Below Terrence Lavin, *Prototype C*, 2013. Glass, brass, bronze, copper, 10" × 15" × 8".
Photo: Derek Dudek.

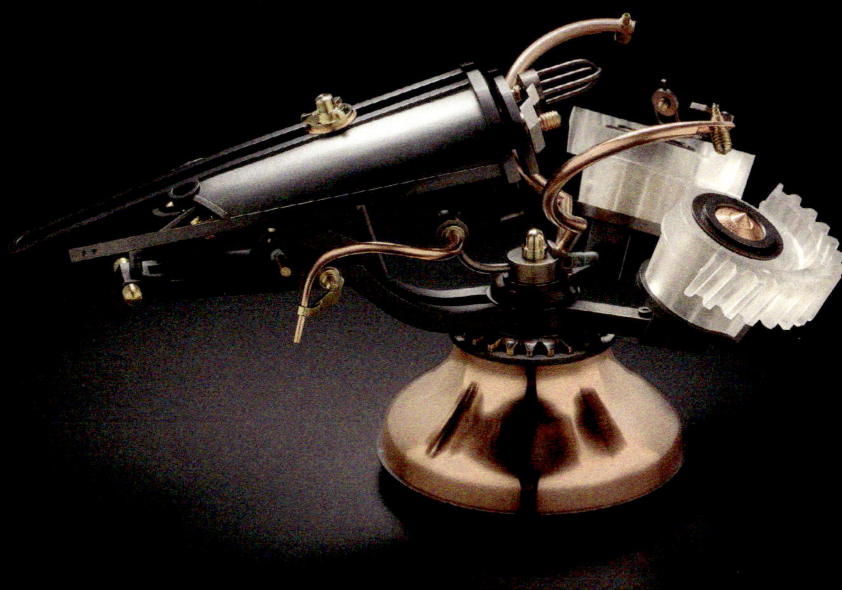

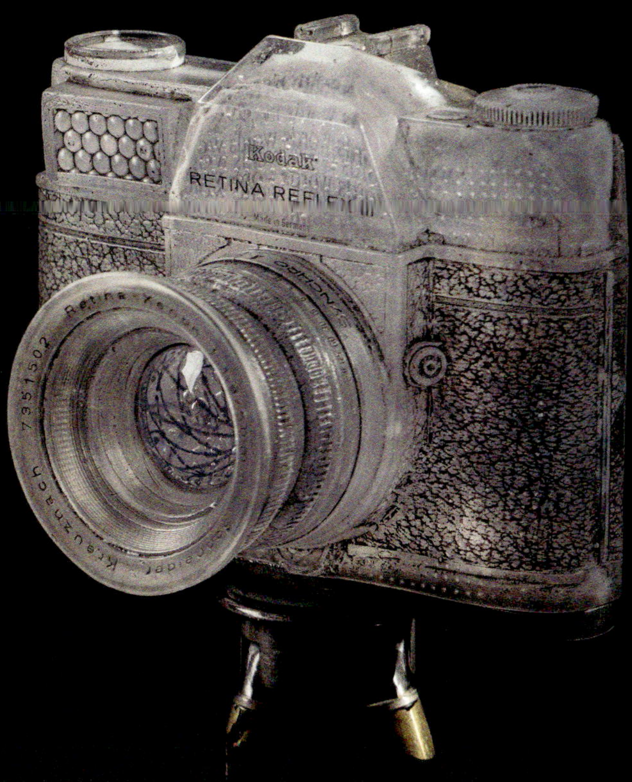

Josh Hershman, *Recollection*, 2013, Lost wax cast glass, polished, several enameled images inlaid, oil paint, antique tripod, 6.5" × 2.25" × 6" (camera only).
Photo: Ben Corda.

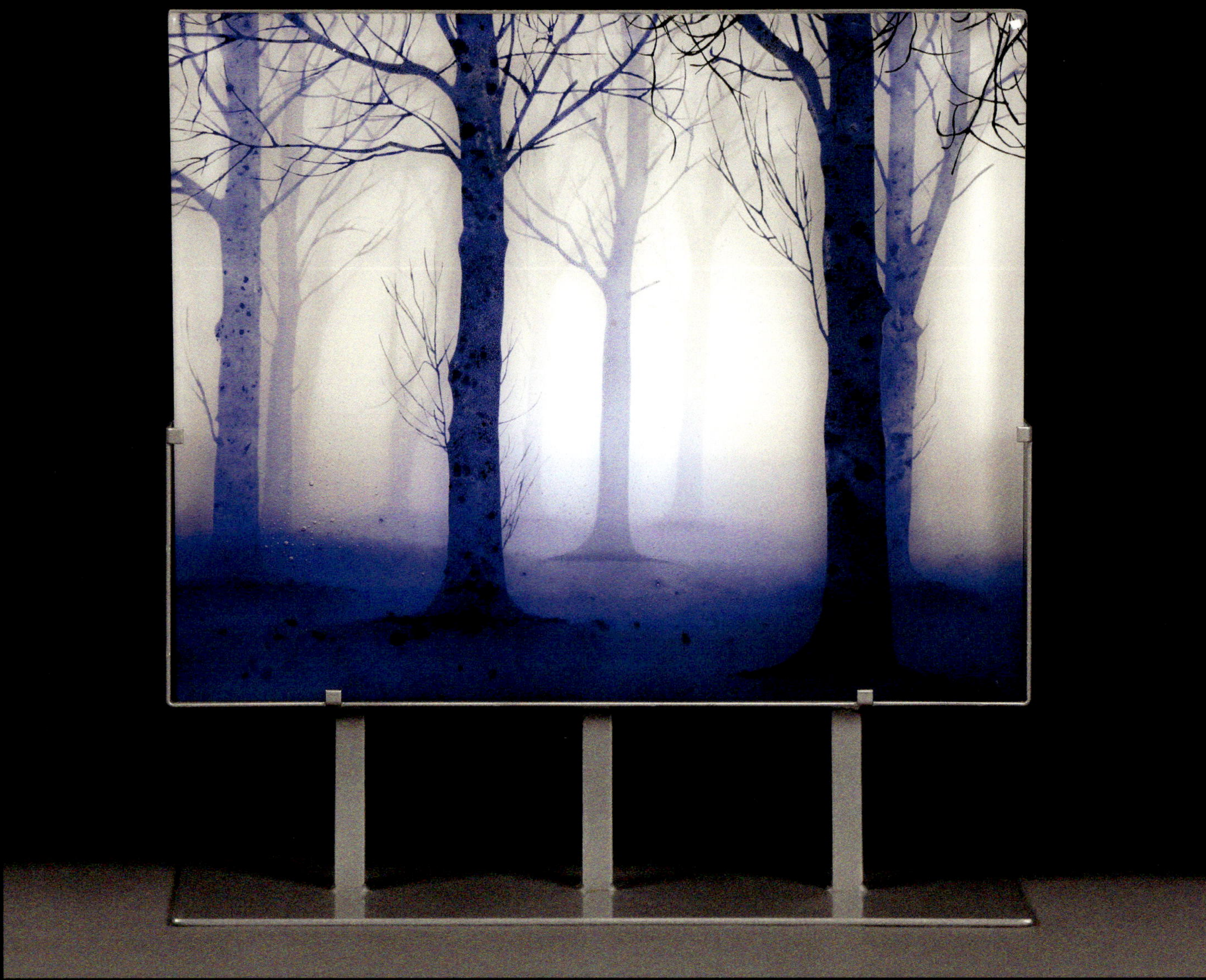

Paul Messink, *Trees in Twilight*, 2014. Enamel, glass, 14" × 13.5" × 1.25".
Photo: Eric Bladholm.

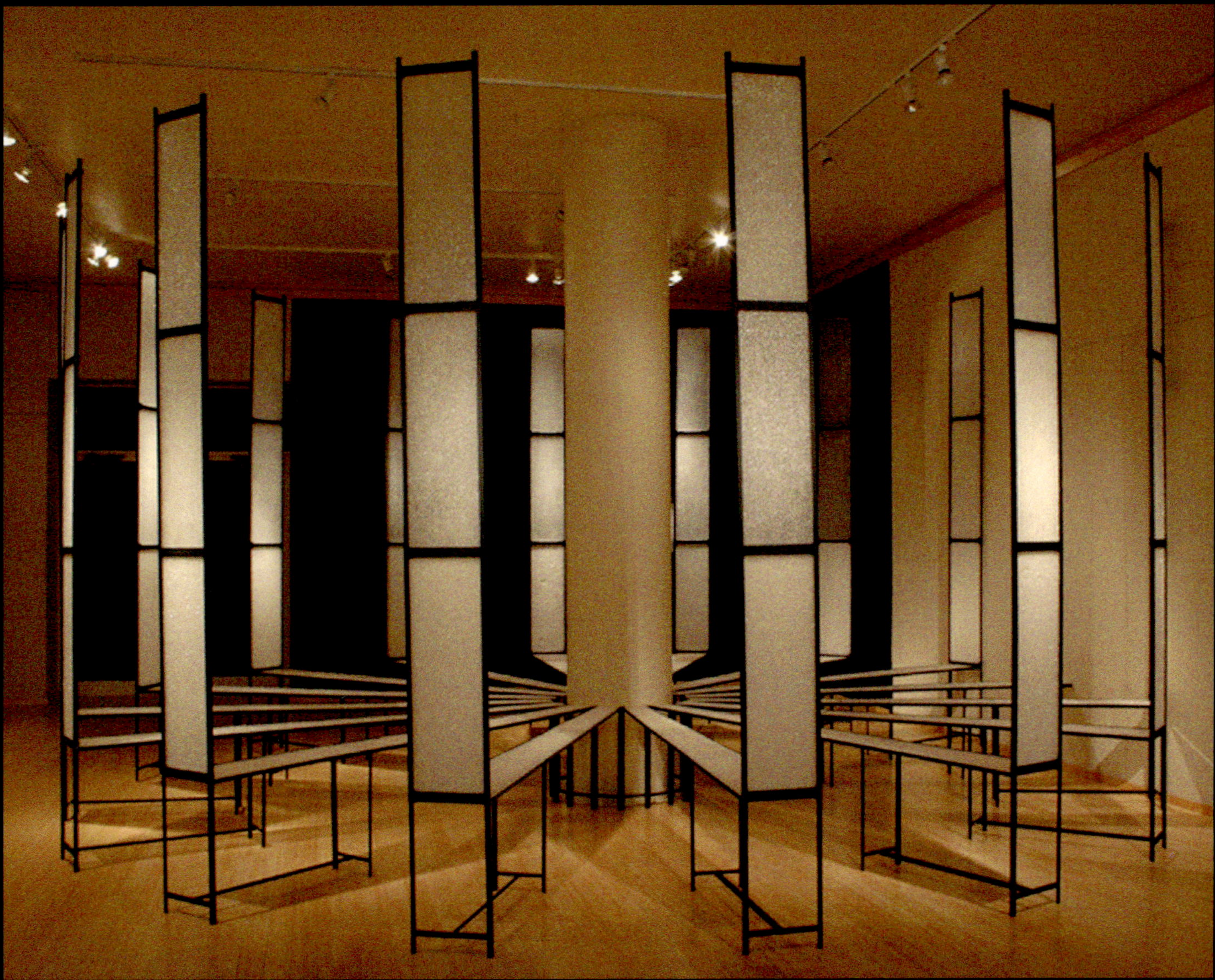

Benjamin Johnson, *Circular Experience*, 2012. Glass, steel, 10'6" × 20' × 20'.
Photo: Benjamin Johnson.

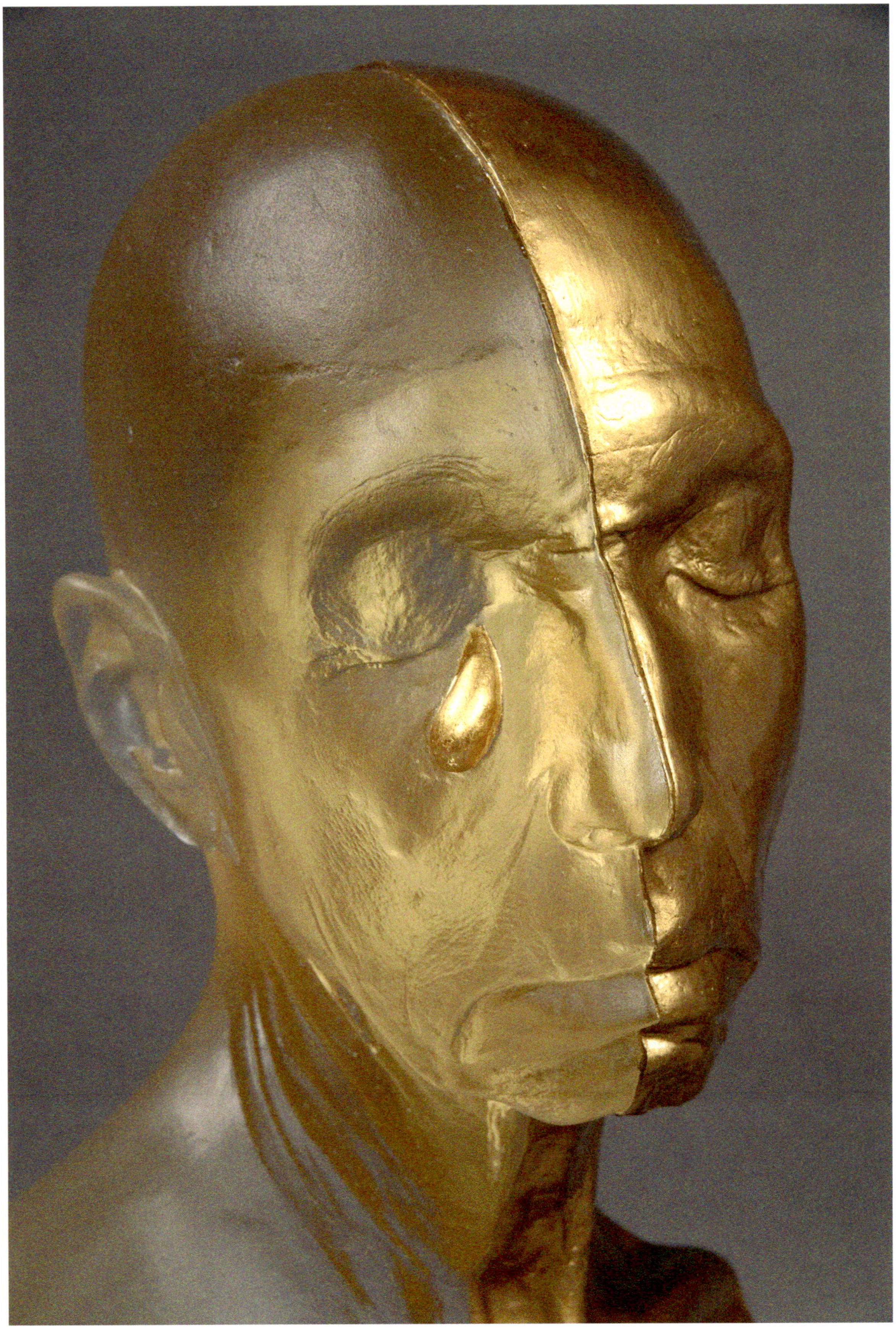

Judy Chicago, *Toby Head with Golden Tear,* 2009. Cast glass, gold leaf, and gilding, 18" × 16.5" × 16.5". Photo: © Donald Woodman, © Judy Chicago.

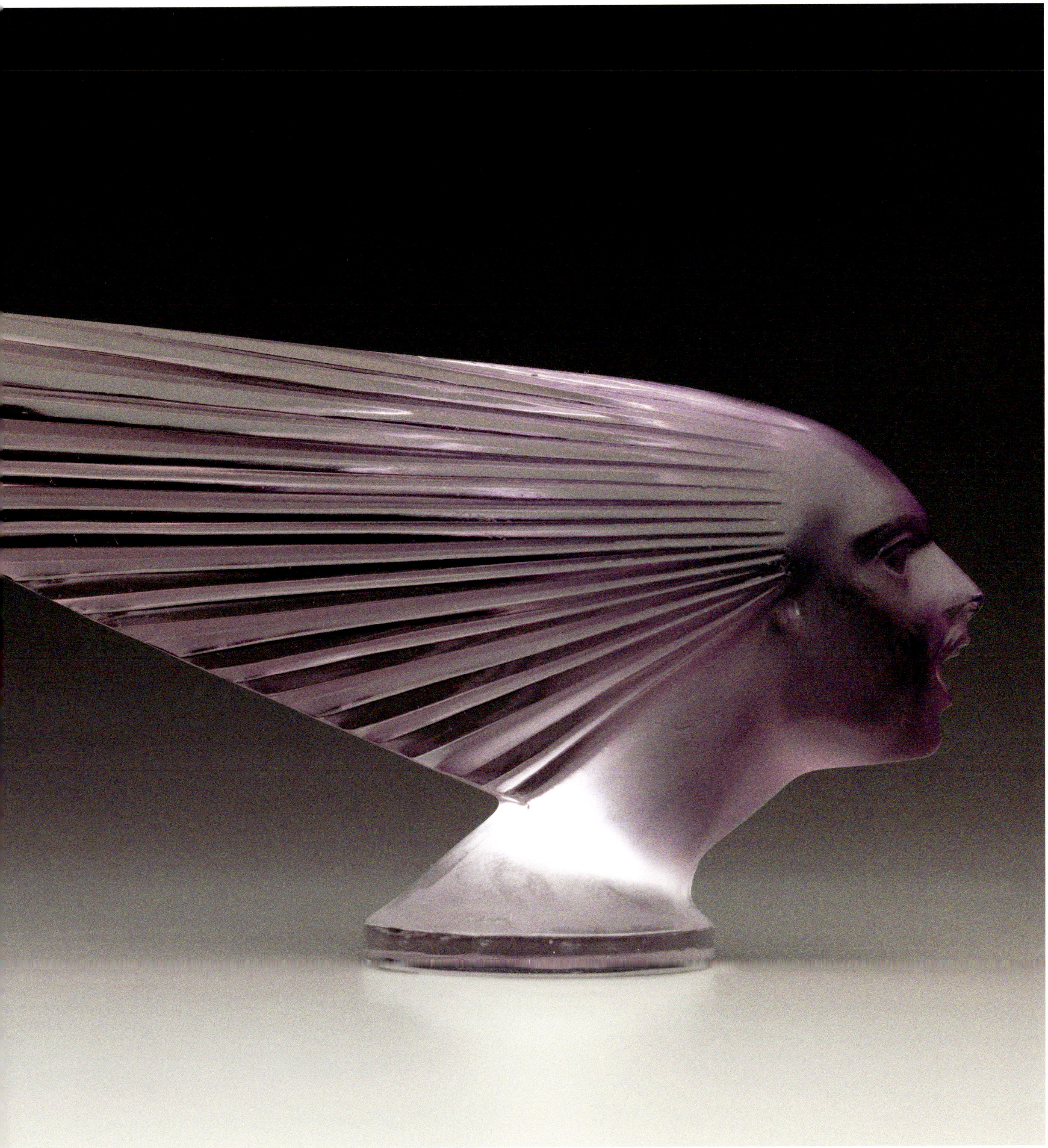

René Jules Lalique, *Victoire* hood ornament, 1928. France, glass, 4.1" × 9" × 2". Courtesy of Bridgeman Images, Gift of Mr. and Mrs. Michael Steinberg, Dallas Museum of Art.

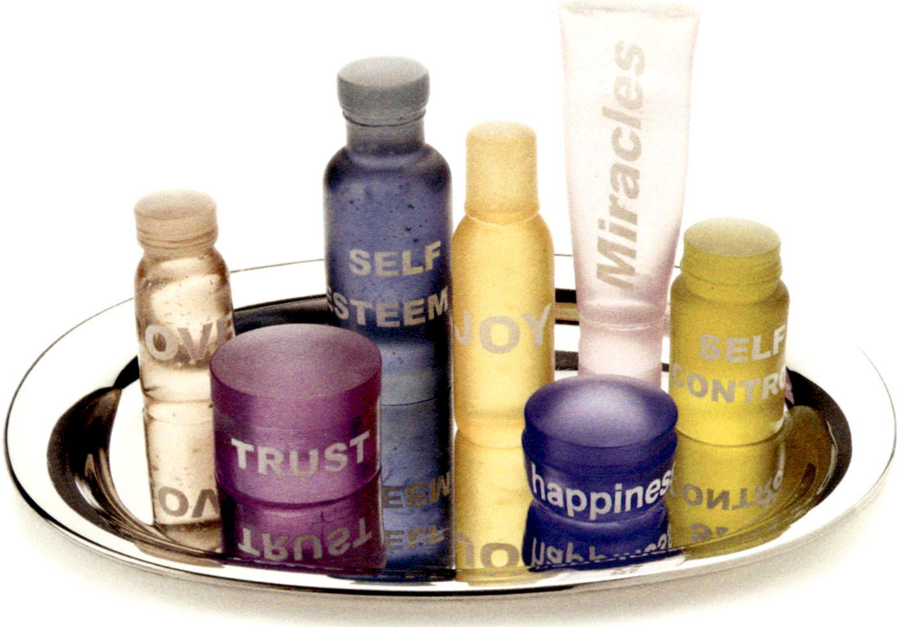
SELF
JOY
Miracles
SELF
TRUST

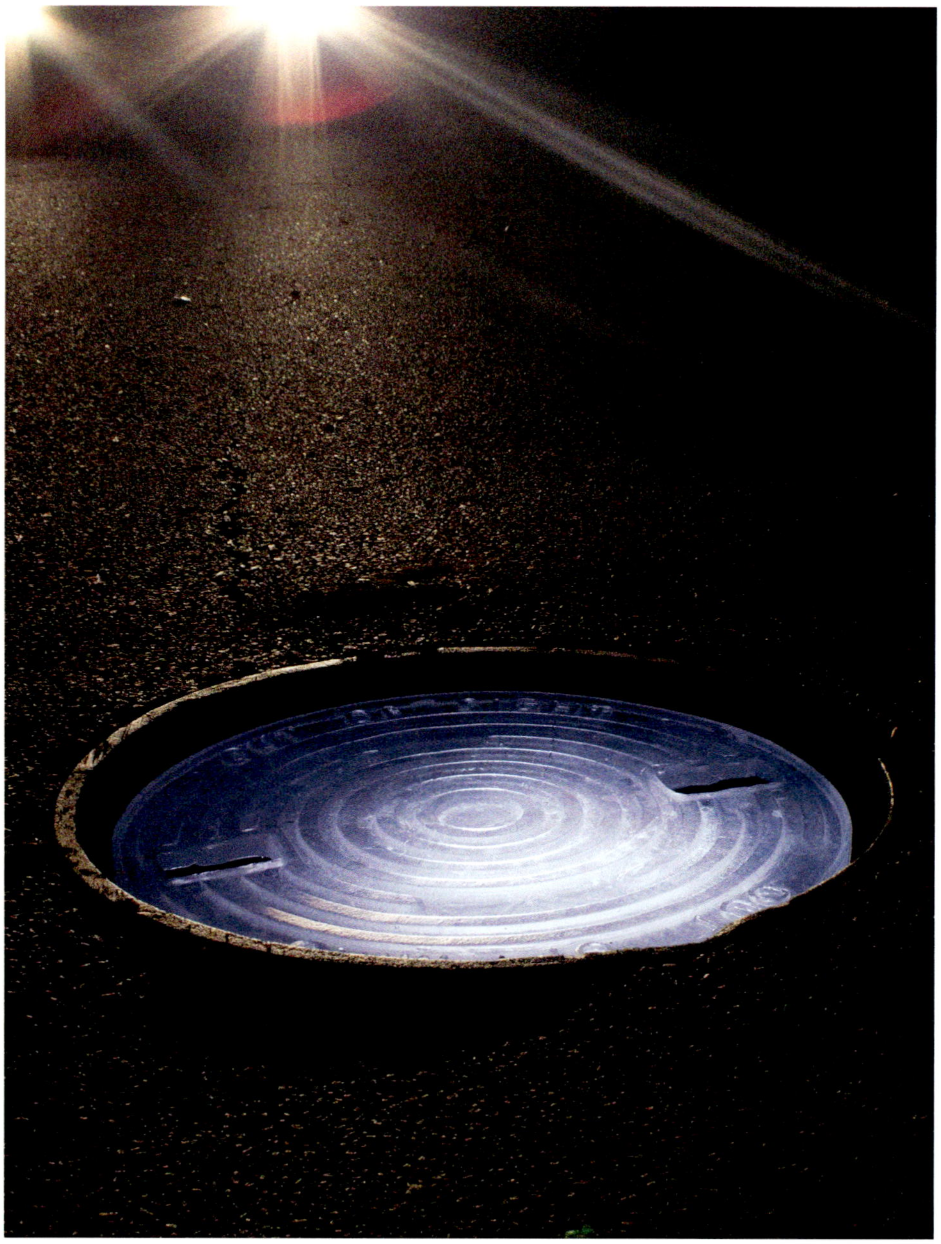

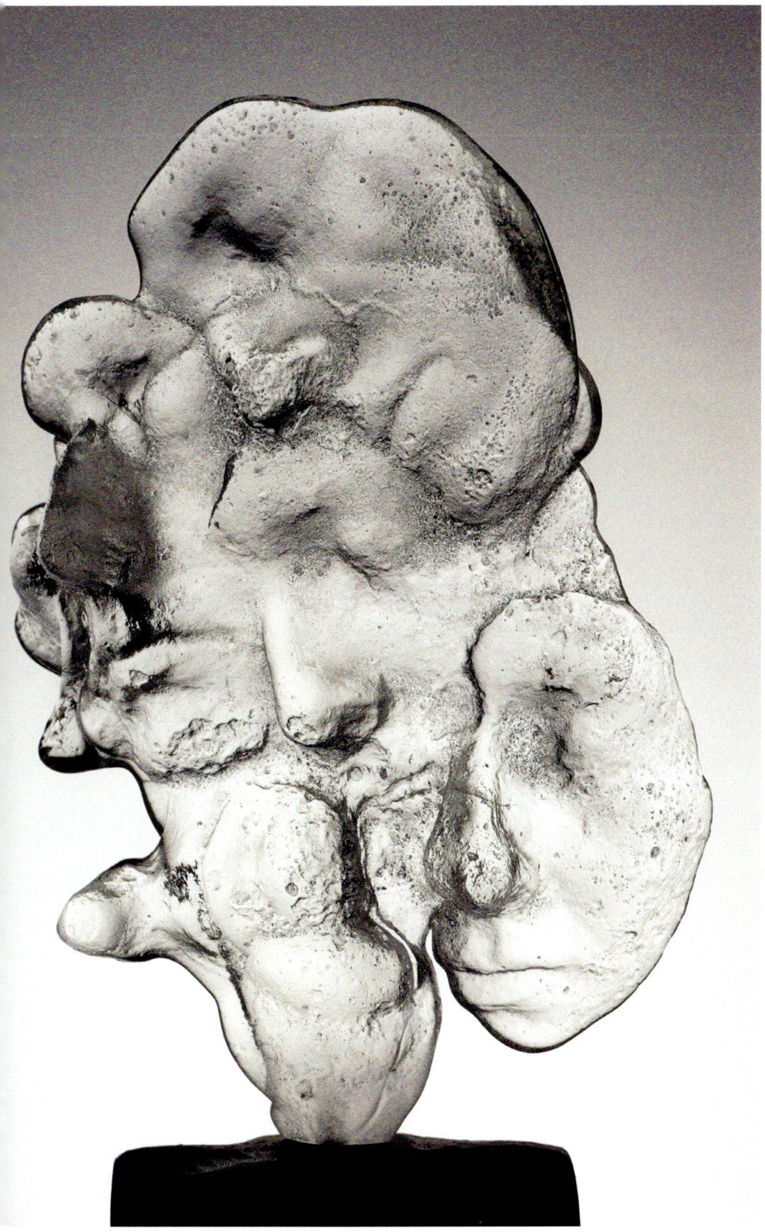

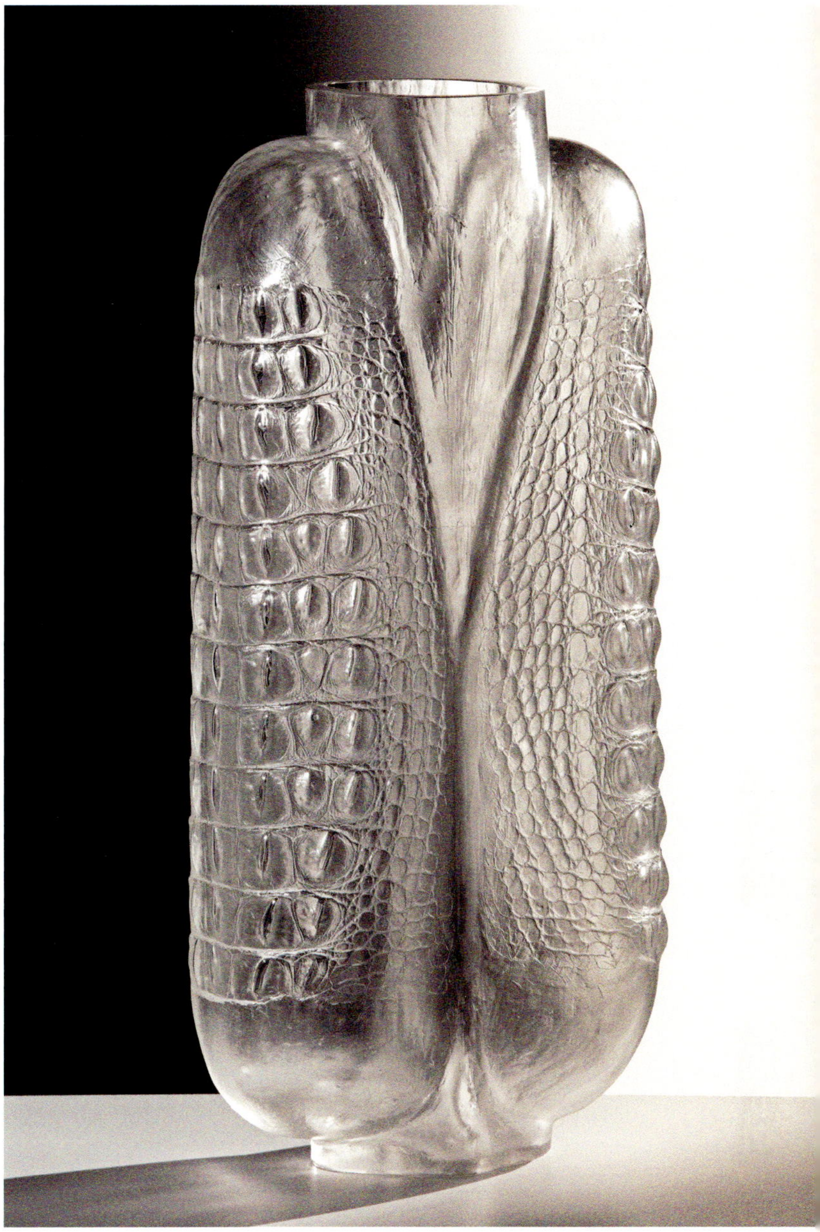

Left Alva Gallagher, *Souls*, 2012. Glass, 19.7" × 7" × 4.7".
Photo: Sylvain Deleu.

Right Amie McNeel, *Gator Vase 5*, 2014. Hot mold blown glass, cire perdue process, 8" × 6" × 17.75".
Photo: Russell Johnson.

OPPOSITE PAGE

Top Silvia Levenson, *Life Strategies*, 2010. Glass, metal, 7.9" × 15.75" × 11.8".
Photo: Natalia Saurin.

Bottom Neal Paustian, *Out of Sight, Out of Mind*, 2014. Glass, 1" × 23" × 23".
Photo: Neal Paustian.

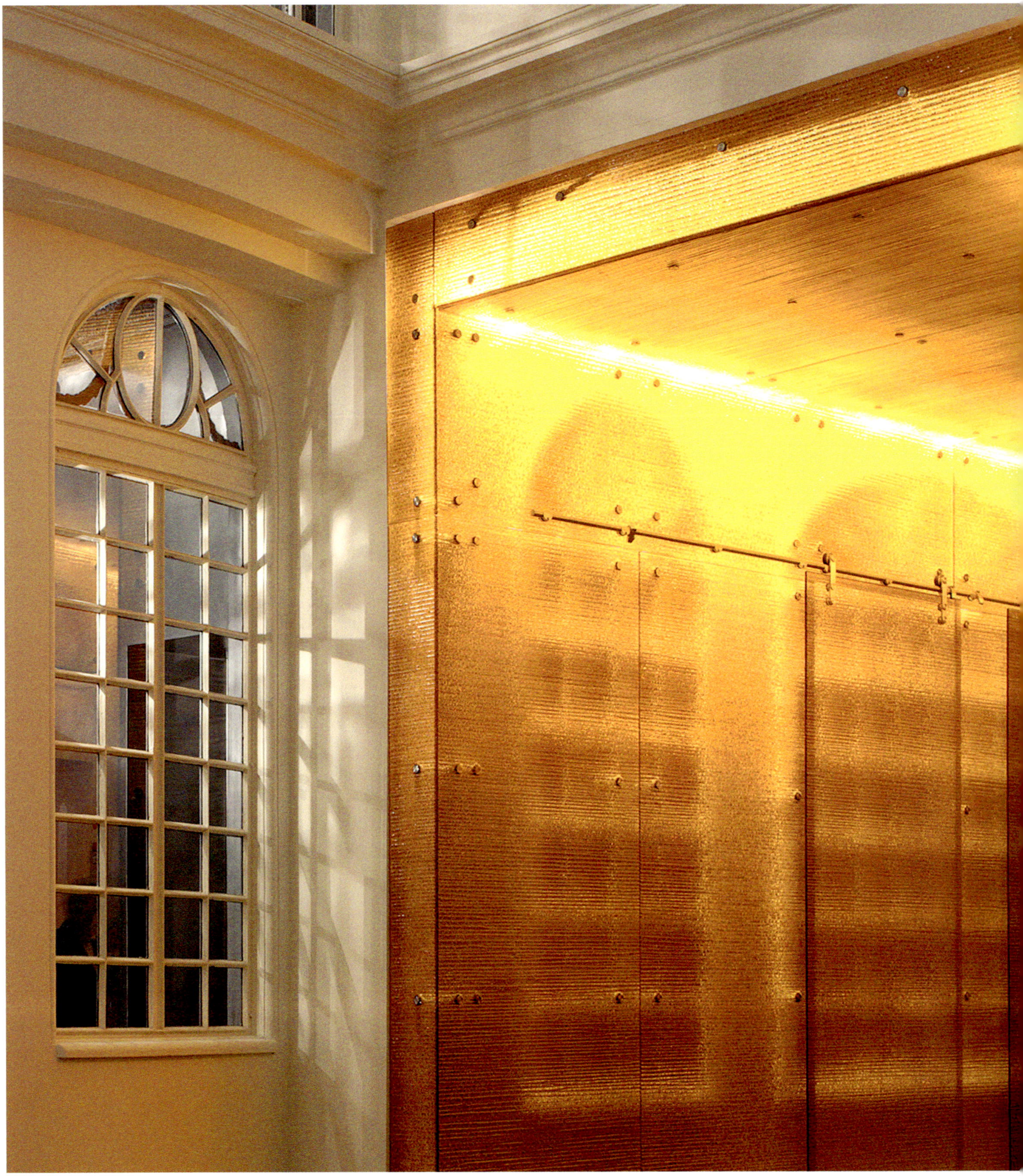

Fendi's prominent glass tunnel was created with Crystal Cognac glass with Rain texture, chosen by Peter Marino, Architect. Using glass in this way creates space full of drama and mystery.

Nathan Allan Glass Studios, *Fendi Glass Tunnel*, 2013. Clear glass, 3" thick.
Photos: © Fendi, Terminal City, BC.

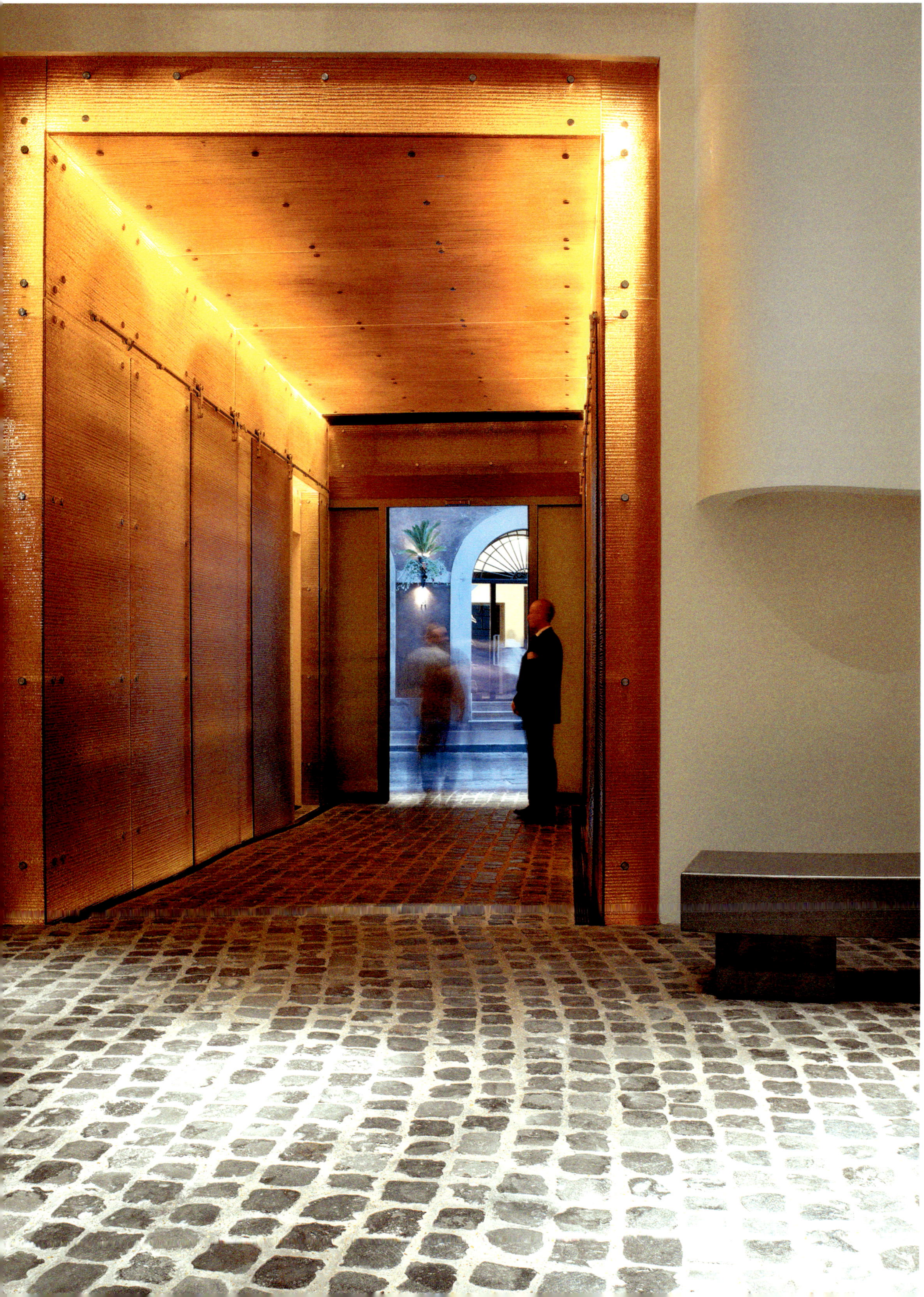

JEWELRY

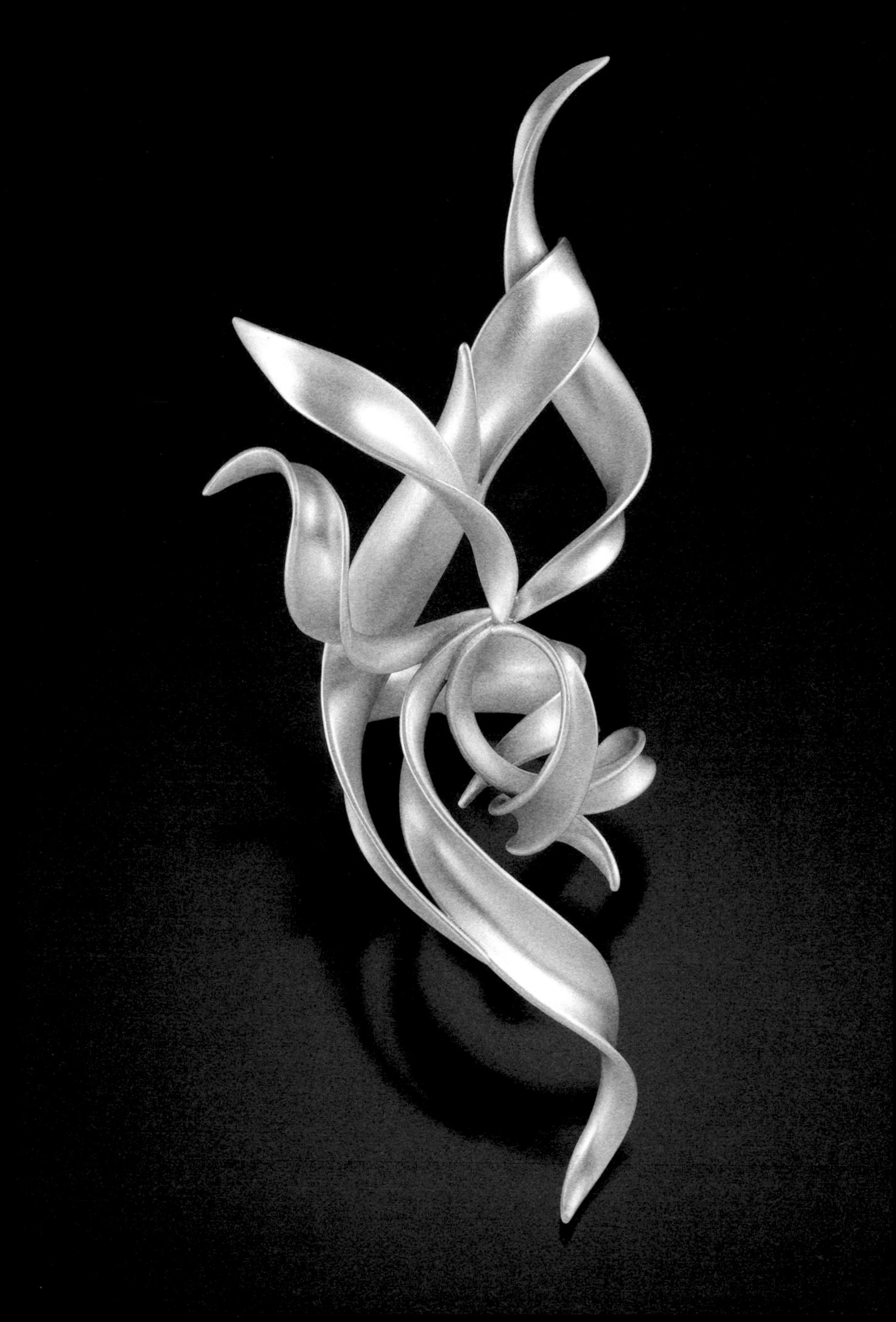

Charles Arnold. *Fire Ring*, 2014. Silver, gold, 7" × 2" × 1".
Photo: Charles Arnold.

Opposite page Donna Veverka. *Rings* from the *Personal Architecture* series, 2004–2014. Sterling silver, variable dimensions.
Photo: James Hull.

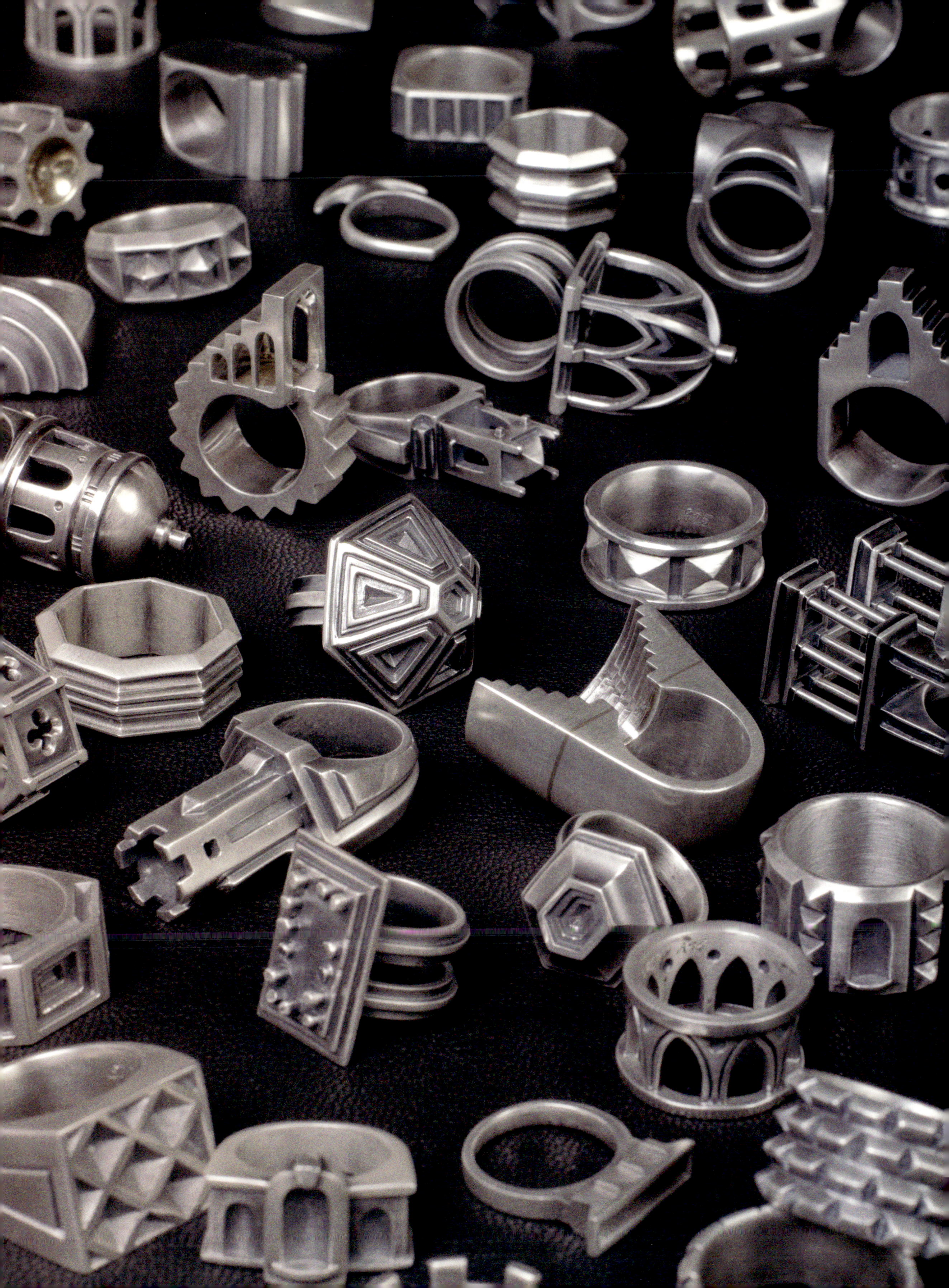

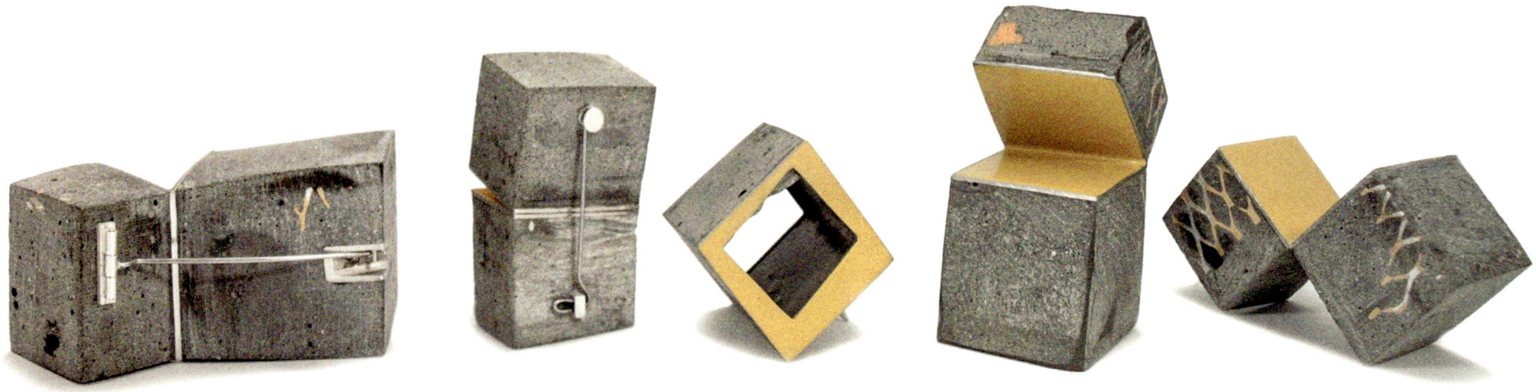

Demitra Thomloudis, *Space Units Stacked*, 2014. Cast cement, powder coat, resin, sterling silver, nickel silver, 2" × 2" × 2". Photo: Demitra Thomloudis.

A little more than a hundred years ago, René Lalique was pioneering modern approaches to jewelry as well. His work could not look more different than Thomloudis's—hers references the built environment while his is an interpretation of the natural world—but both of these artists have created work that speaks to their moment in time, and both makers utilize casting in innovative and exciting ways.

Lalique's dynamic compositions combined with his mastery of design, materials, and process helped define the Art Nouveau movement. The characteristic whiplash curves of Art Nouveau mixed with his approach to the human figure, flora, and fauna combine to result in some of the most sophisticated jewelry ever made.

Manifesting organic and undulating forms is one of casting's greatest strengths; it was through this process that Lalique's vision came to fruition. He cast both metal and glass and was a master of mixed media. He deftly combined various woods and metals, bone and horn, stones and pearls, as well as enamel and glass in a variety of combinations. He was able to incorporate cast, carved, and fabricated elements so seamlessly that it can be difficult to fathom how certain pieces were made.

While cast glass had a well-established place in jewelry for thousands of years, Lalique reinvented it as a modern genre. His cast forms, their scale, and the way he used these elements marked a significant step toward modernism and away from the Victorian style that dominated jewelry for so many years. These cast glass elements were finely articulated narrative forms that often played the leading roles in his pieces, rather than being in supporting roles as simple glass beads or gemstone imitations were in most other jewelry of the time.

Many artists use multiple materials, but often, the treatment of those materials is inconsistent—the handling of one may be clumsier than another and when they are paired, a piece can feel a bit like an awkward marriage. Lalique was equally masterful in every material and process he employed. His work is still easy to identify because his point of view is so distinctive and the execution of his works is consistently superb. He designed almost in spite of the properties and commonly understood confines of his materials, pushing their limits in service of his vision (probably to the chagrin of the master craftsmen populating his atelier). Whether a cluster of cast glass roses or gold and silver clematis pods, Lalique's casting is at the pinnacle of the process.

Few contemporary makers can rival what Lalique achieved, but Lin Stanionis certainly does. Clearly, her work is influenced by his, but hers is distinctly modern in her material choice and her wonderfully contemporary and macabre use of found objects. Casting, her primary process, allows this work to take shape. She says, "My work in large part is made up of natural materials that I collect outside my studio in rural Kansas. These materials—plants, insects, and animals—are found in various states of growth decay, living and dead. I rely on the processes of molding and casting to permanently capture the shapes and details of these life forms. In my finished works I hope to show nature's dark beauty as it relates to the human experience of living in this world. The making of these jewelry pieces involves many steps and stages to reach completion. Through the molding process, I am able to make waxes of the different natural materials that can then be reworked, manipulated, and recombined to create new forms. There are times when I will mold the manipulated forms in order to cast multiples for use in a

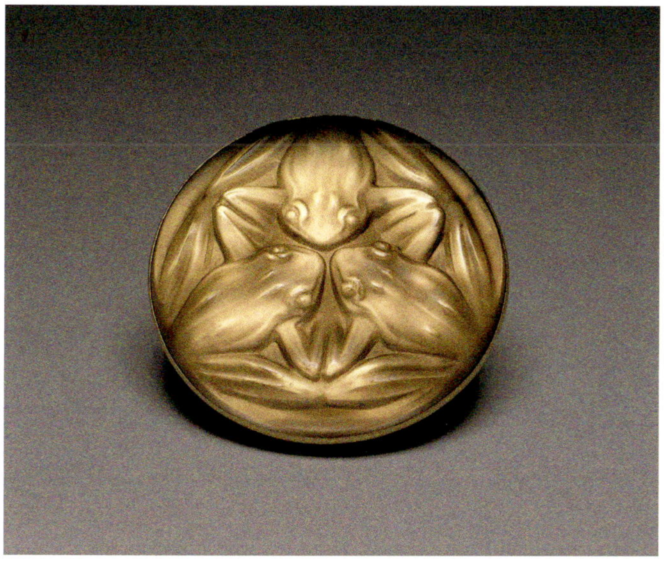

work. The casting process is the only method that will allow me to recreate the natural forms in more permanent materials. I work primarily in metal, resins, and glass though I have used the process for concrete, plaster, and chocolate. A process known to the ancients, casting was thought to be transformative. I still find this to be true; that even with all our technologies there is still magic in the process."[3]

Our primal attraction to casting, the mystery of the materials' transformation, and the plasticity of form it makes possible bring us a dynamic and thrilling way of working, whether in platinum or in plastic.

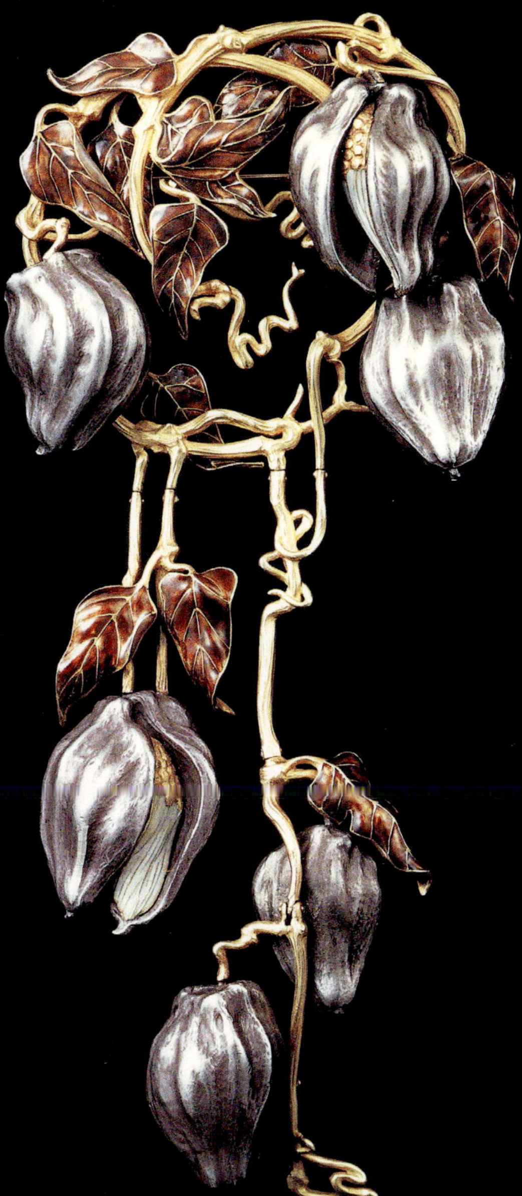

Top René Lalique, *Frog Brooch*, 1911. Amber colored glass, gilded brass (copper and zinc), 1.75".
Courtesy of The Walters Art Museum, Gift of Dr. and Mrs. Edward F. Lewison in 2002, CC0, GNU FDL.

Bottom René Lalique, *Brooch with Clematis*, 1902. Gold, oxidized silver, white and brown enamel.
Private Collection / De Agostini Picture Library / Bridgeman Images.

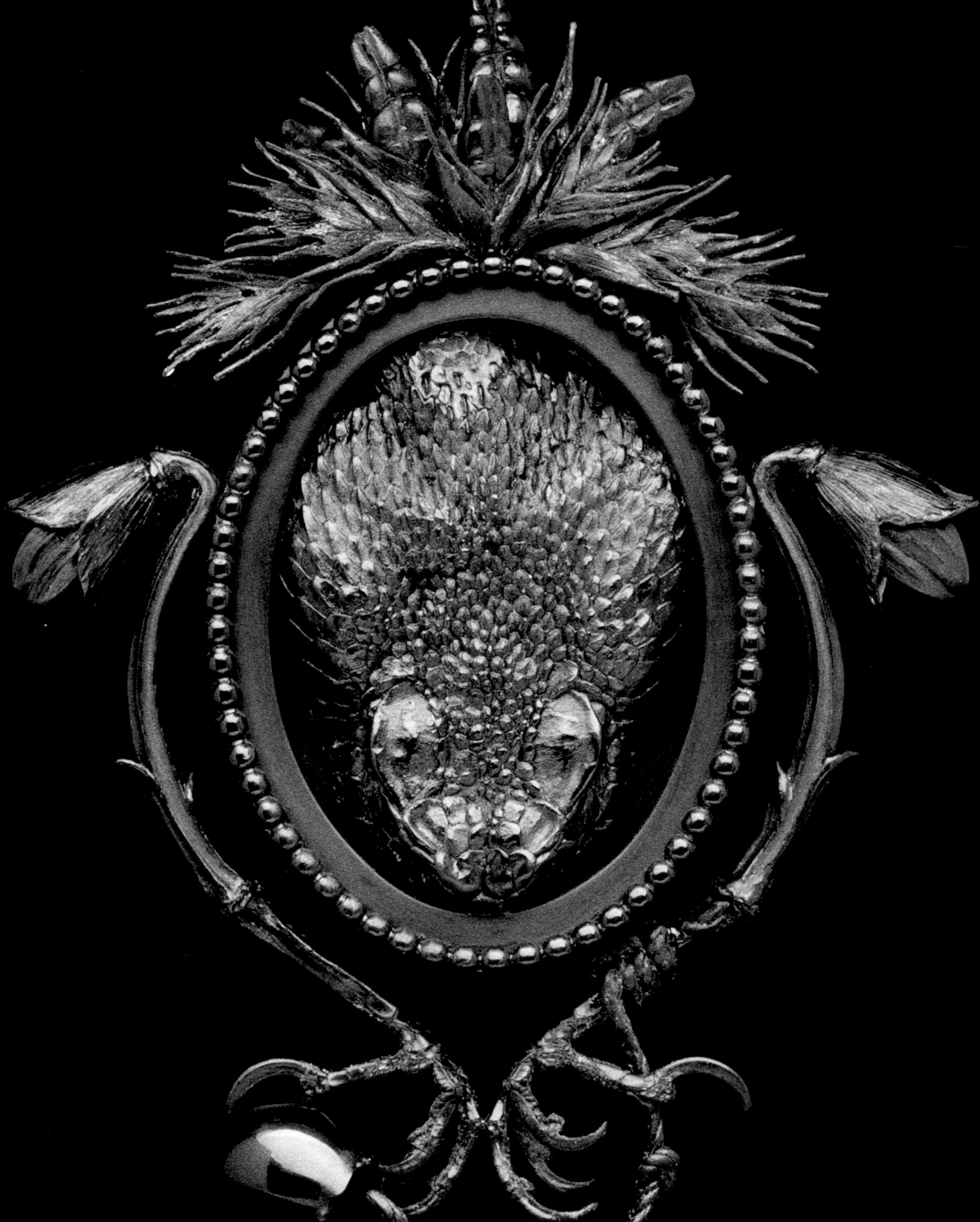

Lin Stanionis, *Rapture*, 2011.
Urethane resin, 4" × 3" × 0.75".
Photo: Jon Blumb.

FUNCTION

From Wearable to Conceptual

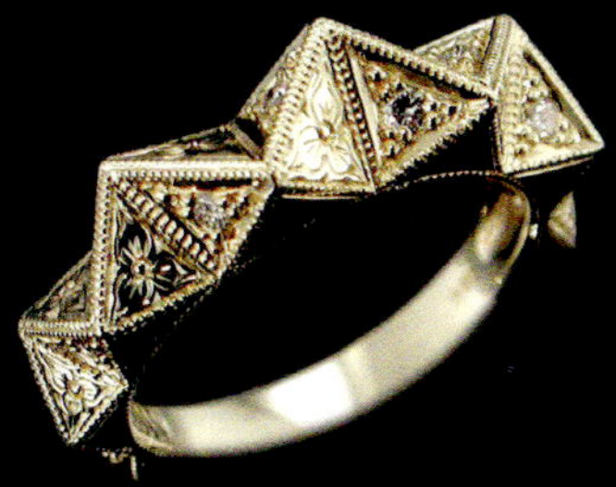

Jewelry ranges from the simple bead or wedding band to ceremonial, editorial, and conceptual wearable sculpture. This scope from basic to elaborate has always existed—since it is a fundamental human impulse to adorn ourselves, every possible permutation has been explored. Casting lends itself to every part of jewelry's spectrum because of its two main strengths: its reproducibility and its capacity for sculptural form.

Casting allows for the most basic commercial wedding bands to be cranked out by the thousands, but it also facilitates thoughtful one-of-a-kind pieces for everyday wear. Jewelers who make traditional wearable work often use casting because they can carve wax forms and cast pieces that would be impractical (or sometimes impossible) to fabricate.

Mark McNown's *ZigZag* ring is a great example of the utility of casting in wearable jewelry. This ring flaunts a complex but delicate form. Fabricating this piece would be possible, but nothing would be gained by doing so. After casting, McNown bead-set the diamonds and engraved the flowers on the angular planes of the ring. He then finished the piece with millegrain edges, giving it an elegant feel.

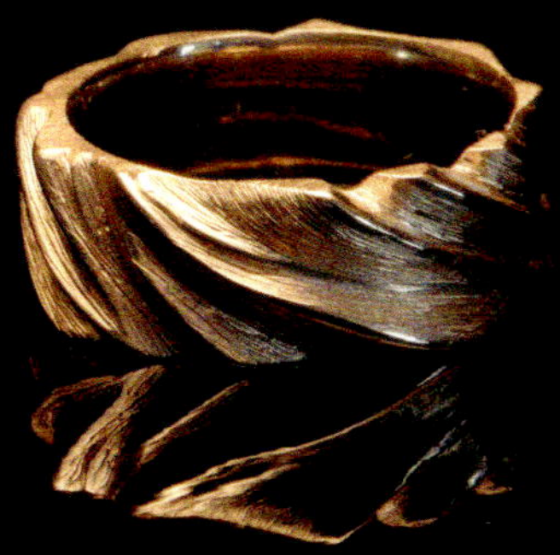

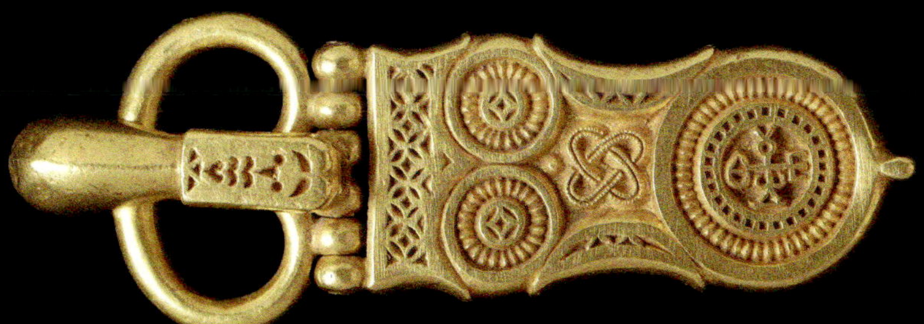

Top to bottom

Mark McNown, *ZigZag Ring*, 2000. 18K royal yellow gold, diamonds, 1" × 1" × 0.3". Photo: Mark McNown.

Jim Dailing, *The Rekindled Flame*, 2013. 14K red gold (alloyed from inherited 18K yellow jewelry and straight copper). Photo: Jim Dailing.

Belt Buckle, late sixth to seventh century (Late Antique). Byzantine, Hamas, Syria (Place of Discovery), gold, 1.12" × 3.12" × 0.4". Courtesy of The Walters Art Museum, Acquired by Henry Walters, CC0, GNU FDL.

Sculptural jewelry is also well-served by casting. Narrative elements, especially naturalistic ones, can be handily and convincingly carved in the round and cast. This Edwardian brooch depicting two swallows carrying a diamond swag in their beaks is a beautiful example; the fullness of the swallows' heads and bodies and the gentle compound curves in their wings and tails were elegantly executed using traditional lost wax casting.

Jewelers can keep a record of pieces they've made and revisit them in the future by taking rubber molds and creating wax replicas, either reproducing pieces exactly or altering them in a variety of ways. Molds are also a form of insurance that protect against a bad casting. Complex and time-consuming models like the Edwardian swallows can just disappear if a casting goes wrong, so it's prudent, even if the piece is to be a one-of-a-kind, to take a mold. Some makers develop a robust library of molds and their work is defined by reconfiguring molded and cast elements in new ways. Casting is indispensable to the bench jeweler for so many reasons, but the benefits of molds are many and varied. Whether a jeweler is simply molding a one-of-a-kind model as a way of saving their work, molding a piece for future production, or molding a single element that will be recontextualized in complex pieces to be made at a later date, a good mold is the jeweler's best friend.

Other studio jewelry is made without the security of rubber molds and the original model is directly burned out in a kiln as a part of the lost wax casting process, as is the case with Carmen Liu's *Origami Rings*. These pieces were made by reinforcing actual paper origami models with wax and burning them out. Working in this way, Liu retains the fine texture of the paper, the subtlety of the folds, and the undercuts from the original origami in a way that wax carving never could.

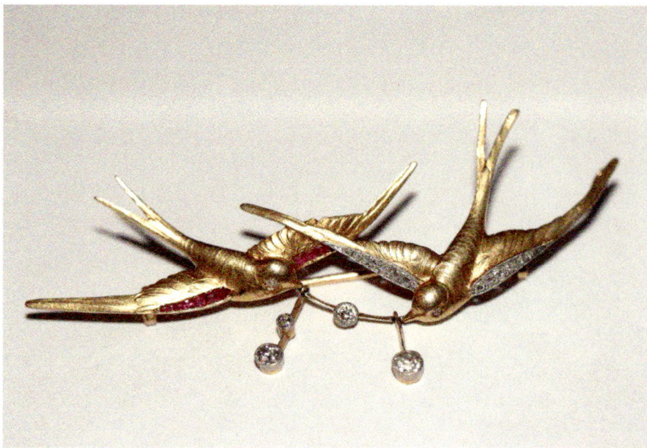

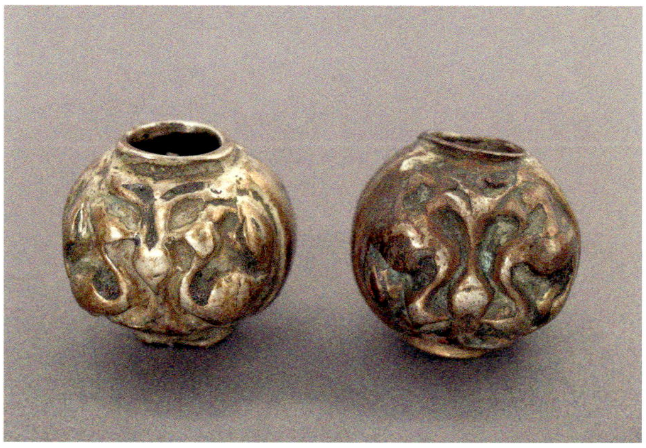

Top Edwardian Double Swallow Brooch, ca. 1910. 18K gold, diamond, synthetic ruby, 2.98" × 1.56".
Courtesy of Elizabeth Rose Antiques.

Center Beads, eleventh to twelfth century. Afghanistan, silver, cast in two halves, decorated with gilding, and niello inlay, 0.75".
Courtesy of LACMA, www.lacma.org, LACMA public domain.

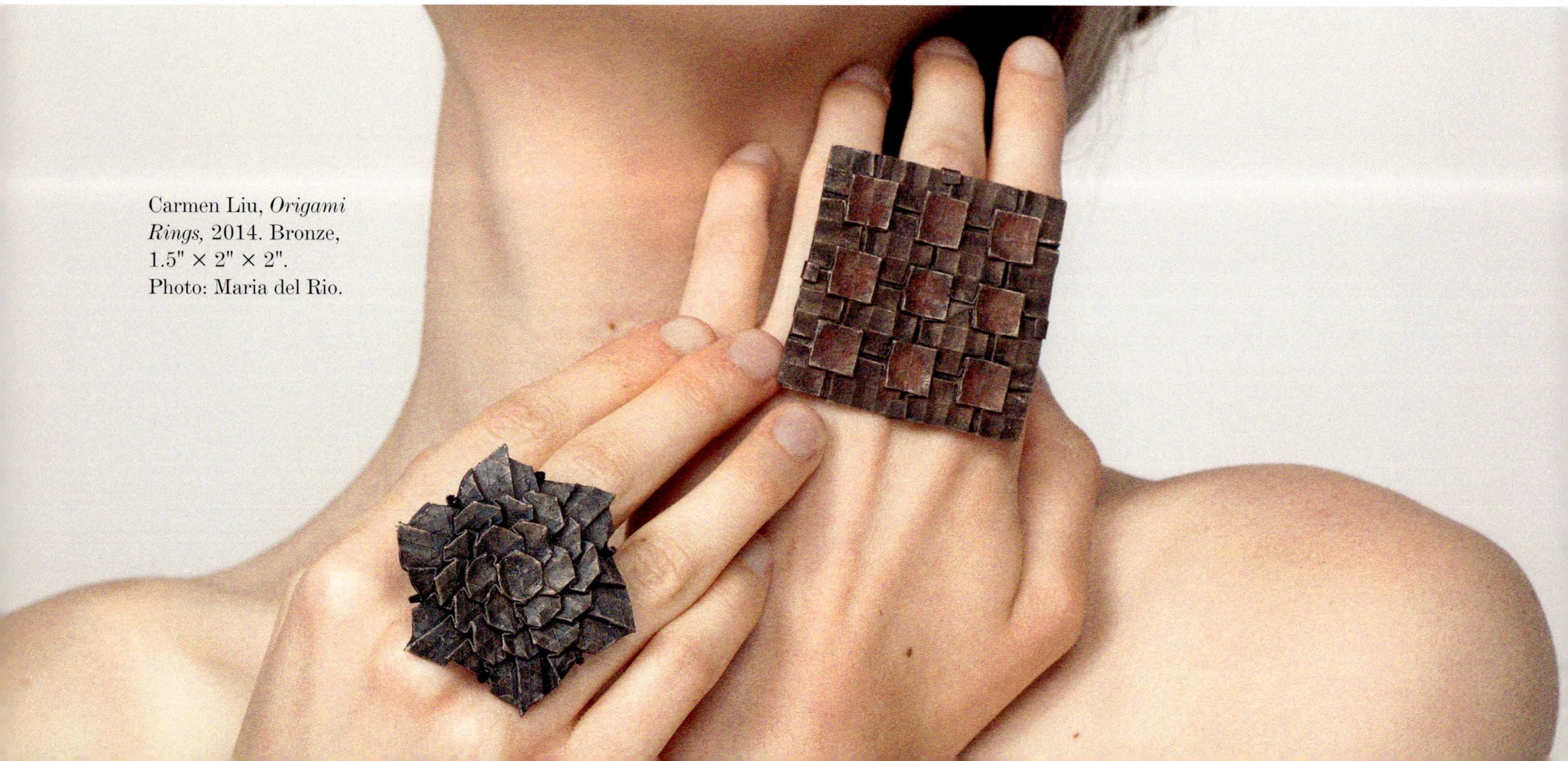

Carmen Liu, *Origami Rings*, 2014. Bronze, 1.5" × 2" × 2".
Photo: Maria del Rio.

Many contemporary jewelers work with nontraditional model materials. Found objects are frequently used, but few artists do it as effectively as Vernon Bowden and Ruudt Peters. They take different approaches and the pieces have their own unique conceptual underpinnings, but both use everything that casting has to offer in these works.

Vernon Bowden's *Barb-aryan* is a necklace of Barbie® heads, sand cast from sixty-two Diet Coke® cans. While wearable, Bowden's work is not about its wearability. It is conceptually layered; the titles, recognizable forms, and his casting mediums all carry associations and preconceptions. Bowden uses Barbie® as a symbol of suburban life and commercialism; casting them in Diet Coke® cans is his nod to the consumer culture of our age. Casting perfectly supports this approach to making. It allows for this mélange of meaning, materials, and the clever transformation of mass-produced objects into this compelling necklace.

CORPUS "ovarium," by Ruudt Peters, is another thought-provoking cast piece that hinges on its identifiable imagery. Peters has taken a Christ figure from its cross, dissected it, and rejoined its two parts at the hip. This image is quite provocative... it suggests a sort of religious vivisection—an attempt to look inside; but what is revealed is a flat, abstract plane that is reminiscent of a Rorschach inkblot. Rather than making a statement, this piece seems to ask questions. Like Vernon Bowden's *Barb-aryan*, Peters's astute approach to *CORPUS "ovarium"* layers ideas and associations resulting in a loaded object that becomes even more potent when worn on the body.

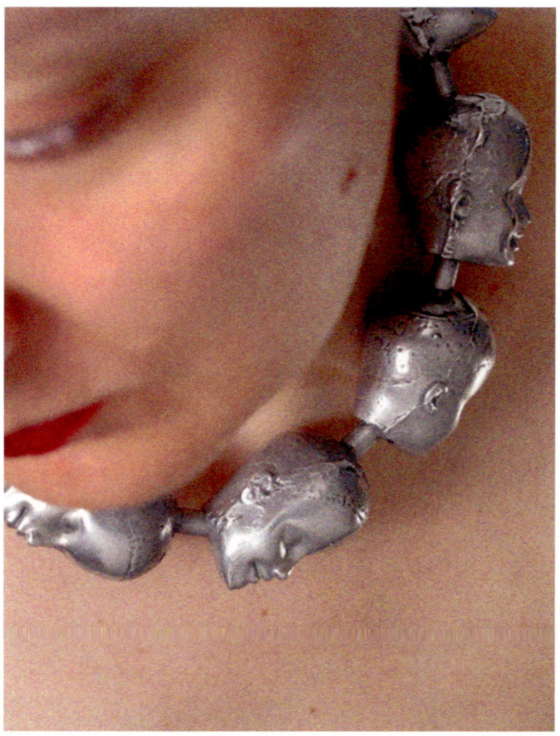

Vernon Bowden, *Barb-aryan*, 2006. Sixty-two Diet Coke® cans, 9.8" × 9.8" × 1.6".
Photo: Vernon Bowden.

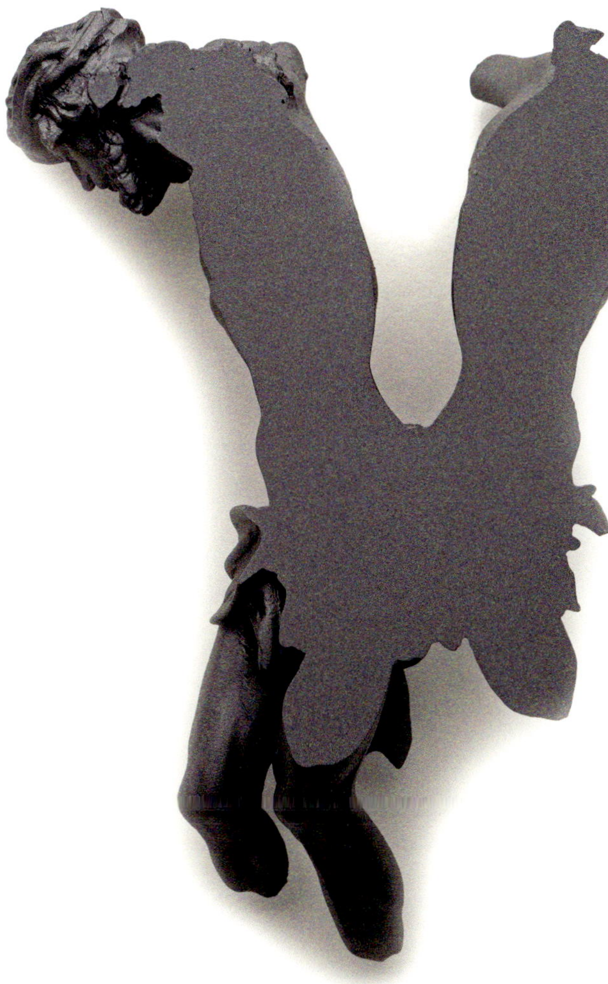

Ruudt Peters, *CORPUS "ovarium,"* 2011. Cast polyurethane, silver, 3.9" × 6.3" × 1.2".
Photo: Rob Versluys, Collection of the CODA Museum, Apeldoorn.

Soohye Park makes jewelry about jewelry, and in part, about casting itself. *Silver Roots* is a ring crowned with a casting tree of commercial rings. Here we see commercial jewelry recontextualized into a piece of conceptual jewelry. The practical becomes impractical, the wearable, unwearable. Park's title, a play on words, seems to suggest that this ring, made up of a casting "tree," has its roots in the hand; or perhaps she is suggesting that the root of conceptual jewelry is in the tradition of wearable jewelry. Either way, casting is central to the ideas that fuel this piece.

In the piece *Touches*, Park is once again making jewelry about jewelry, but this time, she is memorializing a moment. Park used lifecasting to capture the fleeting point in time when one person helps another with the clasp on her necklace. This is a common occurrence for many jewelry wearers. The title, *Touches*, implies a tenderness in this familiar interaction—in Park's bothering to memorialize it, she seems to be saying something profound about domesticity, companionship, and intimacy.

The quality of the castings in this piece allows process to get out of the way of the concept. The fidelity in her lifecasts takes us right to this tender everyday event, allowing it to shine.

Entering the realm of the truly editorial, Yucen Tang's piece, *Mitosis*, is clearly not about wearability in an everyday sense. Its concept, however, revolves around its relationship with the body and this powerful image implies many things about the cell division that is mitosis. The cast resin "cells" are round to roundish with blue elements inside and approximate the look of real cells. They float in front of the model's face by way of silver branch-like wires, perhaps implying a family tree or genetic linkages. The wearer's vision would be filled with the cells if her eyes were open and we can imagine what wearing this piece would do to a field of vision. Is this mitosis of an egg becoming a fetus and the hopes and dreams of having a healthy child, or is it the universally feared mitosis of cancer? Tang doesn't seem to answer this question, but the irregularity in the cells hints at the latter—and more sinister—form of cell division. Either way, this piece creates a lens through which the world is viewed—an all-encompassing cage surrounding thought and vision. In the case of a piece like this, wearability is central to its concept, but clearly is not approached in a traditional sense.

Casting can provide solutions for every kind of jewelry making. It affords practicality for jewelers concerned with the mass production of wearable commercial jewelry, it offers the ability for studio jewelers to make one-of-a-kind, bespoke pieces rich in form and surface, and it provides the sculptural range required for conceptual makers.

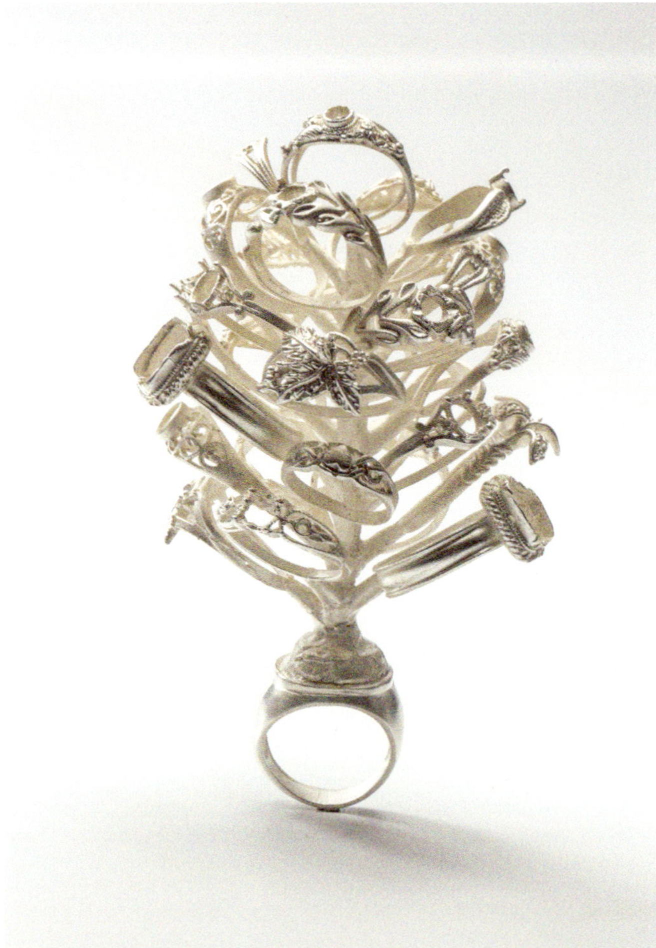

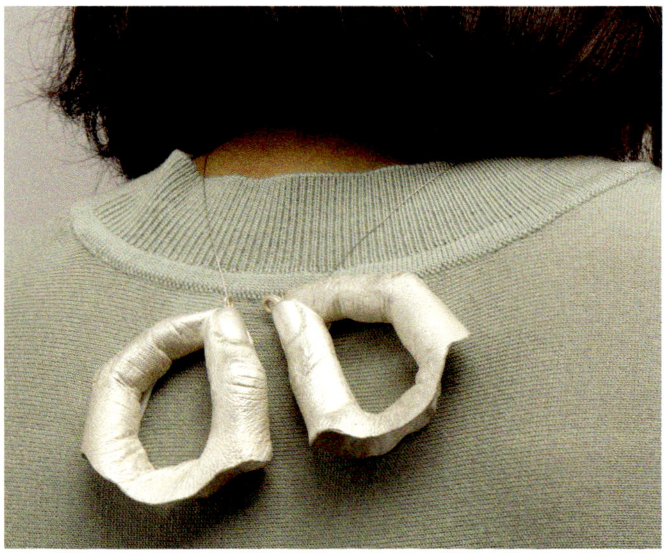

Top Soohye Park, *Silver Roots*, 2015. Sterling silver, 5" × 3" × 3".
Photo: Jim Escalante.

Bottom Soohye Park, *Touches*, 2014. Sterling silver, 4.5" × 8" × 2.5".
Photo: Jim Escalante.

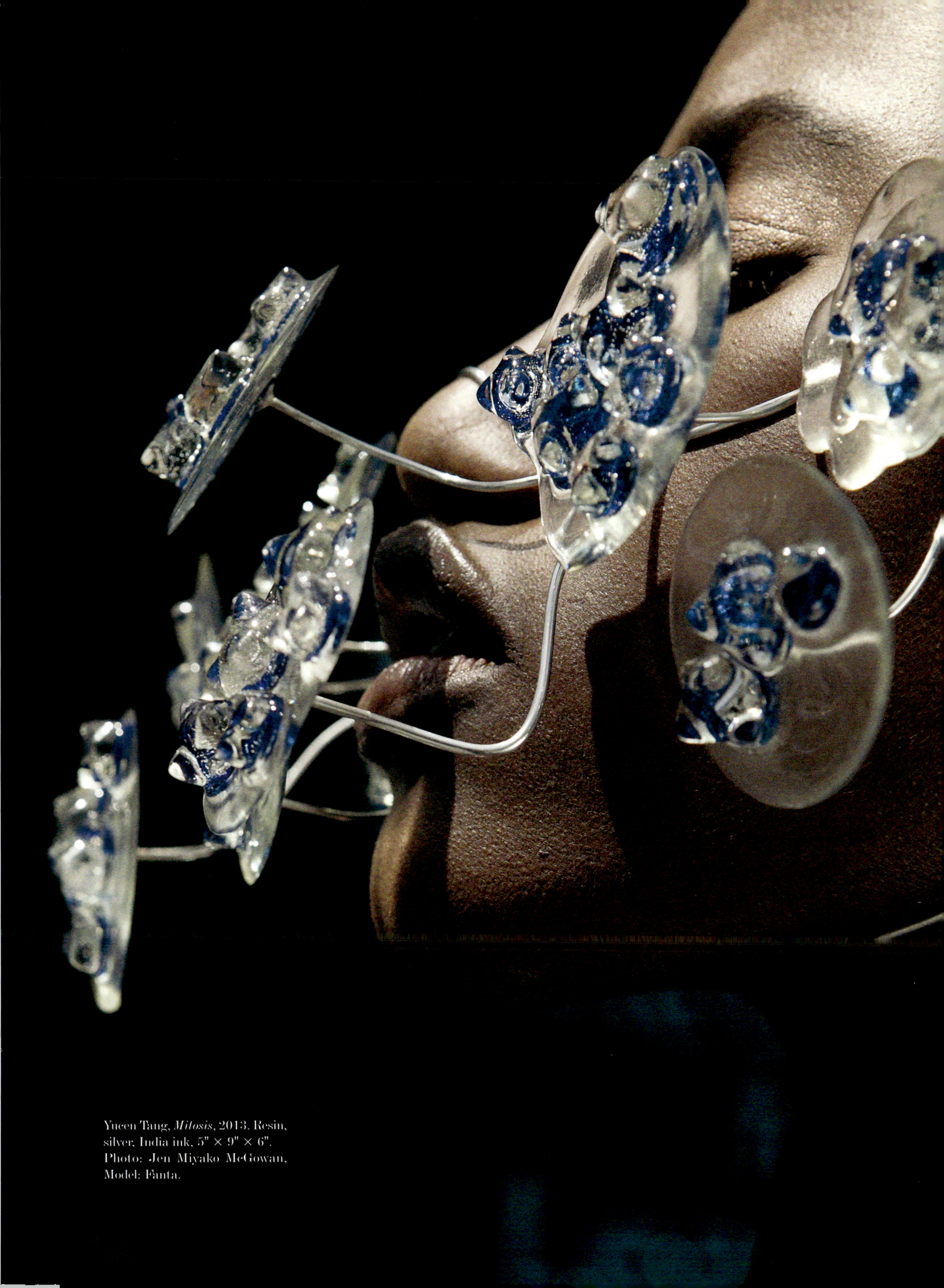

Yuen Tang, *Mitosis*, 2013. Resin, silver, India ink, 5" × 9" × 6". Photo: Jen Miyako McGowan, Model: Fanta.

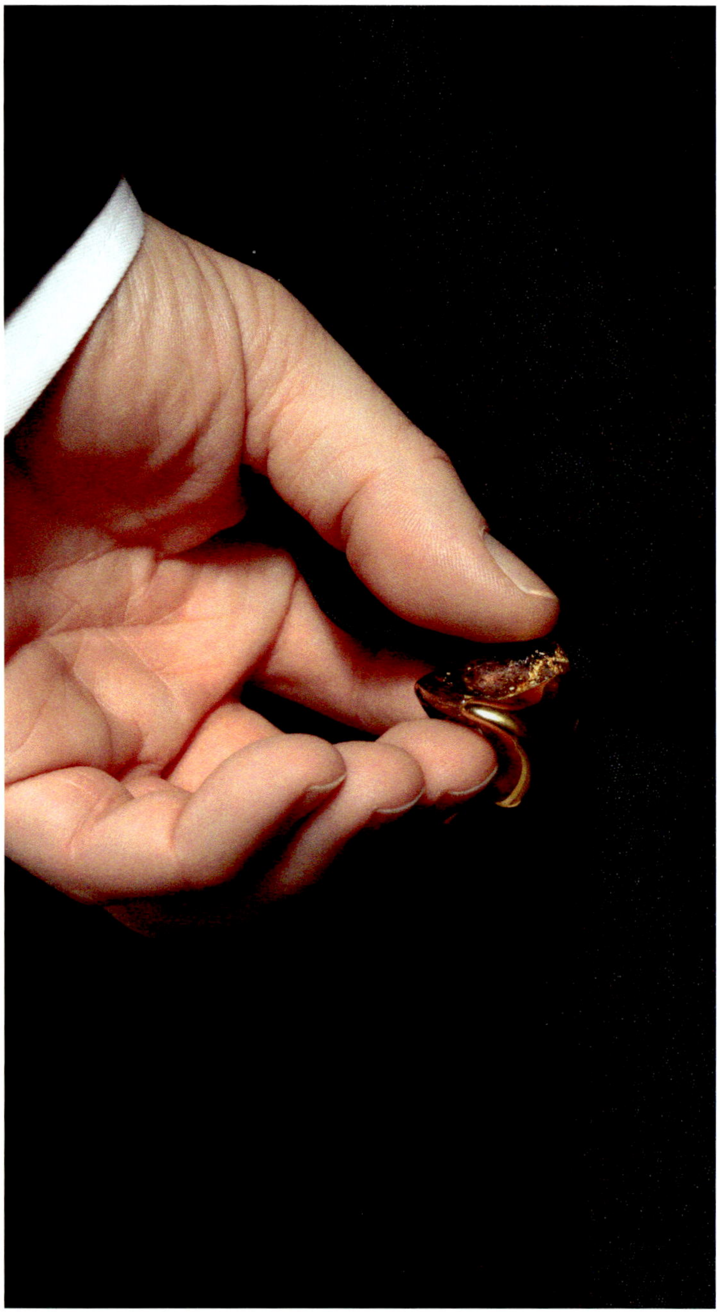

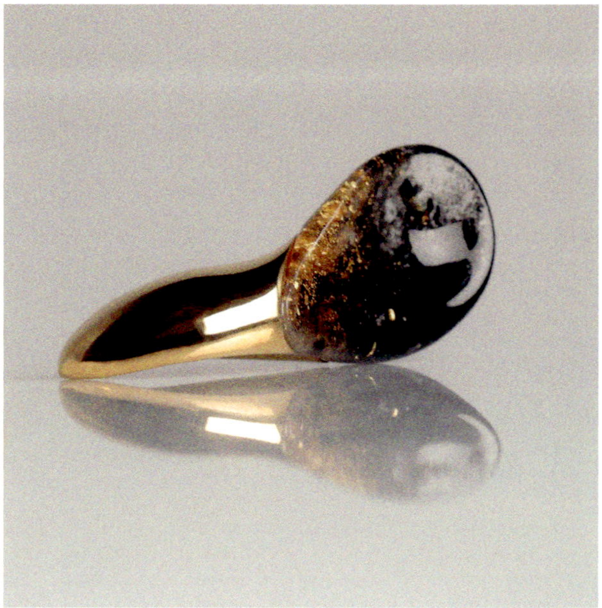

Above Charlotte Garnett, *Pinch* from *Cure for the Itch* collection, 2015. Highly polished clear resin (encapsulating cigarette ash and 23K gold leaf) and 18K gold plated gilding metal, which is 95% copper and 5% zinc, 1.1" × 0.86" × 1.2". Photo: Charlotte Garnett.

Left Charlotte Garnett, *Spoons* from *Cure for the Itch* collection, 2015. Highly polished clear resin (encapsulating cigarette ash and 23K gold leaf) and 18K gold plated gilding metal, which is 95% copper and 5% zinc, 1.75" × 0.7" × 0.9". Photo: Charlotte Garnett.

Opposite page Jessica Todd, *Suburban Growth, Series 5*, 2014. Silicone, digital print (cast silicone on body); house component 3-D modeled in Rhino, 3-D printed in SLS, molded in silicone, and cast in silicone, 24" × 16" (digital print), 1.75" × 1.5" × 2.5" (cast silicone component), 2.25" × 0.25" × 0.5". Photo: Jessica Todd.

Charlotte Garnett makes pieces that aren't quite jewelry, but defy other classification as well…they do have many of the characteristics of jewelry. Garnett casts "pocket pebbles" that are intended to live in the hand (and the pocket). They are seductive little objects made from highly polished clear resin and gilded metal. The resin encapsulates cigarette ash—a reminder for the owner not to smoke. These pieces are made for anxious fiddling and to encourage mindfulness. They feature an ergonomic shape that fits naturally in the hand and between the fingers. Many jewelers have made pieces that satisfy our urge to fidget, but these are especially interesting because they don't actually attach to the body in any way. They are clearly treated like jewelry in their production—fastidiously carved, cast, and finished using traditional methods—but they aren't rings or bracelets, rather they're private ornaments that grace the inside of pockets.

Jewelry that doesn't really attach to the body or jewelry that *literally* attaches to the body, like Jessica Todd's *Suburban Growth*, can be made to exist in another place: the photo. *Suburban Growth* is a cast silicone element tinted to match the model's skin tone. It is applied to the body using a silicone adhesive and blended with makeup, but these technical details are immaterial to the value of this piece. Works like this are made to be editorial; they function similarly to haute couture clothing in the staged, styled, and digitally enhanced photos in fashion magazines—the image *is* the final outcome. Conceptually based jewelry often has nothing to do with practicality, wearability, or any of the other concerns that weigh on the minds of bench jewelers; rather, these pieces are all about the efficacy of the photo.

The gabled window that seamlessly emerges from the model's chest has many possible interpretations. Does this piece address our deep connection to the idea of home? Is it to do with contemporary housing developments that spring up like mushroom colonies? Does it reference the body as a place where emotions, memories and other attachments live? Perhaps there is no answer and the point of conceptual work is in the questions rather than the answers.

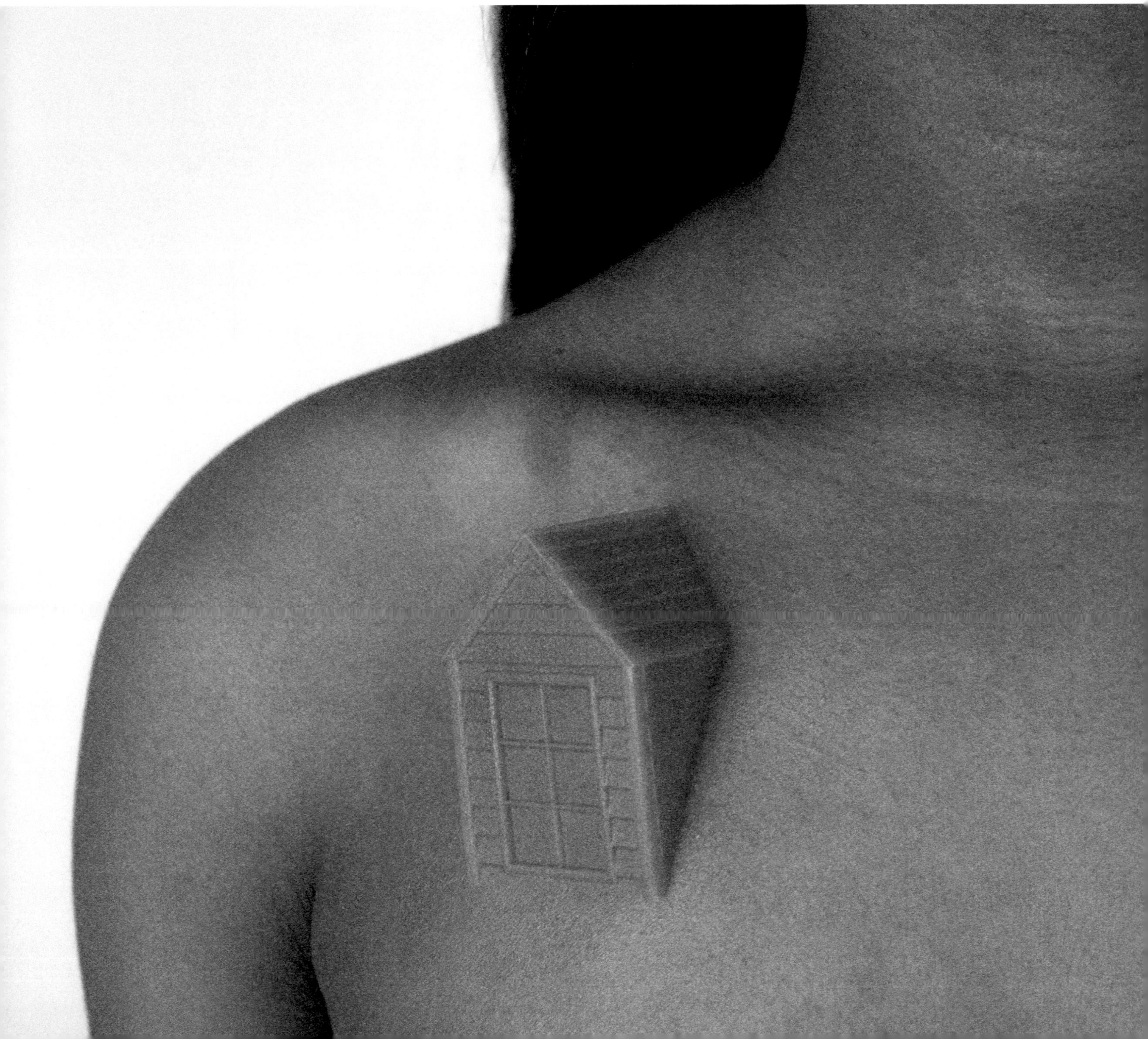

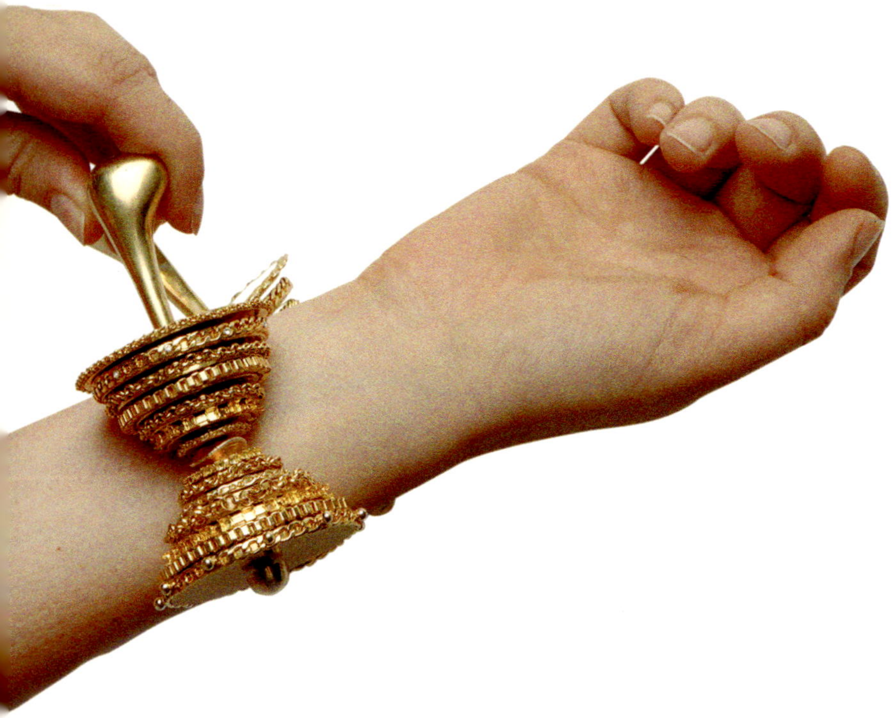

Tamar Navama's pieces, *Jewel Yourself No. 1* and *No. 2*, take another approach to jewelry altogether. Like Jessica Todd's work, the photos are quite important, but unlike Todd's, they are not the final product. These works are about body adornment, although not in a traditional sense—there is no actual jewelry, rather it's the indentations left on skin by these silver and gold contraptions that are important.

Anyone who has ever fallen asleep on a braceleted arm knows what their skin looks like when the jewelry is pulled away; these pieces create a similar result. The silver and gold devices, covered with wavy ridges and rows of chain, are made to be rolled up and down arms leaving patterns on the skin. Fleeting impressions, corporeal memories, a pinch of masochism, and the act of ornamenting one's self are the ideas central to this work.

Whether jewelry takes up residence on a wall, in a museum under glass, in a photo, or as an impression on our skin, it has the power to implant itself firmly in our minds. When jewelry is taken off the body and is recontextualized, it can show us our history, make us think, make us laugh, and most importantly, act as a reflection of our humanity.

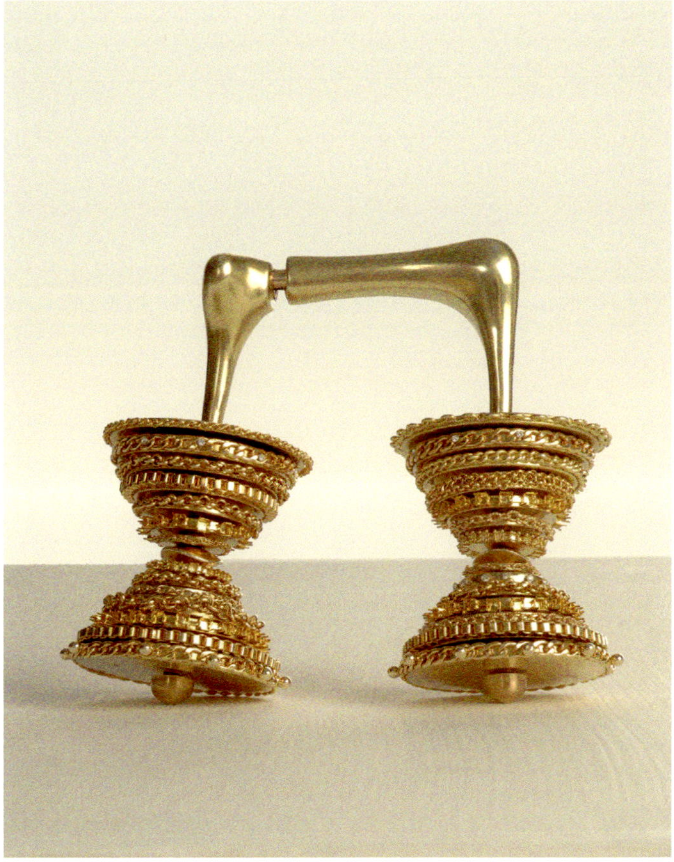

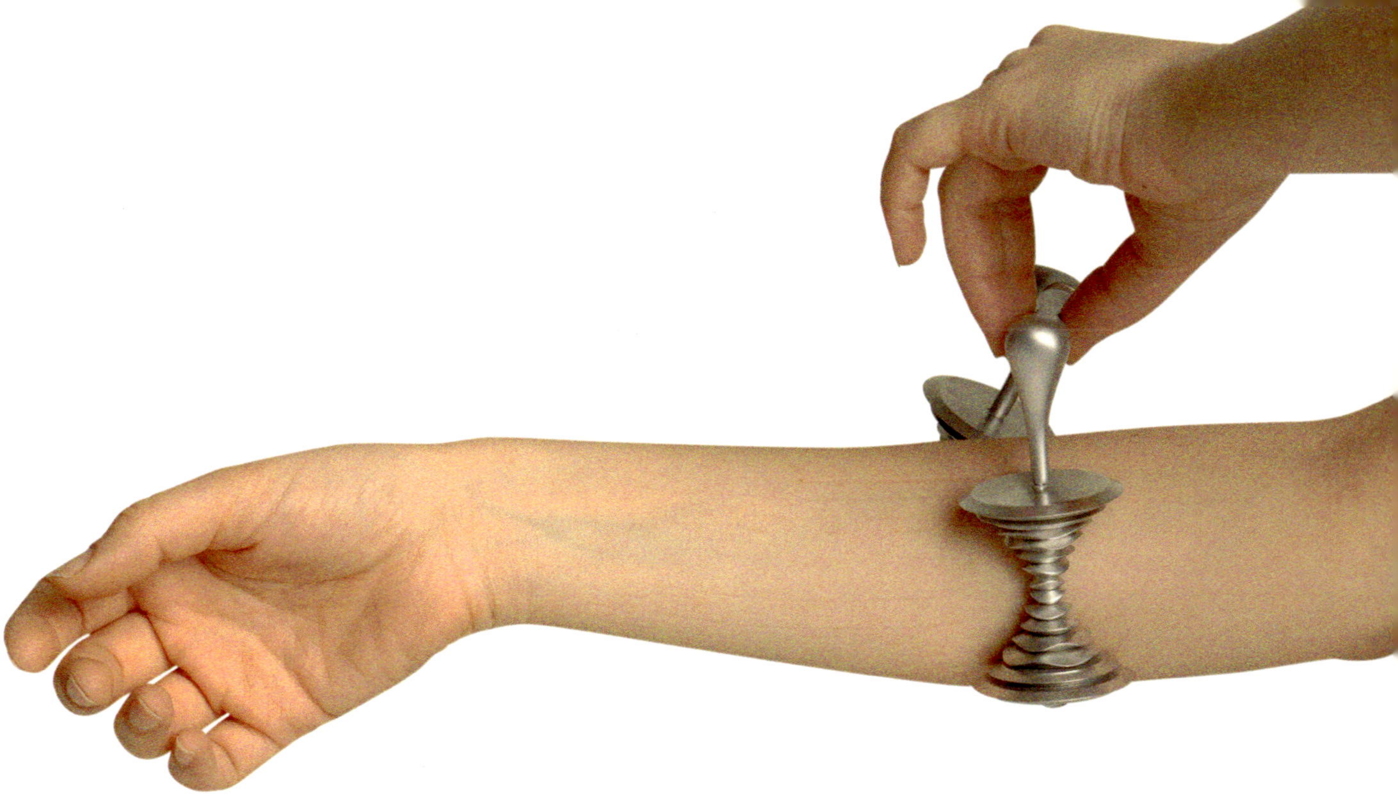

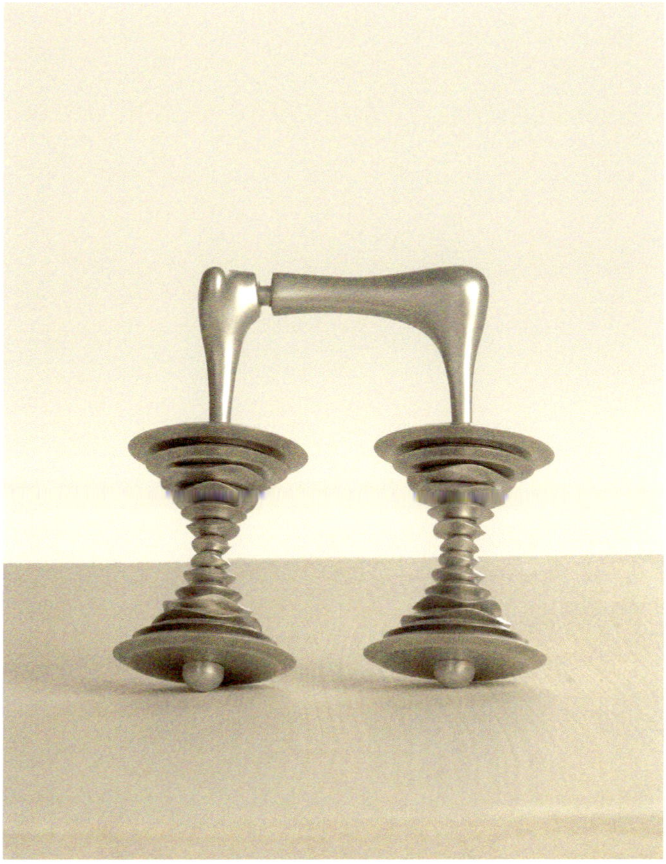

This page Tamar Navama, *Jewel Yourself No. 1*, 2008. Aluminum, brass, spring steel, gold brass, 5.1" × 3.5" × 1.75". Photo: Miri Davidovich, Courtesy of Bezalel Academy.

Opposite page Tamar Navama, *Jewel Yourself No. 2*, 2008. Aluminum, brass, spring steel, gold brass, 5.1" × 3.5" × 1.75". Photo: Miri Davidovich, Courtesy of Bezalel Academy.

FOCUS on MIXED MEDIA

This book was originally intended to be solely about cast jewelry, but it was the prevalence of mixed media in jewelry that inspired its expansion to include all forms of casting in every possible medium. There is no question that metal dominates when it comes to jewelry. That being said, cast jewelry is magnificently and unexpectedly diverse; the constant influx of new materials is inspiring jewelers everywhere to experiment with everything from plastic to paper.

The most traditional use of mixed media in jewelry is via gemstones. For thousands of years, jewelers have included precious and semi-precious stones and pearls in their work. German Kabirski has mastered the use of traditional materials, but applies them in new and cutting-edge ways. This gold and tourmaline ring is a perfect example; it appears to be a standard signet-style ring with a stone, but it is overlaid with cast gold fabric. It's a gutsy move for a jeweler to cover so much of the tourmaline as stones are usually sacrosanct. The texture in the fabric is wonderfully rich and its effortless draping makes the ring both dynamic and surprising. Kabirski's work is an exciting combination of reverence and irreverence; his pieces are always impeccably made, but he throws all of the rote jewelry conventions out the window in the best possible way.

Jade Mellor is clearly in touch with the mineralogical traditions in jewelry as well, but is breaking new ground by creating her own casting mediums that include minerals. Mellor cast her *Mineral Hewn Rings* using a mixture of resin, pyrite, and carnelian, creating a unique aggregate that mimics the look of natural mineral formations. This fascinating approach to casting results in rings that look like they came right out of the earth; when one of them reveals a chunk of carnelian embedded in it, there is a sense of discovery, like something precious is just under the surface.

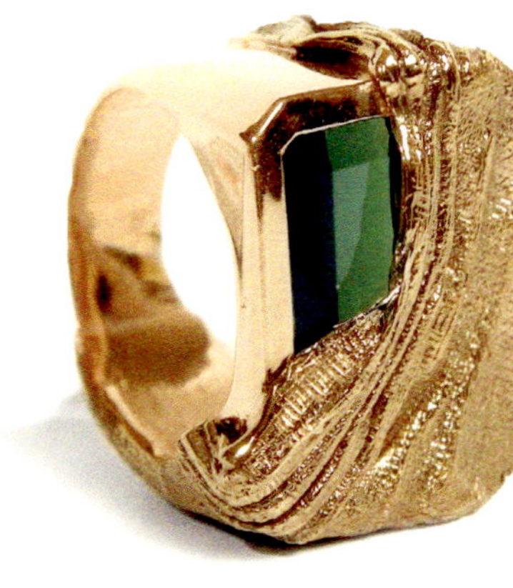

German Kabirski, *Untitled*, 2008. 18K gold, tourmaline, 0.9" × 0.86" × 0.98". Photo: German Kabirski.

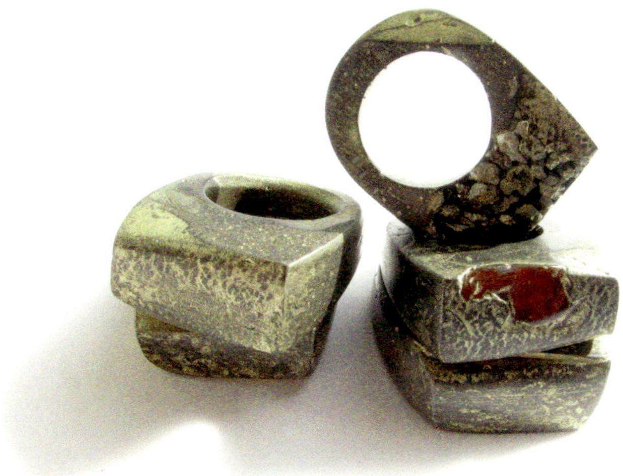

Minerals have always held an exalted position in the jewelry world, but when glass was first discovered and casting technology was being developed, glass was considered to be just as rare and valuable as stones were. Ancient Egyptian jewelers were some of the first to use cast glass in their jewelry and their pieces were often a mix of metal, stone, and cast glass. Many cultures since then have used glass in jewelry, although primarily as a substitute for gemstones. Only a handful of jewelers have really taken advantage of everything glass has to offer as a casting medium for jewelry.

Donald Friedlich is using glass in very savvy ways—he's taking advantage of color, transparency, light refraction, and the plasticity of form offered by glass to create seductive brooches in a range of styles. Friedlich's brooch, *Temptation*, features the distilled form of an apple, elegant and luminous in a way no other material could duplicate.

Top Jade Mellor, *Mineral Encrusted Hewn Rings*, 2009. Resin, pyrite, carnelian, cast with a mold then hand shaped,1.4" × 1.14" × 0.5" each. Photo: Jade Mellor.

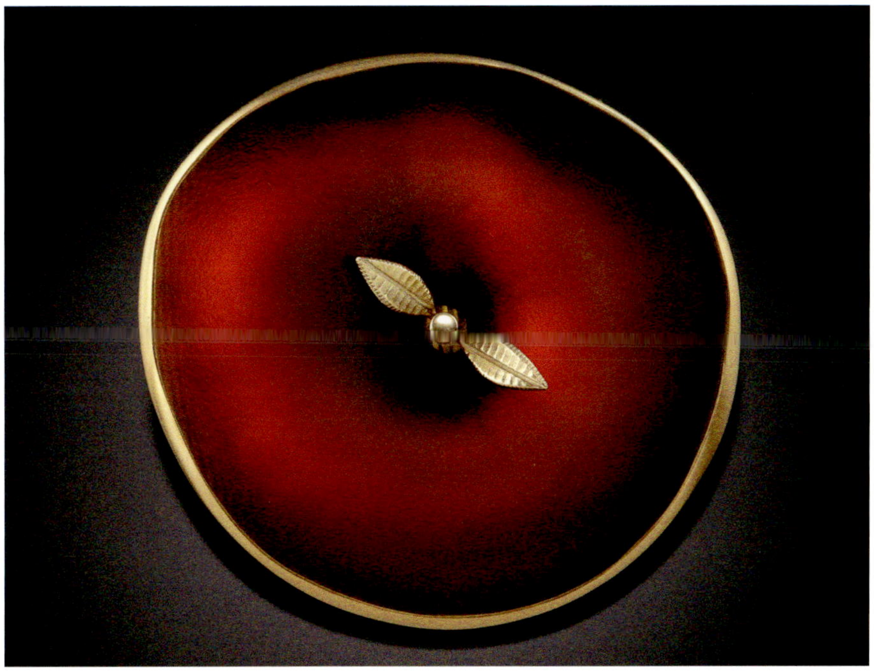

Right Donald Friedlich, *Temptation Brooch*, 2013. Glass, 22K gold, sterling, niobium, 18K gold, 2.5" × 2.7" × 0.65". Photo: Larry Sanders, Collection of the Cooper Hewitt Museum, NYC.

There are, however, many other materials that provide color, transparency, and plasticity of form and can be used in jewelry. Resin, polyurethane, and plastics of all kinds offer interesting possibilities.

Phil Renato's *Grammar* is cast in polyurethane with rubber, silver, and aluminum accents. The polyurethane elements seem to mimic the form of breasts (a conclusion that's fairly obvious considering the photo) but the patriotic color palette of red, white, and blue in conjunction with the title begins to raise questions.

The definition of the word *grammar* always mentions rules, language, and syntax, so what is Renato saying with this piece? The transparent red form over the model's right breast and the blue over the left suggest that this piece is a political statement. Is it about American political discourse? The breakdown of rules held for generations? America as mother? Is it about the overt sexuality of the photo, or were these simply formal choices with no deeper meaning? What does it mean that the piece is polyurethane? *Grammar* is flawlessly crafted, and thoughtfully assembled, but its intention is unclear—red, white, and blue synthetic breast-like forms on a wet, amply endowed model make the piece both attractive and repellant, but maybe being provocative and bold is exactly the point.

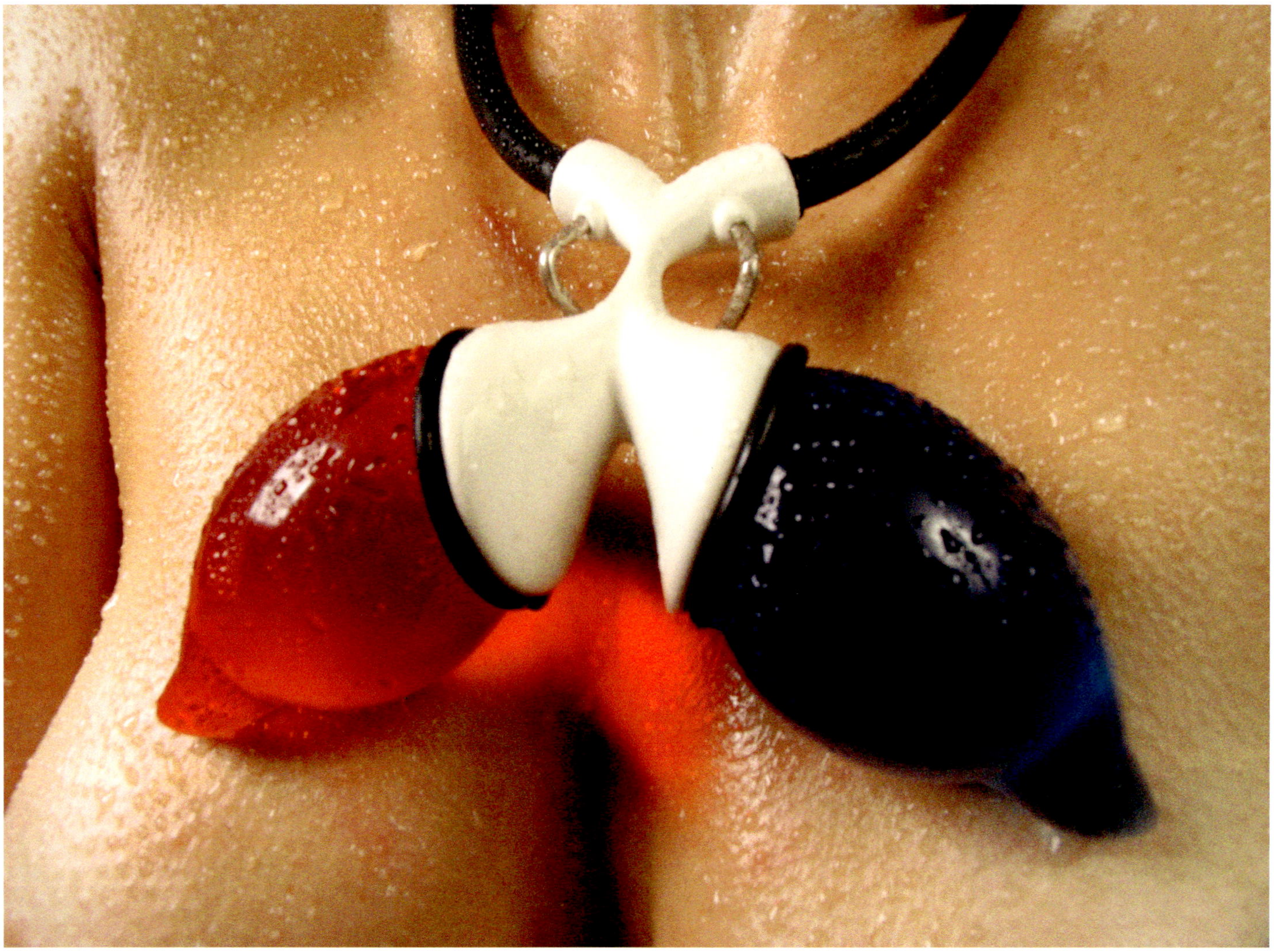

Phillip Renato, *Grammar*, 2007. Polyurethane, aluminum, silver, rubber, 4" × 6" × 2".
Photo: Phil Renato.

Opposite page emiko oye, *Dawning* from the *Musée Series* (neckpiece), 2010. Repurposed LEGO®, silk cord, coated copper wire, fine silver label, 12" × 8.8" × 2".
Photo: Christine Dhein.

Boldness is one of the benefits of mixed media jewelry. Economics and weight make metal a challenge if a maker wants to work in a large scale, and cast metal can be very heavy unless a lot of time is spent carving an intentionally lightweight model.

Plastics allow makers to work expressively at a large scale and maintain both wearability and reasonable material cost. There are several prominent studio jewelers working today whose work hinges on cast plastic elements, although rather than casting it themselves, they're upcycling well-known cast products.

Repurposed LEGO® is what emiko oye uses in her work. LEGO® makes its internationally beloved building blocks out of acrylonitrile butadiene styrene (ABS) through an industrial injection molding process. LEGO® molds have a tolerance of 0.002 millimeters, which is what makes the toy blocks infinitely interchangeable.

She has cleverly used the modular toys to create art jewelry about everything from David Bowie's passing to historic royal jewels. *Dawning*, a piece unmistakably inspired by Louise Nevelson's work, is as large as a neckpiece can be before it enters the realm of costume. This piece references Nevelson's monochromatic assemblages, artfully mimicking the variety in form and surfaces Nevelson is known for.

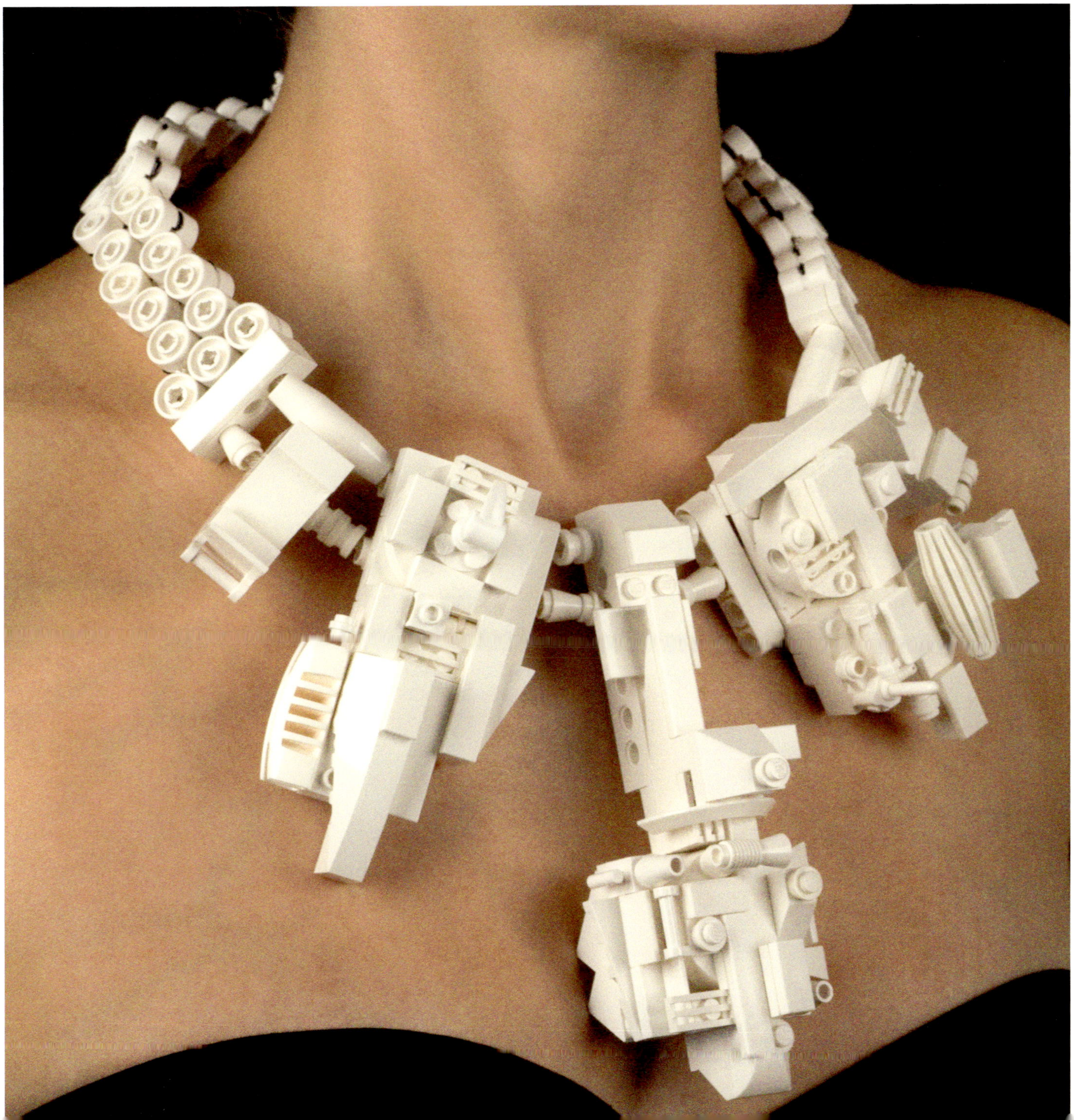

Like Vernon Bowden, Margaux Lange also uses Barbie® components in her work. She cuts apart and recontextualizes the well-known dolls, creating thought-provoking pieces that speak to a range of topics from race to sexuality. In aspirational pieces like *Giggles*, Lange presents a vision of racial equality and harmony, but like oye, her chosen vernacular is one of light childhood associations, thus she avoids going head first into these sticky topics. Lange's shrewd recontouring of the familiar doll allows her to make provocative yet playful work. Both oye and Lange adroitly delve into a variety of weighty subjects without the work feeling heavy, pedantic, or self-important.

While oye and Lange repurpose well-known products that carry many associations, other makers are recontextualizing materials with strong and often disparate associations—materials definitely not associated with jewelry. Concrete is one of those materials.

Tom McCarthy casts earthy concrete compositions that feel both raw and refined at the same time. His approach to this industrial material in the *Caitlin Brooch* is both sensitive and painterly. Embedded steel creates rust that bleeds and drips; ink adds to the range of colors in the casting. The pillowy diamond forms on the left hand side of the piece add dimension and the overall effect is one of a rich abstract composition.

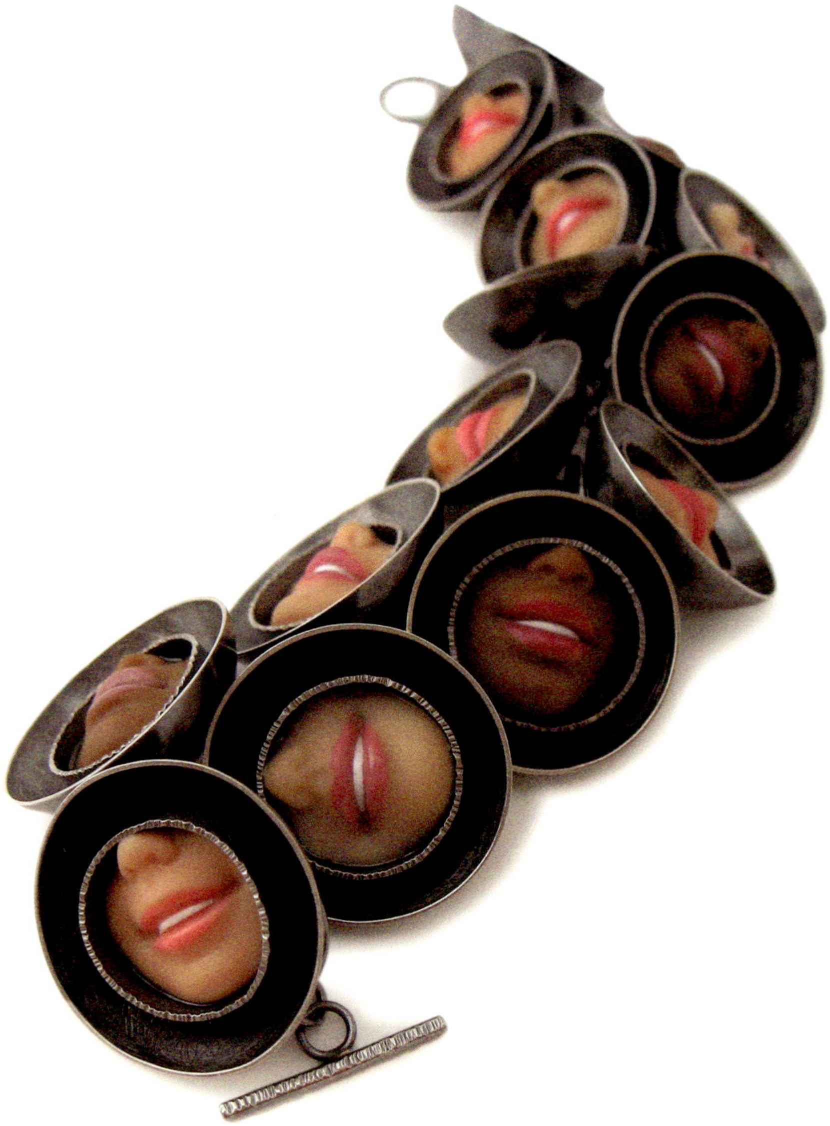

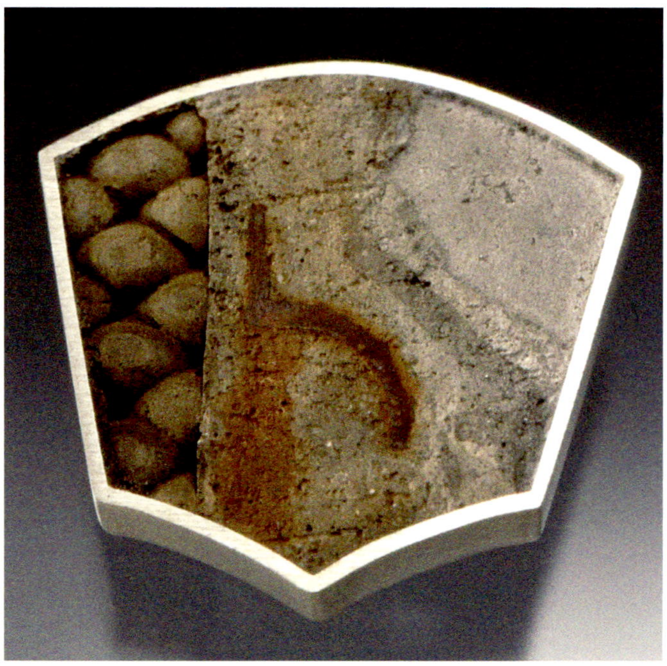

Top Margaux Lange, *14 Giggles Bracelet*, 2008. Sterling silver, plastic Barbie® doll parts, resin, hand-fabrication, 7.5" × 1.5" × 1.5".
Photo: © Margaux Lange.

Right Tom McCarthy, *Caitlin Brooch*, 2005. Silver, concrete, steel, rust, ink, 1.37" × 1.37" × 0.25". Photo Tom McCarthy.

Paper is another unusual material being used today. The cast paper in *Tarped*, by Laura Wood, is such an unexpected material that it's hard to identify when looking at the piece. Wood casts paper pulp and allows it to dry over costume jewelry components. When the pulp dries taut, it leaves ghostly impressions of gems pressing through the surfaces of the cabochon-like castings. Wood holds the castings in place using a traditional prong setting technique.

Precious stones were the original mixed media in jewelry; costume jewelry followed with glass stones made to emulate the gems in fine jewelry. Wood is taking this progression one step further by merely implying gemstones. *Tarped* is a truly modern interpretation of the history of jewelry and brings this exploration of mixed media in jewelry full-circle.

Below Laura Wood, *Tarped Necklace*, 2013. Handmade paper, sterling silver, found object, 10" × 10" × 0.25".
Photo: Ansen Seale Photography.

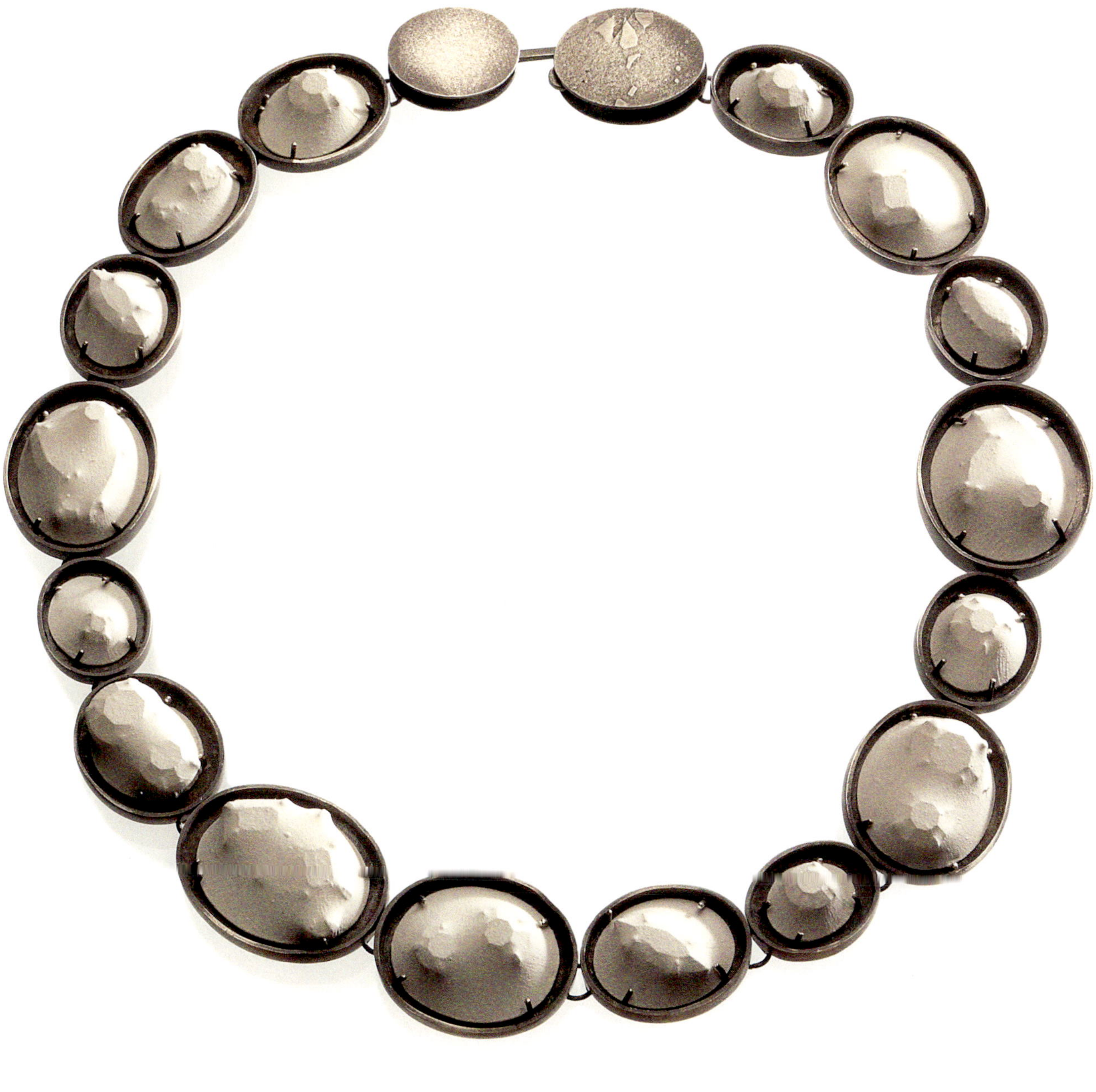

Notes

1. Nancy Hynes, "Oldest Known Jewellery Discovered" (Nature Publishing Group, 22 June 2006), Web, accessed July 20, 2016, http://www.nature.com/news/2006/060619/full/news060619-10.html.
2. Jan Mitchell, *The Art of Precolumbian Gold: The Jan Mitchell Collection* (New York: New York Graphic Society, May 1985), 86.
3. Email from Lin Stanionis, Sept. 7, 2015.

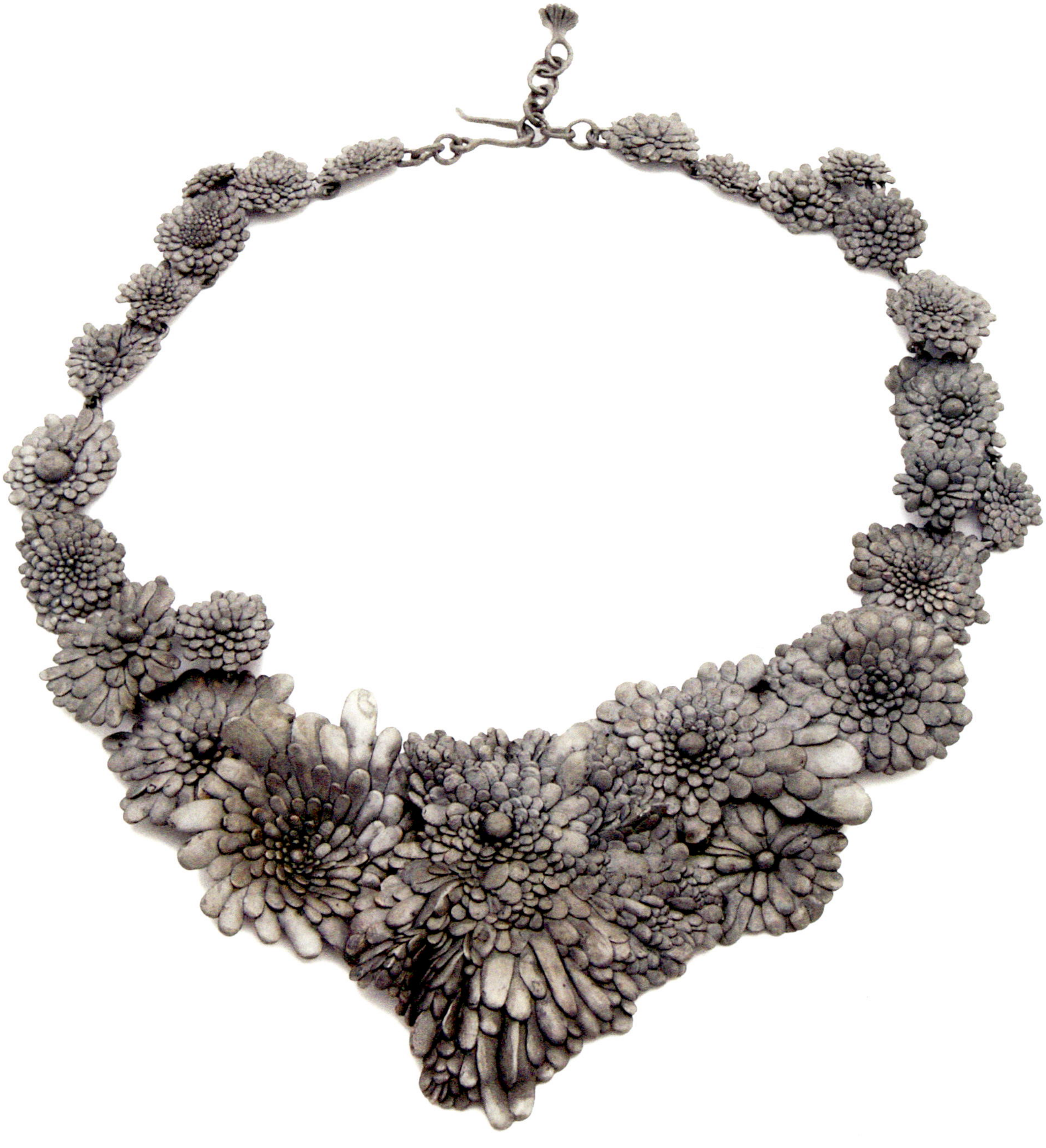

Nora Rochel, *Dandelions*, 2011, fair trade silver (blackened), 17.7" × 2.75" × 0.4".
Photo: Nora Rochel.

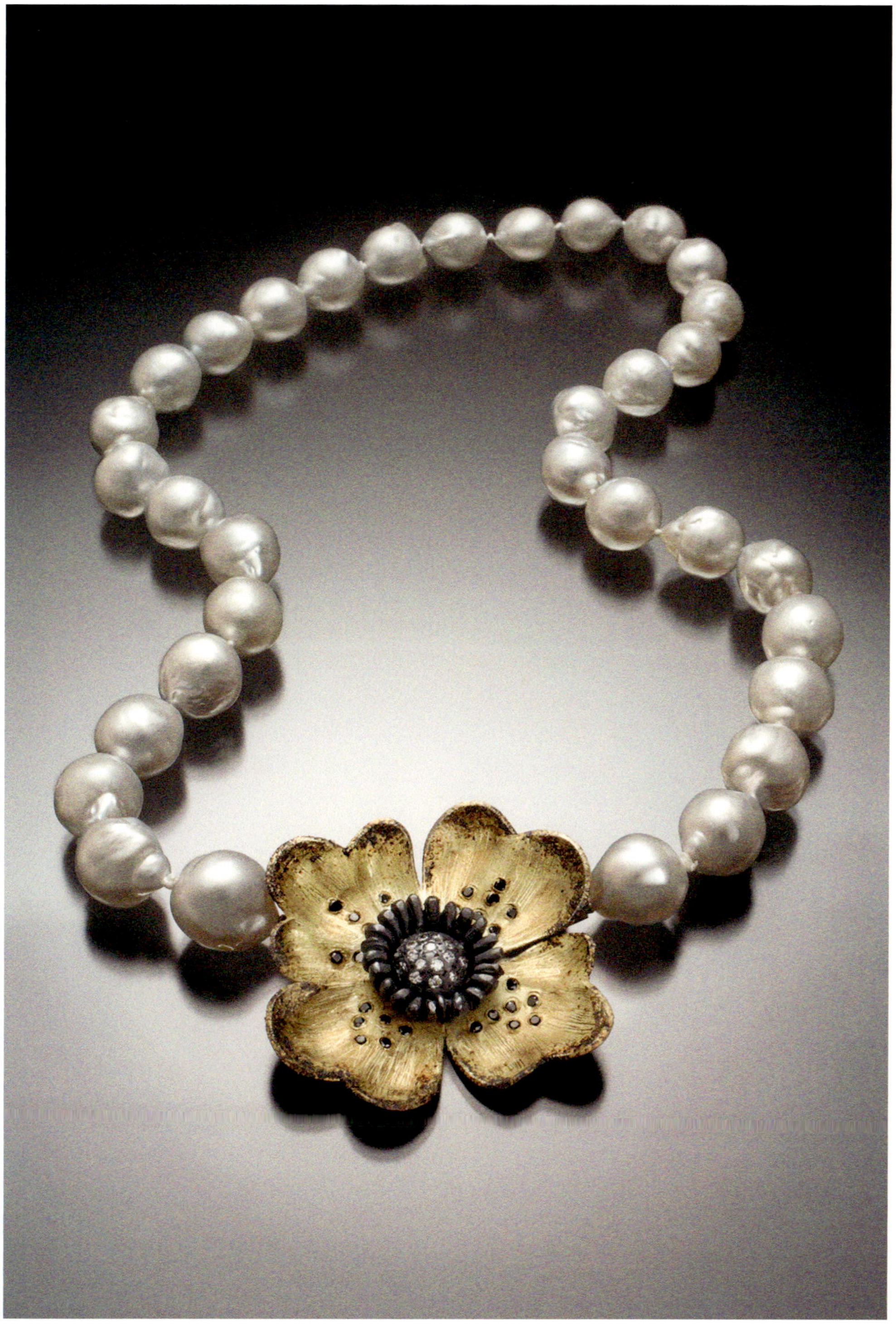

Rebecca Myers, *Dogwood Pendant*, 2008. 18K gold, oxidized silver, pearls, and diamonds, 2.5" × 2.5" × 0.5". Photo: Ralph Gabriner.

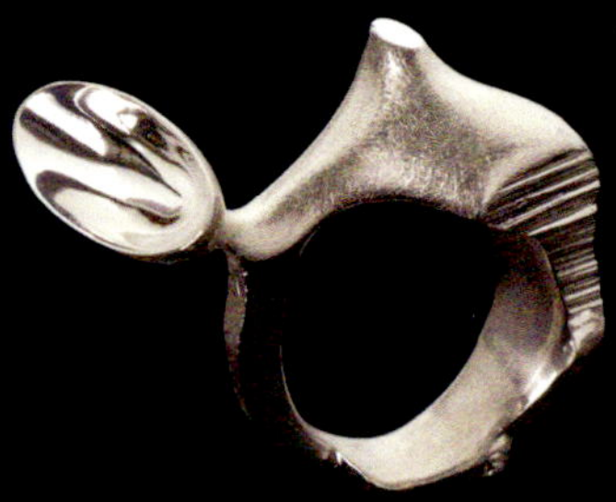

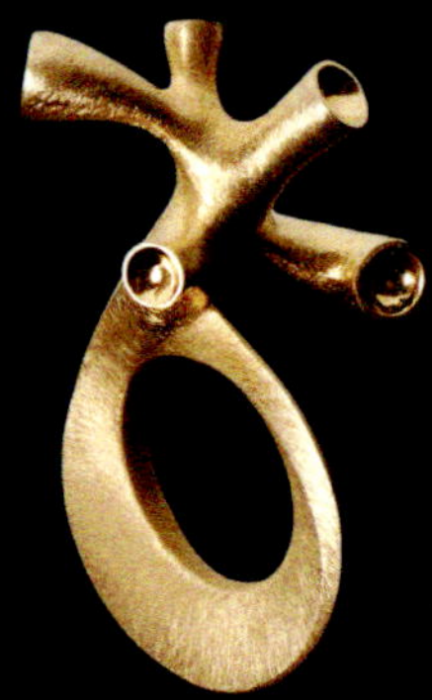

Left Masako Hamaguchi, *Lump*, 2007. Sterling silver, 1.6" × 1.2" × 0.8". Photo: Sylvain Deleu.

Right Masako Hamaguchi, *Deerhorn*, 2010. Sterling silver, gold plated, 2" × 1.2" × 1.2". Photo: Sylvain Deleu.

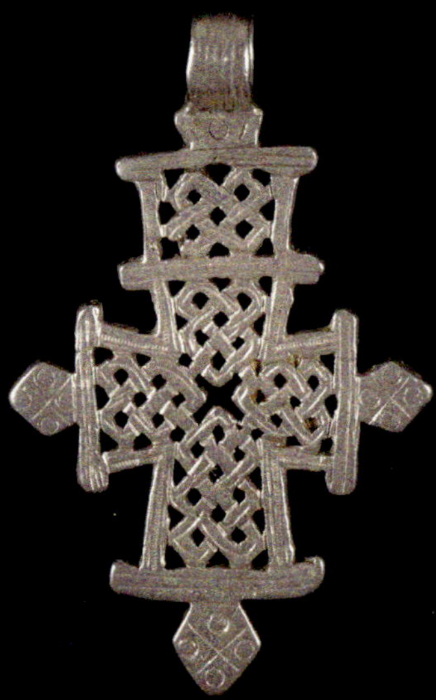

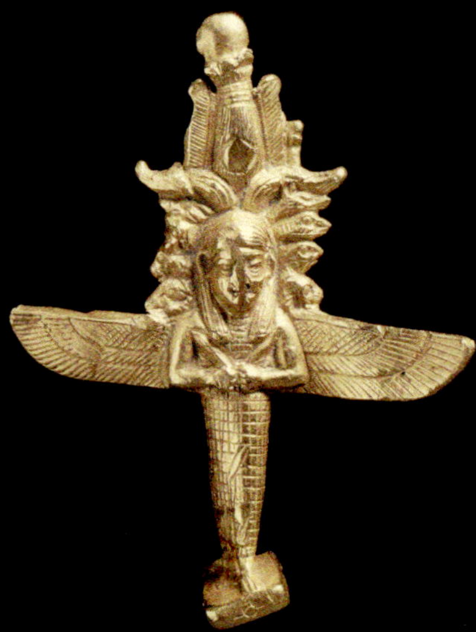

Left Pendant Cross, eighteenth to nineteenth century (Early Modern). Ethiopia, cast, stamped, and engraved silver alloy, 2.38" × 1.43" × 0.38". Courtesy of The Walters Art Museum, Gift of Mrs. Lily Mobille-Pappas in 1997, CC0, GNU FDL.

Right Pantheistic Deity Pendant, late fourth to mid-third century BCE. (Greco-Roman) Egypt, gold, 1.68" × 1.3" × 0.43". Courtesy of The Walters Art Museum, Acquired by Henry Walters, CC0, GNU FDL.

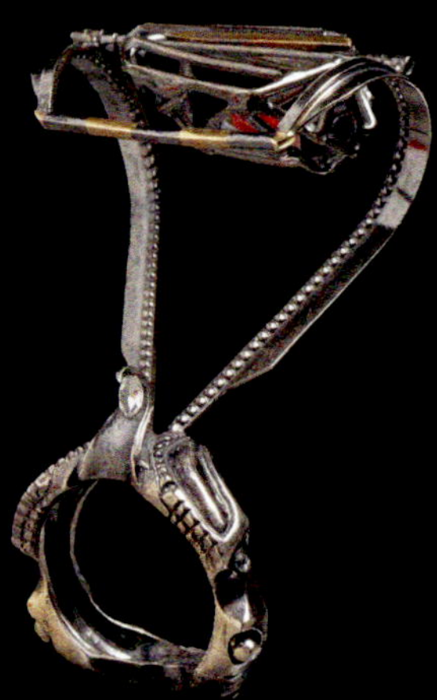

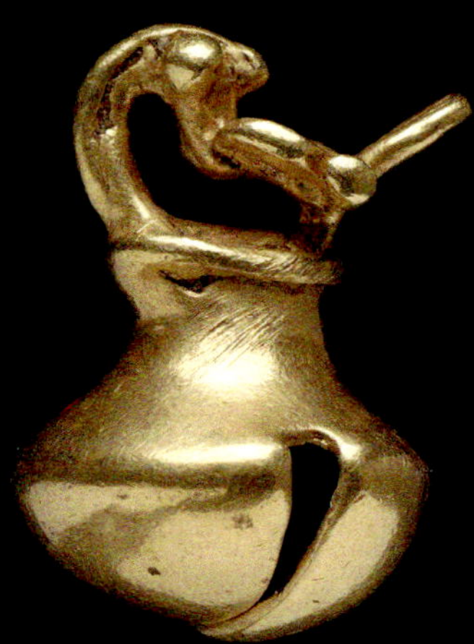

Left Cuong Sy, *Empowerment Ring 2*, 2014. Sterling Silver, 14K, raw ruby crystal, 2.3" × 1" 0.4". Photo: Henry Daniel Galin.

Right Bell Pendant of a Parrot, 800–1521 (Pre-Early Conquest). Veraguas-Gran Chiriquí (in present-day Panama, place of origin), gold and copper alloy, 1" × 0.8". Courtesy of The Walters Art Museum, Acquired by Henry Walters in 1911 from Tiffany & Co., CC0, GNU FDL.

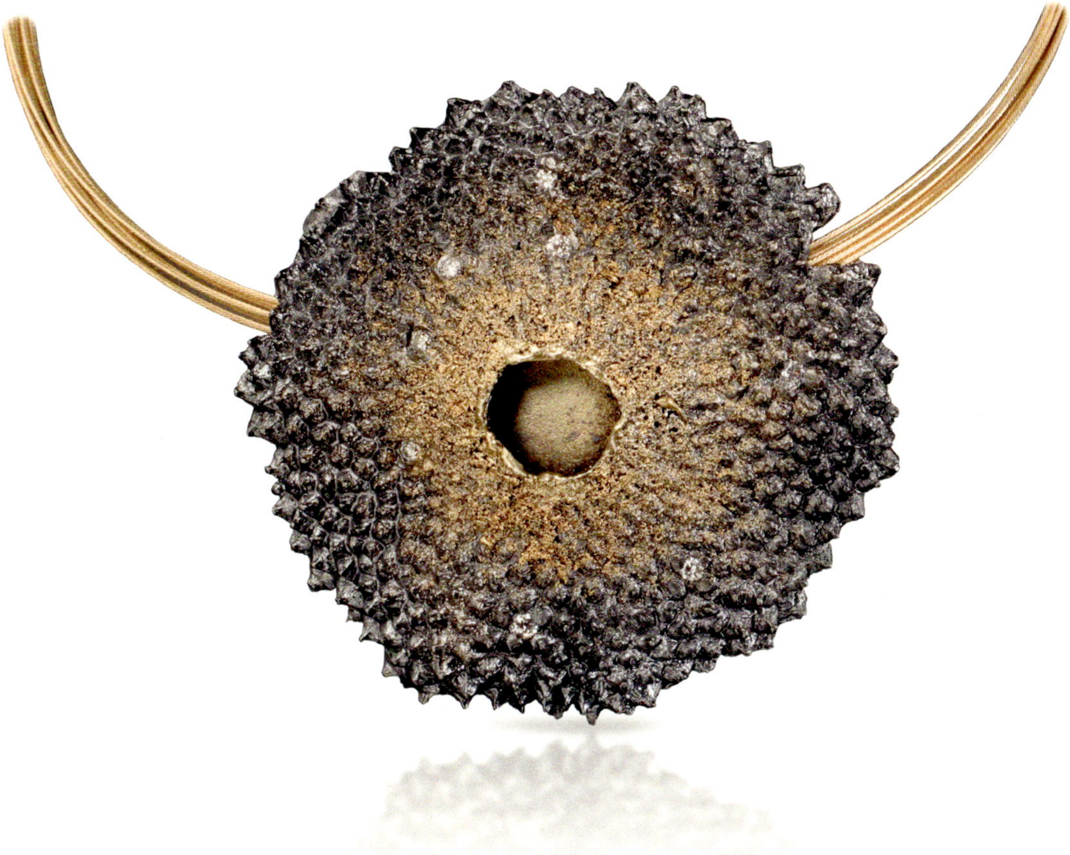

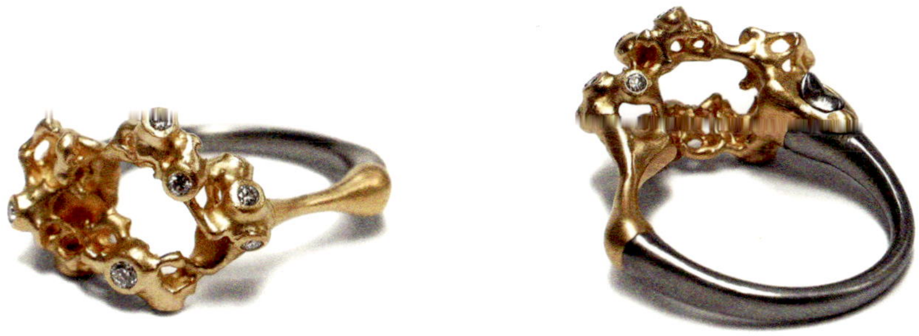

Top center Rebecca Myers, *Seed Pod Pendant*, 2007. Oxidized silver, 22K gold diamonds, 3" × 3" × 0.25". Photo: Ralph Gabriner.

Bottom Johnny C. Ninos, *Industrial Decay Series (reef)*, 2013 (two views). 18K gold, diamonds, 1" × 1" × 1". Photos: Johnny Ninos.

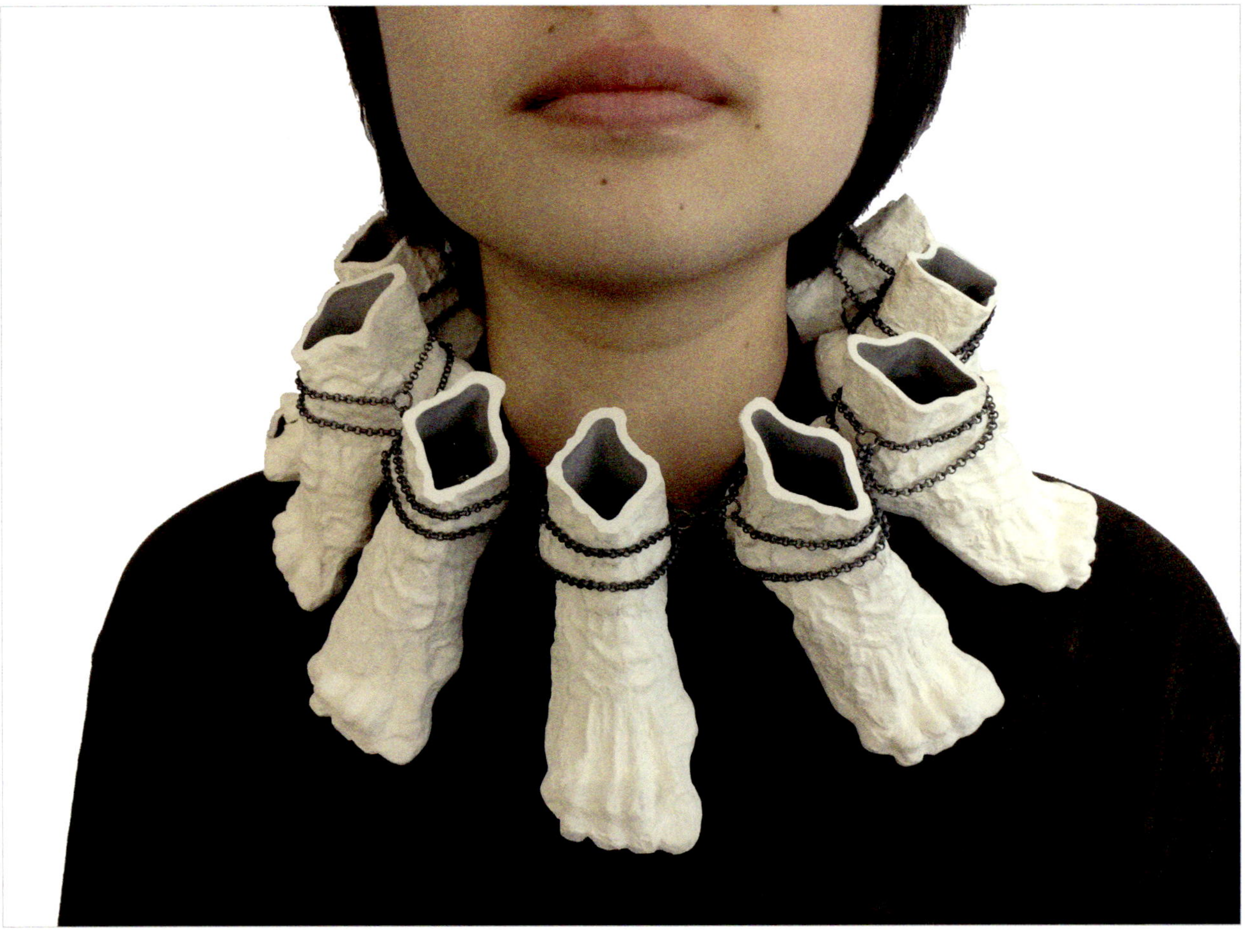

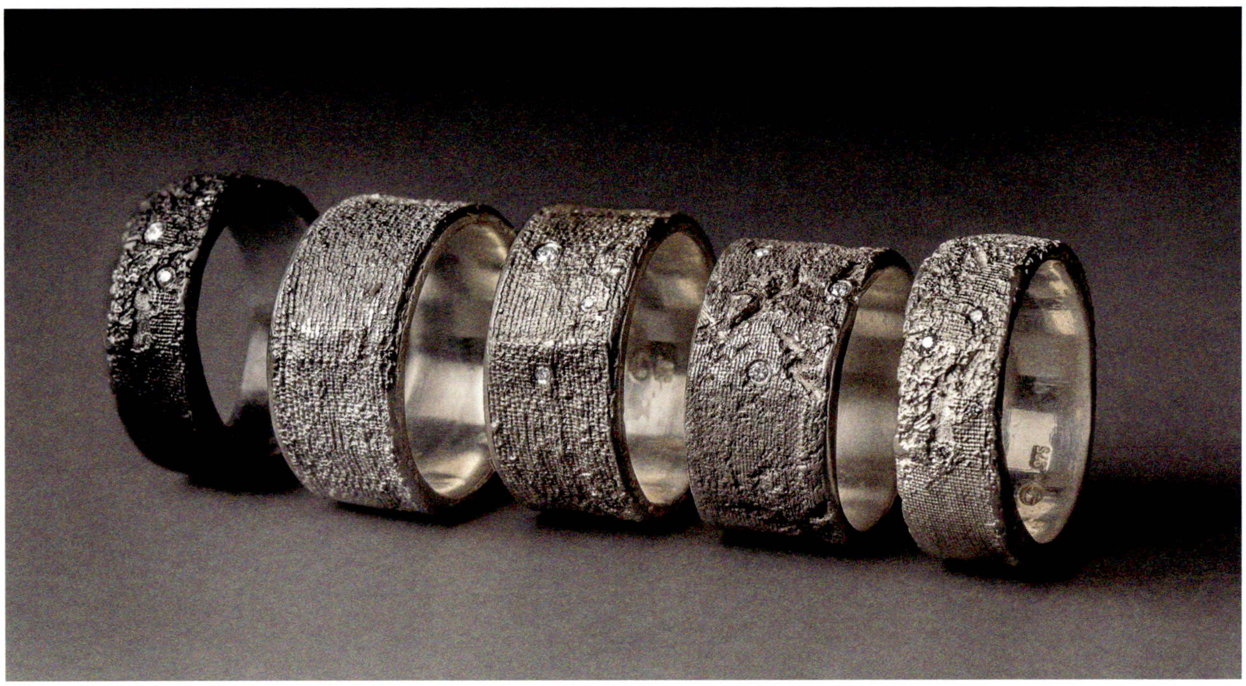

Top Ruudt Peters, *TERRAM "di,"* 2015. Cast porcelain, silver, 9" × 4.3" × 2.75".
Photo: Ruudt Peters.

Bottom Maya Kini, Rings from *The Moths*, 2013. Silver, brilliant cut diamonds, 1" × 1" × 1".
Photo: Rik Keller.

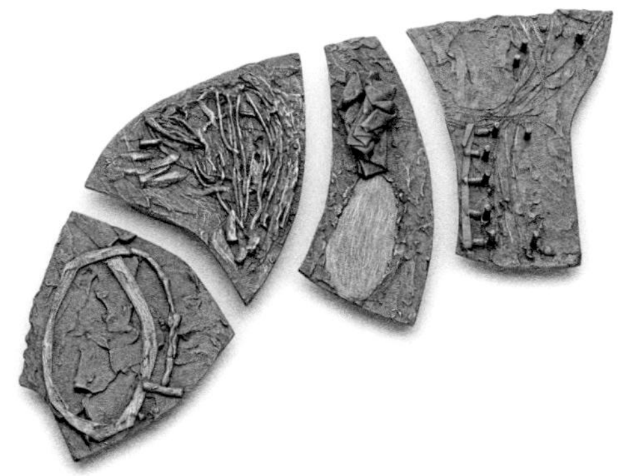

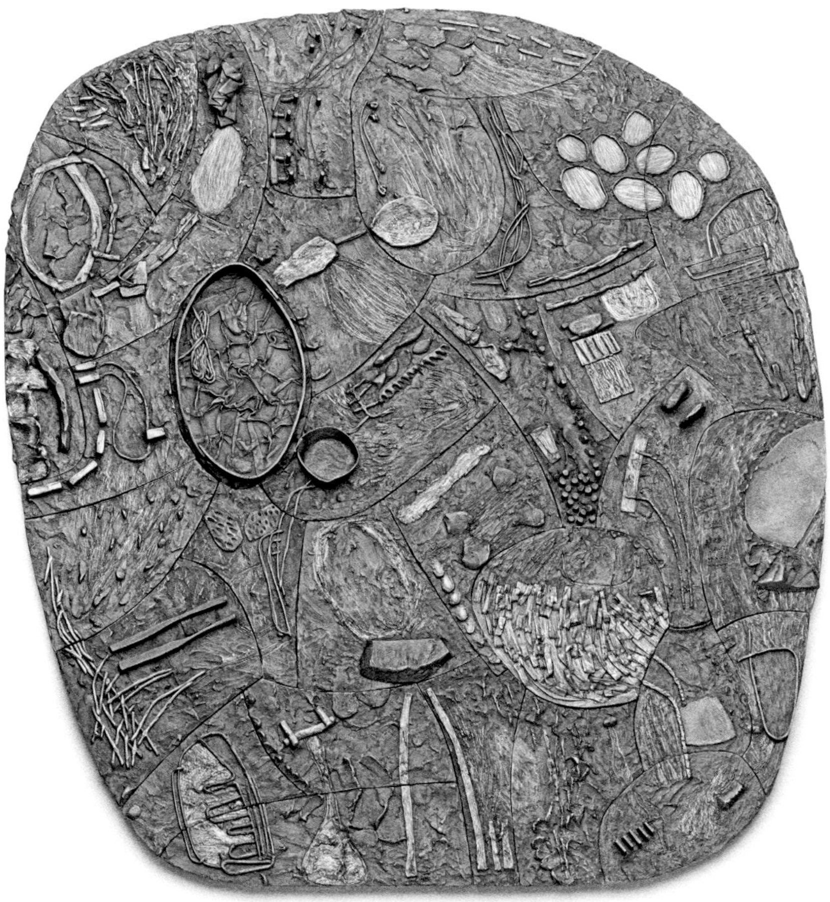

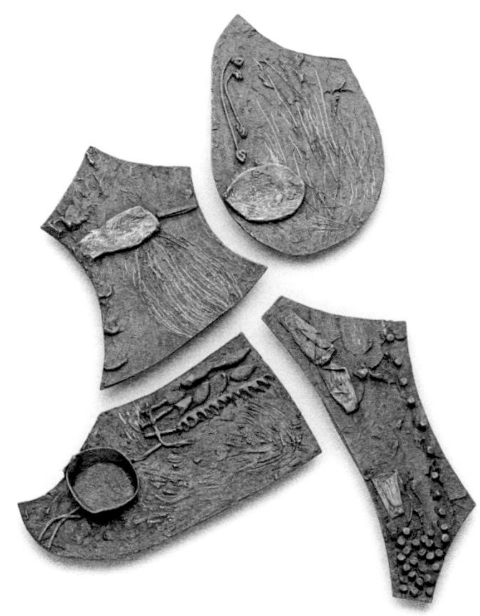

These 34 brooches fit together to form a painterly composition.

Iris Bodemer, *Relief II*, 2013. Silver, 18.9" × 15.75" × 0.6". Photos: Julian Kirschler.

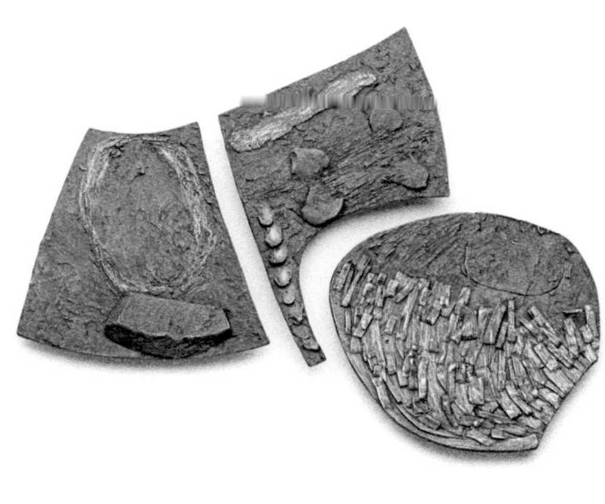

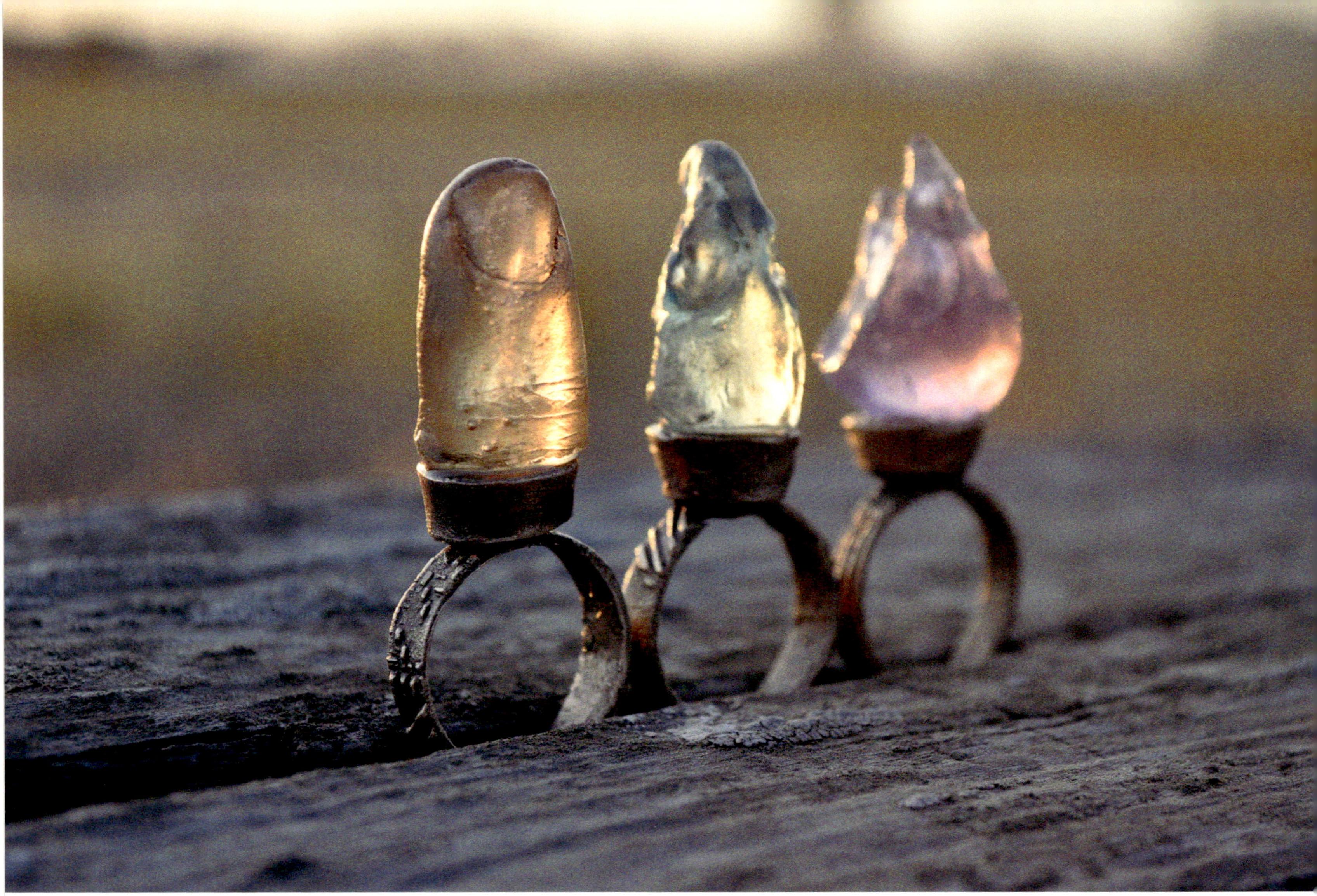

Top Mira F. Woodworth, *Three Rings*, 2014, Cast glass, bronze metal clay, 1.5" × 0.5" × 0.25". Photo: Mira F. Woodworth.

Right Al Maier, *Sharing Air*, 1975. Sterling silver, Owyhee jasper, surgical tubing, coconut shell beads, 2.5" × 3" × 0.25". Photo: Larry Berman.

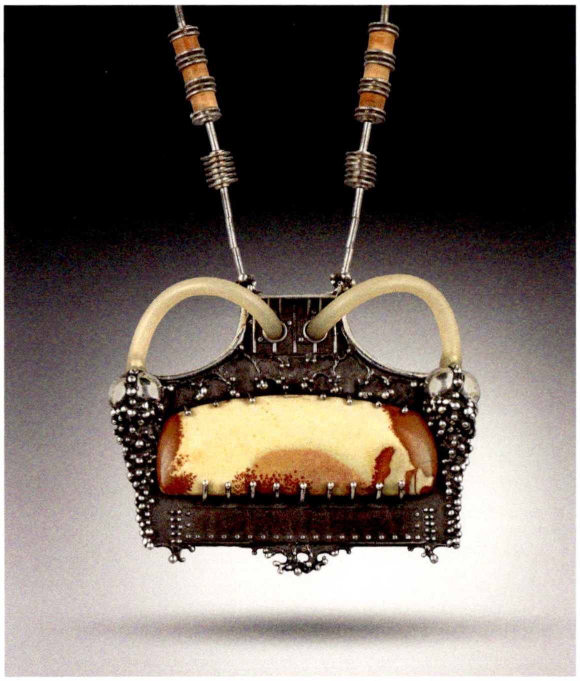

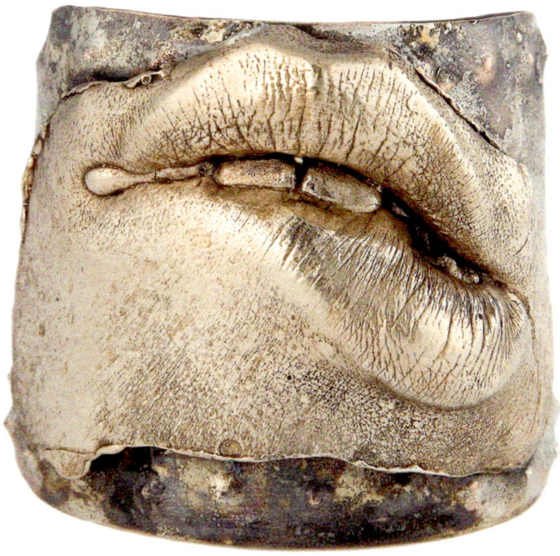

Left Simone De Bernard Mas, *Lip Service*, 2011. Bronze, sterling silver, 6.5" W × 2.5". Photo: Tabatha Mudra.

Below Jim Cotter, *Here's Looking at You Elton*, 1980. Sterling silver, found objects, 7" × 5" × 1". Photo: J. Cotter Studio.

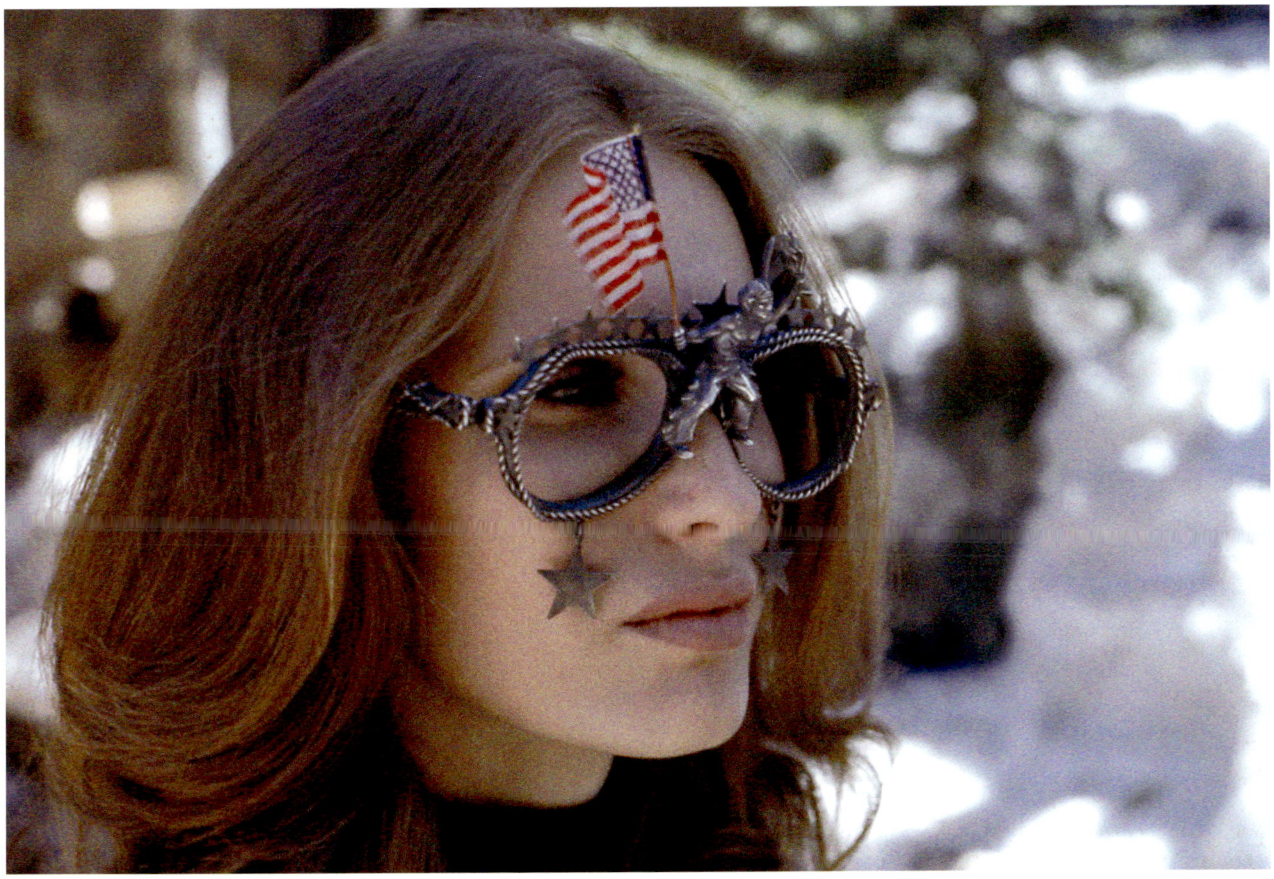

Right Marion Delarue, *Agate Bracelet*, 2012. Porcelain, glass, glaze, chamotte, 4" × 4" × 0.25". Photo: Marion Delarue.

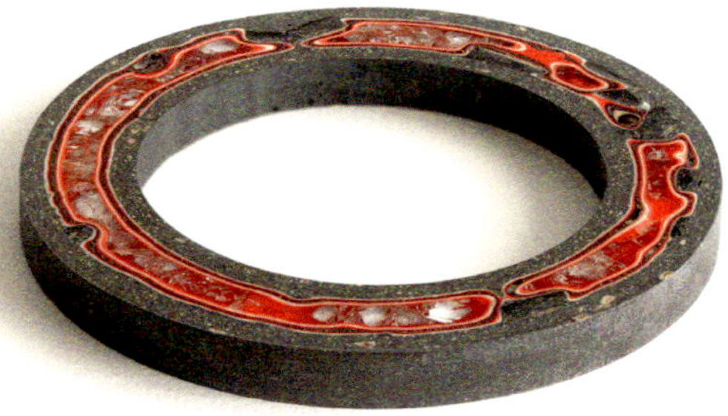

Below Michael Gayk, *Sucker, Tread, Squares,* 2011. Cast silicone, cobalt plated photopolymer, 3.7" × 5.5" × 2". Photo: Zan Maddox.

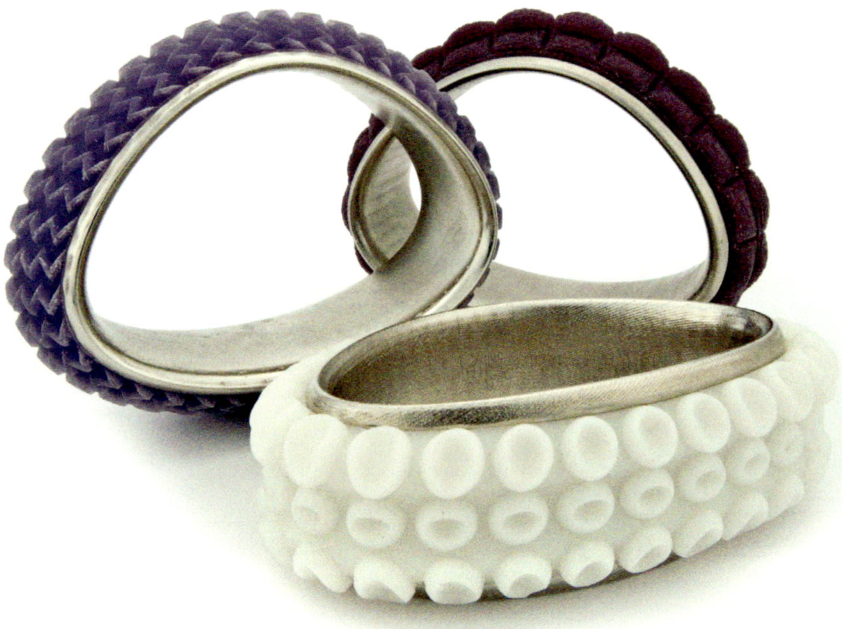

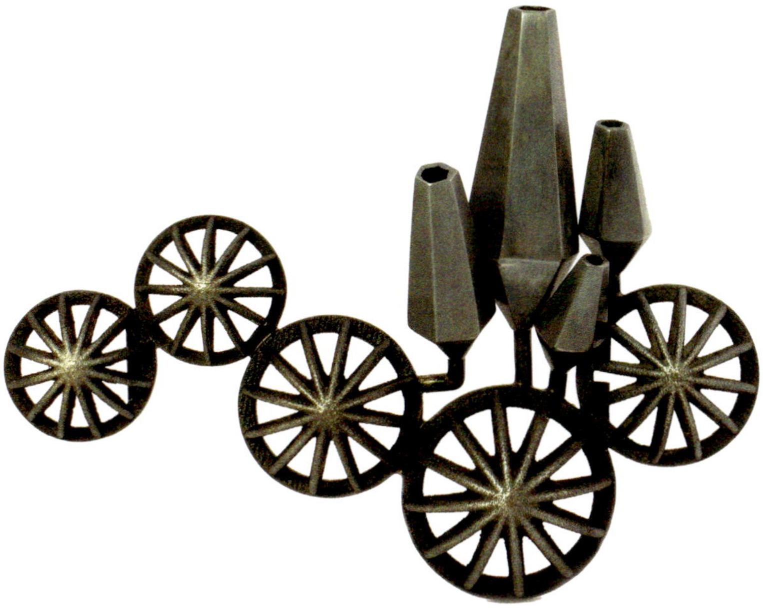

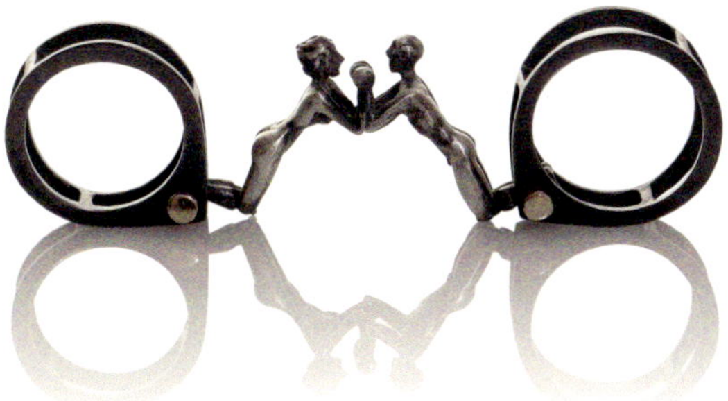

Top Nicole Schuster, *Transformation II*, brooch, 2014. .925 silver oxidized, steel, parts 3-D-designed and printed, cast; parts cut, folded, cast, soldered, 3.2" × 0.66" × 2.5". Photo: Nicole Schuster.

Left Selda Okutan, *Qu'est-ce Que C'est*, 2012. Oxidized sterling silver.
Photo: Mehmet ARDA.

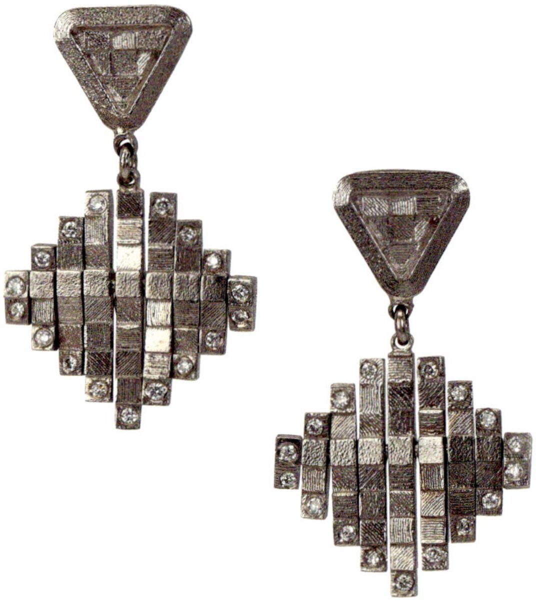

Left Jo Hayes Ward, *Medium Stratus Rain Drop Earrings*, 2015. 18K palladium white gold with 17 brilliant cut diamonds.
Photo: Jeremy Johns.

Below David Giulietti, *Cherry Blossoms on the Water*, 2012. Sterling silver, 18K, diamonds, 1.38" × 6" × 0.12".
Photo: Hap Sakwa.

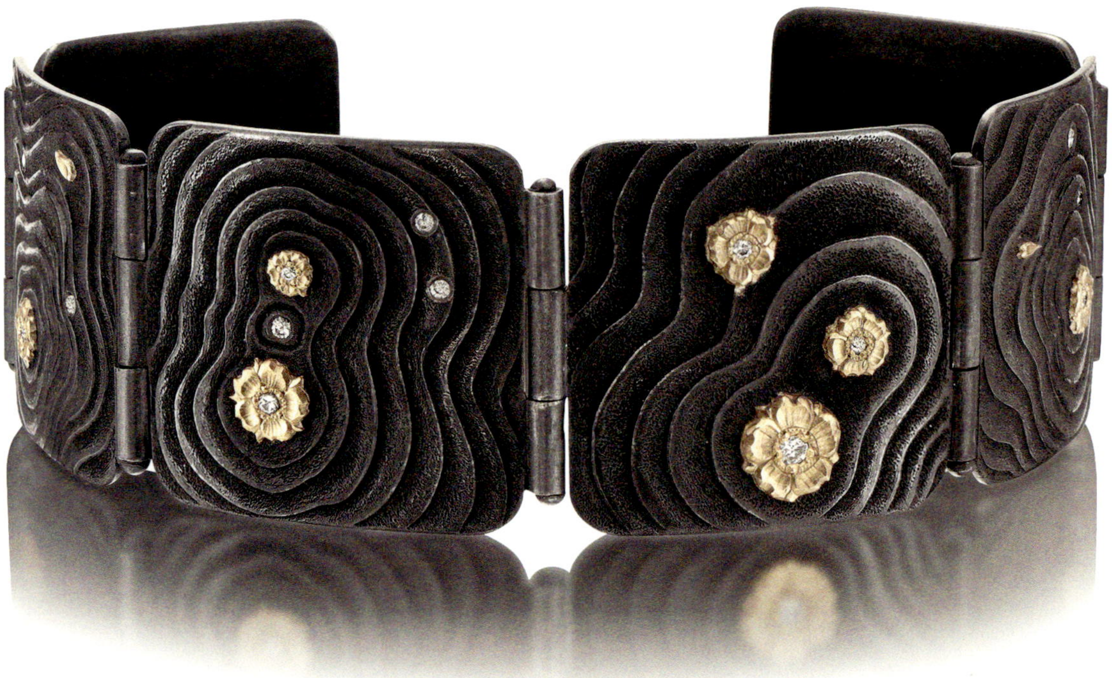

Right German Kabirski, *Untitled Ring*, 2009. Silver, gold, tourmaline, 1.18" × 1.18" × 1.37".
Photo: German Kabirski.

Opposite page Lorraine Kolasa, *Arachne I'm Sorry*, 2015. Bronze, 12" × 6".
Courtesy of Lorraine Kolasa.

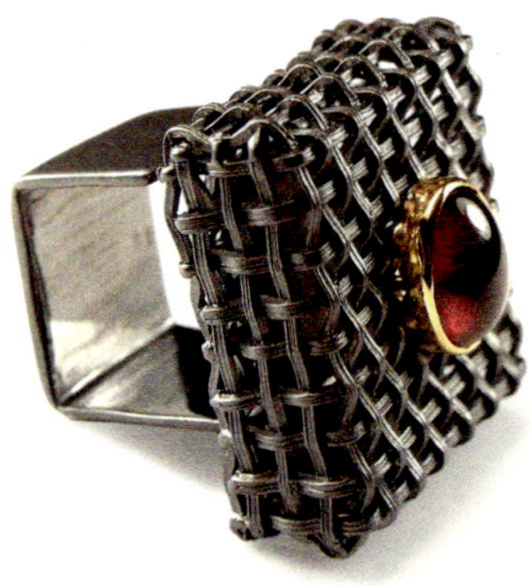

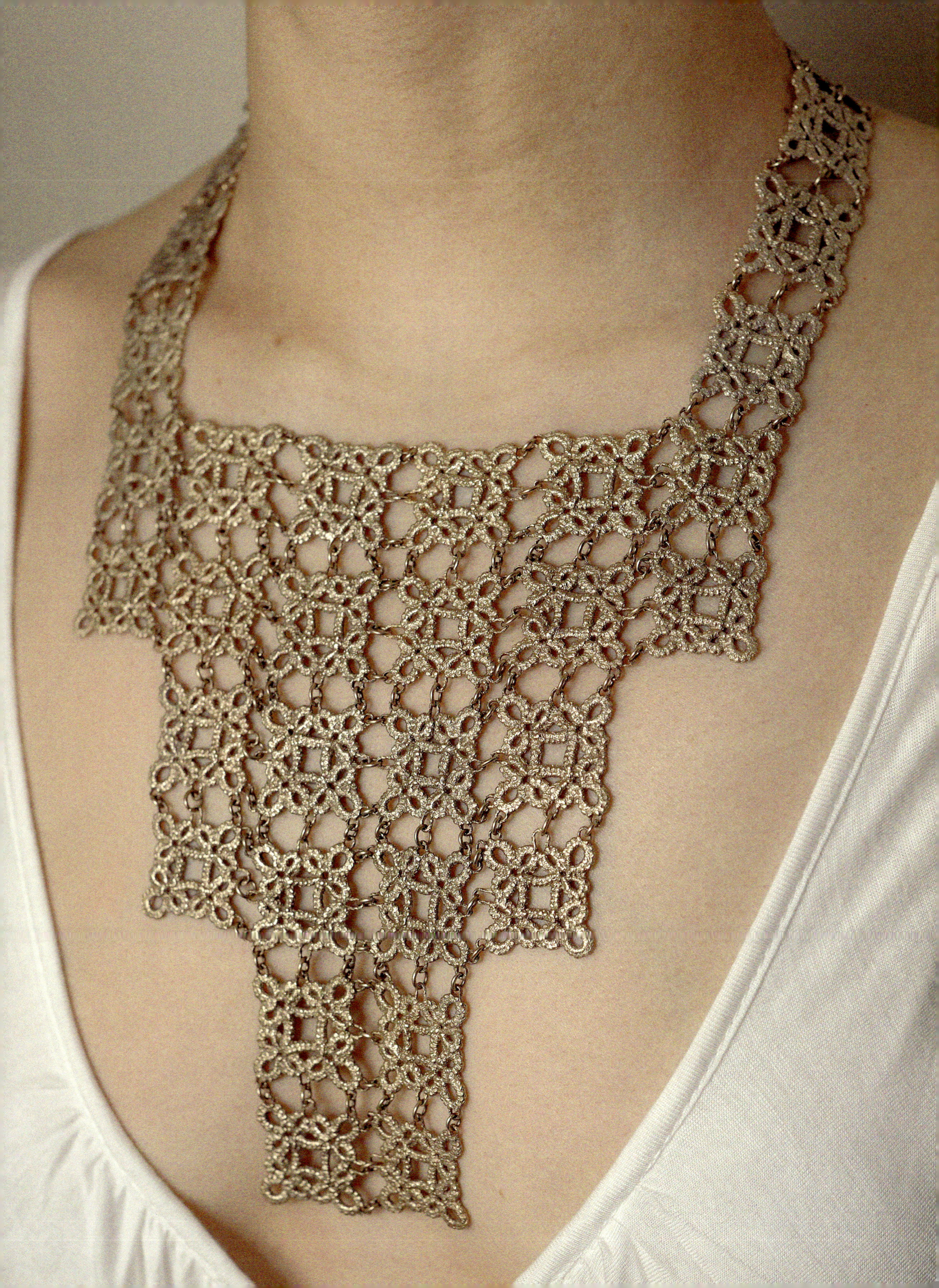

Left Cast iron brooch, nineteenth century. Cast iron, 2" × 1.5". Courtesy of Pink Steel Jewels/Karen Beuning.

Below Vernon Bowden, *Father*, 2006. Action figure, cast silicone rubber, Barb-aryan (2006), sixty-two Diet Coke® cans (hidden inside), 8.6" × 8.6" × 2.4".
Photo: Vernon Bowden.

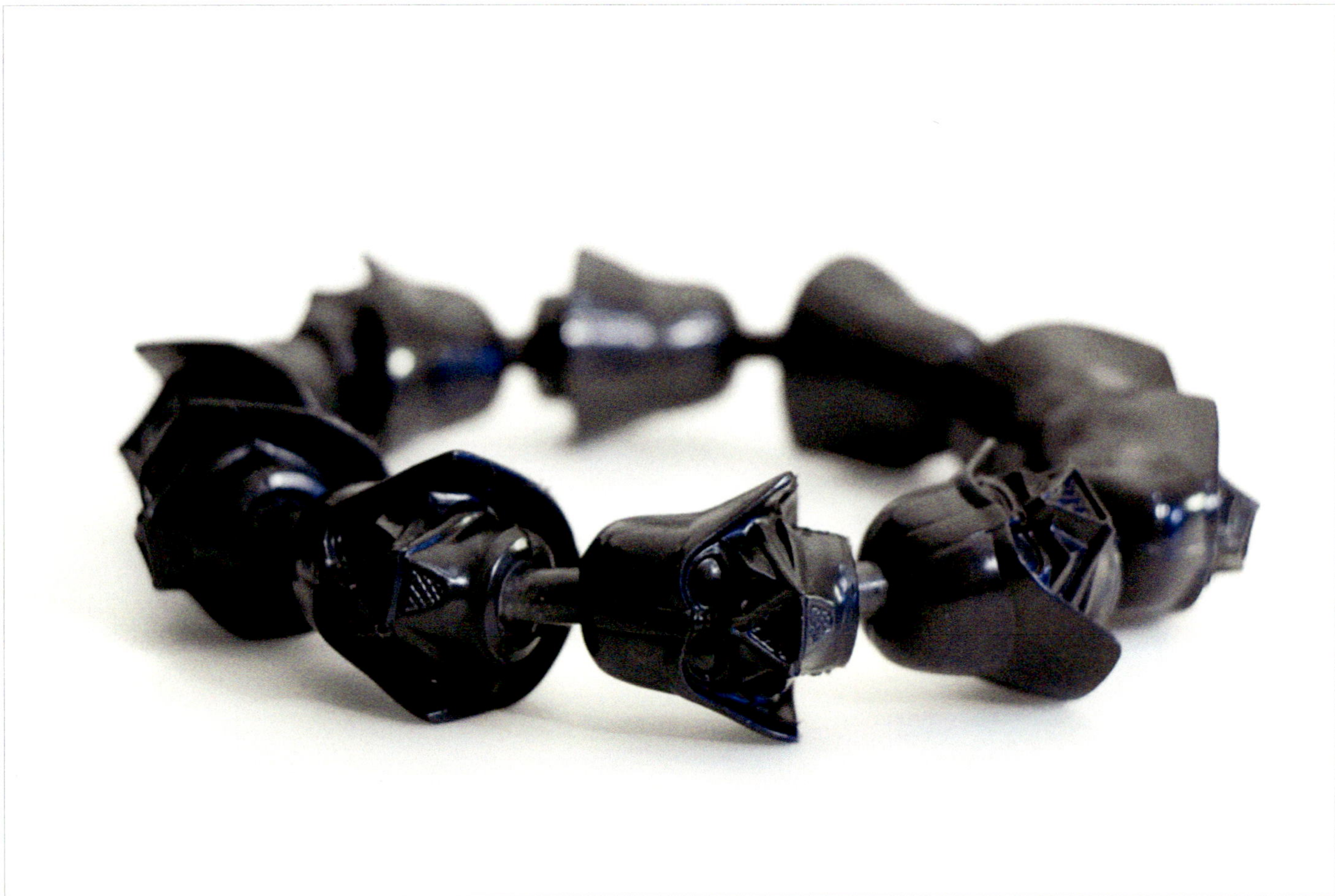

Right Pendant, Iron Age I–II, ca. 1350 to 800 BCE. Northwestern Iran, bronze, 3.25" × 3.75".
Courtesy of LACMA, Gift of The Ahmanson Foundation, www.lacma.org, LACMA public domain.

Below Rona Pondick, *Worry Beads*, 1999–2001. Bronze, edition of 6 + 1 AP, 1.5" × 2" × 24".
Courtesy of Galerie Thaddaeus Ropac, London/Paris/Salzburg and artist.

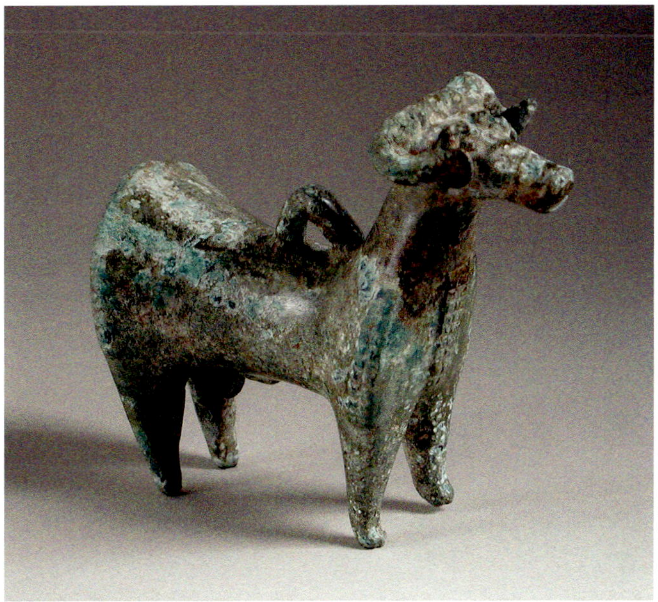

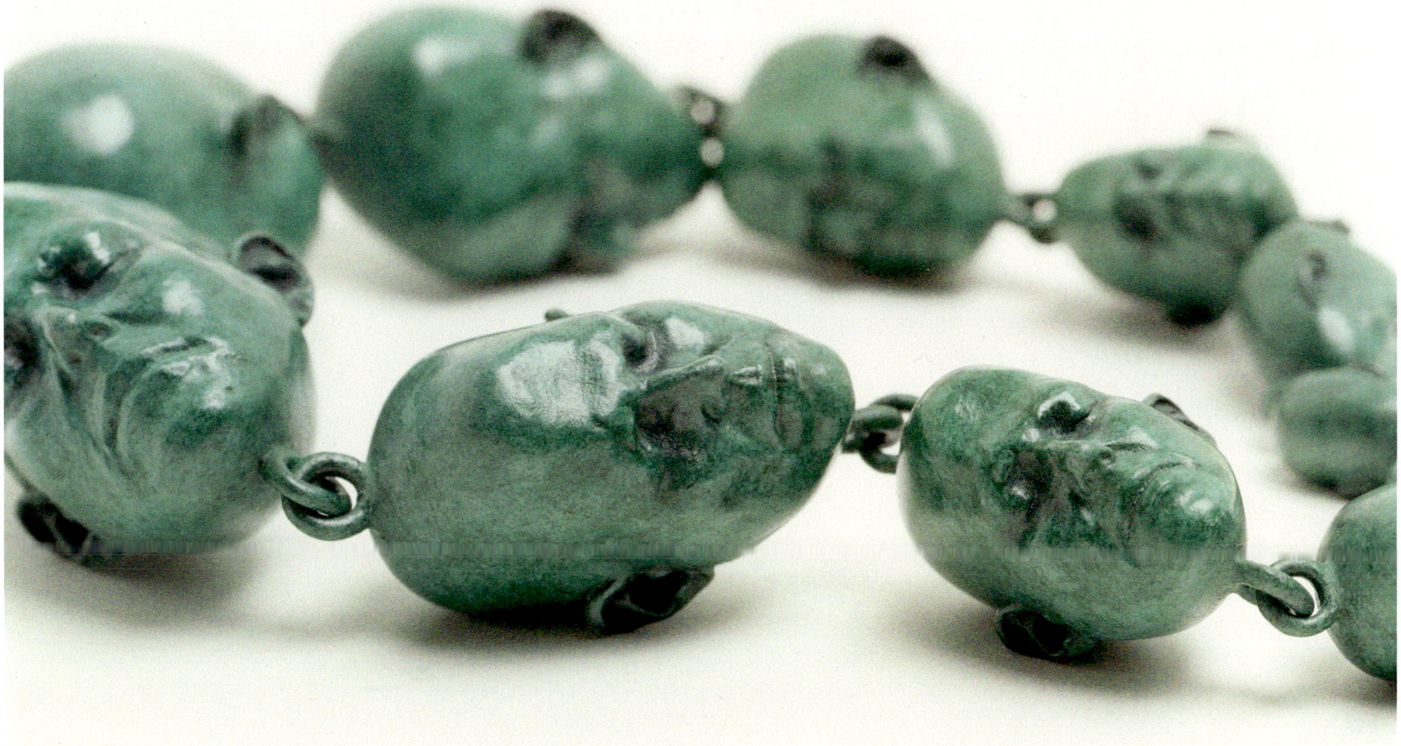

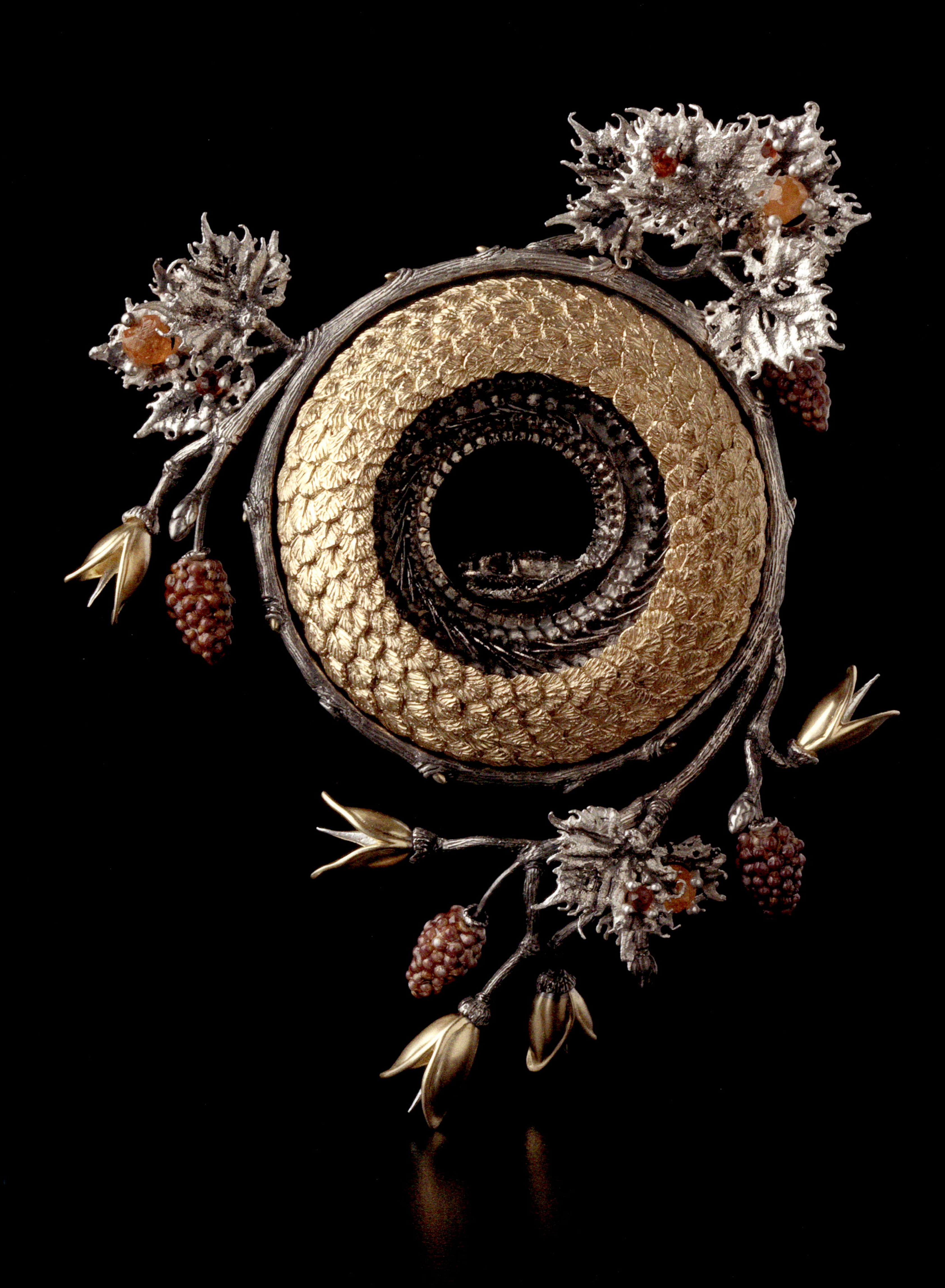

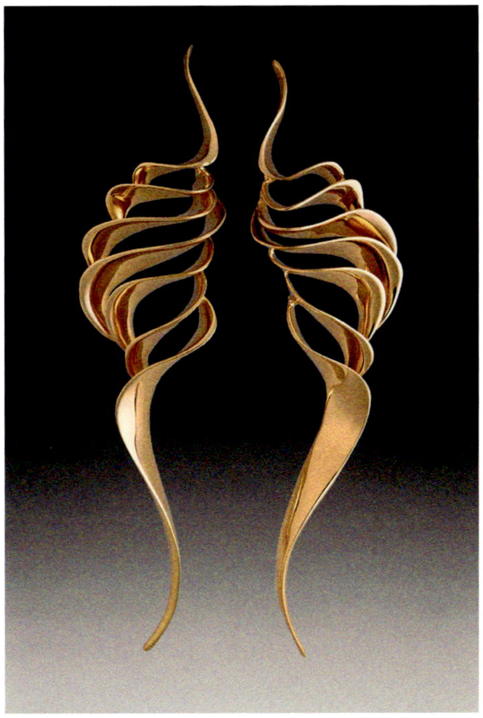

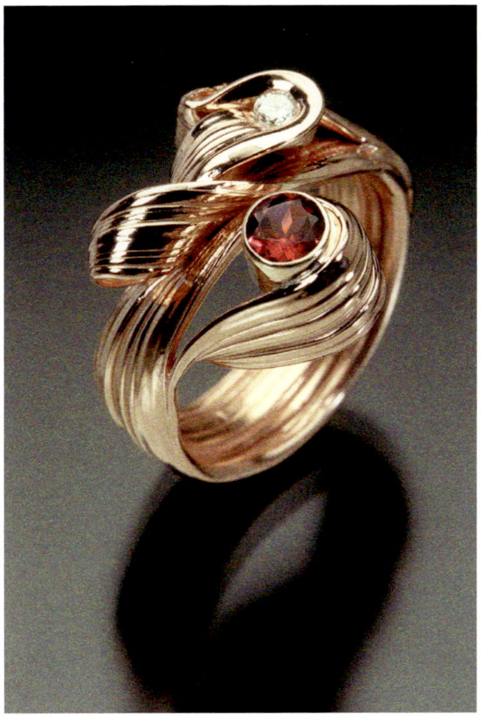

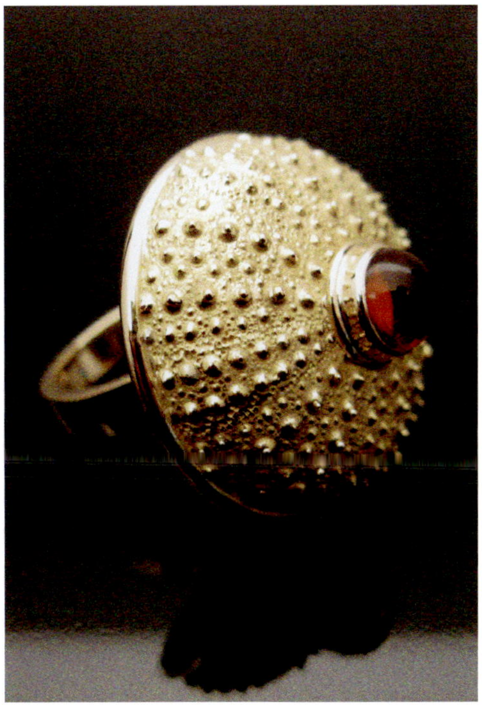

Top left Charles Arnold, *Nautilus*, 2014. Gold, 4" × 0.50" × 0.25".
Photo: Charles Arnold.

Top right Susan Zeiss, *Swirl*, 2010. 18K diamond, rhodolite garnet, 1.25" × 1" × 0.75".
Photo: Susan Zeiss.

Left Hratch Babikian, *Red Sea Urchin*, 2013. 14K gold, dark citrine, 1" × 1" × 1".
Photo: Hratch Babikian.

Opposite page Lin Stanionis, *Awakening*, 2013. 18K gold, sterling, garnets, enamel, snake skeleton, 4" × 3" × 1".
Photo: Jon Blumb.

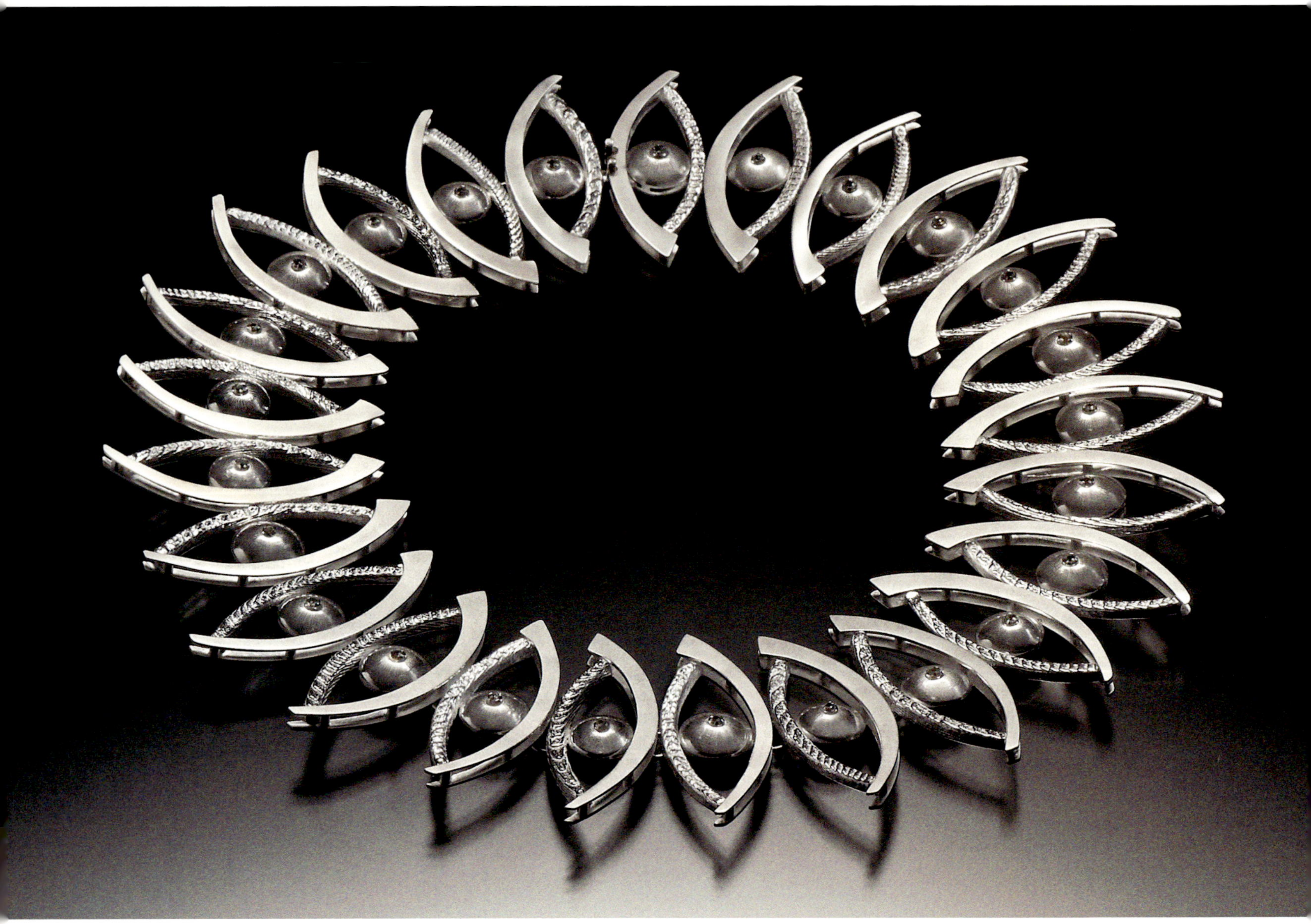

Top Yuka Okane Inoue, *Every Eye Is Unique*, 2012. Silver, sapphire, white gold, 0.3" × 7.7" × 7.7". Photos: Guy Nicol.

Right Yuka Okane Inoue, *Every Eye Is Unique* (detail), 2012.

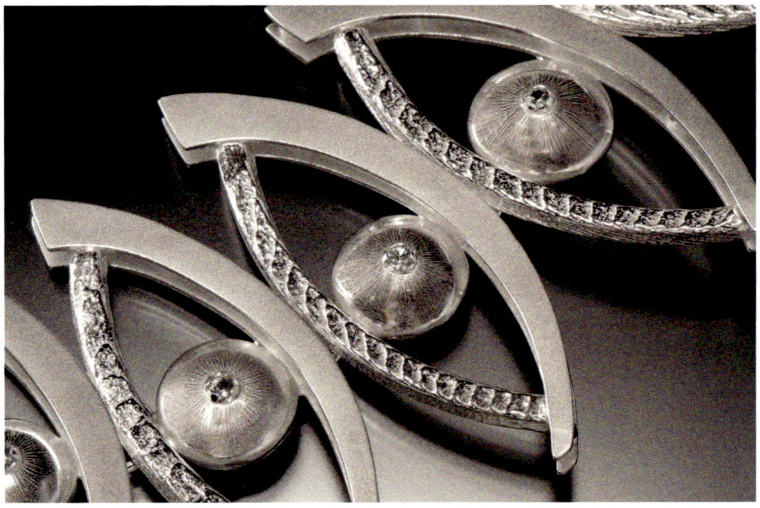

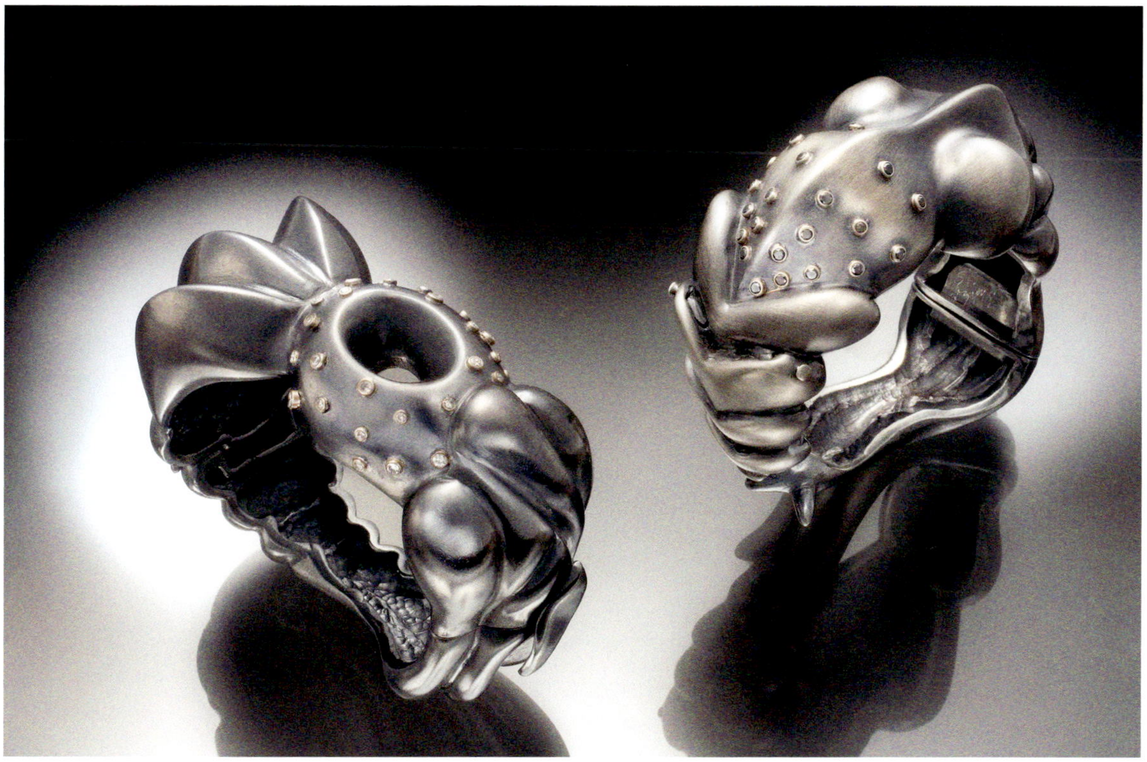

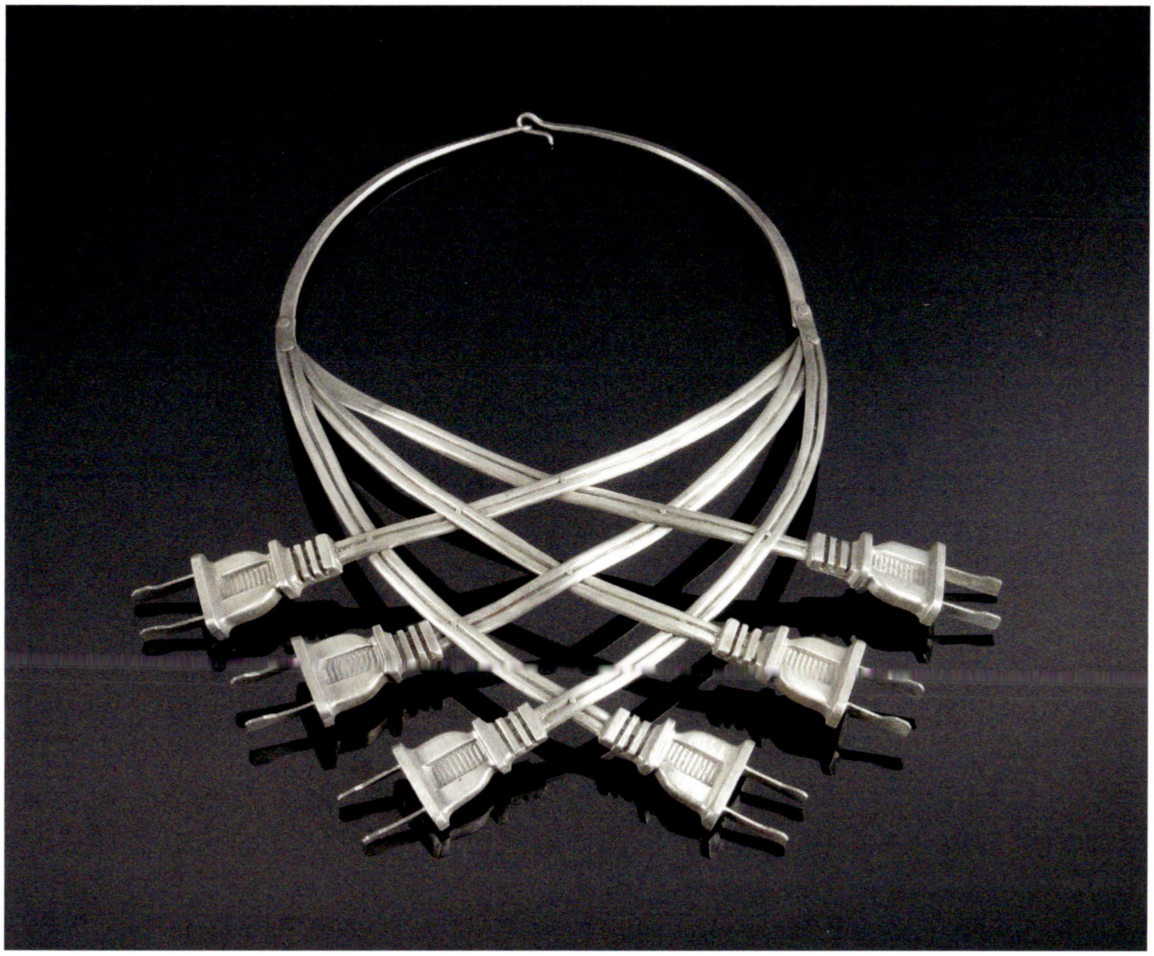

Top Shana Kroiz, *Black and White Sculptural Cuff Bracelets*, 2010. Sterling silver, 14K box clasp, diamonds and black diamonds set in 14K gold, 3" × 1.5" × 2.5". Photo: Ralph Gabriner.

Bottom Aimee Howard, *Untitled Necklace* from the *Consumer Report* series, 2006. Silver, 11" × 10" × 0.75". Photo: Aimee M. Howard.

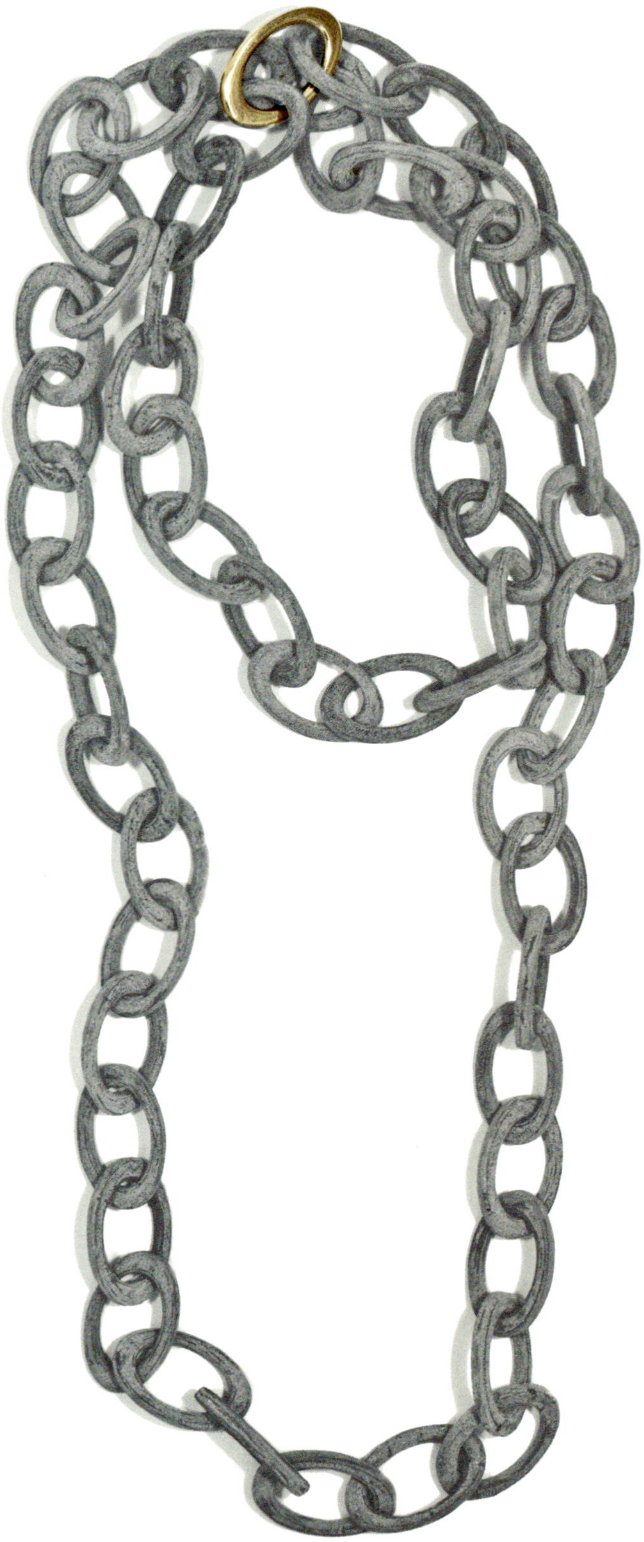

Demitra Thomloudis, *Linked*, 2014. Cement, 14K gold, cast bronze, resin, 30" × 3" × 1.2". Photo: Demitra Thomloudis.

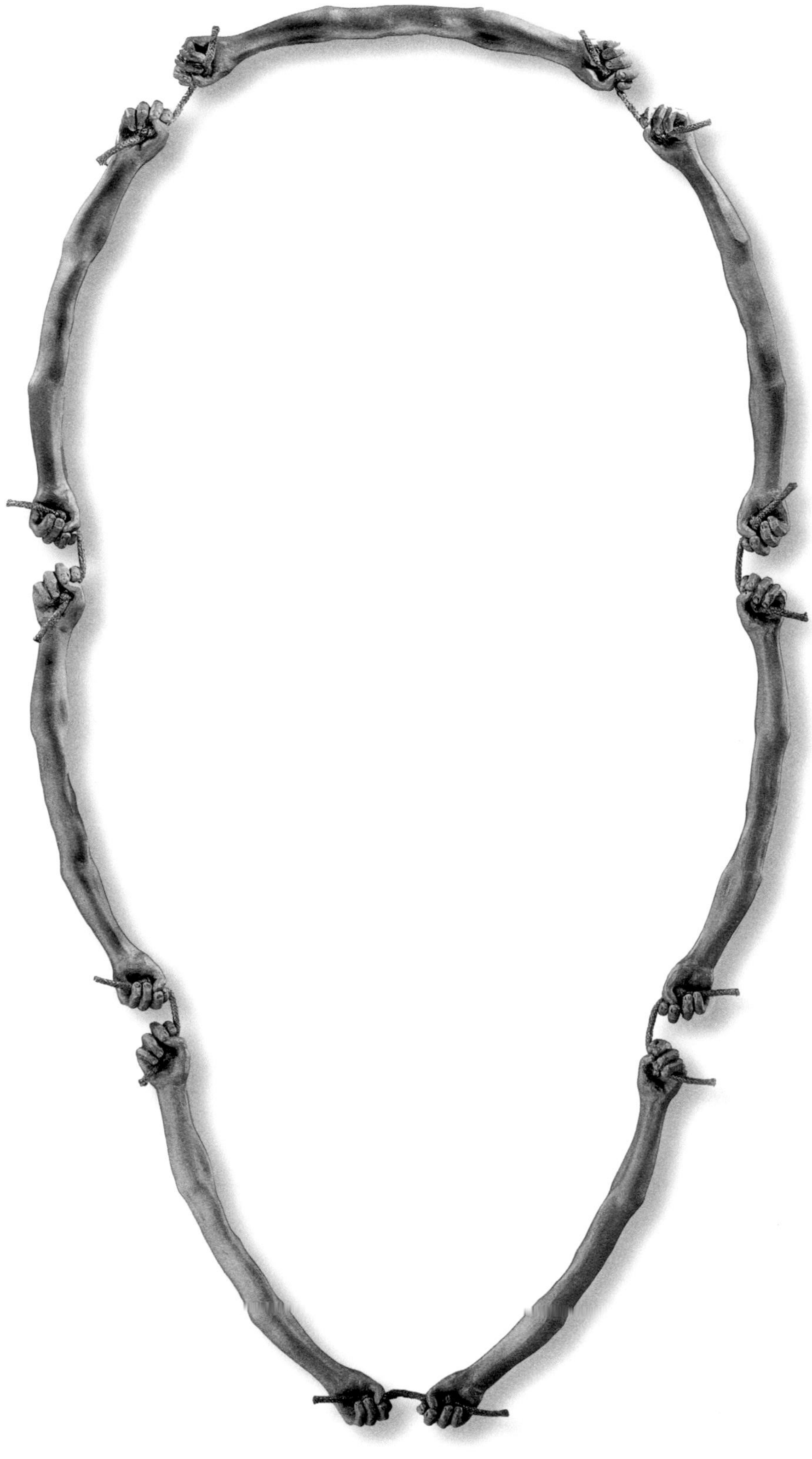

Ruudt Peters, *CORPUS "ossa,"* 2011. Cast aluminum, cotton, 9.8" × 6" × 0.5". Photo: Conor Vella.

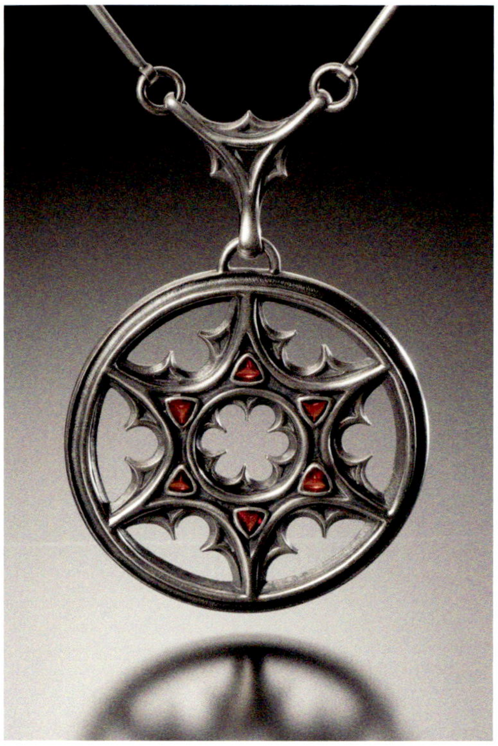

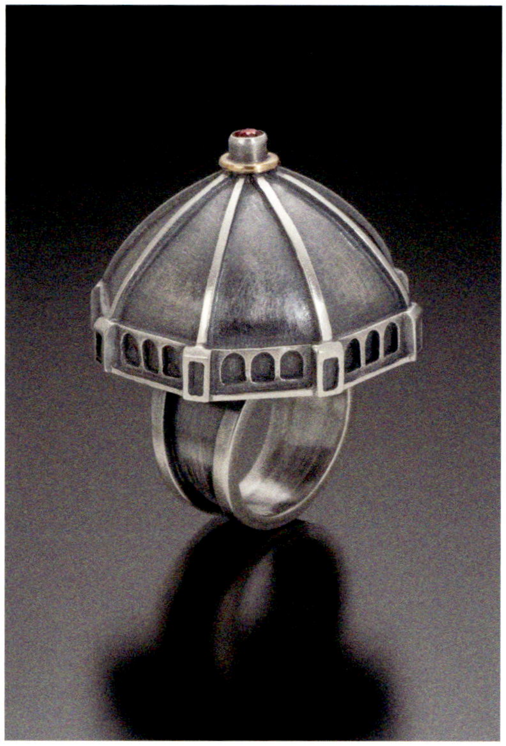

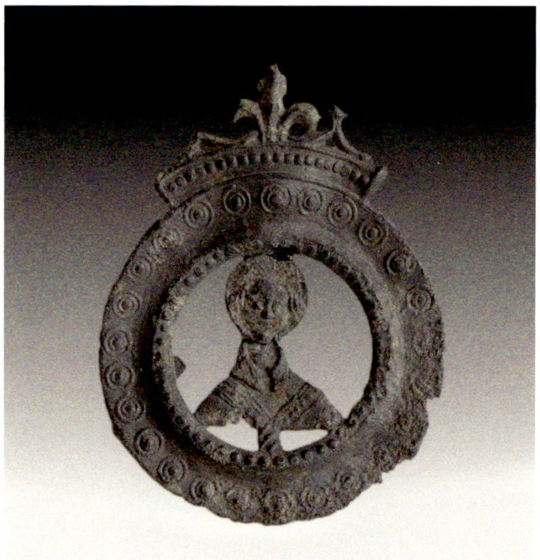

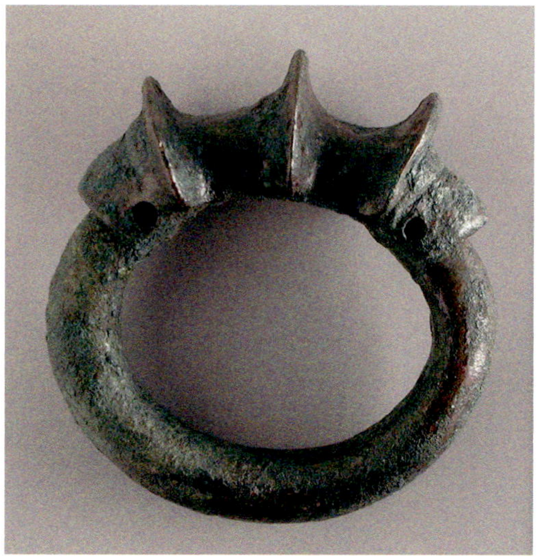

Top left Donna Veverka, *Gothic Pendant*, 2012. Sterling silver, garnet, 3.5" × 2.75" × 0.25". Photo: Robert Diamante.

Top right Donna Veverka, *Duomo Ring*, 2013. Sterling silver, 18K gold details, garnet, 1.5" × 1.25" × 1.25". Photo: Robert Diamante.

Bottom left Venetian (Italian) Artist, Pilgrim Badge, thirteenth to fourteenth century (Medieval). Torcello, Italy (Place of Discovery), lead, 1.12" × 0.8" × 0.12". Courtesy of The Walters Art Museum, Museum purchase with funds provided by the S. & A.P. Fund in 1945, CC0, GNU FDL.

Bottom right Bracelet, ca. 1000 to 650 BCE. Western Iran, bronze, 3.5". Courtesy of LACMA, Gift of Nasli M. Heeramaneck, www.lacma.org, LACMA public domain.

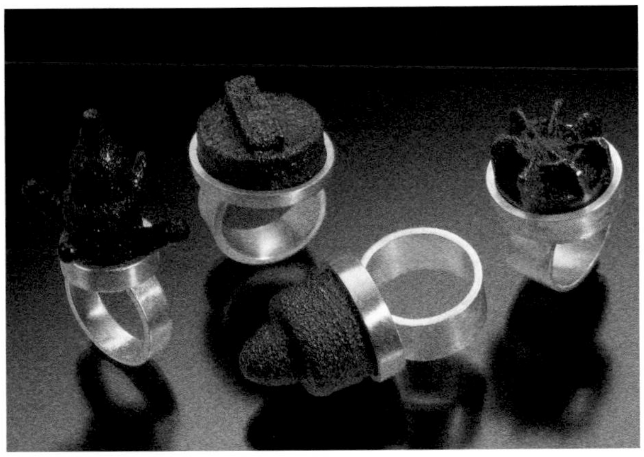

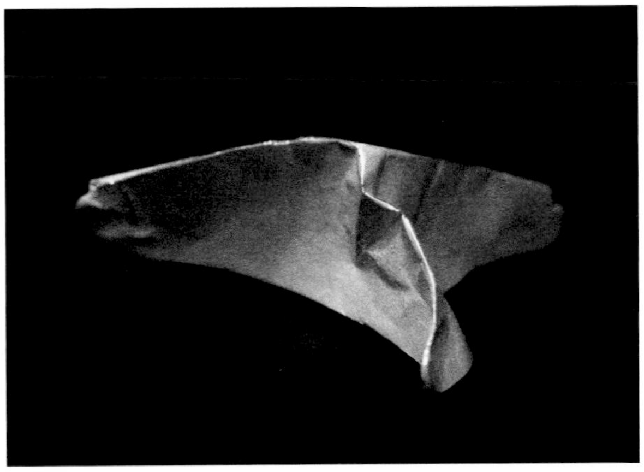

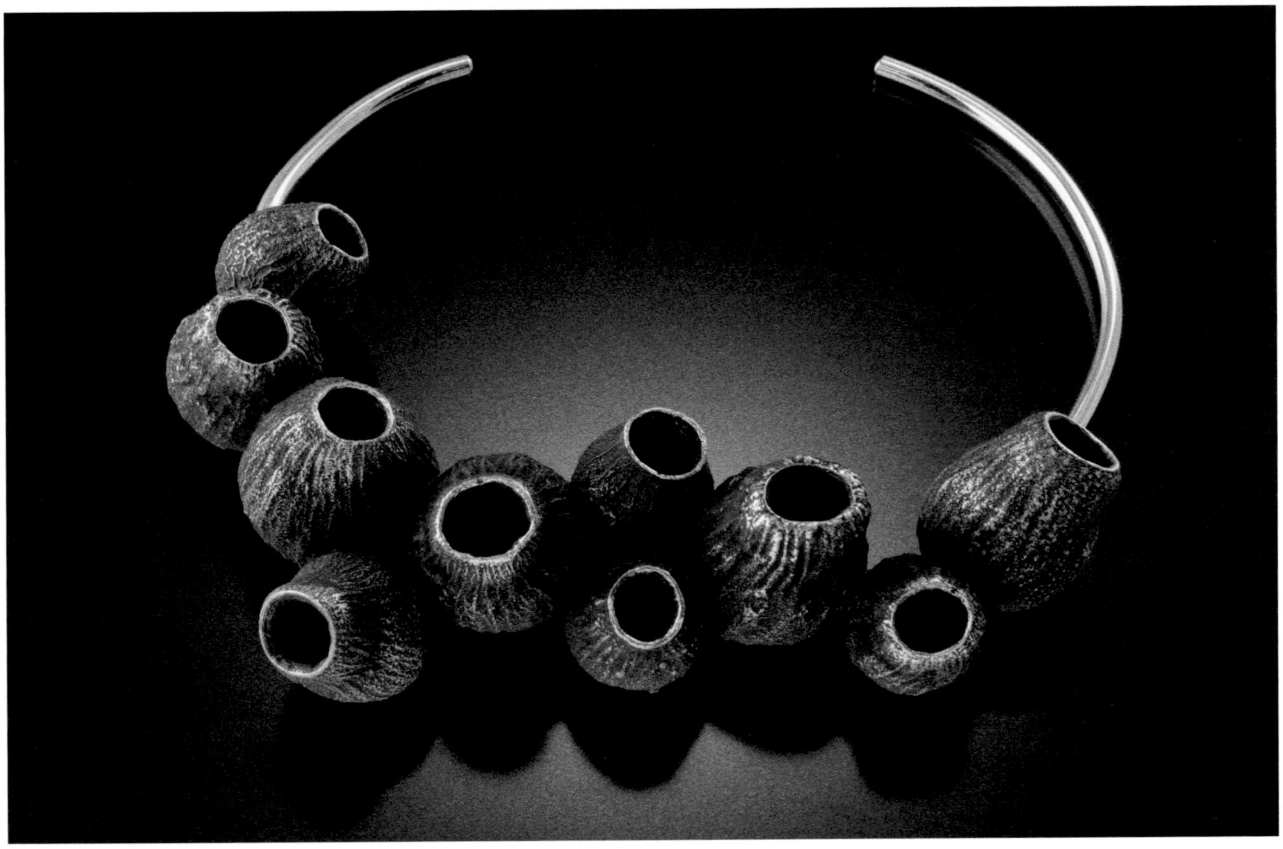

Top left Jana Brevick, *Transmitter Series, Group 1*, 1999, still in production. Lacquered sandcast iron and fabricated sterling silver rings, 1" × 1" × 2.5" (tallest).
Photo: Douglas Yaple.

Bottom Amy Beeler, *Slow Strangulation's Release*, 2013. Sterling silver, 8" × 8" × 2".
Photo: Larry Sanders.

Top right Fie Bodenhoff, *Paperwork 2*, 2015. Sterling silver, 1.5" × 1.4" × 0.6".
Photo: T. Sand.

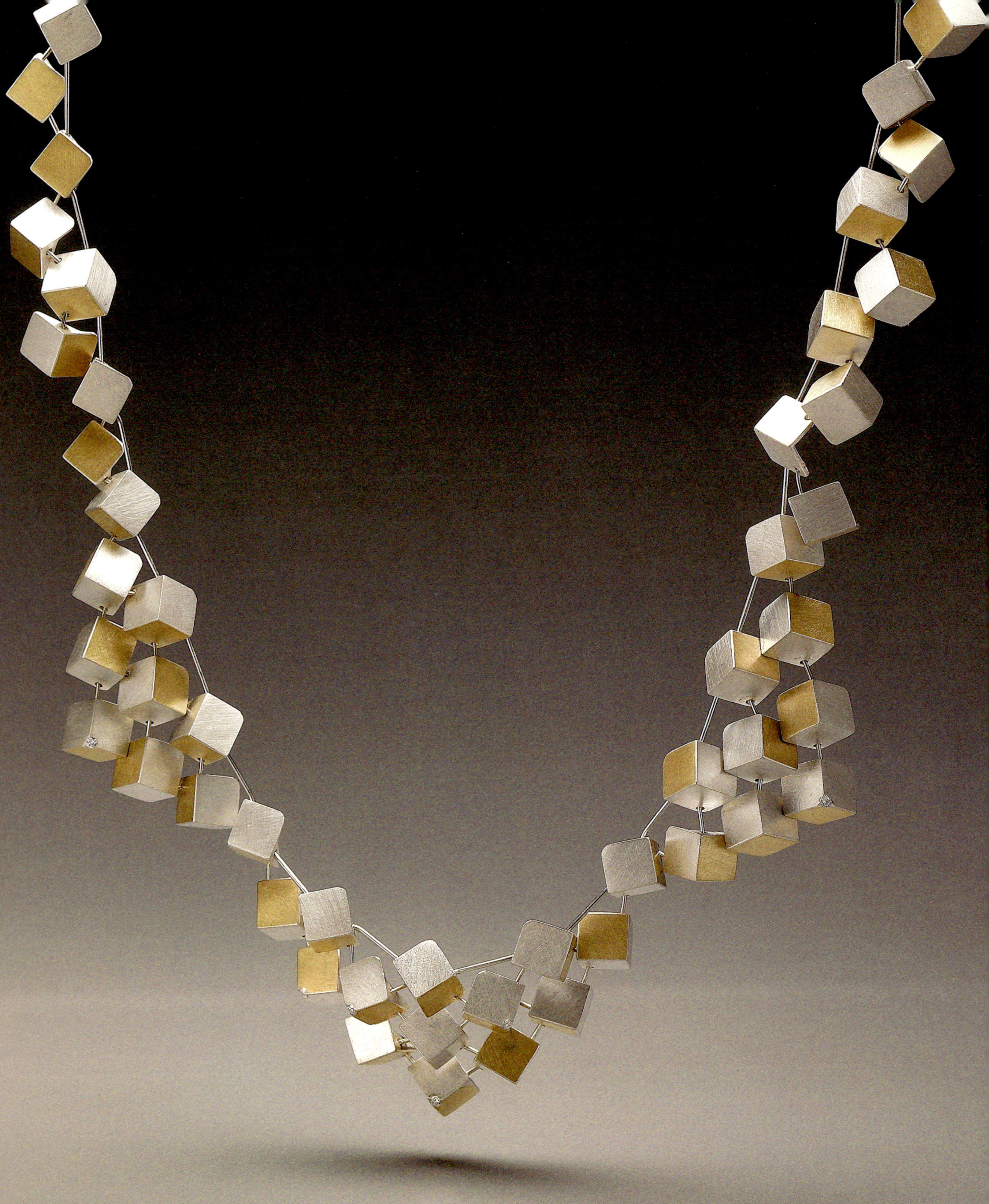

Hongsoek Lee, *Cube Necklace*, 2010. Sterling silver, keum-boo (24K gold), diamonds, 14K gold wire, 0.25" × 16" × 16". Photo: Myoungwook Hu.

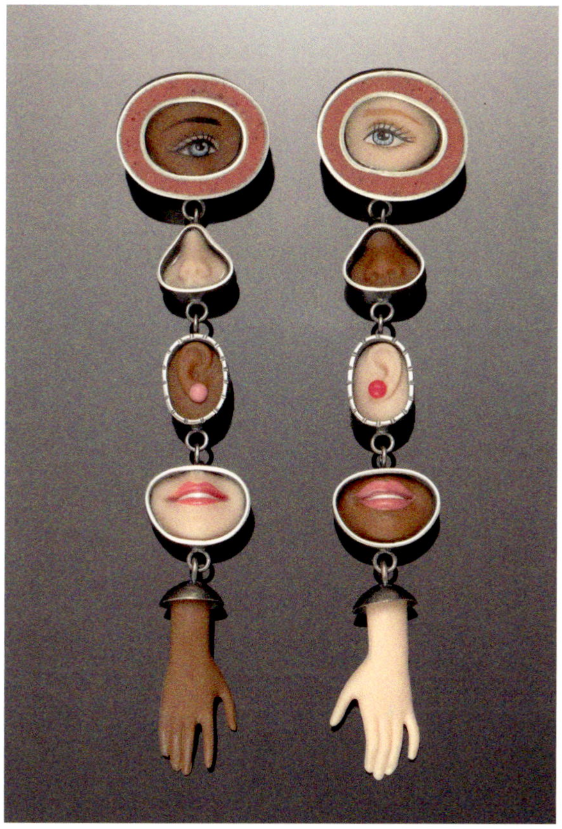

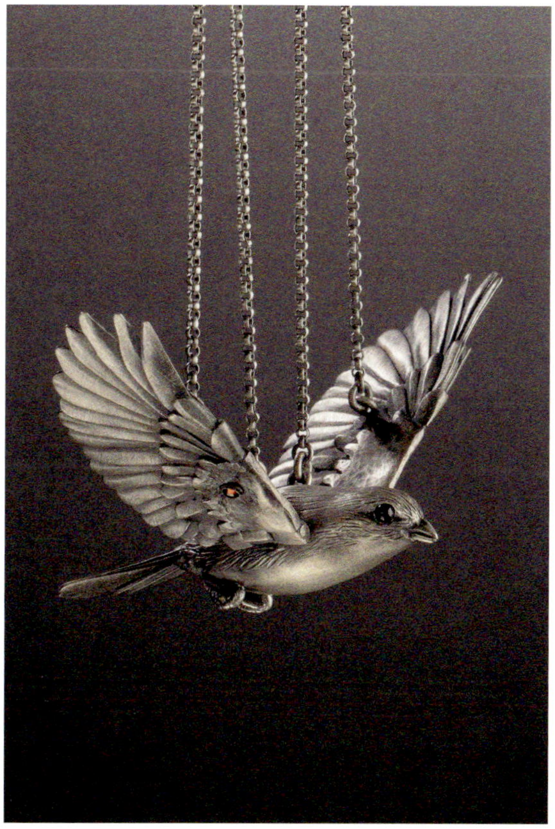

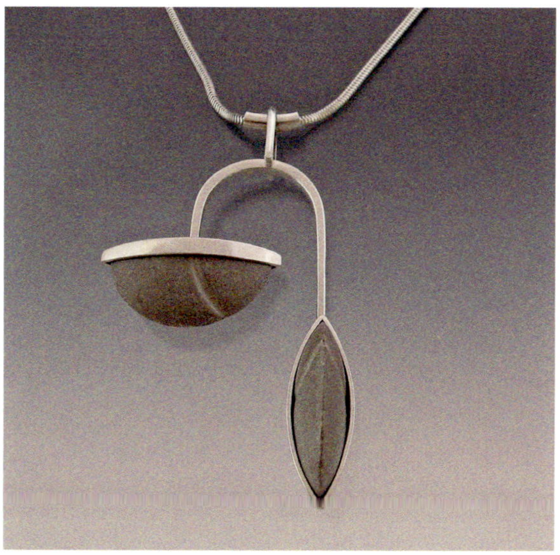

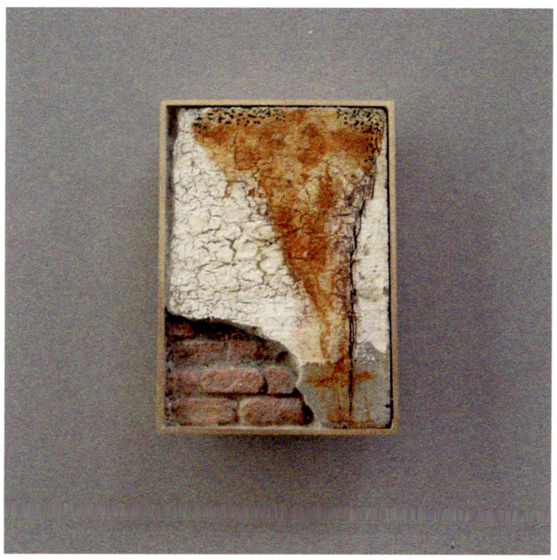

Top left Margaux Lange, *Five Senses Earrings*, 2007. Sterling silver, plastic Barbie® doll parts, pigmented resin, hand-fabrication, 4" × 1" × 0.25".
Photo: © Quad Graphics.

Top right Yuri Tozuka, *Companion*, 2011. Sterling silver, fine silver, brain coral, garnet, onyx, 10" × 5" × 1".
Photo: Dean Powell.

Bottom left Tom McCarthy, *Conversation*, 2010. Sterling silver, concrete, 3.9" × 2.4" × 1".
Photo: Tom McCarthy, Private Collection.

Bottom right Tom McCarthy, *Bucks County Brooch*, 2011. Bronze, concrete, gold leaf, paint, rust, India ink, 1.8" × 1.9" × 0.25".
Photo: Jan Spencley.

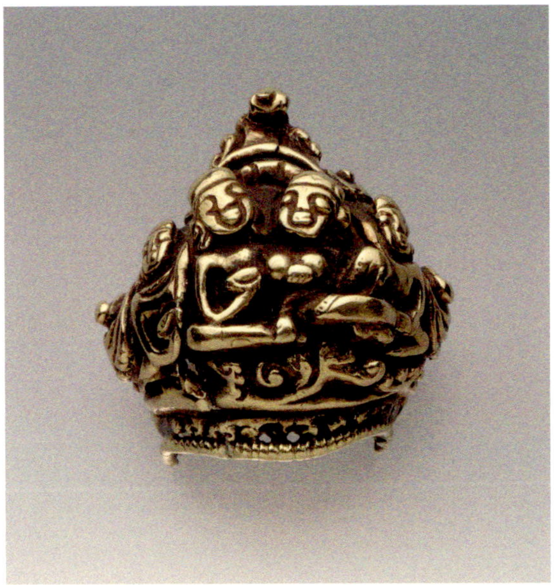

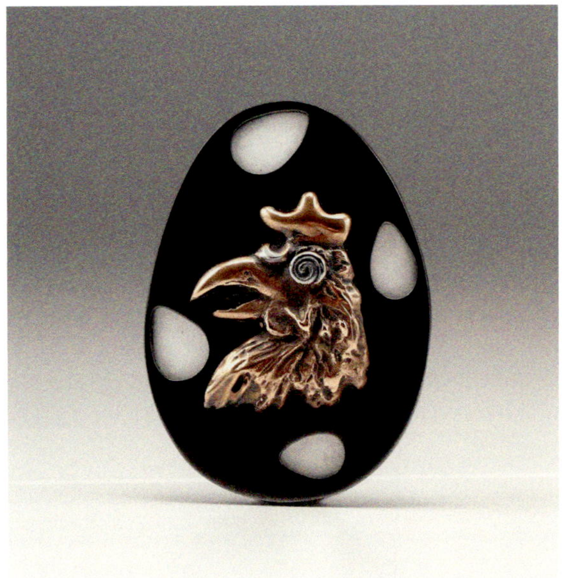

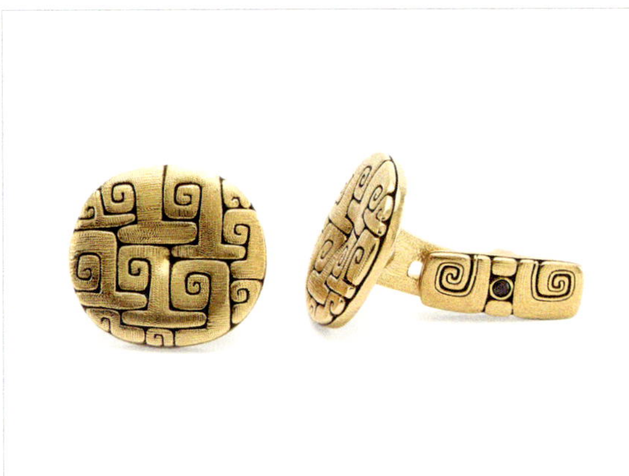

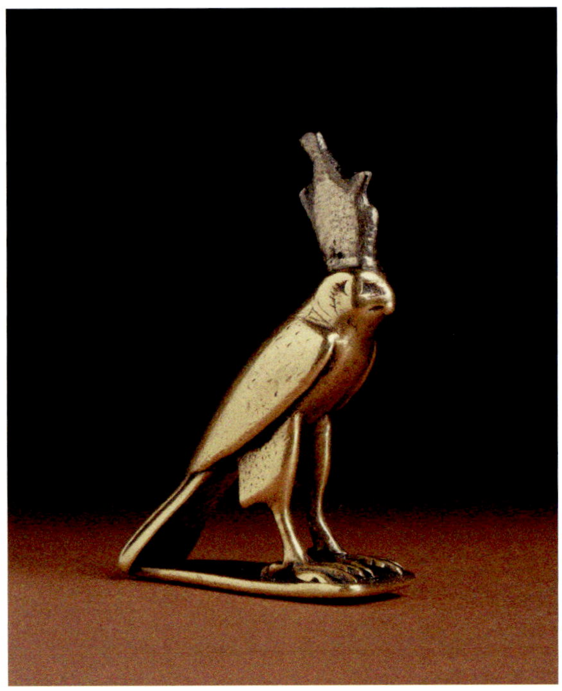

Top left Earring with Figures, ninth to tenth century. Indonesia, Central Java, gold, 0.9" × 0.9" × 0.6". Courtesy of LACMA, Gift of Robert Stolper, www.lacma.org, LACMA public domain.

Top right Jim Bové, *Hypnotizing Chickens*, 2012. Bronze, copper, enamel, 2" × 1.25" × 0.25". Photo: Jim Bové.

Bottom left Alex Šepkus, *J's Garden Cufflinks (C-18)*, 2000. 18K gold, black diamonds, 0.86" × 0.69" × 0.65". Photo: Arunas Kulikauskas.

Above Falcon Shaped Horus, ca. 1070 to 525 BCE (Third Intermediate-Late Period). Egypt, gold, silver, 1.12" × 0.5" × 1.7". Courtesy of The Walters Art Museum, Acquired by Henry Walters, CC0, GNU FDL.

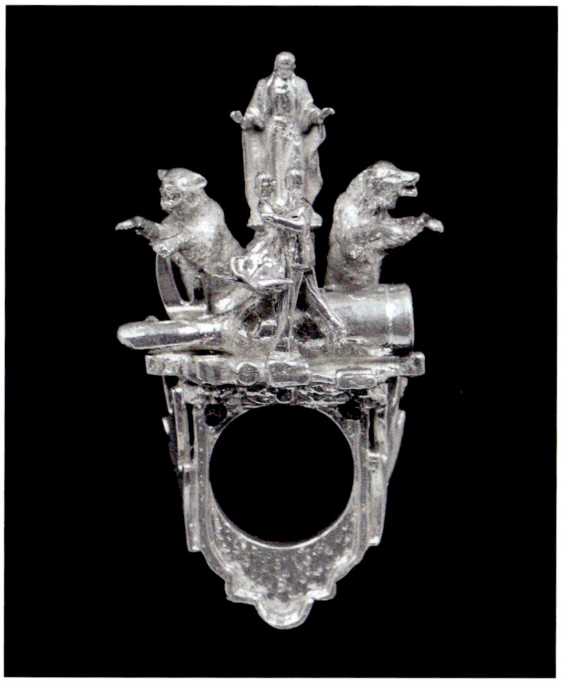

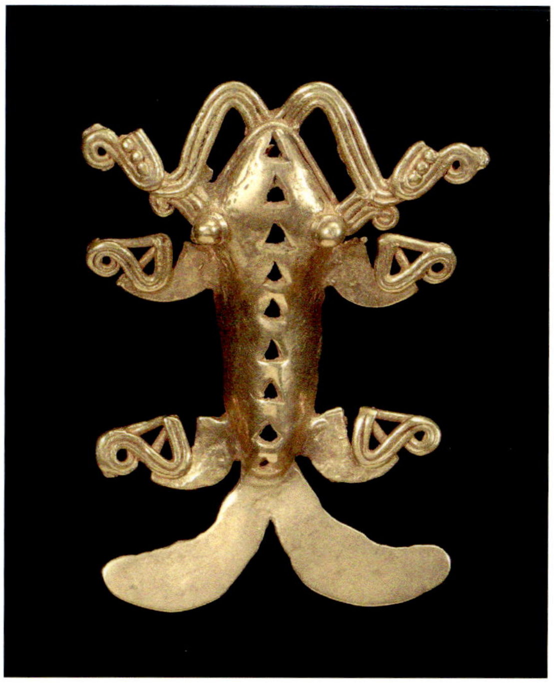

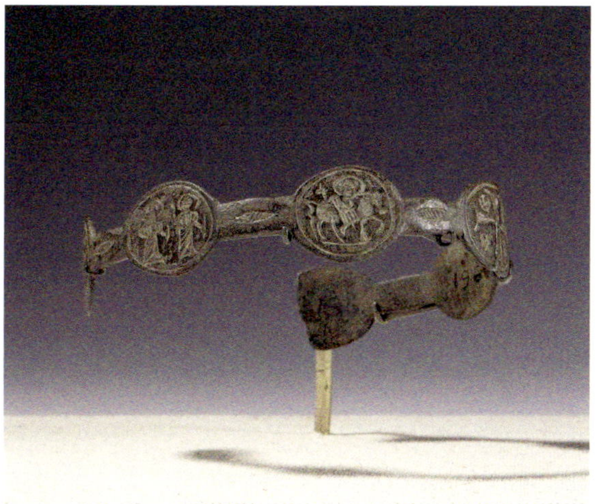

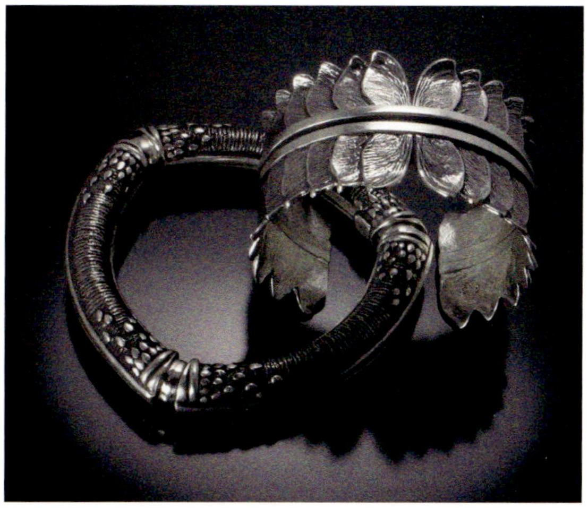

Top left Rebecca Rose, *Embittering*, 2014. Sterling silver, 3.25" × 2" × 1.25".
Photo: Aracely Rosales, Philadelphia, PA.

Bottom left Amuletic Armband with Holy Rider, Saints, and Magical Symbols, Sixth to seventh century (Late Antique). Byzantine Syria or Egypt, bronze, fragment A: 1.45" × 3.18" × 0.06"; fragment B: 0.62" × 2.25" × 0.06"; fragment C: 0.68" × 2.25" × 0.06".
Courtesy of The Walters Art Museum, Museum purchase in 1990, CC0, GNU FDL.

Top right Amphibian Pendant, ca. 800 to 1500 (Pre-Columbian). Veraguas, Panama (place of origin), gold, 3.3" × 2.4" × 0.7".
Courtesy of The Walters Art Museum, Acquired by Henry Walters in 1911 from Tiffany & Co., CC0, GNU FDL.

Bottom right Pam Fox, *Sesame Bangle and Maple Seed Cuff*, 2014. Silver, 0.5" × 4" × 4".
Photo: Pam Fox.

Right Manya Tessler, *Night Chameleon Ring*, 2007. Sterling silver, purple sapphires, diamonds, 1.5" × 1" × 1".
Photo: Chris Trayer.

Bottom left Human Effigy Pendant (from Northwestern Colombia). Gold alloy, 5.1" × 2.75" × 1.12".
Courtesy of The Walters Art Museum, Gift of John Bourne in 2009, CC0, GNU FDL.

Bottom right Barbara Mann, *Monkey Ring*, 2012. Sterling silver, brass, 14K gold, 1.25" × 1.25" × 1.25".
Photo: Walker Montgomery.

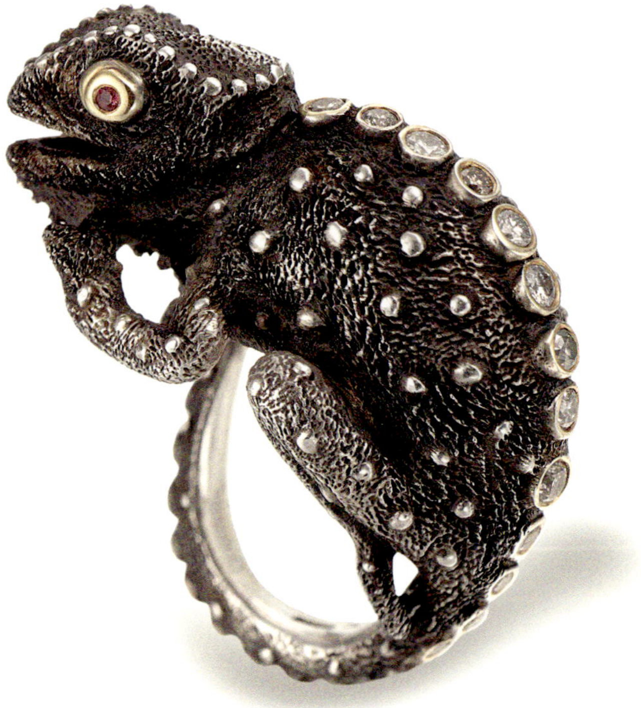

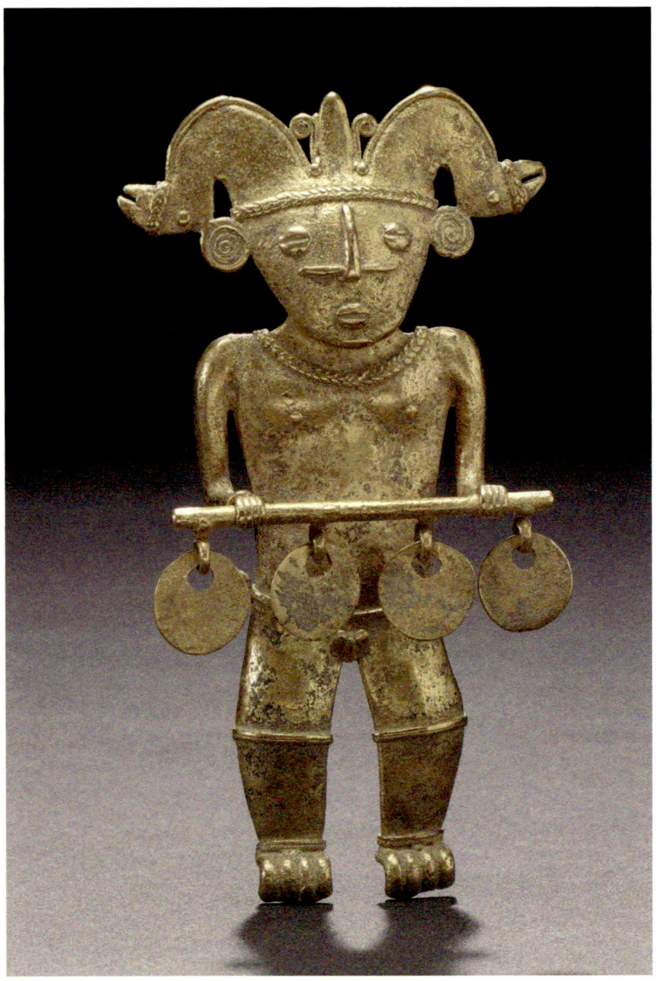

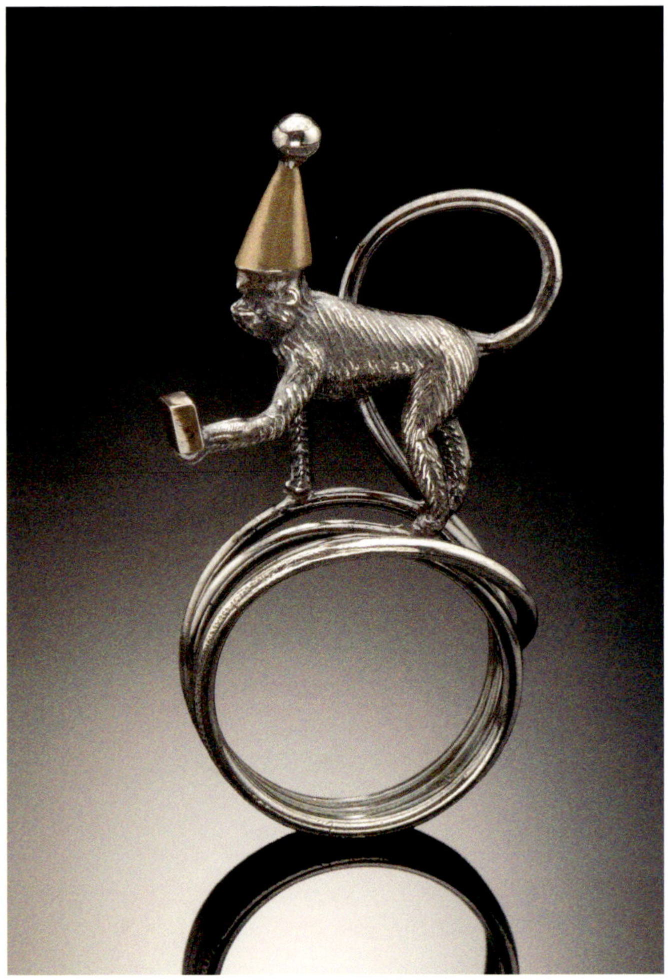

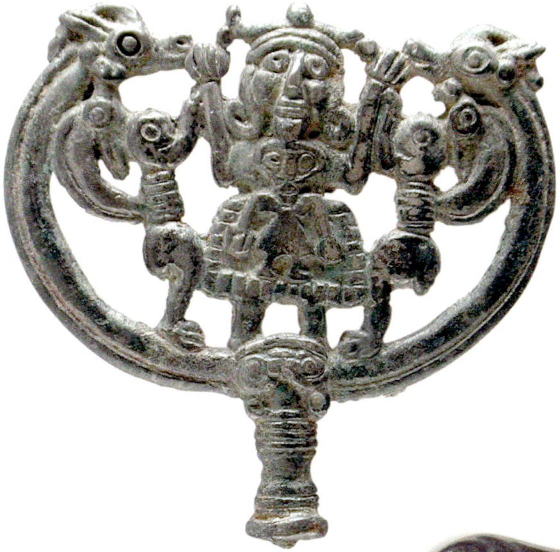

Left Openwork Pinhead, ca. 1000 to 650 BCE. Luristan, Iran, bronze, 2.75" × 2.75".
Courtesy of LACMA, Gift of The Ahmanson Foundation, www.lacma.org, LACMA public domain.

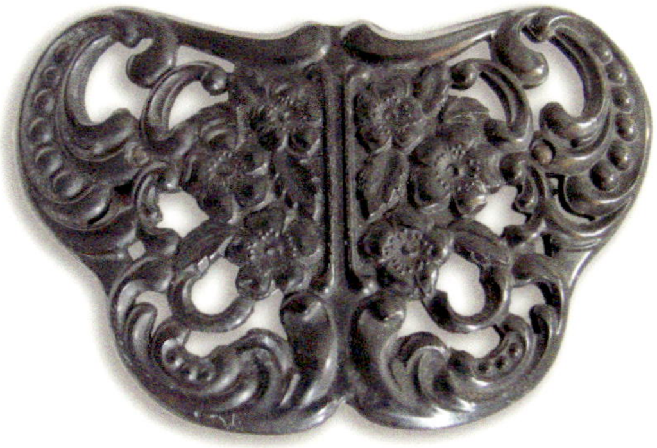

Center Victorian pierced flower brooch, gutta percha, 4" × 2.25".
Courtesy of Colleen Newell.

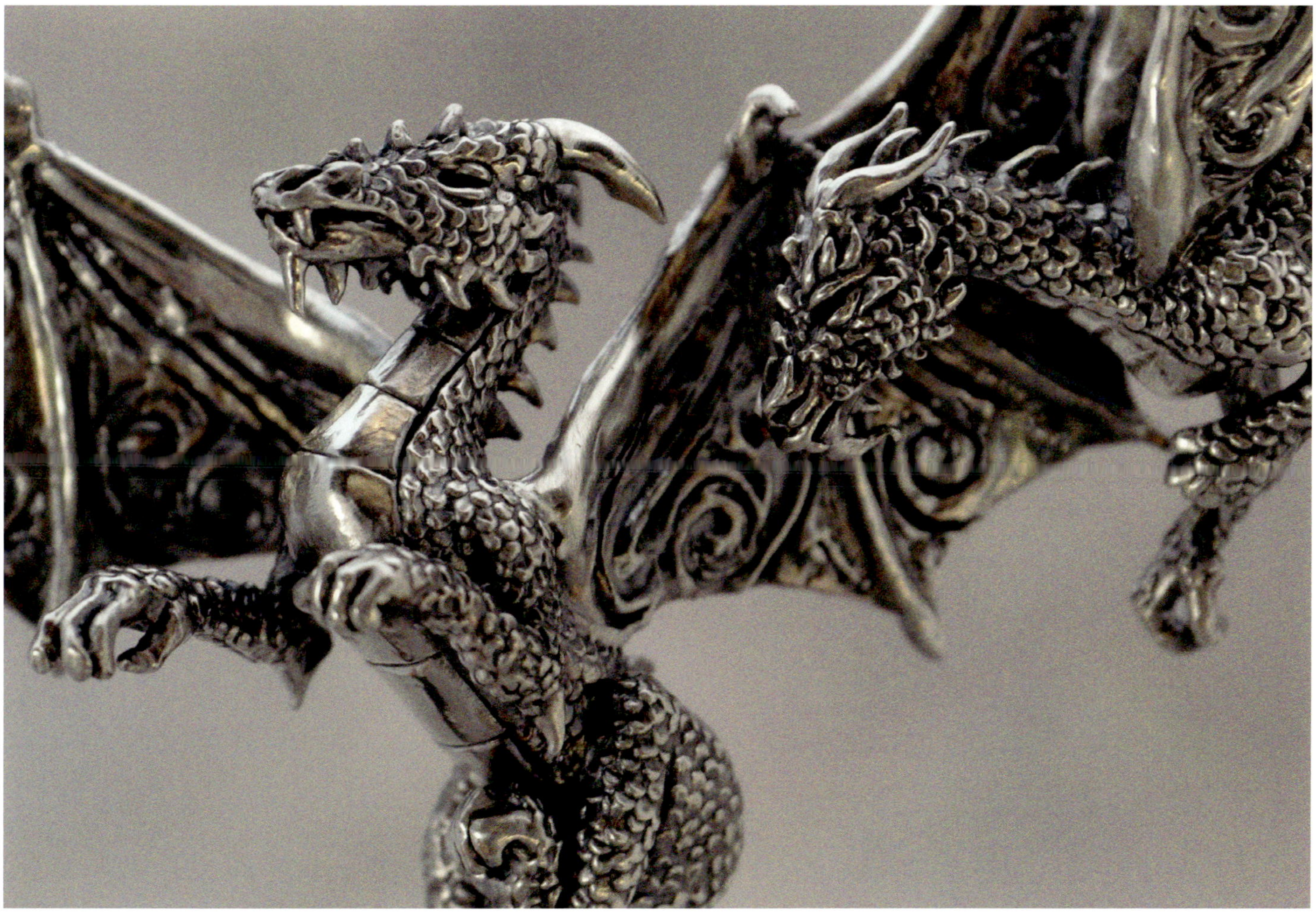

Bottom Sharon K. Berkan-Dent, *Dragon Trio* (detail), 2012. Sterling silver, 1.25" × 1.75" × 0.5".
Photo: Sharon K. Berkan-Dent, Mystic Swan Jewelry.

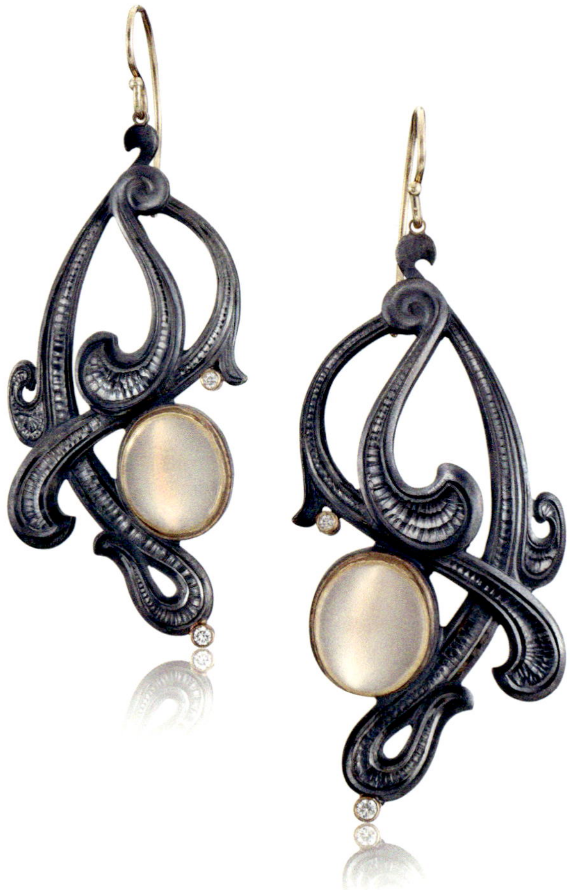

David Giulietti, *Moon Curve Earrings*, 2014. Sterling silver, 18K, 22K, diamonds, moonstone, 3" × 1.4" × 0.38". Photo: Hap Sakwa.

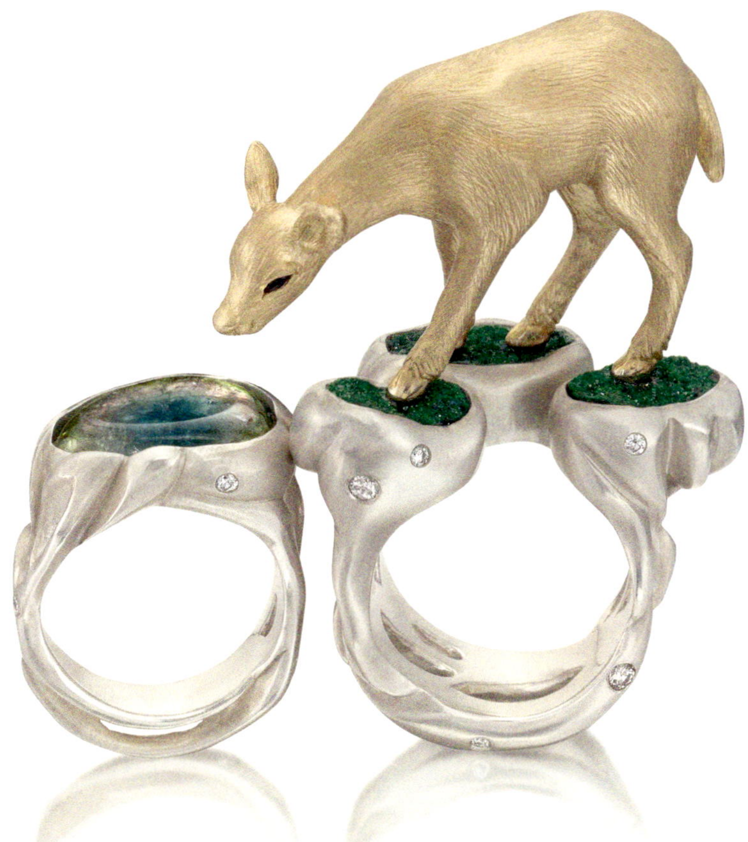

Manya Tessler, *Deer in the Forest* (ring set), 2012. 18K gold, sterling silver, uvarovite druzy, garnet, diamonds, onyxes, tri-color tourmaline, 2" × 2.5" × 0.75".
Photo: Hap Sakwa.

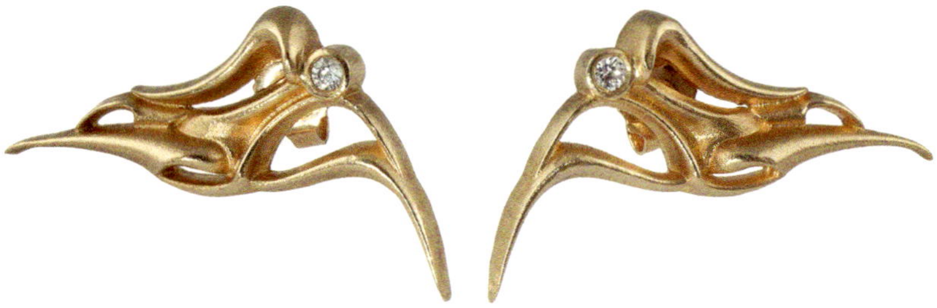

Left Anastasia Tabarovsky, *Ornament Earrings,* 2013.18K gold, diamonds, 0.7" × 0.8" × 0.07". Photo: Anastasia Tabarovsky.

Below Michael Gayk, *Walkers*, 2011. Cast silicone, dyed photopolymer, 5" diameter. Photo: Zan Maddox.

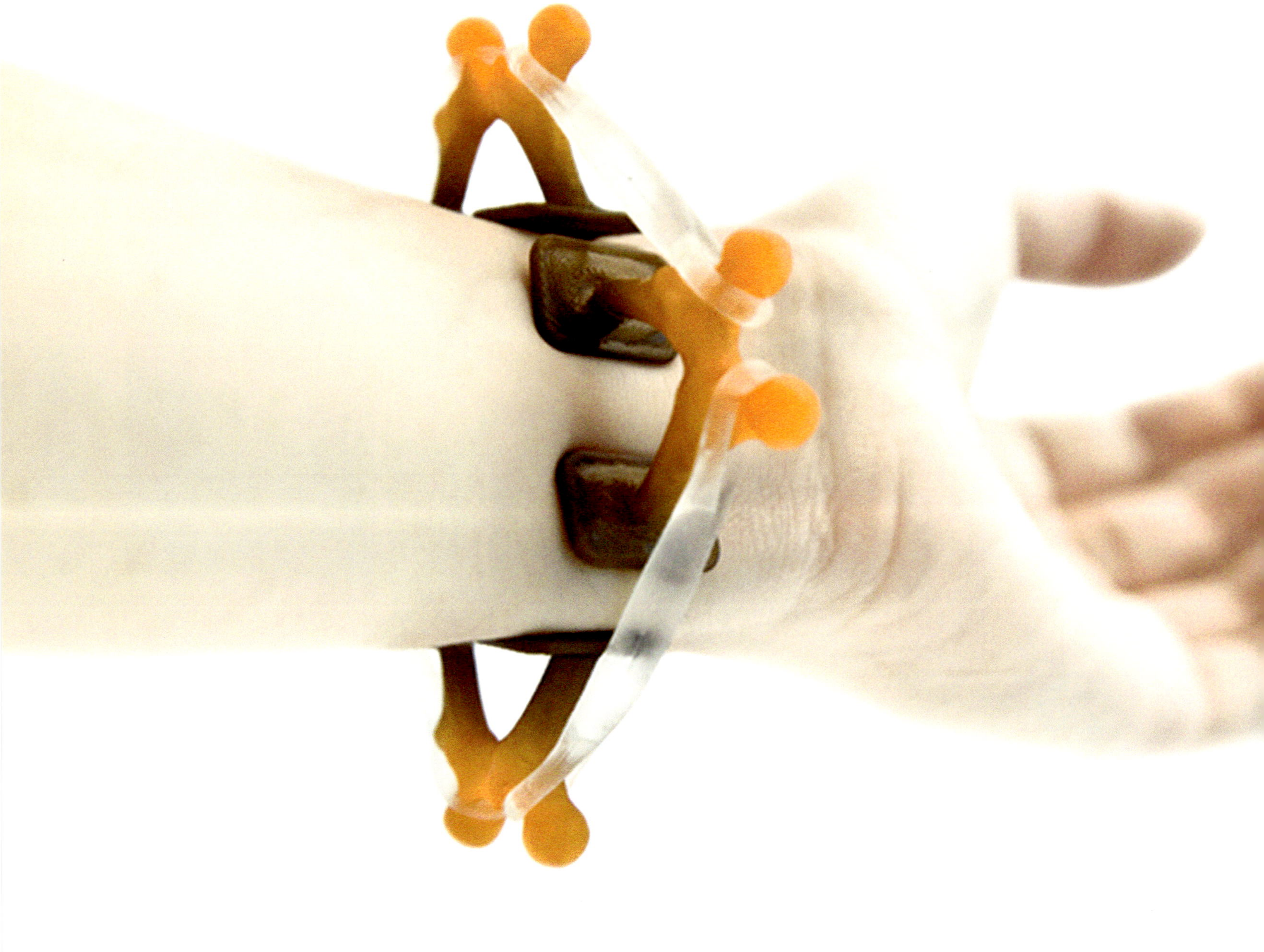

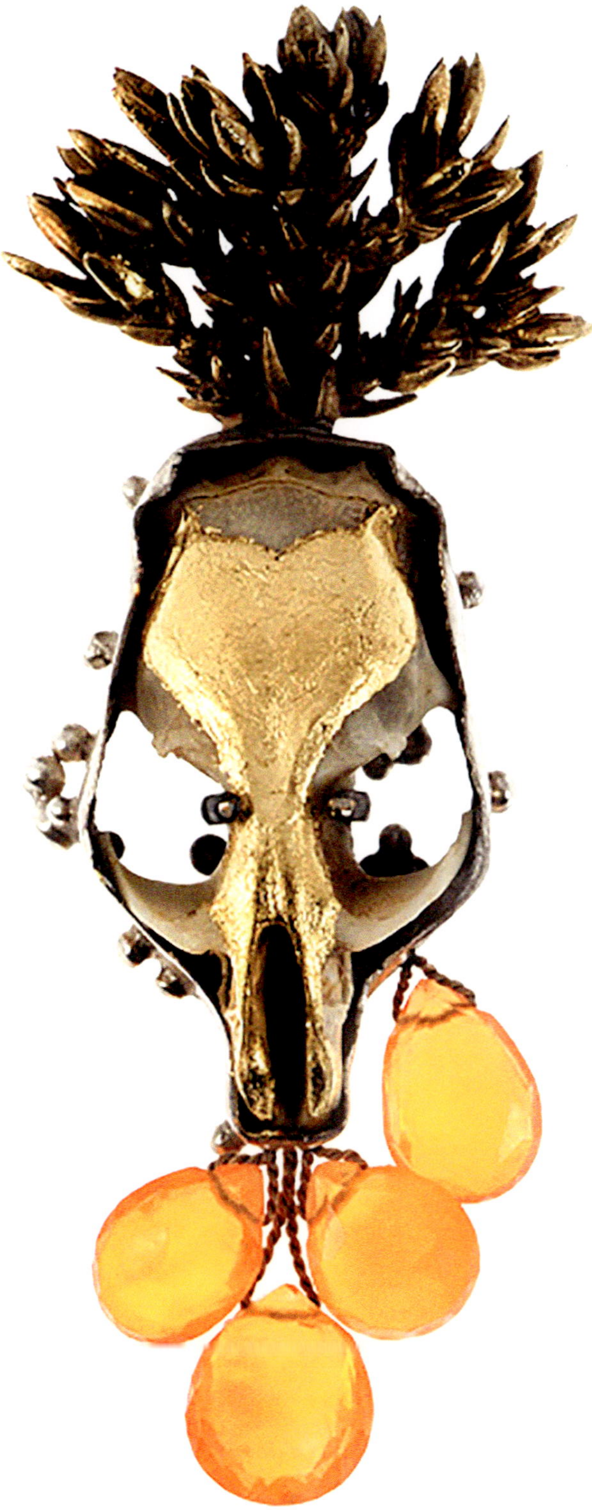

Anna Johnson, *Solstice Brooch*, 2014, Rodent skull, fire opals, cast ancient bronze, ground cedar, 23K gold leaf, fine silver, sterling silver, 3" × 1.5" × 0.75". Photo: Mary Vogel.

(IM)MATERIAL

Art is an engagement with the material world and its continuous transformative energy as well as the immaterial world of emotions and ideas...

—Marc Quinn*

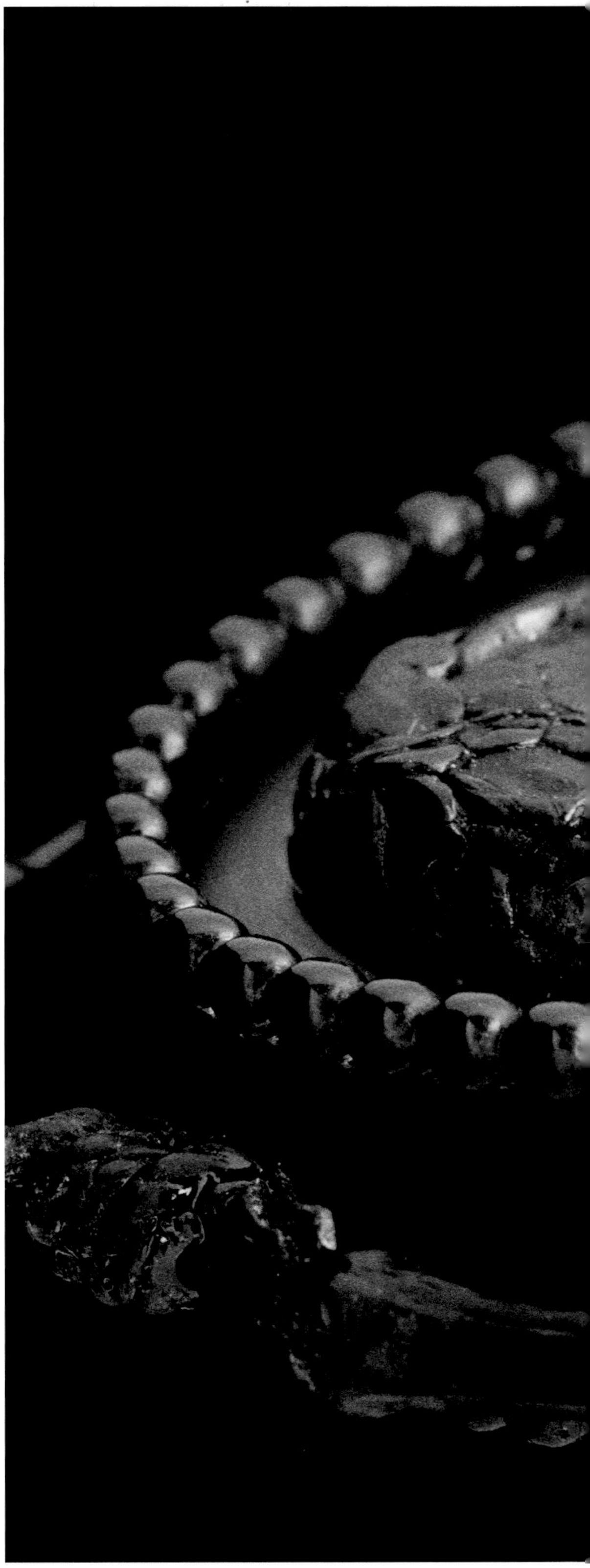

*"Artist's Body" Series 1990–present," Web, accessed July 29, 2016, http://marcquinn.com/artworks/artists-body.

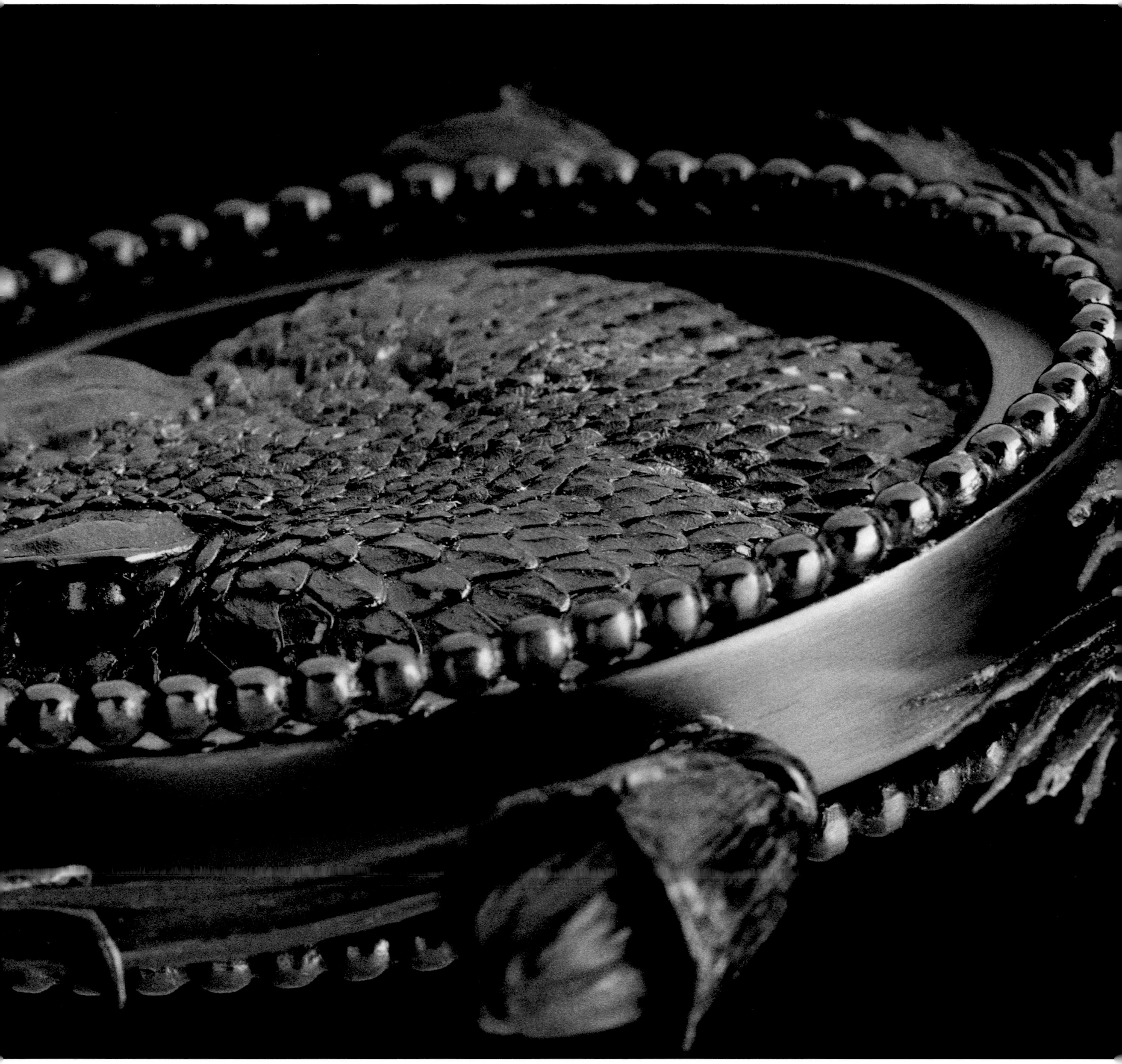

Lin Stanionis, *Rapture*, 2011. Urethane resin, 4" × 3" × 0.75".
Photo: Jon Blumb.

The name of this chapter indicates two contrasting ideas. Much of this book has been organized by medium—metal, ceramic, and glass represent the three major castable mediums used in the fine and decorative arts for thousands of years, but this is hardly the whole picture. There are other major players like plaster, plastic, and concrete, just to name a few. On one hand, there are unique traits to the castings of each of the major mediums, making their materiality worthy of discussion.

On the other hand, there is value to looking at casting holistically. There are shared traits in cast objects: they involve some kind of mold, they are typically difficult (or at least impractical) to produce via other methods, and they often display a plasticity of form. In this way, materials become immaterial. As the artist Marc Quinn states, emotions and ideas also live in the realm of the immaterial…what common conceptual threads run through the world of casting? How are these ideas related to the nature of the process?

Elaine A. King's essay Casting: Unbounded in an Expanded Age introduces this chapter. The two galleries that follow explore the realm of the cast object. The first, Gallery: Spectrum, is a sampling of fascinating objects cast in a variety of materials.

The second gallery, Unconventional Approach, spotlights work that is pushing boundaries in casting. From salt to blood, artists are using any and every material that can be solidified in a mold. This gallery gives us a glimpse at the avant-garde and the experimental. Some of the works are conceptual and weighty, some play with senses other than sight and touch, and some are ephemeral, but they are all thought provoking.

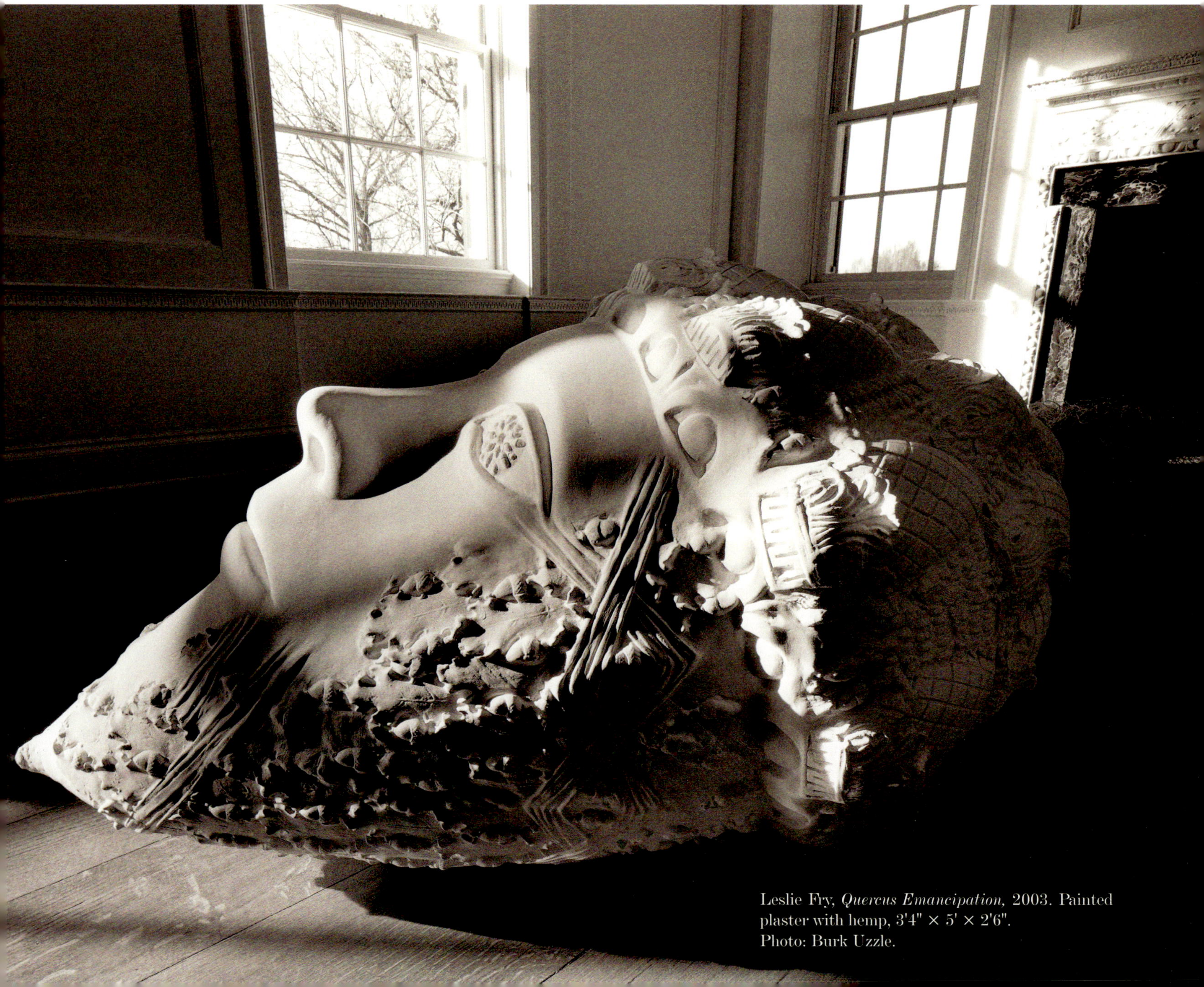

Leslie Fry, *Quercus Emancipation*, 2003. Painted plaster with hemp, 3'4" × 5' × 2'6". Photo: Burk Uzzle.

CASTING
Unbounded in an Expanded Age

by Elaine A. King

Rosalind Krauss was one of the first critics to map the shifts occurring in Postmodern art and what would continue to evolve into this era. In her seminal essay, *Sculpture in the Expanded Field* (1979), she theorized that in this expanded field, the critical operations that have accompanied postwar American art have largely worked in the service of this alteration. The categories of sculpture, painting, and craft have been transformed in past decades in an astonishing manner, demonstrating that art as a cultural manifestation can be just about whatsoever and made from anything.

The tolerance for multi-faceted approaches and media is expressed by Dipti Desai: "Many contemporary artists, who have blurred the boundaries between not only high art and low art, but also art and other disciplines in the social sciences and humanities, have challenged both artistic process and artistic authority."[1] This recognition of mixing art forms was demonstrated in the exhibition *Cast From Life* held at the Skarsted Gallery in the fall of 2014. It showcased works by an assortment of artists including George Condo, Peter Fischli and David Weiss, Isa Genzken, Robert Gober, Mark Grotjahn, Rachel Harrison, Mike Kelley, Martin Kippenberger, Jeff Koons, Juan Muñoz, Cady Noland, Ugo Rondinone, Thomas Schütte, Rebecca Warren, and Franz West. Its organizers stated, "*Cast From Life* loosely interprets the process of casting and examines the ways that sculptors use various forms to mold meaning. Delving into the materiality of contemporary sculpture, the exhibition creates juxtapositions between a wide range of today's sculptors, drawing comparisons in their methodology."[2]

Today, artists make ambitious works through a combination of materials, methods, concepts, and subjects that test conventional boundaries and defy easy definition. Varied and wide-ranging, contemporary art is famed for its very lack of a homogeneous unifying principle or dogma. In our diverse world, artists impart ideas through whatever techniques they deem appropriate; this includes video projections, text, found objects, debris, blood, wrecked furniture, live body performances, multi-media casting, community intervention, and beyond.

Nevertheless, many artists continue to embrace and employ methods connected to time-honored materials and techniques. Many working today share a crucial link to traditional modern art that can be traced back to Auguste Rodin and the technique of bronze casting. In the late nineteenth century, he brought sculpture into the modern age and "restored an ancient role of sculpture—to capture the physical and intellectual force of the human subject."[3]

Rodin preferred working with the foundry established by Alexis Rudier to sand cast his work rather than carrying out that process in his own studio. Today, many artists use independent foundries to materialize their ideas. Two prominent facilities are Polich Tallix in Rock Tavern, New York, and the Modern Art Foundry in Astoria, Queens. Both have become hubs for artists to create ambitious cast objects or works fabricated in metal. The works of Matthew Barney, Ashley Bickerton, Frank Benson, Louise Bourgeois, Eric Fischl, Rona Pondick, and Ursula von Rydingsvard illustrate the varied spectrum of what these foundries produce. Although their works are produced at the foundry, artists play a hands-on role in the production of their sculptures, working closely with superb craftspeople who understand not only material and process, but the artists' concepts as well.

GALLERY
Spectrum

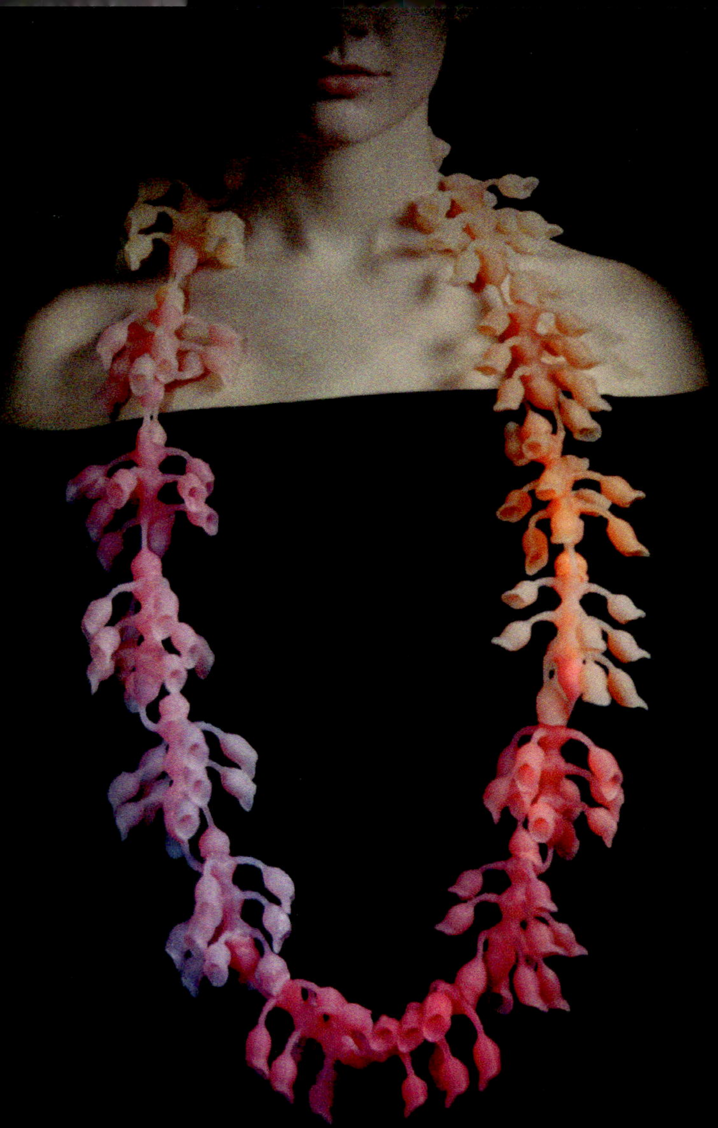

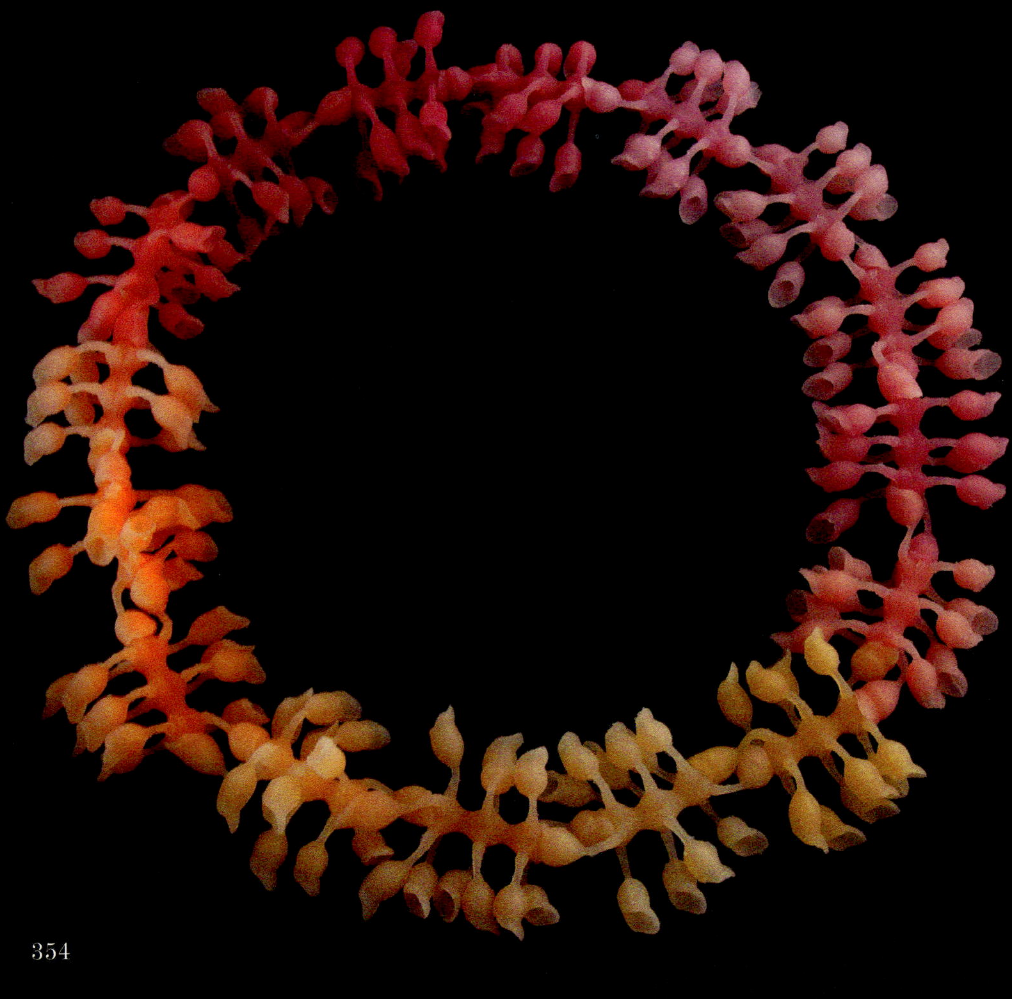

Thomas Anthony Lauria, *Heatseeking Network*, 2014 (two views). Pigmented silicone rubber. 1.5" × 15" × 15". Photos: Thomas Anthony Lauria.

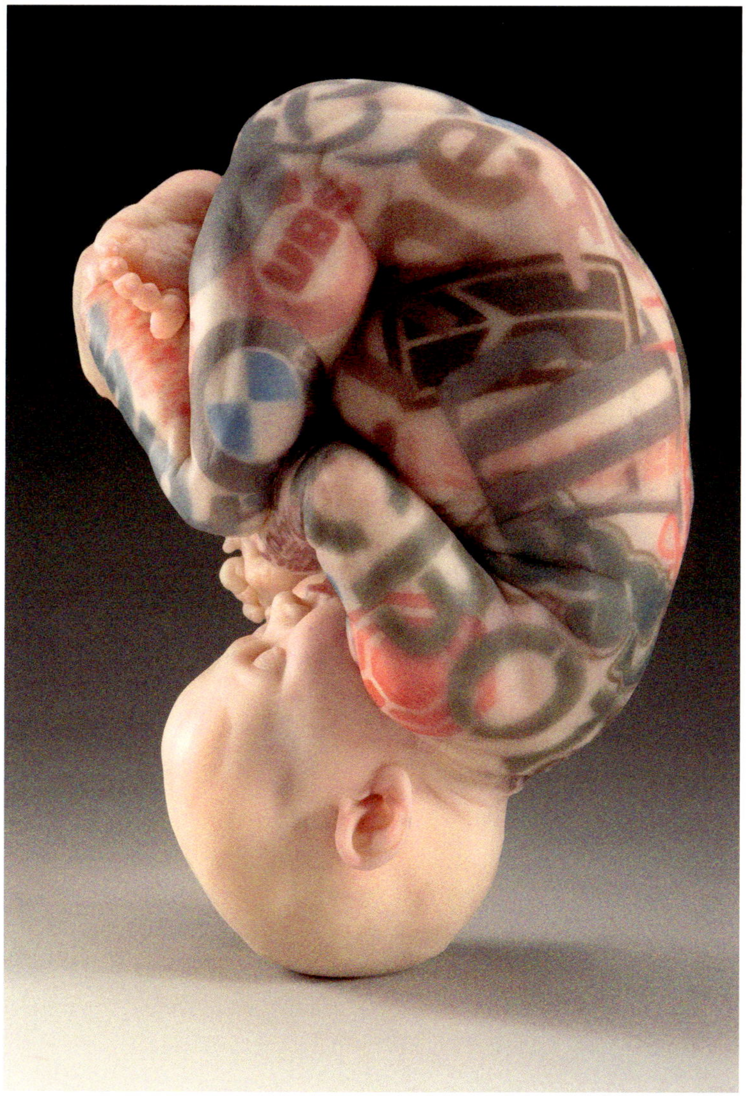

Dietrich Wegner, *Cumulous Brand, Bill*, 2008.
Silicone, urethane foam, 8" × 8" × 12".
Photo: Dietrich Wegner.

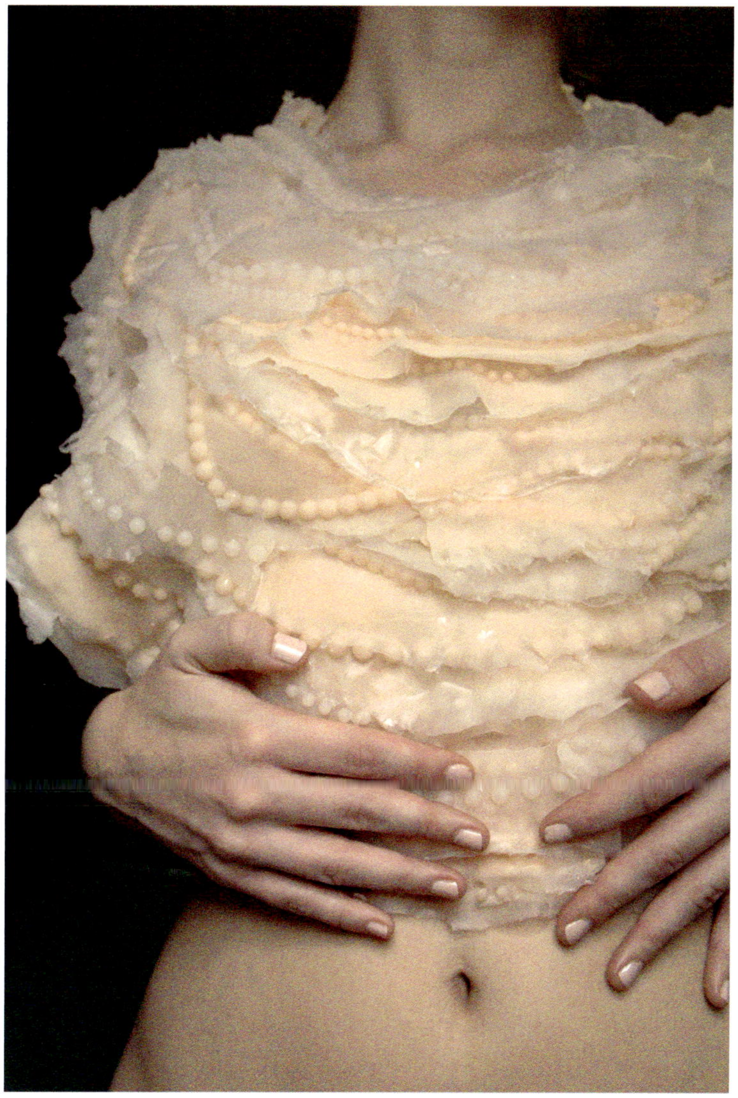

Stephanie Voegele, *Pearls and Ruffles*, 2012.
Silicone rubber, pigment, fresh water pearls, 12" × 3" × 24".
Photo: Sarah Laurentius Ellis.

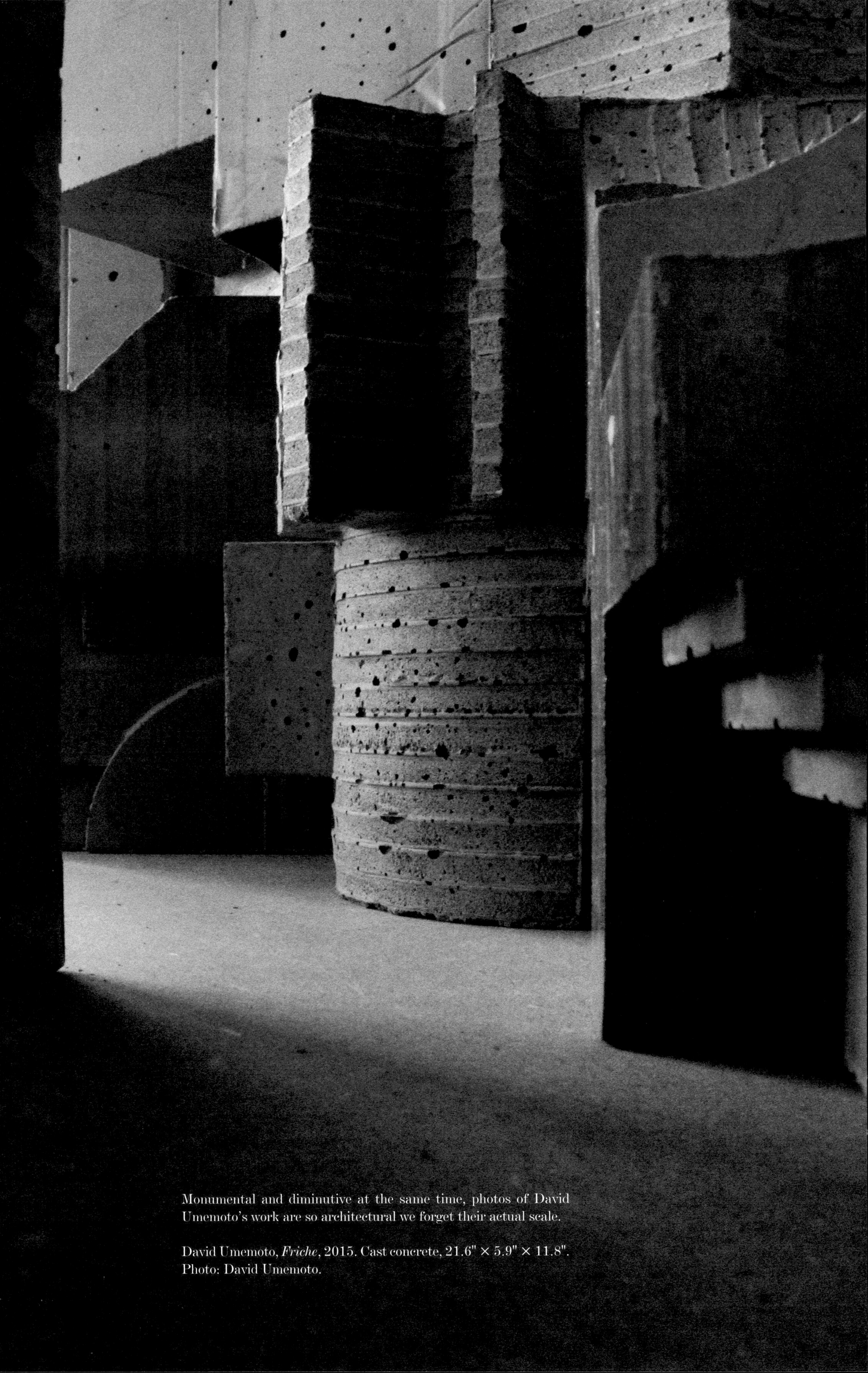

Monumental and diminutive at the same time, photos of David Umemoto's work are so architectural we forget their actual scale.

David Umemoto, *Friche*, 2015. Cast concrete, 21.6" × 5.9" × 11.8". Photo: David Umemoto.

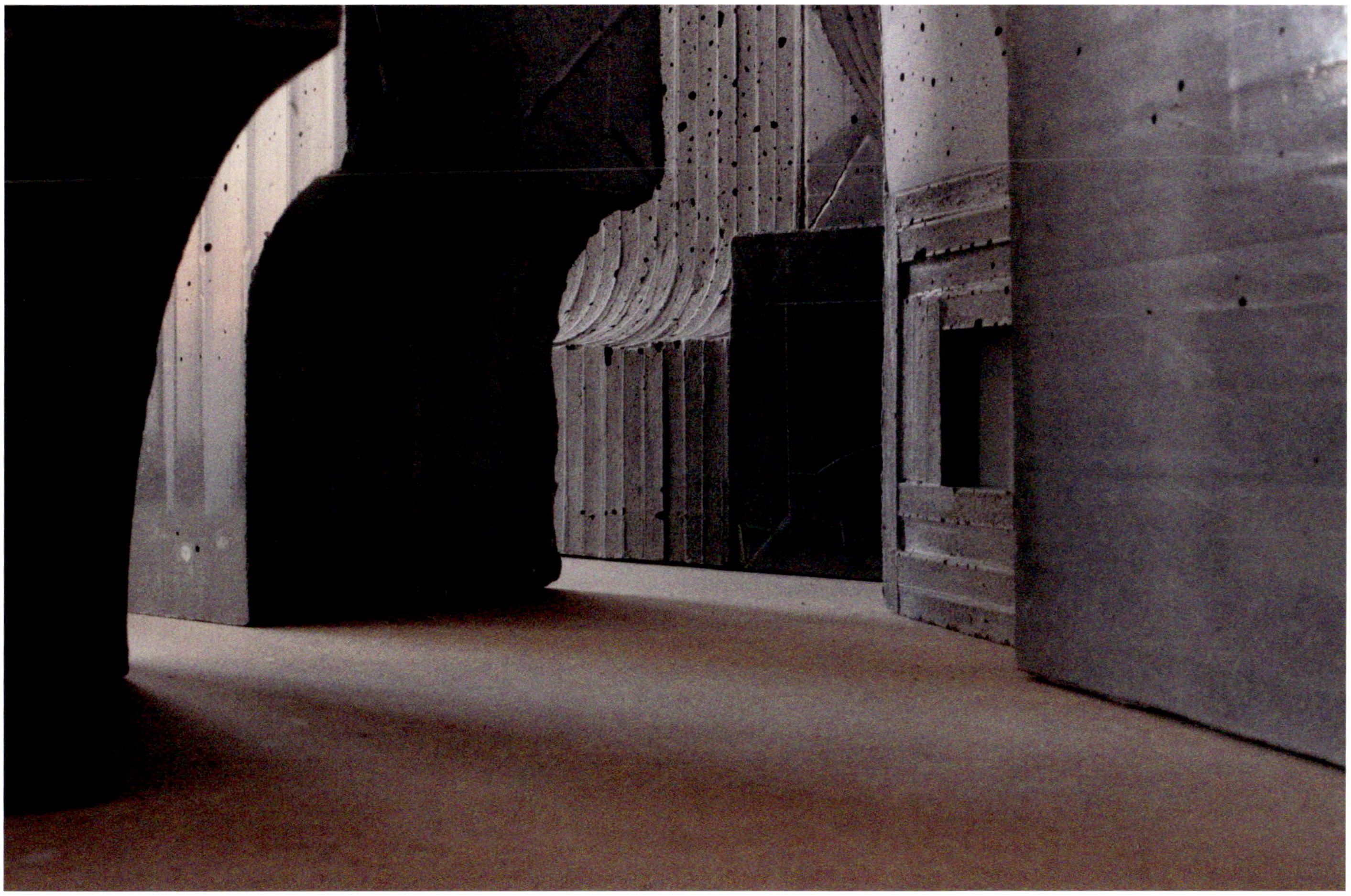

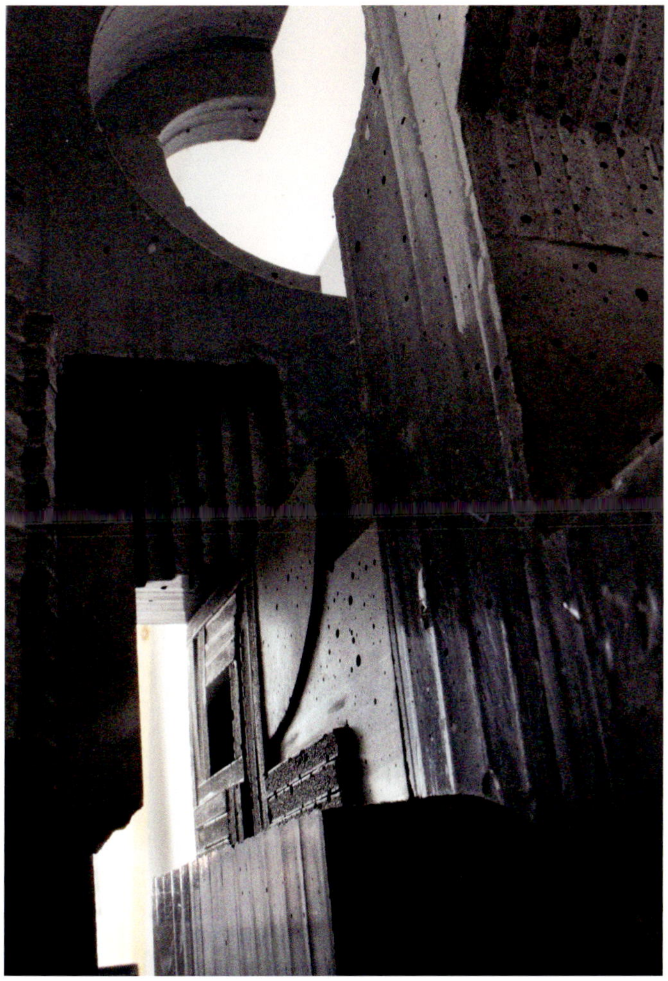

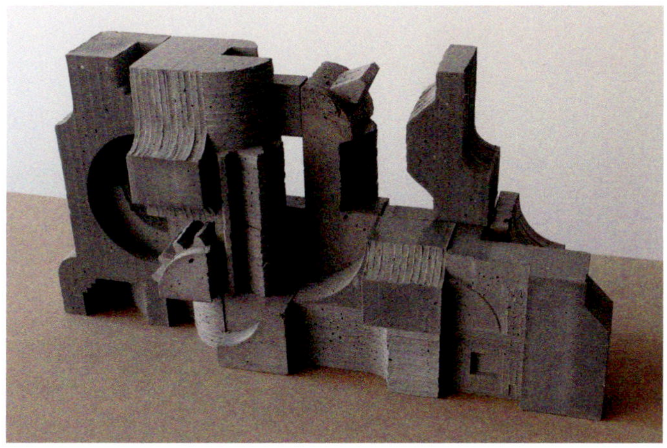

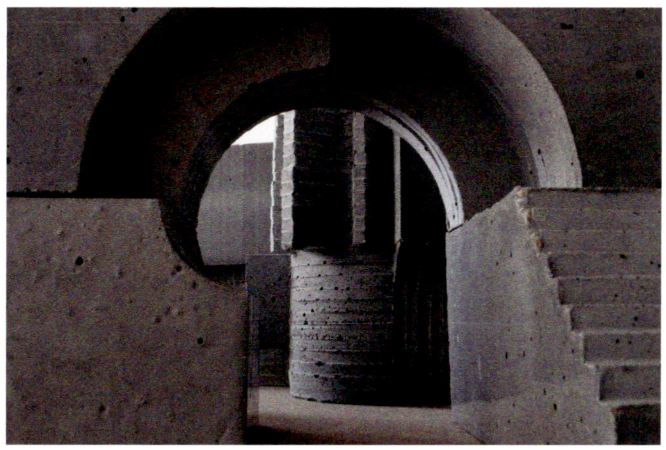

Details David Umemoto, *Friche*, 2015.
Photos: David Umemoto.

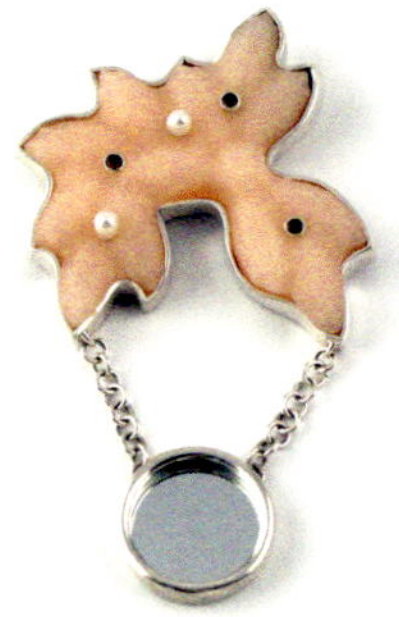

Left Jill Baker Gower, *Fleshgem Brooch #6*, 2015. Silicone rubber, sterling silver, mirror, garnets, pearls, stainless steel, 3.25" × 2" × 0.375". Photo: Jill Baker Gower.

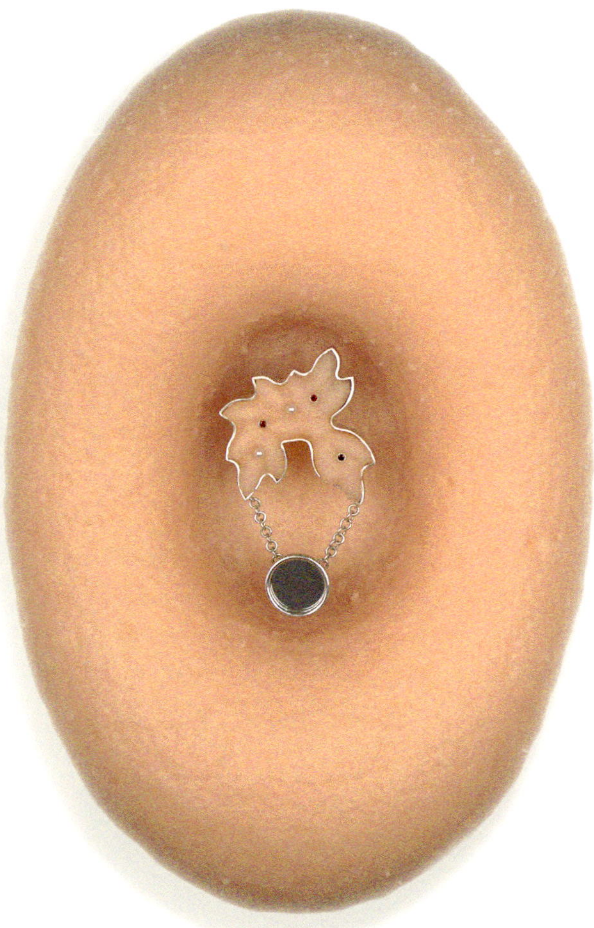

Center left Jill Baker Gower, *Fleshgem Brooch #6* (with wall mount), 2015. Silicone rubber, sterling silver, mirrors, garnets, pearls, stainless steel, wood. 12" × 7.75" × 4.5". Photo: Jill Baker Gower.

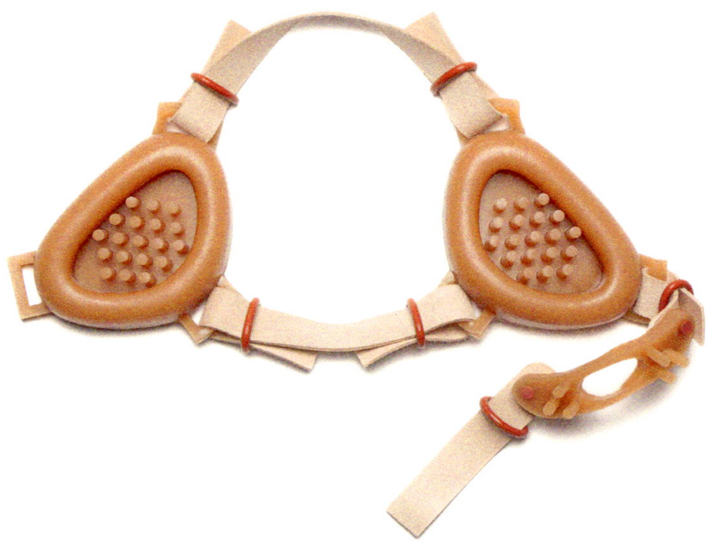

Center right Emily Culver, *Knismesis Headgear*, 2014. Urethane, leather, rubber, rigid photopolymer, 6.5" × 4.25" × 1". Photo: Emily Culver.

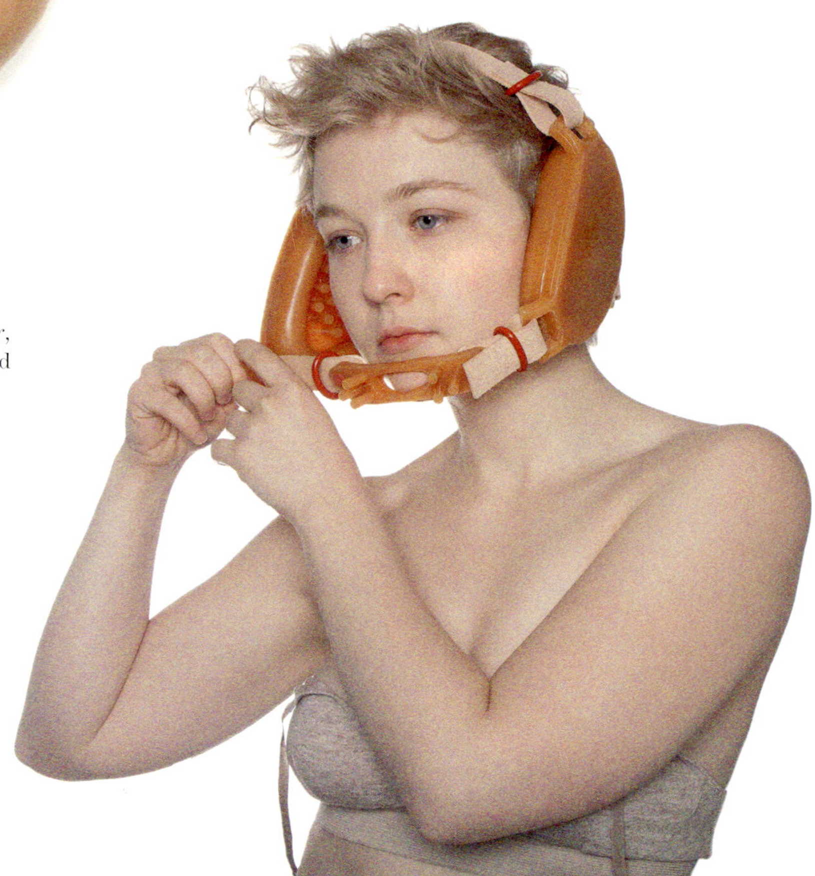

Right Emily Culver, *Knismesis Headgear*, 2014. Urethane, leather, rubber, rigid photopolymer, 6.5" × 4.25" × 1". Photo: Emily Culver.

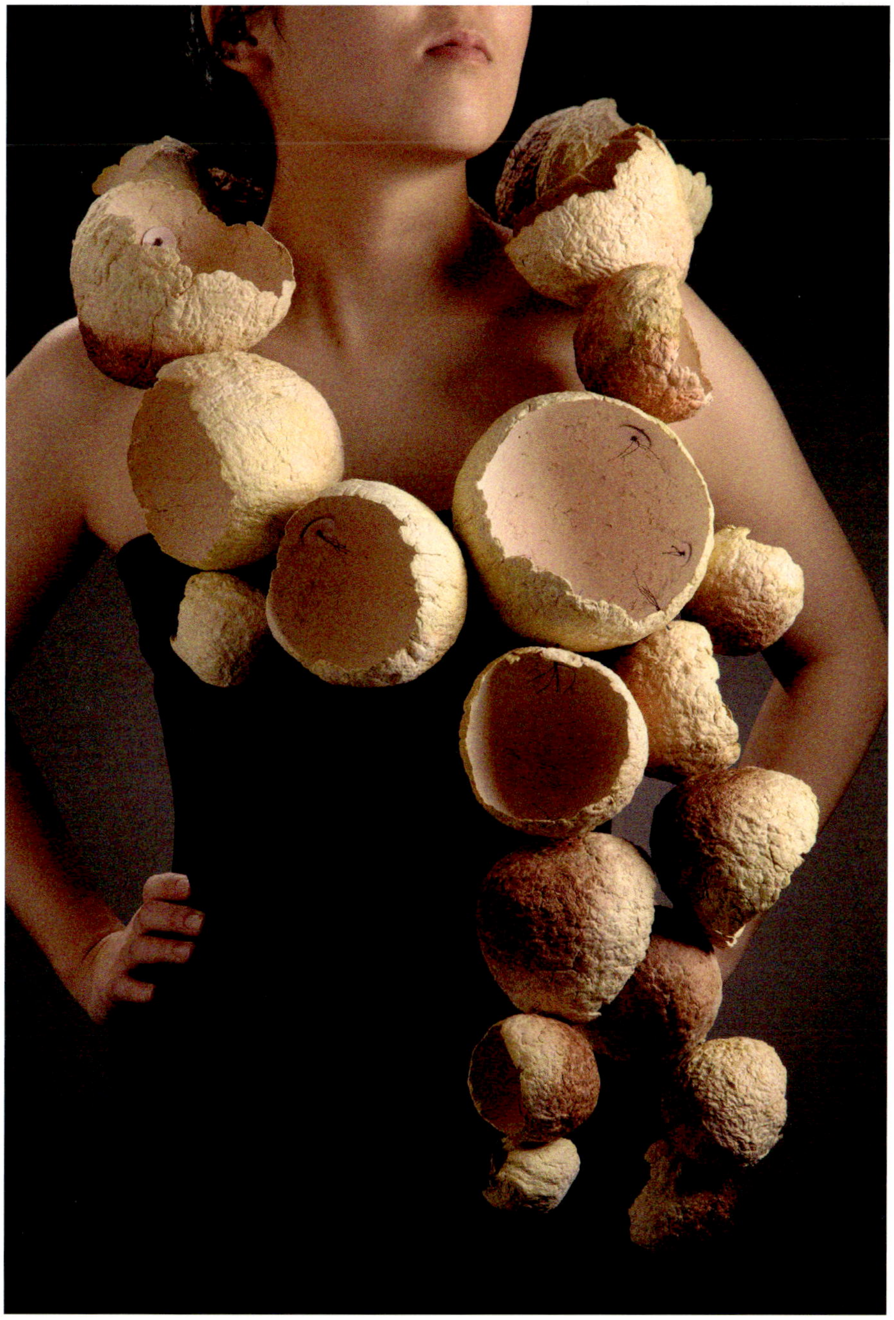

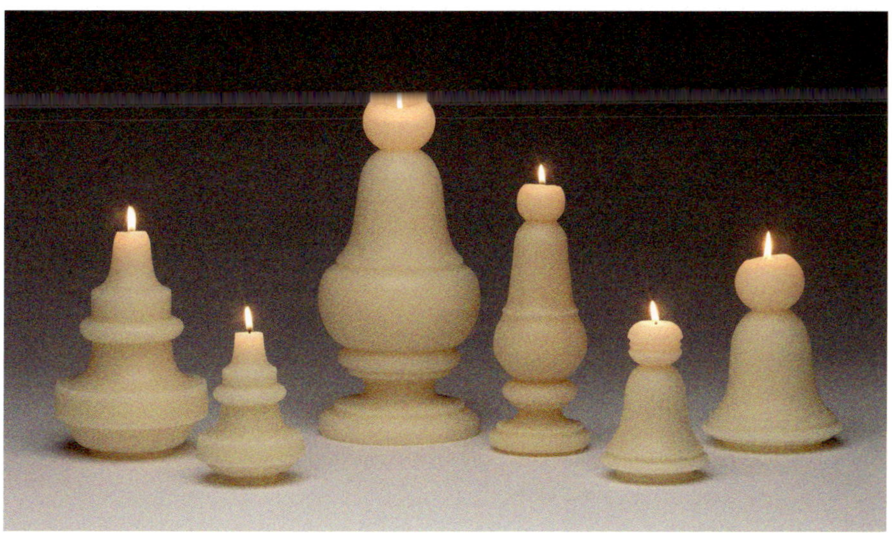

Top Damia Smith, *Apothecia*, 2009. Cast paper pulp, aluminum, copper, 25" × 15" × 18". Photo: Aaron Paden.

Left Niho Kozuru, *NIHO Candles*, 2013. 100% pure beeswax, variable: 3.5" × 3.5" × 5.5" to 6" × 6" × 12.5". Photo: Stewart Clements.

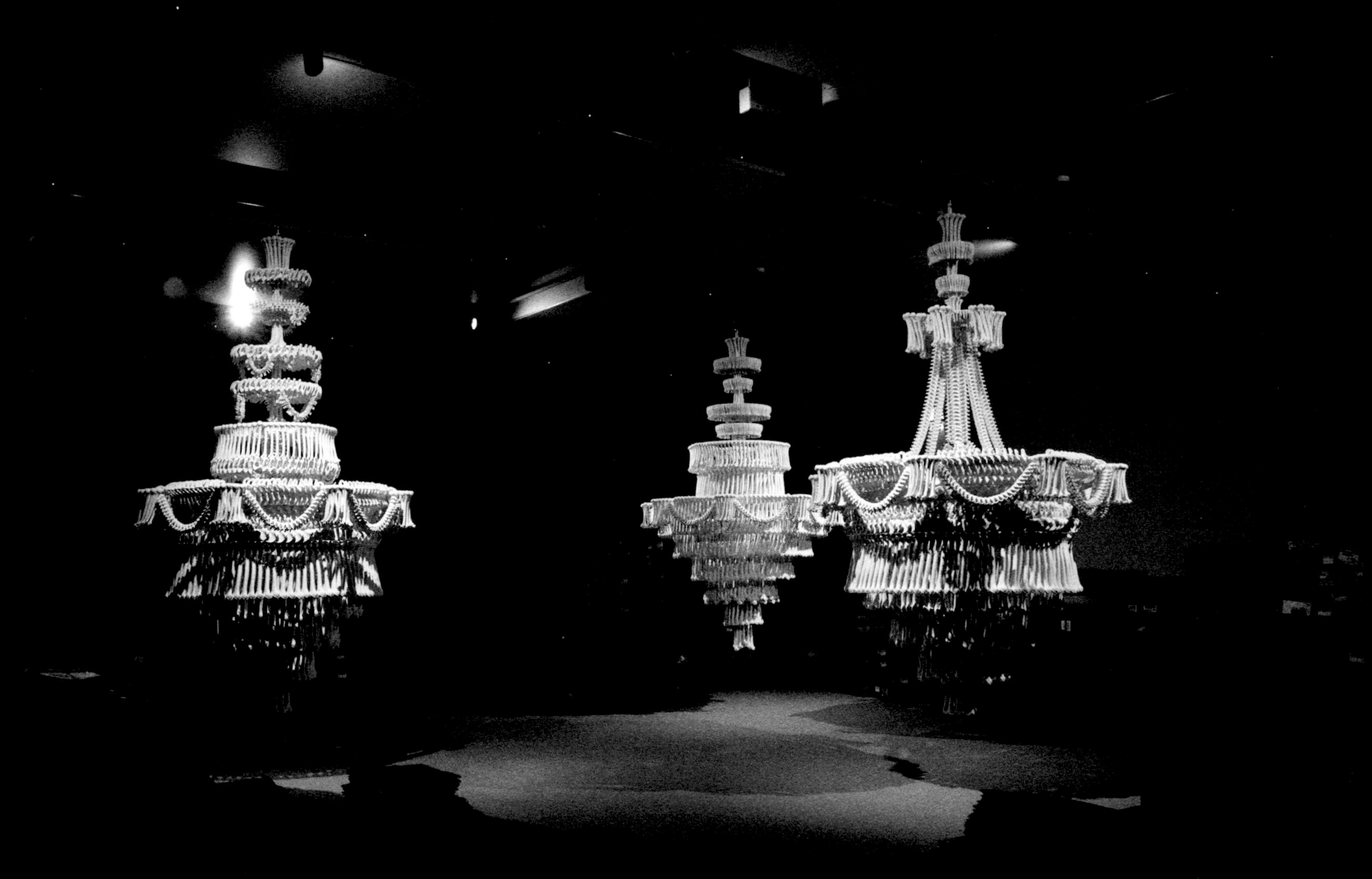

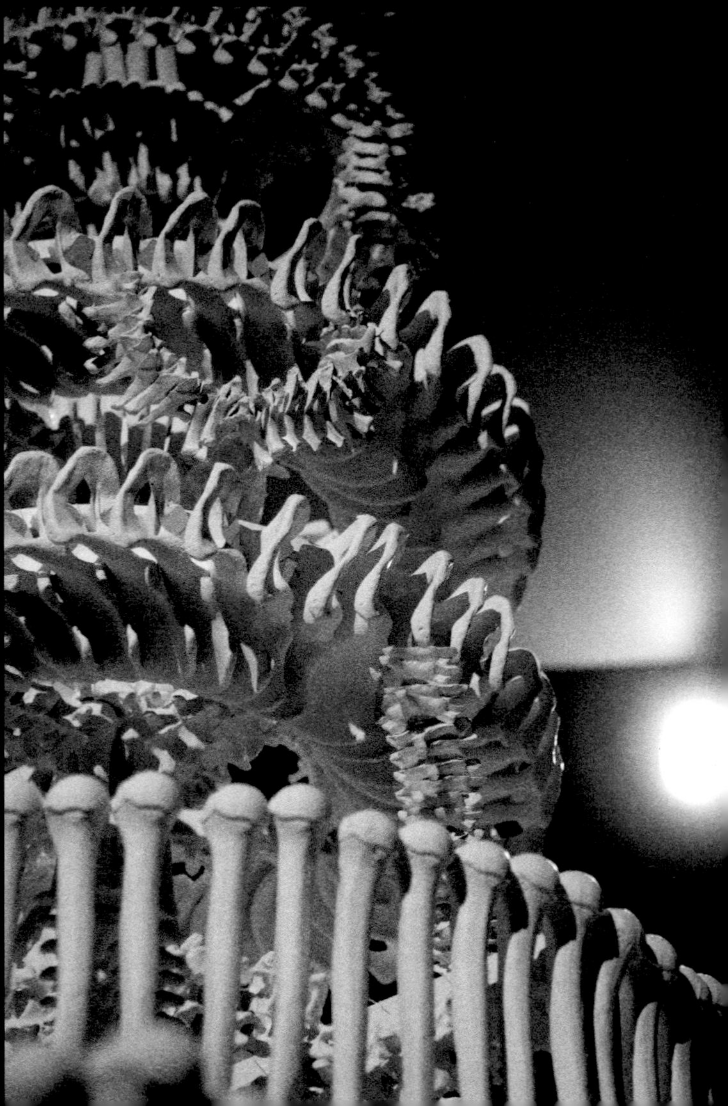

Hung at eye level, these chandeliers are a shocking encounter with our own mortality. The chandeliers are made up of 9,000 hand cast plaster bones total, and weigh more than one ton each.

Jodie Carey, *In the Eyes of Others*, 2009. Plaster, steel, wire, 13'1" × 9'10" × 9'10".
Courtesy of Towner Gallery, UK.

Grace Nickel, *Host* (Installation at the Canadian Clay and Glass Gallery Waterloo Ontario), 2015. Jingdezhen porcelain, metal base, 8'10" × 1'8" × 1'8".
Photos: Michael Zajac.

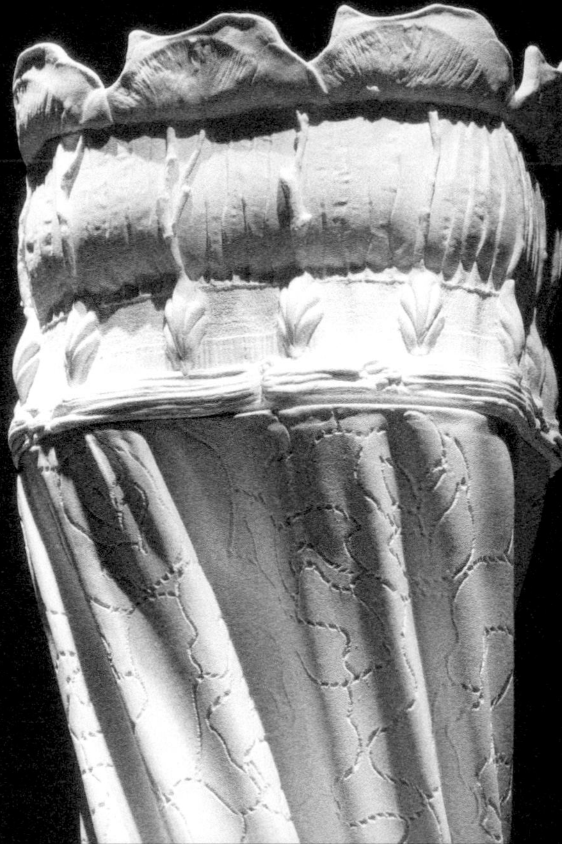

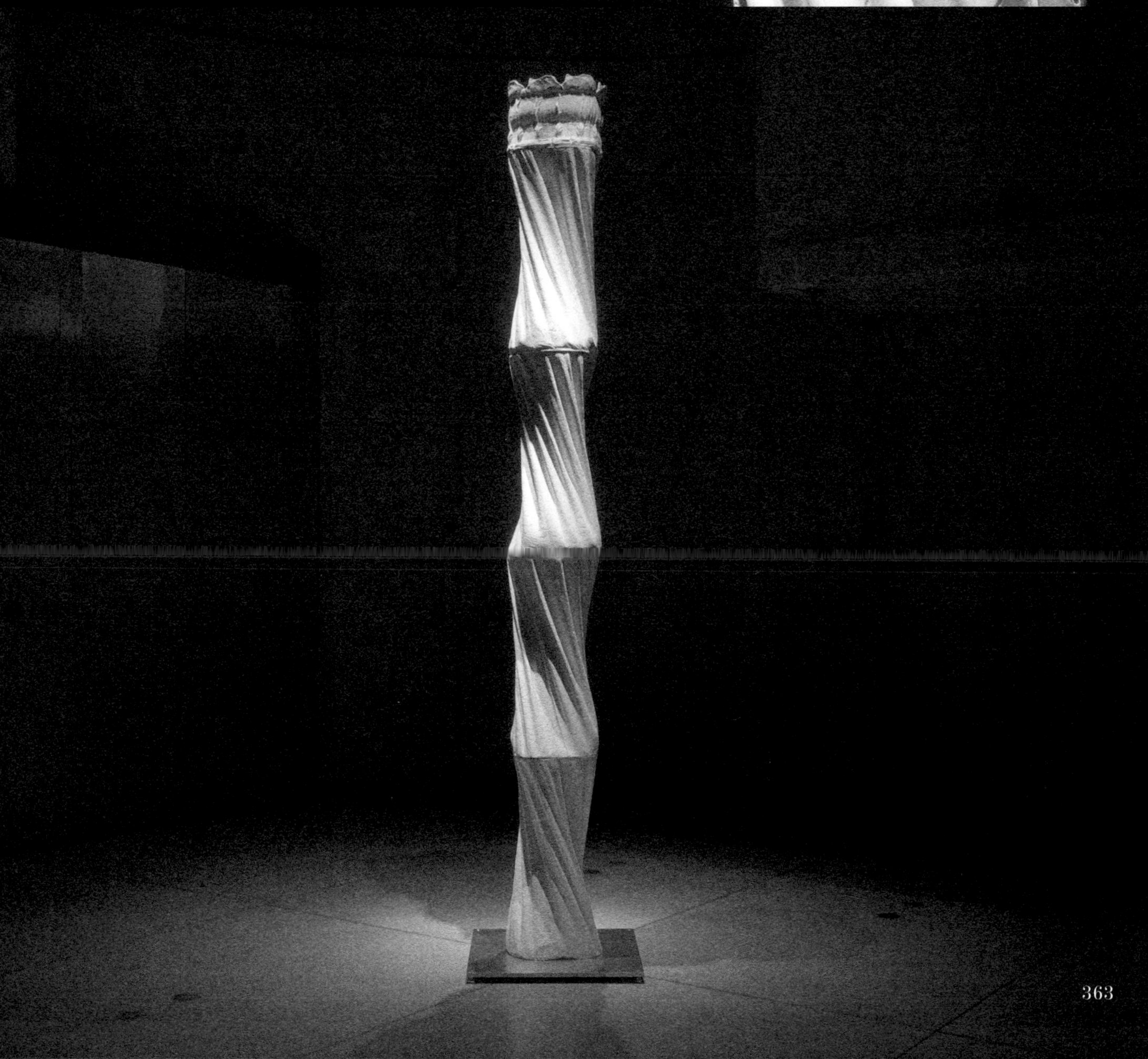

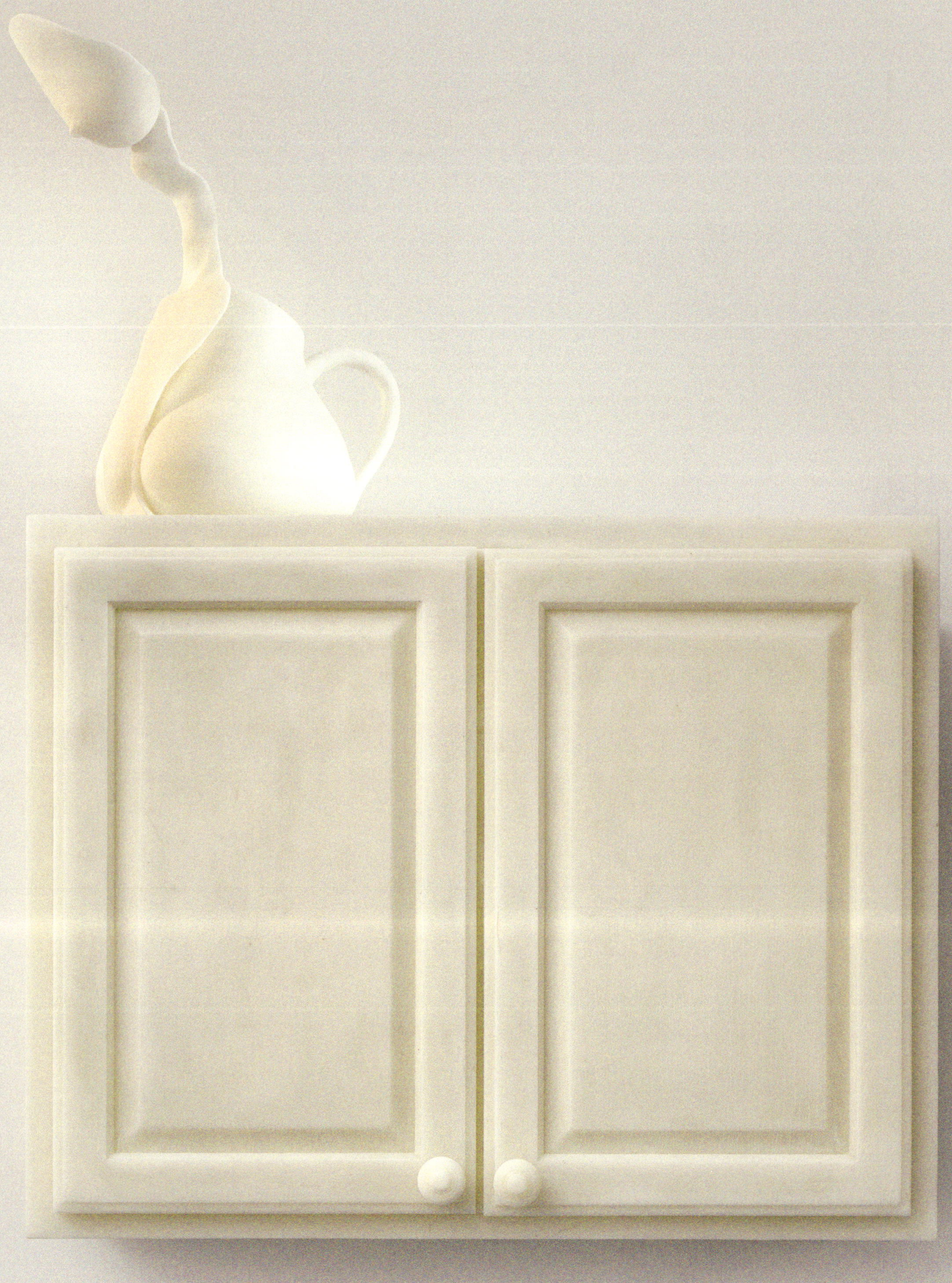

Janine Antoni, *to quench*, 2015. Polyurethane resin, 43" × 30.5" × 14.5". Photo: © Janine Antoni, Courtesy of the artist and Luhring Augustine, New York.

Baldizar artfully uses seams as a design element in her finished works, adding extra dimension and interest.

Right Deborah Baldizar, *Hooded*, 2006. Cast paper, 20" × 8" × 4". Photo: Chuck Pappas.

Bottom Deborah Baldizar, *Edge*, 2006. Cast paper, 4'6" × 2'9" × 4'10". Photos: Chuck Pappas.

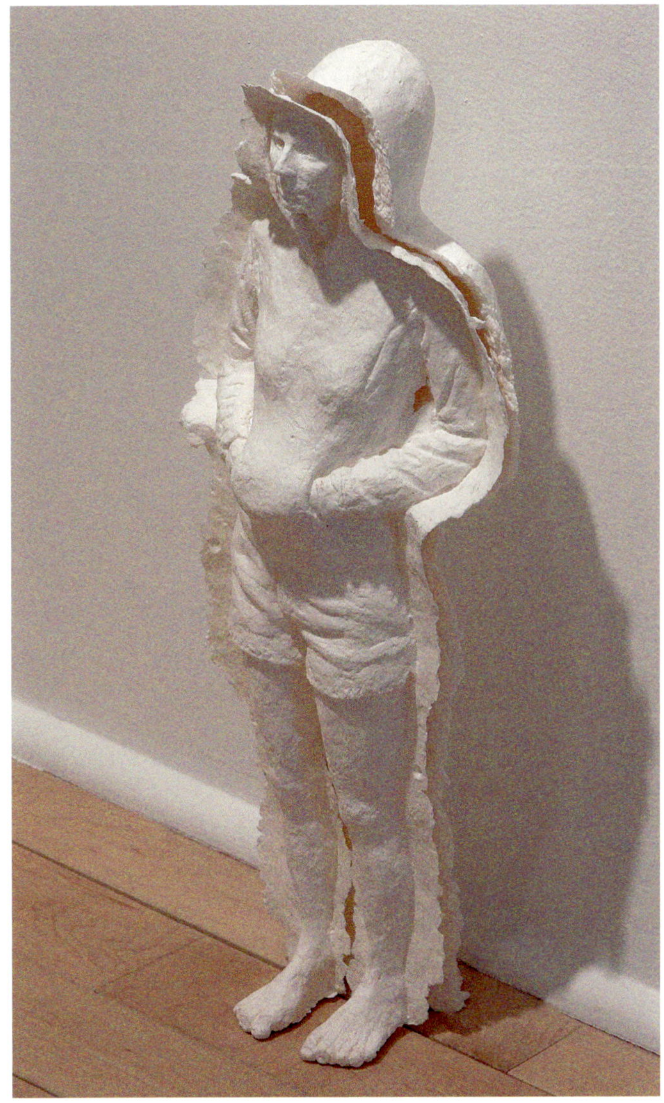

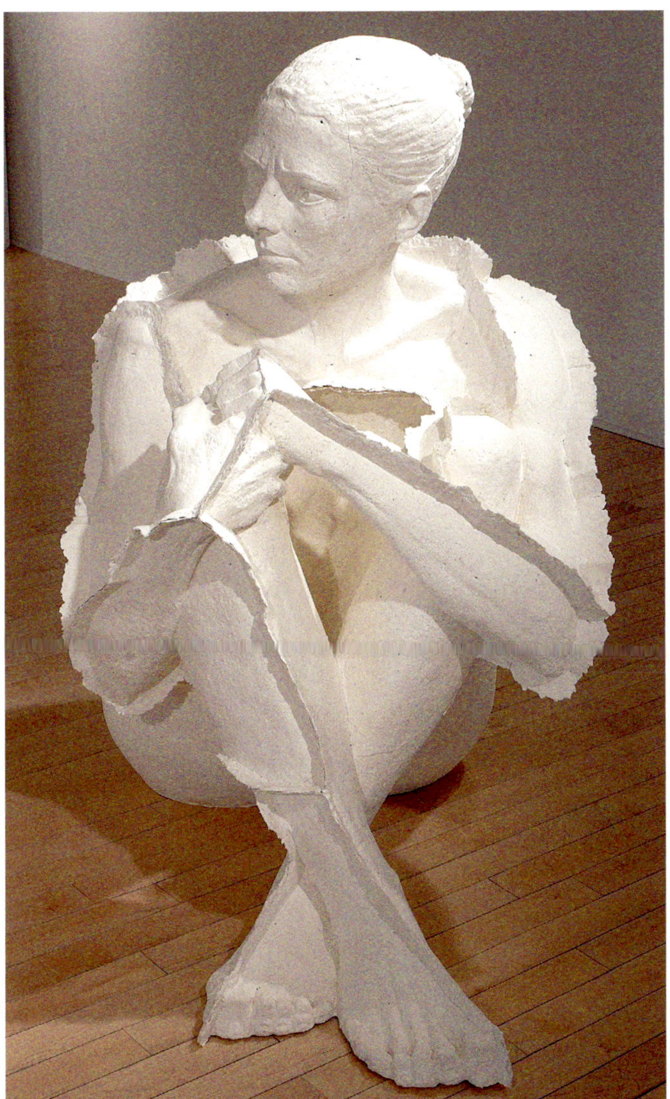

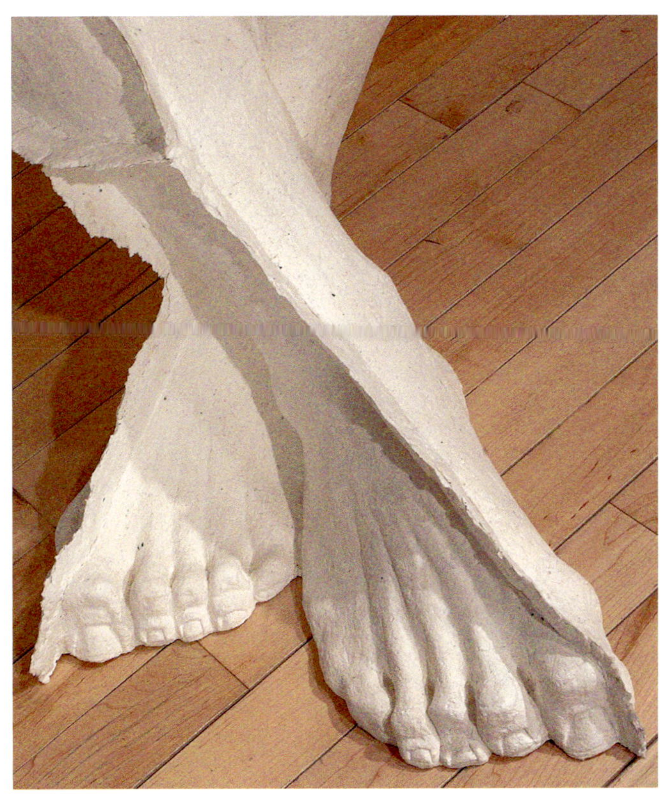

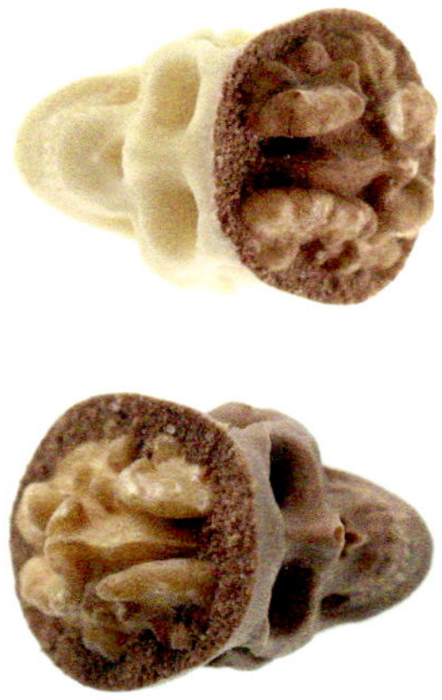

Left Ruth & Sira Garcia (Sparganum), *Chocolate Skulls Gone Nuts*, 2012. Chocolate, walnuts, 2" × 0.55" × 1.25". Photo: Sira García Trigueros.

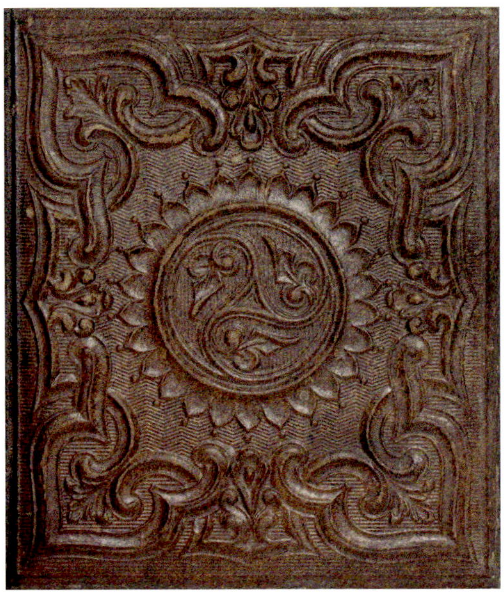

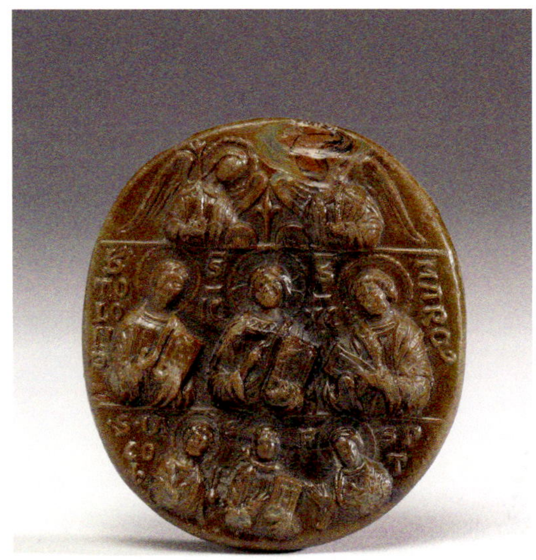

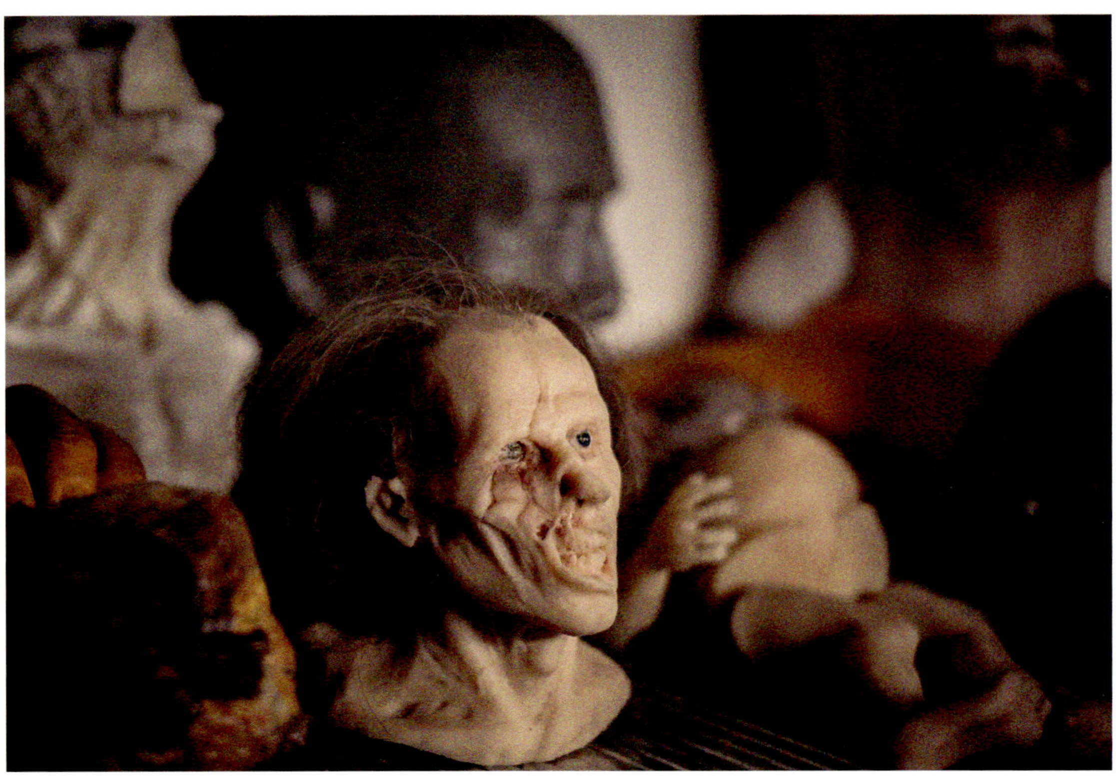

Top Kevin Champeny, *What Remains*, 2012. 35,000+ hand cast urethane flowers, 5' × 4', 75 lbs.
Photos: Kevin Champeny.

Bottom Kevin Champeny, *What Remains* (detail), 2012.

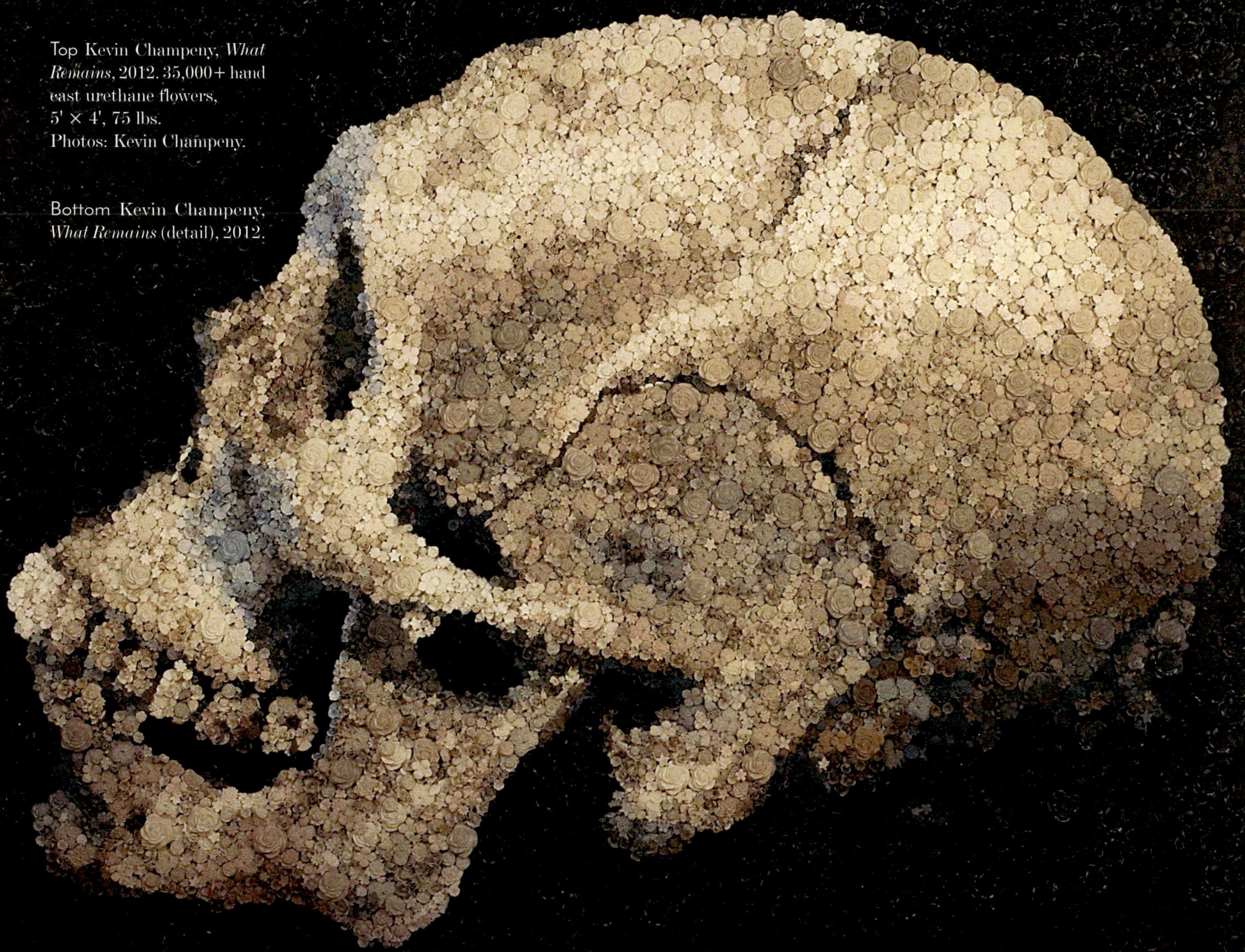

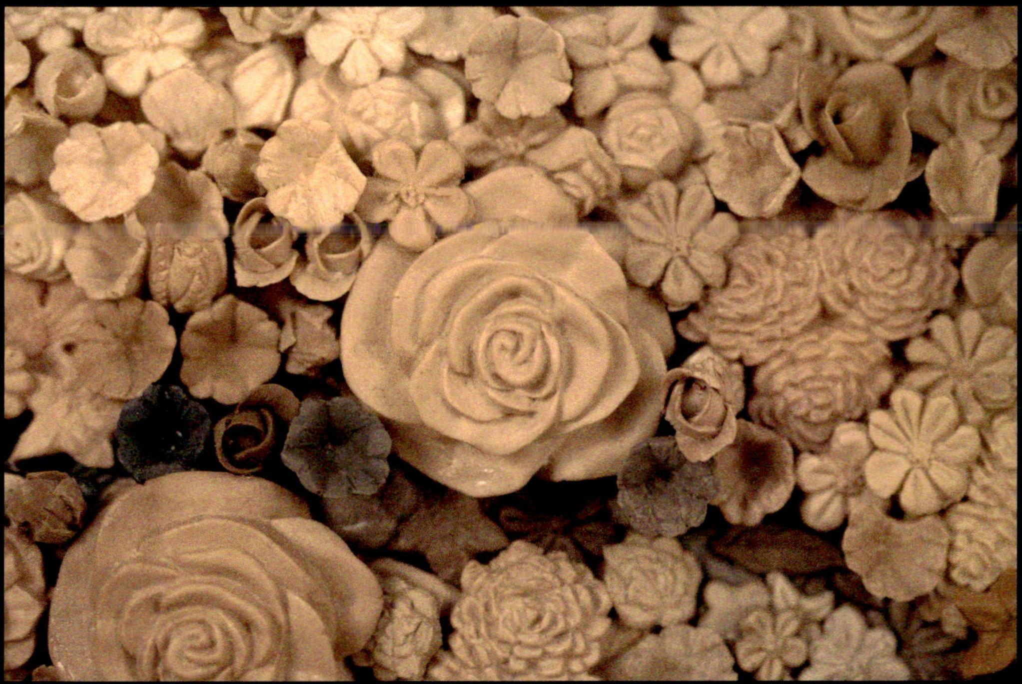

OPPOSITE PAGE

Center left Stamped gutta percha panel, nineteenth century, Gutta percha.
Photo: © Marek Uliasz, Depositzphotos.

Center right *Pilgrim Medallion with Christ, Angels, Apostles and Saints*, thirteenth century (Medieval). Venice, Italy, glass, 2.6" × 2.3" × 0.4".
Courtesy of The Walters Art Museum, Acquired by Henry Walters, CC0, GNU FDL.

Bottom Masks and dummies on a shelf in prosthetic special fx workshop, mixed media.
Photo: © Andrejs Pidjass, Depositphotos.

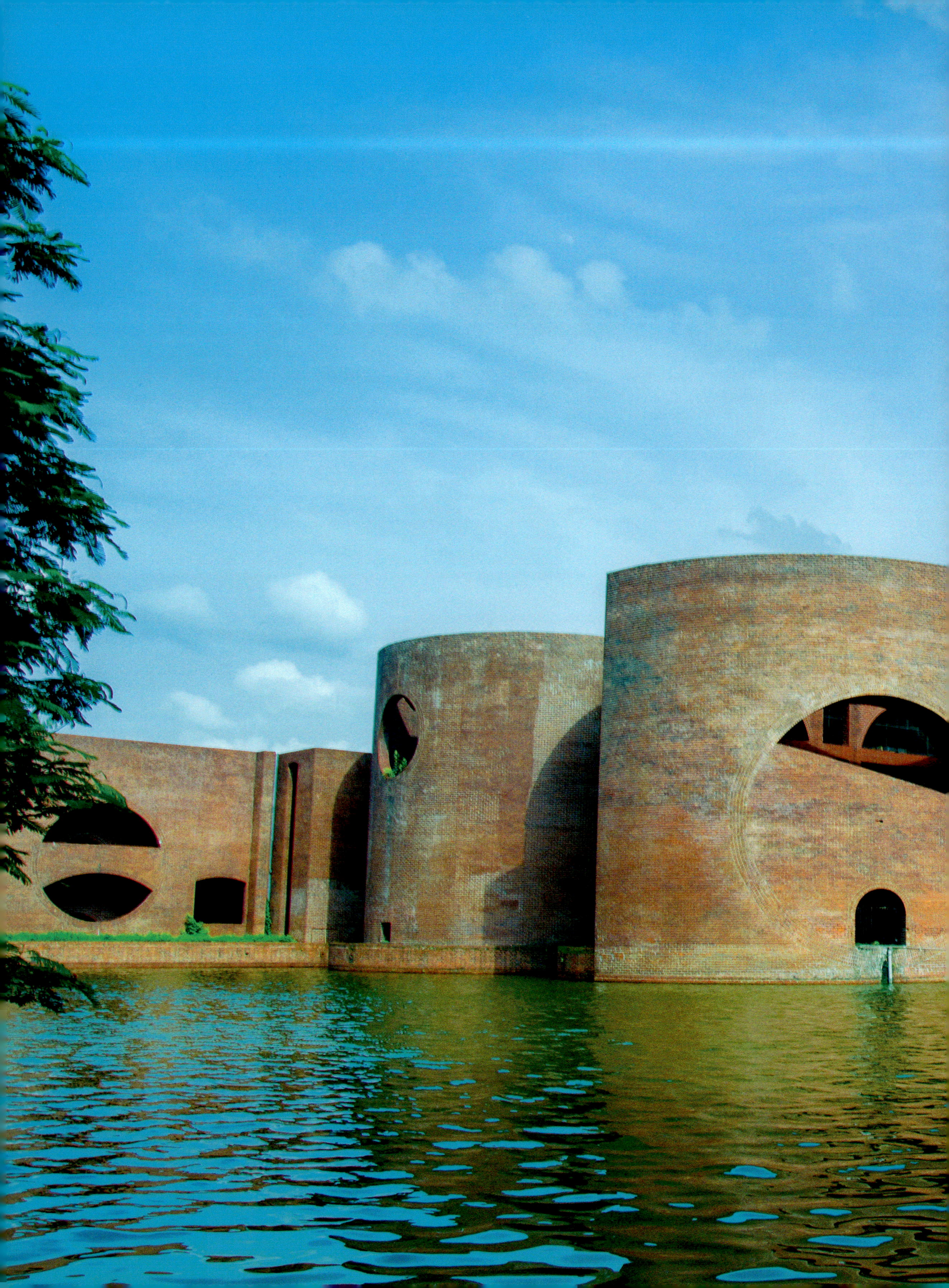

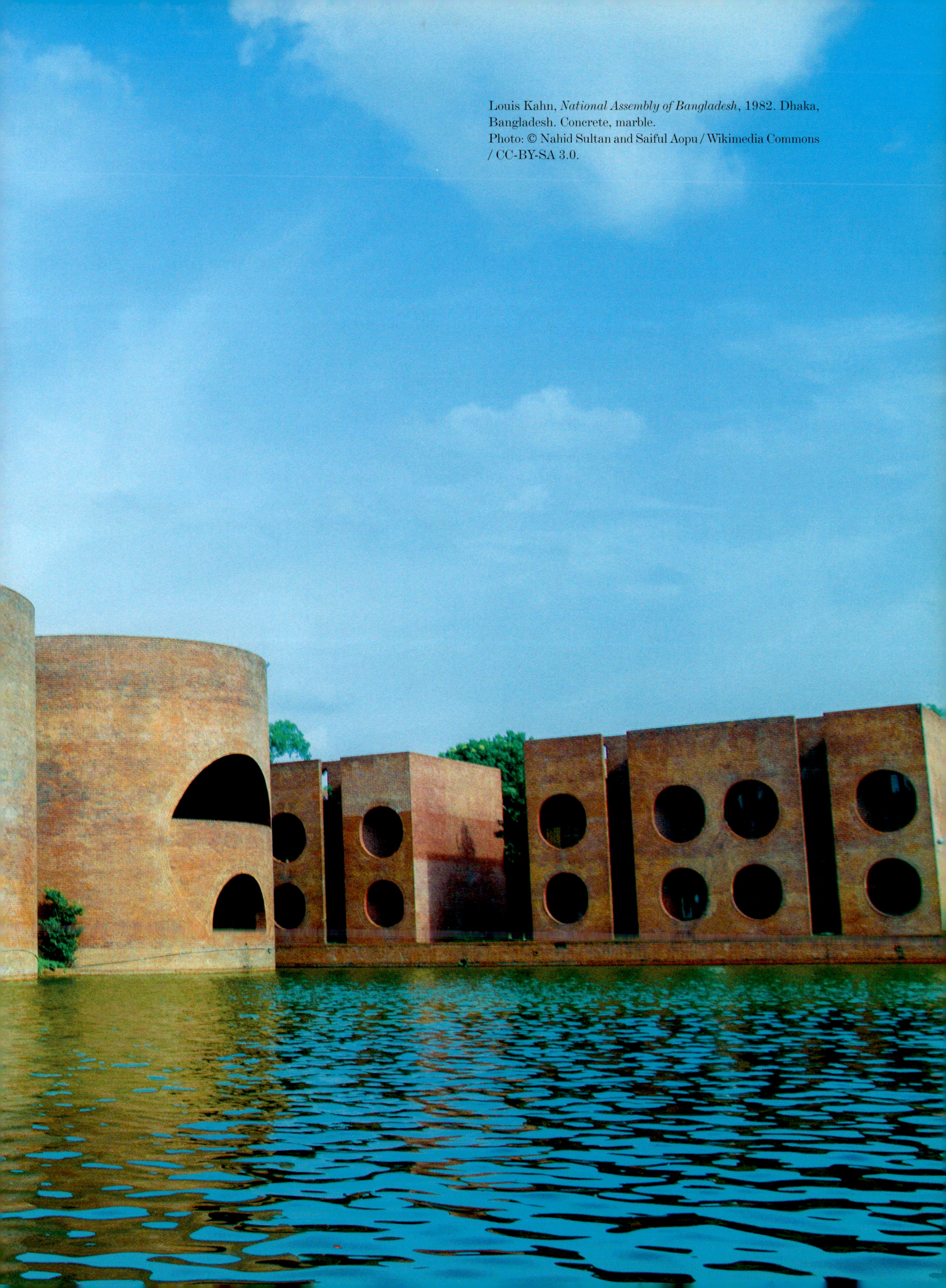

Louis Kahn, *National Assembly of Bangladesh*, 1982. Dhaka, Bangladesh. Concrete, marble.
Photo: © Nahid Sultan and Saiful Aopu / Wikimedia Commons / CC-BY-SA 3.0.

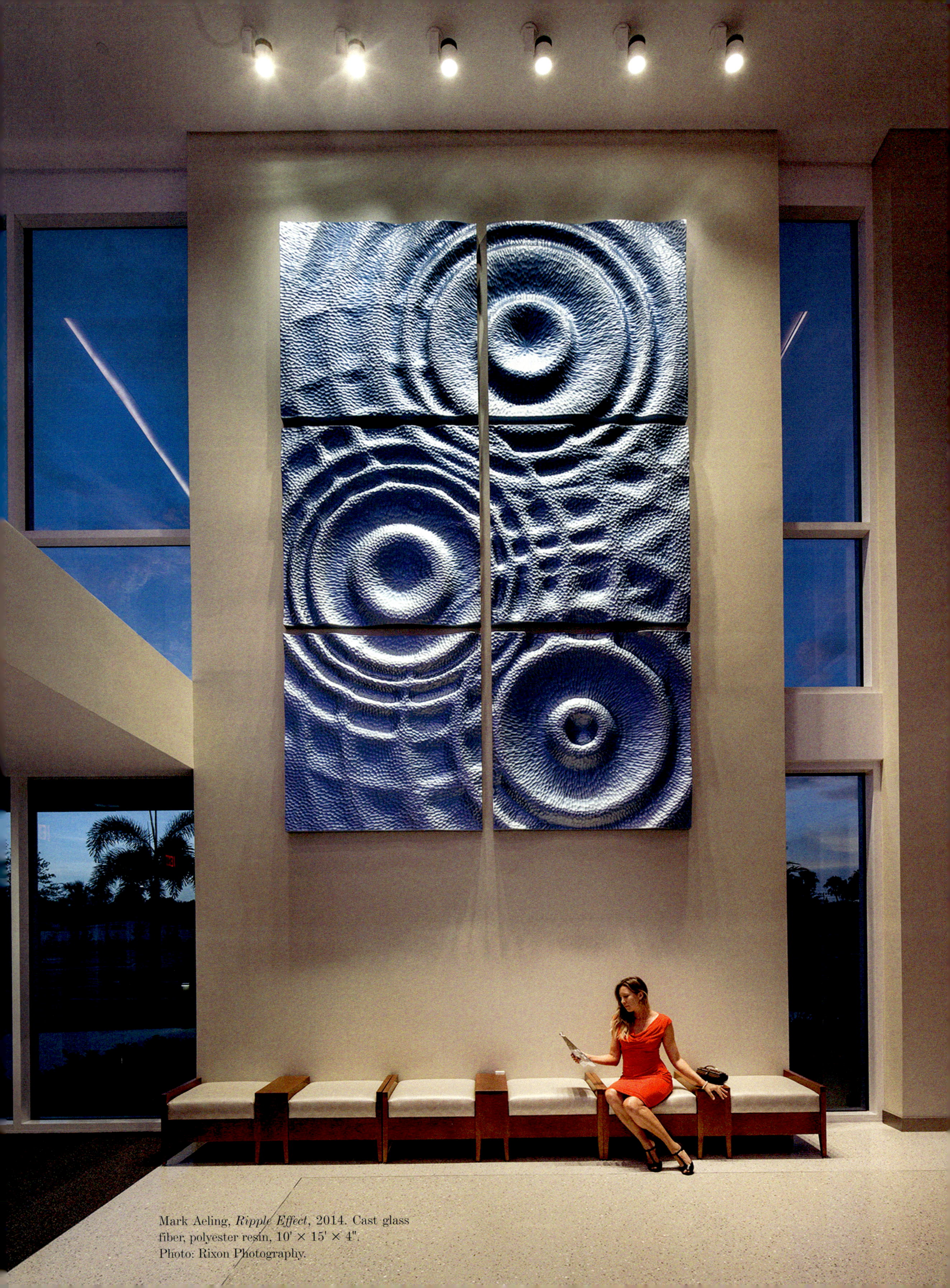

Mark Aeling, *Ripple Effect*, 2014. Cast glass fiber, polyester resin, 10' × 15' × 4".
Photo: Rixon Photography.

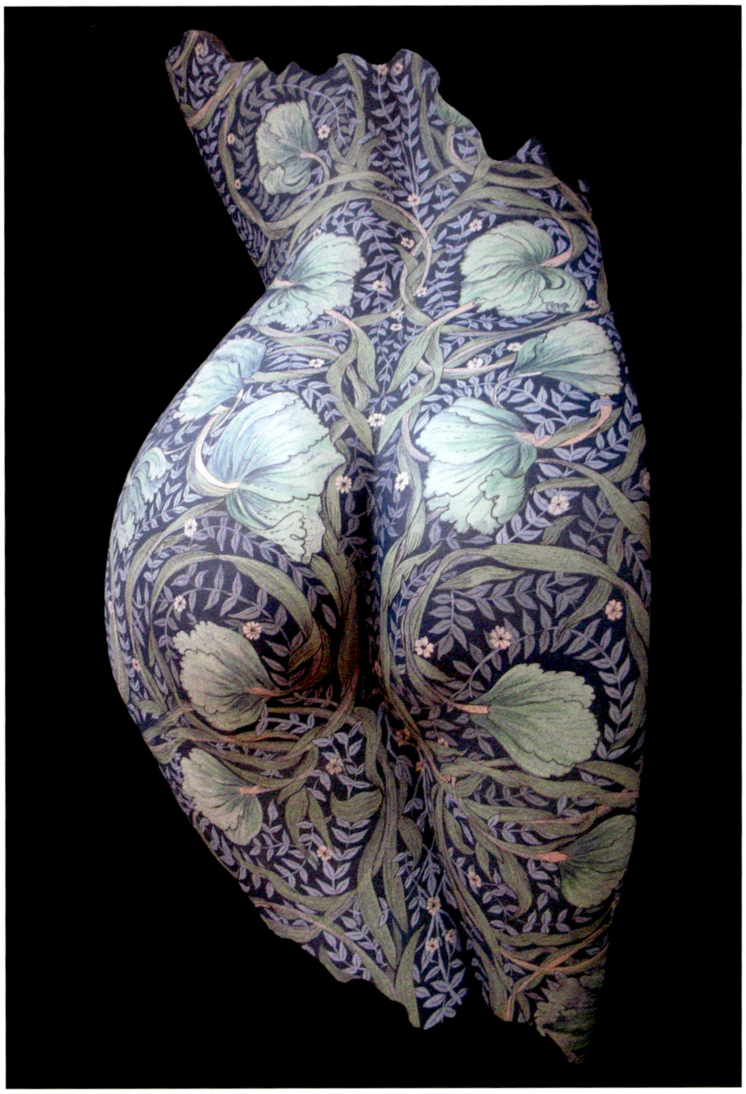

CJ Munn of Masters & Munn, *Marguerite*, from the William Morris Collection, original lifecast from live model taken with alginate supported by a mother mold made with plaster bandages. Cast in Crystacal R® plaster, hand-painted with acrylics and a traditional cow's milk glaze.
Photo: CJ Munn.

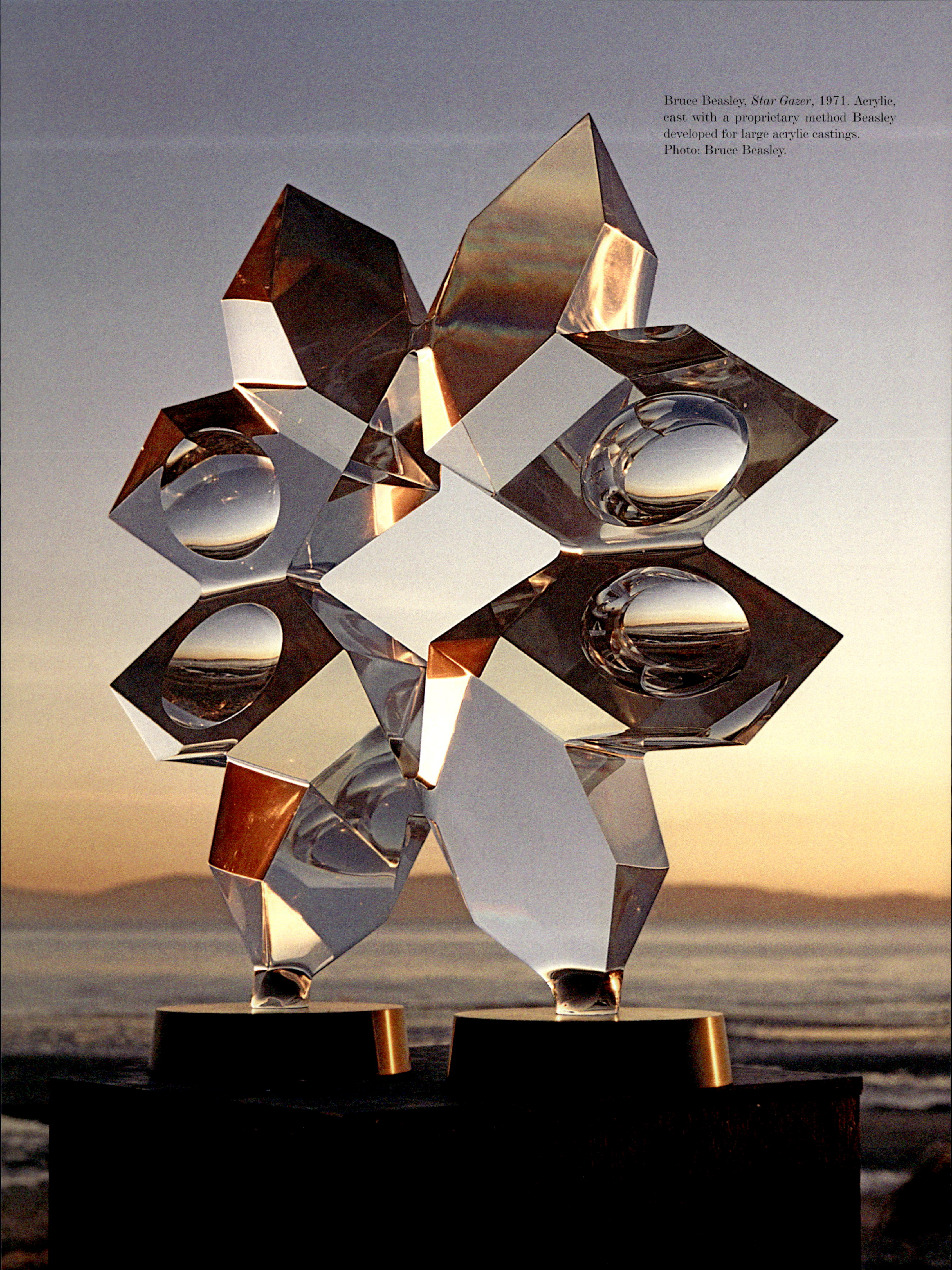

Bruce Beasley, *Star Gazer*, 1971. Acrylic, cast with a proprietary method Beasley developed for large acrylic castings. Photo: Bruce Beasley.

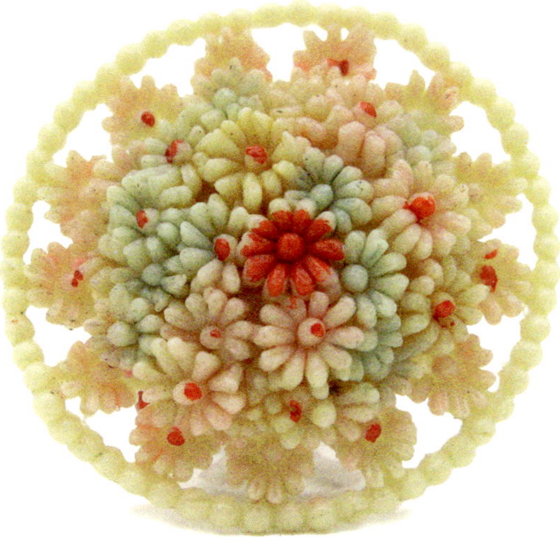

Top Vintage celluloid brooch, 1945–1952. Occupied Japan, celluloid, 1.25" × 0.25". Photo: Renée Zettle-Sterling, Collection of Renée Zettle-Sterling.

Right Courtney Starrett, *Strawberry Kiwi Jello*, 2010. Silicone rubber, sintered nylon, and stainless steel, 3.5" × 3.5" × 1.25". Photo: Zan Maddox.

Bottom Luba Bakicova, *Fragile Support*, 2012. Glass and castable, 2'1" × 7.5" × 6.3". Photo: Ilja Bilek.

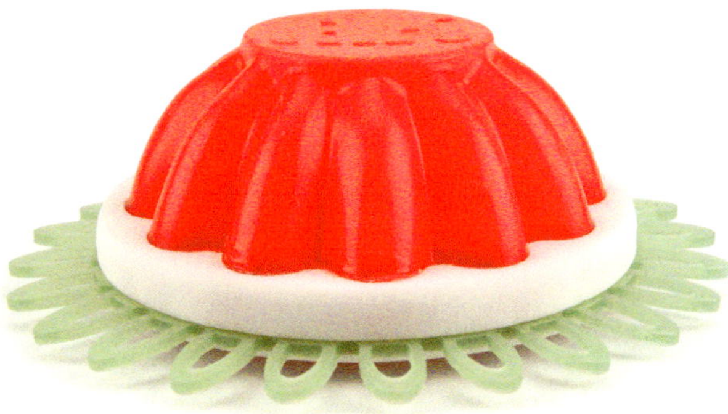

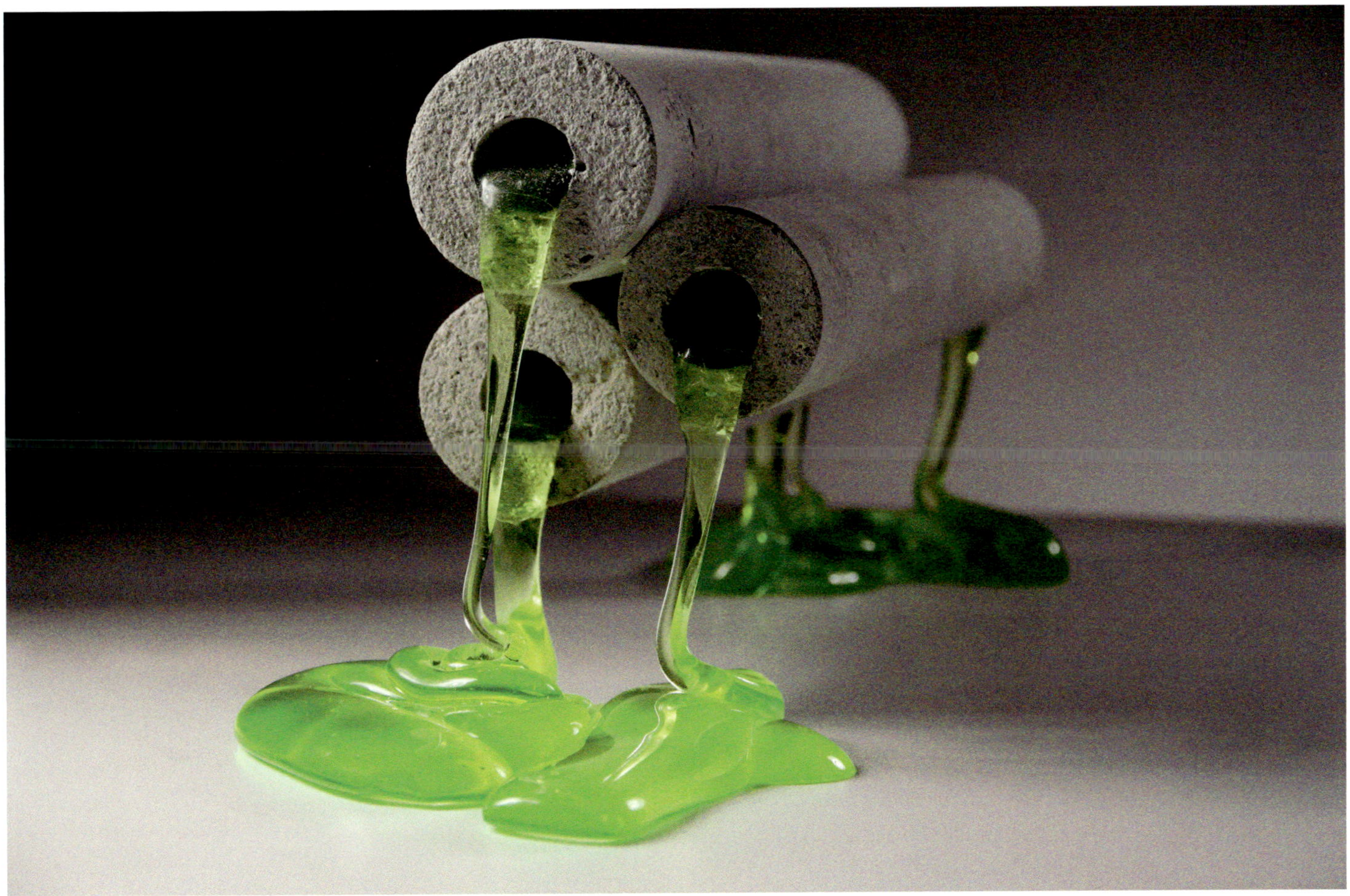

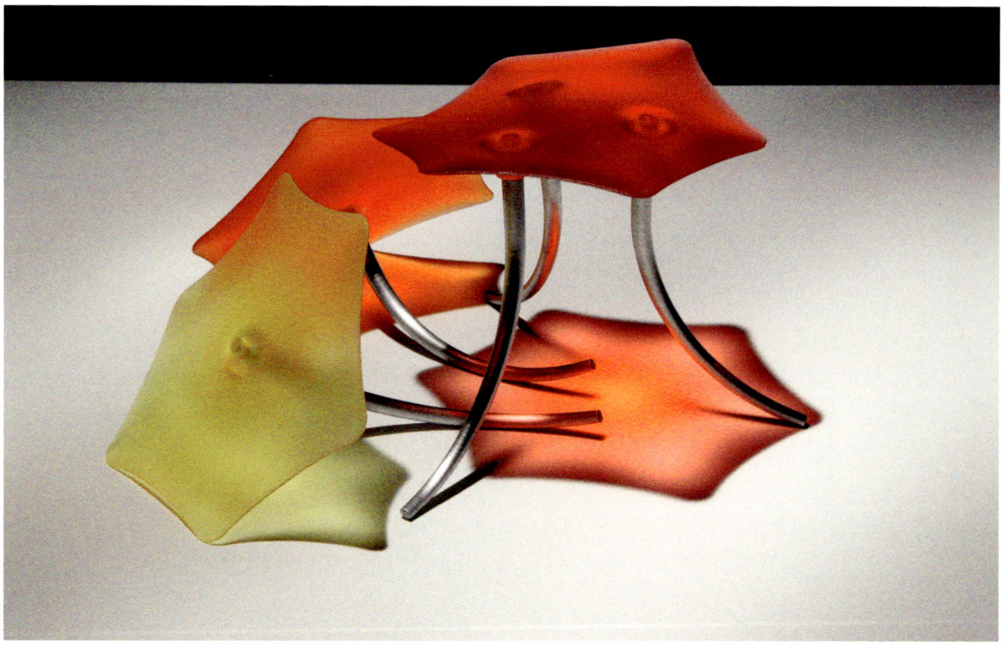

Top Eric H. Troffkin, *Lens Flare Casts*, 2013. Pressure cast urethane plastic, steel, light, 6" × 5" × 5". Photo: Eric Troffkin.

Right Niho Kozuru, *Longfellow Balustrade Columns*, 2004. Cast rubber, 2'6" × 2'6" × 5'. Photo: Robert Perachio.

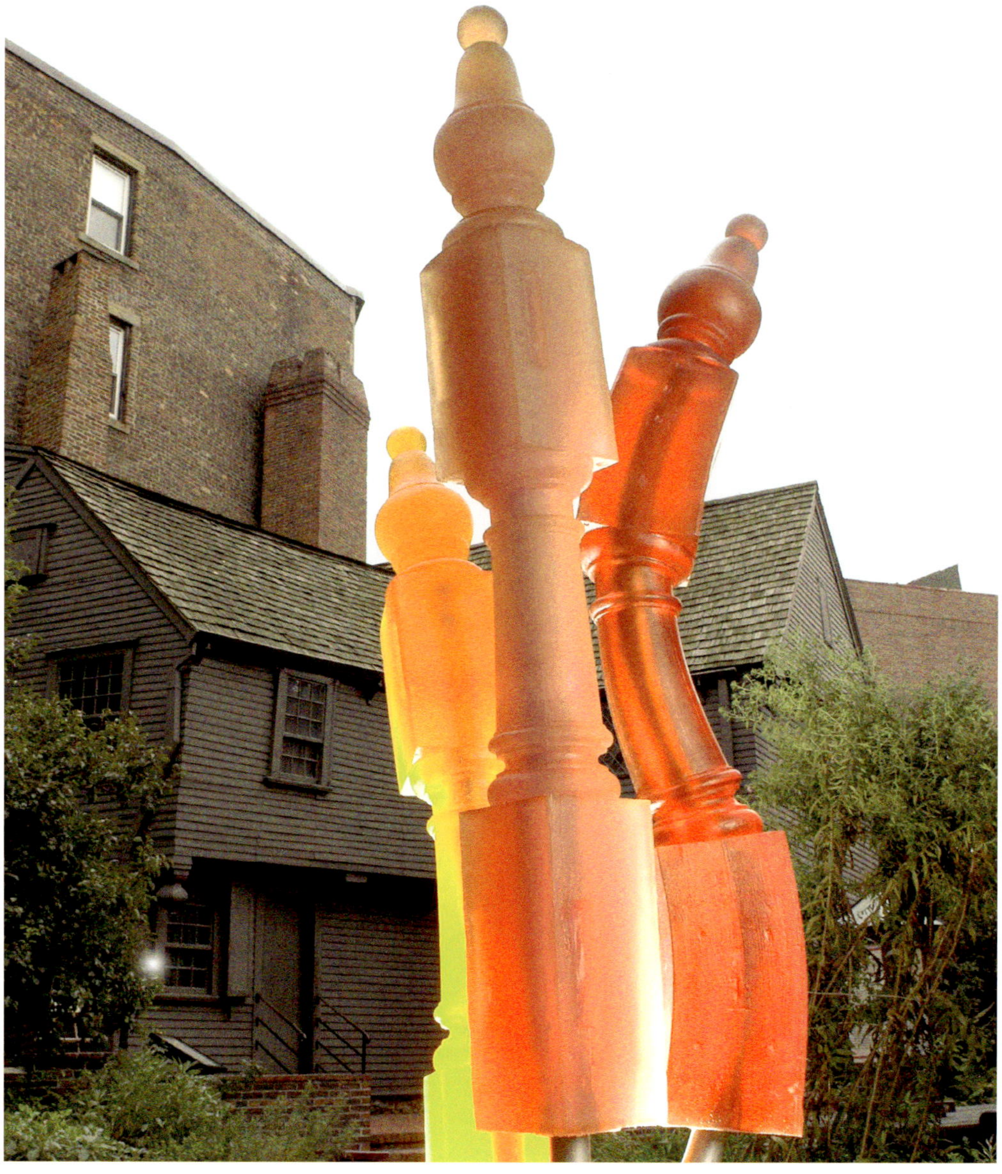

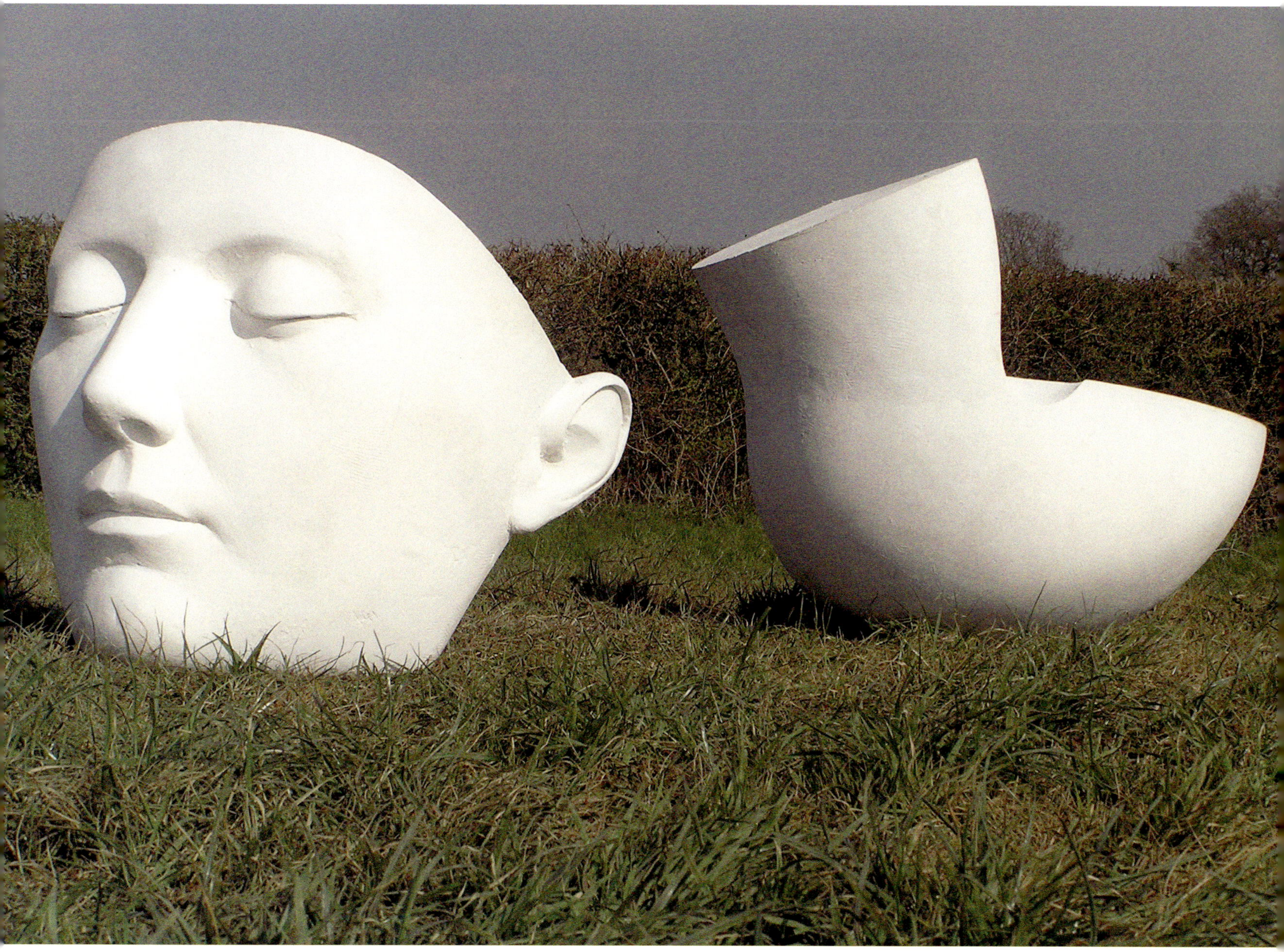

André Wallace, *Two Piece Landscape Head*, 2007. Resin, fiberglass, 26" × 23.6" × 34.6".
Photo: André Wallace.

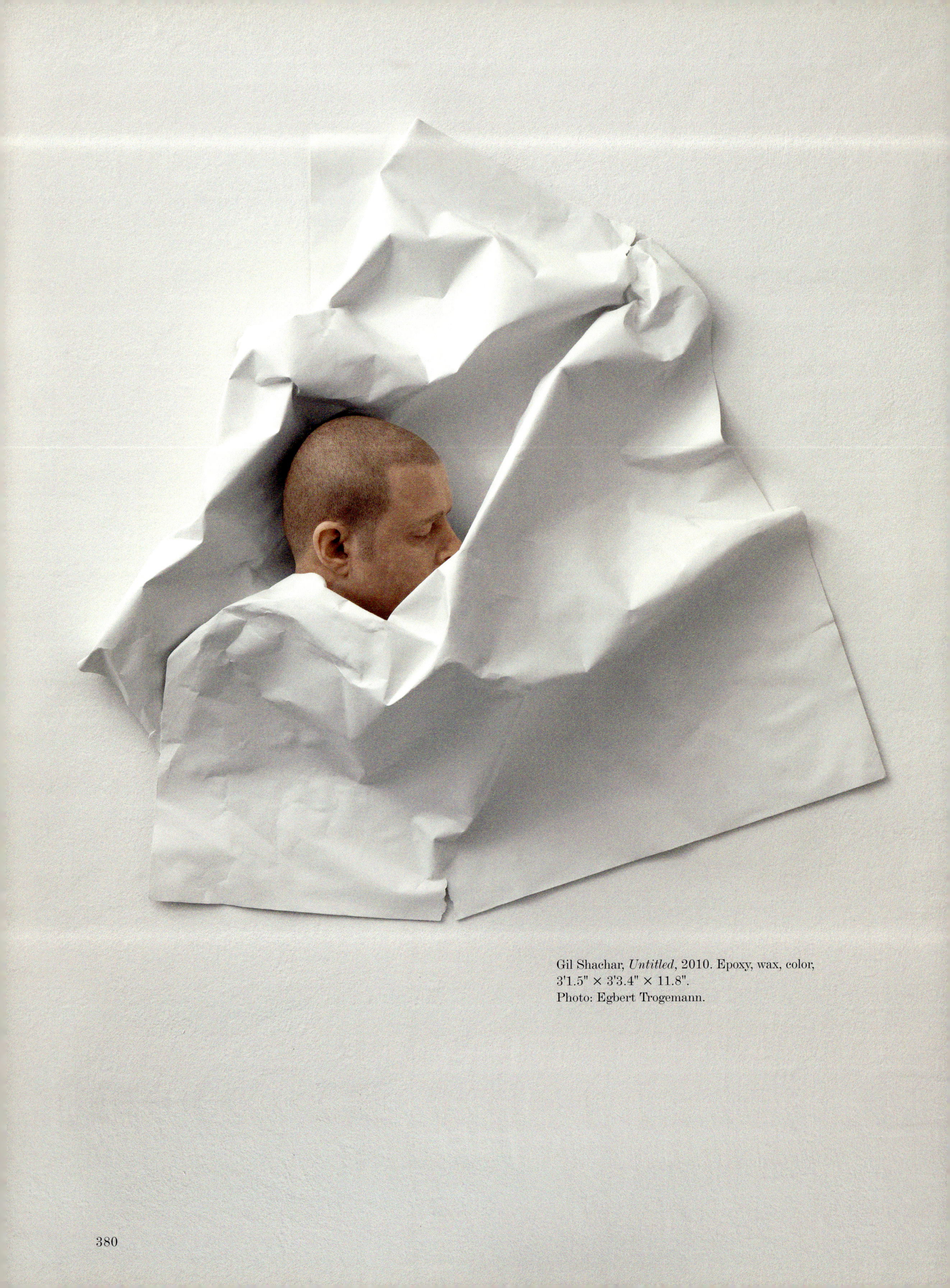

Gil Shachar, *Untitled*, 2010. Epoxy, wax, color, 3'1.5" × 3'3.4" × 11.8".
Photo: Egbert Trogemann.

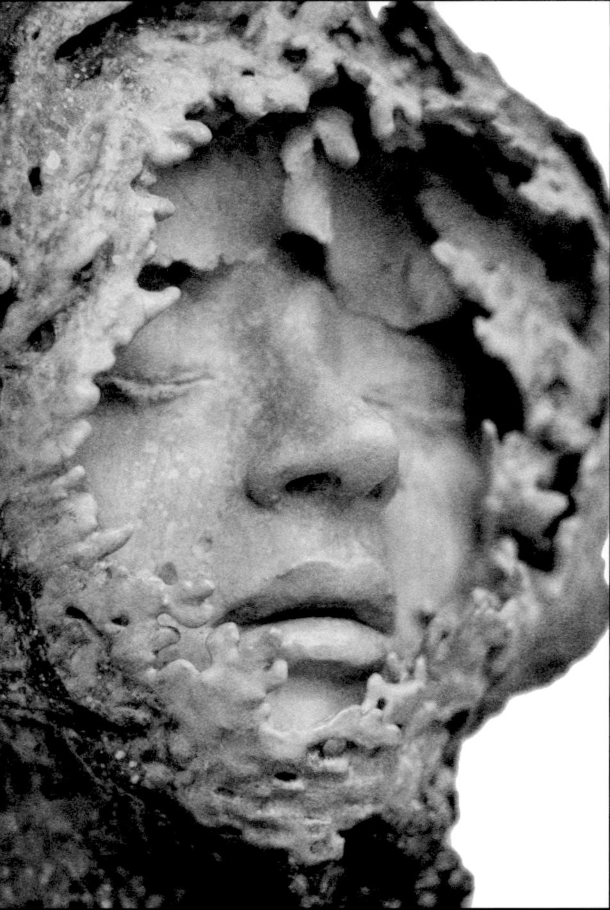

Gosia, *Evening Glow*, 2013. Polymer gypsum, part cast, 12" × 8.5" × 8".

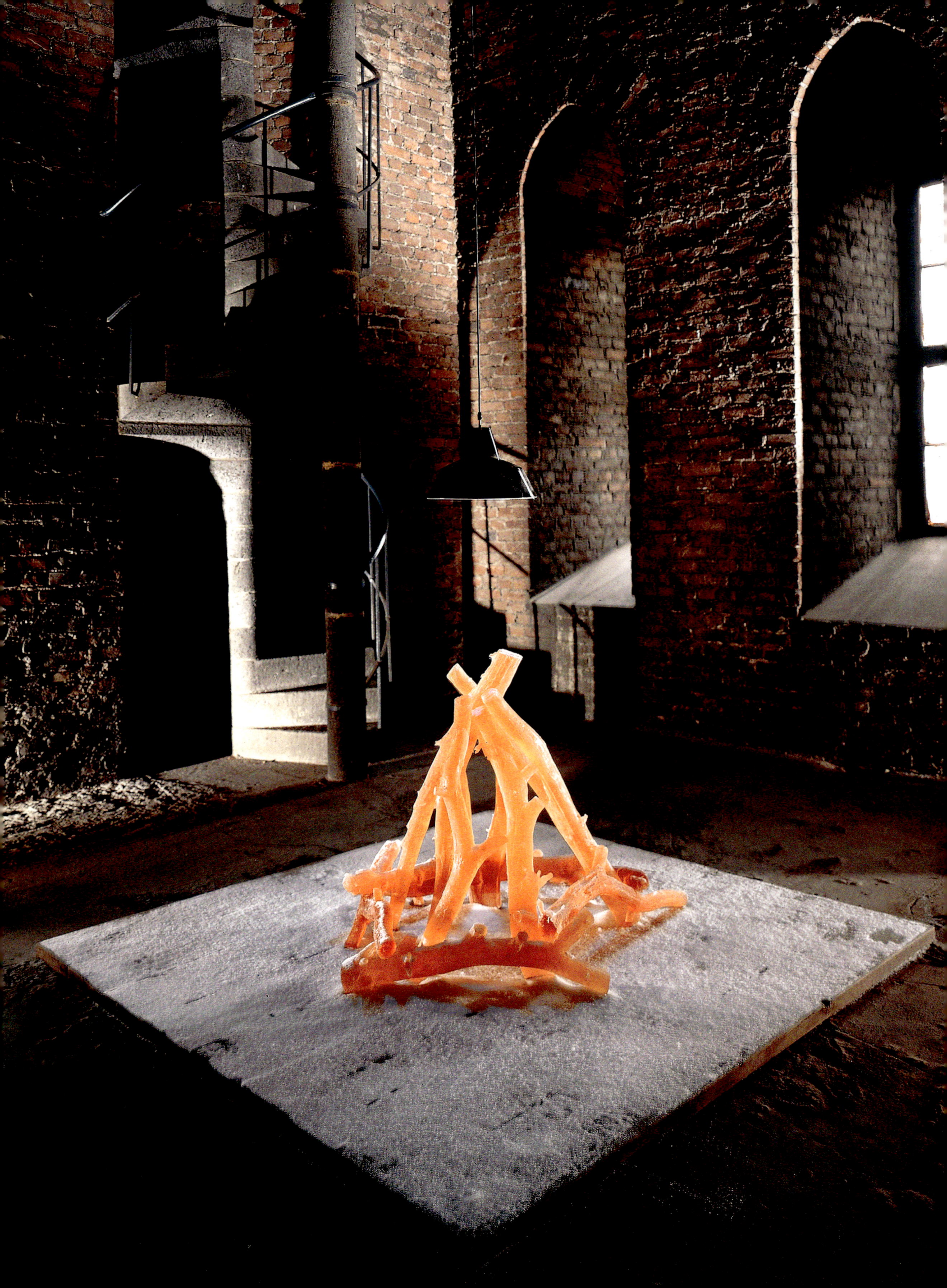

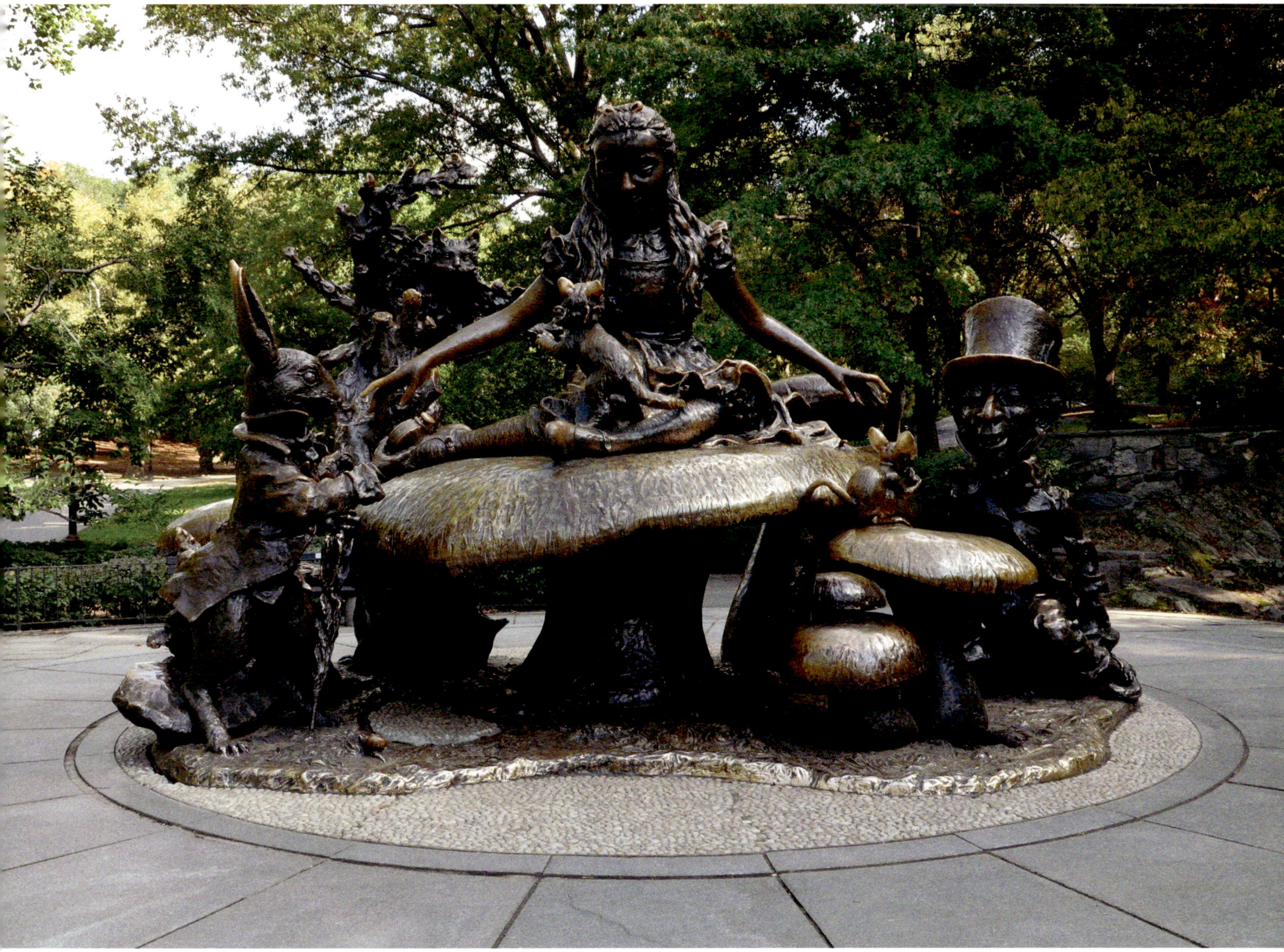

Opposite page Jeffrey Mongrain, *Northern Tinder* (Christus Church, Cologne, Germany), 2014. Resin, Styrofoam, and light, 8' × 6'2" × 6'2".
Photo: Jeffrey Mongrain.

"This campfire configuration rests on a thin layer of artificial snow. It is located in the bell tower of the Christus Church in Cologne, Germany. Positioned above these resin branches is a 1940s German streetlight. During World War II, this room was used as a classroom for children's initiation into the Hitler Youth Party. Their last demonstration of allegiance and courage was to leap over a small fire."
—Jeffrey Mongrain

Above José de Creeft, *Alice in Wonderland*, 1959, Central Park, New York, NY, USA, bronze, group height 9'7", diameter 14'6"; plinth height 1', diameter 34'; esplanade diameter 67'; seven bronze tablets each 1'9" H, 4' W. Gift of George and Margarita Delacorte.
Photo: © Gzzz / Wikimedia Commons / CC-by-3.0.

GALLERY
Unconventional Approach

The Last Supper, Ken and Julia Yonetani

Ken and Julia Yonetani have pioneered casting with salt. After collaborating with scientists at the Murray Darling Freshwater Research Centre (and years of subsequent research) they have mastered the ability to pour salt into molds and solidify it. Their interest in salt is multi-faceted—it has ecological significance related to rising salinity levels and unsustainable agricultural practices used in Australia (it is impossible to grow most plants in salted earth); it is symbolic of trust and friendship; and it has religious significance.

Pairing these meanings with the importance of the Last Supper both as a biblical event and in art history results in a piece layered with meanings and many possible interpretations.

"For us, the connotations of the last supper were of something that was consumed on the eve of death. Here, this death can also mean the death of the environment and the ecosystems that sustain life. The image is one of a large banquet, a feast before entering the afterlife. In this way, notions of food, consumption, death and the spiritual all merge together."[1]

The Last Supper is cast from one metric tonne of Murray River Salt. When asked how they cast the salt, the Yonetanis reply, "Salt + artist magic."

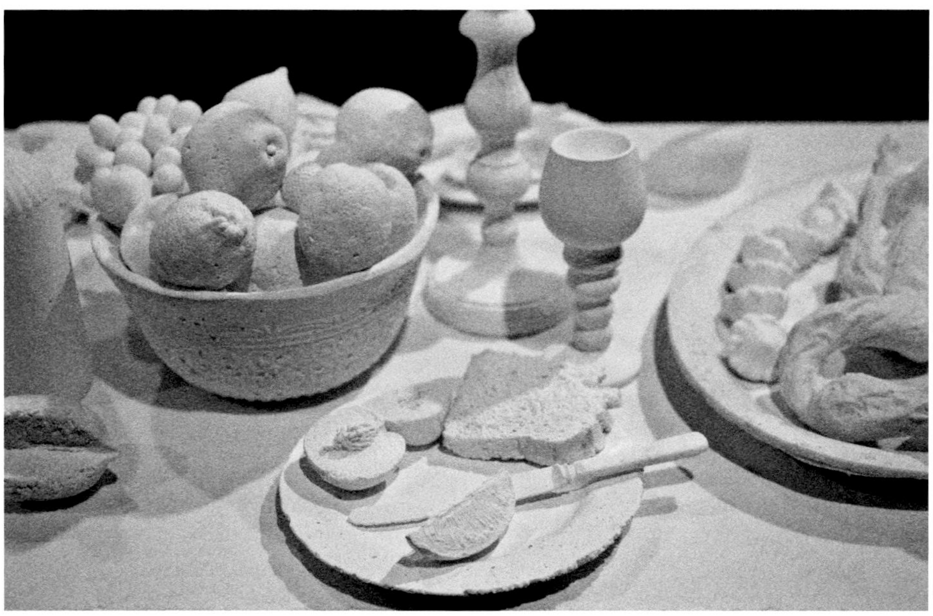

Ken and Julia Yonetani, *The Last Supper* (detail), 2014. Murray River Salt, 29'6" × 1'4" × 4'. Photo: Julia Yonetani.

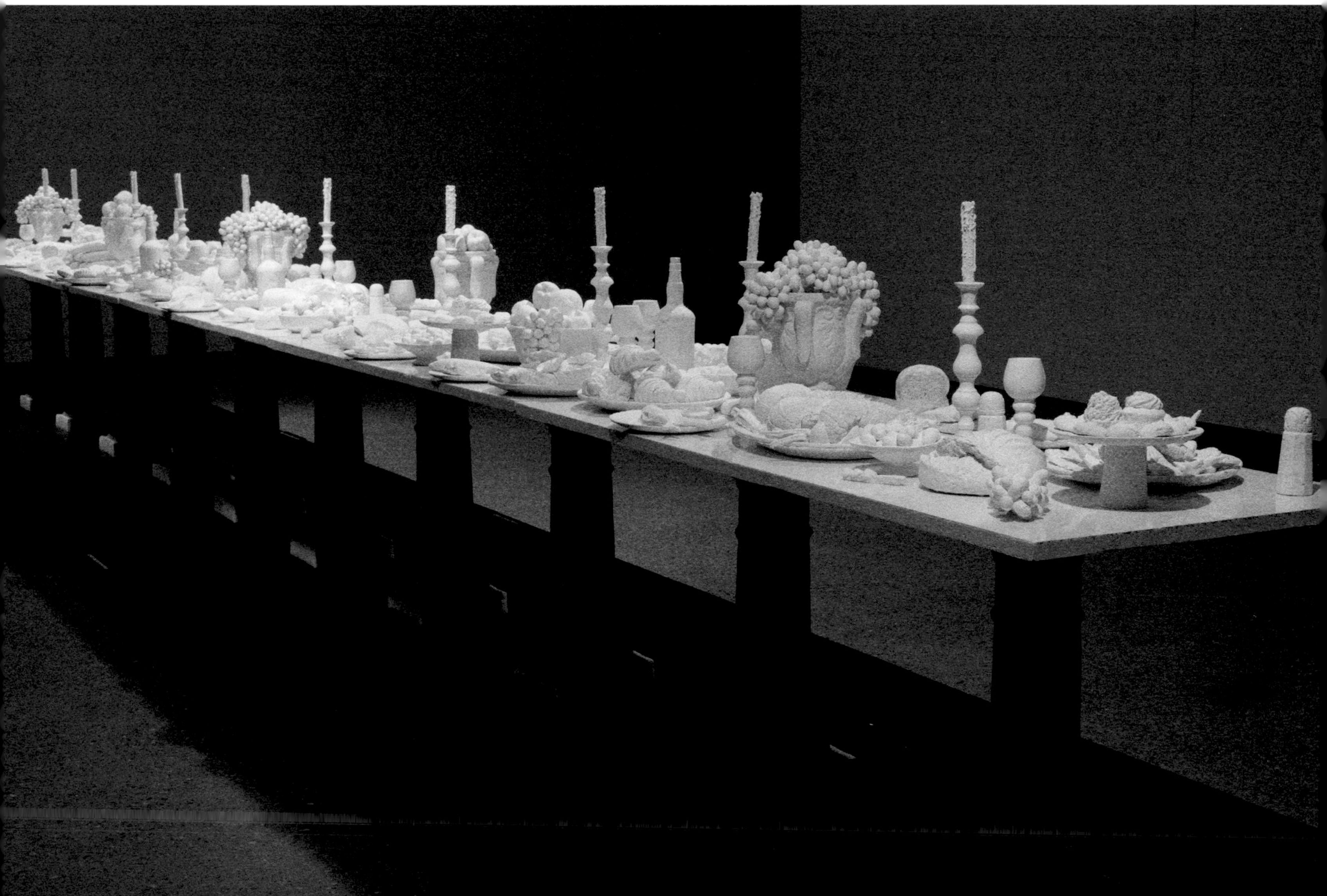

Ken and Julia Yonetani, *The Last Supper,* 2014. Murray River Salt, 29'6" × 1'4" W × 4' H. Photo: Julia Yonetani.

The Nameless Library, Rachel Whiteread

Rachel Whiteread's Holocaust memorial in Vienna, *The Nameless Library*, more commonly known as *Judenplatz* (after the public square where it's located), is unconventional in its imagery and in its long journey coming to fruition. Whiteread won the competition to create this important memorial in 1996, but it wasn't opened to the public until 2000 after what was, by all accounts, a bumpy road to completion.[2]

The Holocaust, an event still brimming with passions more than half a century after the liberation of the concentration camps, was going to be a tricky subject no matter what, but the archaeological excavation of a thirteenth-century synagogue underneath the plaza added extra complications to the process. Whiteread's design, which was never intended to be traditionally beautiful, was controversial as well. The whole project was examined, reexamined, and used politically, but eventually, the powerful piece became a reality.

The memorial is essentially a library turned inside out. We see endless identical volumes of what appear to be the same book, but we see them from the page-side (rather than the spines) rendering their titles unreadable. The repetition of the books almost reads as a texture—individual volumes fade away as the mass begins to blend into a singular surface. The shelves that ought to hold up the books are nonexistent, as is any hardware on the doors to the monument, creating an impenetrable edifice. The austere, blocky mass is commanding in its presence and stands in sharp contrast to the ornate surroundings of Vienna. *The Nameless Library* is cast in concrete and has a bunker-like feeling to it, which is both haunting and apropos. The only words on the memorial are a short dedication to the seventy-seven thousand Austrian Jews[3] who died in the Holocaust and the names of the concentration camps where they were killed. The dedication on the base of the monument outside the doors is written in German, Hebrew, and English.

Placing a solid concrete bunker in the heart of one of the most ornate cities in Europe was a gutsy and unconventional move, but it was the perfect choice for this poignant monument. As Robert Storrs said, "Whiteread's work is the solid shape of an intangible absence—of a gap in a nation's identity, and a hollow at a city's heart."[4] *The Nameless Library* reminds us of not only the individuals who were murdered in the Holocaust, but also of the cost of bigotry, xenophobia, and hate. A more traditional monument might have been overlooked, but Whiteread's savvy choices created a powerful memorial that commands attention and perfectly elucidates the deep wounds of World War II. Simon Wiesenthal, a Holocaust survivor and Nazi hunter, said it best at the unveiling: "This monument shouldn't be beautiful, it must hurt."[5]

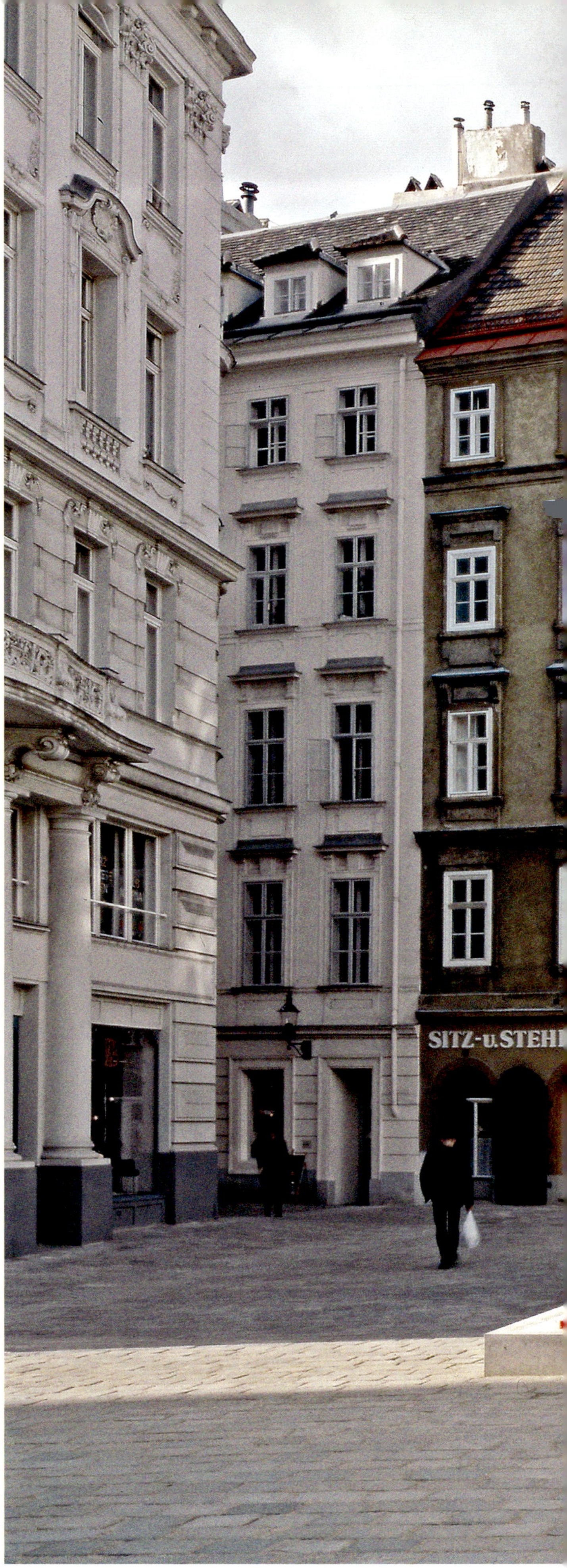

Rachel Whiteread, *The Nameless Library* (more commonly known as *Judenplatz*), 2000. Steel, concrete, 32'7" × 23' × 12'6". Photo: © 2000 Werner Kaligofsky, © Rachel Whiteread, Courtesy of the artist and Gagosian Gallery, London.

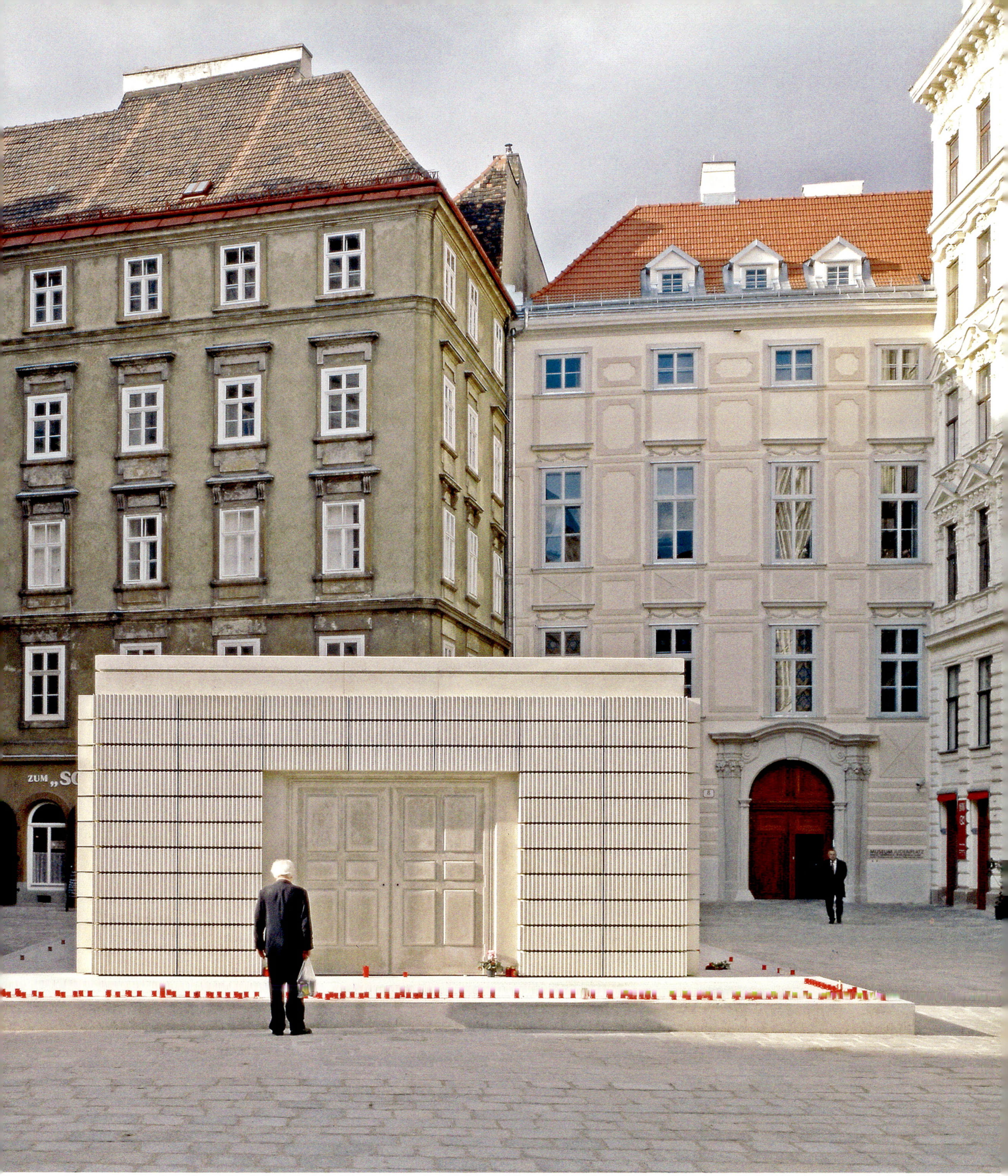
ZUM „SC

Greener Greens (The White Rabbit), Jeremy Brooks

While he used traditional slipcasting methods to produce this charming rabbit hiding in greenery, Jeremy Brooks utilized an unconventional approach in its surface treatment. The paraffin wax coating on the piece is infused with the aroma of freshly cut grass. Scent, the most primitive and arguably the most powerful of our senses, is often overlooked in art. Scent can transport us to places deep in our memories and can evoke visceral associations impossible to achieve by other means. The perfume of freshly cut grass unquestionably adds a special dimension to this piece.

Jeremy R. Brooks, *Greener Greens (The White Rabbit)*, 2006. Slipcast porcelain, underglaze, textured spray paint, fragrant paraffin wax (this figurine is infused with the aroma of freshly cut grass), 8" × 8" × 4".
Photo: Jeremy R. Brooks.

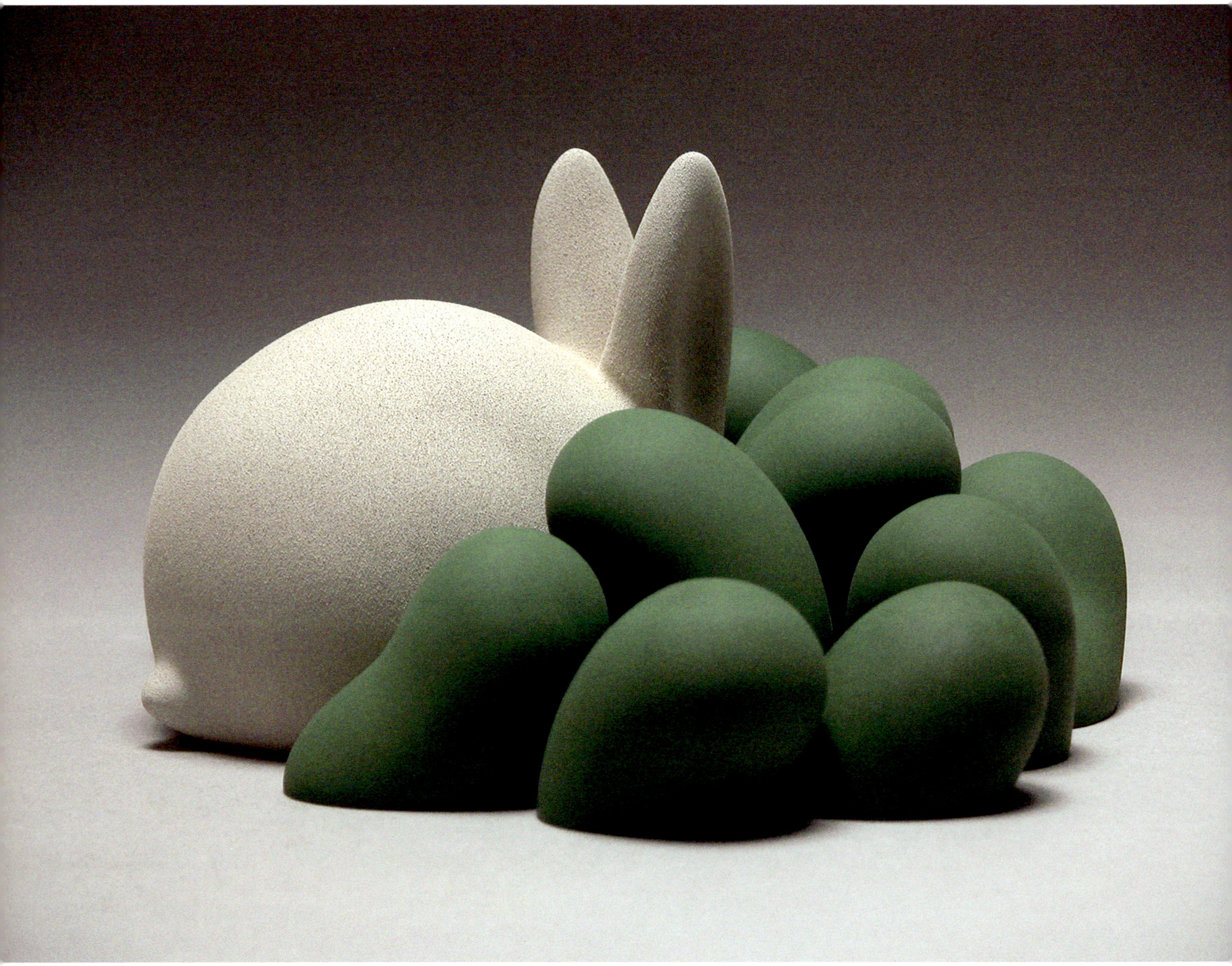

Hive Consciousness, Allen Peterson

Allen Peterson collaborated with a hive of bees to produce this piece.

He explains, "The cast beeswax items that I introduce to the hive have a smell and feel that is familiar to the bees. They choose their own places and ways to build comb onto what I have started—in this way, the process is like having a conversation with the hive. I am careful to respect the bees that I work with. Although the hives that I work with are cared for and tended by humans, the bees are not tame in any sense of the word. I couldn't and wouldn't make them do anything that they didn't want to do. With this in mind, I time my projects to coincide with when they would be building honeycomb anyway, so that I work with the bees, rather than them working 'for' me."[6] *Hive Consciousness* is an exciting example of collaboration and allowing nature to take its course. Whether an artist is collaborating with another artist or with a hive of bees, the piece is the result of sensitivity to others, patience, and allowing/encouraging unexpected outcomes; these are virtues that can inspire makers of all kinds.

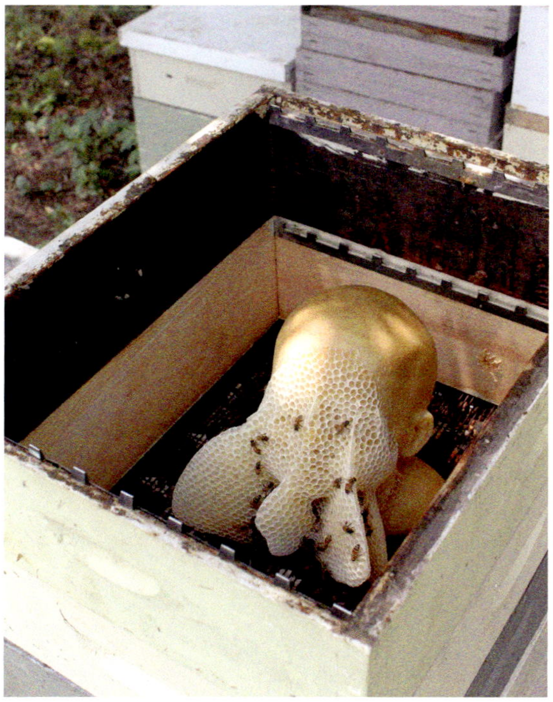

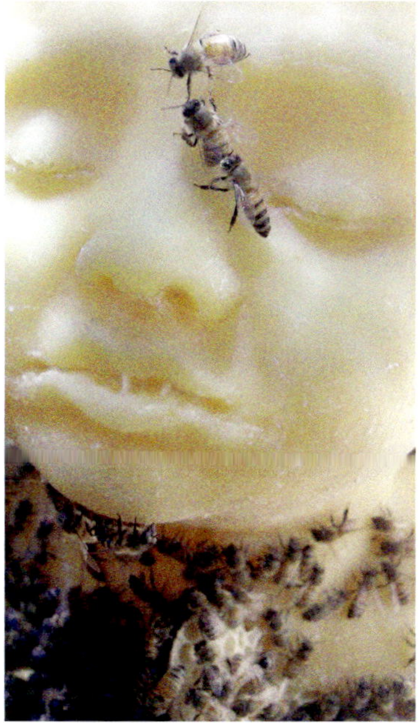

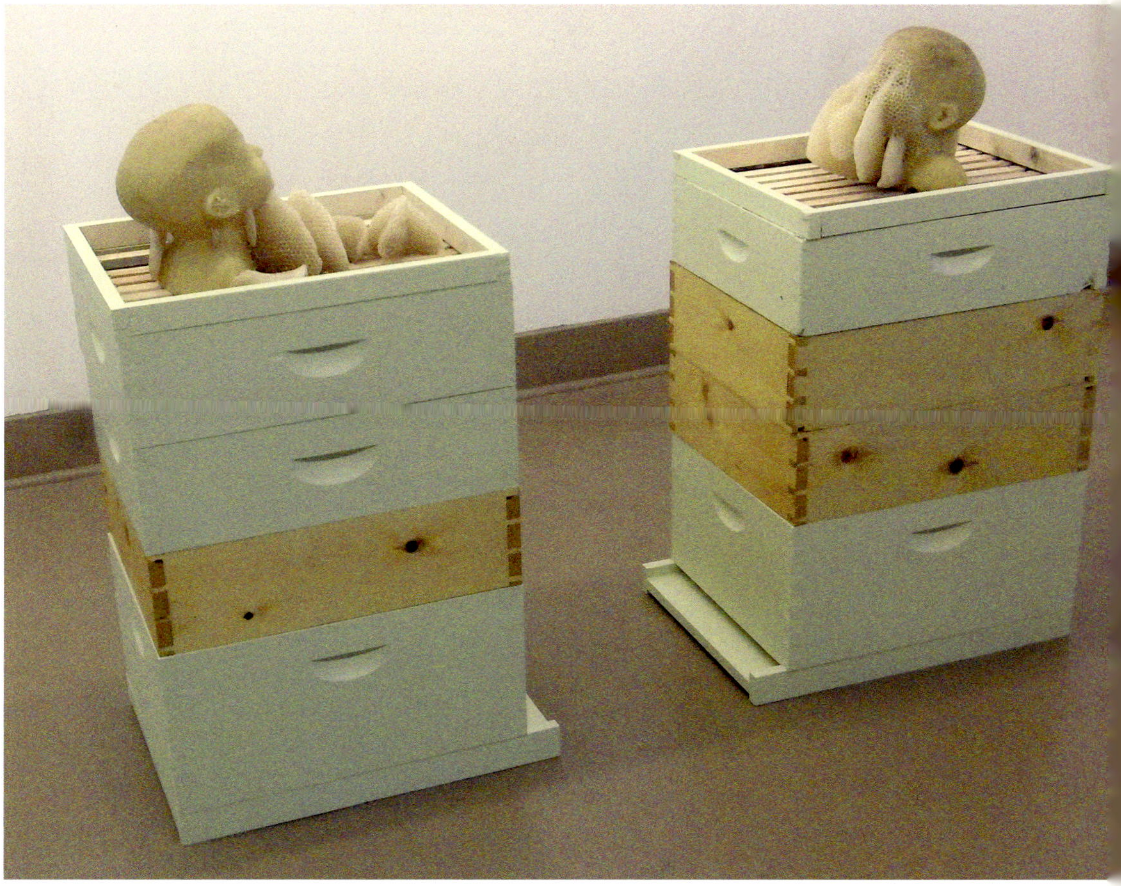

Allen Peterson, *Hive Consciousness*, 2012. Honeycomb built by bees onto cast beeswax, wood, varnish, latex paint, dimensions variable: 5' × 16.5" × 33". Photos: Allen Peterson.

Untitled, Chad Gunderson

Rather than using glaze as a surface treatment on a ceramic object, Chad Gunderson uses it as a casting medium. He developed this material through rigorous testing and in-depth research. His method enables the glaze to bubble, froth, and grow to fill the mold. The resulting objects are reminiscent of mineral specimens or coral, full of irregular bubbles and pits. Gunderson then selectively paints parts of the castings creating eye-popping color schemes. The result is a piece that evokes the natural world, but doesn't look quite natural, nor does it look entirely handmade.

Gunderson explains, "Arbitrarily sculpted by wind, sand, and water, the allure of rocks is converse to the clean aesthetics and deliberate choices of manmade items. With playful naiveté, my work is an attempt to highlight and understand the fringe between these two seemingly opposed objects.

"I see my work as a modern day counterpart to ancient Chinese scholars' rocks. These rocks were carved, distorted, and used as tools for the imagination. My work, though, also merges elements from contemporary culture by using plastic veneers and saturated colors. Ultimately, my creations become part of a collection of fabricated and fetishized objects which represent how the human imagination can mutate geologic and fossilized specimens into engineered designer relics."[7]

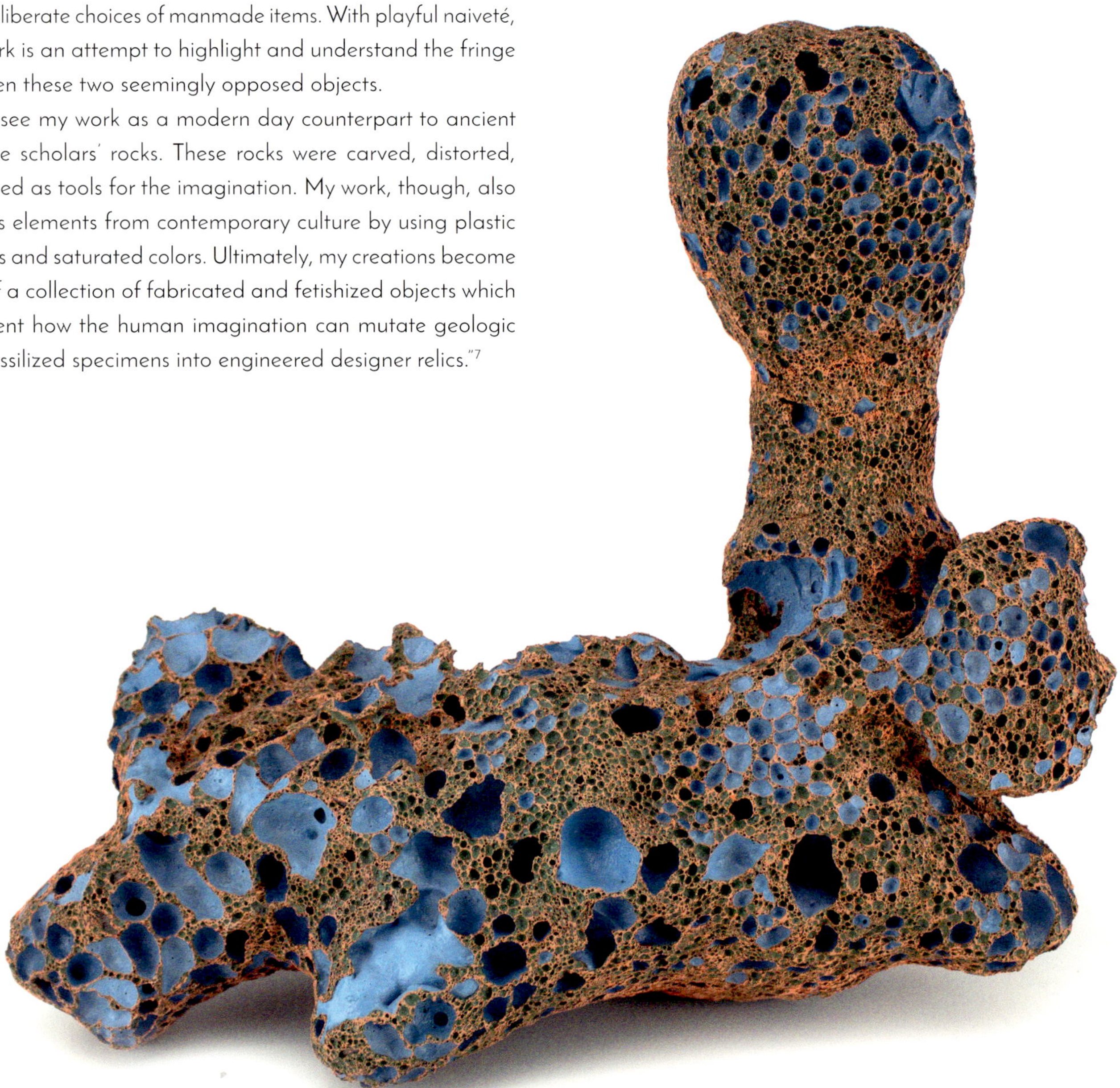

Chad Gunderson, *Untitled*, 2014. Solid cast ceramic glaze, acrylic, 16" × 20" × 8".
Photo: Chad Gunderson.

Island Caldera II, Stephen Hatcher

Largely known as a woodworker, Stephen Hatcher has begun to introduce casting into his practice. He casts burls from different tree species along with mineral crystals such as calcite, gypsum, and mica. The minerals are crushed and sorted to remove fine powder, then are sandwiched in layers of water-clear resin surrounding the burl wood. The resulting casting is then turned on a lathe using traditional woodworking techniques. The final form is then airbrushed with colorfast dyes and finished with twenty coats of hand-rubbed lacquer.

This piece exemplifies the marriage of history and innovation—the tight craftsmanship of a traditional woodworker with the twist of modern material and color.

Stephen Hatcher, *Island Caldera II*, 2014. Yellow box burl (Australia) with gypsum and mica embedded in cast resin, 16" × 16" × 1.5". Photo: Stephen Hatcher.

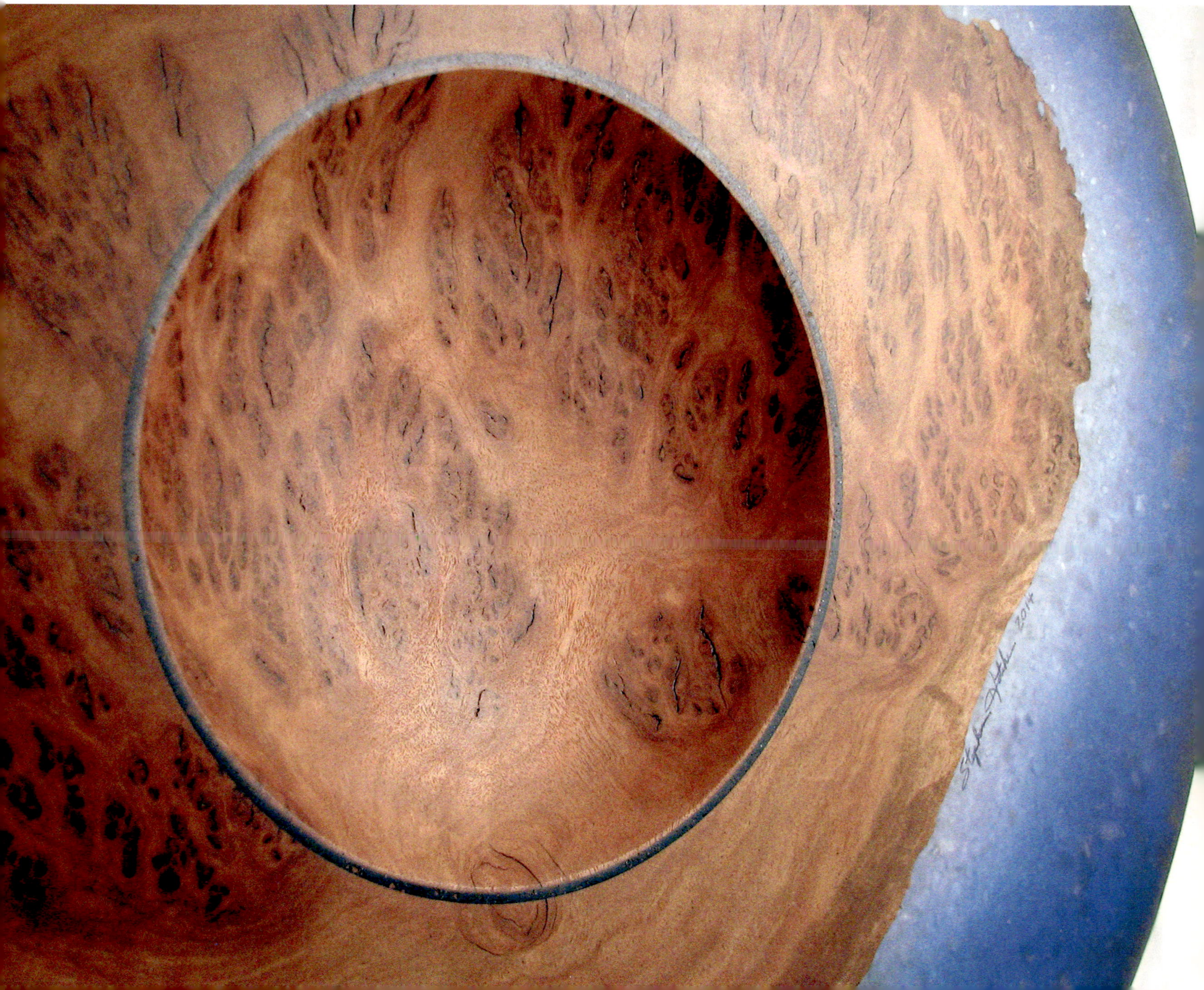

Pellucid and *Jardinière*, Janice Jakielski

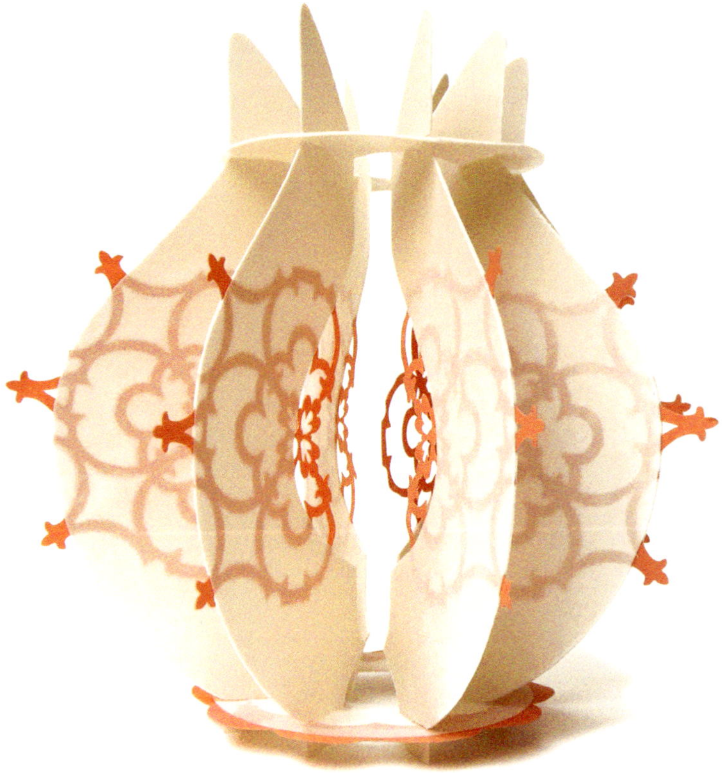

Janice Jakielski's cast porcelain work is shockingly delicate. She is pushing the limits of traditional ceramic casting, both in the thinness of the material and in her chosen application of it. *Pellucid* is made up of fins so eggshell-thin that the soft orange tracery running through them seems to float, backlit in the translucent panels. The crisply cut opaque tracery loses none of its doily-like details by being bonded to the translucent porcelain, rather it's perfectly embedded and exquisite in its precision.

Jakielski and her husband, Joshua Persky, recently started a company called Meld Materials that focuses on combining their art and science backgrounds to create new materials for artists. "Tape casting" is one technique they're exploring. By adapting industrial ceramic techniques for use in the artist's studio, Jakielski and Persky are casting 0.5-millimeter-thick ceramic sheets (traditionally used in microelectronics and solid oxide fuel cells). This recontextualized technique has many exciting applications for anyone interested in working with gossamer-thin porcelain.

Janice Jakielski, *Pellucid #1*, 2016.
Cut porcelain, tapecast, fired to cone 6, 5" × 7".
Photos: Janice Jakielski.

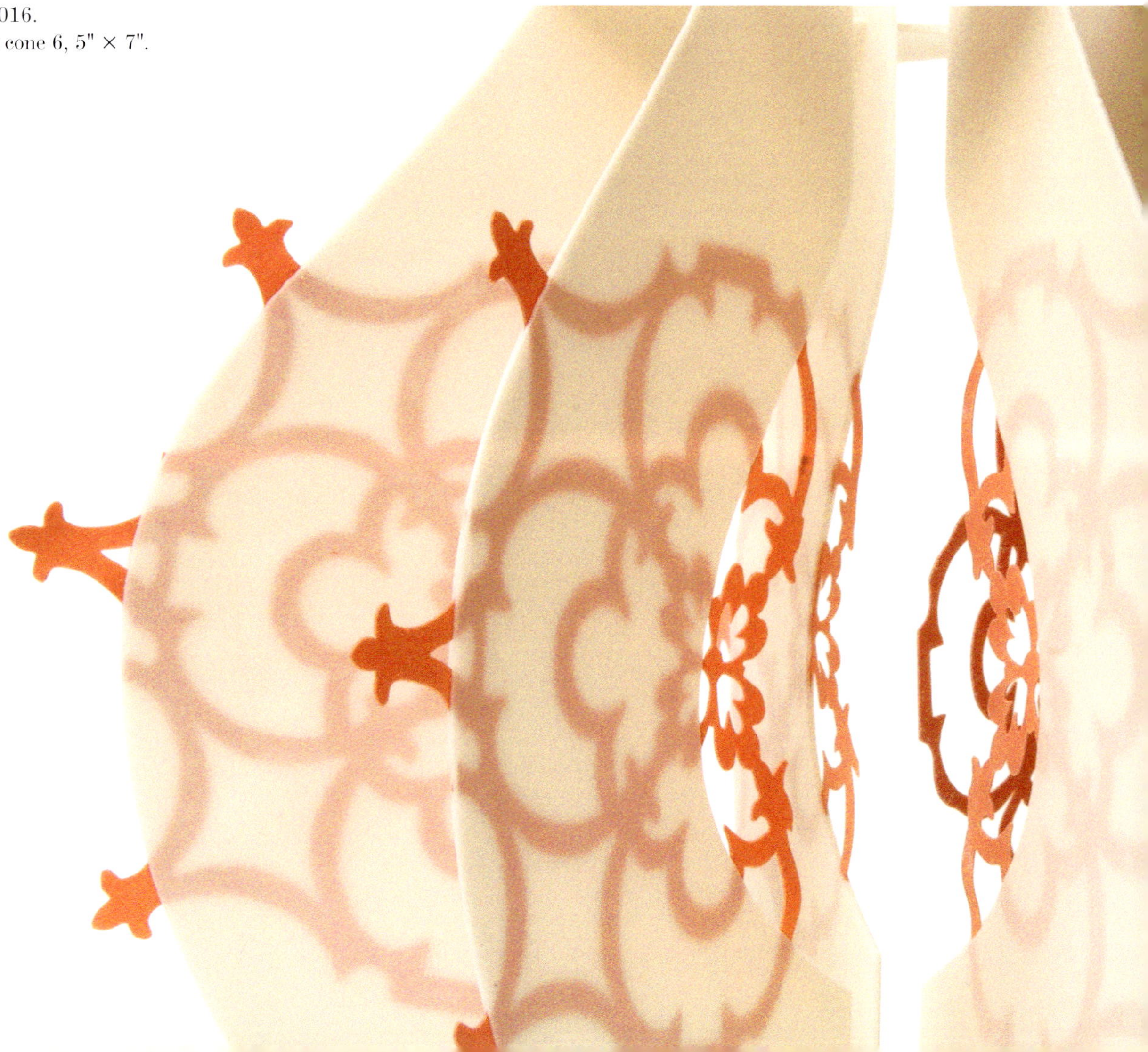

Jardinière was created using traditional paper quilling techniques, made possible by the ultra-thin cast porcelain tape, which is so thin, in fact, that it more closely resembles paper than anything that would be identified as cast ceramic. Paper techniques, therefore, seem to fit as an appropriate application for these filmy castings. Jakielski's vessel forms are a clear nod to the history of ceramics, but she turns that history on its head with her surprising approach.

"Making ceramic tape involves mixing a ceramic slurry or slip. *Pellucid* and *Jardinière* are made from thin sheets of cast porcelain tape. The dry ceramic materials are mixed with a combination of water, plasticizer (to add flexibility), and binder (to hold everything together)," Jakielski says. "This ceramic slurry is cast onto a long, flat, heated surface. The dry tape can be rolled, cut, die cut, etcetera. In its green state, the ceramic tape sheets are very flexible, will not dry out, crack, etcetera, making it possible to work much thinner than with traditional porcelain and without the drying worries that normally go along with clay. When tape cast pieces are fired, all of the binders and plasticizers burn away, leaving normal, vitrified porcelain."[8]

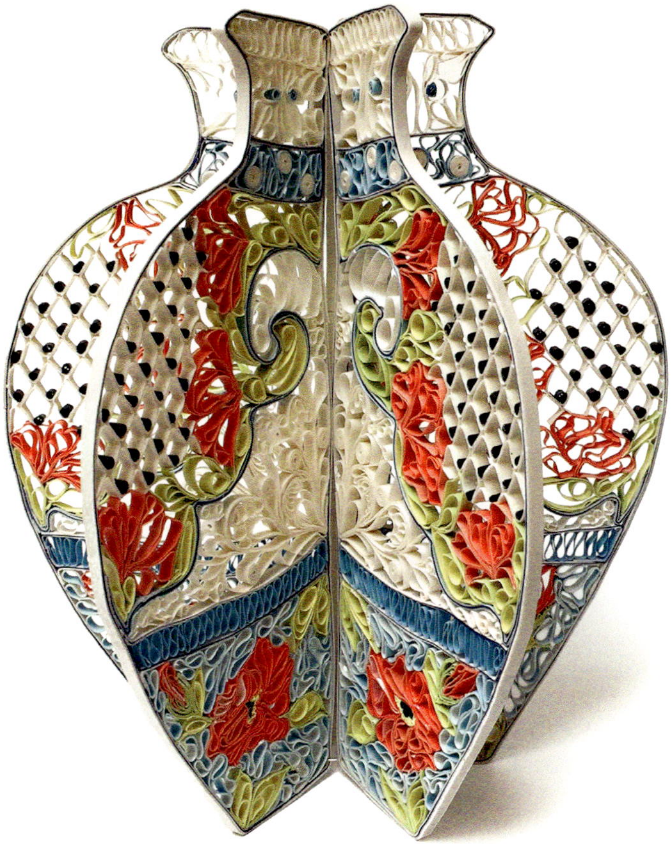

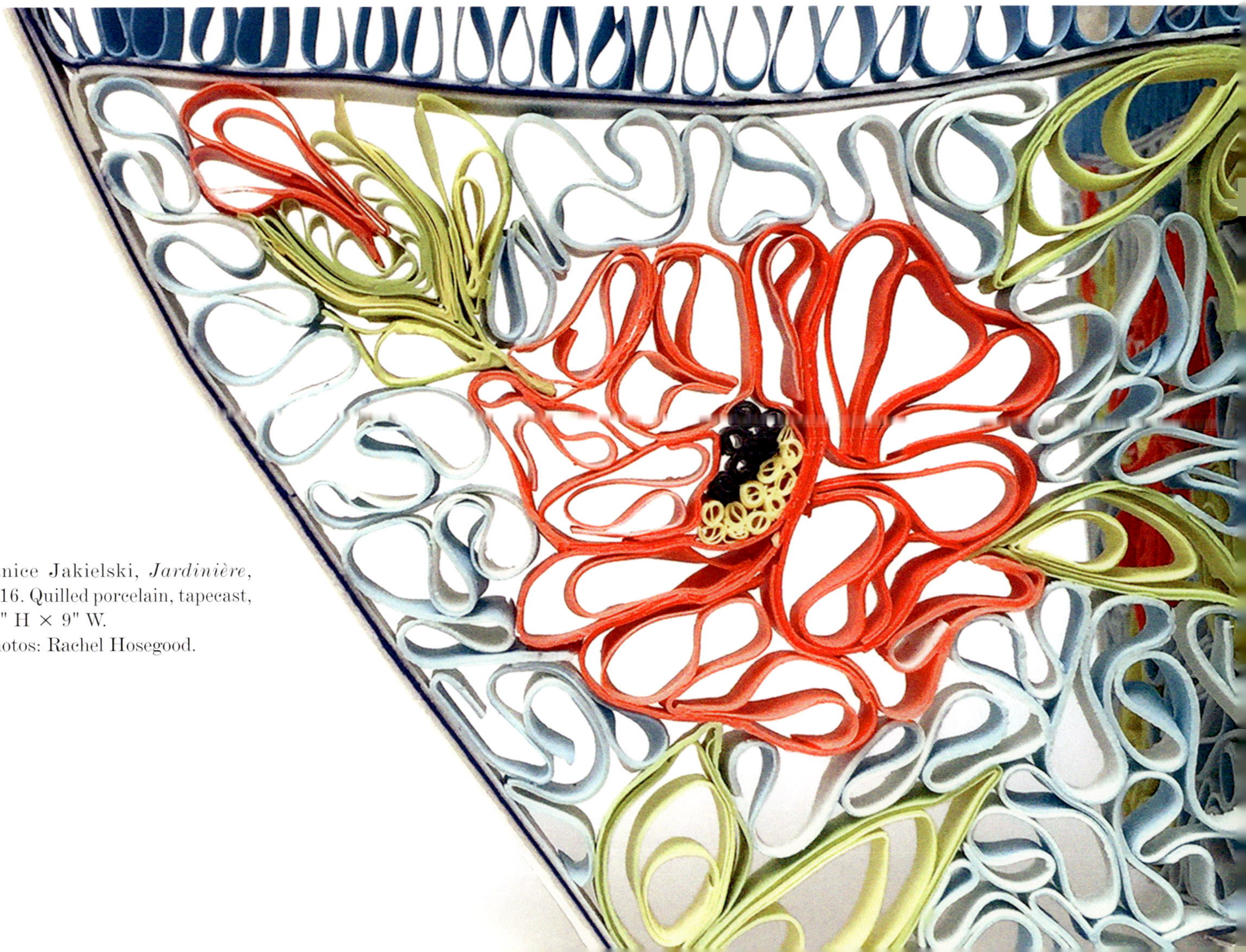

Janice Jakielski, *Jardinière*, 2016. Quilled porcelain, tapecast, 14" H × 9" W. Photos: Rachel Hosegood.

Cascade, Ian McMahon

Ian McMahon's *Cascade* is an impressive 40-by-21-foot curtain made of freestanding plaster. McMahon has pioneered a unique approach to casting thin plaster by spraying it into rubber molds, building up multiple layers until the casting becomes rigid enough to bear its own weight. Plaster can be very fragile, so part of what is jaw-dropping about this work, especially those who have worked in plaster, is that such a piece is even possible. The combination of its overall scale and its relative eggshell-thinness gives the impression of something towering, but so tenuous that it could shatter at any moment.

Monumentality and fragility are funny bedfellows. *Cascade* brings to mind the *Wizard of Oz*, an allegory for the Gilded Age, which somehow feels timely again today...it elicits the question, what lies behind that giant, fragile curtain?

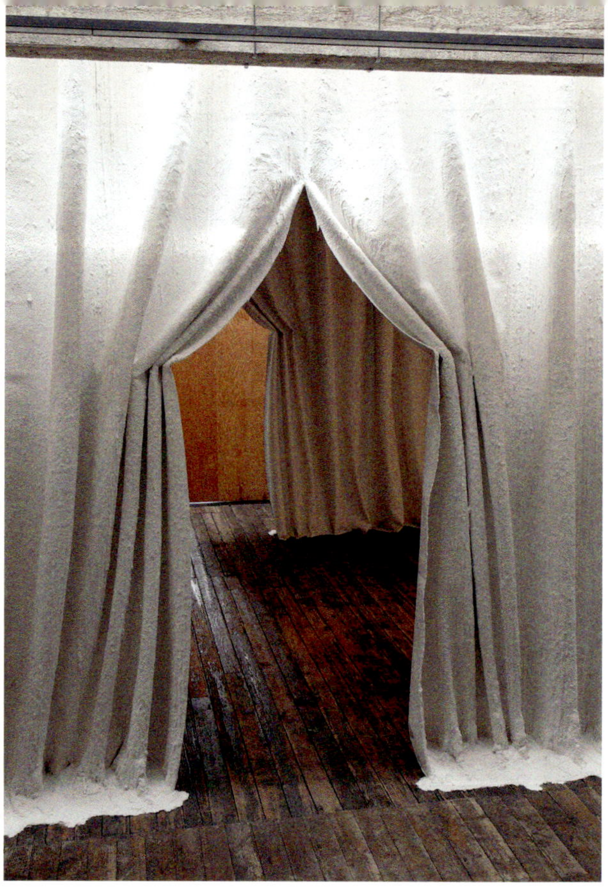

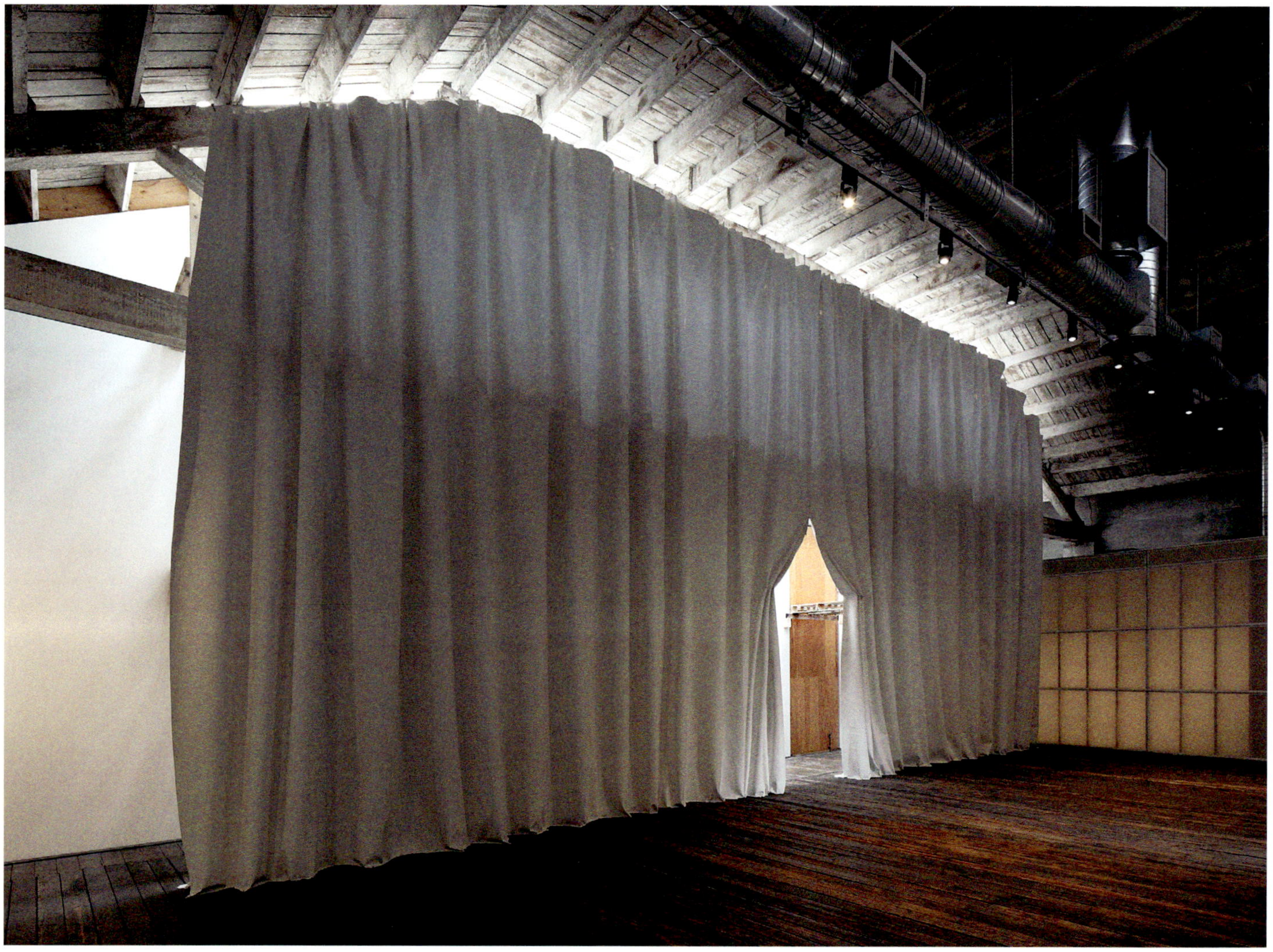

Ian McMahon, *Cascade*, 2014. Freestanding cast plaster, used pallets, each side: 40' × 6' × 21'. Photos: Mark Woods.

Shrouded, David Clarke

Pewter melts between 338° and 446°F (170° and 230°C); this is a surprisingly low melting point compared to most other metals, but this made David Clarke's *Shrouded* possible. Clarke cast pewter into an aluminum foil mold to create this piece. For metalsmiths who are used to casting iron, bronze, and even silver or gold, the thought of pouring molten metal into aluminum foil is nothing short of shocking. However, aluminum has a melting point of 1,221°F (660.5°C) and was thus not in danger of melting from the heat of the molten pewter.

Clarke made a conscious choice to use aluminum foil as a kind of "skin" in this piece. *Shrouded* is Clarke's response to his mother's death. The original model for this piece, one of her well-used teapots, was wrapped in foil, removed, and then Clarke cast pewter into the cavity, creating a teapot that has lost its function. This poignant work is touching in its sensitivity and its deeply personal content.

He says, "I had some keywords in mind to anchor me: void, skin, and loss. These led me toward casting pewter and lead; which was a surprisingly quiet and intimate process as you handle the molds, moving the metal until it sets... Literally, waiting for movement to cease, a skin to form and a void to be present. Stillness."[9]

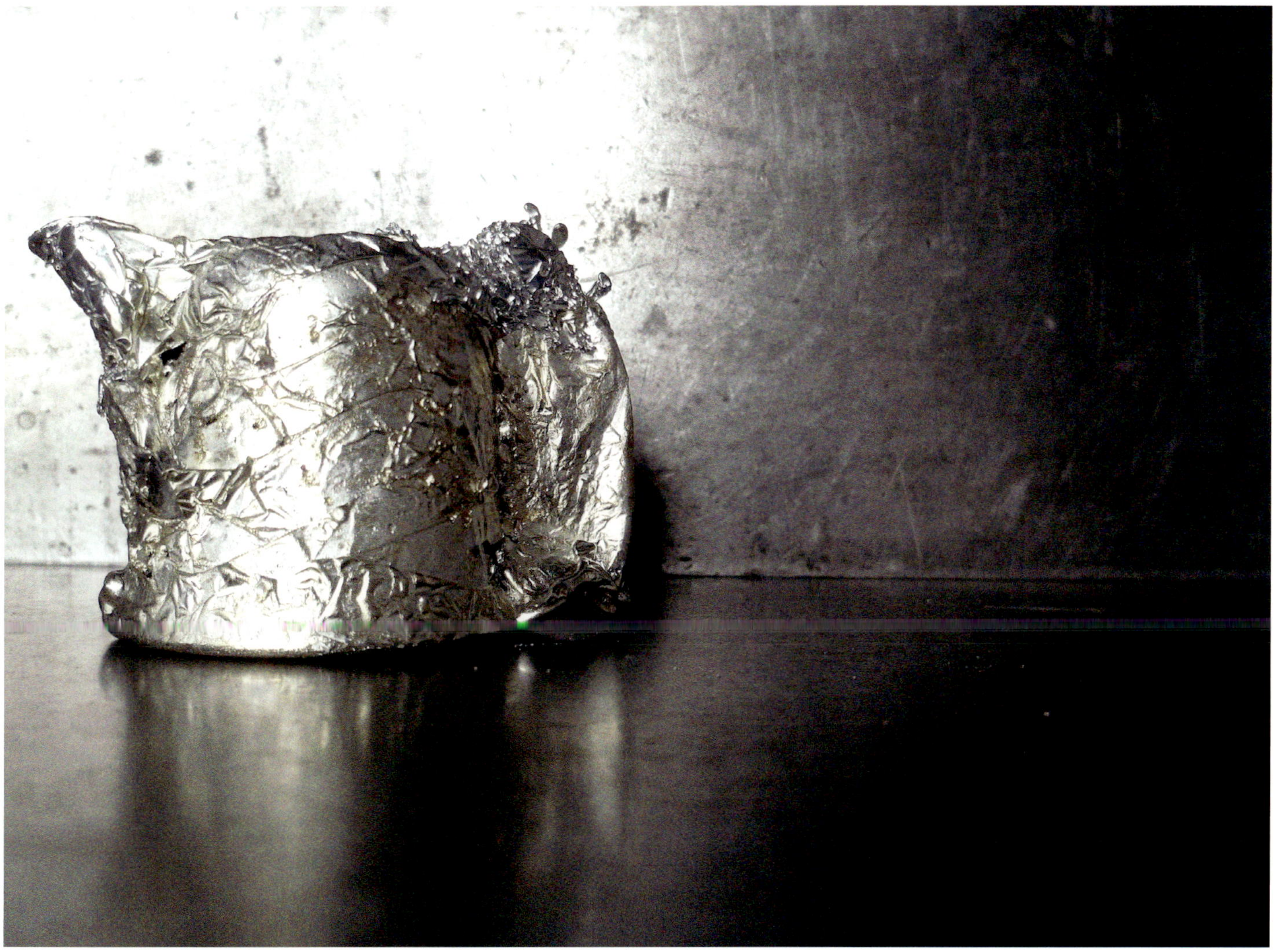

David Clarke, *Shrouded*, 2015. Pewter cast into aluminum foil, 4" × 3" × 6". Photo: David Clarke.

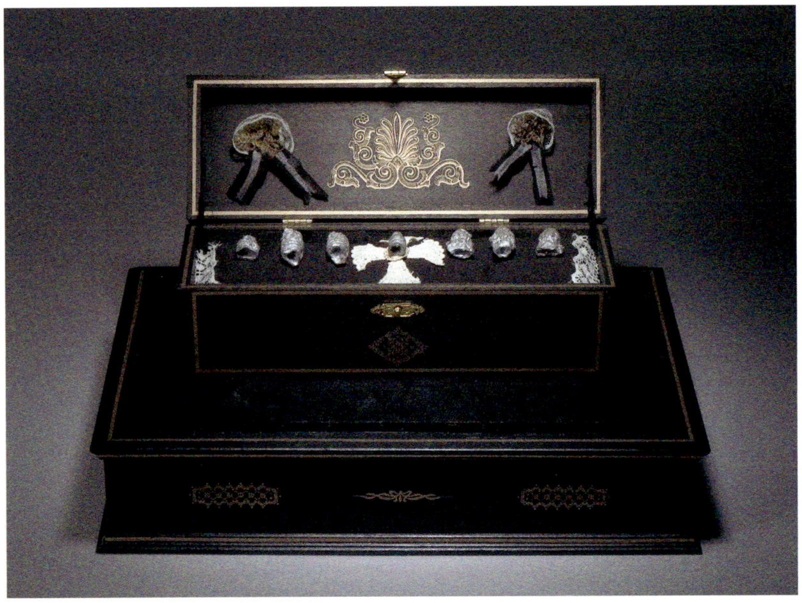

Dario Robleto, *A Sadness Silence Can't Touch*, 2005. Casts of Civil War era "pain bullets" (bullets used by soldiers to bite on during surgery in the absence of painkillers) made from dissolved audio tape recordings of poets' voices, lace and fabric from mourning dresses, ink dyed yellowheart and ash, nickel, silver, milk paint, typeset, 9.5" × 14.5" × 9.5".
Photos: Ansen Seale, Courtesy of Inman Gallery.

A Sadness Silence Can't Touch, Dario Robleto

Dario Robleto alloys his own, utterly unique, casting mediums made up of unusual materials often associated with sound. While the artworks don't emit sound themselves, Robleto's casting mediums evoke the recordings he used to cast his forms.

This piece features casts of Civil War era "pain bullets" (bullets used by soldiers to bite on during surgery in the absence of painkillers) made from dissolved audio tape recordings of poets' voices. The pain bullets are displayed in a Victorian-looking curio box festooned with bits of fabric and lace from mourning dresses, creating a piece that feels like a sacred relic oozing human suffering.

The thought of Walt Whitman, Lord Alfred Tennyson, T.S. Eliot, and other poets echoing from pain bullets is a shiver-inducing and eerie reminder of the costs of war.

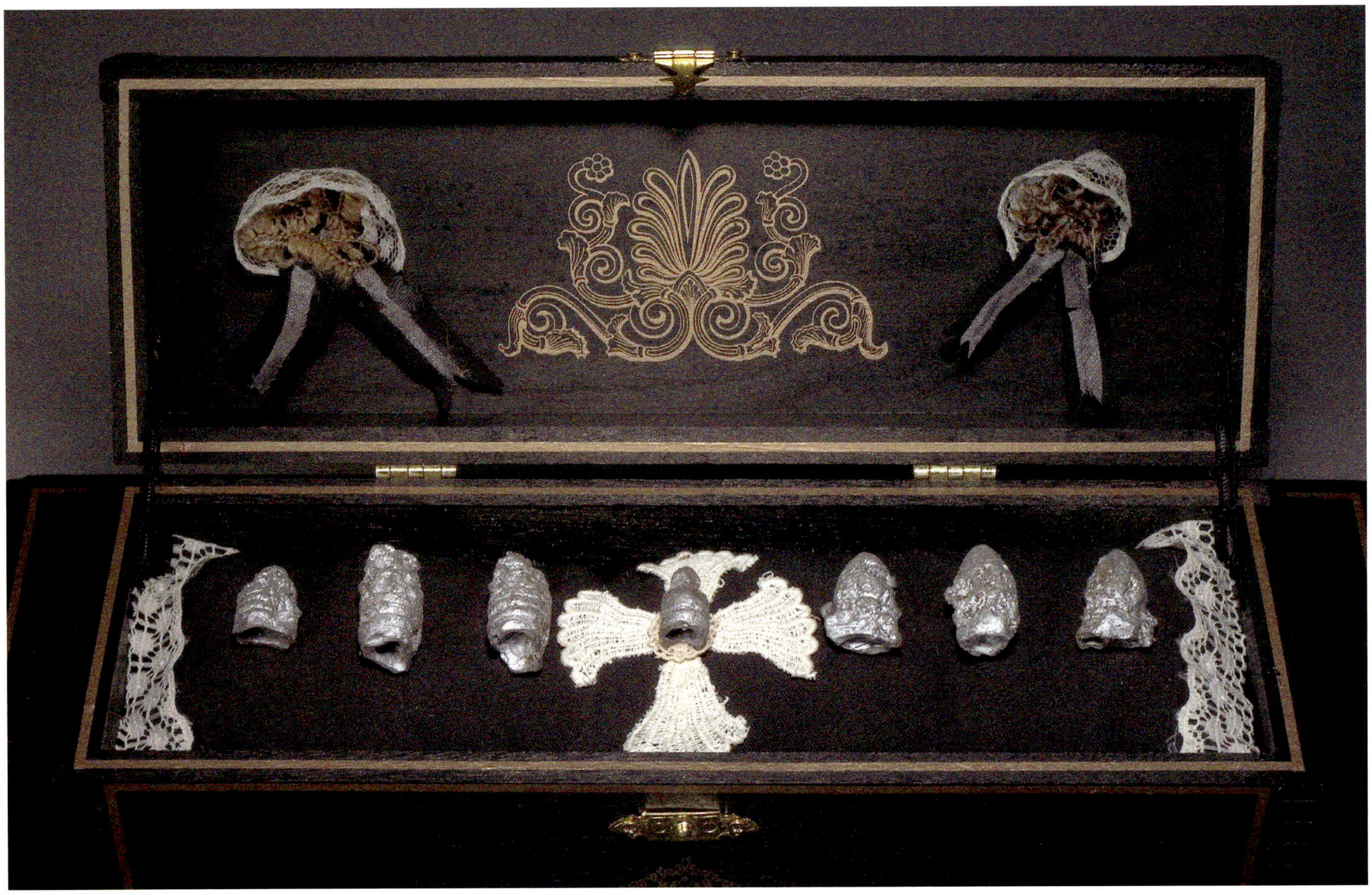

Archaeology of a Fox, Maree Santilla

Maree Santilla approaches casting in a manner completely different than any other artist in this book, yet in a way familiar to most of us. She casts roadkill using clay bandages just like a doctor would cast a broken limb. While casting with bandages is common in medicine, it is unusual in art as a final product. In the context of this book, Santilla's interpretation of the word "cast" breaks the mold, so to speak.

This piece began when Santilla found a dead fox on the side of the road. She collected the body, froze it, and "cast" it in clay bandages. After that, she wrapped the bandaged animal in a fabric shroud. Then, she fired the mummified form in a raku style, essentially cremating the body of the fox, leaving an empty ceramic shell behind as a reminder of the fragile, broken body.

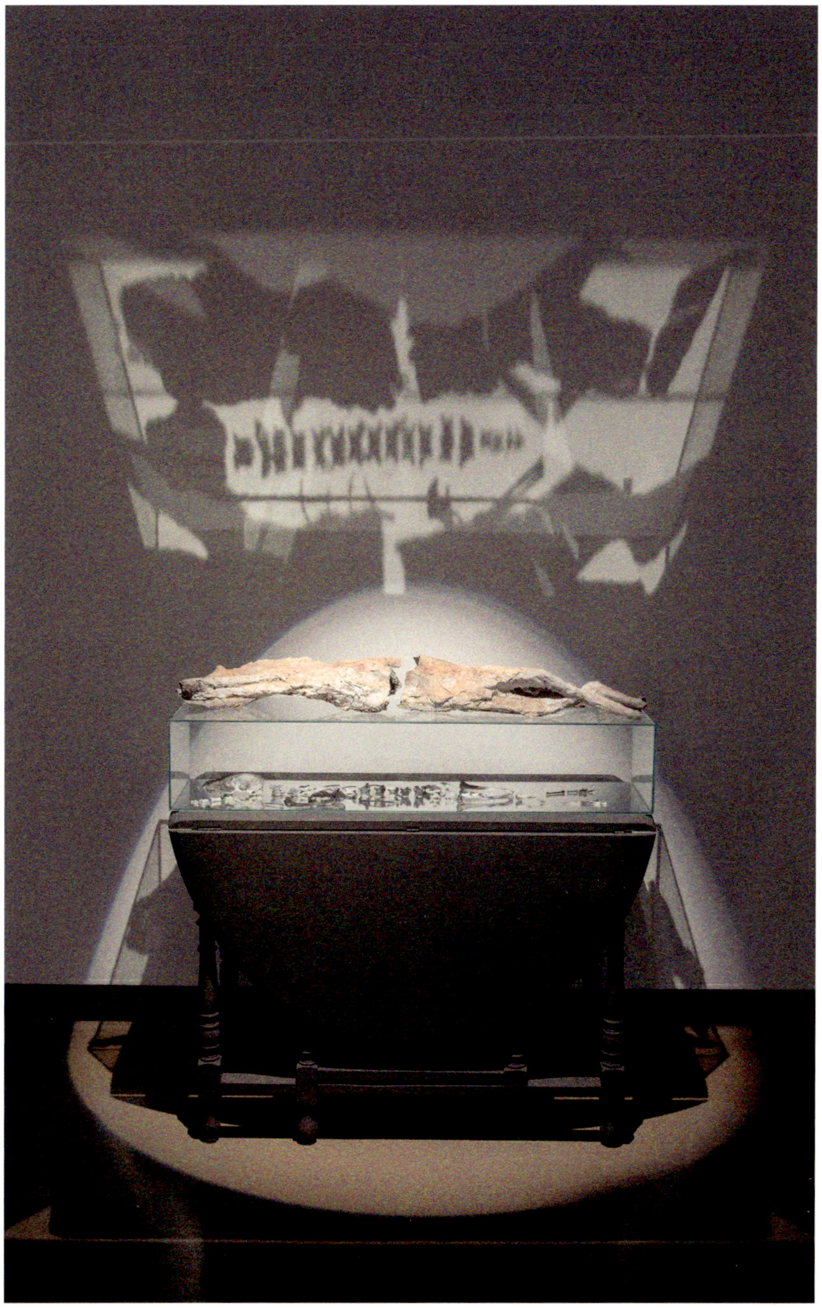

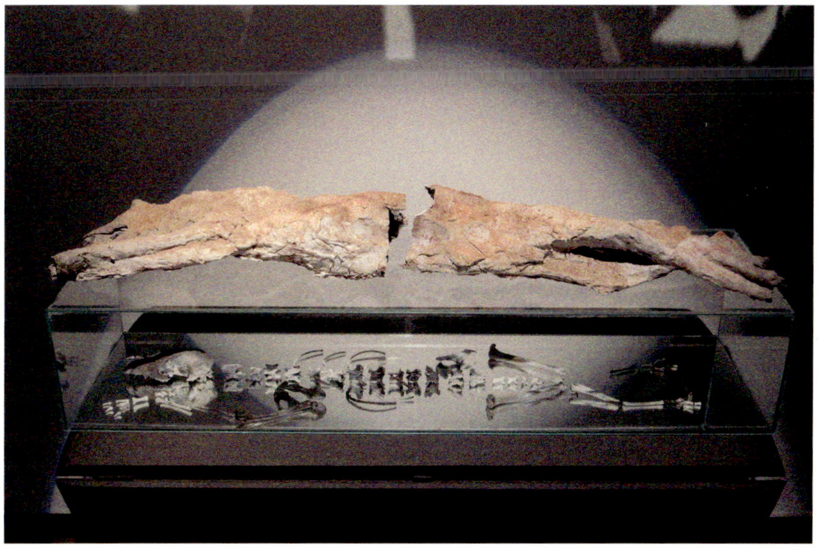

Maree Santilla, *Archaeology of a Fox*, 2014. Raku, paper clay, oxides, 3'11.25" × 1'5.7" × 6". Photos: Ian Hill.

Self, Marc Quinn

When considering unconventional approaches to casting, Marc Quinn's *Self* is a superb example. This groundbreaking work, inspired by Rembrandt's lifelong commitment to self-portraits, is actually a series of lifecasts of Quinn's head cast in his own blood. The potency of blood as a material combined with its ties to science, medicine, religion, fragility, mortality, and life itself, makes it a powerful medium for self-portraiture.

Every five years, over a course of months, Quinn has eight pints of his own blood drawn to cast the next iteration of *Self*. There are five of the "blood heads" so far and that number will continue to grow into a gang of aging lookalikes. When we consider this piece in the context of cloning research, *Self* is quite thought provoking.

This temporal work could not exist without modern medicine and technology. Perfectly calibrated refrigeration technology to keep the casting in a state of suspended animation and the ability to draw and store blood without coagulation are undeniable signs of our time. *Self* explores our relationship to and dependency on science and technology, but more importantly, Quinn is exploring his complex relationship to his own mortality. This ongoing series of works begs the question, how many will be made?

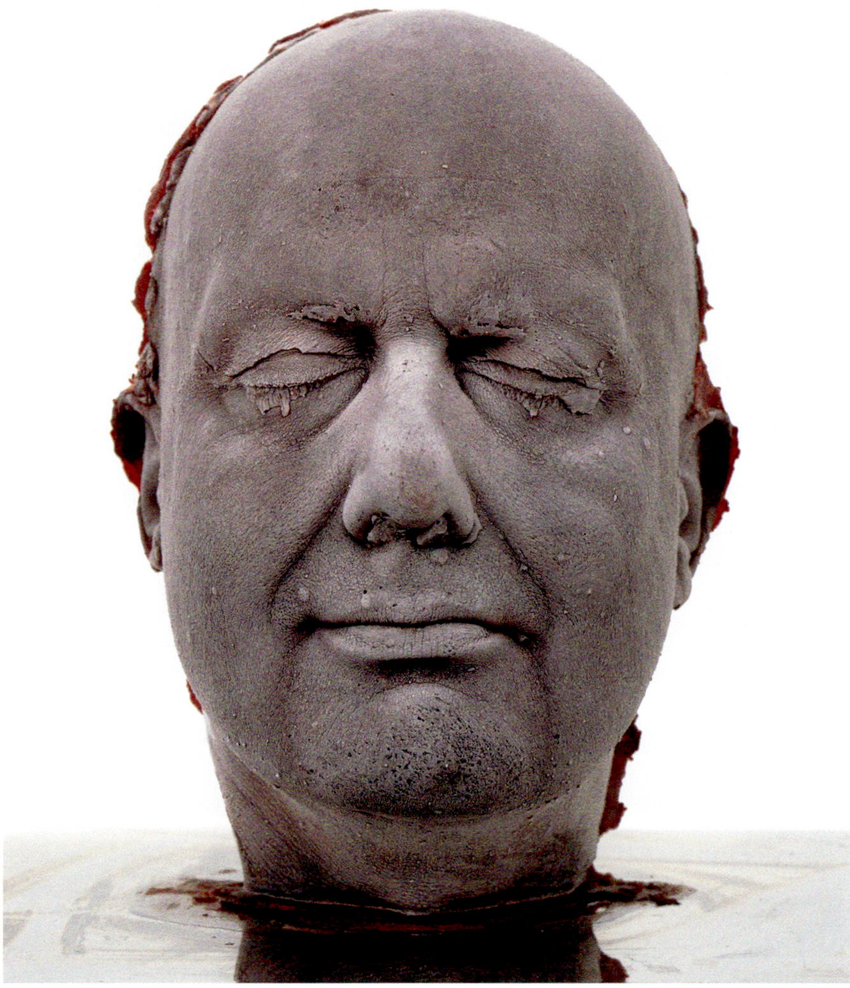

Marc Quinn, *Self*, 2006. Blood (artist's), stainless steel, Perspex and refrigeration equipment, 6'10" × 2' × 2'.
Courtesy of Marc Quinn Studio.

Janine Antoni also used lifecasting to create cutting edge self-portraits in her series, *Lick and Lather*. Inspired by classical sculpture, this work comprises fourteen busts, seven cast in chocolate and seven in soap. After these pieces were cast, Antoni washed herself with the soap busts and licked the chocolate ones until their features began to fade. Each one was organically eroded, so while they were all cast from the same mold, some are rendered completely unrecognizable while others retain most of Antoni's features.

She says, "Both the licking and the bathing are quite gentle and loving acts, but what's interesting is that I'm slowly erasing myself through the process. So for me it's about that conflict, that love/hate relationship we have with our physical appearance, and the problem I have with looking in the mirror and thinking, 'Is that who I am?'"[10]

Light and darkness, self-care and self-destruction, indulgence, intimacy, erosion, and our corporeality...*Lick and Lather* probes all of these ideas and presents us with our own multiplicity through the many faces of Janine Antoni.

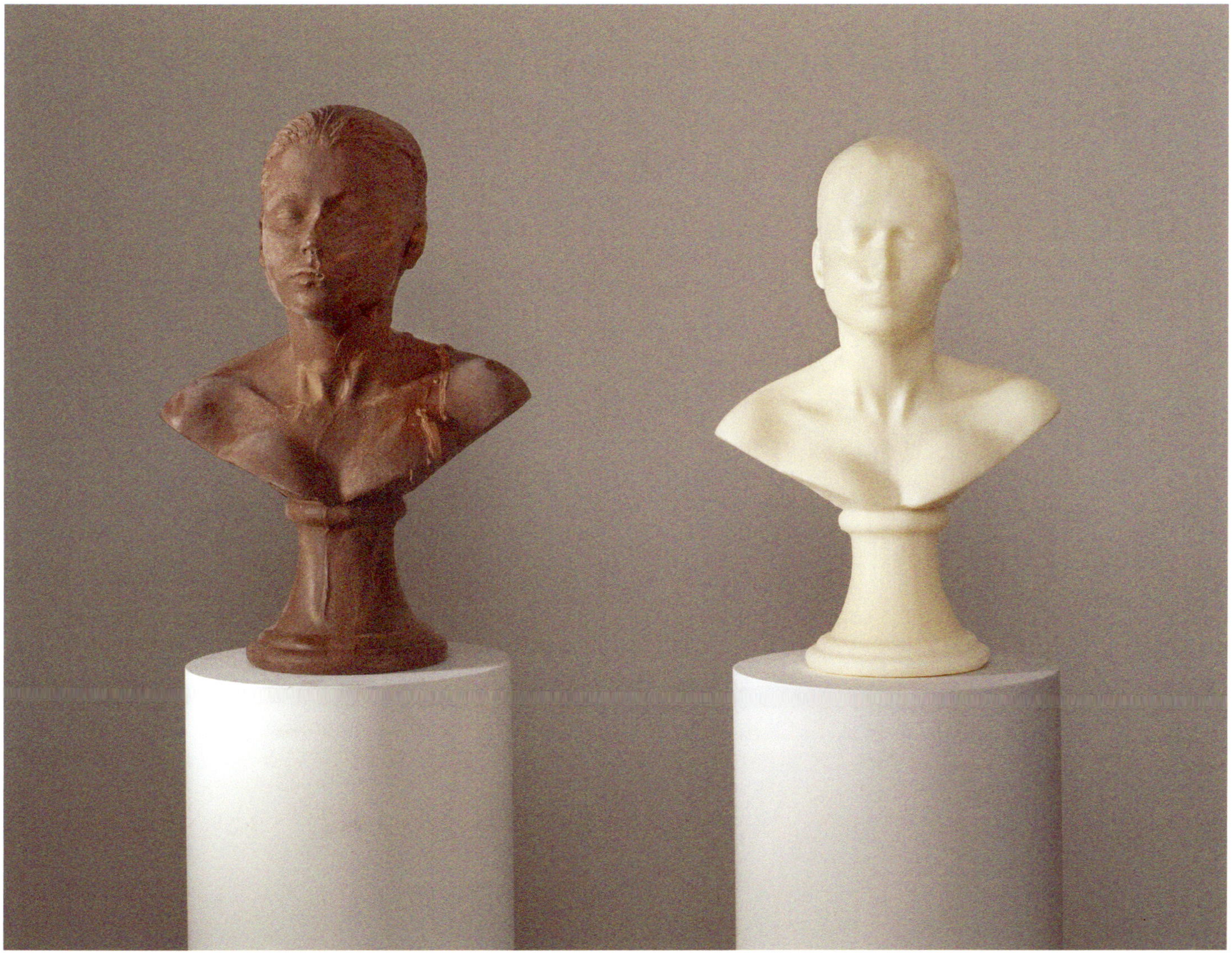

Janine Antoni, *Lick and Lather*, 1993. Two self-portrait busts: one chocolate and one soap, 24" × 16" × 13". Pedestals 45.875" × 16". Photo: Ben Blackwell, © Janine Antoni, Courtesy of the artist and Luhring Augustine, New York.

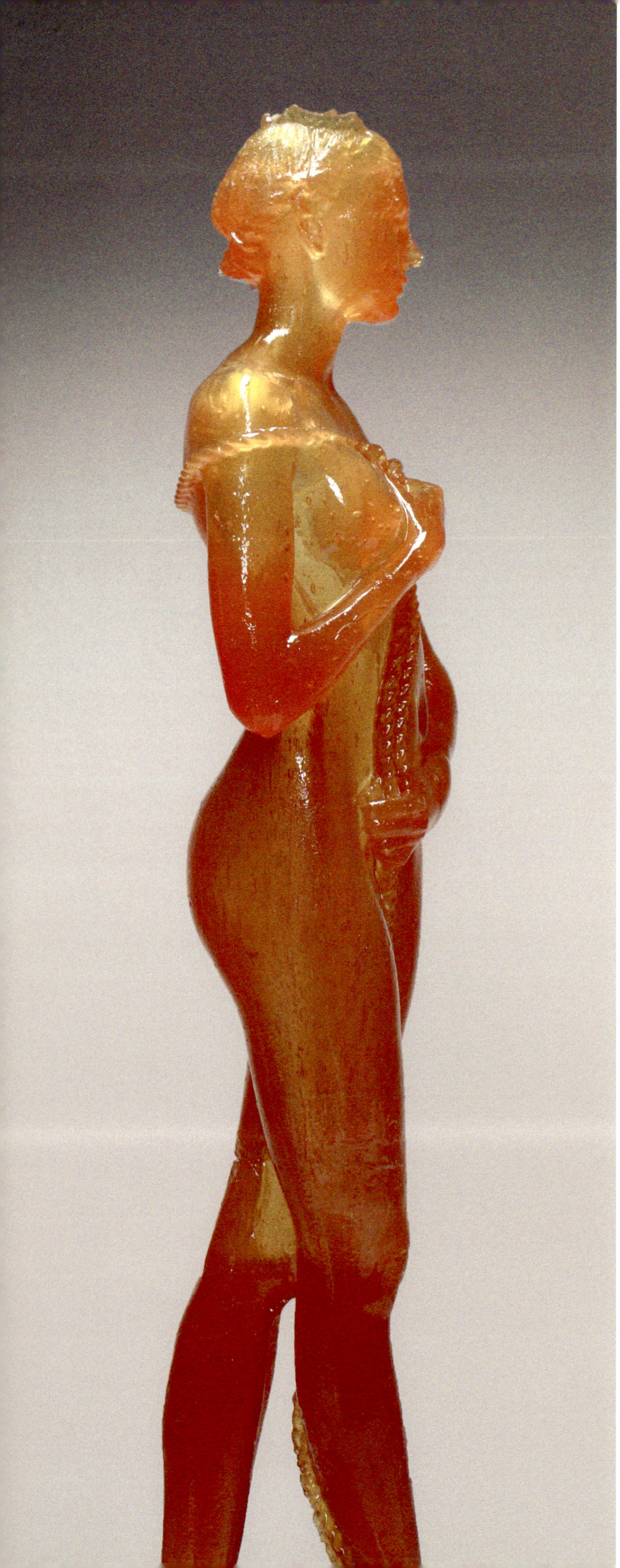

Fanta Laura, Joseph Marr

Joseph Marr's *Fanta Laura*, from the series *Desire is a trap*, is cast in passionfruit flavored sugar preserved in resin. Casting with sugar, especially on a large scale, is technically very challenging, but is an exciting (if dangerous) sensory explosion. The smell, color, heat, and the fluidity of the molten candy is as sexy as the figure itself. The glass-like shards of shattered sugar and potential for burns from the liquefied syrup, are every bit as painful as desire gone wrong....

Marr explains, "There is a statement in Hindu Sufism, 'desire is a trap,' which was the first impulse for me to work with sugar. The glistening beautiful colors of the different sugary flavors are the sticky trap, the object/form being the bait."[11]

Joseph Marr, *Fanta Laura*, 2011. Passionfruit flavored sugar preserved in resin, 43.3" × 12.6" × 9.45". Photo: © Atelier Joseph Marr, Private collections in Australia, Germany and America.

Sugar Liquor Capsules, Fernando LaPosse

Fernando LaPosse's cast sugar "gems" are made to be injected with fresh cocktails. They can either be taken as a shot or dropped into a glass to dissolve before being consumed.

Because the gems are filled with liquid, even a pinhole flaw can ruin the effect. Although they are ingested and only exist for a brief moment in time, these pieces require just as much fine craft as any other casting. They arouse all our senses—their precision of form and beautiful color, their smell, taste, tactility, and the sound as they crack apart. They create an elevated and memorable experience for anyone lucky enough to have one. It's easy to imagine one of these jewels breaking in an explosion of fresh negroni and crisp sugar and the intoxicating thrill of devouring something so precious.

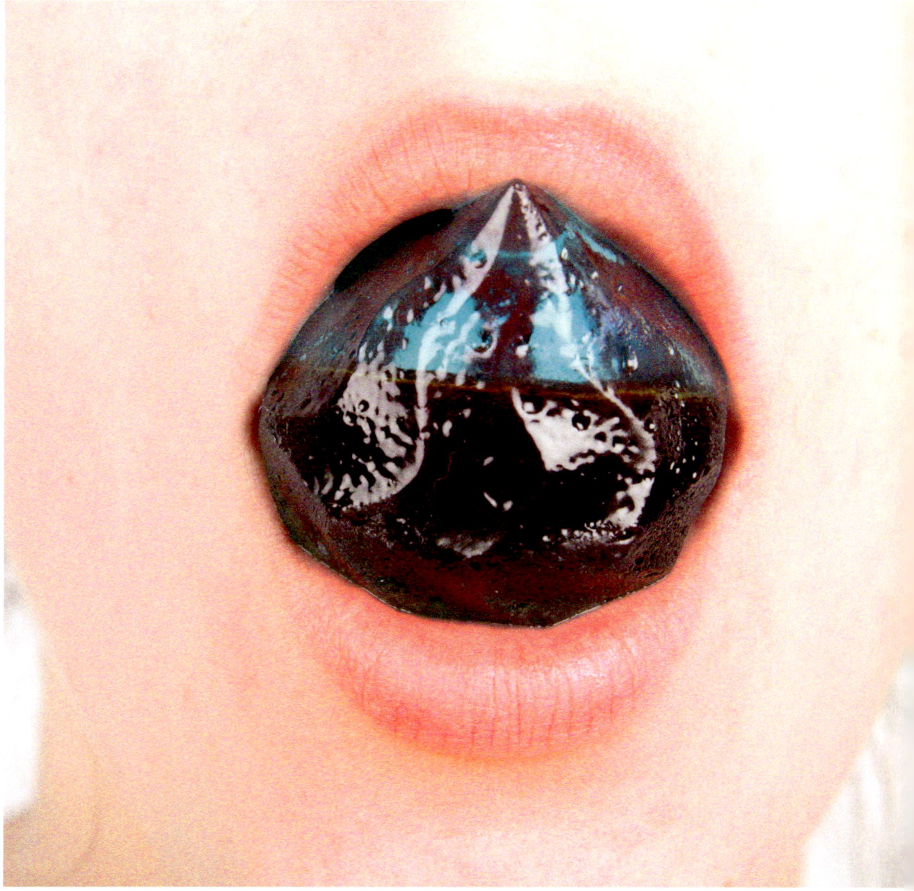

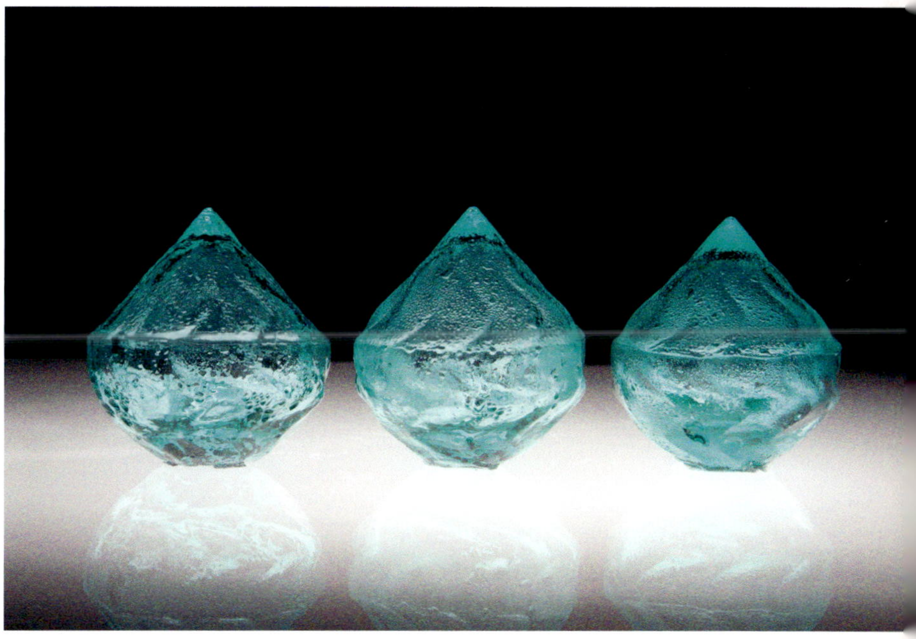

Top Fernando LaPosse, *Sugar Liquor Capsule* taken as a shot, 2014. Sugar, liquor. Photo: Fernando LaPosse.

Bottom Fernando LaPosse, *Sugar Liquor Capsules*, 2014. Sugar, liquor. Photo: Fernando LaPosse.

Monumento Minimo, Néle Azevedo

Néle Azevedo casts in ice, creating installations that slowly melt and fade away to nothing but puddles. *Monumento Minimo* stands in stark contrast to the permanent sculptures that adorn public spaces worldwide. While most monuments celebrate leaders and heroes, Azevedo's work honors the anonymous masses, the everyday, and the ephemeral.

Monumento Minimo has been installed in cities around the world. Azevedo casts thousands of figures in ice and transports them in freezer chests to her installation sites. Members of the public help put them in place, a fitting way to install this egalitarian work. Azevedo's installations offer a quiet, poignant, and contemplative moment in the heart of bustling cities reminding us of the ephemeral nature of our lives.

Néle Azevedo, *Monumento Minimo* (Santiago de Chile Agosto). © 2012. Ice, dimensions variable.
Photo: © Néle Azevedo.

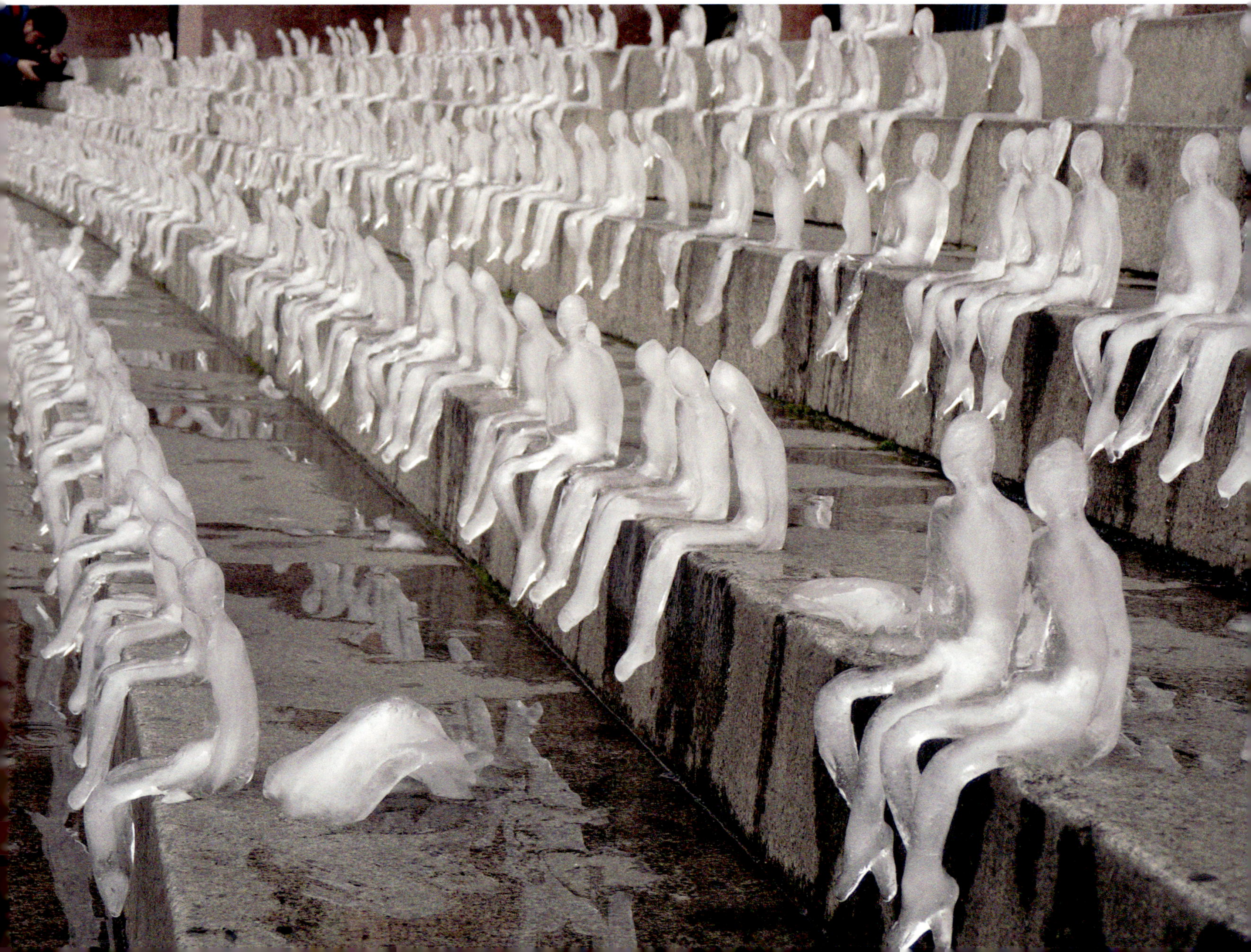

Vicissitudes, Jason deCaires Taylor

This underwater work by Jason deCaires Taylor is made up of a circle of lifecasts of children holding hands. This concrete piece permanently lives on the seafloor off the coast of Grenada and is intended to become a part of the oceanic ecosystem. While it was not originally made as a statement on the Middle Passage and slave trade, the ethnicity of the children in the work as well as its physical location make it easy to draw connections and many have done so.

Jason deCaires Taylor faces huge technical challenges in creating his underwater concrete sculptures. Using just the right kind of concrete, getting the work into place, and assuring its stability are all hurdles on top of the standard complications of making large sculpture. Beyond its technical bravado, this haunting work is both touching and provocative, even as it's being quietly engulfed by the sea.

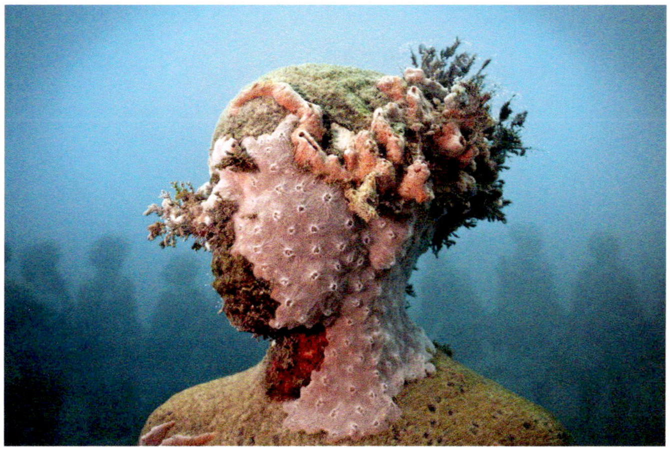

Top Jason deCaires Taylor, *Vicissitudes* (detail), 2007.

Photo: Jason deCaires Taylor.

Bottom Jason deCaires Taylor, *Vicissitudes*, 2007. Grenada Underwater Sculpture Park, cement.

Photo: Jason deCaires Taylor.

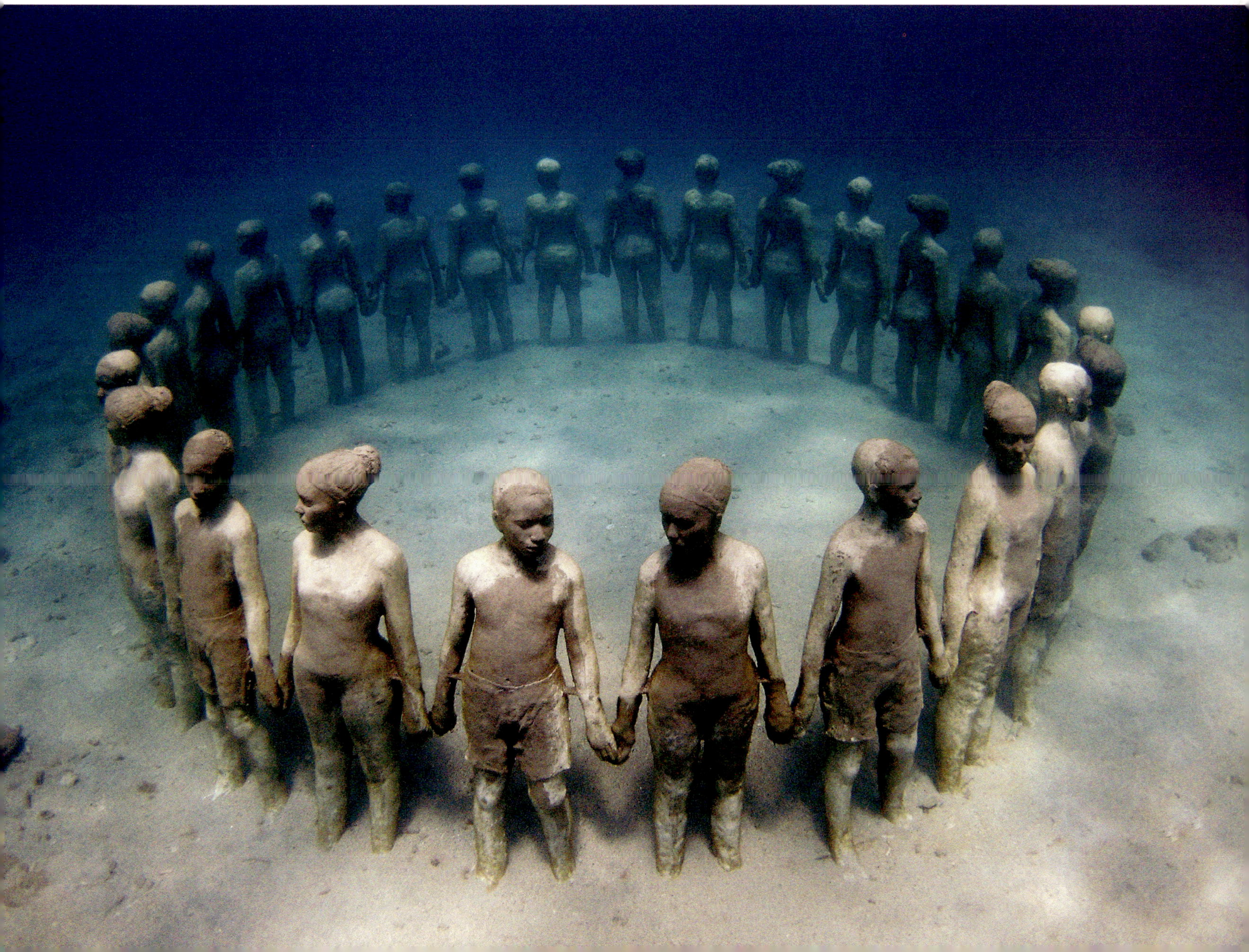

My Soul Is Rarely with Me Anymore, Dario Robleto

Casting is a mourner's plea, a not-yet-ready reckoning with loss on the threshold. This plea turns to an irrational rebellion and refusal, insisting on the exactitude of the replica as a form of cheating death: the cast as after-life. But in this solemn effort, this casting-as-rebellion-to-death, one must come to terms with the futility of the gesture. Something changed—through the uneasy mix of earth, chemicals, and emotional expectations; in the trembling of the pour; the breaking of the inexact mold that now passes on its impurities like an heirloom; the materials struggling to embody the impossibility now asked of them; and disappointment that the grief still persists. In the cast's inevitable mutations, fluctuations, and inadequacies is the mourner's chance for a reluctant but open-eyed honesty: it can never be the same. Instead of the cast as after-life—as forever-preserved replication—the cast becomes a form of healing in the present life, a truce with the transient.[12]

—Dario Robleto

Opposite page Dario Robleto, *My Soul Is Rarely with Me Anymore*, 2003. Cast dehydrated bone calcium, bone dust from every bone in the body, swamp salt, swamp root, beeswax, ironized yeast, brimstone, witch hazel leaf, spirit of ammonia, diatomaceous earth, sawdust, leeches made from Muddy Waters's "Rollin' Stone" and The Rolling Stones' "Sympathy for the Devil" melted vinyl records, water extendable resin, typeset, 19" × 9" × 7".
Photo: Ansen Seale, Inman Gallery.

Notes

1. Email from Ken and Julia Yonetani, October 18, 2015.
2. Bethany Bell, "Austria's Delayed Holocaust Memorial," BBC News Europe, Oct. 25, 2000, Web, accessed July 20, 2016, http://news.bbc.co.uk/2/hi/europe/989255.stm.
3. Jane Perlez, "Vienna to Go Ahead with Holocaust Shrine," *New York Times,* Mar. 3, 1998, Web, accessed July 20, 2016, http://www.nytimes.com/1998/03/04/world/vienna-to-go-ahead-with-holocaust-shrine.html.
4. James E. Young, "Memory and Counter-Memory," editorial, *Harvard Design Magazine,* 1999, Web, accessed July 20, 2016, http://www.harvarddesignmagazine.org/issues/9/memory-and-counter-memory.
5. Hugh Honour and John Fleming, *A World History of Art* (London: Laurence King Publishing, 2005), 215.
6. Email from Allen Peterson, December 17, 2015.
7. Email from Chad Gunderson, September 7, 2015.
8. Email from Janice Jakielski, June 28, 2016.
9. Email from David Clarke, June 5, 2016.
10. "Lick and Lather, Detail (1993)," PBS, Web, accessed July 29, 2016, http://www.pbs.org/art21/images/janine-antoni/lick-and-lather-detail-1993.
11. Email from Joseph Marr, April 25, 2016.
12. Email from Dario Robleto, March 19, 2016.

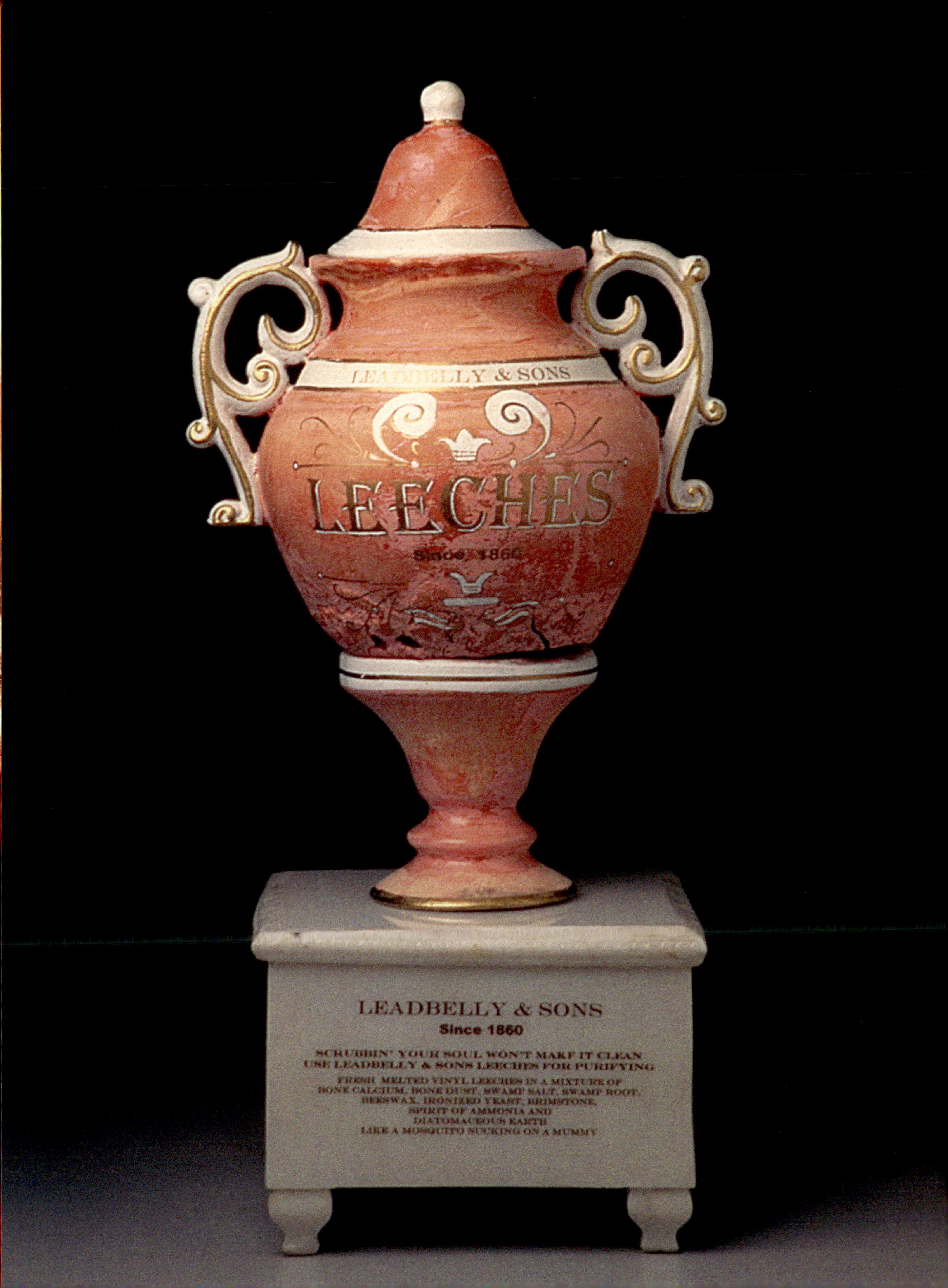

LEADBELLY & SONS
LEECHES
Since 1860
LEADBELLY & SONS
Since 1860
SCRUBBIN' YOUR SOUL WON'T MAKE IT CLEAN
USE LEADBELLY & SONS LEECHES FOR PURIFYING
FRESH MELTED VINYL LEECHES IN A MIXTURE OF
BONE CALCIUM, BONE DUST, SWAMP SALT, SWAMP ROOT,
BEESWAX, IRONIZED YEAST, BRIMSTONE,
SPIRIT OF AMMONIA AND
DIATOMACEOUS EARTH
LIKE A MOSQUITO SUCKING ON A MUMMY

TRANSFORMATION

One definition of alchemy is: "any magical power or process of transmuting a common substance, usually of little value, into a substance of great value."[1] Another source defines it as "a seemingly magical process of transformation, creation, or combination."[2]

Casting can have a magical feeling to it and it's easy to see how comparisons can be drawn between the process and the ancient tradition. Alchemy was practiced in ancient Egypt, the Islamic world, Asia, and Europe; each region had its own alchemical practices, but they shared some common goals. Turning base metals into gold, the creation of a universal solvent, and curing illness with panaceas were some of the common ambitions of alchemists everywhere. There was also a spiritual component to alchemy in the form of purifying, perfecting, and transforming the state of body and soul, with immortality being the ultimate goal. Alchemy was pursued with a religious fervor for centuries in many parts of the world. Each tradition was unique, but transmutation and transformation were at the core of all of them. These concepts can be viewed literally or symbolically.

Casting can also be viewed through those two lenses: the interest in the technical and the interest in casting as metaphor.

Casting (like alchemy) is a lifelong fascination for many practitioners—both the practical and the esoteric traditions can bewitch devotees, with fascinating results. The obsession with transformation is something many casters talk about... the drama of the smoke, the flash of molten metal, and the magic of de-molding a casting are moments that are hard to explain, but the instant a casting is revealed is one all its own. Like the alchemists' quest to transmute base metals into gold, casters voraciously pursue their own mission to transform, relishing every development and success.

The "practical" side of casting is the process itself. It includes design, model and moldmaking, the pour, and the finishing (completing the transformative process). Many makers use the process of casting as a driving force in their work.

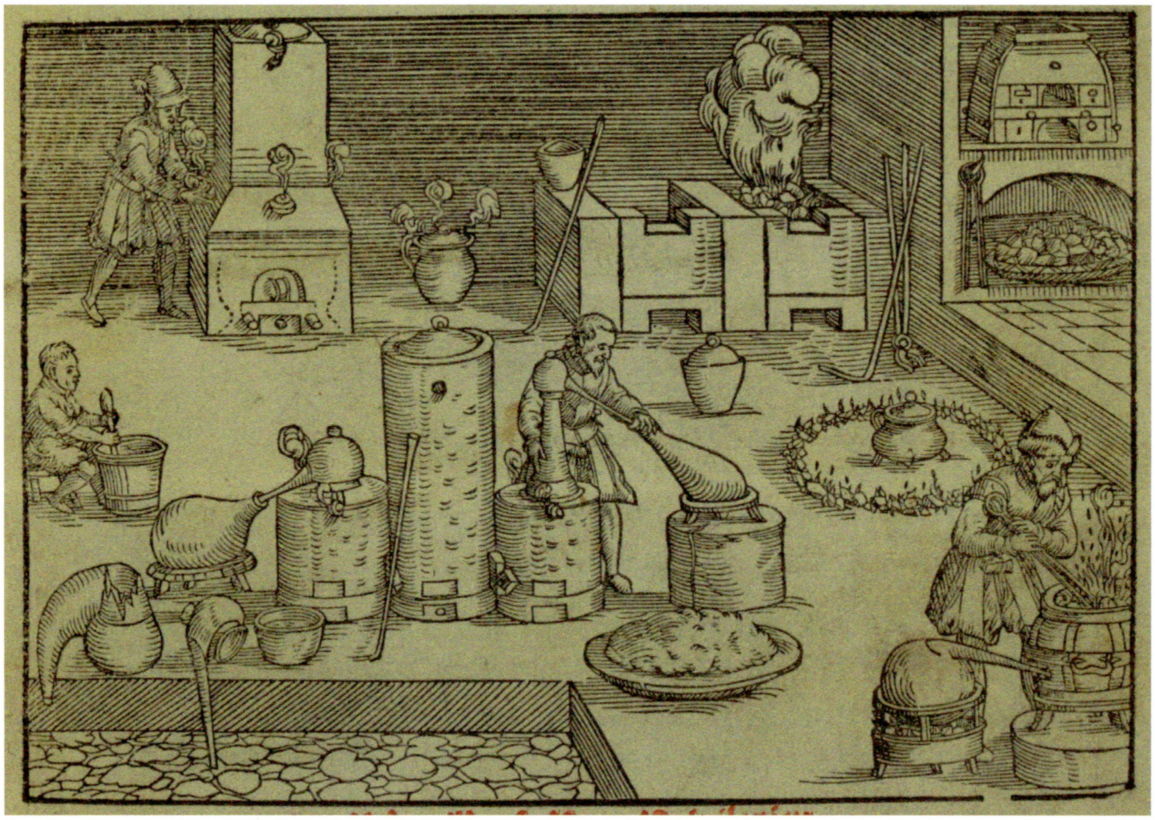

Lazarus Ercker, *The Alchemist's Workshop*, 1580. Engraving.
Courtesy of the Chemical Heritage Foundation, Roy G. Neville Historical Chemical Library, Chemical Heritage Foundation, public domain.

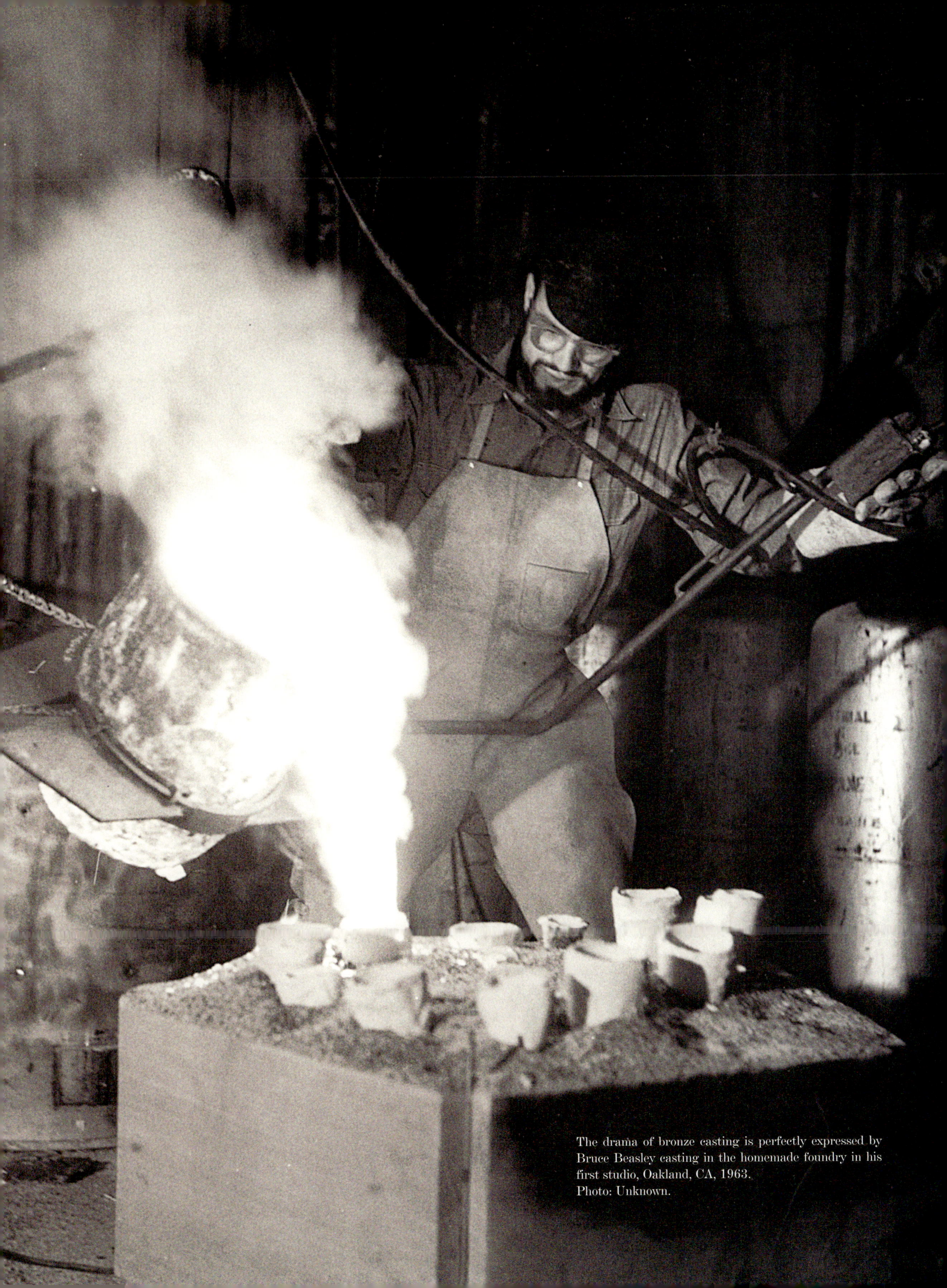

The drama of bronze casting is perfectly expressed by Bruce Beasley casting in the homemade foundry in his first studio, Oakland, CA, 1963.
Photo: Unknown.

Zsolt Simon, for example, transforms negative space into positive forms. Unlike most casters, Simon begins not with a model, but with making a mold. His molds, made up of countless hand-carved plaster parts, fit together like a three-dimensional puzzle. Using tiny porcelain shims, each piece of the mold is ever-so-slightly separated from the next, creating a maze of narrow passageways. The porcelain slip then fills these passageways creating forms that feel both natural and mysterious.

Simon intuitively transforms his materials; starting with a block of plaster and a bucket of liquid porcelain slip, he gives form to the void. His unique connection to the casting process allows him to transform matter and create seemingly effortless and endlessly fascinating work.

Natalie Tyler is also a purveyor of transformation, but hers is making the ephemeral durable. *I Girasoli* is a relatively straightforward piece of sculpture but its creation was surely not simple. The practical, process-based hurdles of scale, changes in thickness, burning out the original sunflowers properly, and achieving just the right temperature (in both the mold and the bronze) for the molten metal to fill every nook and cranny all have to be mastered to pull off this deceptively simple piece. The instant cognizance of these transmuted sunflowers conjures a deep connection…such is the power of transformation.

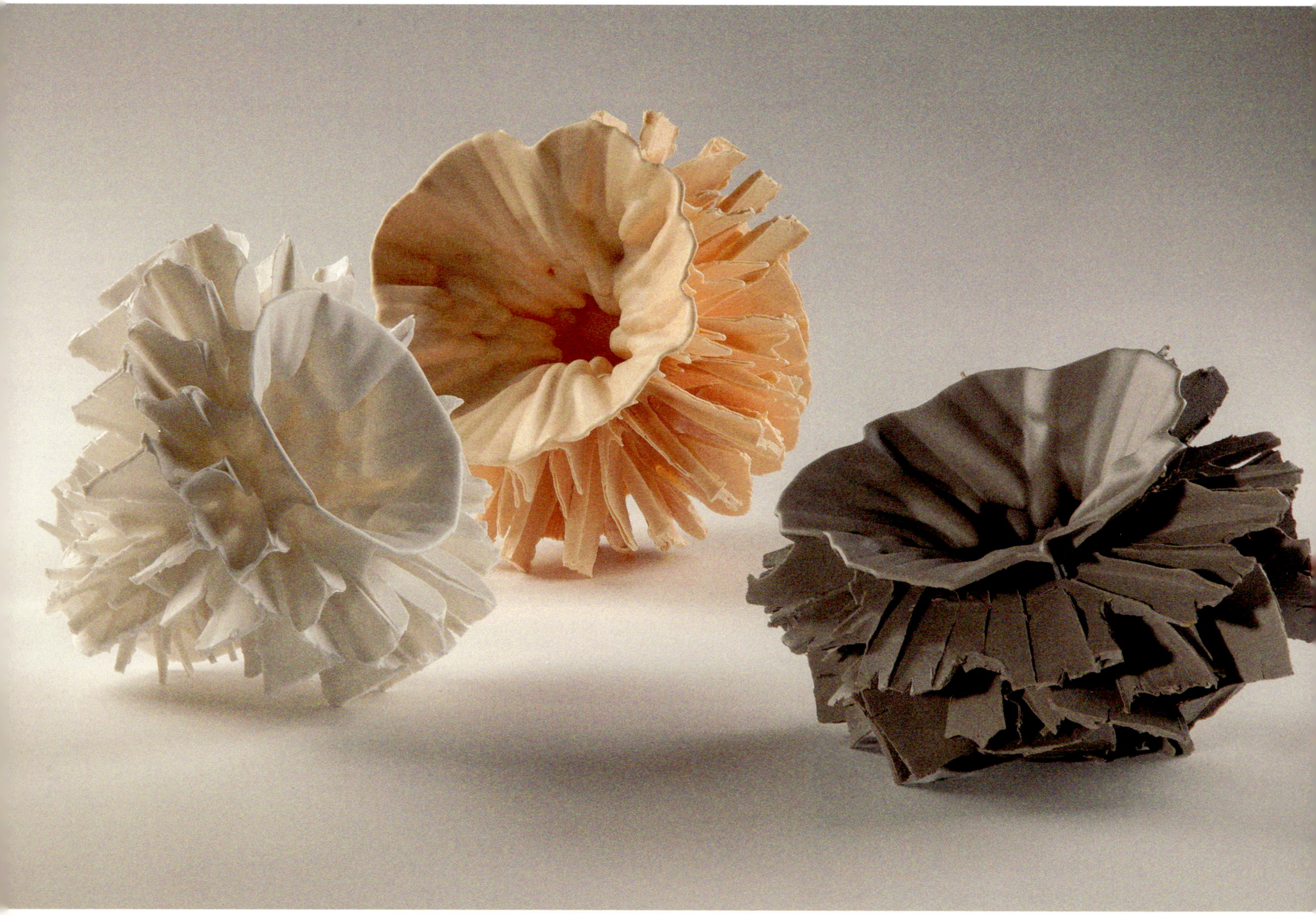

Tyler's brand of transformation is different than Zsolt Simon's though—hers reaches for immortality. The alchemist's obsession with immortality is matched by many who cast; whether an army of terracotta soldiers or bronze sunflowers, many castings fascinate us because they bring durability to ephemera. Tyler's sunflowers, with their nodding seed heads and withering petals, are frozen at the end of their lives in a permanent autumn. This poignant piece seems to conjure lifecycles, the beauty of age, mortality and immortality. Freezing a moment in time has an otherworldly feeling to it and for the caster, there is a deep satisfaction in being able to stop time and bring about this type of wondrous transformation.

I Girasoli is successful in both the practical and technical as well as the esoteric and conceptual aspects. Some makers are interested simply in the technical aspects of the process and these concerns are important, especially for industrial casting. A kitchen sink need not have a conceptual underpinning, but "simply" requires technical perfection and to be reproducible on a mass scale, which is no small feat. Casting, a quintessentially useful and practical process, also has some baked-in conceptual implications and opportunities, many of which go hand-in-hand with the ideas of alchemy.

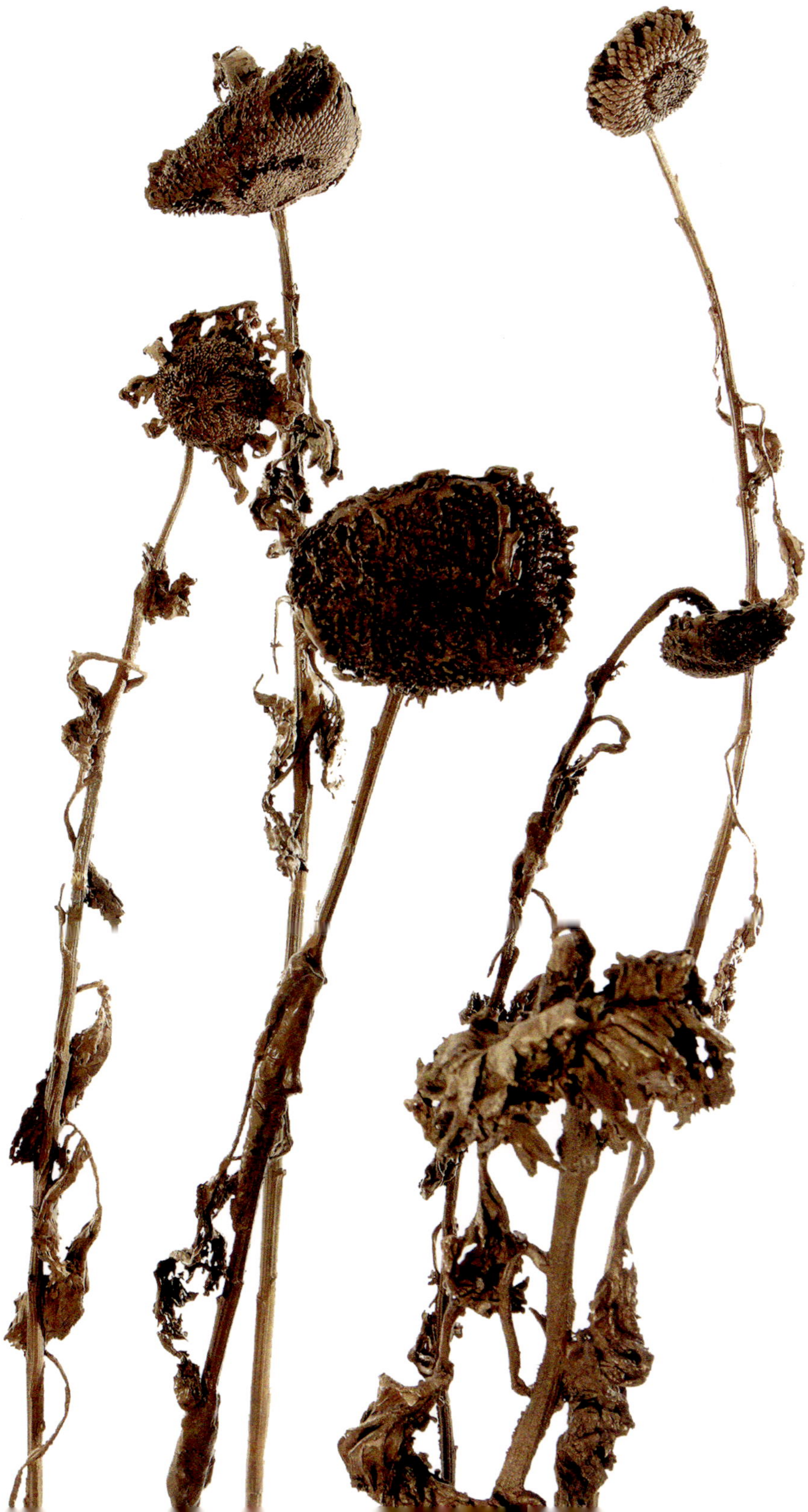

Natalie Tyler, *I Girasoli*, 2002. Bronze, 7' × 15' × 7'. Photo: Natalie Tyler.

Opposite page Zsolt József Simon, *Untitled Trio*, 2014. Porcelain, pigment, no glaze, (*left to right*) 10.25" × 10.25" × 7.9"; 10.25" × 10.25" × 7.9"; 10.6" × 10.6" × 4.25". Photo: János Bánáti.

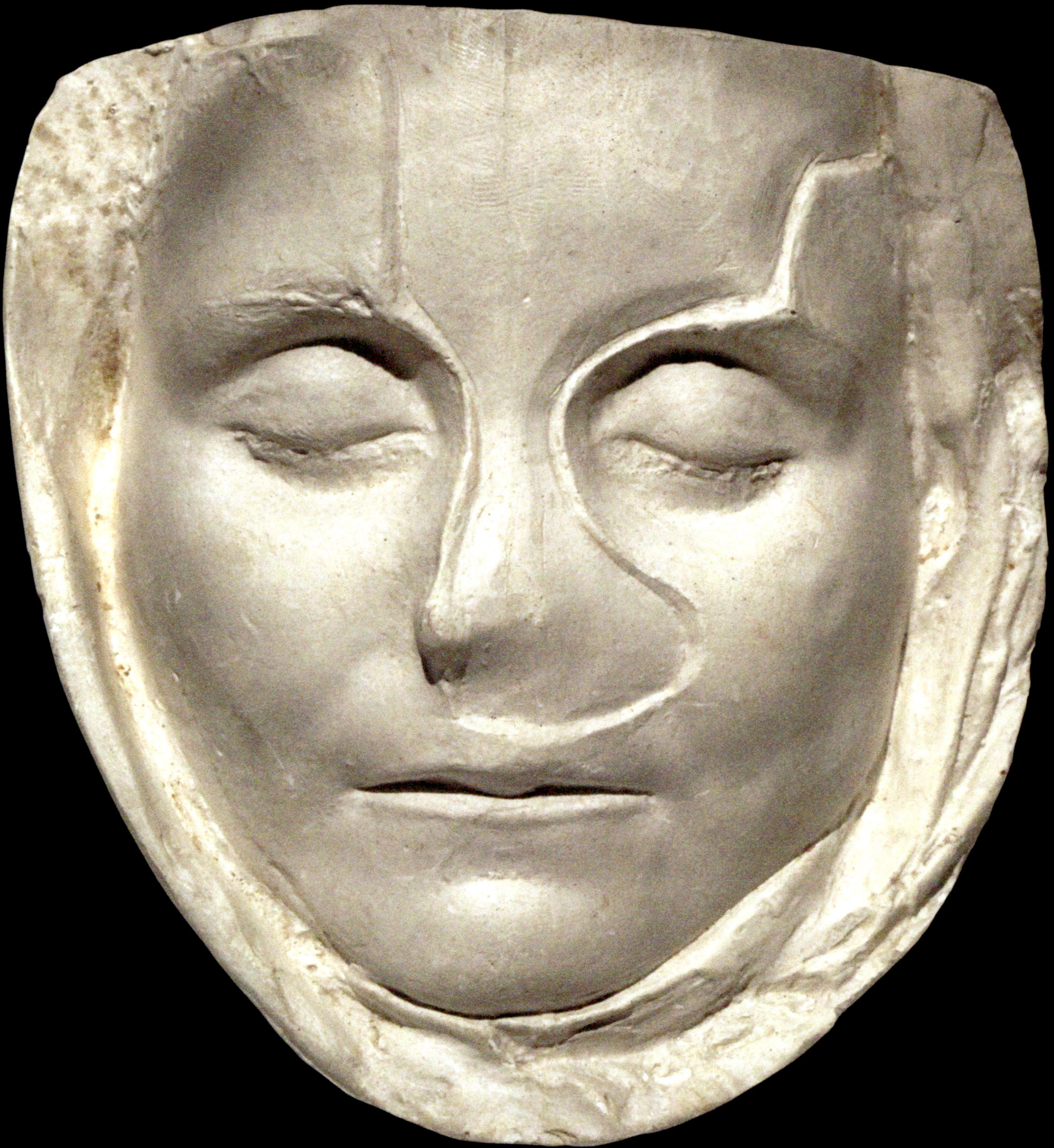

One of casting's most powerful aspects is its ability to make fleeting moments in time permanent. In a pre-photographic world, lifecasting and death masks offered unfading realism—an immortality of sorts. The ability to keep a loved one's likeness was surely a comfort to countless grieving people across millennia. This death mask of a young girl from Gaul is a heartbreaking example of the power in making the ephemeral durable.

Another way casting relates to immortality is through its archaeological applications. Nothing illustrates this better than the castings made at Pompeii. These people, smothered in ash in one horrible moment and lost to history, were resurrected in a sense and their images made immortal.

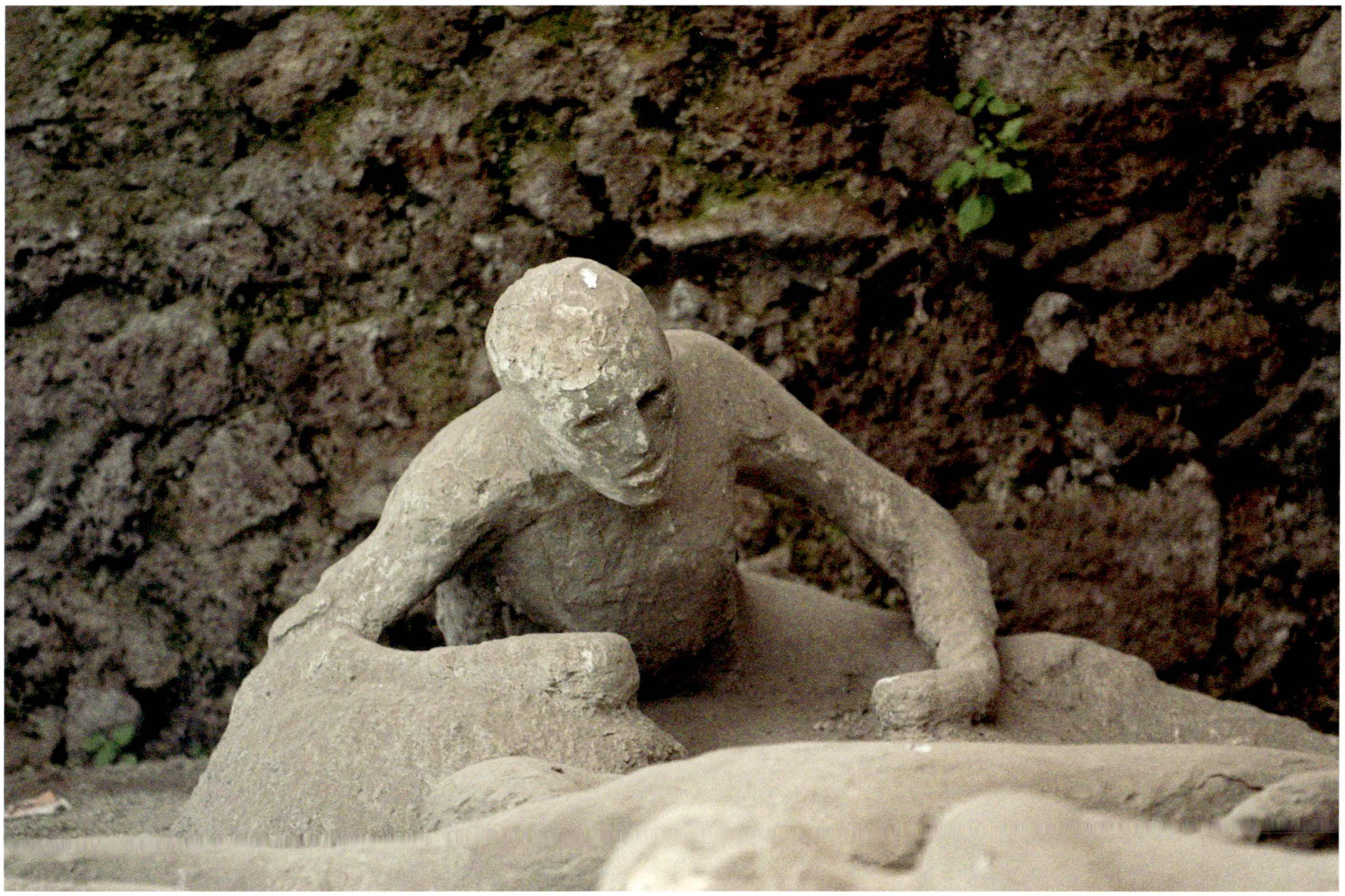

Crawling victim from Pompeii, Eruption of Mt. Vesuvius, 79 CE, discovery of the site late sixteenth century, excavation began in 1748. Pompeii, Italy, cement.
Photo: © Floriano Rescigno, Depositphotos.

Opposite page Face of a child from Gaul (death mask), plaster.
Photo: © Rama at the Museum of Gallo-Roman Civilization / Wikimedia Commons / Cc-by-sa-2.0-fr.

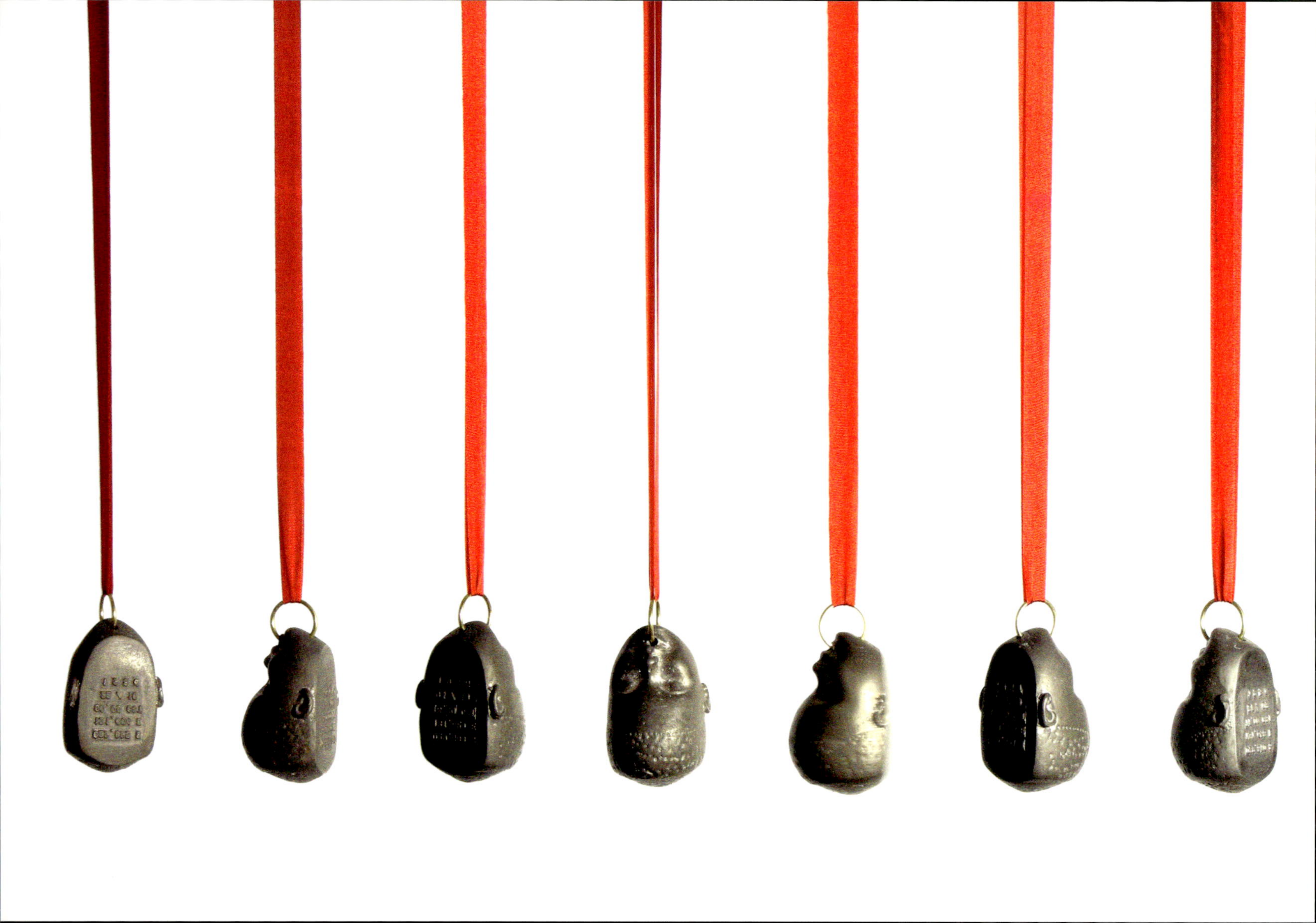

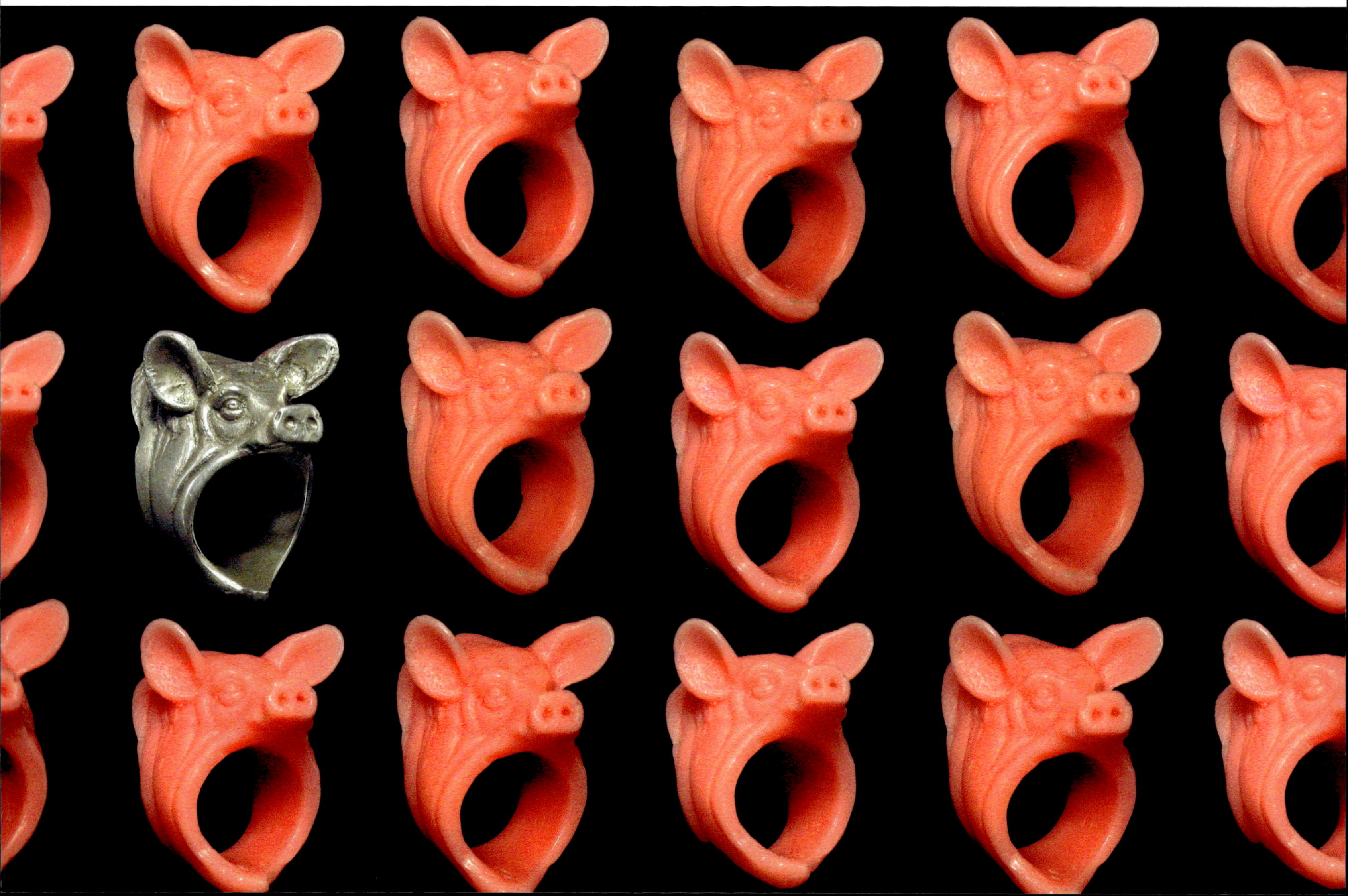

MEANINGFUL MATERIALS

Since just about any material that solidifies can be cast, artists have endless conceptual opportunities when choosing a medium. Materials can carry powerful messages and those who thoughtfully consider them can add a conceptual foundation to their work that's impossible to achieve any other way.

Vernon Bowden is a perfect example of someone whose casting mediums pack a conceptual punch. In fact, the meaning of his work can only be fully understood through a careful inspection of his materials.

Banker's Ring benefits from an ideologically weighty material. This ring, in the shape of a silver pig with a huge gluttonous mouth, references consumer culture, capitalism, wealth, and greed. Bowden drives his intent home by casting the ring in American silver dollars. The imagery in this piece communicates its message, so Bowden is simply reinforcing his concept through the transmuted silver dollars. Concocting with narrative heft, Bowden takes his work to a deeper place when he creates his own casting mediums.

Combining materials can be a powerful act. The quest to transmute base metals enraptured generations of alchemists. The first person to mix copper and tin created a new metal that started a worldwide paradigm shift, the Bronze Age. Alloys like bronze and steel are important for different reasons: bronze for its superior castings, and steel for its strength and malleability. Alloys can be made for working characteristics, for color (rose gold is a good example), or, in the case of the artist, for symbolic reasons. The thoughtful mixing of materials can lead to unique and potent casting mediums that have the power to conjure war, mortality, and love by their very essence.

Bowden's piece *Empire Series: One Tree, Marcoo, Kite, Breakaway, Tadje, Biak, Taranaki* is made up of metal Barbie® heads that hang from red silk ribbons. Bowden uses Barbie® as a common cipher in his work to represent mainstream culture, but the famous face can take on a variety of additional meanings depending on its casting medium.

"Cast from lead musket balls originating from a Victorian era British armory in India, these stylized scarab/Barbie® forms each commemorate a nuclear test that was carried out in Australia by the British army in the early 1950s," says Bowden. "The date and location of the test is inscribed on the back in reverse as a stamp/seal. Encased in each scarab is a piece of trinitite, reputedly from the first atomic test, *Trinity*, in 1945.

OPPOSITE PAGE

Top Vernon Bowden, *Empire Series: One Tree, Marcoo, Kite, Breakaway, Tadje, Biak, Taranaki*, 2007. Trinitite glass 1945 (nuclear blast fused sand), 18K gold, 0.75cal British musket balls ca. 1800, Chinese silk, dimensions variable. Photo: Vernon Bowden.

Bottom Vernon Bowden, *Banker's Ring*, 2008. American silver liberty dollar, 1.6" × 1.2" × 1.2". Photo: Vernon Bowden.

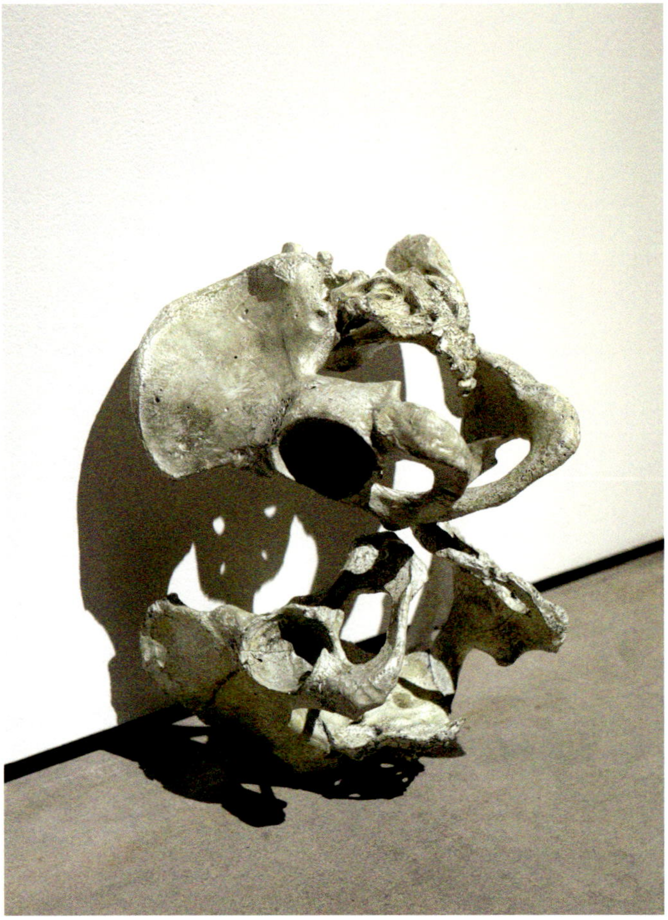

Dario Robleto, *Our Sin Was In Our Hips*, 2001–2002. Hand-ground and powderized vinyl records, melted vinyl records, male and female pelvic bone dust, polyester resin, spray paint, pigments, dirt, concert spotlight, 12" × 9" × 9".
Photo: Pablo Mason and Thomas R. Dubrock, Inman Gallery.

The work references nuclear stupidity, colonialism, and race in Australia; many indigenous people were displaced or ignored when the tests occurred on their lands. The land is still unoccupied to this day."[3]

Dario Robleto is a master of compelling and evocative casting mediums. *Our Sin Was in Our Hips*, realistic looking human pelvises, one male and one female, stacked one on top of the other, are cast in an alloy of hand-ground and powderized vinyl records, melted vinyl records, male and female pelvic bone dust, and polyester resin. Robleto handily summons love, sex, and mortality in this piece simply through his shrewd choice of ingredients.

It's impossible to understand *Our Sin Was in Our Hips* without knowing what it's made of. When combined, pelvic bone and vinyl records imply the erotic. The impassioned corporeal bodies are long gone, though—their moment of rapture evaporated and the music ceased.....

Combining materials in this way allows Robleto to alloy ideas; he mixes associations, memories, and emotions and we find ourselves falling down a rabbit hole of his making into a surreal world where music lives in our bones. Robleto's approach to casting and materials is powerful in this way. He inextricably links his objects and his concepts by creating unexpected and evocative casting mediums. The object is made of the ideas.

Marc Quinn has an almost preternatural sense for materials. From his own blood to lead, Quinn's materials carry expansive associations with them; when these materials are cast into his powerful forms, there is almost a whiff of alchemy. His genius for joining powerful images and potent materials with impeccable craft leads to arresting works he then crowns in astute titles.

Quinn's series *Chemical Life Support* is composed of recumbent figures, each cast in polymer wax with a unique combination of pharmaceuticals mixed in. Each figure is cast in the drug that preserves the life of its real-world model by managing a disease or keeping a condition at bay. From *Innoscience*, a likeness of Quinn's infant son cast with the lab-developed milk substitute that saved his life, to *Nicholas Grogan – Insulin (Diabetes)*, the series has a sobering and touching—yet hopeful—feeling.

These works would be a shadow of themselves without the inclusion of the pharmaceuticals in the castings. *Chemical Life Support* expresses not only the miracle of life-saving drugs but the human aspects of being sick: the vulnerability, the dependency, and the reminder of our own mortality.

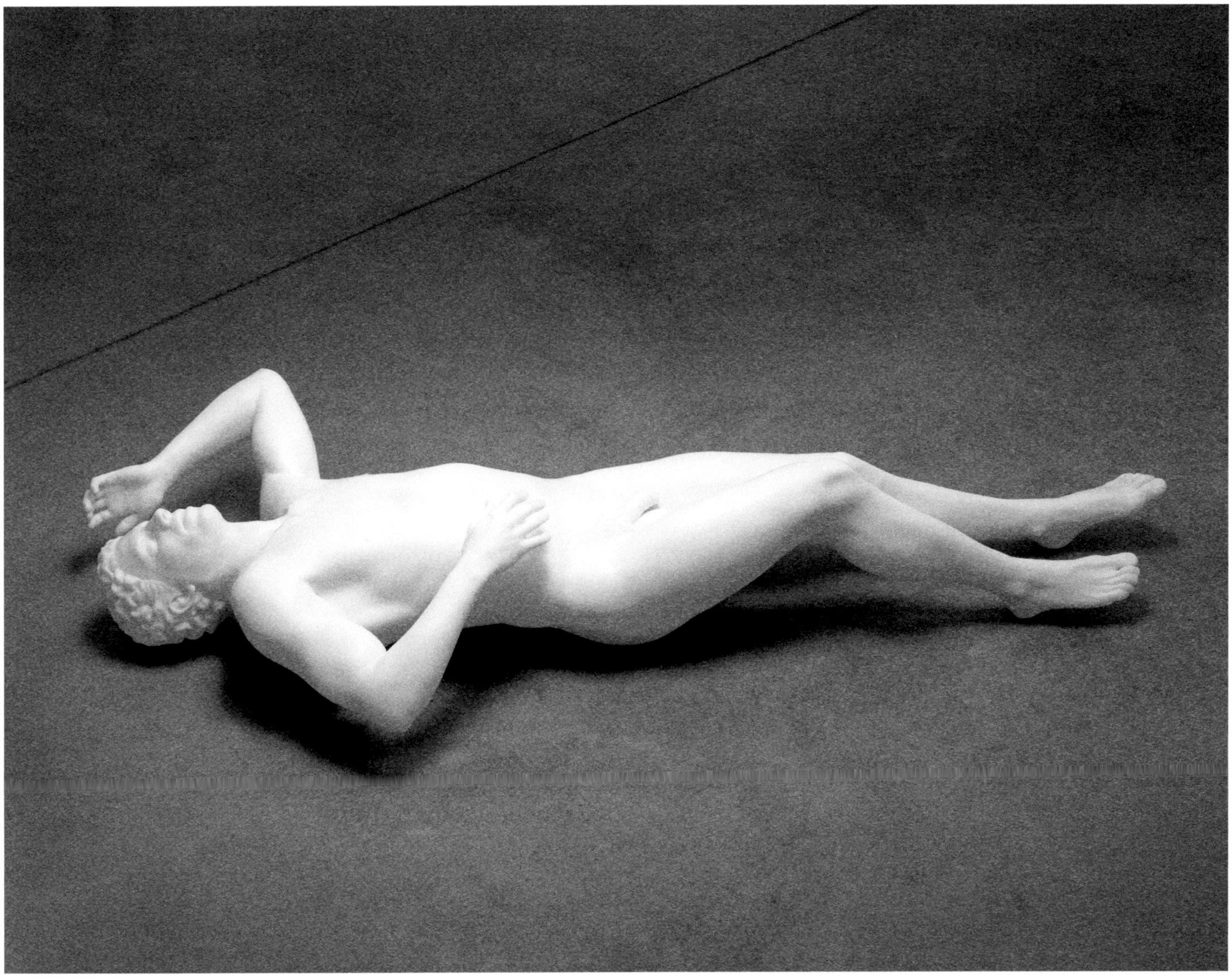

Marc Quinn, *Nicholas Grogan – Insulin (Diabetes)*, 2005. Polymer wax and drugs, 14" × 74.4" × 33". Photo: Stephen White, Courtesy of White Cube Gallery and Marc Quinn Studio.

Eternal Reefs, Inc., a company based in Sarasota, Florida, approaches mortality and casting from a different vantage point. The company "provides a creative, environmentally enhancing means to memorialize the cremated remains of a loved one."[4]

The company works with the Reef Ball Foundation, the leader in scientifically designed artificial reefs. The foundation is dedicated to casting environmentally safe concrete "reef balls" to replace our vanishing coral ecosystems. The foundation's patented mold system and environmentally friendly concrete formula has led to more than seven hundred thousand of the forms having been placed in oceans worldwide.

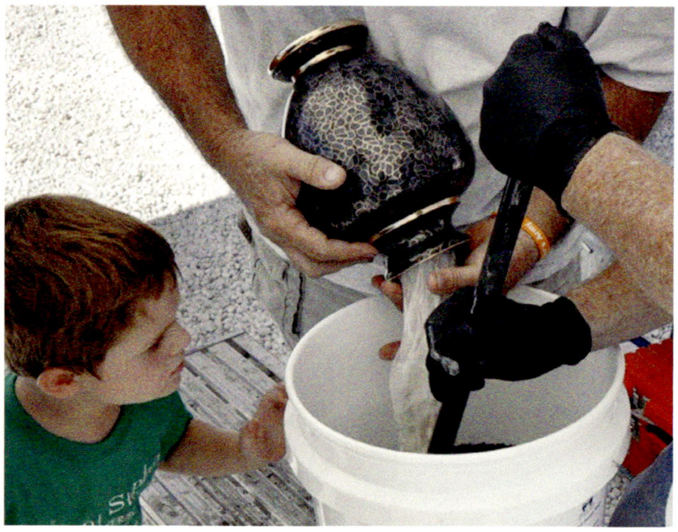

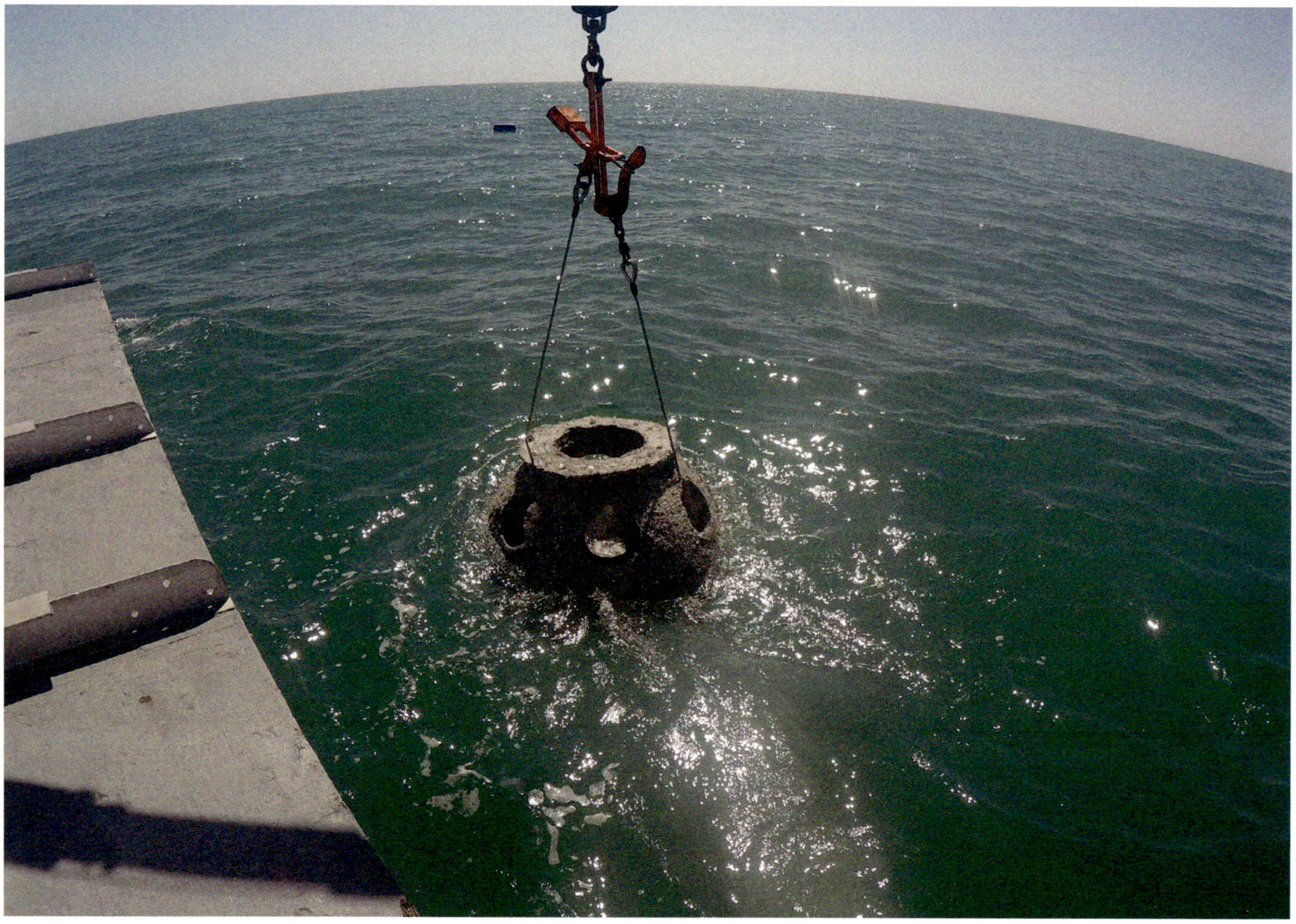

Top right Family members mixing cremation ashes into the specially formulated concrete. Reef balls take one day to cure and then they are unveiled. If appropriate, memorial activities and military honors are given.

Eternal Reefs, Inc., 2015.
Photo: George Frankel.

Above Placing the reef balls. Family members can go out on a small private boat and watch as their loved one's reef ball is lowered into its final resting place

Eternal Reefs, Inc., 2015. Specialized concrete formula with a neutral pH content and cremains. Reef ball dimensions: 2' × 3' and 4' × 5'.
Photo: George Frankel.

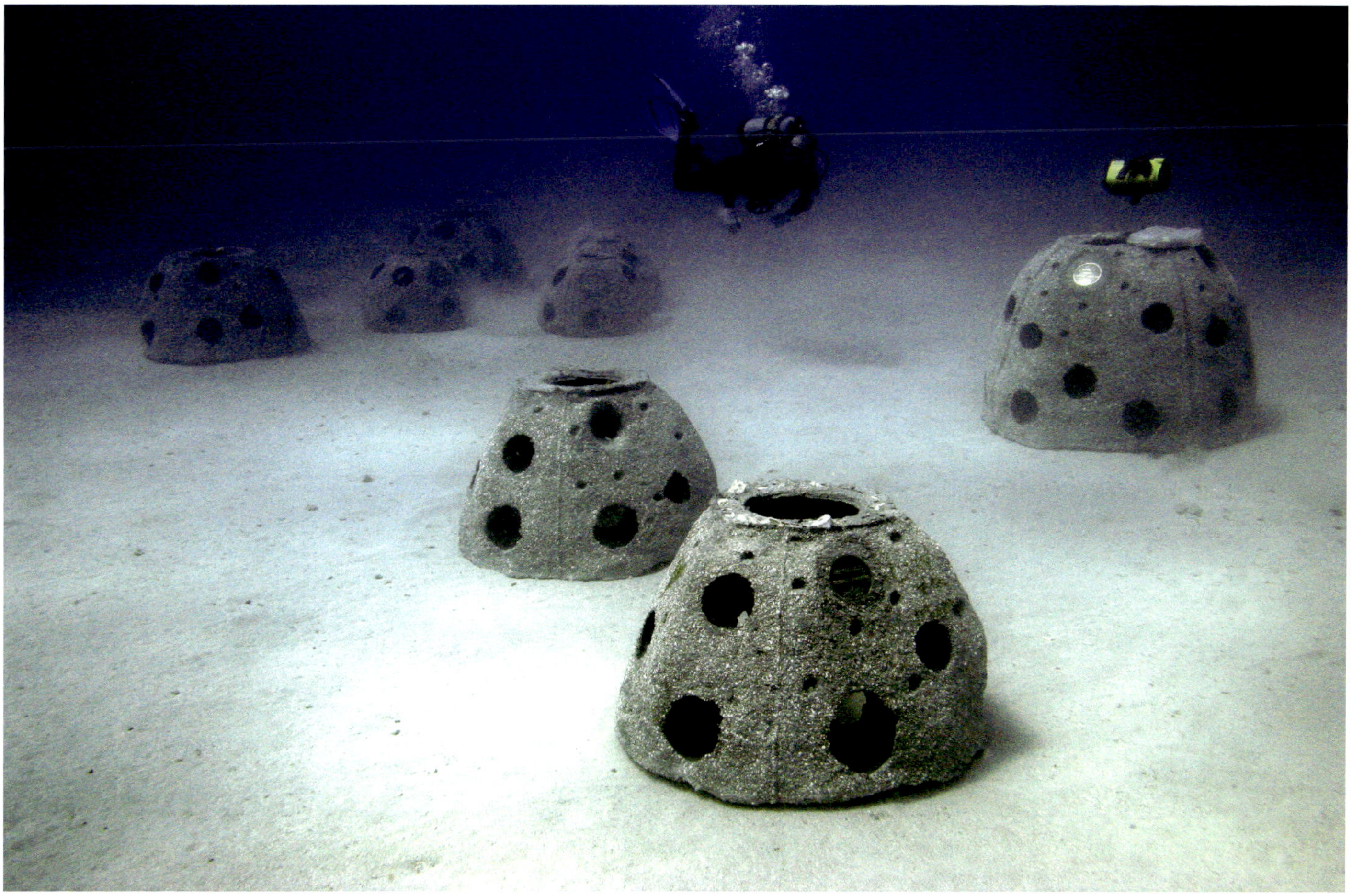

Eternal Reefs, Inc., *Memorial Reef*, 2015. Specialized concrete formula and cremains, reef ball dimensions: 2' × 3' and 4' × 5'. Photo: George Frankel.

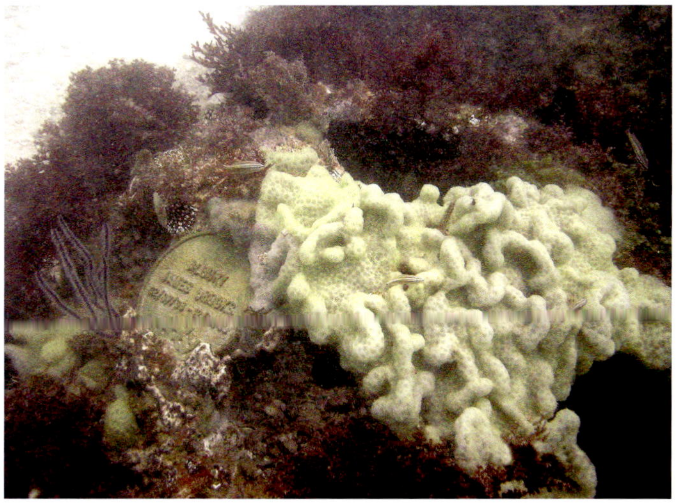

Eternal Reefs, Inc., *Living Monument*, 2015, Specialized concrete formula and cremains, reef ball dimensions: 2' × 3' and 4' × 5'. Photo: George Frankel.

Eternal Reefs applies this casting technology and offers people the opportunity to memorialize their loved ones by adding cremains to the concrete structures.

The company typically creates a group of reef balls over the course of a weekend. The casting is done with families present to add their loved one's cremains to the concrete. They are also invited to add their handprints or bronze plaques to their freshly cast reef ball. "Memorial Reefs," a collection of these forms, become new marine habitats, a significant contribution toward the health of our planet. This celebration of the cycle of life and death is emotionally touching and conjures an almost alchemical type of immortality, a reincarnation of sorts in which a loved one lives on in contribution to an ecosystem in deep need of healing.

Tom Joyce's *Reliquary, September 11, 2001* also centers on life and death with a gentle reverence and sense of hope. These handheld pieces are cast in an alloy of World Trade Center ash, soil from New Mexico's Sanctuario de Chimayo (a Catholic pilgrimage site famous for the healing properties of its earth), sand from a Tibetan Buddhist mandala made by monks as a prayer for peace just after 9/11, and ductile iron.

This metal, steeped in tragedy, fear, and horror, is also imbued with faith, peace, hope, and healing. Joyce cast the forms to incorporate the universal geometry of interlocking vesicae piscis that relate to Christian and Islamic depictions of sacred iconography, but they also closely resemble two seeds. The smooth, innocent, and hopeful symbol of the seed is a poignant choice for such a piece. Their splitting open (presumably to germinate) feels like a true rebirth and a much-needed reminder that life continues, even in the wake of a traumatic event.

Joyce has recast tragedy into hope and death into new life with this piece. His sensitive and nuanced metalwork is always beautifully designed and executed, but when he uses meaningful materials, his work becomes something more...his approach to this piece is rooted both in its materiality and in the transformative nature of casting. When makers transmute meaningful materials and access the alchemy of casting, magic happens.

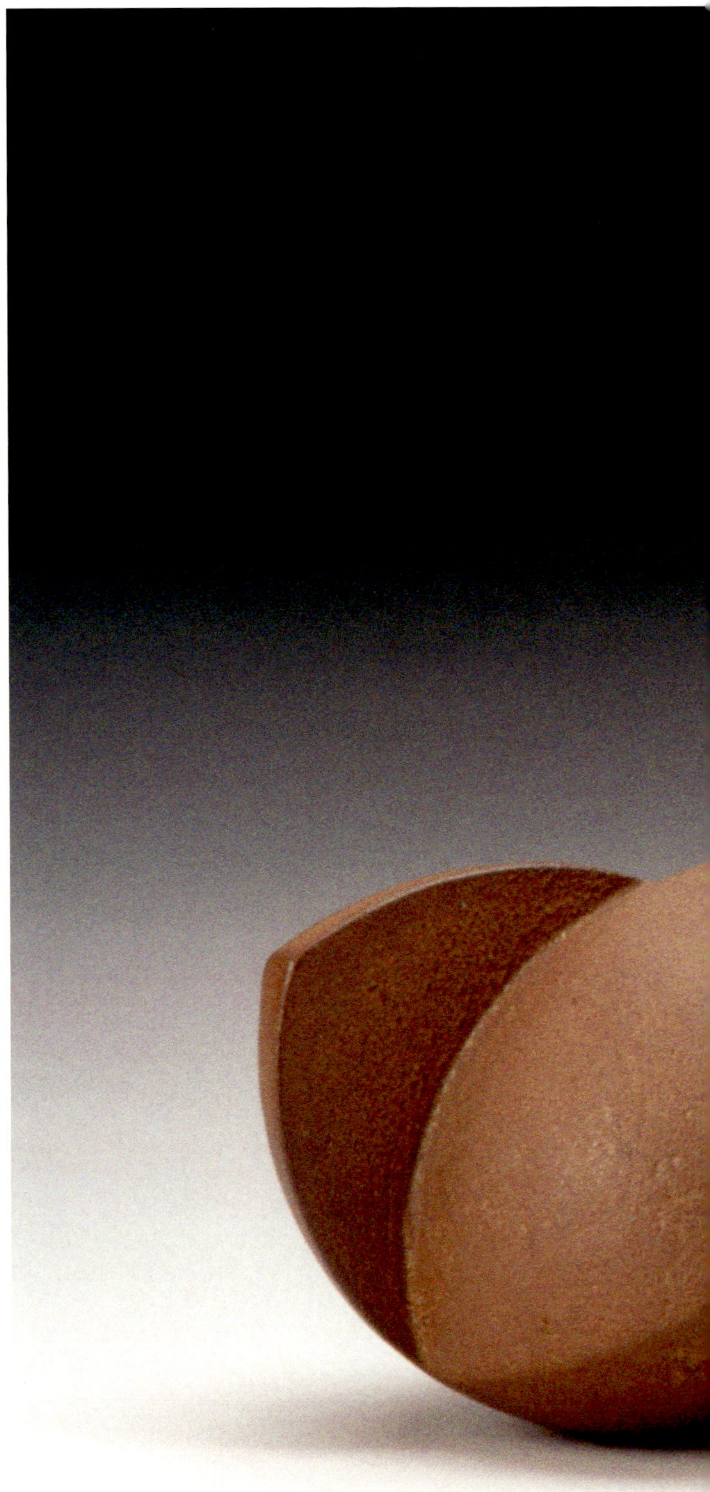

Notes

1. Dictionary.com, s.v. "alchemy," accessed July 14, 2016, http://www.dictionary.com/browse/alchemically.
2. Oxford Dictionaries, s.v. "alchemy," accessed July 14, 2016, http://www.oxforddictionaries.com/us/definition/american_english/alchemy.
3. Email from Vernon Bowden, February 10, 2016.
4. Email from Becky Peterson, Eternal Reefs, January 19, 2016.

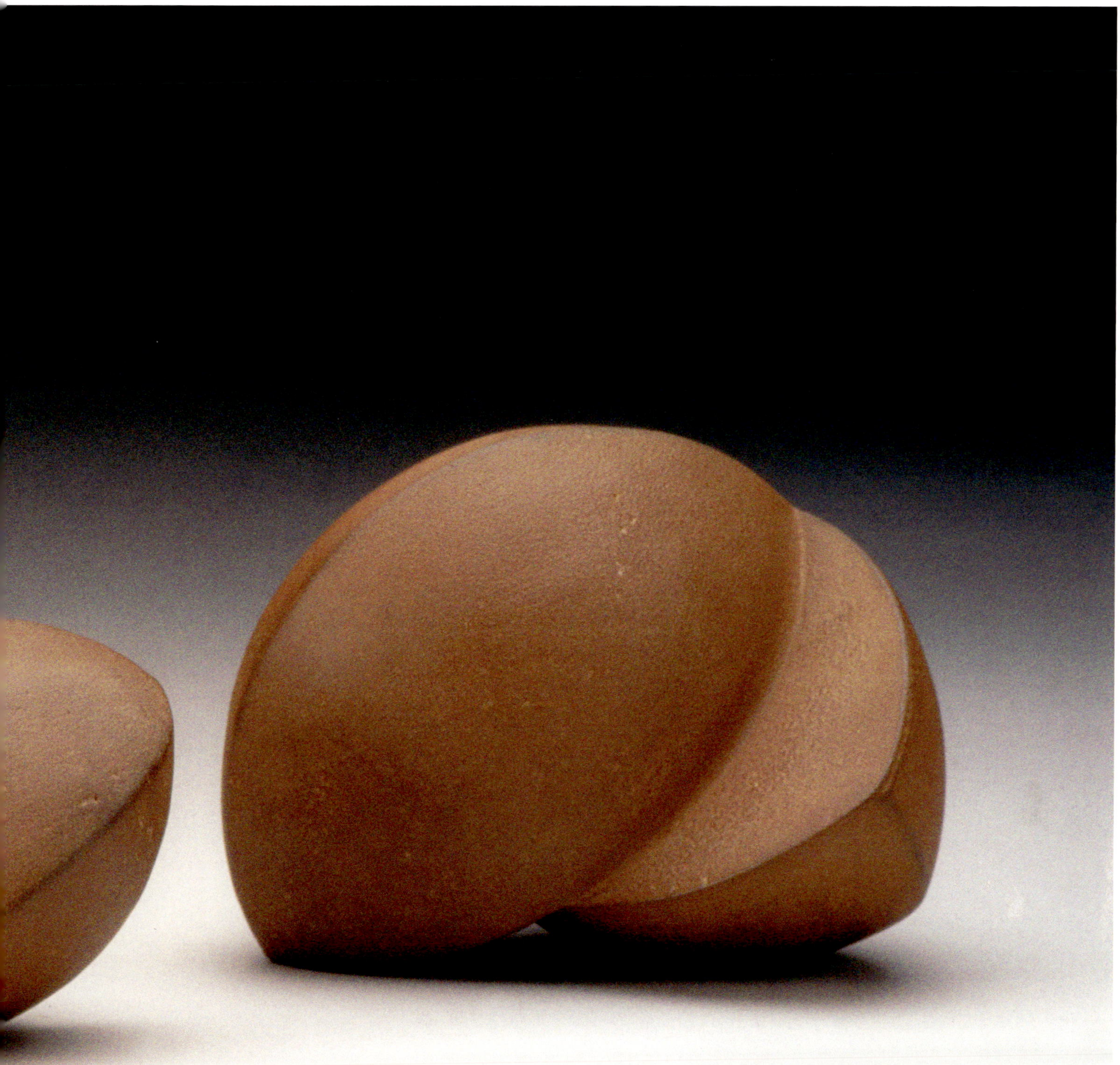

Tom Joyce, *Reliquary, September 11, 2001*, made on September 11, 2002. Ductile iron with World Trade Center ash, soil from New Mexico's Sanctuario de Chimayo, and sand from a Tibetan Buddhist mandala made by monks as a prayer for peace just after the 9/11, (left) 6" × 6" × 6"; (right) 8" × 8" × 8". Created in Arts/Industry, a long-term residency program of the John Michael Kohler Arts Center in Sheboygan, Wisconsin. Arts/Industry takes place at Kohler Co.
Photo: © 2002 John Michael Kohler Art/Industry, Collection of the National Metal Museum, Memphis, Tennessee.

CONCLUSION

Our world is, quite literally, built on casting. Its omnipresence in everyday life is reason enough to study it, but beyond that, cast objects are endlessly fascinating. The common thread of process unites everything from an injection-molded plastic toy to a poignant and heartbreaking work of sculpture. Casting offers a tool for every purpose and reminds us how much there is to learn. The mind-boggling array of applications for casting makes it an unwieldy topic, but a variety of angles and a chorus of voices provide a balanced perspective and broad look at this diverse process.

Examining casting from so many different vantage points creates a kind of kaleidoscopic perspective. This book has explored casting's history and impact, its countless mediums, how it's used in industry and the arts, common and uncommon approaches, its many functions, its conceptual possibilities, and casters themselves. Casting is a process impossible to know in its entirety—it's simply too vast—but our ever-expanding and shifting perspective that results from studying it in this way creates a holistic look at the process, its objects, and its impact.

One of the goals of *Cast* is to promote integrative thinking, especially among makers. Isolating designations like "sculptor," "designer," and "craftsperson" will ideally fall away in the face of a common process, leading to new ideas, unexpected conversations, and new approaches to making.

Contemporary works of fine and decorative art are the backbone of *Cast*, but peppering the galleries with works from antiquity (often every bit as sophisticated as their contemporary counterparts) is a reminder that we are a part of the continuum of making that stretches over millennia. History has much to offer us, and while casting has been revolutionized by technology just like every other process, it still boils down to the same basic parts. Whether an object was produced using low-tech, ancient materials and processes or through the exciting high-tech offerings of computer aided design, 3-D printing, and injection molding, the same core considerations apply.

Thinking three-dimensionally, model creation, moldmaking, and the cast itself are at the heart of the process, regardless of medium. These are also topics capable of inspiring lifelong obsession, not to mention spurring countless hours of "shop talk." However, it's the intangible magic of casting that seems to get into the blood as much as the fascination with technique. This dual attraction means the process appeals to both the left brain and the right, which explains the fervor it can inspire.

Casting has indelibly altered human existence for thousands of years and shows no signs of abating. Its utility is unparalleled and its applications are as diverse as creation itself, serving industry as well as individual artists. Again, we encounter duality—the logical and the emotional, the technical and the philosophical.

These two aspects of the process hold great promise for the future. What will the next generation do with new casting technologies and materials? Will philosophies guiding the casting industry shift to reflect the spirit of conservation? We know that cast plastic contributes to waste and environmental injury, but can this discarded material be recycled or repurposed, perhaps through recasting? The creativity, technical innovation, and pioneering spirit expressed by casters from every era and discipline are not only inspiring, but also hold great potential for transforming our world once again.

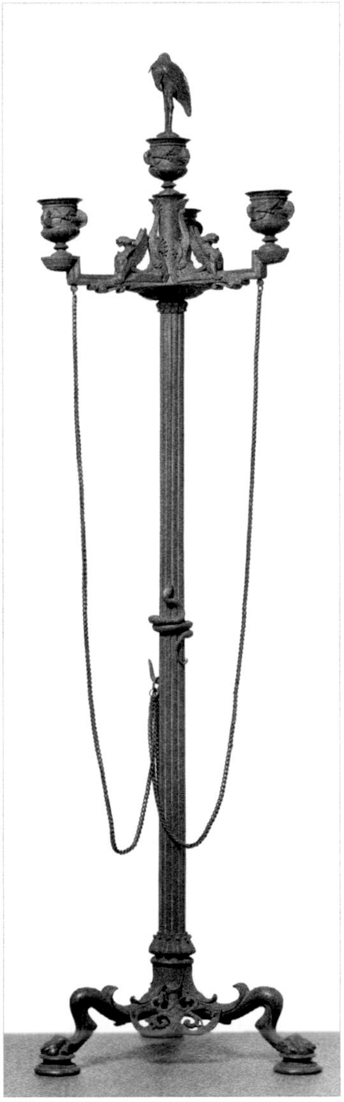

Antoine-Louis Barye, *Candelabra with Three Branches, Chains, and a Stork Finial*, nineteenth century. Paris, France, bronze, 31.8" tall. Courtesy of The Walters Art Museum, Acquired by William T. or Henry Walters, CC-by-0, GNU FDL.

ACKNOWLEDGMENTS

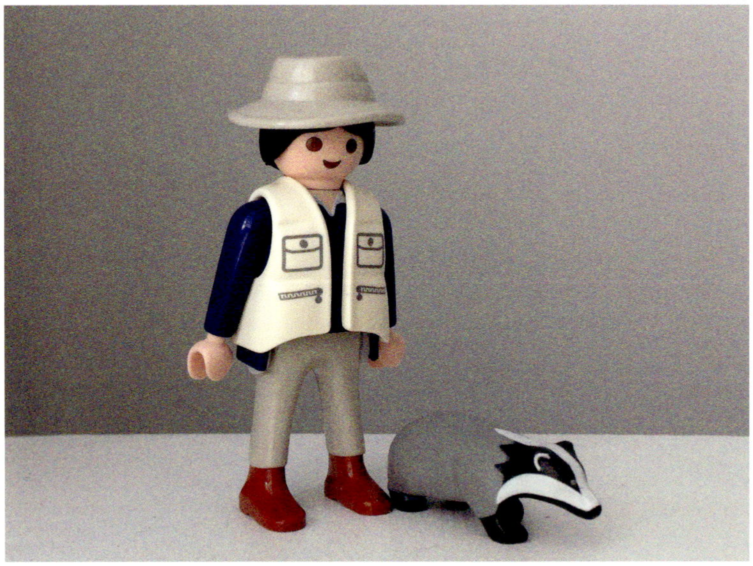

Playmobil Adventure Figure with Honey Badger, plastic.
Photo: Monica Hampton.

SPECIAL THANKS TO:

Ed Vates
John Sterling
Tommy Sterling
Participating artists
Susie J. Silbert
Suzanne Ramljak
Joseph Becherer
Elaine A. King
Ezra Shales
Jim Bové
Sandra Korinchak
Peter Schiffer
Nicci Townsend
Monica Hampton
Dr. Kim Theriault
Connie Binder
Studio Us
Brigitte Martin
Donald Friedlich
Norwood Viviano
Natalie Sinisgalli Photography
The Brooklyn Museum
Frederik Meijer Gardens and Sculpture Park
The Corning Museum of Glass
Grand Rapids Art Museum
The Wege Foundation
Grand Valley State University
John Michael Kohler Arts Center
Los Angeles County Museum of Art
Rijksmuseum
The Walters Art Museum
Wikimedia Commons

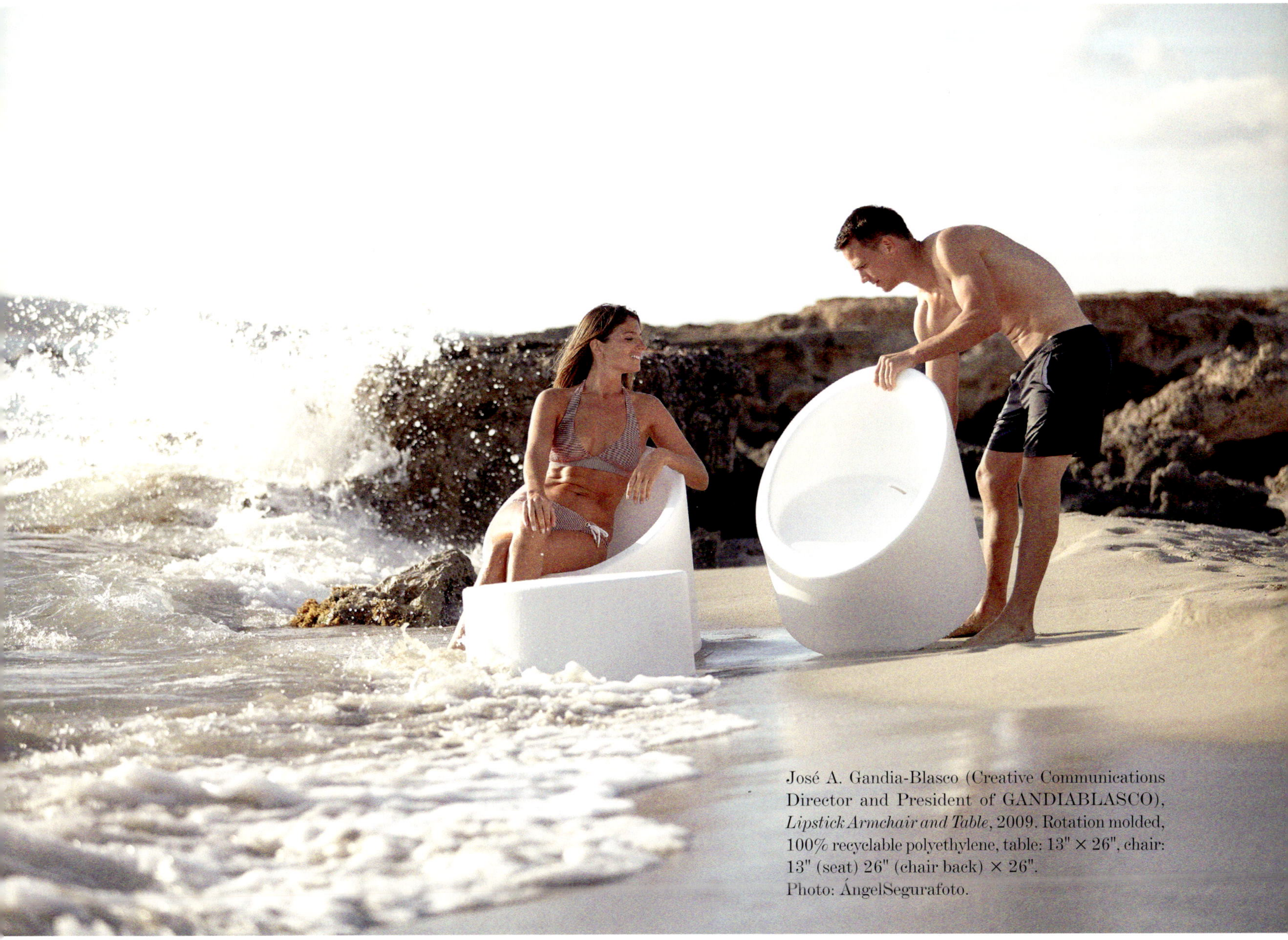

José A. Gandia-Blasco (Creative Communications Director and President of GANDIABLASCO), *Lipstick Armchair and Table*, 2009. Rotation molded, 100% recyclable polyethylene, table: 13" × 26", chair: 13" (seat) 26" (chair back) × 26".
Photo: ÁngelSegurafoto.

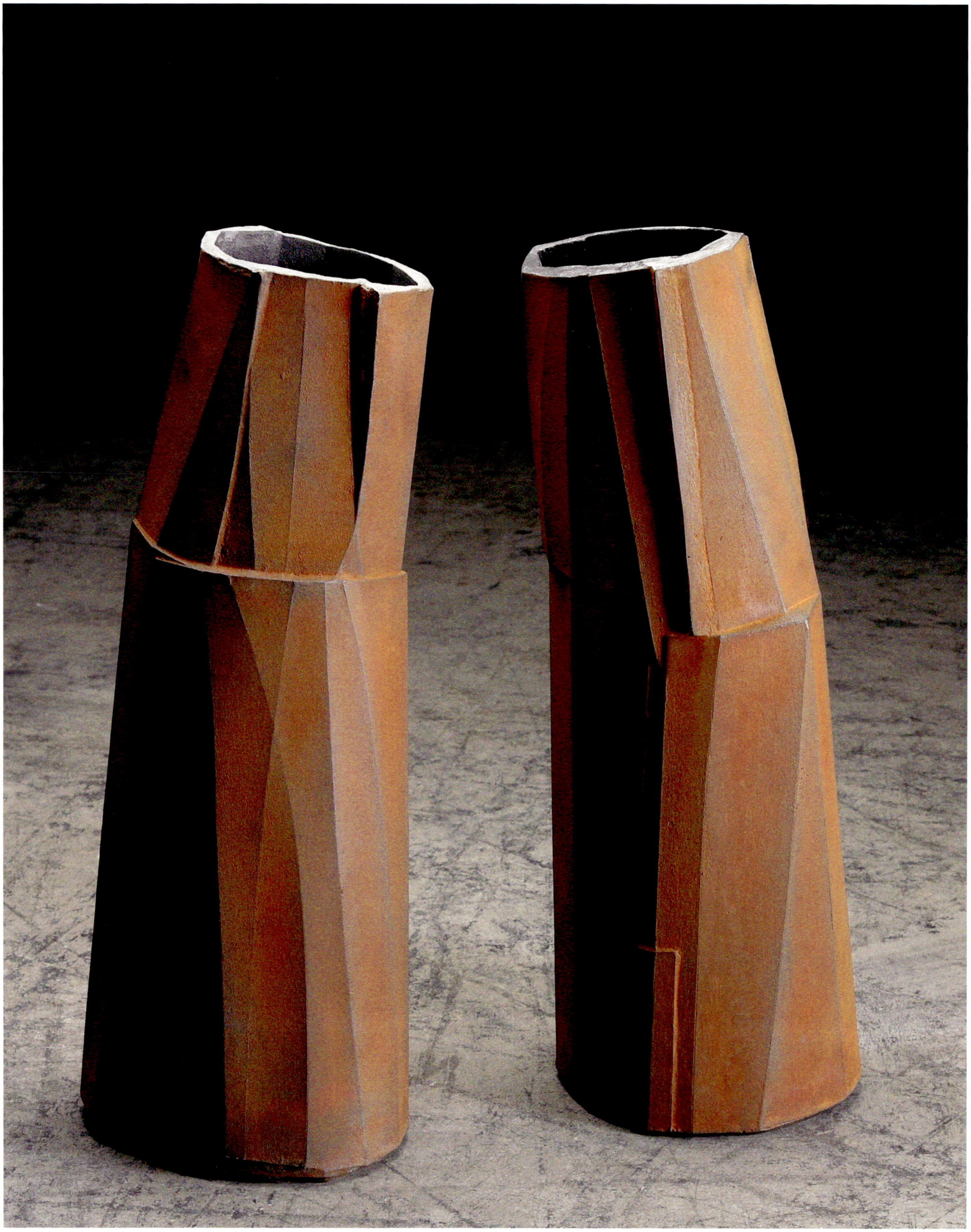

Myra Mimlitsch-Gray, *Limbs*, 2007. Ductile iron cast in a resin-bonded sand mold from a chipboard pattern, 2' × 10" × 9". Created in Arts/Industry, a long-term residency program of the John Michael Kohler Arts Center in Sheboygan, Wisconsin. Arts/Industry takes place at Kohler Co.
Photo: Rich Maciejewski.

GLOSSARY

by Jim Bové

3-D printing: Also known as additive manufacturing, it is the process of creating a three-dimensional object from a digital file. Material is built up in layers in order to create the object. A wide range of materials, from plastics to ceramics and metal, can be 3-D printed. Some 3-D printed materials can be directly burned out in the lost wax casting process.

Additive process: Introducing small amounts of material to build up or alter a form. Often used in wax, clay, and plasticene modeling to add to a structure.

Alginate: An organic polymer derived from the stems of seaweed, often used in lifecasting and dentistry to create molds.

Annealing: ***Glass:*** slowly cooling a glass piece in a kiln so as to avoid thermal cracking and stresses. ***Metal:*** heating metal above its recrystallization temperature in order to make it more pliable and workable.

Bas relief: A shallow sculpture that is raised a few inches from a flat background to give a three-dimensional effect. Undercuts are minimal or nonexistent. Coins and medallions are examples of bas relief.

Billets: A solid ingot of glass used in glass casting.

Burnout cycle: Burnout cycles consist of stepped increases of temperature in a kiln over a period of time as part of the lost wax casting process. Cycles depend on mold size and material being burned out.

CAD: Computer-Aided Design.

Carving wax: A hard wax designed to be carved, drilled, filed, and machined without clogging tools. Often made of a mixture of waxes and plastics. Comes in various degrees of hardness.

Casting: Forming a material into a three-dimensional shape using a mold.

Casting furnace: An enclosed structure usually made of refractory brick in which glass or metal is melted and heated. A glass furnace may be used for melting batches of glass or for the reheating of glass at the glory hole during the glass blowing process. A cupola is a vertical furnace in which the heated material is poured out through a bottom opening into a mold or crucible.

Casting tree: A main sprue with multiple castings branching off of it. Often used in industry to cast multiple objects, such as rings.

Centrifugal casting: A casting method that utilizes centrifugal force to push molten metal into a mold.

Ceramic shell: A ceramic mold made by dipping or coating a wax model with a slurry (liquid ceramic). The shell is built up around the form creating the mold that is broken away after the piece has been cast.

Cleanup: Removing investment or other mold materials, sprue removal, and correcting imperfections from the casting process using a variety of tools.

Cold working: Working with glass at room temperature, including sandblasting, grinding, drilling, polishing, and cutting.

Crucible: A vessel made of refractory material, such as ceramic, silicon carbide, or graphite, in which metal is melted prior to casting.

Cuttlefish casting: An early form of casting in which the calcareous internal structural bone of the cuttlefish is carved and used as a waste mold. Molten metal is poured directly into the carved cavity.

De-spruing: The removal of the sprue from the final casting; one of the stages of cleanup.

Devesting: The removal of the mold material from a cast object; the first stage of cleanup.

Die: A mold, usually made of tool steel, or a jig in which objects are formed. When a material is forced into a metal mold under pressure, it is called die-casting.

Fabrication: Manipulating a material through bending, cutting, joining, and shaping, generally through the use of mill stock like sheet metal, wire, rod, and tubing.

Faience: A moldable paste composed of silica with small amounts of sodium and calcium. Upon firing, the paste hardens and the silica forms a glassy surface.

Ferrous: A metal that contains iron.

Flask: A container or form that supports a mold.

Flex shaft: A tool composed of a small motor that provides rotational power to an interchangeable hand piece through a flexible shaft. It is used in jewelry to drill, grind, carve, cut, and polish.

Forging: Forming or shaping metal using hammers. Forging may be done while the metal is in a hot or cold state.

Found object casting: Using a waste mold, organic and/or plastic objects are directly burned out in a kiln.

Foundry or **Casting house:** A company or studio that specializes in casting metal and moldmaking, and can also assist designers with model creation.

Frit: Small granules of glass, often colored, which can be used to color clear glass.

Glass pressing, Pressed glass: Glass that has been formed by pressing it into a mold through the use of a plunger.

Developed in the 1820s to create glass knobs and handles. The most widely known examples are Carnival glass.

Hand-building: Using only the hands and simple tools for creating objects out of clay. Generally refers to any ceramic work that is not made on a potter's wheel or cast.

Hot casting or Ladle casting: The material is ladled out of a melt pot or furnace and poured directly into a mold. Used in both glass and metal.

Ingot: A block of metal cast in a standard shape for easy stacking and storage.

Injection molding: A process by which the material to be molded is heated and injected into a mold. A variety of materials may be cast in this way, from plastics to waxes and metals.

Investment: A fine-grade plaster that can withstand high temperatures. Used in the lost wax casting process to create a waste mold around the model.

Kiln: A furnace or oven capable of producing the high temperatures needed for the lost wax casting method, cast or slumped glass, or firing ceramics or faience.

Kiln cast: Any one of a number of methods for shaping glass in a kiln, typically by filling a mold with molten glass.

Lead bearing glass: Glass with lead added to it to improve the appearance and to make it easier to melt.

Leather-hard: Clay that has been allowed to partly dry and is firm enough to be tooled and worked.

Lifecasting: Taking a mold of a body with the intention of casting a replica.

Lost wax casting: The process by which a model (often wax) is encased in a waste mold then melted away leaving a void in the exact shape of the model. Molten metal is then poured or injected into the negative space thereby creating a metal copy of the wax. Both the wax model and the mold are "lost" in the creation of the finished piece.

Microcrystalline wax: Wax that is created through the de-oiling of petroleum and has fine crystalline structures, that are dense, elastic, and flexible. It is a preferred model-making material for sculpture.

Model: An object (carved, modeled, constructed, or found) in the image of the final casting, used to create the negative space in a mold.

Mold: A cavity into which a substance is poured, injected, or pushed.

Mold-blown glass: Glass that is shaped by having a molten bubble of glass blown under air pressure into a mold made of multiple parts. This technique was used by the Romans and is still in use by artisans and in industry to create hollow glass objects. A common example is the Mason jar.

Mother mold: A structural outer mold that reinforces a lighter, flexible inner mold. Also used to hold multiple pieces of a mold together in their proper relationship. Usually made from plaster or plastic.

Non-ferrous: A metal or metal alloy that does not contain iron. Examples are silver, copper, gold, platinum, and bronze.

Open-faced mold: A one-sided mold that allows materials to be poured or pushed directly into it without the use of sprues or gates, resulting in a bas relief form.

Parting lines: The seam on a cast object formed where the parts of the mold meet. It can be a raised ridge or an indentation.

Pâte de verre: A material made by grinding glass into a fine powder and adding a binder to create a paste. The paste is then brushed or tamped into a mold, dried, and fused by firing.

Patina: Oxidation on the surface of metal. Patina can be intentional or can happen naturally over time. Various chemicals and/or heat treatments are used to create color on metal.

Plasticity of form: A three-dimensional quality, a state of being plastic, and the capacity to be formed or molded

Polyurethane: A liquid polymer material that is flexible when dry, and can be used as a casting medium or can be brushed on or sprayed to seal a surface.

Press mold: A mold that allows the caster to push material directly into the mold.

Reductive process: Material is carved or ground away to achieve the desired form in the model or artwork. Also called the subtractive process.

Refining: The removal of impurities through the heating and separation of metals based on melting temperatures or through the use of an acid that dissolves the material, allowing it to be precipitated and reclaimed.

Refractory: Heat resistant materials in the form of bricks, castable cement, insulation, or moldable clay bodies. Often used to make kilns, furnaces, and crucibles.

Resin: Natural or synthetic, often in the form of a two-part mixture, resins are liquids that harden over time or through chemical reaction.

Rotocast or Rotomolding: A slow-speed, multidirectional casting in which the mold is turned on at least two axes, resulting in a hollow casting with relatively thin walls. Examples are a chocolate Easter bunny or a rotomolded kayak.

RTV mold: Room Temperature Vulcanizing (RTV) molds are made from a class of flexible rubber mold materials that are able to cure without heat, allowing the molding of heat-sensitive mold materials like some waxes or organic materials.

Sand casting: Sand held together with a binder used as a mold material and compressed around a model. Commonly used for metal or glass casting.

Silicone mold: A flexible rubber mold that can withstand high temperatures. Wax or low-melting-point metal alloys may be poured or injected directly into the mold.

Slipcasting: A ceramic process in which a partially liquefied clay body is poured into a plaster mold. The slip is allowed to set for a short time, then poured back out. The process is repeated until the desired thickness is achieved. Once the clay is leather hard, the mold is opened and the slipcast piece is cleaned of any parting-lines or flashing. The same process is also used to create hollow waxes for lost wax casting using hot wax instead of slip.

Sprig-molding: The use of press-molds (sprig-molds) for creating decorative accents to pottery, such as handles or appliques. Sprigs are formed in a mold of bisque-fired clay or plaster, then adhered to the ceramic piece using slip (a paste-like material made of the same clay body that has been watered down).

Sprues: Wax sprues are attached to a model to create channels for wax to flow out and for metal to flow into the mold. Sprues are burned out along with the model.

Undercut: A cut that goes around and behind the form to create greater shadows or the impression that the form continues, thereby giving the sculpture a greater sense of dimension and depth. An undercut on a form may prevent the form's withdrawal from a mold, therefore consideration is required in the design of the mold and/or the model to ensure successful de-molding.

Vacuum casting: The use of a vacuum to pull molten metal into an investment waste mold.

Vulcanized mold: A mold made with natural rubber that is formed around a model through heat and pressure; cures at 300°F (149°C). The model must be able to withstand heat and pressure.

Waste mold: A single-use mold designed to be dissolved, peeled, chipped, or cut away from the casting, leaving no opportunity for further reproduction. Example materials are investment, ceramic shell, and alginate.

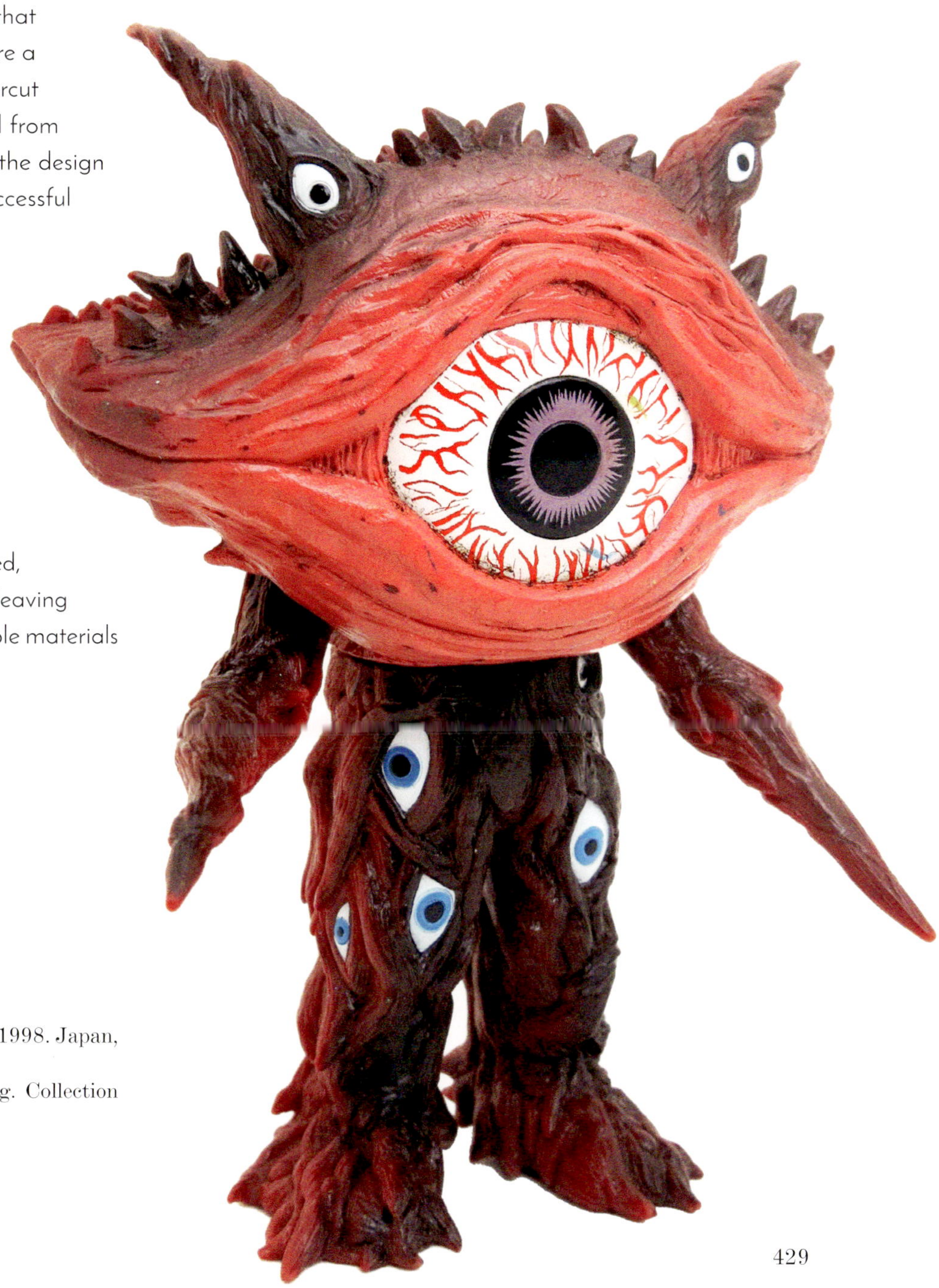

Kaiju Monster Action Figure, 1998. Japan, vinyl, 6.25" × 0.5" × 2.5". Photo: Renée Zettle-Sterling. Collection of Tommy Sterling.

BIBLIOGRAPHY

Akker, Robin van den, and Timotheus Vermeulen. "Notes on Metamodernism." *Journal of Aesthetics and Culture,* 2010. http://aestheticsandculture.net/index.php/jac/article/view/5677/6304.

Allington, Edward. Unpublished interview with author, September 23, 1998.

Benson, Sarah. "Reproduction, Fragmentation, and Collection: Rome and the Origin of Souvenirs." In *Architecture and Tourism: Perception, Performance and Place,* edited by Medina Lasansky and Brian McLaren. Oxford: Berg Publishers, 2004.

Brunhammer, Yvonne. *The Jewels of Lalique.* Paris: Flammarion, 2002.

Burt, Nathaniel. *Palaces for the People: A Social History of the American Museum.* New York: Little Brown & Co., 1977.

Ciarla, Roberto, Lionello Lanciotti, Maurizio Scapari, and Alexandra Wetzel, photographs by Araldo de Luca. *The Eternal Army: The Terracotta Soldiers of the First Emperor.* Novara, Italy: White Star Publishers, 2012.

Cole, Henry. *Memorandum of the International Exchange of Copies of Works of Fine Arts.* London: South Kensington Museum, 1864.

Cole, Ina. "Mapping Traces: A Conversation with Rachel Whiteread." International Sculpture Center. Accessed July 20, 2016. http://www.sculpture.org/documents/scmag04/april04/WebSpecials/whiteread.shtml.

Corces, Arturo, MD. "Metallic Alloys." *Medscape,* edited by Jason H. Calhoun (Dec. 3, 2015). Accessed July 20, 2016. http://emedicine.medscape.com/article/1230554-overview.

Cummings, Keith. *Contemporary Kiln-Formed Glass.* London and Philadelphia: A&C Black; University of Pennsylvania Press, 2009.

Desai, Dipti. "The Ethnographic Move in Contemporary Art: What Does It Mean for Art Education?" *Studies in Art Education,* Summer 2002, 320.

Dowson, Thomas. "Under the Acropolis Museum—Archaeology Travel." *Archaeology Travel.* April 29, 2014. Accessed July 20, 2016. http://archaeology-travel.com/museums/under-the-acropolis-museum/.

Faurschou Foundation. "Ai Weiwei Sunflower Seeds." Accessed July 20, 2016. http://www.aiweiweiseeds.com/about-ai-weiweis-sunflower-seeds.

Garcia, Brittany. "Ishtar Gate." *Ancient History Encyclopedia.* August 23, 2013. Accessed July 20, 2016. http://www.ancient.eu/Ishtar_Gate/.

Gardner, Albert Ten Eyck. "The Hand of Rodin." *The Metropolitan Museum of Art Bulletin, New Series* 15 (9), May 1957, 200–204.

George, Sean. "A Brief History of Antique Pressed Glass." Accessed July 21, 2016. http://www.pressedglassandgoblets.com/articles/history.shtml.

Goldstein, Carl. *Teaching Art: Academies and Schools from Vasari to Albers.* Cambridge: Cambridge University Press, 1996.

Grimmer, Anne E., and Paul K. Williams. "Preservation Brief 30: The Preservation and Repair of Historic Clay Tile Roofs." National Parks Service, U.S. Department of the Interior. Accessed July 19, 2016. https://www.nps.gov/tps/how-to-preserve/briefs/30-clay-tile-roofs.htm.

Honour, Hugh, and John Fleming. *A World History of Art.* London: Laurence King Publishing, 2009.

Hynes, Nancy. "Oldest Known Jewellery Discovered." June 22, 2006. Accessed July 20, 2016. http://www.nature.com/news/2006/060619/full/news060619-10.html.

"Iron Bridge and Tollhouse." Accessed July 19, 2016. http://www.ironbridge.org.uk/our-attractions/the-iron-bridge-and-tollhouse/.

Jobey, Liz. "Cornelia Parker." *Financial Times,* Life & Arts section, May 17, 2013.

Kohler, Ruth, Ezra Shales, and Glen Brown. *Arts/Industry: Collaboration and Revelation.* Sheboygan, WI: John Michael Kohler Arts Center, 2014.

"Lick and Lather, Detail (1993)." Accessed July 29, 2016. http://www.pbs.org/art21/images/janine-antoni/lick-and-lather-detail-1993.

Lindsey, Bill. "Bottle Finishes (aka 'Lips') & Closures." Historic Glass Bottle Identification & Information Website, Society for Historical Archaeology. February 14, 2016. Accessed July 21, 2016. https://sha.org/bottle/closures.htm#External%20Threaded%20screw%20cap.

Macky, Ian. "Deck Lights/Deck Prisms." Accessed July 20, 2016. http://glassian.org/Prism/Deck/index.html.

McIntosh, Jane. *The Ancient Indus Valley: New Perspectives.* Santa Barbara, CA: ABC-CLIO, 2008.

Mills, John W. *Technique of Casting for Sculpture*. London: B T Batsford Ltd, 1990.

Mitchell, Jan. *The Art of Precolumbian Gold: The Jan Mitchell Collection*. New York: New York Graphic Society, 1985.

Munger, Jeffrey, and Alice Cooney Frelinghuysen. *East and West: Chinese Export Porcelain*. In Heilbrunn Timeline of Art History. New York: The Metropolitan Museum of Art, 2000–. http://www.metmuseum.org/toah/hd/ewpor/hd_ewpor.htm.

Murphy-Gnatz, Mary. "The Porcelain Trade." Minneapolis: University of Minnesota Libraries. Accessed July 20, 2016. https://www.lib.umn.edu/bell/tradeproducts/porcelain.

Ochota, Mary-Ann. *Britain's Secret Treasures*. London: Headline Publishing Group, 2013.

Osterman, Matthias. *The Ceramic Narrative*. Philadelphia: University of Pennsylvania Press, 2006.

Peiser, Mark. *Looking Within: Mark Peiser, the Art of Glass: Works from 1967-2003*. Asheville, NC: Asheville Art Museum, 2003.

Pendergrast, Mark. *Mirror Mirror: A History of the Human Love Affair with Reflection*. New York: Basic Books, 2009.

Perlez, Jane. "Vienna to Go Ahead With Holocaust Shrine." *New York Times*, March 3, 1998. Accessed July 20, 2016. http://www.nytimes.com/1998/03/04/world/vienna-to-go-ahead-with-holocaust-shrine.html.

Rachlin, Samuel. "Stalin's Long Shadow." *New York Times*, March 4, 2013. Accessed July 28, 2016. http://www.nytimes.com/2013/03/05/opinion/global/stalins-long-shadow.html?_r=0.

Ravi, B. *METAL CastING: Computer-Aided Design and Analysis*. New Delhi: PHI Learning Private Limited, 2005.

Roach, John. "Emperor Qin's Terra Cotta Army." National Geographic. Accessed July 20, 2016. http://science.nationalgeographic.com/science/archaeology/emperor-qin/.

Shortland, Andrew J. *Lapis Lazuli from the Kiln: Glass and Glassmaking in the Late Bronze Age*. Leuven: Leuven University Press, 2012.

Sias, Fred R. Jr. *Lost-Wax Casting: Old, New, and Inexpensive Methods*. Pendleton, SC: Woodsmere Press, LLC, 2006.

Skarstedt Gallery, New York. Press release for the exhibition *Cast From Life*, September 18–October 25, 2014.

Smyth, Jonathan Blood. "English Pressed Glass of the Victorian Era." Accessed July 20, 2016. http://englishpressedglass.com.

Stern, E. M., and Birgit Schlick-Nolte. *Early Glass of the Ancient World: 1600 B.C.-A.D. 50: Ernesto Wolf Collection*. English edition. New York: Verlag Gerd Hatje, 1994.

Townsend, Chris. *The Art of Rachel Whiteread*. New York: Thames & Hudson, 2004.

Tsuda, Noritake, and Patricia Graham. *A History of Japanese Art: From Prehistory to the Taisho Period* (Tuttle Classics). North Clarendon, VT: Tuttle Publishing, 2015.

Vermeulen, Robin van den Akker, and Timotheus Vermeulen. "Notes on Metamodernism." *Journal of Aesthetics and Culture*, 2010. http://aestheticsandculture.net/index.php/jac/article/view/5677/6304.

Wong, Peter L. "Casting Call: Spatial Impressions in the Work of Rachel Whiteread." January 27, 2016. Accessed July 20, 2016. http://www.theplanjournal.com/article/casting-call-spatial-impressions-work-rachel-whiteread.

Young, James E. "Memory and Counter-Memory." *Harvard Design Magazine*, 1999, editorial. Accessed July 20, 2016. http://www.harvarddesignmagazine.org/issues/9/memory-and-counter-memory.

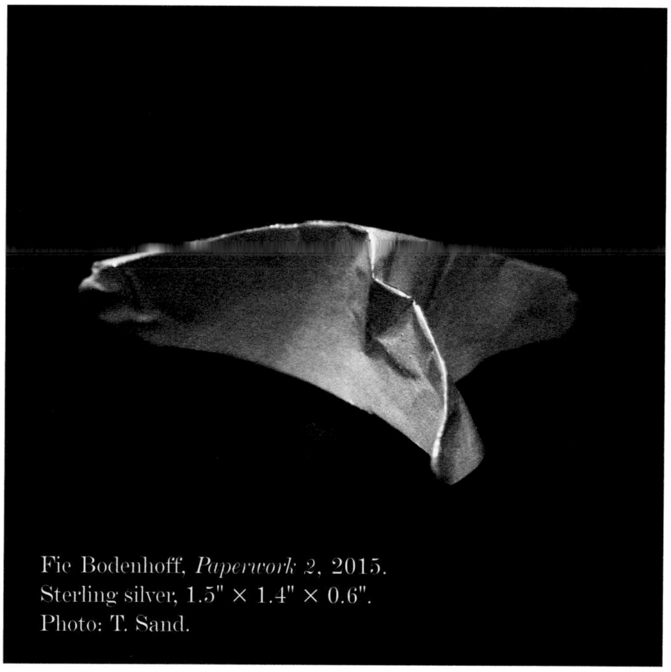

Fie Bodenhoff, *Paperwork 2*, 2015. Sterling silver, 1.5" × 1.4" × 0.6". Photo: T. Sand.

INDEX

Page numbers in *italics* indicate illustrations.